Plant Nutrition and Food Security in the Era of Climate Change

Plant Nutrition and Food Security in the Era of Climate Change

Edited by

Vinay Kumar

Ashish Kumar Srivastava

Penna Suprasanna

ELSEVIER

ACADEMIC PRESS

An imprint of Elsevier

Academic Press is an imprint of Elsevier
125 London Wall, London EC2Y 5AS, United Kingdom
525 B Street, Suite 1650, San Diego, CA 92101, United States
50 Hampshire Street, 5th Floor, Cambridge, MA 02139, United States
The Boulevard, Langford Lane, Kidlington, Oxford OX5 1GB, United Kingdom

Notices
Knowledge and best practice in this field are constantly changing. As new research and experience broaden our understanding, changes in research methods, professional practices, or medical treatment may become necessary.

Practitioners and researchers must always rely on their own experience and knowledge in evaluating and using any information, methods, compounds, or experiments described herein. In using such information or methods they should be mindful of their own safety and the safety of others, including parties for whom they have a professional responsibility.

To the fullest extent of the law, neither the Publisher nor the authors, contributors, or editors, assume any liability for any injury and/or damage to persons or property as a matter of products liability, negligence or otherwise, or from any use or operation of any methods, products, instructions, or ideas contained in the material herein.

British Library Cataloguing-in-Publication Data
A catalogue record for this book is available from the British Library

Library of Congress Cataloging-in-Publication Data
A catalog record for this book is available from the Library of Congress

ISBN: 978-0-12-822916-3

Publisher: Charlotte Cockle
Editorial Project Manager: Sara Valentino
Production Project Manager: Swapna Srinivasan
Cover Designer: Christian J. Bilbow

Typeset by MPS Limited, Chennai, India

Contents

List of contributors

Muna A. Ali Department of Applied Biology, College of Sciences, University of Sharjah, Sharjah, United Arab Emirates

Surabhi Awasthi Plant Stress Biology Laboratory, Institute of Environment and Sustainable Development, Banaras Hindu University, Varanasi, India

Rahul Badhwar Reliance Research and Development Centre, Reliance Industries Ltd., Navi Mumbai, India

Arun Banerjee Reliance Research and Development Centre, Reliance Industries Ltd., Navi Mumbai, India

Adalberto Benavides-Mendoza Horticulture Department, Autonomus Agricultural University Antonio Narro, Saltillo, Mexico

Debarati Bhaduri ICAR-National Rice Research Institute, Cuttack, India

S.M.P. Carvalho GreenUPorto – Sustainable Agrifood Production Research Centre / Inov4Agro & DGAOT, Faculty of Sciences, University of Porto, Vairão, Portugal

Koushik Chakraborty ICAR-National Rice Research Institute, Cuttack, India

Swarnavo Chakraborty Department of Biotechnology, St. Xavier's College (Autonomous), Kolkata, India

Reshu Chauhan Plant Stress Biology Laboratory, Institute of Environment and Sustainable Development, Banaras Hindu University, Varanasi, India

P.S. John Daniel International Crops Research Institute for the Semi-Arid Tropics, Patancheru, India

Santanu Dasgupta Reliance Research and Development Centre, Reliance Industries Ltd., Navi Mumbai, India

Sonia Elthury ICAR-Indian Institute of Oilseeds Research, Hyderabad, India

A.P.G. Fernandes GreenUPorto – Sustainable Agrifood Production Research Centre / Inov4Agro & DGAOT, Faculty of Sciences, University of Porto, Vairão, Portugal

T.R. Fernandes GreenUPorto – Sustainable Agrifood Production Research Centre / Inov4Agro & DGAOT, Faculty of Sciences, University of Porto, Vairão, Portugal

Komal Goel CSIR-Institute of Himalayan Bioresource Technology (IHBT), Palampur, India; Academy of Scientific and Innovative Research (AcSIR), CSIR-HRDC Campus, Ghaziabad, India

Jorge González-Villagra Department of Agricultural Sciences and Aquaculture, Faculty of Natural Resources, Universidad Católica de Temuco, Temuco, Chile; Nucleus of Research in Food Production, Faculty of Natural Resources, Universidad Católica de Temuco, Temuco, Chile

K. Gopika ICAR-Indian Institute of Oilseeds Research, Hyderabad, India

Kathleen Hefferon Department of Cell & Systems Biology, University of Toronto, Toronto, ON, Canada

E. Heuvelink Horticulture and Product Physiology Group, Department of Plant Sciences, Wageningen University, Wageningen, The Netherlands

Claudio Inostroza-Blancheteau Department of Agricultural Sciences and Aquaculture, Faculty of Natural Resources, Universidad Católica de Temuco, Temuco, Chile; Nucleus of Research in Food Production, Faculty of Natural Resources, Universidad Católica de Temuco, Temuco, Chile

Ahmed Ismail Department of Biotechnology, Faculty of Agriculture, Al-Azhar University, Cairo, Egypt; Public Health Department, College of Applied Medical Science, Majmaah University, Al Majmáah, Saudi Arabia

Ayushi Jaiswal Genetics and Plant Breeding, Rani Lakshmi Bai Central Agricultural University, Jhansi, India

Stanislav Kopriva Institute for Plant Sciences, Cluster of Excellence on Plant Sciences, University of Cologne, Cologne, Germany

Anna Koprivova Institute for Plant Sciences, Cluster of Excellence on Plant Sciences, University of Cologne, Cologne, Germany

Ramesh Kulasekaran ICAR-Indian Institute of Oilseeds Research, Hyderabad, India

Anil Kumar Rani Lakshmi Bai Central Agricultural University, Jhansi, India

Pravesh Kundu CSIR-Institute of Himalayan Bioresource Technology (IHBT), Palampur, India; Academy of Scientific and Innovative Research (AcSIR), CSIR-HRDC Campus, Ghaziabad, India

J. Machado GreenUPorto – Sustainable Agrifood Production Research Centre / Inov4Agro & DGAOT, Faculty of Sciences, University of Porto, Vairão, Portugal; Centro de Biotecnologia e Química Fina (CBQF) – Laboratório Associado, Escola Superior de Biotecnologia, Universidade Católica Portuguesa, Porto, Portugal; Horticulture and Product Physiology Group, Department of Plant Sciences, Wageningen University, Wageningen, The Netherlands

Bhumika Madan School of Biotechnology, Guru Gobind Singh Indraprastha University, Dwarka, India

Swati Mahiwal Department of Plant Molecular Biology, University of Delhi South Campus, New Delhi, India

Aakansha Malik School of Biotechnology, Guru Gobind Singh Indraprastha University, Dwarka, India

C.H.L.N. Manikanta ICAR-Indian Institute of Oilseeds Research, Hyderabad, India

Samriti Mankotia Department of Biological Sciences, Indian Institute of Science Education and Research (IISER) Mohali, SAS Nagar, India

Ankita Mohanty ICAR-National Rice Research Institute, Cuttack, India

Subhankar Mondal ICAR-National Rice Research Institute, Cuttack, India

Kareem A. Mosa Department of Applied Biology, College of Sciences, University of Sharjah, Sharjah, United Arab Emirates; Department of Biotechnology, Faculty of Agriculture, Al-Azhar University, Cairo, Egypt

Rajesh Nandru Reliance Research and Development Centre, Reliance Industries Ltd., Navi Mumbai, India

Brij Bihari Pandey ICAR-Indian Institute of Oilseeds Research, Hyderabad, India

Girdhar K. Pandey Department of Plant Molecular Biology, University of Delhi South Campus, New Delhi, India

Ratnakumar Pasala ICAR-Indian Institute of Oilseeds Research, Hyderabad, India

Alivia Paul Ballygunge Science College, University of Calcutta, Kolkata, India

Kenny Paul Reliance Research and Development Centre, Reliance Industries Ltd., Navi Mumbai, India

Nandula Raghuram School of Biotechnology, Guru Gobind Singh Indraprastha University, Dwarka, India

Kalidoss Ramamoorthy Department of Applied Biology, College of Sciences, University of Sharjah, Sharjah, United Arab Emirates

Zed Rengel UWA School of Agriculture and Environment, University of Western Australia, Perth, WA, Australia

Marjorie Reyes-Díaz Department of Chemical Sciences and Natural Resources, Faculty of Engineering and Sciences, Universidad de La Frontera, Temuco, Chile; Center of Plant, Soil Interaction and Natural Resources Biotechnology, Scientific and Technological Bioresource Nucleus (BIOREN-UFRO), Universidad de La Frontera, Temuco, Chile

Aryadeep Roychoudhury Department of Biotechnology, St. Xavier's College (Autonomous), Kolkata, India

Richa Salwan College of Horticulture and Forestry, Dr YS Parmar University of Horticulture and Forestry, Neri, Hamirpur, India

Santanu Samanta Department of Biotechnology, St. Xavier's College (Autonomous), Kolkata, India

Ajit Sapre Reliance Research and Development Centre, Reliance Industries Ltd., Navi Mumbai, India

Santosh B. Satbhai Department of Biological Sciences, Indian Institute of Science Education and Research (IISER) Mohali, SAS Nagar, India

Sergey Shabala Tasmanian Institute of Agriculture, College of Science and Engineering, University of Tasmania, Hobart, TAS, Australia

Vivek Sharma University Centre for Research and Development, Chandigarh University, Gharuan, Mohali, India

Erum Shoeb Department of Cell & Systems Biology, University of Toronto, Toronto, ON, Canada

Anshuman Singh Genetics and Plant Breeding, Rani Lakshmi Bai Central Agricultural University, Jhansi, India

Ashutosh Singh Biotechnology and Crop Improvement, Rani Lakshmi Bai Central Agricultural University, Jhansi, India

Sudhakar Srivastava Plant Stress Biology Laboratory, Institute of Environment and Sustainable Development, Banaras Hindu University, Varanasi, India

Jagannath Swain Department of Biological Sciences, Indian Institute of Science Education and Research (IISER) Mohali, SAS Nagar, India

Mohsin Tanveer Tasmanian Institute of Agriculture, College of Science and Engineering, University of Tasmania, Hobart, TAS, Australia

Ram Sewak Tomar Biotechnology, Rani Lakshmi Bai Central Agricultural University, Jhansi, India

Alok Varshney Reliance Research and Development Centre, Reliance Industries Ltd., Navi Mumbai, India

M.W. Vasconcelos Centro de Biotecnologia e Química Fina (CBQF) – Laboratório Associado, Escola Superior de Biotecnologia, Universidade Católica Portuguesa, Porto, Portugal

Praduman Yadav ICAR-Indian Institute of Oilseeds Research, Hyderabad, India

Gaurav Zinta CSIR-Institute of Himalayan Bioresource Technology (IHBT), Palampur, India; Academy of Scientific and Innovative Research (AcSIR), CSIR-HRDC Campus, Ghaziabad, India

About the editors

Dr. Vinay Kumar is an Associate Professor at the Department of Biotechnology, Modern College of Arts, Science and Commerce, Savitribai Phule Pune University, Pune, India. He completed his PhD in Biotechnology from Savitribai Phule Pune University in 2009 and has published more than 50 peer-reviewed research/review articles. He is a recipient of the Government of India's Science and Engineering Board, Department of Science and Technology (SERB-DST) Young Scientist Award in 2011. His current research interests include elucidating molecular mechanisms underlying environmental stress responses and tolerance in plants. He has been a guest editor for *Physiologia Plantarum* and *Current Topics in Medicinal Chemistry*, has reviewed several highly regarded journals, and has edited six books from various publishers including Springer, Wiley, and Elsevier. He has received several grants from funding agencies, including DBT (Department of Biotechnology), DST, and SERB (Government of India).

Dr. Ashish Kumar Srivastava is a Scientific Officer at Bhabha Atomic Research Centre, Mumbai. His research is focused on developing strategies for enhancing crop resilience toward different abiotic stresses. The stimulatory potential of thiourea has been demonstrated for enhancing stress tolerance and crop productivity through laboratory and small-scale field experiments, and by conducting multilocation field trials at salt-, drought-, and arsenic-affected fields. The molecular mechanism of stress perception and tolerance has been elucidated using a systems biology–based approach. He has more than 50 research and review articles and book chapters to his credit and has been awarded the Young Scientist Award of National Academy of Sciences, India (NASI), Allahabad in 2018; the Young Scientist Medal from Indian National Science

Academy (INSA) in 2014; and the Young Scientist Award from the Department of Atomic Energy (DAE) in 2014. He has also received the Newton-Bhabha International Grant from DBT-BBSRC in 2018; President International Fellowship from Chinese Academy of Sciences, China in 2016; and an EMBO Short-Term Fellowship in 2011.

Penna Suprasanna is a Professor of Homi Bhabha National Institute, Mumbai, and former Head of Nuclear Agriculture and Biotechnology Division, Bhabha Atomic Research Centre, Mumbai, India. His research contributions are in crop biotechnology, radiation-induced mutagenesis, plant genomics, and abiotic stress tolerance. He has made concerted efforts to apply radiation mutagenesis techniques in vegetatively propagated plants through collaborative research projects with several national and international bodies (IAEA, Vienna). He has more than 300 published works and has edited 3 Springer books: *Salinity Responses and Tolerance in Plants, Volume 1: Targeting Sensory, Transport and Signaling Mechanisms* (2018); *Salinity Responses and Tolerance in Plants, Volume 2: Exploring RNAi, Genome Editing and Systems Biology* (2018); and *Plant–Metal Interactions* (2019). In addition to serving as the editor of several journals of repute, he has been a Guest Editor for special issues of Elsevier, Wiley, and Springer journals.

Preface

Food security and sustainable agriculture are global priorities in the face of the rising world human population and emerging global climate changes. Both biotic and abiotic stresses impose severe restrictions on global crop production and quality. In view of this, there is general consensus on improving crop productivity and climate resilience so as to develop better crops with optimal plant nutrient management. Plant nutrients are key to ensuring food and nutritional security and maintaining nutritional levels in human food and animal feeds. Nutrient stewardship needs to be in place to ensure plant nutrient availability, nutrient acquisition, nutrient assimilation, and nutrient allocation under environmentally challenging conditions. Extensive research and development is thus a necessity to integrate genetic and agronomic factors to fine-tune the "capacitance" of crop plants to ensure global food security.

This comprehensive, unique book aims to present the current knowledge and updates on plant nutrition and nutrients, including their uptake, transport, and roles and regulation during plant growth, development, and stress responses. The chapters are written by highly acclaimed experts of international repute working on plant nutrition and in allied fields. The wide spectrum of topics includes the importance of plant nutrition and nutrients, interactions between essential and nonessential nutrients, their implications on global food security, how nutritional homeostasis holds the key for plants as well as consumers, and how the global effects of abiotic stressors, stressful conditions, and rising atmospheric carbon dioxide levels impact plant nutrition. This book also discusses modern tools used for plant nutritional research, including high-throughput ionomics as well as biotechnological tools. New dimensions of exploring nanomaterials and beneficial microbes in improving plant nutrition and alleviating plant stress are also covered. Further, the key priorities of crop nutrient management and better crops with better nutrient efficiency have also been discussed. We believe this collection to be useful as a platform for integrating agronomy, crop biology, biotechnology, genetic engineering, and plant health. This is an attempt to promote research about the importance of mineral nutrients in plant biology and how to produce plants with the desired levels of nutrients even amidst changing climatic conditions. The book provides an integrated and comprehensive mix of basic and advanced information for students, scholars, and scientists interested in, or already engaged in, research involving plant nutrients. We firmly believe that this book will meet the needs of a wide spectrum of scientific readers including students, research scholars, scientists, and academics.

We highly appreciate the excellent and timely contributions by the authors presenting the relevant information on different aspects of plant nutrition, nutrients, their biology, and biotechnology in the purview of food security in challenging environments and climate change. We gratefully acknowledge the reviewers for their valuable comments which helped in the quality improvement of the chapters. We are happy of this collaborative undertaking comprising of very valuable contributions of 52 scientists from 11 countries.

We thank the Elsevier team comprising Ms. Charlotte Cockle, Publisher, and Ms. Sara Valentino, Editorial Project Manager, and the entire Elsevier production team for their consistent hard work in the publication of this book.

Vinay Kumar,
Ashish Kumar Srivastava and
Penna Suprasanna

Chapter 1

Entangling the interaction between essential and nonessential nutrients: implications for global food security

Mohsin Tanveer and Sergey Shabala
Tasmanian Institute of Agriculture, College of Science and Engineering, University of Tasmania, Hobart, TAS, Australia

1.1 Introduction

The world population stands currently at ~7.77 billion and is predicted to exceed 10 billion by 2050 (FAO, 2017). This increase in population growth must be accompanied by a massive (at least 60%; Springmann et al., 2018) increase in food production. However, because of the increasing rate of urbanization and land degradation caused by various environmental constraints, per capita availability of arable land has decreased by about twofold over the last 50 years (Liu et al., 2020). Global food security is further threatened by current climate trends, with expected increases in several stress events (e.g., high temperatures, cold, frost, drought, soil salinization, and flooding) that drastically limit local plant production.

Until now, the backbone of modern agriculture was a broad use of agrochemicals (and, specifically, chemical fertilizers) and a heavy reliance on irrigation (Fita et al., 2015). Starting in the 1960s as a part of the Green Revolution, this trend resulted in the development of high-yielding semi-dwarf plant varieties that were highly demanding of synthetic fertilizers and needed large amounts of irrigation water. However, the current agricultural practices appear to be unsustainable under future scenarios. For example, the major source for production of inorganic phosphorus (P) fertilizers is a rock phosphate (RP). Since the Green Revolution, the use of RP has increased fourfold (Childers et al., 2011; Daneshgar et al., 2018), and the demand for

Plant Nutrition and Food Security in the Era of Climate Change. DOI: https://doi.org/10.1016/B978-0-12-822916-3.00019-6

P fertilizers is increasing along with population growth. However, RP is a nonrenewable source and may be completely exhausted in 50–200 years (Baker et al., 2015). Paradoxically, while global reserves of RP are being depleted, much of P fertilizers that are produced and used in agricultural systems are not used by plants but lost in runoff (Bouwman et al., 2013), resulting in massive environmental damage through eutrophication. In addition, while staple crops were improved and the total amounts of protein and energy available to people were increased, the nutritional value of food was largely ignored, leading to an increase in micronutrient malnutrition, which is known as "hidden hunger" (Welch and Graham, 2002; Fita et al., 2015). Also, varieties that were responsive to fertilization practices and produced high yield will not necessarily perform as well when grown under abiotic stress conditions (Palmgren et al., 2015). The increased reliance on irrigation and a lack of quality freshwater result in progressive land salinization (Liu et al., 2020), thus affecting plants' ability to take up essential nutrients from the soil. The same is true for other abiotic stresses such as drought, flooding, and temperature extremes.

Increased anthropogenic activities, rapid industrialization, and modern agricultural practices have resulted in increased heavy metal (HM) contamination in the environment (Eapen and D'Souza, 2005; Kavamura and Esposito, 2010; Miransari, 2011). The HMs most commonly found in contaminated soils are arsenic (As), cadmium (Cd), chromium, and lead (Pb). All pose significant risks for human health when they enter a food chain. Importantly, while climate-driven abiotic stresses reduce the availability of some essential nutrients, the availability of other (nonessential to plants) elements may increase. The latter further reduce the biological availability of essential nutrients (via interference with their uptake). The current management practices are of a little help to entangle this issue. Thus the sustainability of the agricultural production system in the 21st century is critically dependent on understanding the molecular basis of plant nutrient uptake and entanglement of interaction between essential and nonessential nutrients, with the goals of increasing plant nutrient use efficiency, reducing the resource (fertilizer) input, and managing the environmental footprint. This chapter undertakes three case studies [using sodium (Na), Cd, and As as examples] to summarize the current knowledge of mechanisms of acquisition of some essential macronutrients and micronutrients under adverse environmental conditions.

1.2 Potassium/sodium interaction

Potassium (K) is the most abundant inorganic cation in plant cells, with an average content being between 2% and 10% of plant dry matter (Marschner, 1995). Due to potassium's broad range of metabolic and physiological functions, plants maintain rather constant cytosolic K^+ content, typically within the 100–200 mM range (Britto and Kronzucker, 2008). The failure to maintain cytosolic K^+ homeostasis comes with a significant disturbance to a plant's operation; because of this, K is recognized as a rate-limiting factor for crop yield and quality (Dreyer and Uozumi, 2011). A plant's inability to maintain optimal cytosolic K^+ levels also comes with a danger of inducing program cell death (thus a complete loss of cell viability) by unblocking activities of caspase-like proteases and endonucleases (Shabala, 2009; Demidchik et al., 2010). Stress-induced K^+ deficiency is a well-reported phenomenon observed under adverse environmental conditions such as drought (Hu and Schmidhalter, 2005; Hu et al., 2013), salinity (Aziz and Khan, 2001, Chen et al., 2005, Tavakkoli et al., 2011), or flooding (Close and Davidson, 2003; Smethurst et al., 2005; Board, 2008), and the ability of various plant tissues to retain potassium under stress conditions has been recognized as a novel and essential mechanism of abiotic stress tolerance in plants (Shabala and Pottosin, 2014; Shabala et al., 2016a; Wu et al., 2018).

Potassium homeostasis in plants is regulated by a large number of transporter proteins. In model *Arabidopsis* plants, seven major families of cation transporters (75 genes in total) mediate K^+ transport across cellular membranes. These include (Mäser et al., 2001; Véry and Sentenac, 2002, 2003) (1) Shaker-type K^+ channels (9 genes), (2) two-pore K^+ channels (6 genes), (3) putative K^+/H^+ antiporters (6 genes), (4) KUP/HAK/KT transporters (13 genes), (5) HKT transporters (1 gene), (6) GLR glutamate receptors (20 genes), and (7) CNGC cyclic nucleotide gated channels (20 genes). Some of these transporters are specifically selective for K^+ (e.g., Shaker-type channels), while others are nonselective and permeate other cations including Na (e.g., CNGC or GLR). They also differ in their gating properties and control modes and are expressed in a highly tissue-specific manner. In addition, low-affinity K^+-permeable transporters have been identified in other species (e.g., LCT1 in wheat; Schachtman et al., 1997).

At the physiological level, the K^+ transporting proteins previously mentioned operate either in a unidirectional K^+ uptake (e.g., high-affinity K^+ transporters from the HAK family or inward-rectifying AKT1 channels) or K^+ release (e.g., GORK channel in the root epidermis or SKOR channels in the xylem parenchyma cells) or could mediate a bidirectional transport of K^+ along its electrochemical gradient (e.g., nonselective GLR and CNGC channels). AKT1 and/or AtKC1/AKT1 heteromers constitute the major channel-mediated pathway for K^+ uptake from the soil (Nieves-Cordones et al., 2016), while HAK5 is the only system that operates at external K^+ concentrations below 10 μM (Rubio et al., 2020). At concentrations between 10 and 200 μM, both AtHAK5 and AKT1 contribute to K^+ supply.

Nearly 10% of land surface and 50% of all irrigated land in the world is affected by salinity (Ruan et al., 2010; Flowers et al., 2010), which is characterized by the presence of high amounts of Na in the rhizosphere. Most of this salinity is natural; however, both land clearing and extensive irrigation have caused extensive salinization and land degradation in many countries (Ghassemi et al., 1995; Liu et al., 2020). On average, between 3 and 6 tons of salt are added to every hectare of irrigated land every year (Liu et al., 2020). As all modern crops were selected to be salt excluders, a constant salinity buildup occurs. The latter affects plants' ability to acquire essential nutrients, specifically K^+, thus resulting in massive yield losses (estimated at \$27.3 billion per annum; Qadir et al., 2014).

The detrimental effects of salinity on K^+ acquisition can be explained by the combined effect of at least nine confounding factors:

1. When positively charged Na^+ crosses the plasma membrane (PM), the PM is depolarized by as much as 60–80 mV (depending on NaCl concentration and species; Shabala et al., 2003, 2005; Chen et al., 2007; Cuin et al., 2008; Bose et al., 2015; Hamam et al., 2019). This depolarization is independent of the plant's ability to exclude Na^+ via the SOS pathway (Zhao et al., 2020), as SOS1 Na^+/H^+ exchanger is electrically neutral and does not contribute to MP maintenance. Such significant PM depolarization opens outward-rectifying depolarization-activated GORK K^+ efflux channels, resulting in a massive K^+ loss from the cytosol (Wu et al., 2018). As the vacuolar K^+ buffering capacity is limited (ca. 4–6 h; Rubio et al., 2020), unless balanced by increased K^+ uptake, a prolonged depolarization results in substantial K^+ loss from the roots, leading to K^+ deficiency observed under saline conditions.
2. To prevent depolarization-induced K^+ leakage, plants attempt to restore the membrane potential by more active H^+ ATPase pumping (Falhof et al., 2016). This comes with a significant ATP cost and may compromise the plant's ability to adapt and grow. Naturally salt-tolerant halophyte species possess constitutively more negative MP values (Bose et al., 2015); the same is true for salt-tolerant genotypes (Chen et al., 2007). This comes, however, at the expense of yield (e.g., a negative correlation between productivity and tolerance; Shabala et al., 2016a).
3. This PM depolarization also makes a passive K^+ uptake through inward-rectifying K^+ channels (such as AKT or KAT) thermodynamically impossible. Hence, K^+ uptake under saline conditions relies mainly on high-affinity (HAK-mediated) transporters that are much less efficient and require ATP to operate (i.e., need to be energized by H^+-ATPases for a symport operation). This depletes the ATP pool and comes with penalties for growth and yield. In quantitative terms, accumulation of 1 mol of K^+ requires 1 mol of ATP for uniport (channel-mediated) mechanism but 2 mol of ATP for a K^+/H^+ symport (Britto and Kronzucker, 2008).
4. In addition to reducing the K^+ electrochemical gradient, Na^+ can also inhibit K^+ uptake mediated by the HAK and AKT transport systems (Bertl et al., 1997; Santa-María et al., 1997). Both HAK and AKT

transport systems exhibit a similar (about ~1/1000) K^+/Na^+ selectivity when expressed in heterologous systems (Rubio et al., 2000; Véry et al., 2014).

5. Salinity stress also results in a significant accumulation of reactive oxygen species (ROS) in root and leaf tissues. These ROS are mainly produced in the apoplast, chloroplasts, mitochondria, and peroxisomes (Miller et al., 2010). Plant cells harbor a large number of ROS-activated K^+-permeable NSCC cation channels (Wu et al., 2018). Most of these channels are only weakly voltage dependent (Demidchik and Maathuis, 2007). As cytosolic K^+ concentrations are typically at around 100 mM, ROS-induced NSCC opening favors passive K^+ efflux. The differential sensitivity of K^+-permeable NSCC to various ROS (e.g., H_2O_2 and •OH) explains the intraspecific (Bose et al., 2015; Chakraborty et al., 2016), genotypic (Chen et al., 2007; Wang et al., 2018), and tissue-specific (Demidchik et al., 2007; Shabala et al., 2016b) differences in salinity stress tolerance.
6. Even if the electrochemical gradient favors passive K^+ uptake via NSCC, excessive Na^+ will compete with K^+ for uptake sites at the PM. This is also true for high-affinity transporters (e.g., HKT).
7. High levels of salt in the soil reduce its water potential and result in an osmotic stress. Plants respond to this stress by closing their stomata, thus reducing transpiration rate and, hence, delivery of K^+ to the shoot. Reduced transpiration rate also slows down the bulk flow of the water in the soil and, hence, delivery of K^+ to the root surface via this mechanism.
8. Salinity-induced changes in soil water potential require plants to osmotically adjust (Zhao et al., 2020), and most plants achieve this by increased de novo synthesis of various compatible solutes. This comes with a huge carbon cost (Munns et al., 2020) and severely reduces the available ATP pool, making high-affinity K^+ uptake even more problematic.
9. Finally, high concentrations of Na^+ in the soil substantially reduce the activity of many essential nutrients (including K^+), making them less available for plants (Shabala and Cuin, 2008). This is a purely "passive" physical process that is independent of genotype and explained by the soil chemistry.

The possible strategies of improving K^+ nutrition under saline conditions (in the presence of interfering Na^+) could be as follows. First, plants need to prevent salinity-induced K^+ leak to avoid a futile cycle. This could be done by both better control of the voltage gating (e.g., reduce depolarization) and desensitization of K^+-permeable channels to ROS (Shabala et al., 2016b). The first avenue is achievable by a more efficient operation of electrogenic H^+ pumps (Palmgren and Nissen, 2011). Stress-tolerant species and genotypes have either constitutively higher H^+-ATPase activity (Chen et al., 2007;

Bose et al., 2015) or are capable of rapidly upregulating it upon salinity exposure (Bose et al., 2015). However, because the operation of H^+-ATPase comes with an energy cost, such genotypes and species need to allocate bigger ATP pools to fuel such pumps, at the expense of other metabolic processes. The above dilemma could be potentially overcome by increasing the affinity for the substrate (lower K_m) to achieve a maximum H^+ extrusion (higher V_{max}) upon sensing Na^+ ions (Wu and Seliskar, 1998). Another (and, arguably, the more promising) approach is to make transgenic plants that would be able to use pyrophosphate instead of ATP for membrane potential maintenance (Shabala et al., 2016a). The classical view is that H^+-PPases are located at the intracellular membranes (tonoplast and Golgi; Maeshima, 2000). However, both structural (microscopy; Ratajczak et al., 1999; Langhans et al., 2001) and functional (pharmacological experiments; Bose et al., 2014) evidence suggests that type I H^+-PPase may have dual membrane localization and be also present at the PM, at least in some tissues. Thus exploring the possibility of substituting H^+-ATPases by the functional H^+-PPases at the PM represents a previously untapped resource for dealing with salinity stress and its negative impact on K^+ nutrition.

Desensitization of ROS-inducible K^+ efflux channels is an equally important task, as a failure to achieve this may result in a futile cycle and a complete depletion of energy resources in an attempt to restore an MP via increased H^+ pumping. It was shown that a higher sensitivity of the root apex to salinity compared to the mature zone was causally related to a larger density of ROS-activated nonselective currents in this tissue that resulted in a 10-fold larger K^+ efflux from epidermal root cells (Shabala et al., 2016b). While the mechanisms of ion channels activation by ROS are still understood poorly, reactive cysteine residues are considered to be the major targets (Alansary et al., 2016). Garcia-Mata et al. (2010) showed that the K^+ outward-rectifying SKOR channel was activated by H_2O_2 via targeted oxidation of Cys168 at the S3 α-helix within the channel's voltage sensor, and the higher K^+ retention ability and salinity tolerance in polyploid plants (Liu et al., 2019) have been explained by the desensitization of the appropriate transport system to H_2O_2 resulting from modification of thiol groups in the sensory domain (Shabala, 2019).

1.3 Zinc/cadmium interaction

Zinc (Zn) is an essential micronutrient with multiples roles in plant growth and development. Zn acts as a catalyst for more than 300 enzymes (Frassinetti et al., 2006) that are critical for a broad range of metabolic and developmental processes. Zn is also essential for humans, with the recommended dietary allowance ranging between 8 and 11 mg/day, depending on gender. Zn deficiency is ranked fifth among nutritional causes of disease in developing countries (Cakmak and Kutman, 2018). Insufficient micronutrient intake is also widespread in developed countries (Gibson et al., 2002), and many diseases associated with aging (cancer, neural deterioration causing mental disorders, etc.) are linked to micronutrient insufficiency. The international research program Harvest Plus has met with some success over the last decade, releasing biofortified cereal genotypes (rice, wheat, and maize) with enhanced capacity to accumulate Zn in grains (Bouis and Saltzman, 2017). However, a drawback of biofortified genotypes is that they frequently accumulate more cadmium (Cd)—a toxic HM taken up by Zn transporters—than the standard genotypes (Huang et al., 2020). So, can uptake of these nutrients be entangled?

Cd is a nonessential HM that significantly reduces plant growth and yield (Anjum et al., 2016). When Cd is accumulated in plants, it can also enter a food chain and cause a broad range of toxic effects to humans (Prankel et al., 2005). Cd is first taken by roots and is then translocated to the shoot, where it is destined for either vacuolar sequestration or detoxification (Mendoza-Cózatl et al., 2011). These processes are highly dependent on the efficiency and operation of different membrane transport systems. However, most of them have a high affinity for essential micronutrients and, specifically, Zn. As a result, all attempts to reduce Cd uptake and accumulation in plants come with the risk of reducing uptake of Zn and other micronutrients.

The key transport systems for Zn and Cd in plants include: the zinc-regulated transporter protein (ZIP) family, the HM ATPase (HMA) family, the cation diffusion facilitator (CDF) family, the natural resistance-associated macrophage protein (NRAMP) family, and the low cation transporter family 1 (LCT1) (Pilon et al., 2009; Lin et al., 2012). These transporters are highly tissue-specific and work in collaboration to govern Zn/Cd homeostasis. The functional roles of these transporters in regulating Zn/Cd homeostasis are briefly summarized in Fig. 1.1.

ZIP is a metalloprotein that is responsible for Zn transport across the cellular membranes in living organisms (Eide, 2005). ZIP proteins have eight transmembrane domains and facilitate intracellular Zn homeostasis at both the transcriptional and posttranscriptional levels (Do et al., 2016). ZIP also transports Cd across the PM, for instance in *Brassica campestris* (Wu et al., 2019a,b). In *Arabidopsis*, AtZIP5 and AtZIP9 are tandem duplicates and act synergistically in Zn/Cd uptake.

HMAs are P 1B-type transporters, which are highly diverse in terms of metal specificity, subcellular localization, and tissue distribution (Tan et al., 2013). Nine members of HMA transporters are functionally characterized in rice, while eight members are present in *Arabidopsis*. Based on their functional characterization, HMAs can be divided into two groups: (1) copper (Cu)/silver (Ag) (P1B1-ATPase transporters), and (2) Zn/cobalt (Co)/Cd/Pb (P1B2-ATPase transporters). In *Arabidopsis*, AtHMA1-AtHMA4, and in rice, OsHMA1-OsHMA3 belong to the P1B2-ATPases, whereas AtHMA5-AtHMA8 and OsHMA4-OsHMA9 belong to the P_{1B1}-ATPases (Takahashi et al., 2012a;

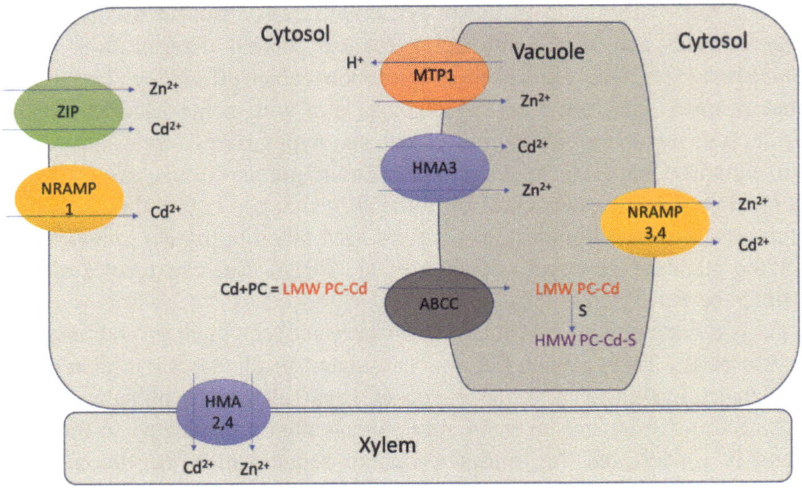

FIGURE 1.1 Uptake and intracellular compartmentation of zinc (Zn) and cadmium (Cd) in plant roots. Both Zn and Cd are taken up by ZIP1 (zinc transporter-like protein), located at the PM of the root epidermis. Cd can also be transported into cells through NRAMP1/5/6 (natural resistance-associated macrophage protein) transporters. HMA3 (metal-transporting ATPases) play an important role in the vacuolar sequestration of Zn and Cd. Chelated Cd is stored in vacuoles via ABCC (ATP-binding cassette transporters). MTP1 (metal-tolerance protein) also involves the vacuolar sequestration of Zn in response to cytosolic Zn toxicity, while NRAMP3,4 transfer vacuolar Zn to cytosol in response to Zn deficiency. NRAMP3,4 mediate Cd transport and sequestration in the vacuole. Both Zn and Cd are translocated from root to shoot via the HMA2 located at the xylem parenchyma.

Guo et al., 2019). HMAs play an important role in maintaining Zn/Cd homeostasis. For example, the expression of OsHMA1 in rice was 7.4 times higher under Zn-deficient conditions than under Zn sufficiency, indicating the role of HMA1 in Zn uptake (Suzuki et al., 2012). HMA2 plays a critical role in translocation of Zn and Cd to the shoot (Wong and Cobbett, 2009; Takahashi et al., 2012b; Guo et al., 2019), while HMA3 plays a role in vacuole sequestration of Cd and Zn in roots under Cd and Zn (Ueno et al., 2010, 2011; Cai et al., 2019). In a Cd hyperaccumulator *Arabidopsis halleri*, AhHMA4 plays an important role in the xylem Cd and Zn loading (Courbot et al., 2007; Hanikenne et al., 2008).

CDF (also known as metal-tolerance protein 1; **MTP1**) is one of the most important Zn transporters for its vacuolar sequestration during Zn toxicity. Functionally, CDF/MTP1 is a tonoplast localized Zn^{2+}/H^{+} antiporter (Kobae et al., 2004). The expression of CDF/MTP1 was higher in the shoots of HM hyperaccumulating plant species as compared with nonhyperaccumulators (Assunção et al., 2001; Gustin et al., 2009; Zhang et al., 2011). In *Arabidopsis*, MTP1 is responsible for the vacuolar Zn sequestration and a knockdown of MTP1 resulted in decreased Zn tolerance (Kawachi et al., 2009). CDF/MTP1 transporters act as efflux transporters, and in cucumber, *CsMTP9*, known as an efflux transporter for manganese (Mn), rescued the Cd-hypersensitive phenotype when expressed in yeast, resulting from higher Cd^{2+} efflux activity (Migocka et al., 2015). However, in rice, OsMTP9 does not have the capacity to transport cadmium, suggesting a different unidentified transporter involved in this pathway (Ueno et al., 2015). In barley, MTP1 encodes for Zn^{2+}/Co^{2+} transport and showed high affinity to Zn (Podar et al., 2012). Other members of the CDF/MTP1 family, such as MTP8 and MTP11, are also important for HM homeostasis (Hammond et al., 2006; Van de Mortel et al., 2006; Krämer et al., 2007). These MTPs are very diverse in nature in context of their functionality, characterization, and ion selectivity. The latter is determined by some specific amino acid sequence that could potentially be targeted in breeding programs. For example, in *Arabidopsis*, a selectivity of AtMTP1 towards Zn was shown to be increased by modifying a five-residue sequence within the MTP1 *N*-segment of the His-rich intracytoplasmic loop (Kawachi et al., 2012; Huang et al., 2020).

NRAMP proteins represent another pathway for HM and micronutrient transport across cellular membranes (Zhang et al., 2020). Six NRAMP transporters have been identified in a model plant species *Arabidopsis thaliana*. These can be divided into two subgroups: the first subgroup includes AtNRAMP1 and AtNRAMP6, and the second includes AtNRAMP2, AtNRAMP3, AtNRAMP4, and AtNRAMP5 (Mäser et al., 2001; Song et al., 2014). In *Arachis hypogaea*, AhNRAMP1 operates in Zn uptake (Wang et al., 2019) and also encodes for Cd transport (Chen et al., 2017a,b,c; Zhang et al., 2020).

Other than the transporters mentioned previously, there are ATP-binding cassette (**ABCC**) transporters that play an important role in HM vacuolar sequestration. In *Arabidopsis*, two ABCC-type transporters, AtABCC1 and AtABCC2, enhanced Cd tolerance by conducting vacuolar sequestration of Cd (Song et al., 2010). Upon the sequestration of Cd into cells, low-molecular-weight (LMW) phytochelatins (PCs) make complexes with Cd and then sequestration of Cd is done in PC–Cd form via ABCC transporters. After entering the vacuole, LMW PC–Cd complexes bind sulfides to form a stable, high-molecular-weight (HMW) PC–Cd complex, which is stored in the vacuole (Clemens 2006; Lin et al., 2012). Members of the MTP/ABCC belonging to the superfamily of the CDF protein are involved in Zn loading in the vacuoles (Dräger et al., 2004). In the shoot, the vacuolar protein MTP1 and other MTP8 and MTP11 play key roles in Zn sequestration and detoxification (Gustin et al., 2009; Hanikenne et al., 2011).

Given the diversity and complexity of metal transport systems, controversial reports are available in the literature about the impact of elevated concentrations of one ion on uptake of another one. For example, many studies demonstrated that Zn at low concentration increases Cd uptake (Hussain et al., 2016; Huang et al., 2019), while Zn at high concentration decreases

Cd uptake and accumulation (He et al., 2004; Rizwan et al., 2019). This antagonistic behavior can be easily explained by the chemical structure of these metals: Both Zn and Cd are divalent ions and share the same group number in the periodic table. Their similar characteristics may lead to competition for binding sites in the soil, which would affect their uptake in plants (Ming et al., 2016) and on the root surface (Zare et al., 2018). Besides this, a synergistic relation of Zn and Cd has also been observed when they have been applied at low concentration under field conditions (Nan et al., 2002; Köleli et al., 2004; Sharma and Agrawal, 2006). A typical dose-dependent curve model is presented in Fig. 1.2 and consists of two submodels that depict the synergistic and antagonistic interaction of Zn and Cd. For Zn, there are four phases, Zn deficiency, Zn at low concentration, optimal Zn, and Zn toxicity, while Cd has two phases, Cd tolerance and Cd toxicity. In the first submodel (Fig. 1.2A), Zn deficiency results in a high Cd uptake that tends to decline when Zn is available at low concentrations, and Cd concentration becomes negligible under optimal Zn concentration. However, when Zn concentration exceeds the optimal requirement and enters into Zn toxicity range, plants may show Cd toxicity symptoms, but the latter are less severe than Zn toxicity symptoms. In the second submodel (Fig. 1.2B), Cd uptake increases when Zn concentration increases from Zn deficiency to Zn at low concentration and then Cd declines continuously from optimal Zn concentration to the Zn toxicity phase. This putative model requires further validation.

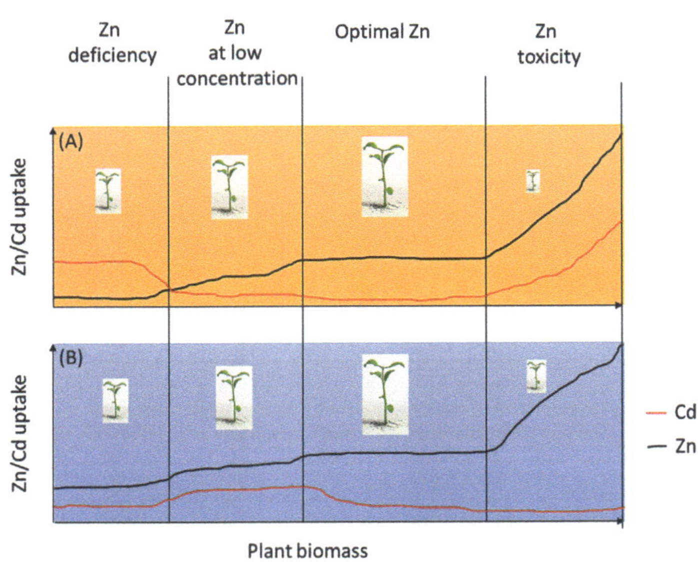

FIGURE 1.2 A suggested model depicting (A) an antagonistic and (B) a synergistic behavior of Zn and Cd homeostasis in the root cell under different Zn and Cd application conditions. See text for details.

The diversity and functionality, spatial and temporal activation, and genetic potential of different plant species should also be reflected in a proposed model. For instance, downregulation of SpHMA3 by RNA interference in *Sedum plumbizincicola* resulted in a marked loss of Cd hypertolerance (Liu et al., 2017). HvHMA1 from barley may also be a broad specificity exporter of Zn^{2+} from the chloroplasts (Mikkelsen et al., 2012). In *Arabidopsis*, AtMTP2 and AtHMA2 act additively in the partitioning and translocation of Zn from roots to shoots and can further enhance the transcriptional activation of ZIP genes; thus Zn concentration of shoots is regulated systemically by HMA2, MTPs, and ZIPs (Sinclair et al., 2018). Other factors such as timing of exposure and plant species (Feng et al., 2013; Hussain et al., 2019); the presence of other competing elements, such as iron (Fe), Mn, or Pb in the soil (Sipos et al., 2008); and the forms and placement of Zn in soil (Zhang and Song, 2006; Sipos et al., 2008; Forster et al., 2018) could also affect interactions between Zn and Cd uptake.

1.4 Arsenic/nitrogen/phosphorus interaction

Arsenic (As) is the 20th most abundant component in the earth's crust and a highly toxic metalloid. As exists either in organic form such as arsenolipids and arsenosugars, or in inorganic form as arsenate (As^{V}) and arsenite (As^{III}); the latter are the most phytoavailable forms of As in the soil solution (Meharg and Hartley-Whitaker, 2002; Ali et al., 2009). In plants, As^{V} is taken up by phosphate transporters (PHT1 and PTH4) (Shin et al., 2004), while As^{III} is taken up by noduline-like intrinsic aquaporin protein (NIP) (Xu et al., 2015). In *Arabidopsis*, the NIP subfamily consists of nine members in the genome. In yeast and oocyte expression systems, NIP5;1, NIP6;1, NIP7;1, and NIP1;2 are capable of As^{III} transport (Bienert et al., 2008; Kamiya et al., 2009). Furthermore, a strong root-growth As^{III}-tolerant phenotype has been found in NIP1;1 loss-of-function mutants (Kamiya et al., 2009). Under anaerobic conditions, plants absorbed mostly As^{III}, and As^{V} is reduced to As^{III} by arsenate reductase using glutathione (GSH) as a reductant and As^{III} can form complexes with thiol groups from glutathione and PCs to lower its cytotoxicity (Fig. 1.1B; Liu et al., 2010; Xu et al., 2015). The complexed As^{III} and inorganic As^{III} are believed to be mostly sequestered into the central vacuole via the arsenite efflux transporter (arsenical compound resistance3 [ACR3]) gene (Chen et al., 2017a,b,c); however, such vacuolar sequestration activity of the ACR3 gene is reported in the As-hyperaccumulating fern, *Pteris vittata* (Indriolo et al., 2010). In As non-hyperaccumulating *Arabidopsis*, PC transporters AtABCC1 and AtABCC2 are responsible for transporting As^{III}-PC2 into vacuoles (Song et al., 2010). Nonetheless the overexpression of PvACR3;1 gene in rice (*Oryza sativa* L.) decreases inorganic As content in rice grains (Chen et al., 2019). The complexation of As^{III} with PCs contributes to As sequestration in roots, whereas

As^{III} is extruded out of cells into the external medium and even into the xylem for root-to-shoot translocation (Liu et al., 2010). As^{III} is then exported to the xylem by the silicon/As^{III} efflux carrier Lsi2 and transported to the shoot (Ma et al., 2008).

Apart from causing direct harm to animal and human health, As exposure also causes detrimental effects on plant growth by interference with uptake of some key mineral nutrients such as nitrogen (N) and P. N is an essential macronutrient for plants and a major limiting factor for plant growth and crop production. Nitrate (NO_3^-) is one of the most abundant sources of N in nature (O'Brien et al., 2016; Hussain et al., 2020). Besides its role as a nutrient, NO_3^- can also act as a signaling molecule that regulates numerous growth-related processes such as root architecture, leaf development (Rahayu et al., 2005), seed germination and dormancy (Alboresi et al., 2005), and fruiting time (Marin et al., 2011). In plants three transport systems exist for NO_3^- transport from soil to plants: (1) a constitutive high-affinity transport system; (2) an inducible high-affinity transport system that works up to the concentration of 500 μM NO_3^-; and (3) a low-affinity transport system that works at concentrations of NO_3^- higher than 500 μM (Glass 2002, 2009; Dechorgnat et al., 2011). Five protein families mediate NO_3^- transport. These include nitrate the transporter 1/peptide transporter family (NPF); nitrate transporter 2 (NRT2); chloride channels (CLC); slow anion channel-associated 1 (SLAC1); and aluminum-activated malate transporter (ALMT) (Léran et al., 2015; Krapp et al., 2014). However, their relative contribution differs dramatically, and the majority of NO_3^- transport is conducted by NPF and NRT2, which encode for the low-affinity transport system (LATS) and high-affinity transport system (HATS), respectively (Nacry et al., 2013). An exception to NPF transporters is NPF6.3 (a 59-amino-acid protein), which acts as a dual-affinity NO_3^- transporter depending on the phosphorylation of threonine residue, Thr101 (Ho et al., 2009; Sun and Zheng, 2015). The functionality of NRT2 largely depends on the presence of protein NAR2 (also known as NRT3.1), which directly interacts with NPT2 proteins (Orsel et al., 2006; Kotur et al., 2012). In *Arabidopsis*, two NAR2 genes are present; however, NAR2.1 is more critical for the activation of NRT2 (Li et al., 2007). In maize, NAR2.1 also interacts with H^+-ATPase activity along with NO_3^- acquisition (Lupini et al., 2016).

Little is known about the interaction of NO_3^- and As^{III}. However, it was observed that low doses of N application under As^{III} stress resulted in the reduction of As concentration in roots (Srivastava et al., 2019). Likewise, in As hyperaccumulator *P. vittata*, As^{III} application resulted in significant decline in NO_3^- levels in roots, due to a reduction in the enzymes involved in N metabolism such as nitrate reductase (Singh et al., 2009) and reduced expression levels of nitrate transporters (Srivastava et al., 2019). At the same time, exogenous application of NO_3^- reduced As^{III} accumulation and increased nitrate reductase activity (Agnihotri and Seth, 2016). Because sulfhydryl (SH) groups are required for NADH binding and catalytic activity of nitrate reductase enzyme (Solomonson and Barber, 1990), arsenic might affect the nitrate reductase activity by binding to functional −SH groups present in the active sites of this enzyme (Sharma and Dubey, 2005; Xiong et al., 2006). As^{III} has been shown to interact with functional SH groups (Schat and Kalff, 2002), thus indicating a possible causal relationship between As and NO_3^-. Another indirect factor that can play a critical role in their interaction is calcium (Ca^{2+}). Upon stress, elevation in the cytosolic Ca^{2+} leads to the activation of calcineurin B-like proteins (CBLs), and CBL-interacting protein kinases (CIPKs) (Tang et al., 2020). These proteins (CBL and CIPK) interact with the transport of NO_3^- and As. For example, in *Arabidopsis* CIPK31 acts as an interacting protein of a NIP1;1 protein and regulates As uptake (Ji et al., 2017). Moreover, the promoter-β-glucuronidase and quantitative real-time PCR analysis revealed CIPK31 displayed overlapping expression profiles with NIP1;1 in the roots, suggesting that these two transporters might operate together in roots (Ji et al., 2017). Moreover, CIPK31 contains a myristoylation site at its N-terminus and this site can recruit CIPK31 by anchoring target NIP1;1 (Cheng et al., 2002; Kamiya et al., 2009). On the other hand, CIPK23 has been shown to phosphorylate Thr101 and allows NRT1.1 to function as a high-affinity NO_3^- transporter under limited NO_3^- supply (Ho et al., 2009). CIPK8 in *Arabidopsis* is involved in glucose sensing and regulates the low-affinity phase of nitrate response involved in long-term nitrate-regulated root growth (Hu et al., 2009). Thus it is plausible to suggest that CBL and CIPK proteins interact with NO_3^- and As transport via a phosphorylation mechanism.

Besides NO_3^-, ammonium (NH_4^+) and ammonia (NH_3) are two other important forms of nitrogen. There is some evidence that the cellular uptake of both NH_4^+ and NH_3 could be mediated by ammonium transporters of the AMT/MEP/Rh (AMT) protein family (Maurel et al., 2015; Ludewig et al., 2007). The crystal structure analysis of AMT protein in *E. coli* showed that the presence of two phenylalanine side chains blocks the entry of NH_4^+ into the pore of AMT protein from the periplasmic side, while presence of two highly conserved histidine (H168 and H318) residues may facilitate the deprotonation of an ammonium ion entering the pore, and the electrostatic barrier between H168 and H318 hinders the permeation of cations but not that of the uncharged NH_3, suggesting a role of the AMT proteins as NH_3 gas channels (Zheng et al., 2004). Along with the AMT family, aquaporins also mediate NH_4^+/NH_3 transport (Holm et al., 2005; Maurel et al., 2015), just like NIPs do for As^{III} uptake. Aquaporins may also transport urea (Soto et al., 2008). TIPs from wheat (TaTIP2;1), maize (ZmTIP1;1 and ZmTIP1;2), and *A. thaliana* (AtTIP2;1 and AtTIP2;3) were shown to facilitate transport of NH_3 across membranes and therefore mediate the loading and acid-trapping of the protonated form (NH_4^+) in the vacuole (Loqué et al., 2005; Bárzana et al., 2014). Expression of AtTIP2;1 and AtTIP2;3 was upregulated by NH_4^+ in *Arabidopsis* roots (Bertl et al., 2007).

In plants, the transportation of substrates across the aquaporin is regulated by two-pore constrictions: (1) asparagine–proline–alanine boxes (which are highly conserved) and (2) aromatic/arginine (ar/R). The first one is involved in proton exclusion (Tajkhorshid et al., 2002), while the second one acts as a selectivity filter (Azad et al., 2012). All NIPs have different pore structures at ar/R selective filters (Abbas et al., 2018). Mutations at this ar/R selectivity filter are thought to determine the broad spectrum of substrate conductance in plant aquaporins (Azad et al., 2012). The extended selectivity filter capacity of TIP was increased by an additional histidine in a different part of AtTIP2;1, which fortifies the position of the arginine and interacts directly with the substrate in the channel (Kirscht et al., 2016). It was also suggested that a water-filled side pore, next to the substrate-binding histidine, participates in deprotonating NH_4^+ ions, which could increase the net permeation of NH_3 (Kirscht et al., 2016). Besides NH_3 transport, mutation in the ar/R filter also impairs the selectivity filter ability of aquaporins to permeate As^{III} across membranes. For instance, similar to As hyperaccumulator *P. vittata*, the permeability to As^{III} uptake and translocation were confirmed for PvTIP4;1 in *A. thaliana*, while substitution of argenine with cysteine in ar/R selectivity filter of PvTIP4;1 made it impermeable to As^{III}. Several aquaporins permeable to As^{III} have been identified. For example, AtNIP1;1, AtNIP1;2, AtNIP3;1, AtNIP5;1, AtNIP6;1, and AtNIP7;1 from *Arabidopsis*, and OsNIP1;1, OsNIP2;1, OsNIP2;2, OsNIP3;1, OsNIP3;2, OsNIP3;3, OsPIP2;4, OsPIP2;6, and OsPIP2;7 from rice, have been reported to show transport activities for As^{III} (Bienert et al., 2008; Ma et al., 2008; Kamiya et al., 2009; Mosa et al., 2012; Katsuhara et al., 2014; Xu et al., 2015; Sun et al., 2018). Thus it appears that the ability of plants to uncouple transport of NH_4^+/NH_3 and As^{III} is causally related to the mutation in the ar/R selectivity filter.

Other than N compounds, P is another important essential mineral nutrient and its deficiency results in tremendous yield loss (Singh et al., 2018; Stangoulis, 2019). One key factor inducing P deficiency in As^V-contaminated soil is the shared transport pathway of P and As^V via phosphate transporters (PHT1 and PHT4). In *Arabidopsis*, AtPht1;1 and AtPht1;4 are known as high- and medium-affinity P transporters, respectively, which also have been shown to uptake As^V (Shin et al., 2004). Moreover, AtPht1;5 is responsible for the transport of P and As^V from source to sink (Nagarajan et al., 2011). A double mutation in AtPHT1;1 resulted in the downregulation of P related genes but led to upregulation of As^V-activated genes (Catarecha et al., 2007), suggesting an antagonistic transport behavior of P and As^V via the same transport system. The overexpression of AtPht1;7 and AtPht1;9 showed higher As^V accumulation in *Arabidopsis* (Remy et al., 2012; LeBlanc et al., 2013). Additionally, the involvement of AtPht1;8 in As^V uptake along with P has been demonstrated (Remy et al., 2012). Cloning of the PvPht1;4 gene from *P. vittata* in transgenic tobacco lines showed that PvPht1;4 potentially detoxifies As by reducing As^{V} to As^{III}, which then makes complex with GSH and stores in the root vacuoles, thereby reducing As translocation in transgenic tobacco (Sun et al., 2019).

1.5 Concluding remarks

Modern agriculture relies heavily on soil fertilization, with 14 inorganic nutrients considered to be essential to plants. However, other minerals are also present in agricultural soils that are considered as nonessential but can, nonetheless, interfere with uptake and translocation of essential nutrients. Due to similarities in their physical and chemical properties, these nonessential nutrients are often transported by the same proteins as essential minerals. This comes with a caveat of either reducing availability of essential nutrients (due to competition for uptake and/or interference), or accumulation of undesirable elements in plant tissues (that could be toxic to both plants and humans). Until now, plant breeders and agronomists have been trying to resolve this conundrum and entangle this interference via empirical means, by experimenting with various ratios of fertilizers and exploring the genetic diversity of modern crops. However, it is becoming increasingly obvious that this approach has some severe limitations and is unlikely to be able to resolve the issue of increasing plant nutrient use efficiency and reduce the resource (fertilizer) input and its environmental footprint to match the required growth in food quantity and quality. Accordingly, these empirical studies should be replaced by a holistic and targeted approach combining electrophysiological studies that enable quantification of transport kinetics of the minerals of interest via transporting proteins with molecular genetics tools allowing the side-specific mutagenesis and/or gene editing of critical amino acids that determine the selectivity and gating properties of transporters.

References

Abbas, G., Murtaza, B., Bibi, I., Shahid, M., Niazi, N.K., Khan, M.I., et al., 2018. Arsenic uptake, toxicity, detoxification, and speciation in plants: physiological, biochemical, and molecular aspects. Int. J. Environ. Res. Public Health 15, 59. Available from: https://doi.org/10.3390/ijerph15010059.

Agnihotri, A., Seth, C.S., 2016. Exogenously applied nitrate improves the photosynthetic performance and nitrogen metabolism in tomato (*Solanum lycopersicum* L. cv Pusa Rohini) under arsenic (V) toxicity. Physiol. Mol. Biol. Plant. 22, 341–349.

Alansary, D., Schmidt, B., Dörr, K., Bogeski, I., Rieger, H., Kless, A., et al., 2016. Thiol dependent intramolecular locking of Orai1 channels. Sci. Rep. 6, 1–11.

Alboresi, A., Gestin, C., Leydecker, M.T., Bedu, M., Meyer, C., Truong, H.N., 2005. Nitrate, a signal relieving seed dormancy in *Arabidopsis*. Plant Cell Environ. 28, 500–512.

Ali, W., Isayenkov, S.V., Zhao, F.J., Maathuis, F.J., 2009. Arsenite transport in plants. Cell. Mol. Life Sci. 66, 2329–2339.

Anjum, S.A., Tanveer, M., Hussain, S., Shahzad, B., Ashraf, U., et al., 2016. Osmoregulation and antioxidant production in maize under combined cadmium and arsenic stress. Environ. Sci. Pollut. Res. 23, 11864–11875.

Assunção, A.G.L., Martins, P.D.C., De Folter, S., Vooijs, R., Schat, H., Aarts, M.G.M., 2001. Elevated expression of metal transporter genes in three accessions of the metal hyperaccumulator *Thlaspi caerulescens*. Plant Cell Environ. 24, 217–226.

Azad, A.K., Yoshikawa, N., Ishikawa, T., Sawa, Y., Shibata, H., 2012. Substitution of a single amino acid residue in the aromatic/arginine selectivity filter alters the transport profiles of tonoplast aquaporin homologs. Biochim. Biophys. Acta 1818, 1–11.

Aziz, I., Khan, M.A., 2001. Effect of seawater on the growth, ion content and water potential of *Rhizophora mucronata* Lam. J. Plant Res. 114, 369–373.

Baker, A., Ceasar, S.A., Palmer, A.J., Paterson, J.B., Qi, W., Muench, S.P., et al., 2015. Replace, reuse, recycle: improving the sustainable use of phosphorus by plants. J. Exp. Bot. 66, 3523–3540.

Bárzana, G., Aroca, R., Bienert, G.P., Chaumont, F., Ruiz-Lozano, J.M., 2014. New insights into the regulation of aquaporins by the arbuscular mycorrhizal symbiosis in maize plants under drought stress and possible implications for plant performance. Mol. Plant Microbe Interact. 27, 349–363.

Bertl, A., Kaldenhoff, R., 2007. Function of a separate NH_3-pore in aquaporin TIP2; 2 from wheat. FEBS Lett. 581, 5413–5417.

Bertl, A., Reid, J.D., Sentenac, H., Slayman, C.L., 1997. Functional comparison of plant inward-rectifier channels expressed in yeast. J. Exp. Bot. 48, 405–413.

Bienert, G.P., Thorsen, M., Schüssler, M.D., Nilsson, H.R., Wagner, A., Tamás, M.J., et al., 2008. A subgroup of plant aquaporins facilitate the bi-directional diffusion of $As(OH)_3$ and $Sb(OH)_3$ across membranes. BMC Biol. 6, 1–15.

Board, J.E., 2008. Waterlogging effects on plant nutrient concentrations in soybean. J. Plant Nutr. 31, 828–838.

Bose, J., Rodrigo-Moreno, A., Shabala, S., 2014. ROS homeostasis in halophytes in the context of salinity stress tolerance. J. Exp. Bot. 65, 1241–1257.

Bose, J., Rodrigo-Moreno, A., Lai, D., Xie, Y., Shen, W., Shabala, S., 2015. Rapid regulation of the plasma membrane H^+-ATPase activity is essential to salinity tolerance in two halophyte species, *Atriplex lentiformis* and *Chenopodium quinoa*. Ann. Bot. 115, 481–494.

Bouis, H.E., Saltzman, A., 2017. Improving nutrition through biofortification: a review of evidence from HarvestPlus, 2003 through 2016. Glob. Food Sec. 12, 49–58.

Bouwman, L., Goldewijk, K.K., Van Der Hoek, K.W., Beusen, A.H., Van Vuuren, D.P., Willems, J., et al., 2013. Exploring global changes in nitrogen and phosphorus cycles in agriculture induced by livestock production over the 1900–2050 period. Proc. Natl. Acad. Sci. U.S.A. 110, 20882–20887.

Britto, D.T., Kronzucker, H.J., 2008. Cellular mechanisms of potassium transport in plants. Physiol. Plant. 133, 637–650.

Cai, H., Huang, S., Che, J., Yamaji, N., Ma, J.F., 2019. The tonoplast-localized transporter OsHMA3 plays an important role in maintaining Zn homeostasis in rice. J. Exp. Bot. 70, 2717–2725.

Cakmak, I., Kutman, U.Á., 2018. Agronomic biofortification of cereals with zinc: a review. Eur. J. Soil Sci. 69, 172–180.

Catarecha, P., Segura, M.D., Franco-Zorrilla, J.M., García-Ponce, B., Lanza, M., Solano, R., et al., 2007. A mutant of the *Arabidopsis* phosphate transporter PHT1; 1 displays enhanced arsenic accumulation. Plant Cell 19, 1123–1133.

Chakraborty, K., Bose, J., Shabala, L., Shabala, S., 2016. Difference in root K^+ retention ability and reduced sensitivity of K^+-permeable channels to reactive oxygen species confer differential salt tolerance in three *Brassica* species. J. Exp. Bot. 67, 4611–4625.

Chen, Z., Newman, I., Zhou, M., Mendham, N., Zhang, G., Shabala, S., 2005. Screening plants for salt tolerance by measuring K^+ flux: a case study for barley. Plant Cell Environ. 28 (10), 1230–1246.

Chen, Z., Pottosin, I.I., Cuin, T.A., Fuglsang, A.T., Tester, M., Jha, D., et al., 2007. Root plasma membrane transporters controlling K^+/Na^+ homeostasis in salt-stressed barley. Plant Physiol. 145, 1714–1725.

Chen, C., Xia, S., Deng, R., Liu, C., Shi, G., 2017a. *AhIRT1* and *AhNRAMP1* metal transporter expression correlates with Cd uptake in peanuts under iron deficiency. PLoS One 12, e0185144.

Chen, Y., Hua, C.Y., Jia, M.R., Fu, J.W., Liu, X., Han, Y.H., et al., 2017b. Heterologous expression of *Pteris vittata* arsenite antiporter PvACR3; 1 reduces arsenic accumulation in plant shoots. Environ. Sci. Technol. 51, 10387–10395.

Chen, Y., Sun, S.K., Tang, Z., Liu, G., Moore, K.L., Maathuis, F.J., et al., 2017c. The Nodulin 26-like intrinsic membrane protein OsNIP3; 2 is involved in arsenite uptake by lateral roots in rice. J. Exp. Bot. 68, 3007–3016.

Chen, Y., Hua, C.Y., Chen, J.X., Rathinasabapathi, B., Cao, Y., Ma, L.Q., 2019. Expressing arsenite antiporter PvACR3; 1 in rice (*Oryza sativa* L.) decreases inorganic arsenic content in rice grains. Environ. Sci. Technol. 53, 10062–10069.

Cheng, S.H., Willmann, M.R., Chen, H.C., Sheen, J., 2002. Calcium signaling through protein kinases. The *Arabidopsis* calcium-dependent protein kinase gene family. Plant Physiol. 129, 469–485.

Childers, D.L., Corman, J., Edwards, M., Elser, J.J., 2011. Sustainability challenges of phosphorus and food: solutions from closing the human phosphorus cycle. Bioscience 61, 117–124.

Clemens, S., 2006. Toxic metal accumulation, responses to exposure and mechanisms of tolerance in plants. Biochimie 88, 1707–1719.

Close, D.C., Davidson, N.J., 2003. Long-term waterlogging: nutrient, gas exchange, photochemical and pigment characteristics of *Eucalyptus nitens* saplings. Russ. J. Plant Physiol. 50, 843–847.

Courbot, M., Willems, G., Motte, P., Arvidsson, S., Roosens, N., Saumitou-Laprade, P., et al., 2007. A major quantitative trait locus for cadmium tolerance in *Arabidopsis halleri* colocalizes with HMA4, a gene encoding a heavy metal ATPase. Plant Physiol. 144, 1052–1065.

Cuin, T.A., Betts, S.A., Chalmandrier, R., Shabala, S., 2008. A root's ability to retain K^+ correlates with salt tolerance in wheat. J. Exp. Bot. 59, 2697–2706.

Daneshgar, S., Buttafava, A., Capsoni, D., Callegari, A., Capodaglio, A.G., 2018. Impact of pH and ionic molar ratios on phosphorous forms precipitation and recovery from different wastewater sludges. Resources 7, 71. Available from: https://doi.org/10.3390/resources7040071.

Dechorgnat, J., Nguyen, C.T., Armengaud, P., Jossier, M., Diatloff, E., Filleur, S., et al., 2011. From the soil to the seeds: the long journey of nitrate in plants. J. Exp. Bot. 62, 1349–1359.

Demidchik, V., Maathuis, F.J., 2007. Physiological roles of nonselective cation channels in plants: from salt stress to signalling and development. New Phytol. 175, 387–404.

Demidchik, V., Shabala, S.N., Davies, J.M., 2007. Spatial variation in H_2O_2 response of *Arabidopsis thaliana* root epidermal Ca^{2+} flux and plasma membrane Ca^{2+} channels. Plant J. 49 (3), 377–386.

Demidchik, V., Cuin, T.A., Svistunenko, D., Smith, S.J., Miller, A.J., Shabala, S., et al., 2010. *Arabidopsis* root K^+-efflux conductance activated by hydroxyl radicals: single-channel properties, genetic basis and involvement in stress-induced cell death. J. Cell Sci. 123, 1468–1479.

Do, E., Hu, G., Caza, M., Kronstad, J.W., Jung, W.H., 2016. The ZIP family zinc transporters support the virulence of *Cryptococcus neoformans*. Sabouraudia 54, 605–615.

Dräger, D.B., Desbrosses-Fonrouge, A.G., Krach, C., Chardonnens, A.N., Meyer, R.C., Saumitou-Laprade, P., et al., 2004. Two genes encoding *Arabidopsis halleri* MTP1 metal transport proteins co-segregate with zinc tolerance and account for high MTP1 transcript levels. Plant J. 39, 425–439.

Dreyer, I., Uozumi, N., 2011. Potassium channels in plant cells. FEBS J. 278, 4293–4303.

Eapen, S., D'souza, S.F., 2005. Prospects of genetic engineering of plants for phytoremediation of toxic metals. Biotechnol. Adv. 23, 97–114.

Eide, D.J., 2005. The zip family of zinc transporters. In: Iuchi, S., Kuldell, N. (Eds.), Zinc Finger Proteins. Molecular Biology Intelligence Unit. Boston, MA, Springer. Available from: https://doi.org/10.1007/0-387-27421-9_35.

Falhof, J., Pedersen, J.T., Fuglsang, A.T., Palmgren, M., 2016. Plasma membrane H^+-ATPase regulation in the center of plant physiology. Mol. Plant 9, 323–337.

FAO, 2017. The Future of Food and Agriculture: Trends and Challenges. FAO, Rome.

Feng, R., Qiu, W., Lian, F., Yu, Z., Yang, Y., Song, Z., 2013. Field evaluation of in situ remediation of Cd-contaminated soil using four additives, two foliar fertilisers and two varieties of pakchoi. J. Environ. Manage. 124, 17–24.

Fita, A., Rodríguez-Burruezo, A., Boscaiu, M., Prohens, J., Vicente, O., 2015. Breeding and domesticating crops adapted to drought and salinity: a new paradigm for increasing food production. Front. Plant Sci. 6, 978. Available from: https://doi.org/10.3389/fpls.2015.00978.

Flowers, T.J., Galal, H.K., Bromham, L., 2010. Evolution of halophytes: multiple origins of salt tolerance in land plants. Funct. Plant Biol. 37, 604–612.

Forster, S.M., Rickertsen, J.R., Mehring, G.H., Ransom, J.K., 2018. Type and placement of zinc fertilizer impacts cadmium content of harvested durum wheat grain. J. Plant Nutr. 41, 1471–1481.

Frassinetti, S., Bronzetti, G.L., Caltavuturo, L., Cini, M., Della Croce, C., 2006. The role of zinc in life: a review. J. Environ. Pathol. Toxicol. Oncol. 25, 597–610.

Garcia-Mata, C., Wang, J., Gajdanowicz, P., Gonzalez, W., Hills, A., Donald, N., et al., 2010. A minimal cysteine motif required to activate the SKOR K^+ channel of *Arabidopsis* by the reactive oxygen species H_2O_2. J. Biol. Chem. 285, 29286–29294.

Ghassemi, F., Jakeman, A.J., Nix, H.A., 1995. Salinisation of Land and Water Resources: Human Causes, Extent, Management, and Case Studies. CAB International, Wallingford.

Gibson, R.S., Heath, A.L., Ferguson, E.L., 2002. Risk of suboptimal iron and zinc nutriture among adolescent girls in Australia and New Zealand: causes, consequences, and solutions. Asia Pac. J. Clin. Nutr. 11, 543–552.

Glass, A.D., 2009. Nitrate uptake by plant roots. Botany 87, 659–667.

Glass, A.D., Britto, D.T., Kaiser, B.N., Kinghorn, J.R., Kronzucker, H.J., Kumar, A., et al., 2002. The regulation of nitrate and ammonium transport systems in plants. J. Exp. Bot. 53, 855–864.

Guo, Q., Tian, X., Mao, P., Meng, L., 2019. Functional characterization of IlHMA2, a P_{1B2}-ATPase in *Iris lactea* response to Cd. Environ. Exp. Bot. 157, 131–139.

Gustin, J.L., Loureiro, M.E., Kim, D., Na, G., Tikhonova, M., Salt, D.E., 2009. MTP1-dependent Zn sequestration into shoot vacuoles suggests dual roles in Zn tolerance and accumulation in Zn-hyperaccumulating plants. Plant J. 57, 1116–1127.

Hamam, A.M., Coskun, D., Britto, D.T., Plett, D., Kronzucker, H.J., 2019.). Plasma-membrane electrical responses to salt and osmotic gradients contradict radiotracer kinetics and reveal Na^+-transport dynamics in rice (*Oryza sativa* L.). Planta 249, 1037–1051.

Hammond, J.P., Bowen, H.C., White, P.J., Mills, V., Pyke, K.A., Baker, A.J., et al., 2006. A comparison of the *Thlaspi caerulescens* and *Thlaspi arvense* shoot transcriptomes. New Phytol. 170, 239–260.

Hanikenne, M., Nouet, C., 2011. Metal hyperaccumulation and hypertolerance: a model for plant evolutionary genomics. Curr. Opin. Plant Biol. 14, 252–259.

Hanikenne, M., Talke, I.N., Haydon, M.J., Lanz, C., Nolte, A., Motte, P., et al., 2008. Evolution of metal hyperaccumulation required cis-regulatory changes and triplication of HMA_4. Nature 453, 391–395.

He, P.P., Lv, X.Z., Wang, G.Y., 2004. Effects of Se and Zn supplementation on the antagonism against Pb and Cd in vegetables. Environ. Int. 30, 167–172.

Holm, L.M., Jahn, T.P., Møller, A.L., Schjoerring, J.K., Ferri, D., Klaerke, D.A., et al., 2005. NH_3 and NH_4^+ permeability in aquaporin-expressing Xenopus oocytes. Pflügers Arch. 450, 415–428.

Ho, C.H., Lin, S.H., Hu, H.C., Tsay, Y.F., 2009. CHL1 functions as a nitrate sensor in plants. Cell 138, 1184–1194.

Hu, Y., Schmidhalter, U., 2005. Drought and salinity: a comparison of their effects on mineral nutrition of plants. J. Plant Nutr. Soil Sci. 168, 541–549.

Hu, L., Wang, Z., Huang, B., 2013. Effects of cytokinin and potassium on stomatal and photosynthetic recovery of Kentucky bluegrass from drought stress. Crop Sci. 53, 221–231.

Hu, H.C., Wang, Y.Y., Tsay, Y.F., 2009. AtCIPK8, a CBL-interacting protein kinase, regulates the low-affinity phase of the primary nitrate response. Plant J. 57, 264–278.

Huang, G., Ding, C., Zhou, Z., Zhang, T., Wang, X., 2019.). A tillering application of zinc fertilizer based on basal stabilization reduces Cd accumulation in rice (*Oryza sativa* L.). Ecotoxicol. Environ. Saf. 167, 338–344.

Huang, X., Duan, S., Wu, Q., Yu, M., Shabala, S., 2020. Reducing cadmium accumulation in plants: structure–function relations and tissue-specific operation of transporters in the spotlight. Plants 9, 223. Available from: https://doi.org/10.3390/plants9020223.

Hussain, T., Murtaza, G., Ghafoor, A., Cheema, M.A., 2016. The Cd: Zn ratio in a soil affects Cd toxicity in spinach (*Spinacea oleracea* L.). Pak. J. Agric. Sci. 53, 419–424.

Hussain, S., Khan, A.M., Rengel, Z., 2019. Zinc-biofortified wheat accumulates more cadmium in grains than standard wheat when grown on cadmium-contaminated soil regardless of soil and foliar zinc application. Sci. Total Environ. 654, 402–408.

Hussain, S., Khaliq, A., Noor, M.A., Tanveer, M., Hussain, H.A., Hussain, S., et al., 2020. Metal toxicity and nitrogen metabolism in plants: an overview. In: Datta, R., Meena, R., Pathan, S., Ceccherini, M. (Eds.), Carbon and Nitrogen Cycling in Soil. Springer, Singapore, pp. 221–248. Available from: https://doi.org/10.1007/978-981-13-7264-3_7.

Indriolo, E., Na, G., Ellis, D., Salt, D.E., Banks, J.A., 2010. A vacuolar arsenite transporter necessary for arsenic tolerance in the arsenic hyperaccumulating fern *Pteris vittata* is missing in flowering plants. Plant Cell 22, 2045–2057.

Ji, R., Zhou, L., Liu, J., Wang, Y., Yang, L., Zheng, Q., et al., 2017. Calcium-dependent protein kinase CPK31 interacts with arsenic transporter AtNIP1; 1 and regulates arsenite uptake in *Arabidopsis thaliana*. PLoS One 12, e0173681.

Kamiya, T., Tanaka, M., Mitani, N., Ma, J.F., Maeshima, M., Fujiwara, T., 2009. NIP1; 1, an aquaporin homolog, determines the arsenite sensitivity of *Arabidopsis thaliana*. J. Biol. Chem. 284, 2114–2120.

Katsuhara, M., Sasano, S., Horie, T., Matsumoto, T., Rhee, J., Shibasaka, M., 2014. Functional and molecular characteristics of rice and barley NIP aquaporins transporting water, hydrogen peroxide and arsenite. Plant Biotechnol. 31, 213–219.

Kavamura, V.N., Esposito, E., 2010. Biotechnological strategies applied to the decontamination of soils polluted with heavy metals. Biotechnol. Adv. 28, 61–69.

Kawachi, M., Kobae, Y., Kogawa, S., Mimura, T., Krämer, U., Maeshima, M., 2012. Amino acid screening based on structural modelling identifies critical residues for the function, ion selectivity and structure of *Arabidopsis* MTP1. FEBS J. 279, 2339–2356.

Kawachi, M., Kobae, Y., Mori, H., Tomioka, R., Lee, Y., Maeshima, M., 2009. A mutant strain *Arabidopsis thaliana* that lacks vacuolar membrane zinc transporter MTP1 revealed the latent tolerance to excessive zinc. Plant Cell Physiol. 50, 1156–1170.

Kirscht, A., Kaptan, S.S., Bienert, G.P., Chaumont, F., Nissen, P., de Groot, B.L., et al., 2016. Crystal structure of an ammonia-permeable aquaporin. PLoS Biol. 14, e1002411.

Kobae, Y., Uemura, T., Sato, M.H., Ohnishi, M., Mimura, T., Nakagawa, T., et al., 2004. Zinc transporter of *Arabidopsis thaliana* AtMTP1 is localized to vacuolar membranes and implicated in zinc homeostasis. Plant Cell Physiol. 45, 1749–1758.

Köleli, N., Eker, S., Cakmak, I., 2004. Effect of zinc fertilization on cadmium toxicity in durum and bread wheat grown in zinc-deficient soil. Environ. Pollut. 131, 453–459.

Kotur, Z., Mackenzie, N., Ramesh, S., Tyerman, S.D., Kaiser, B.N., Glass, A.D., 2012. Nitrate transport capacity of the *Arabidopsis thaliana* NRT2 family members and their interactions with AtNAR2. 1. New Phytol. 194, 724–731.

Krämer, U., Talke, I.N., Hanikenne, M., 2007. Transition metal transport. FEBS Lett. 581, 2263–2272.

Krapp, A., David, L.C., Chardin, C., Girin, T., Marmagne, A., Leprince, A.S., et al., 2014. Nitrate transport and signalling in *Arabidopsis*. J. Exp. Bot. 65, 789–798.

Langhans, M., Ratajczak, R., Lützelschwab, M., Michalke, W., Wächter, R., Fischer-Schliebs, E., et al., 2001. Immunolocalization of plasma-membrane H^+-ATPase and tonoplast-type pyrophosphatase in the plasma membrane of the sieve element-companion cell complex in the stem of *Ricinus communis* L. Planta 213, 11–19.

LeBlanc, M.S., McKinney, E.C., Meagher, R.B., Smith, A.P., 2013. Hijacking membrane transporters for arsenic phytoextraction. J. Biotechnol. 163, 1–9.

Léran, S., Garg, B., Boursiac, Y., Corratgé-Faillie, C., Brachet, C., Tillard, P., et al., 2015. AtNPF5. 5, a nitrate transporter affecting nitrogen accumulation in Arabidopsis embryo. Sci. Rep. 5, 1–7.

Li, W., Wang, Y., Okamoto, M., Crawford, N.M., Siddiqi, M.Y., Glass, A.D., 2007. Dissection of the AtNRT2. 1: AtNRT2. 2 inducible high-affinity nitrate transporter gene cluster. Plant Physiol. 143, 425–433.

Lin, Y.F., Aarts, M.G., 2012. The molecular mechanism of zinc and cadmium stress response in plants. Cell. Mol. Life Sci. 69, 3187–3206.

Liu, W.J., Wood, B.A., Raab, A., McGrath, S.P., Zhao, F.J., Feldmann, J., 2010. Complexation of arsenite with phytochelatins reduces arsenite efflux and translocation from roots to shoots in *Arabidopsis*. Plant Physiol. 152, 2211–2221.

Liu, H., Zhao, H.X., Wu, L.H., Xu, W.Z., 2017. A genetic transformation method for cadmium hyperaccumulator *Sedum plumbizincicola* and non-hyperaccumulating ecotype of *Sedum alfredii*. Front. Plant Sci. 8, 1047. Available from: https://doi.org/10.3389/fpls.2017.01047.

Liu, Y., Yu, Y., Sun, J., Cao, Q., Tang, Z., Liu, M., et al., 2019. Root-zone-specific sensitivity of K^+-and Ca^{2+}-permeable channels to H_2O_2 determines ion homeostasis in salinized diploid and hexaploid *Ipomoea trifida*. J. Exp. Bot. 70, 1389–1405.

Liu, M., Pan, T., Allakhverdiev, S.I., Yu, M., Shabala, S., 2020. Crop halophytism: an environmentally sustainable solution for global food security. Trends Plant Sci. 25, 630–634.

Loqué, D., Ludewig, U., Yuan, L., von Wirén, N., 2005. Tonoplast intrinsic proteins AtTIP2; 1 and AtTIP2; 3 facilitate NH_3 transport into the vacuole. Plant Physiol. 137, 671–680.

Ludewig, U., Neuhäuser, B., Dynowski, M., 2007. Molecular mechanisms of ammonium transport and accumulation in plants. FEBS Lett. 581, 2301–2308.

Lupini, A., Mercati, F., Araniti, F., Miller, A.J., Sunseri, F., Abenavoli, M.R., 2016. NAR2. 1/ NRT2. 1 functional interaction with NO_3-and H^+ fluxes in high-affinity nitrate transport in maize root regions. Plant Physiol. Biochem. 102, 107–114.

Ma, J.F., Yamaji, N., Mitani, N., Xu, X.Y., Su, Y.H., McGrath, S.P., et al., 2008. Transporters of arsenite in rice and their role in arsenic accumulation in rice grain. Proc. Natl. Acad. Sci. U.S.A. 105, 9931–9935.

Maeshima, M., 2000. Vacuolar H^+-pyrophosphatase. Biochim. Biophys. Acta 1465, 37–51.

Marin, I.C., Loef, I., Bartetzko, L., Searle, I., Coupland, G., Stitt, M., et al., 2011. Nitrate regulates floral induction in *Arabidopsis*, acting independently of light, gibberellin and autonomous pathways. Planta 233, 539–552.

Mäser, P., Thomine, S., Schroeder, J.I., Ward, J.M., Hirschi, K., Sze, H., et al., 2001. Phylogenetic relationships within cation transporter families of *Arabidopsis*. Plant Physiol. 126, 1646–1667.

Maurel, C., Boursiac, Y., Luu, D.T., Santoni, V., Shahzad, Z., Verdoucq, L., 2015. Aquaporins in plants. Physiol. Rev. 95, 1321–1358.

Meharg, A.A., Hartley-Whitaker, J., 2002. Arsenic uptake and metabolism in arsenic resistant and non-resistant plant species. New Phytol. 154, 29–43.

Mendoza-Cózatl, D.G., Jobe, T.O., Hauser, F., Schroeder, J.I., 2011. Long-distance transport, vacuolar sequestration, tolerance, and transcriptional responses induced by cadmium and arsenic. Curr. Opin. Plant Biol. 14, 554–562.

Migocka, M., Papierniak, A., Kosieradzka, A., Posyniak, E., Maciaszczyk-Dziubinska, E., Biskup, R., et al., 2015. Retracted: Cucumber metal tolerance protein *CsMTP9* is a plasma membrane H^+-coupled antiporter involved in the Mn^{2+} and Cd^{2+} efflux from root cells. Plant J. 84, 1045–1058.

Mikkelsen, M.D., Pedas, P., Schiller, M., Vincze, E., Mills, R.F., Borg, S., et al., 2012. Barley HvHMA1 is a heavy metal pump involved in mobilizing organellar Zn and Cu and plays a role in metal loading into grains. PLoS One 7, e49027. Available from: https://doi.org/10.1371/journal.pone.0049027.

Miller, G., Suzuki, N., Ciftci-Yilmaz, S., Mittler, R., 2010. Reactive oxygen species homeostasis and signalling during drought and salinity stresses. Plant Cell Environ. 33, 453–467.

Ming, H., Naidu, R., Sarkar, B., Lamb, D.T., Liu, Y., Megharaj, M., et al., 2016. Competitive sorption of cadmium and zinc in contrasting soils. Geoderma 268, 60–68.

Miransari, M., 2011. Hyperaccumulators, arbuscular mycorrhizal fungi and stress of heavy metals. Biotechnol. Adv. 29, 645–653.

Mosa, K.A., Kumar, K., Chhikara, S., Mcdermott, J., Liu, Z., Musante, C., et al., 2012. Members of rice plasma membrane intrinsic proteins subfamily are involved in arsenite permeability and tolerance in plants. Transgenic Res. 21, 1265–1277.

Munns, R., Day, D.A., Fricke, W., Watt, M., Arsova, B., Barkla, B.J., et al., 2020. Energy costs of salt tolerance in crop plants. New Phytol. 225, 1072–1090.

Nacry, P., Bouguyon, E., Gojon, A., 2013. Nitrogen acquisition by roots: physiological and developmental mechanisms ensuring plant adaptation to a fluctuating resource. Plant Soil 370, 1–29.

Nagarajan, V.K., Jain, A., Poling, M.D., Lewis, A.J., Raghothama, K.G., Smith, A.P., 2011. *Arabidopsis* Pht1; 5 mobilizes phosphate between source and sink organs and influences the interaction between phosphate homeostasis and ethylene signalling. Plant Physiol. 156, 1149–1163.

Nan, Z., Li, J., Zhang, J., Cheng, G., 2002. Cadmium and zinc interactions and their transfer in soil-crop system under actual field conditions. Sci. Total Environ. 285, 187–195.

Nieves-Cordones, M., Martinez, V., Benito, B., Rubio, F., 2016. Comparison between *Arabidopsis* and rice for main pathways of K^+ and Na^+ uptake by roots. Front. Plant Sci. 7, 992. Available from: https://doi.org/10.3389/fpls.2016.00992.

O'Brien, J.A., Vega, A., Bouguyon, E., Krouk, G., Gojon, A., Coruzzi, G., et al., 2016. Nitrate transport, sensing, and responses in plants. Mol. Plant 9, 837–856.

Orsel, M., Chopin, F., Leleu, O., Smith, S.J., Krapp, A., Daniel-Vedele, F., et al., 2006. Characterization of a two-component high-affinity nitrate uptake system in *Arabidopsis*. physiology and protein–protein interaction. Plant Physiol. 142, 1304–1317.

Palmgren, M.G., Nissen, P., 2011. P-type ATPases. Annu. Rev. Biophys. 40, 243–266.

Palmgren, M.G., Edenbrandt, A.K., Vedel, S.E., Andersen, M.M., Landes, X., Østerberg, J.T., et al., 2015. Are we ready for back-to-nature crop breeding? Trends Plant Sci. 20, 155–164.

Pilon, M., Cohu, C.M., Ravet, K., Abdel-Ghany, S.E., Gaymard, F., 2009. Essential transition metal homeostasis in plants. Curr. Opin. Plant Biol. 12, 347–357.

Podar, D., Scherer, J., Noordally, Z., Herzyk, P., Nies, D., Sanders, D., 2012. Metal selectivity determinants in a family of transition metal transporters. J. Biol. Chem. 287, 3185–3196.

Prankel, S.H., Nixon, R.M., Phillips, C., 2005. Implications for the human food chain of models of cadmium accumulation in sheep. Environ. Res. 97, 348–358.

Qadir, M., Quillerou, E., Nangia, V., Murtaza, G., Singh, M., Thomas, R.J., et al., 2014. Economics of salt-induced land degradation and restoration. Nat. Resour. Forum 38, 282–295.

Rahayu, Y.S., Walch-Liu, P., Neumann, G., Romheld, V., von Wiren, N., Bangerth, F., 2005. Root-derived cytokinins as long-distance signals for NO_3-induced stimulation of leaf growth. J. Exp. Bot. 56, 1143–1152.

Ratajczak, R., Hinz, G., Robinson, D.G., 1999. Localization of pyrophosphatase in membranes of cauliflower inflorescence cells. Planta 208, 205–211.

Remy, E., Cabrito, T.R., Batista, R.A., Teixeira, M.C., Sá-Correia, I., Duque, P., 2012. The Pht1;9 and Pht1;8 transporters mediate inorganic phosphate acquisition by the *Arabidopsis thaliana* root during phosphorus starvation. New Phytol. 195, 356–371.

Rizwan, M., Ali, S., Ur Rehman, M.Z., Maqbool, A., 2019. A critical review on the effects of zinc at toxic levels of cadmium in plants. Environ. Sci. Pollut. Res. 26, 6279–6289.

Ruan, C.J., da Silva, J.A.T., Mopper, S., Qin, P., Lutts, S., 2010. Halophyte improvement for a salinized world. Crit. Rev. Plant Sci. 29, 329–359.

Rubio, F., Santa-María, G.E., Rodríguez-Navarro, A., 2000. Cloning of *Arabidopsis* and barley cDNAs encoding HAK potassium transporters in root and shoot cells. Physiol. Plant. 109, 34–43.

Rubio, F., Nieves-Cordones, M., Horie, T., Shabala, S., 2020. Doing 'business as usual' comes with a cost: evaluating energy cost of maintaining plant intracellular K^+ homeostasis under saline conditions. New Phytol. 225, 1097–1104.

Santa-María, G.E., Rubio, F., Dubcovsky, J., Rodríguez-Navarro, A., 1997. The HAK1 gene of barley is a member of a large gene family and encodes a high-affinity potassium transporter. Plant Cell 9, 2281–2289.

Schachtman, D.P., Kumar, R., Schroeder, J.I., Marsh, E.L., 1997. Molecular and functional characterization of a novel low-affinity cation transporter (LCT1) in higher plants. Proc. Natl. Acad. Sci. U.S.A. 94, 11079–11084.

Schat, H., Kalff, M.M.A., 2002. Are phytochelatins involved in differential metal tolerance or do they merely reflect metal-imposed strains? Plant Physiol. 99, 1475–1480.

Shabala, S., 2009. Salinity and programmed cell death: unravelling mechanisms for ion specific signalling. J. Exp. Bot. 60, 709–712.

Shabala, S., 2019. Linking ploidy level with salinity tolerance: NADPH-dependent 'ROS–Ca^{2+} hub' in the spotlight. J. Exp. Bot. 70, 1063–1067.

Shabala, S., Cuin, T.A., 2008. Potassium transport and plant salt tolerance. Physiol. Plant. 133, 651–669.

Shabala, S., Pottosin, I., 2014. Regulation of potassium transport in plants under hostile conditions: implications for abiotic and biotic stress tolerance. Physiol. Plant. 151, 257–279.

Shabala, S., Shabala, L., Van Volkenburgh, E., 2003. Effect of calcium on root development and root ion fluxes in salinised barley seedlings. Funct. Plant Biol. 30, 507–514.

Shabala, L., Cuin, T.A., Newman, I.A., Shabala, S., 2005. Salinity-induced ion flux patterns from the excised roots of *Arabidopsis* SOS mutants. Planta 222, 1041–1050.

Shabala, L., Zhang, J., Pottosin, I., Bose, J., Zhu, M., Fuglsang, A.T., et al., 2016a. Cell-type-specific H^{+}-ATPase activity in root tissues enables K^{+} retention and mediates acclimation of barley (*Hordeum vulgare*) to salinity stress. Plant Physiol. 172 (4), 2445–2458.

Shabala, S., Bose, J., Fuglsang, A.T., Pottosin, I., 2016b. On a quest for stress tolerance genes: membrane transporters in sensing and adapting to hostile soils. J. Exp. Bot. 67, 1015–1031.

Sharma, P., Dubey, R.S., 2005. Modulation of nitrate reductase activity in rice seedlings under aluminium toxicity and water stress: role of osmolytes as enzyme protectant. J. Plant Physiol. 162, 854–864.

Sharma, R.K., Agrawal, M., 2006. Single and combined effects of cadmium and zinc on carrots: uptake and bioaccumulation. J. Plant Nutr. 29, 1791–1804.

Shin, H., Shin, H.S., Dewbre, G.R., Harrison, M.J., 2004. Phosphate transport in *Arabidopsis*: Pht1;1 and Pht1;4 play a major role in phosphate acquisition from both low- and high-phosphate environments. Plant J. 39, 629–642.

Sinclair, S.A., Senger, T., Talke, I.N., Cobbett, C.S., Haydon, M.J., Krämer, U., 2018. Systemic upregulation of MTP2-and HMA2-mediated Zn partitioning to the shoot supplements local Zn deficiency responses. Plant Cell 30, 2463–2479.

Singh, N., Ma, L.Q., Vu, J.C., Raj, A., 2009. Effects of arsenic on nitrate metabolism in arsenic hyperaccumulating and non-hyperaccumulating ferns. Environ. Pollut. 157, 2300–2305.

Jaiswal, S., Sheokand, S., Duhan, S., 2018. Morpho-physiological and oxidative responses of nitrogen and phosphorus deficiency in wheat (*Triticum aestivum* L.). Indian J. Agric. Res. 52, 40–45.

Sipos, P., Németh, T., Kis, V.K., Mohai, I., 2008. Sorption of copper, zinc and lead on soil mineral phases. Chemosphere 73, 461–469.

Smethurst, C.F., Garnett, T., Shabala, S., 2005. Nutritional and chlorophyll fluorescence responses of lucerne (*Medicago sativa*) to waterlogging and subsequent recovery. Plant Soil 270, 31–45.

Solomonson, L.P., Barber, M.J., 1990. Assimilatory nitrate reductase; functional properties and regulation. Annu. Rev. Plant Physiol. Mol. Biol. 41, 225–253.

Song, W.Y., Park, J., Mendoza-Cózatl, D.G., Suter-Grotemeyer, M., Shim, D., Hörtensteiner, S., et al., 2010. Arsenic tolerance in *Arabidopsis* is mediated by two ABCC-type phytochelatin transporters. Proc. Natl. Acad. Sci. U.S.A. 107, 21187–21192.

Song, Y., Hudek, L., Freestone, D., Puhui, J., Michalczyk, A.A., Senlin, Z., et al., 2014. Comparative analyses of cadmium and zinc uptake correlated with changes in natural resistance-associated macrophage protein (NRAMP) expression in *Solanum nigrum* L. and *Brassica rapa*. Environ. Chem. 11, 653–660.

Soto, G., Alleva, K., Mazzella, M.A., Amodeo, G., Muschietti, J.P., 2008. AtTIP1; 3 and AtTIP5; 1, the only highly expressed *Arabidopsis* pollen-specific aquaporins, transport water and urea. FEBS Lett. 582, 4077–4082.

Springmann, M., Clark, M., Mason-D'Croz, D., Wiebe, K., Bodirsky, B.L., Lassaletta, L., et al., 2018. Options for keeping the food system within environmental limits. Nature 562, 519–525.

Srivastava, S., Pathare, V.S., Sounderajan, S., Suprasanna, P., 2019. Nitrogen supply influences arsenic accumulation and stress responses of rice (*Oryza sativa* L.) seedlings. J. Hazard. Mater. 367, 599–606.

Stangoulis, J., 2019. Variation in root system architecture and morphology of two wheat genotypes is a predictor of their tolerance to phosphorus deficiency. Acta Physiol. Plant. 41, 1–13.

Sun, J., Zheng, N., 2015. Molecular mechanism underlying the plant NRT1. 1 dual-affinity nitrate transporter. Front. Physiol. 6, 386. Available from: https://doi.org/10.3389/fphys.2015.00386.

Sun, S.K., Chen, Y., Che, J., Konishi, N., Tang, Z., Miller, A.J., et al., 2018. Decreasing arsenic accumulation in rice by overexpressing OsNIP 1; 1 and OsNIP 3; 3 through disrupting arsenite radial transport in roots. New Phytol. 219, 641–653.

Sun, D., Feng, H., Li, X., Ai, H., Sun, S., Chen, Y., et al., 2019. Expression of new *Pteris vittata* phosphate transporter PvPht1; 4 reduces arsenic translocation from the roots to shoots in tobacco plants. Environ. Sci. Technol. 54, 1045–1053.

Suzuki, M., Bashir, K., Inoue, H., Takahashi, M., Nakanishi, H., Nishizawa, N.K., 2012. Accumulation of starch in Zn-deficient rice. Rice 5, 1–8.

Tajkhorshid, E., Nollert, P., Jensen, M.Ø., Miercke, L.J., O'Connell, J., Stroud, R.M., et al., 2002. Control of the selectivity of the aquaporin water channel family by global orientational tuning. Science 296, 525–530.

Takahashi, R., Bashir, K., Ishimaru, Y., Nishizawa, N.K., Nakanishi, H., 2012a. The role of heavy-metal ATPases, HMAs, in zinc and cadmium transport in rice. Plant Signal. Behav. 7, 1605–1607.

Takahashi, R., Ishimaru, Y., Shimo, H., Ogo, Y., Senoura, T., Nishizawa, N.K., et al., 2012b. The OsHMA2 transporter is involved in root-to-shoot translocation of Zn and Cd in rice. Plant Cell Environ. 35, 1948–1957.

Tan, J., Wang, J., Chai, T., Zhang, Y., Feng, S., Li, Y., et al., 2013. Functional analyses of TaHMA2, a P 1B-type ATPase in wheat. Plant Biotechnol. J. 11, 420–431.

Tang, R.J., Wang, C., Li, K., Luan, S., 2020. The CBL–CIPK calcium signalling network: unified paradigm from 20 years of discoveries. Trends Plant Sci. 25, 604–617.

Tavakkoli, E., Fatehi, F., Coventry, S., Rengasamy, P., McDonald, G.K., 2011. Additive effects of Na^+ and Cl^- ions on barley growth under salinity stress. J. Exp. Bot. 62, 2189–2203.

Ueno, D., Yamaji, N., Kono, I., Huang, C.F., Ando, T., Yano, M., et al., 2010. Gene limiting cadmium accumulation in rice. Proc. Natl. Acad. Sci. U.S.A. 107, 16500–16505.

Ueno, D., Koyama, E., Yamaji, N., Ma, J.F., 2011. Physiological, genetic, and molecular characterization of a high-Cd-accumulating rice cultivar, Jarjan. J. Exp. Bot. 62, 2265–2272.

Ueno, D., Sasaki, A., Yamaji, N., Miyaji, T., Fujii, Y., Takemoto, Y., et al., 2015. A polarly localized transporter for efficient manganese uptake in rice. Nat. Plants. 1, 15170.

Van de Mortel, J.E., Villanueva, L.A., Schat, H., Kwekkeboom, J., Coughlan, S., Moerland, P. D., et al., 2006. Large expression differences in genes for iron and zinc homeostasis, stress response, and lignin biosynthesis distinguish roots of *Arabidopsis thaliana* and the related metal hyperaccumulator *Thlaspi caerulescens*. Plant Physiol. 142, 1127–1147.

Véry, A.A., Sentenac, H., 2002. Cation channels in the *Arabidopsis* plasma membrane. Trends Plant Sci. 7, 168–175.

Very, A.A., Sentenac, H., 2003. Molecular mechanisms and regulation of K^+ transport in higher plants. Annu. Rev. Plant Biol. 54, 575–603.

Véry, A.A., Nieves-Cordones, M., Daly, M., Khan, I., Fizames, C., Sentenac, H., 2014. Molecular biology of K^+ transport across the plant cell membrane: what do we learn from comparison between plant species? J. Plant Physiol. 171, 748–769.

Wang, H., Shabala, L., Zhou, M., Shabala, S., 2018. Hydrogen peroxide-induced root Ca^{2+} and K^+ fluxes correlate with salt tolerance in cereals: towards the cell-based phenotyping. Int. J. Mol. Sci. 19, 702.

Wang, N., Qiu, W., Dai, J., Guo, X., Lu, Q., Wang, T., et al., 2019. AhNRAMP1 enhances manganese and zinc uptake in plants. Front. Plant Sci. 10, 415. Available from: https://doi.org/10.3389/fpls.2019.00415.

Welch, R.M., Graham, R.D., 2002. Breeding crops for enhanced micronutrient content. In: Adu-Gyamfi, J.J. (Ed.), Food Security in Nutrient-Stressed Environments: Exploiting Plants' Genetic Capabilities. Springer, Dordrecht, pp. 267–276.

Wong, C.K.E., Cobbett, C.S., 2009. HMA P-type ATPases are the major mechanism for root-to-shoot Cd translocation in *Arabidopsis thaliana*. New Phytol. 181, 71–78.

Wu, J., Seliskar, D.M., 1998. Salinity adaptation of plasma membrane H^+-ATPase in the salt marsh plant Spartina patens: ATP hydrolysis and enzyme kinetics. J. Exp. Bot. 49, 1005–1013.

Wu, H., Zhang, X., Giraldo, J.P., Shabala, S., 2018. It is not all about sodium: revealing tissue specificity and signalling roles of potassium in plant responses to salt stress. Plant Soil 431, 1–17.

Wu, X., Chen, J., Yue, X., Wei, X., Zou, J., Chen, Y., et al., 2019a. The zinc-regulated protein (ZIP) family genes and glutathione s-transferase (GST) family genes play roles in Cd resistance and accumulation of pak choi (*Brassica campestris* ssp. chinensis). Ecotoxicol. Environ. Saf. 183, 109571.

Wu, X., Zhu, Z.B., Chen, J.H., Huang, Y.F., Liu, Z.L., Zou, J.W., et al., 2019b. Transcriptome analysis revealed pivotal transporters involved in the reduction of cadmium accumulation in pak choi (*Brassica chinensis* L.) by exogenous hydrogen-rich water. Chemosphere 216, 684–697.

Xiong, Z.-T., Liu, C., Geng, B., 2006. Phytotoxic effects of copper on nitrogen metabolism and plant growth in Brassica pekinensis Rupr. Ecotoxicol. Environ. Saf. 64, 273–280.

Xu, W., Dai, W., Yan, H., Li, S., Shen, H., Chen, Y., et al., 2015. *Arabidopsis* NIP3; 1 plays an important role in arsenic uptake and root-to-shoot translocation under arsenite stress conditions. Mol. Plant 8, 722–733.

Zare, A.A., Khoshgoftarmanesh, A.H., Malakouti, M.J., Bahrami, H.A., Chaney, R.L., 2018. Root uptake and shoot accumulation of cadmium by lettuce at various Cd: Zn ratios in nutrient solution. Ecotoxicol. Environ. Saf. 148, 441–446.

Zhang, L., Song, F.B., 2006. Effects of forms and rates of zinc fertilizers on cadmium concentrations in two cultivars of maize. Commun. Soil Sci. Plant Anal. 37, 1905–1916.

Zhang, M., Senoura, T., Yang, X., Nishizawa, N.K., 2011. Functional analysis of metal tolerance proteins isolated from Zn/Cd hyperaccumulating ecotype and non-hyperaccumulating ecotype of *Sedum alfredii* Hance. FEBS Lett. 585, 2604–2609.

Zhang, J., Zhang, M., Song, H., Zhao, J., Shabala, S., Tian, S., et al., 2020. A novel plasma membrane-based NRAMP transporter contributes to Cd and Zn hyperaccumulation in *Sedum alfredii* Hance. Environ. Exp. Bot. 176, 104121.

Zhao, C., Zhang, H., Song, C., Zhu, J.K., Shabala, S., 2020. Mechanisms of plant responses and adaptation to soil salinity. Innovation 1, 100017. Available from: https://doi.org/10.1016/j.xinn.2020.100017.

Zheng, L., Kostrewa, D., Bernèche, S., Winkler, F.K., Li, X.D., 2004. The mechanism of ammonia transport based on the crystal structure of AmtB of *Escherichia coli*. Proc. Natl. Acad. Sci. U.S.A. 101, 17090–17095.

Chapter 2

The importance of beneficial and essential trace and ultratrace elements in plant nutrition, growth, and stress tolerance

Surabhi Awasthi*, Reshu Chauhan* and Sudhakar Srivastava
Plant Stress Biology Laboratory, Institute of Environment and Sustainable Development, Banaras Hindu University, Varanasi, India

2.1 Introduction

Environmental stresses (biotic and abiotic) are a primary concern nowadays due to their impact on plant growth and development. Environmental stresses result in great losses in crop yields and productivity. For plant/agricultural science researchers, it is a great challenge to maintain the yield and nutritional status of a crop under stressful conditions. Plant nutrition plays a vital role in the maintenance of healthy growth as well as stress tolerance. Several studies have reported the importance of micronutrients for stress tolerance and plant growth (Maathuis and Diatloff, 2013; Bradacova et al., 2016). According to Marschner (2011), out of 92 known elements, 17 elements are considered to be essential and are classified into macroelements *viz*. C, H, O, Ca, K, Mg, N, S, and P (requirement >1000 mg kg^{-1} dry weight) and micro/trace elements *viz*. B, Cl, Cu, Fe, Mn, Mo, Ni, and Zn (required for growth but may not be required for species to survive).

In the early 1900s, scientists were able to quantitatively detect small amounts of several elements in living organisms, and these elements were present in trace amounts. Thus within a short period of time these elements became known as the trace elements. In 1980, the category of ultratrace elements was established to include elements whose dietary requirements for animals were less than 1.0 mg g^{-1}. Several elements of the transition metals

* Authors contributed equally.

Plant Nutrition and Food Security in the Era of Climate Change. DOI: https://doi.org/10.1016/B978-0-12-822916-3.00001-9

are involved in metabolism and thus are essential for plant growth and development. Ultratrace metals such as Cu, Fe, Mn, Mo, Ni, and Zn are generally accepted to be essential for all plants. In recent years, there has been increasing awareness about the beneficial and ultratrace elements.

Beneficial and ultratrace elements play several roles in plant growth, with the main five as follows:

1. Some trace and ultratrace elements are integral parts of catalytic centers in close association with enzymes.
2. For the chemical transformation of molecules, oxidation and reduction reactions take place to generate and utilize metabolic energy. Trace and ultratrace elements donate or accept electrons in reactions of oxidation and reduction.
3. Trace elements are involved in binding, transporting, and releasing oxygen for maintaining homeostasis.
4. Some trace and ultratrace elements are involved in the structural roles of biological molecules.
5. Some trace and ultratrace elements are involved in various biological processes such as activating hormones, facilitating the binding of molecules at receptor sites on cell membranes, altering the structure of membranes to facilitate entry of specific molecules into a cell, and inducing gene expression.

Trace metals are transported to biological target sites by movement of ions against a concentration gradient; thus the transport of the trace metal ions requires energy (Williams et al., 2000). This energy can be provided by hydrolysis of Adenosine triphosphate (ATP), in which case the transport is called *primary active transport*. For optimum nutrition, dissolved metal ion concentrations required inside the plant are usually higher than those in the soil solution. The weathering of parent rocks in a field is the major source of trace metals; however, deposition through wind, water, and glaciations also contributes trace metals to soil (Thornton and Webb, 1980). Further, a number of anthropogenic sources such as mining and smelting, agricultural activities, coal burning, various industrial manufacturing units, and sewage dumping may add metals to the soil. However, it is not only the concentration of trace metals but also their bioavailability that affects trace metal accumulation in plants. A number of physico-chemical factors like pH, redox potential (Eh), soil moisture, presence of oxides/hydroxides of Fe/Mn/Al, and temperature as well as biological factors *viz.* community of bacteria and fungi influence metal concentration and bioavailability to plants (Majumdar et al., 2020).

2.2 Beneficial and essential elements

A number of elements have been characterized as essential and perform various roles in plants (Table 2.1). There are some trace metals that are beneficial to plant growth and stress tolerance but are not essential. The following discussion presents an account of various beneficial and essential trace and

TABLE 2.1 Effect of micronutrients on plants.

Micronutrient	Effect on plants
Aluminum	Promotes plant growth and increases the uptake of other nutrient elements
Boron	Increases root growth and elongation, and improves yield
Cobalt	Increases growth, yield, and nutrient levels (N, P, K, Ca, Mg, Zn, and Cu)
Copper	Plays an important role in plant growth as well as stress tolerance
Iodine	Improves the productivity and yield of the plant
Iron	Improves the yield and nutritional quality of crops
Manganese	Important in photosynthesis, scavenging of reactive oxygen species (ROS), respiration, hormone signaling, and pathogen defense
Molybdenum	Provides tolerance against various environmental stresses, such as drought, salt stress, and cold.
Nickel	Stimulates the germination of seeds and regulates enzyme activity, mineral metabolism, and several other metabolic processes
Selenium	Improves plant growth by improving water use efficiency, CO_2 assimilation, stomatal conductance, and transpiration, which leads to improved photosynthetic efficiency
Silica	Provides strength to the plant by making the plant tissues stronger and rigid and increases resistance to disease
Zinc	Involved in the regulation and maintenance of the gene expression required for the tolerance of several environmental stresses

ultratrace metals required for and supportive of plant growth and development.

2.2.1 Aluminum

Aluminum (Al) is an abundant element in soil; however, it is not an essential nutrient for plants. Aluminum can cause toxicity to plants upon high accumulation and this is particularly important in acid soils (Muhammad et al., 2018). However, low concentrations of Al can promote growth for a number of cereal crops and citrus plants (Foy et al., 1973), especially those plants adapted to acidic soil or native to such environments (Pilon-Smits et al., 2009). The growth of both Al accumulator plants (*Melastoma malabathricum*) and nonaccumulator plants (*Miscanthus sinensis)* has been found to improve upon Al supply to some extent (Watanabe et al., 2005; Kayama, 2001). Even cash crops like rice and maize have shown improvement in growth with better Al supplies (Famoso et al., 2011; Wang et al., 2015). The application of Al can increase the uptake of other nutrient elements, such as nitrogen (N), phosphorus (P), and potassium (K) to bring out positive effects. Normally, Al precipitates with phosphorus and makes Al−P precipitates and reduces P availability. However, Al is also known to induce the secretion of organic acids (OAs), which free P from Al−P precipitated over root surfaces and in the rhizospheric zone. These effects enhance P concentrations around roots and lead to increased uptake of P (Clegg and Gobran, 1995). Further, the application of Al promotes root growth (Clune and Copeland, 1999) and consequently increases plants' potential for P uptake (Ligaba et al., 2004; Watanabe et al., 2006; Liang et al., 2013). Transgenic barley plants expressing aluminum-activated malate transporter from wheat (TaALMT1) showed enhanced root exudation associated with increases in grain yield when grown in P-deficient conditions as compared with wild-type plants (Delhaize et al., 2004).

Acidic soils can cause toxicity to plants due to an excess of ion protons (H^+); this has been shown to be alleviated by Al supply. Aluminum treatment induces H^+ extrusion and reduces H^+ toxicity (Kinraide, 1993). Al application is also demonstrated to alleviate Mn and Fe toxicity in plants (Watanabe et al., 2006; Yang et al., 2009; Wang et al., 2015; Muhammad et al., 2016) through reduction in excess Fe and Mn accumulation in plants (Alia et al., 2015). The formation of Fe−OAs and Mn−OAs complexes in the presence of Al, which induces secretion of OAs, is suggested as a possible mechanism of Fe/Mn toxicity alleviation (Ghanati et al., 2005; Muhammad et al., 2016). Aluminum treatment affects the expression of several genes involved in various biological functions including scavenging of reactive oxygen species (ROS; Li et al., 2016), low-phosphate responsive genes, genes for OA secretion (Muhammad et al., 2016), cold tolerance (Duressa et al., 2010), and heat-shock proteins (Karuppanapandian et al., 2012; Li et al., 2016). Aluminum-induced OA secretion has also been found to stimulate colonization of plant growth-promoting rhizobacteria (PGPRs) (Rudrappa et al., 2008). Thus an optimum Al supply to plants, especially in acidic soil conditions, can help plants avoid toxicity of excess ions (Fe, Mn) and protons and deficiency of P and also stimulate plant growth through changes in a number of genes and PGPR activities.

2.2.2 Boron

The main sources of boron (B) in soil are volcanic eruptions, natural rock and soil weathering, and industrial activities (mining, biomass burning, fossil fuel combustion) (Kot, 2009). Boron is an essential element for the optimum growth and development of plants and also influences crop yields and crop produce quality (Brown et al., 2002). Boron is a unique ultratrace element among the nutrients in which the range between its deficiency and toxicity is very narrow (Yau and Ryan, 2008). Boron promotes plant growth and the production of pectin. The synthesis of pectin is important for the functions of the cell wall (O'Neill et al., 2004) because pectin plays a key role in the porosity and elasticity of the cell wall. Up to about 98% of B may be localized in the cell walls of plants, where it remains cross-linked to glycosyl inositol phosphoryl ceramides (GIPCs) and rhamnoglacturonan-II (RG-II) (O'Neill et al., 2004). Boron is also curial for the functional integrity of cell membranes, inducing membrane hyperpolarization, and maintaining optimum HATPase activity for ionic flux (Goldbach et al., 2001, 2007).

An optimum B supply is important for root growth and elongation owing to the importance of B in cell division (Gupta and Solanki, 2013). In cases of B deficiency, formation of B-sugar cis-diol complex is affected, which leads to excessive phenol accumulation. Thus B is involved in the synthesis and accumulation of phenolics. Further, oxidative processes are increased in the presence of excess phenol, which can lead to accumulation of quinine in toxic levels (Atique-ur-Rahman et al., 2018). Thus B deficiency induces oxidative damage to plants that is also aggravated by increased generation of ROS. The role of B in N metabolism and symbiotic N fixation has also been demonstrated. Boron-mediated effects on chloroplastic membranes and thylakoid integrity and functions influence the photosynthetic efficiency of plants (Goldbach and Wimmer, 2007). The optimum supply of B has been linked to seed setting, yield enhancement in barley, and produce quality in tomato (El-Feky et al., 2012). Boron supply also affects plant growth and development via its impact on major (N, P, K) and trace (Fe, Cu, Zn) element uptake and accumulation in various plant tissues. The effects on nutrient uptake and accumulation are demonstrated to be mediated through changes in expression of transporters (Quiles-Pando et al., 2013).

2.2.3 Cobalt

Cobalt (Co) is a transition metal and is categorized as a beneficial element for plant growth and development (Gad, 2012). The beneficial impacts of Co are particularly visible in leguminous plants because of the essentiality of Co for nitrogen fixation by root nodule bacteria (Gad and El–Metwally, 2015). Cobalt is a cofactor for enzymes and coenzymes including vitamin B_{12} (cobalamin) (Gad and El–Metwally, 2015). Cobalamin is required for the activity of several enzymes of nitrogen fixation by rhizobacteria and cyanobacteria in root nodules of leguminous plants. Low concentrations of Co allow proper nodulation and promote the growth and yield of crop plants. The role of Co in the protection against salinity stress in tomato and barley plants has been reported (Gad, 2005; Gad et al., 2011); the application of Co provides tolerance against salinity in tomato plants. The application of Co has also been found to increase the growth, yield, and nutrient levels (N, P, K, Ca, Mg, Zn, and Cu) in wheat plants (Kandil and El-Maghraby, 2017). In pea (*Pisum sativum*) and sweet potato (*Ipomoea batatas*), Co increased growth, nodule numbers, yield, and nutrient content. Gad (2012) reported that in groundnut (*Arachis hypogaea*), supplementation of Co significantly increased nitrogenase activity, number of root nodules, enhanced growth and yield, and also improved the quality of pods and oil yield. Similarly, low levels of Co have shown positive effects on the growth and yield characteristics of soybean (*Glycine max*; Kandil et al., 2013), lentil (*Lens culinaris*), chickpea (*Cicer arietinum*; Khan and Khan, 2010), cowpea (*Vigna unguiculata*), and broad bean (*Vicia faba*). However, at higher levels, Co caused toxicity to various growth parameters and yield in different plants.

2.2.4 Copper

Copper (Cu) is a well-known essential element for normal plant growth and development. It is a redox-active element due to its existence in different forms (Cu^{2+} and Cu^{+}) in cellular environments. Due to its redox-active behavior, it can participate in several enzymatic reactions that involve oxidation and reduction of molecules and electron transport. Copper is an essential cofactor as well as a toxic element. Copper regulates a complex network of metal-trafficking pathways, and plants have evolved different strategies to regulate homeostasis of Cu as a function of the environmental copper level. Copper chaperones are soluble, low-molecular-weight (LMW) receptor proteins, which regulate intracellular metal ions. Copper chaperones have an important role in transport as Cu chaperones bind to Cu ions and deliver them into the active sites of Cu-dependent enzymes. Copper is associated with several enzymes and metalloproteins as a cofactor or as a structural constituent. In plants, copper-containing proteins and enzymes include Cu−Zn superoxide dismutase (Cu/Zn-SOD), cytochrome c oxidase, plastocyanin, amino oxidase, laccase, and polyphenol oxidase (Yruela, 2005). Copper also plays an important role in photosynthetic electron transport, cell wall metabolism, oxidative stress response, and hormone signaling (Raven et al., 1999). Thus Cu is an essential element for plant growth and development. Due to Cu participation in such a wide range of enzymatic reactions, Cu deficiency affects crucial processes such as ROS regulation, photosynthetic and mitochondrial electron transport, cell wall metabolism, and hormone signaling (Raven et al., 1999). Copper homeostasis plays a very important role in the response to oxidative stress due to environmental stress. Copper also plays an important role in transcription, trafficking of proteins, and Fe mobilization (Yruela, 2005). The *Arabidopsis* proteome comprises 105 "copper proteins" and 21 "copper-binding proteins", which play an important role in plant growth as well as stress tolerance.

The involvement of Cu in ROS detoxification via Cu/Zn-SOD is one of its major functions. However, the interconversion of Cu^{2+} and Cu^{+} is itself a reaction leading to production of hydroxyl radicals, which may itself produce other ROS and cause oxidative damage to cellular biomolecules. Hence, the margin of essential and toxic Cu levels is very minimal (Srivastava et al., 2006); therefore, plants have a complex system of metalloproteins and chelators to complex extra Cu ions and sequester or store them for later use. In conditions of Cu deficiency, root and leaf architecture are modified so as to ensure more Cu uptake. The level of SOD and plastocyanin, two major Cu-containing proteins, is also altered based on the Cu content of plants. To avoid reduction of photosystem I electron transport, the SOD level may decline in order to make Cu available for plastocyanin (Yamasaki et al., 2007). Copper deficiency leads to changes in the thylakoid membrane; negative effect on photosystem I and II leads to reduced photosynthesis and plant growth (Yruela, 2005).

At high concentration, Cu is toxic and causes necrosis, chlorosis, stunting, and inhibition of root growth. Copper inhibits enzymatic activity by binding at the sulfhydryl group in proteins and causes oxidative damage (Meharg, 1994). Copper toxicity causes oxidative stress due to the formation of the hydroxyl ion by Haber-Weiss reaction (Halliwell and Gutteridge, 1984). Copper toxicity also induces changes in the activity of antioxidant enzymes such as glutathione reductase, superoxide dismutase, ascorbate peroxidase, dehydroascorbate reductase, and monodehydroascorbate reductase (Wang et al., 2004).

2.2.5 Iodine

For human nutrition, iodine (I) is an essential trace element. Currently, iodine deficiency is present almost all over the world. Iodine is irregularly distributed in the earth's crust, and some areas face the problem of decreasing iodine concentrations in soil through accelerated leaching, deforestation, and erosion (Li et al., 2017). Insufficient iodine intake causes various iodine-deficiency disorders (IDD) such as cretinism, goiter, miscarriages, reduced IQ, birth defects, and higher neonatal mortality (Medrano-Macías et al., 2016).

Iodine is not considered as an essential element for plants, particularly land plants. However, iodine plays a central role in the antioxidant system of some aquatic plants. Iodine plays an important role as a beneficial element for plants, as it improves plant productivity and yields (Medrano-Macías et al., 2016). The application of iodine in different chemical forms has been studied in several crops, and positive impacts have been found in terms of growth and yield and tolerance to a number of stresses. Iodine application has also been studied from the perspective of biofortification of iodine in crop produce. Iodine metabolism in seaweeds of the genus *Laminaria* volatilizes iodine by producing molecular iodine (I_2) and organoiodine compounds (CH_3I and CH_2I_2 (Leblanc et al., 2006; Jones et al., 2010)), which ultimately enter the process of antioxidant metabolism to reduce oxidative stress (Küpper et al., 2008). It has been reported that the application of iodine in crop plants increases the antioxidant potential and consequently decreases the oxidative damage induced by several stresses (Gupta et al., 2015). Plants utilize iodine as I^- via ionic channels and chloride transporters, which are regulated by proton pumps (Barbier-Brygoo et al., 2011). The application of iodine increases the productivity of plants in stressed environments (Leyva et al., 2011; Blasco et al., 2013). An increase in plant biomass, growth, as well as fruit quality was observed upon iodine supplementation in spinach, strawberry, and tomato (Dai et al., 2004; Li et al., 2016). Iodine supplementation also increased the level of nutrients and free amino acids in radish (Strzetelski et al., 2010). Weng et al. (Weng et al., 2008) reported that upon iodine supply at lower concentrations, iodine content in the plant was improved whereas supply of iodine at high concentration was found to be toxic to the plant. Supplementation of soil with potassium iodide improved the concentration of iodine in *Telfairia occidentalis* and *Zea mays* (Ujowundu et al., 2009). Zhu et al. (Zhu et al., 2003) also reported that high concentration of iodine is toxic to plants and reduces plant yield. The application of low concentrations of IO_{3-} improved the response to salinity stress in lettuce (Leyva et al., 2011). Application of iodine in soil increased the dietary iodine content of leafy vegetables such as spinach, lettuce, and cabbage (Mao et al., 2014). Li et al. (2016) reported that application of iodine improved fruit quality by increasing the ascorbic acid and soluble sugar contents in *Capsicum annuum.*

2.2.6 Iron

Iron (Fe) is an essential element for optimal growth and development of plants due to its crucial involvement in various metabolic processes including respiration, photosynthesis, and DNA synthesis. The Fe availability to the plant ultimately affects the crop yield and the concentration of Fe in edible tissues. Iron is predominantly present in the Fe($^{3+}$) form as a constituent of oxyhydroxide polymers with extremely low solubility in aerobic soils, and therefore it is not easily bioavailable to plants. Iron is a micronutrient and also highly reactive and toxic to the plant. Iron is an important part of electron transport chains and a cofactor of several enzymes. The Fe concentration in soils improves the yield and nutritional quality of crops. Plants have devised several strategies to enhance Fe bioavailability in soil for eventual uptake. Nongraminaceous plants are able to reduce Fe^{3+} to Fe^{2+} by releasing protons and secreting phenolics to take up Fe (Jeong and Guerinot, 2009; Jeong and Connolly, 2009; Cesco et al., 2010). Some graminaceous plants take up Fe after its chelation with metabolites secreted by them such as mugineic acid (MA; Ishimaru et al., 2010). Iron is the main constituent of many enzymes and is involved in the activation of several metabolic processes. In plants, Fe plays an important role in the synthesis of chlorophyll, and it is also essential for the maintenance of chloroplast structure and function. Iron is also an active cofactor of several enzymes that are necessary for plant hormone synthesis, such as lipoxygenase, ethylene, abscisic acid, and 1-aminocyclopropane acid-1-carboxylic oxidase (Rai et al., 2021). In addition, Fe has a structural role in the prosthetic groups of enzymes, such as cytochromes, catalases, and peroxidases. Iron is also part of SOD (Fe-SOD) and participates in the quenching of superoxide radicals. Therefore, Fe plays an important role in plant defense systems by activating the antioxidant enzymes, thus reducing the oxidative stresses under normal and stressed conditions and improving plant growth.

Iron deficiency causes retarded growth, interveinal chlorosis, and reduced fitness. Iron deficiency significantly alters root morphology (Marschner, 1995) and also modulates the expression of genes involved in Fe uptake (Colangelo and Guerinot, 2004). Dinneny et al. (2008) reported that several genes are differentially regulated by Fe in the roots of *Arabidopsis thaliana*. Ward et al. (2008) reported that Fe toxicity reduces the root growth of *Arabidopsis*; the reason behind this was that high concentration of Fe was no longer complexed with P. Non-graminaceous plants in Fe-deficient conditions secrete phenolic compounds in the rhizosphere and uptake the apoplastic Fe. In Fe-deficient conditions, within 24 hours genes of barley involved in PS synthesis, methionine synthesis, and sulfur uptake were found upregulated (Nagasaka et al., 2009). Similarly in rice during 5 days of Fe-deficient conditions, transcription factor OsIRO2 increased and regulated the activation of genes involved in Fe uptake and PS synthesis (Ogo et al., 2006). The upregulation of *OsIRT1* and *OsIRT2* in rice mutant increased the other micronutrients such as Zn, Cu, and Mn in shoots (Cheng et al., 2007). Cadmium can interfere with Fe movement, as exposure of Cd to *Brassica napus* increases the accumulation of Fe in roots, while the concentration of Fe was decreased in the shoot (Mendoza-Cózatl et al., 2008).

2.2.7 Manganese

Manganese (Mn) is a micronutrient that plays a vital role in plant growth and development. Manganese availability to plants depends on various soil properties including pH and Eh. White and Greenwood (White and Greenwood, 2013) reported that if soil pH is correct then application of Mn fertilizer can alleviate Mn deficiency. Manganese plays an important role in photosynthesis, as the photosystem II-water oxidizing system needs an absolute Mn requirement (Hakala et al., 2006). Manganese plays an important role in various processes of a plant's life cycle such as photosynthesis, scavenging of ROS, respiration, hormone signaling, and pathogen defense. The most important role of Mn is in the function of oxygen-evolving complex (OEC) of the photosynthetic machinery. Manganese is an essential cofactor in the complex and catalyzes the water-splitting reaction in photosystem II (PSII) (Andresen et al., 2018). Manganese can exist in four oxidation states that allow transfer of four electrons from PSII during water-splitting reaction. Manganese participates in the production of phenolic compounds and other plant defense mechanisms (Fernando et al., 2009). Regarding antioxidant functions, Mn is a cofactor of MnSODs located in mitochondria and peroxisomes through which Mn plays an essential role in the quenching of superoxide radicals (Corpas and Barroso, 2017).

In plants, Mn has low phloem mobility, which leads to Mn deficiency first developed in younger leaves. Manganese deficiency affects lignin content as it is a cofactor of the enzyme involved in lignin biosynthesis (Marschner, 2011). Application of Mn provides resistance against several soil-borne diseases such as common scab in potato, root rot in cotton, and take-all in wheat, as well as fungal leaf diseases including powdery mildew in grape, black leaf mold in tomato, and tan spot in wheat (Brennan, 1992; Graham and Webb, 1991; Heine et al., 2011). Heine et al. (2011) reported that Mn is a cofactor of peroxidase. Application of Mn provides stress tolerance by regulating the superoxide dismutase enzymes, which detoxify the destructive free radicals. Increased activity of Mn-superoxide dismutase improves the plant tolerance against different environmental stress factors such as ozone stress, salinity, and drought stress. Manganese toxicity is a major limiting factor on growth in acidic soils, as high concentrations of Mn reduce photosynthesis, resulting in reduced growth. Brown spots appear on mature leaves and chlorotic dots appear at the tips of young leaves (Millaleo et al., 2010). The application of Mn increased the yield of potato (Mousavi et al., 2007). Manganese facilitates the production of carbohydrates and is required for optimum utilization of macronutrients in plants.

2.2.8 Molybdenum

Molybdenum (Mo) is a trace element required for the growth of most biological organisms including plants. The availability of Mo for plants is strongly dependent on the soil pH, extent of water drainage, concentration of adsorbing oxides (e.g., Fe oxides), and organic compounds found in the soil. Molybdenum can exist in several oxidation states ranging from zero to VI, and VI is the most common form of Mo found in agricultural soils. Molybdenum is utilized by specific plant enzymes and participates in oxidation-reduction reactions. Molybdenum is predominantly found as an integral part of organic pterin, which is known as molybdenum cofactor (Moco). Molybdenum itself is not biologically active, and, therefore, in the form of Moco it binds to Mo-requiring enzymes in plants (Williams and Da Silva, 2002). In plants five Mo-enzymes have been identified: (1) aldehyde oxidase (AO), (2) xanthine dehydrogenase (XDH), (3) mitochondrial amidoxime reductase (mARC), (4) nitrate reductase (NR), and (5) sulfite oxidase (SO). These enzymes catalyze significant reactions in degradation of purine, synthesis of phytohormone, nitrogen assimilation, and detoxification of sulfite. Molybdenum also provides tolerance against various environmental stresses, such as drought, salt stress, and cold (Al-Issawi et al., 2013). Yu and Wang (Yu and Wang, 2006) reported that Mo improved photosynthesis and promoted plant growth as well as provided tolerance against cold stress in wheat. Several studies report that application of Mo improved the stomatal conductance and photosynthetic rate and decreased the transpiration rate in winter wheat under low-temperature stress (Al-Issawi et al., 2013). The mode of action of Mo has been found to be through increased activity of antioxidant enzymes such as CAT (catalase), SOD (superoxide dismutase), APX (ascorbate peroxidase), and POD (peroxidase) or nonenzymatic antioxidants, such as glutathione, ascorbic acid, and carotenoid.

2.2.9 Nickel

Nickel (Ni) is an essential plant micronutrient. Nickel stimulates the germination of seeds and regulates enzyme activity, mineral metabolism, and several other metabolic processes in plants. Ni is an integral part of the active site of the enzyme urease, which plays a role in hydrolysis of urea in plant tissues and also regulates N metabolism (Küpper and Kroneck, 2007). Two forms of urease are present in plants, one in seeds and another in vegetative tissues. Polacco et al. (2013) reported that the seed form of urease is highly active, while the activity of the urease present in vegetative tissues is low. Despite the low activity, it plays an important role in N recycling in plants (Hogan et al., 1983). Ubiquitous urease metabolizes the toxic urea into ammonia and the ammonia can be reused in other metabolic pathways, such as synthesis of polyamines, amino acids, and other nitrogen compounds (Gerendas and Sattelmacher, 1997). Nickel also has protective properties so it is used as a fungicide to control cereal rusts (Mishra and Kar, 1974). Several studies have reported that Ni may activate glyoxalase I, which plays an important role in the degradation of methylglyoxal (MG), which is a potent cytotoxic compound naturally produced by cellular metabolism. It is reported that Ni may activate glyoxalase I in rice (*Oryza sativa;* Mustafiz et al., 2014). Ni also plays an important role in the regeneration of GSH and therefore in redox homeostasis. Thus Ni also plays an important role in plants' tolerance against various stresses. It is suggested that Ni promotes plant growth by increasing the nutrient contents (Freitas et al., 2018).

Nickel deficiency decreases plant growth, plant senescence, and metabolism. Ni-deficient plants develop chlorosis in the youngest leaves and ultimately suffer meristematic necrosis (Bai et al., 2006). Nickel is essential for plants in low concentrations, but high concentrations cause deleterious effects on plant growth and can be toxic. Ni toxicity to plants results in reduced root and shoot growth, deformation of plant parts and abnormal flower shape, reduced biomass, leaf spotting, disturbed mitotic root tips, inhibition of germination, and chlorosis that leads to foliar necrosis (Pandey and Sharma, 2002; Nakazawa et al., 2004; Rahman et al., 2005). Moraes et al. (2009) reported that application of Ni in low concentration improved the growth and yield of rice. A study by Tan et al. (2000) reported that application of Ni affects the nitrogen metabolism in *Lycopersicon esculentum*.

2.2.10 Selenium

Required in trace amounts, selenium is a beneficial trace element that exerts a positive effect on crop growth and stress tolerance at low concentrations. The essentiality of Se for plants is still dubious and needs further research (Chauhan et al., 2019). Selenium may be present in four different oxidation states: Se(IV), Se(VI), Se(0), and inorganic and organic selenide (−2) in the soil. The bioavailability of Se in soil depends upon the soil pH and redox potential. Several studies reported that at low concentration Se increases the growth and antioxidative potential of plants (Chauhan et al., 2020). The growth-promoting response to Se was demonstrated in rice (Chauhan et al., 2017) and soybean (*G. max* L.) (Djanaguiraman et al., 2005). Se can also delay senescence and promote the growth of aging seedlings. At suboptimal temperatures, Se(IV) priming promoted germination of bitter gourd (Chen and Sung, 2001). Foliar application of Se is suggested as the most appropriate method of application. Foliar application of Se stimulates the growth of lettuce ryegrass, green tea leaves, and potato (Gupta and Gupta, 2017). The application of Se provides tolerance against salt (Yao et al., 2009) and cold stresses (Chu et al., 2010). The growth-promoting attribute of Se might be due to an increased accumulation of starch in chloroplasts (Pennanen et al., 2002). Recent studies have reported that Se not only promotes the growth of plants but also increases the antioxidant activity to resist various stresses (Djanaguiraman et al., 2005; Chauhan et al., 2017). Several studies demonstrated that the protective role of Se against the oxidative stress in higher plants is attributable to its role in glutathione peroxidase (GPX) activity (Xue and Hartikainen, 2000; Djanaguiraman et al., 2005). Further, Se supplementation has also been found to improve the nutrient status of plants (Chauhan et al., 2017).

Se supplementation improved plant growth by improving water use efficiency, CO_2 assimilation, stomatal conductance, and transpiration leading to improved photosynthetic efficiency. Selenium is antagonistic to arsenic (As) and reduces the oxidative stress caused by As. Upregulation of photosynthesis-related proteins such as fructose-bisphosphate aldolase and triose phosphate under As + Se exposure would have promoted carbon fixation and also allowed light reaction to operate efficiently in order to produce more reducing equivalents (NADPH). Supplementation of Se modulates the expression of different transporters including ABC transporter G family, V type proton ATPases, and STAR1, and thus reduces As accumulation resulting in reduced oxidative stress. This suggests the Se-mediated As amelioration in rice. Application of Se during As stress also regulates the expression of transcription factors Myb94 and WRKY62, which were upregulated during As stress but downregulated during Se supplementation. This suggested the Se-mediated alleviation of As toxicity.

2.2.11 Silicon

For plant growth and development, silicon (Si) is considered as a nonessential element. However, it is also a beneficial element as the optimum supply of Si. Silica provides various benefits to plants, especially under stress conditions. Zargar et al. (2012) defined Si as a "multitalented" quasi-essential element. Silicon is a nutritive element and is receiving increased attention due to its potential for enhancing plant tolerance against abiotic and biotic stresses such as extreme temperature, UV radiation, metal toxicity, nutrient deficiency, drought, salinity, pathogen, and fungus attack. Silica plays an important role in plant growth, yield and crop quality, photosynthesis, and nitrogen fixation (Epstein, 2009; Zargar et al., 2012). Recently, Si has been declared as a nutritive element for plants by the International Plant Nutrition Institute (http://www.ipni.net/nutrifacts-northamerican) and a beneficial substance by the Association of American Plant Food Control Officials (http://www.aapfco.org/). Silicon provides strength to the plant by making the plant tissues stronger and rigid and increases resistance to disease by modulating several key compounds . In addition to this, the accumulation of Si results in synthesis of phytoalexins and phenolics, which provide tolerance against the various stresses. Ye et al. (2013) reported the role of Si in the jasmonate-mediated defense mechanism in rice. Silica provides tolerance to plants via two major processes: (1) Si provides a physical and mechanical protection afforded by deposition of Si as SiO_2, and (2) Si stimulates a number of biochemical and molecular responses. Silica strengthens the stem of rice, which protects the rice from lodging (Liang et al., 2013). Silica enhances UV tolerance due to deposition of Si on the epidermis of the leaf, which protects the plants from UV radiation and thus reduces the UV-induced membrane damages (Shen et al., 2010). Silica maintains the plant–water relation during drought conditions by forming a double layer of Si cuticle under the epidermis, which reduces water loss by transpiration. Thus Si provides tolerance against drought stress (Gong et al., 2003). Zhu and Gong (2014) also reported that Si also reduces stomatal conductance in relation to turgor loss of guard cells.

The application of Si alleviates many abiotic and biotic stresses. Due to its beneficial role, it has been used as a silicate in many fertilizers. Si is able to exert such a protective effect as it provides a physical barrier and a biochemical defense. The deposition of silica as a physical barrier provides protection against several pathogens (Currie and Perry, 2007). Silica application provides drought tolerance by reducing the transpiration rate in plants (Epstein, 1999). Wang et al. (2005) reported that Si provides tolerance against high temperature as it provides a cooling effect to *Agrostis palustris*. Silica also provide resistance to salt stress by increasing the activity of antioxidant enzymes such as SOD and catalase, preventing membrane oxidative damage in *Z. mays* (Moussa, 2006).

2.2.12 Zinc

Zinc (Zn), a plant micronutrient, is involved in several physiological functions. Zinc is required in various enzymatic reactions, metabolic processes, and oxidation-reduction reactions and is vital to crop nutrition. Zinc is also an essential element for many enzymes that are involved in N metabolism and protein synthesis. Zinc is related to the enzymes that are involved in maintenance of the integrity of cellular membranes, carbohydrate metabolism, and regulation of auxin and protein synthesis (Hafeez et al., 2013). Zinc helps to regulate and maintain the gene expression required for the tolerance of several environmental stresses in plants (Cakmak and Hoffland, 2012). Zinc regulates the water uptake and transport in plants and also reduces the adverse effects of heat and salt stress. Application of Zn improved the nutrient content of the plant as it increased the synthesis of amino acids such as tryptophan, which is a precursor of IAA (Alloway, 2004). The application of Zn also improves the integrity of cellular membranes to preserve the structural orientation of macromolecules and ion transport systems. Its interaction with phospholipids and sulphydryl groups of membrane proteins helps in the maintenance of membranes (Cakmak, 2000; Disante et al., 2010). Zinc also has a positive effect on plant growth and development and stress tolerance as zinc finger proteins regulate plant responses to biotic stress conditions (Noman et al., 2019). Application of Zn also plays an important role in plant defense as it helps protect against salt stress (Aktaş et al., 2006).

References

Aktaş, H., Abak, K., Öztürk, L., Çakmak, I., 2006. The effect of zinc on growth and shoot concentrations of sodium and potassium in pepper plants under salinity stress. Turkish J. Agric. For. 30, 407–412.

Alia, F.J., Shamshuddin, J., Fauziah, C.I., Husni, M.H.A., Panhwar, Q.A., 2015. Effects of aluminum, iron and/or low pH on rice seedlings grown in solution culture. Int. J. Agric. Biol. 17 (4).

Al-Issawi, M., Rihan, H.Z., Woldie, W.A., Burchett, S., Fuller, M.P., 2013. Exogenous application of molybdenum affects the expression of CBF14 and the development of frost tolerance in wheat. Plant Physiol Biochem 63, 77–81.

Alloway, B.J., 2004. Zinc in soils and crop nutrition. International Zinc Association, Brussels, pp. 130–pp.

Andresen, E., Peiter, E., Kupper, H., 2018. Trace metal metabolism in plants. J. Exp. Bot. 69, 909–954.

Atique-ur-RahmanFarooq, M., Rashid, A., Nadeem, F., Stuerz, S., Asch, F., et al., 2018. Boron nutrition of rice in different production systems: A review. Agron. Sustain. Develop. 38, 25.

Bai, C., Reilly, C.C., Wood, B.W., 2006. Nickel deficiency disrupts metabolism of ureides, amino acids, and organic acids of young pecan foliage. Plant Physiol. 140 (2), 433–443.

Barbier-Brygoo, H., De Angeli, A., Filleur, S., Frachisse, J.-M., Gambale, F., Thomine, S., et al., 2011. Anion channels/transporters in plants: from molecular bases to regulatory networks. Annu. Rev. Plant Biol. 62, 25–51.

Blasco, B., Leyva, R., Romero, L., Ruiz, 2013. Iodine effects on phenolic metabolism in lettuce plants under salt stress. J. Agric. Food Chem. 61, 2591–2596.

Bradacova, K., Weber, N.F., Morad-Talab, N., Asim, M., Imran, M., Weinmann, M., et al., 2016. Micronutrients (Zn/Mn), seaweed extracts, and plant growth-promoting bacteria as cold-stress protectants in maize. Chem. Biol. Technol. Agric. 3, 19.

Brennan, R.F., 1992. The role of manganese and nitrogen nutrition in the susceptibility of wheat plants to take-all in Western Australia. Fert. Res. 31 (1), 35–41.

Brown, P.H., Bellaloui, N., Wimmer, M.A., Bassil, E.S., Ruiz, J., Hu, H., et al., 2002. Boron in plant biology. Plant Biol 4, 211–229.

Cakmak, I., 2000. Role of zinc in protecting plant cells from reactive oxygen species. New Phytol. 146, 185–205.

Cakmak, I., Hoffland, E., 2012. Zinc for the improvement of crop production and human health. Plant Soil 361, 1–2.

Cesco, S., Neumann, G., Tomasi, N., Pinton, R., Weisskopf, L., 2010. Release of plant-borne flavonoids into the rhizosphere and their role in plant nutrition. Plant Soil 329, 1–25.

Chauhan, R., Awasthi, S., Tripathi, P., Mishra, S., Dwivedi, S., Niranjan, A., et al., 2017. Selenite modulates the level of phenolics and nutrient element to alleviate the toxicity of arsenite in rice (*Oryza sativa* L.). Ecotoxicol. Environ. Saf. 138, 47–55.

Chauhan, R., Awasthi, S., Srivastava, S., Dwivedi, S., Pilon-Smits, E.A.H., Dhankher, O.P., et al., 2019. Understanding selenium metabolism in plants and its role as a beneficial element. Crit. Rev. Environ. Sci. Technol. 49, 1937–1958.

Chauhan, R., Awasthi, S., Indoliya, Y., Chauhan, A.S., Mishra, S., Agrawal, L., et al., 2020. Transcriptome and proteome analyses reveal selenium mediated amelioration of arsenic toxicity in rice (*Oryza sativa* L.). J. Hazard. Mater. 390, 122122.

Chen, C.C., Sung, J.M., 2001. Priming bitter gourd seeds with selenium solution enhances germinability and antioxidative responses under sub-optimal temperature. Physiol. Plant. 111, 9–16.

Cheng, L., Wang, F., Shou, H., Huang, F., Zheng, L., He, F., et al., 2007. Mutation in nicotianamine aminotransferase stimulated the Fe (II) acquisition system and led to iron accumulation in rice. Plant Physiol. 145 (4), 1647–1657.

Chu, J., Yao, X., Zhang, Z., 2010. Responses of wheat seedlings to exogenous selenium supply under cold stress. Biol. Trace Elem. Res. 136, 355–363.

Clegg, S., Gobran, G.R., 1995. Effects of aluminium on growth and root reactions of phosphorus stressed Betula pendula seedlings. Plant Soil. 168 (1), 173–178.

Clune, T.S., Copeland, L., 1999. Effects of aluminium on canola roots. Plant Soil. 216 (1-2), 27–33.

Colangelo, E.P., Guerinot, M.L., 2004. The essential basic helix-loop-helix protein FIT1 is required for the iron deficiency response. Plant Cell 16 (12), 3400–3412.

Corpas, F.J., Barroso, J.B., 2017. Lead-induced stress, which triggers the production of nitric oxide (NO) and superoxide anion (O2·-) in Arabidopsis peroxisomes, affects catalase activity. Nitric Oxide 68, 103–110.

Currie, H.A., Perry, C.C., 2007. Silica in plants: biological, biochemical and chemical studies. Ann. Bot. 100 (7), 1383–1389.

Dai, J.-L., Zhu, Y.-G., Zhang, M., Huang, Y.-Z., 2004. Selecting iodine-enriched vegetables and the residual effect of iodate application to soil. Biol. Trace Elem. Res. 101, 265–276.

Delhaize, E., Ryan, P.R., Hebb, D.M., Yamamoto, Y., Sasaki, T., Matsumoto, H., 2004. Engineering high-level aluminum tolerance in barley with the *ALMT1* gene. Proc. Nat. Acad. Sci. 101 (42), 15249–15254.

Dinneny, J.R., Long, T.A., Wang, J.Y., Jung, J.W., Mace, D., Pointer, S., et al., 2008. Cell identity mediates the response of *Arabidopsis* roots to abiotic stress. Science. 320 (5878), 942–945.

Disante, K.B., Fuentes, D., Cortina, J., 2010. Sensitivity to zinc of Mediterranean woody species important for restoration. Sci Total Environ 408, 2216–2225.

Djanaguiraman, M., Devi, D.D., Shanker, A.K., Sheeba, J.A., Bangarusamy, U., 2005. Selenium—An Antioxidative Protectant in Soybean during Senescence. Plant Soil 272, 77–86.

Duressa, D., Soliman, K., Chen, D., 2010. Identification of aluminum responsive genes in Al-tolerant soybean line PI 416937. Int. J. Plant Genomics 2010.

El-Feky, S.S., El-Shintinawy, F.A., Shaker, E.M, Shams El-Din, H.A., 2012. Effect of elevated boron concentrations on the growth and yield of barley (*Hordeum vulgare* L.) and alleviation of its toxicity using different plant growth modulators. Austr. J. Crop Sci. 6, 1687–1695.

Epstein, E., 1999. Silicon. Ann. Rev. Plant Biol. 50 (1), 641–664.

Epstein, E., 2009. Silicon: its manyfold roles in plants. Ann. Plant Biol. 155, 155–160.

Famoso, A.N., Zhao, K., Clark, R.T., Tung, C.W., Wright, M.H., Bustamante, C., et al., 2011. Genetic architecture of aluminum tolerance in rice (*Oryza sativa*) determined through genome-wide association analysis and QTL mapping. PLoS Genet 7 (8), e1002221.

Fernando, D.R., Baker, A.J.M., Woodrow, I.E., 2009. Physiological responses in *Macadamia integrifolia* on exposure to manganese treatment. Aust. J. Bot. 57 (5), 406–413.

Freitas, D.S., Rodak, B.W., dos Reis, A.R., Reis, F.D.B., de Carvalho, T.S., Schulze, J., et al., 2018. Hidden nickel deficiency? Nickel fertilization via soil improves nitrogen metabolism and grain yield in soybean genotypes. Front. Plant Sci. 9, 614.

Foy, C.D., Fleming, A.L., Schwartz, J.W., 1973. Opposite alluminum and manganese tolerances of two wheat varieties. Agron. J. 65 (1), 123–126.

Gad, N., 2005. Interactive effect of cobalt and salinity on tomato plants I- growth and mineral composition as affected by cobalt and salinity. Res. J. Agric. Biol. Sci. 1 (3), 261–269.

Gad, N., 2012. Physiological and Chemical Response of Groundnut (Arachis hypogaea) to Cobalt Nutrition. World Appl. Sci. J. 202, 327–335.

Gad, N., Abd El-Moez, M.R., Kandil, H., 2011. Barley response to salt stress at varied levels of cobalt ıı. some physiological and chemical characteristics. Res. J. Appl. Sci. 7 (11), 1447–1453.

Gad, N., El–Metwally, I.M., 2015. Chemical and physiological response of maize to salinity using cobalt supplement. Int. J. Chemtech Res. 8 (10), 45–52.

Gajewska, E., Skłodowska, M., 2005. Antioxidative responses and proline level in leaves and roots of pea plants subjected to nickel stress. Acta Physiol. Plant. 27 (3), 329–340.

Gerendas, J., Sattelmacher, B., 1997. Significance of N source (urea vs. NH_4NO_3) and Ni supply for growth, urease activity and nitrogen metabolism of zucchini (Cucurbita pepo convar. giromontiina). Plant Soil 196 (2), 217–222.

Ghanati, F., Morita, A., Yokota, H., 2005. Effects of aluminum on the growth of tea plant and activation of antioxidant system. Plant Soil 276 (1–2), 133–141.

Goldbach, H., Huang, L., Wimmer, M., 2007. Boron functions in plants and animals: recent advances in boron research and open questions. In: Xu, F, Goldbach, HE, Brown, PH, Bell, RW, Fujiwara, T, Hunt, CD, Goldberg, S, Shi, L (Eds.), Advances in plant and animal boron nutrition. Springer, Dordrecht, pp. 3–25.

Goldbach, H.E., Wimmer, M.A., 2007. Boron in plants and animals: is there a role beyond cell-wall structure? J. Plant Nutr. Soil Sci. 170, 39–48.

Goldbach, H.E., Yu, Q., Wingender, R., Schulz, M., Wimmer, M., Findeklee, P., Baluska, F., 2001. Rapid response reactions of roots to boron deprivation. J Plant Nutr. Soil Sci. 164, 173–181.

Gong, H.J., Chen, K.M., Chen, G.C., Wang, S.M., Zhang, C.L., 2003. Effects of silicon on growth of wheat under drought. J. Plant Nutr. 26, 1055–1063.

Graham, R.D., Webb, M.J., 1991. Micronutrients and plant disease resistance and tolerance in plants. Micronutr. Agric. 329–370.

Gupta, N., Bajpai, M., Majumdar, R., Mishra, P., 2015. Response of iodine on antioxidant levels of Glycine max L. Grown under Cd2+ stress. Adv. Biol. Res. (Rennes). 9, 40–48.

Gupta, M., Gupta, S., 2017. An Overview of Selenium Uptake, Metabolism, and Toxicity in Plants. Front. Plant Sci. 7, 2074.

Gupta, U., Solanki, H., 2013. Impact of boron deficiency on plant growth. Int. J. Bioassays 2, 1048–1050.

Hafeez, B., Khanif, Y.M., Saleem, M., 2013. Role of zinc in plant nutrition- A review. Am. J. Exp. Agric. 3, 374–391.

Hakala, M., Rantamäki, S., Puputti, E.M., Tyystjärvi, T., Tyystjärvi, E., 2006. Photoinhibition of manganese enzymes: insights into the mechanism of photosystem II photoinhibition. J. Exp. Bot. 57 (8), 1809–1816.

Heine, G., Max, J.F., Führs, H., Moran-Puente, D.W., Heintz, D., Horst, W.J., 2011. Effect of manganese on the resistance of tomato to *Pseudocercospora fuligena*. J. Plant Nutr. Soil Sci. 174 (5), 827–836.

Hogan, M.E., Swift, I.E., Done, J., 1983. Urease assay and ammonia release from leaf tissues. Phytochemistry 22, 663–667.

Ishimaru, Y., Masuda, H., Bashir, K., et al., 2010. Rice metal–nicotianamine transporter, OsYSL2, is required for the long-distance transport of iron and manganese. The Plant Journal 62, 379–390.

Jeong, J, Connolly, EL, 2009. Iron uptake mechanisms in plants: functions of the FRO family of ferric reductases. Plant Science 176, 709–714.

Jones, C.E., Hornsby, K.E., Sommariva, R., Dunk, R.M., von Glasow, R., McFiggans, G., et al., 2010. Quantifying the contribution of marine organic gases to atmospheric iodine. Geophys. Res. Lett. 37, 1–6.

Kandil, H., El-Maghraby, A., 2017. Impact of cobalt form and level addition on wheat plants (*Triticum aestivum* L.): II. Yield and nutrient contents. J. Basic Appl. Sci. Res. 7, 1–7.

Kandil, H., Farid, I.M., El-Maghraby, A., 2013. Effect of cobalt level and nitrogen source on quantity and quality of soybean plant. J. Basic. Appl. Sci. Res. 3, 185–192.

Karuppanapandian, T., Rhee, S.J., Kim, E.J., Han, B.K., Hoekenga, O.A., Lee, G.P., 2012. Proteomic analysis of differentially expressed proteins in the roots of Columbia-0 and Landsberg *erecta* ecotypes of *Arabidopsis thaliana* in response to aluminum toxicity. Can. J. Plant Sci. 92 (7), 1267–1282.

Kayama, M., 2001. Comparison of the aluminum tolerance of *Miscanthus sinensis* Anderss. and *Miscanthus sacchariflorus* Bentham in hydroculture. Int. J. Plant Sci. 162 (5), 1025–1031.

Khan, M.R., Khan, M.M., 2010. Effect of varying concentration of nickel and cobalt on the plant growth and yield of chickpea. Austr. J. Basic Appl. Sci. 4, 1036–1046.

Kinraide, T.B., 1993. Aluminum enhancement of plant growth in acid rooting media. A case of reciprocal alleviation of toxicity by two toxic cations. Physiol. Plant. 88 (4), 619–625.

Kot, F.S., 2009. Boron sources, speciation and its potential impact on health. Rev. Environ. Sci. Biotechnol. 8, 3–28.

Küpper, H., Kroneck, P.M.H., 2007. Nickel in the environment and its role in the metabolism of plants and cyanobacteria. In: A. Sigel, H. Sigel, R.K.O. Sigel, (Eds.), Nickel and Its Surprising Impact in Nature, vol. 2. John Wiley & Sons Ltd., Sussex, England. ISBN: 9780470016718.

Küpper, F.C., Carpenter, L.J., McFiggans, G.B., Palmer, C.J., Waite, T.J., Boneberg, E.M., et al., 2008. Iodide accumulation provides kelp with an inorganic antioxidant impacting atmospheric chemistry. Proc Natl Acad Sci 105 (19), 6954–6958.

Leblanc, C., Colin, C., Cosse, A., Delage, L., La Barre, S., Morin, P., et al., 2006. Iodine transfers in the coastal marine environment: the key role of brown algae and of their vanadium-dependent haloperoxidases. Biochimie 88, 1773–1785.

Leyva, R., Sánchez-Rodríguez, E., Ríos, J.J., Rubio-Wilhelmi, M.M., Romero, L., Ruiz, J.M., et al., 2011. Beneficial effects of exogenous iodine in lettuce plants subjected to salinity stress. Plant Sci. 181 (2), 195–202.

Li, H., Yang, L.T., Qi, Y.P., Guo, P., Lu, Y.B., Chen, L.S., 2016. Aluminum toxicity-induced alterations of leaf proteome in two citrus species differing in aluminum tolerance. Int. J. Mol. Sci. 17 (7), 1180.

Li, R., Li, D.W., Liu, H.P., Hong, C.L., Song, M.Y., Dai, Z.X., et al., 2017. Enhancing iodine content and fruit quality of pepper (*Capsicum annuum* L.) through biofortification. Sci. Hortic. 214, 165–173.

Liang, C., Piñeros, M.A., Tian, J., Yao, Z., Sun, L., Liu, J., et al., 2013. Low pH, aluminum, and phosphorus coordinately regulate malate exudation through GmALMT1 to improve soybean adaptation to acid soils. Plant Physiol. 161 (3), 1347–1361.

Ligaba, A., Yamaguchi, M., Shen, H., Sasaki, T., Yamamoto, Y., Matsumoto, H., 2004. Phosphorus deficiency enhances plasma membrane H^+-ATPase activity and citrate exudation in greater purple lupin (Lupinus pilosus). Funct. Plant Biol. 31 (11), 1075–1083.

Maathuis, F.J.M., Diatloff, E., 2013. Roles and functions of plant mineral nutrients. Methods Mol. Biol. 953, 1–21.

Majumdar, A., Kumar, J.S., Bose, S., 2020. Agricultural water management practices and environmental influences on arsenic dynamics in rice field. In: Arsenic in Drinking Water and Food, Springer, Singapore, pp. 425–443.

Mao, H., Wang, J., Wang, Z., Zan, Y., Lyons, G., Zou, C., 2014. Using agronomic biofortification to boost zinc, selenium, and iodine concentrations of food crops grown on the loess plateau in China. J. Soil Sci. Plant Nutr. 14 (2), 459–470.

Marschner, H., 1995. Mineral Nutrition of Higher Plants, second ed. Academic, Great Britain.

Marschner, H., 2011. Marschner's Mineral Nutrition of Higher Plants, third ed. Academic press. ISBN: 9780123849052.

Medrano-Macías, J., Leija-Martínez, P., González-Morales, S., Juárez-Maldonado, A., Benavides-Mendoza, A., 2016. Use of iodine to biofortify and promote growth and stress tolerance in crops. Front. Plant Sci. 7, 1146.

Meharg, A.A., 1994. Integrated tolerance mechanisms: constitutive and adaptive plant responses to elevated metal concentrations in the environment. Plant Cell Environ. 17 (9), 989–993.

Mendoza-Cózatl, D.G., Butko, E., Springer, F., Torpey, J.W., Komives, E.A., Kehr, J., et al., 2008. Identification of high levels of phytochelatins, glutathione and cadmium in the phloem sap of Brassica *napus*. A role for thiol-peptides in the long-distance transport of cadmium and the effect of cadmium on iron translocation. Plant J. 54 (2), 249–259.

Millaleo, R., Reyes-Díaz, M., Ivanov, A.G., Mora, M.L., Alberdi, M., 2010. Manganese as essential and toxic element for plants: transport, accumulation and resistance mechanisms. J. Soil Sci. Plant Nutr. 10 (4), 470–481.

Mishra, D., Kar, M., 1974. Nickel in plant growth and metabolism. Bot. Rev. 40 (4), 395–452.

Moraes, M.F., Reis, A.R., Moraes, L.A.C., Lavres-Junior, J., Vivian, R., Cabral, C.P., et al., 2009. Effects of molybdenum, nickel, and nitrogen sources on the mineral nutrition and growth of rice plants. Commun. Soil Sci. Plant Anal. 40 (21–22), 3238–3251.

Mousavi, S.R., Galavi, M., Ahmadvand, G., 2007. Effect of zinc and manganese foliar application on yield, quality and enrichment on potato (*Solanum tuberosum* L.). Asian J. Plant Sci 6, 1256–1260.

Moussa, H.R., 2006. Influence of exogenous application of silicon on physiological response of salt-stressed maize (*Zea mays* L.). Int. J. Agric. Biol. 8 (3), 293–297.

Muhammad, N., Cai, S., Shah, J.M., Zhang, G., 2016. The combined treatment of Mn and Al alleviates the toxicity of Al or Mn stress alone in barley. Acta Physiol. Plant. 38 (12), 277.

Muhammad, N., Zvobgo, G., Guo-ping, Z., 2018. A review: the beneficial effect of aluminum on plant growth in acid soil and the possible mechanisms. J. Integr. Agric. 17, 60345–60347.

Mustafiz, A., Ghosh, A., Tripathi, A.K., Kaur, C., Ganguly, A.K., Bhavesh, N.S., et al., 2014. A unique Ni^{2+} -dependent and methylglyoxal-inducible rice glyoxalase I possesses a single active site and functions in abiotic stress response. Plant J. 78, 951–963.

Nagasaka, S., Takahashi, M., Nakanishi-Itai, R., Bashir, K., Nakanishi, H., Mori, S., et al., 2009. Time course analysis of gene expression over 24 hours in Fe-deficient barley roots. Plant Mol. Biol. 69 (5), 621–631.

Nakazawa, R., Kameda, Y., Ito, T., Ogita, Y., Michihata, R., Takenaga, H., 2004. Selection and characterization of nickel-tolerant tobacco cells. Biol. Plant. 48 (4), 497–502.

Noman, A., Aqeel, M., Khalid, N., Islam, W., Sanaullah, T., Anwar, M., et al., 2019. Zinc finger protein transcription factors: Integrated line of action for plant antimicrobial activity. Microb. Pathog. 132, 141–149.

Ogo, Y., Itai, R.N., Nakanishi, H., Inoue, H., Kobayashi, T., Suzuki, M., et al., 2006. Isolation and characterization of IRO_2, a novel iron-regulated bHLH transcription factor in graminaceous plants. J. Exp. Bot. 57 (11), 2867–2878.

O'Neill, M.A., Ishii, T., Albersheim, P., Darvill, A.G., 2004. Rhamnogalacturonan II: structure and function of a borate cross-linked cell wall pectic polysaccharide. Annu. Rev. Plant Biol. 55, 109–139.

Pandey, N., Sharma, C.P., 2002. Effect of heavy metals Co^{2+}, Ni^{2+} and Cd^{2+} on growth and metabolism of cabbage. Plant Sci. 163 (4), 753–758.

Pennanen, A., Xue, T., Hartikainen, H., 2002. Protective role of selenium in plant subjected to severe UV irradiation stress. J Appl Botany (1995), 76(1-2), 66–76.

Pilon-Smits, E.A., Quinn, C.F., Tapken, W., Malagoli, M., Schiavon, M., 2009. Physiological functions of beneficial elements. Curr. Opin. Plant Biol. 12 (3), 267–274.

Polacco, J.C., Mazzafera, P., Tezotto, T., 2013. Opinion: nickel and urease in plants: still many knowledge gaps. Plant Sci. 199–200, 79–90.

Quiles-Pando, C., Rexach, J., Navarro-Gochicoa, M.T., Camacho-Cristóbal, J.J., Herrera-Rodríguez, M. B., González-Fontes, A., 2013. Boron deficiency increases the levels of cytosolic Ca2+ and expression of Ca2+ -related genes in Arabidopsis thaliana roots. Plant Physiol Biochem 65, 55–60.

Rahman, H., Sabreen, S., Alam, S., Kawai, S., 2005. Effects of nickel on growth and composition of metal micronutrients in barley plants grown in nutrient solution. J. Plant Nutr. 28 (3), 393–404.

Rai, S., Singh, P.K., Mankotia, S., Swain, J., Satbhai, S.B., 2021. Iron Homeostasis in Plants and its Crosstalk with Copper, Zinc, and Manganese. Plant Stress, p. 100008.

Raven, J.A., Evans, M.C., Korb, R.E., 1999. The role of trace metals in photosynthetic electron transport in O_2-evolving organisms. Photosynth. Res. 60 (2-3), 111–150.

Rudrappa, T., Czymmek, K.J., Paré, P.W., Bais, H.P., 2008. Root-secreted malic acid recruits beneficial soil bacteria. Plant Physiol. 148 (3), 1547–1556.

Shen, X., Zhou, Y., Duan, L., Li, Z., Eneji, A.E., Li, J., 2010. Silicon effects on photosynthesis and antioxidant parameters of soybean seedlings under drought and ultraviolet-B radiation. J. Plant Physiol. 167 (15), 1248–1252.

Srivastava, S., Mishra, S., Tripathi, R.D., Dwivedi, S., Gupta, D.K., 2006. Copper-induced oxidative stress and responses of antioxidants and phytochelatins in Hydrilla verticillata (Lf) Royle. Aquatic Toxicology 80 (4), 405–415.

Strzetelski, P., Smoleń, S., Rożek, S., Sady, W., 2010. The effect of diverse iodine fertilization on nitrate accumulation and content of selected compounds in radish plants (Raphanus sativus L.). Acta Sci. Pol. Hort. Cult 9 (2), 65–73.

Tan, X.W., Ikeda, H., Oda, M., 2000. Effects of nickel concentration in the nutrient solution on the nitrogen assimilation and growth of tomato seedlings in hydroponic culture supplied with urea or nitrate as the sole nitrogen source. Sci. Hortic. 84 (3–4), 265–273.

Thornton, I., Webb, J.S., 1980. Trace elements in soils and plants. In: K.L. Blaxter (Ed.), Food Chains and Human Nutrition. Applied Science Publishers Ltd, London, pp. 273–312.

Ujowundu, C.O., Ukoha, A.I., Agha, N.C., Nwachukwu, N., Igwe, K.O., 2009. Iodine Biofortification of selected plants using potassium iodide. Niger. J. Biochem. Mol. Biol. 24 (2), 17.

Wang, L., Nie, Q., Li, M., Zhang, F., Zhuang, J., Yang, W., et al., 2005. Biosilicified structures for cooling plant leaves: a mechanism of highly efficient midinfrared thermal emission. Appl. Phys. Lett. 87 (19), 194105.

Wang, S.H., Yang, Z.M., Yang, H., Lu, B., Li, S.Q., Lu, Y.P., 2004. Copper-induced stress and antioxidative responses in roots of Brassica juncea L. Bot. Bull. of Acad. Sinica 45.

Wang, W., Zhao, X.Q., Hu, Z.M., Shao, J.F., Che, J., Chen, R.F., et al., 2015. Aluminium alleviates manganese toxicity to rice by decreasing root symplastic Mn uptake and reducing availability to shoots of Mn stored in roots. Ann. Bot. 116 (2), 237–246.

Ward, J.T., Lahner, B., Yakubova, E., Salt, D.E., Raghothama, K.G., 2008. The effect of iron on the primary root elongation of *Arabidopsis* during phosphate deficiency. Plant Physiol. 147 (3), 1181–1191.

Watanabe, T., Jansen, S., Osaki, M., 2005. The beneficial effect of aluminium and the role of citrate in Al accumulation in *Melastoma malabathricum*. New Phytol. 773–780.

Watanabe, T., Jansen, S., Osaki, M., 2006. Al–Fe interactions and growth enhancement in *Melastoma malabathricum* and Miscanthus *sinensis* dominating acid sulphate soils. Plant Cell Environ. 29 (12), 2124–2132.

Weng, H.X., Weng, J.K., Yan, A.L., Hong, C.L., Yong, W.B., Qin, Y.C., 2008. Increment of iodine content in vegetable plants by applying iodized fertilizer and the residual characteristics of iodine in soil. Biol. Trace Elem. Res. 123 (1), 218–228.

White, P.J., Greenwood, D.J., 2013. Properties and management of cationic elements for crop growth. Soil conditions and plant growth 12, 160–194.

Williams, R.J.P., Da Silva, J.F., 2002. The involvement of molybdenum in life. Biochem. Biophys. Res. Commun. 292 (2), 293–299.

Williams, L.E., Pittman, J.K., Hall, J.L., 2000. Emerging mechanisms for heavy metal transport in plants. Biochim Biophys Acta Biomembr BBA-Biomembranes 1465 (1–2), 104–126.

Xue, T., Hartikainen, H., 2000. Association of antioxidative enzymes with the synergistic effect of selenium and UV irradiation in enhancing plant growth. Agric. Food Sci. 9 (2), 177–186.

Yamasaki, H., Abdel-Ghany, S.E., Cohu, C.M., Kobayashi, Y., Shikanai, T., Pilon, M., 2007. Regulation of copper homeostasis by micro-RNA in Arabidopsis. J. Biol. Chem. 282 (22), 16369–16378.

Yang, Z.B., You, J.F., Xu, M.Y., Yang, Z.M., 2009. Interaction between aluminum toxicity and manganese toxicity in soybean (*Glycine max*). Plant Soil 319 (1–2), 277–289.

Yao, X., Chu, J., Wang, G., 2009. Effects of selenium on wheat seedlings under drought stress. Biol. Trace Elem. Res. 130 (3), 283–290.

Yau, S.K., Ryan, J., 2008. Boron toxicity tolerance in crops: a viable alternative to soil amelioration. Crop Science 48 (3), 854–865.

Ye, M., Song, Y., Long, J., Wang, R., Baerson, S.R., Pan, Z., et al., 2013. Priming of jasmonate-mediated antiherbivore defense responses in rice by silicon. Proc. Natl. Acad. Sci. 110 (38), E3631–E3639.

Yruela, I., 2005. Copper in plants. Braz. J. Plant Physiol. 17 (1), 145–156.

Yu, M., Wang, Y.H., 2006. Effects of molybdenum on the precursors of chlorophyll biosynthesis in winter wheat cultivars under low temperature. Sci. Agric. Sin. 39, 702–708.

Zargar, M.S., Macha, A.M., Nazir, M., Kumar A., Rakwal, R., 2012. Silicon: multitalented micronutrient in OMICS perspective: an update. Curr. Proteom. 9 (4), 245–254.

Zhu, Y., Gong, H., 2014. Beneficial effects of silicon on salt and drought tolerance in plants. Agron Sustain Dev 34 (2), 455–472.

Zhu, Y.-G., Huang, Y.-Z., Hu, Y., Liu, Y.-X., 2003. Iodine uptake by spinach (Spinacia oleracea L.) plants grown in solution culture: effects of iodine species and solution concentrations. Environ. Int. 29, 33–37.

Chapter 3

Crop nitrogen use efficiency for sustainable food security and climate change mitigation

Bhumika Madan, Aakansha Malik and Nandula Raghuram
School of Biotechnology, Guru Gobind Singh Indraprastha University, Dwarka, India

3.1 Introduction

With the global rise in population predicted to reach 9 billion by 2050 coupled with net decrease in arable land area, increasing crop productivity is critical for global food security. Agriculture is not only affected by climate change but also contributes to it due to emission of methane, nitrous oxide (N_2O), oxides of nitrogen (NOx), ammonia (NH_3), etc., despite carbon sequestration by crop plants (Fagodiya et al., 2017, 2020).This is because intensive crop production relies heavily upon the exogenous application of nitrogenous fertilizers, as nitrogen (N) is quantitatively the most important nutrient input needed for plants. The deviation of modern monoculture from the traditional practice of having legumes as an essential component of all cropping systems prevented the symbiotic fixation of N_2 present abundantly in the atmosphere and instead promoted dependence on industrially manufactured N fertilizers such as urea, nitrate, and ammonium salts. A large fraction of this applied N is lost into the environment, as globally only 33% of applied N fertilizer is harvested as usable crop output. The loss of the rest as emissions into air as N_2O, NH_3, NOx, or to ground and surface waters as nitrates, nitrites, or ammonium salts represents a huge waste of this resource, negatively impacting the economy, health, and the environment.

3.2 Reactive nitrogen, climate change, and agriculture

Nitrous oxide is mainly produced by soil microbial action on already fixed N and is a greenhouse gas 300 times more potent than CO_2 over a century (Johnson et al., 2007), whereas NOx is 10 times more powerful. Nitrous oxide (N_2O) emission causes global warming and depletes the ozone layer,

Plant Nutrition and Food Security in the Era of Climate Change. DOI: https://doi.org/10.1016/B978-0-12-822916-3.00003-2

whereas the oxides of nitrogen (NO and NO_2) and ammonia (NH_3) cause air pollution and aerosol formation (Malyan et al., 2016). While a thorough global N assessment is still underway, regional N assessment for Europe (Sutton et al., 2011), national N assessment for India (Abrol et al., 2017), and preliminary global estimates have been made (Sutton et al., 2013; Winiwarter et al., 2018) regarding the contribution of agriculture to global N pollution (Fagodiya et al., 2017) and the shared socioeconomic pathways to sustainability (Kanter et al., 2020). The pioneering efforts of the International Nitrogen Initiative (INI) contributed hugely in this regard. The Indian effort was independently pioneered with the establishment of the Indian Nitrogen Group in 2006, and the South Asian Nitrogen Centre for INI in 2007 (Raghuram et al., 2007; Raghuram et al., 2020). The Indian nitrogen assessment in civil society made during the last decade has shown that N losses through nitrification, leaching, and volatilization pathways have already become a serious problem (Abrol et al., 2007, 2008, 2017), as elaborated below.

Over 70% of all N_2O emitted from India is attributed to agricultural soils (Bhattacharya et al., 2017), reflecting global trends (Winiwarter et al., 2018). Out of this, up to 80% is attributed to unused N fertilizers in the fields (Pathak and Bhatia, 2017; Fagodiya et al., 2020). Most of these fertilizers (over 69%) are used in food grain production, indicating their importance for India's food security (Tewatia and Chanda, 2017). The health impacts of nitrogen pollution have also been well documented from lab studies (Gupta et al., 2017), though wider impacts on public health need further assessment at national and global levels. In the meantime, there was growing recognition that reactive nitrogen crossed planetary boundaries of repair (Rockström et al., 2009; Steffen et al., 2015). These global and Indian developments were highlighted (Galloway et al., 2008; Sutton et al., 2017) at several international and intergovernmental forums, including the World Environment Day events hosted by the Indian government in 2018. They culminated in the UN Environment Programme identifying reactive nitrogen as one of the five most important threats to the planet (Sutton et al., 2019) and the adoption of the first-ever UN resolution on "sustainable nitrogen management" led by India (UNEP, 2019).

As agricultural N use is the largest contributor to N pollution, minimization of N losses from agriculture becomes the mainstay of sustainable nitrogen management. This will also result in huge economic savings, considering that the Indian government subsidizes 75% of the market price of urea, the predominant N fertilizer in India, at a cost of around Rs 50,000 crores per annum. Minimizing N losses requires improvement of crop nitrogen use efficiency (NUE), which was neglected along with other input-use efficiencies when yield became the sole criterion for crop improvement during the years of the Green Revolution. It is also likely that some traditional crop varieties that grew well under low-N-input conditions may have been rejected due to low yield despite high NUE. A nationwide low-input screening program is required to identify such germplasm for genetic improvement of crop NUE and other input-use efficiencies such as phosphorus, water, etc., apart from agronomic improvements. They should be aimed at enhancing the crop's N acquisition, utilization, and maximal translocation to the yield components (grain, fruit, leaf, tuber, etc., depending on the crop). In other words, reducing the environmental impact of intensive agricultural systems requires improved grain production technologies to maximize resource-use efficiencies (Nair, 2019).

3.3 Understanding nitrogen use efficiency

NUE is an inherited, multigenic, and quantitative trait comprising uptake, assimilation, and remobilization processes in plants. In the absence of any direct biological measure of NUE so far, it is derived as a ratio of output in the form of total biomass/grain yield or grain N and the input in the form of applied N through fertilizer or available N in the soil. Its two primary components are nitrogen uptake efficiency (NUpE, the efficiency of N uptake from the soil/environment) and nitrogen utilization efficiency (NUtE, the efficiency of utilizing and converting the absorbed N to the grain N). The plant NUE is defined in simpler terms as grain yield per unit of applied N, also defined by the multiplication of these two components (Pathak et al., 2011; Raghuram and Sharma, 2019). One important constituent of NUtE is how efficiently the plants remobilize the absorbed N from the source tissue (leaves) to the sink tissue (developing seeds) at the time of grain filling that contributes to the overall N content of the grains. This is known as the N remobilization efficiency of the plant. Both the estimations of NUE are important in different contexts. For example, in low-N conditions, NUpE contributes more to the overall NUE of the plant whereas in high-N conditions the contributions of both the components vary, as revealed in plant-breeding programs in several important crop species including rice, maize, and barley (Brauer and Shelp, 2010; Good et al., 2004). Increasing the efficiency of either uptake or utilization or both could lead to NUE improvement in crops (Beatty and Good, 2018).

For measuring NUE, several formulas exist in the literature, based on different parameters such as biomass, grain weight, and capacity of N uptake from the soil, as summarized in Table 3.1 (Lochab et al., 2007; Hawkesford and Barraclough, 2011; Raghuram and Sharma, 2019). Researchers often choose the appropriate definition of NUE depending on the crop, its harvest product, and the type of investigation, that is physiological, agronomic, and molecular. Their ultimate focus is on developing genotypes that provide higher yield under lower N (genetic N efficiency) while maintaining the ability to convert applied fertilizer into yield (N-responsiveness) (Han et al., 2015).

TABLE 3.1 Different formulas to calculate nitrogen use efficiency.

Term	Abbreviation	Formula/definition
Fertilizer use efficiency	FUE	(NUp/N applied)*100
Nitrogen use efficiency	NUE	NUpE*NUtE = yield/N available, Sw/N
Usage index	UI	Sw*(Sw/N)
Nitrogen use efficiency grain	NUE_g	Gw/Ns
Uptake efficiency	UpE	Nt/Ns
Utilization efficiency	UtE	Gw/Nt
Agronomic efficiency	AE	$(Gw_f-Gw_c)/N_f$
Apparent nitrogen recovery	AR	$(N_f$ uptake$-N_c$uptake$)/N_f*100$
Physiological efficiency	PE	$(Gw_f-Gw_c)/(N_f$ uptake$-N_c$uptake)
Nitrogen transport efficiency	NTE	Total N transported into the aboveground parts/total N in the whole plant
Nitrogen remobilization efficiency	NRE	N remobilization from senescent leaves/that of developing grains
Harvest index	HI	Mass of harvested fraction (grain)/total aboveground biomass (grain + straw)
Nitrogen harvest index	NHI	G_N/S_N*100

Abbreviations: G_N, grain N; Gw, grain weight; Gw_f, grain weight with fertilizer; Gw_c, grain weight without fertilizer (control); N, total nitrogen content of shoots; N_c uptake, plant nitrogen unfertilized control; N_f, nitrogen fertilizer applied; N_f uptake, plant nitrogen with fertilizer; Ns, nitrogen supplied in gram per plant; Nt, total nitrogen in plant; S_N, shoot N; Sw, shoot weight.
Source: Adapted from Hawkesford, M.J., Barraclough, P., 2011. An overview of nutrient use efficiency and strategies for crop improvement. In: The Molecular and Physiological Basis of Nutrient use Efficiency in Crops, 5–19; Raghuram, N., Sharma, N., 2019. Improving crop nitrogen use efficiency. In: Moo-Young, M. (Ed.), Comprehensive Biotechnology, vol. 4. Elsevier, Pergamon, pp. 211–220.

3.4 Agronomic approaches to improve nitrogen use efficiency

There exists a great variation among NUE in different regions due to the differences in cultivars used and farmers' practices. Yet, for any given cultivar and the prevailing agroclimatic conditions, short-term agronomic improvements are possible by identifying and promoting the right type of N fertilizer, its dosage, timing, frequency, and even the method of application, while balancing all other inputs. Several methods such as soil testing, plant tissue testing, spectral response, vegetative indexes (leaf area index), handheld sensors, and satellite imagery can be used for effective decisions to improve NUE, considering soil organic matter, residual N from the previous crop, and also the N available in irrigation water and manures (Sharma and Bali, 2018). These methods also help in identifying variable nitrogen management zones for formulating appropriate fertilizer recommendations accordingly.

The simplest and most effective and sustainable way to improve NUE is to adopt legume-based cropping systems to fix atmospheric N and minimize other N inputs. This can be done through traditional cereal-legume rotations, mixed cropping, and intercropping of legumes with nonlegumes, cover crops, etc. The different cropping systems can help in better uptake of nutrients from different depths of soil profile and further restoration of nutrient balance (Noor, 2017). Where this is not possible, manure, urine, compost, biochar, crop residue mulch, and other organic N sources can be supplemented with chemical N fertilizer as appropriate. Temporal and spatial management of fertilizer applications before planting and during growing seasons ensures minimum N loss in the form of leaching and volatilization. Similarly, split applications of N in accordance with growing stages helps the photosynthetic apparatus to enhance the grain N deposition with improved grain quality (Zhou et al., 2016). Many other factors, such as environment, soil type, crop type, and cropping system influence NUE.

Other agronomic options include the use of chlorophyll meters, leaf color charts, slow-release fertilizers, neem-coated urea, or fertilizers incorporating nitrification inhibitors that have the potential to reduce N losses and further increase NUE (Singh et al., 2019). Some polymer-coated fertilizers are also available that can control nutrient release rates depending on soil temperature and moisture conditions. Moreover, the use of decision-support tools along with integrated and precision nitrogen management and site-specific nitrogen management could be substantially used to enhance NUE for sustainable crop production (Jat et al., 2012).

3.5 Microbial nitrogen fixation and crop nitrogen use efficiency

The only biological components of the earth's nitrogen cycle were N-fixing microbes, nitrifying and denitrifying bacteria. They maintained the natural balance between nitrification and denitrification, even though anthropogenic legume cultivation tilted the balance slightly in favor of nitrification. This balance was altered heavily with the advent of synthetic nitrogen fertilizers produced from ammonia, a process invented by Haber and Bosch. Today, anthropogenic generation of reactive N is twice that of natural denitrification, leading to undesirable environmental accumulation due to the lack of human-made options for denitrification. Reducing the dependence on synthetic nitrogen fertilizers for sustainable agriculture requires better exploitation of microbial N fixation to deliver nitrogen to cereals or other nonlegume crops (Dent and Cocking, 2017; Pathak et al., 2011; Raghuram and Sharma, 2019). In 1988, a nonnodulating, nonrhizobial, nitrogen-fixing bacterium, *Gluconacetobacter diazotrophicus (Gd)* was isolated from intercellular juice of sugarcane that provided fresh momentum in this area of research. After that, the strains of *Gd* were inoculated in wheat, maize (corn) and rice, potato, tea, oilseed rape, grass, and tomato and the bacterium was successfully colonized in the roots and shoots of these crops. The other benefit was that it works under both presence and absence of synthetic nitrogen fertilizers. In some of the crops such as wheat, maize, oilseed rape, and grasses the significant improvement in yield was achieved attributed to the symbiotic nitrogen fixation, enhanced photosynthetic rates, and the presence of additional plant growth factors (Dent and Cocking, 2017).

Certain microbes have been identified to exhibit plant growth promoting (PGP) traits. Also the presence of various symbiotic N fixers, such as *Rhizobium, Bradyrhizobium, Mesorhizobium, Sinorhizobium*, and *Allorhizobium*, and nonsymbiotic N fixers, such as *Beijerinckia, Azotobacter, Azospirillum, Herbaspirillum, Gluconacetobacter, Burkholderia, Pseudomonas, Clostridium, Methanosarcina*, and *Paenibacillus*, provided us the opportunity to develop microbial biofertilizers. Their efficacy depends on many factors including soil conditions, but one of the important features of these PGP bacteria is the increased nitrogen acquisition by modification of root system architecture. Biofertilizers act synergistically with other measures of improving NUE in plants (Bargaz et al., 2018).

Recently, it has been shown that the mucilage secreted by the aerial roots of one of the indigenous landrace of maize contained diazotrophs and active nitrogenase activity. Field experiments over multiple years have shown that this maize landrace can acquire 29%–82% of its nitrogen from atmospheric dinitrogen (Van Deynze et al., 2018; Bennett et al., 2020). Whether this mechanism is operative in other cereal crops needs to be investigated further.

The idea of making symbiotic nitrogen-fixing cereals has been attracting molecular biologists for the last four decades with little success, but recent advances in synthetic biology are rekindling hope in this idea (Mus et al., 2016; Rosenblueth et al., 2018; Burén and Rubio, 2018; Bloch et al., 2020). In a recent finding, an engineered ammonium-excreting strain of *Azotobacter vinelandii* was shown to support growth of cucumber in low-nitrogen condition (Ambrosio et al., 2017). The recent expression of 16 nitrogenase proteins from the diazotroph *Klebsiella pneuomoniae* in the *Nicotiana benthamiana* mitochondrial matrix is another source of hope for engineering N fixation in higher plants, as mitochondria provide the most suitable prokaryotic ambience for the reconstitution and activity of nitrogenase complex (Allen et al., 2017).

Another important but underexplored area is the exploitation of more efficient microbial nitrate assimilatory systems for improving higher plant NUE. The nitrate-inducible abundant expression of long-lasting and thermostable N-assimilatory enzymes in *Arthrospira* (Spirulina) *platensis* has been shown in comparison with rice (Jha et al., 2007; Ali et al., 2008; Lochab et al., 2009) and their genes have been isolated and characterized (Lochab et al., 2014). However, as cyanobacterial research remained more focused on N fixation rather than nitrate assimilation, the potential of these findings to improve higher plant NUE remains underexplored.

3.6 Plant biological approaches to improve nitrogen use efficiency

NUE is a complex quantitative trait, which can be improved beyond agronomic management through genetic improvement of crops (Ranjan and Yadav, 2019). For several decades, physiological, biochemical, molecular, and genetic studies have been carried out on plant N assimilation (Pathak et al., 2008, 2011). An important aspect of this was the direct role of light as a signal and its indirect role through photosynthesis (Sharma et al., 1994; Raghuram and Sopory, 1995a; Pathak et al., 2020). Another aspect was sensing and signaling (Raghuram et al., 2006; Chakraborty and Raghuram, 2011), including the role of G-proteins, kinases/phosphatases (Chakraborty et al., 2015b; Chakraborty et al., 2019; Jaiswal and Raghuram, 2019; Kumari and Raghuram, 2020), and the segregation of light and nitrate effects on it (Raghuram and Sopory, 1995b; Raghuram et al., 1999; Ali et al., 2007b). The identification of nitrate-response elements has proved elusive for a long time (Das et al., 2007; Pathak et al., 2009), as was the interface between low nitrogen stress, other abiotic stresses, and NUE (Jangam and Raghuram, 2015), even as G-proteins were connecting them (Chakraborty et al., 2015a, b,c; Jangam et al., 2016).

Despite generating considerable knowledge, the complex nature of NUE and the lack of clear phenotype have hampered crop improvement (Plett et al., 2017; Mandal et al., 2018; Raghuram and Sharma, 2019; Sinha et al., 2020). The prediction of phenotype for NUE is difficult because its interaction with the environment can mask genetic gains. For example, different usable N compounds exist in the soil simultaneously, but they have different effects on plant N response (Raghuram and Sopory, 1999; Ali et al., 2007a), which makes field experiments difficult and necessitates carefully designed experiments in controlled and replicable environments (Hansen et al., 2018). This also requires methods for nutrient depletion of soils for the precise determination of plant N needs (Sharma et al., 2019). Using these methods, several components of the NUE phenotype have been recently identified in rice (Sharma et al., submitted).

Identifying the most important trait components of the NUE phenotype and shortlisting those that are amenable for automation can be a major aid in phenomics using proximal sensing technologies (Berry et al., 2018). They will further aid in the identification of quantitative trait locus (QTL) mapping and genome-wide association studies (GWAS) related to NUE (Chietera and Chardon, 2014; Han et al., 2016), though some QTL were found even earlier (Hirel et al., 2011). With the use of automated high-throughput phenotyping techniques several QTL such as Jasmonate Responsive 1 (JR1), Calcium Sensor Receptor, Root System Architecture 1 (RSA1), Phosphate 1, and *PhzC* have been identified in *Arabidopsis* as promising targets for NUE (Table 3.5). Recently, TOND1 and qNGR2/GR4 have been identified in rice (Zhang et al., 2020), as shown in Table 3.5.

3.7 Transgenic and genome-editing approaches for improving nitrogen use efficiency

NUE is a complex trait involving hundreds of genes that extends beyond nitrogen metabolism. Genetic manipulation studies have been concentrated on diverse targets including genes involved in nitrogen metabolic pathways, signaling, regulation and C-metabolism, etc. Recently the genes, microRNAs, and QTL involved in root system architecture are also emerging as prospective targets for NUE. Transgenic studies mainly consist of overexpression of target genes and assessing the effect of overexpression of candidate genes on plant growth, overall yield, and other parameters that are exclusively related with NUE. Few studies involving knockout mutants of these candidate genes have also been reported. Although transgenic studies involving constitutive overexpression of candidate genes using ubiquitous promoters have been successful but for significant increase in NUE, one should aim to achieve organ/cell specific expression using tissue-specific or native promoters of the genes (Pathak et al., 2011). The following paragraphs summarize these studies based on the processes of N metabolic pathways targeted.

3.8 Manipulation of genes involved in nitrogen acquisition

The first step in assimilatory pathway is nitrogen uptake from the soil in the form of water-soluble NO_3^- (most abundant form, preferable source in aerobic soils), NH_4^+ (preferable source in acidic soils), and to a lesser extent in the form of proteins, peptides, or amino acids mediated by various substrate-specific transporters present in the roots of plant cells. In Asian countries, fertilizers are supplied mainly in the form of urea that gets degraded in the soil to ammonium and CO_2 by urease. Nitrate, ammonium, and urea transport takes place in a biphasic manner including the low-affinity transport system (LATS) and high-affinity transport system (HATS), which are further categorized into inducible and constitutive. Nitrate transporters in plants are encoded by five gene families including Nitrate Transporter 1/ Peptide Transporter (NPF) family (NRT1/PTR family) (except NRT1.1 which is a dual-affinity transporter, all other members belong to LATS family), Nitrate Transporter 2 (NRT2) family (members belong to HATS family, some require NAR2 as a partner protein for their activity), the chloride channel family, slow anion channel-associated homologs (SLAC/SLAH), and aluminumaluminium-activated malate transporter1 (ALMT1) (Kant, 2018; Reddy and Ulaganathan, 2015). Ammonium uptake in plants is mediated by ammonium transporter/methyl ammonium permeases/Rhesus (AMT/MEP/ Rh) transporter gene family having several members. Urea uptake is mediated by major intrinsic proteins and DUR3 transporter gene family. N acquisition is a highly regulated process mediated by external signals (endogenous N status, rhizosphere pH, and water availability in rhizosphere), phosphorylation of transporters, transcription factors, and microRNAs (Xuan et al., 2017). These transporters have been used as the candidate genes in overexpression studies in various plants including tobacco, maize, *Arabidopsis*, and rice (Fan et al., 2017; Li et al., 2017; Zhang et al., 2020) and have shown enhanced N uptake in most cases but subsequent increase in NUE is not significant. These transporters are very well characterized in plants, but limited success has been achieved in overexpression studies of these transporter genes.

Various nitrate and ammonium transporters in rice have been overexpressed to study their effects on NUE. It is worth mentioning a recent review in this field (Zhang et al., 2020). Recently, overexpression of *Os*NRT1.1A resulted in enhanced grain yield and decreased maturation time attributed to the regulation of N utilization and flowering time in rice. *Os*NPF7.2 overexpression leads to improved grain yield in rice by promoting tillering. Overexpression of both splicing variants of *Os*NPF7.7 leads to increase in grain yield and enhanced NUE in rice. Similarly, it has been shown that overexpression of *Os*AMT1.1 leads to improved plant growth, grain yield even under suboptimal N fertilizer conditions, and also regulation of ammonium-potassium homeostasis (Zhang et al., 2020; Raghuram and Sharma, 2019).

3.9 Manipulation of genes involved in nitrogen assimilation

Various attempts of overexpression studies involving nitrate assimilation enzymes in different plants have been made as described in Table 3.2, with limited success. After nitrate is absorbed into roots, it is reduced to ammonia via an enzymatic process. First, nitrate (NO_3^-) is reduced to nitrite (NO_2^-) by nitrate reductase (NR/NIA) in a NAD (P) H-dependent manner in the

TABLE 3.2 A list of N-responsive genes (involved in N metabolic processes) with potential roles in nitrogen use efficiency in plants.

Gene name	Gene family	Target plant
Transporter genes		
*Os*NPF2.4	pH-dependent low-affinity nitrate transporter	Rice
*Os*NPF6.3/ *Os*NRT1.1A	Tonoplast-localized dual-affinity nitrate transporter	Rice
*Os*NPF6.5/ *Os*NRT1.1B	Dual-affinity nitrate transporter	Rice
*Os*NPF7.2	Low-affinity nitrate transporter	Rice
*Os*NPF7.3	Organic-nitrogen regulated vacuolar membrane peptide transporter	Rice
*Os*NPF7.7	Nitrate transporter	Rice
*Os*NAR2.1, *Os*NRT2.1	High-affinity nitrate transporter (under native promoter, p*Os*NAR2.1)	Rice
*Os*NRT2.3b	pH-sensitive high-affinity nitrate transporter	Rice
*Os*AMT1.1	Dual-affinity ammonium transporter	Rice
*Os*AMT2.1	Low-affinity ammonium transporter	Rice
*Os*DUR3	High-affinity urea transporter (Beier et al., 2019)	Rice
LHT, *Os*LHT1	Lysine histidine transporter, lysine-histidine-type transporter1 (Guo et al., 2020)	*Arabidopsis*, rice
*Vf*AAP1, *Os*AAP3	Amino acid transporters, amino acid permease 1 (Perchlik and Tegeder, 2017), amino acid permease 3	Pea, rice
STP13	Hexose transporter, sugar transport protein 13	*Arabidopsis*
N assimilation genes		
NR (Nia)	Nitrate reductase isoform 1	*Nicotina tabacum, Lactuca sativa*
NR (Nia2)	Nitrate reductase isoform 2	*Solanum tuberosum*
NiR	Ferredoxin-nitrite reductase	*Nicotiana plumbaginifolia, Arabidopsis*

(Continued)

TABLE 3.2 (Continued)

Amino acid biosynthesis genes		
GS2, *Ta*GS2	Plastidic glutamine synthetase isoform	*N. tabacum, Oryza sativa, Triticum aestivum*
Fd-GOGAT	Ferredoxin-dependent glutamine-oxoglutarate amidotransferase	*N. tabacum, O. sativa, T. aestivum*
GS1	Cytosolic glutamine synthetase	*N. tabacum, T. aestivum*, rice, maize
NADH-GOGAT	NADH-dependent glutamate synthase	*N. tabacum, O. sativa*, alfalfa
gdhA/GDH	NADP-dependent glutamate dehydrogenase	*N. tabacum*, rice, *Zea mays, Lycopersicon esculentum, T. aestivum*
AS1/ASN1	Glutamine-dependent asparagine synthetase 1	*N. tabacum, Arabidopsis*
ASNase	Asparaginase	Rice
AspAT/ASP	Mitochondrial aspartate aminotransferase	*N. tabacum, Arabidopsis*
AlaAT	Alanine aminotransferase (Shrawat et al., 2008)	*Brassica napus, Sorghum bicolor*, rice
C:N metabolism and storage genes		
Rubisco	Rubisco small subunit	Tobacco
FNR	Ferredoxin NADP(H) reductase	*Arabidopsis*, rice
IPT	Isopentenyl transferase	Tobacco, *Arabidopsis*
CIN	GRAIN INCOMPLETE FILLING 1 (GIF1) encodes cell wall invertase	Rice
C_4-PEPC	Phosphoenolpyruvate carboxylase (Tang et al., 2018)	Rice

Source: Adapted from Raghuram, N., Sharma, N. 2019. Improving crop nitrogen use efficiency. In: Comprehensive Biotechnology, vol. 4, Moo-Young, M., (Ed.), Elsevier:Pergamon; pp. 211–220; Lightfoot, D.A. 2009. Genes for use in improving nitrate use efficiency in crops, chapter 7. In: Genes for Plant Abiotic Stress, Wood, A.J., Jenks, M.A. (Eds.), Wiley Blackwell, New York, pp. 167-182; Zhang, Z., Gao, S., Chu, C. 2020. Improvement of nutrient use efficiency in rice: current toolbox and future perspectives. Theor. Appl. Genet., 1-20; Xuan, W., Beeckman, T., Xu, G. 2017. Plant nitrogen nutrition: sensing and signaling. Curr. Opin. Plant Biol., 39, 57-65; Fan, X., Naz, M., Fan, X., Xuan, W., Miller, A.J., Xu, G. 2017. Plant nitrate transporters: from gene function to application. J. Exp. Bot., 68(10), 2463-2475; Han, M., Wong, J., Su, T., Beatty, P.H., Good, A.G. 2016. Identification of nitrogen use efficiency genes in barley: searching for QTLs controlling complex physiological traits. Front. Plant Sci., 7, 1587; Beatty, P.H., Good, A.G. (2018). Improving nitrogen use efficient in crop plants using biotechnology approaches. In: Engineering Nitrogen Utilization in Crop Plants (pp. 15-35). Springer, Cham.

cytosol, which is also the rate-limiting step, and then nitrite (NO_2^-) is reduced further to ammonium (NH_4^+) by nitrite reductase (NiR) in the chloroplast and finally converted to amino acids via GS/GOGAT pathway involving plastidic glutamine synthetase (GS2) and glutamate synthase (Fd-GOGAT). This constitutes the primary nitrate assimilation pathway. The cytosolic isoforms of GS (GS1) and GOGAT (NADH-GOGAT) are involved in secondary ammonia assimilation and remobilization (Reddy and Ulaganathan, 2015). These nitrate assimilation enzymes have been well characterized in different plants.

Nitrate reductase overexpression has already been studied in various plants, mostly dicots, resulting in no improvement of NUE. However, a recent study has shown that overexpression of *Os*NAR2.1 under its native promoter resulted in increased N uptake, grain yield, and NUE in rice, suggesting that this enzyme is more suitable as a promising target for NUE in monocots. Similarly, primary and secondary ammonia assimilation enzymes have already been tried for genetic manipulation studies but the results have not been very conclusive. In a recent study in wheat, overexpression of GS2 has reported an increased grain yield and NUE. Recently, Fd-GOGAT has also been shown to be colocalized with a marker that is a QTL for grain protein content (GPC) in durum wheat. It is also useful to study GPC (NUE-related trait) in other plants. Overexpression studies of GS1 have been carried out in many cereals including rice and have resulted in higher grain yield and improved N utilization efficiency suggesting GS1 as a strong candidate gene for NUE (Lightfoot, 2009; Raghuram and Sharma, 2019).

3.10 Manipulation of genes involved in nitrogen translocation and remobilization

Remobilization of organic N compounds is common in seasonal crops from their old senescing plant parts to growing vegetative parts or to the storage organs during the reproductive phase. It is one of the key steps for NUE improvement in plants as it is the major contributor (60%–92%) to seed nitrogen content in cereals. The amount of N remobilized depends on many factors including N remobilization efficiency, available nitrogen in the plant, environmental factors, and genotype. For remobilization of N from leaves, chloroplast and its proteins along with Rubisco are degraded by autophagy and vacuolar proteases during developmental senescence, making autophagy and senescence-associated genes promising targets for NUE (Reddy and Ulaganathan, 2015). Besides these genes, enzymes such as asparagine synthetase (AS), alanine aminotransferase (AlaAT), aspartate aminotransferases (AspAT), glutamate dehydrogenase (GDH), and amino acid transporters such as amino acid permeases (AAP) and lysine-histidine like transporters (LHT) and transporters such as DUR3 and *Os*NPF7.3 are also reported to play important roles in N translocation, making them potential targets for NUE (Lightfoot, 2009; Han et al., 2016).

Few recent studies in this field have reported that overexpression of a peptide transporter, *Os*NPF7.3 in rice, caused increased N allocation to grains resulting in enhanced grain yield and NUtE. Inhibition of *Os*AAP3 expression by creating a knockout line in rice promoted bud outgrowth, leading to larger numbers of tillers that resulted in increased grain yield (Raghuram and Sharma, 2019). Engineered pea plants overexpressing amino acid permease 1(AAP1) allocated more nitrogen to seeds resulting in more biomass and higher seed yields (Perchlik and Tegeder, 2017). In one study, transgenic canola and rice were created by introducing barley AlaAT cDNA driven by a root-specific promoter (btg26) and a rice tissue-specific promoter (*Os*Ant1) resulted in high biomass, grain yield, and total N content leading to enhanced NUpE (a strong NUE phenotype) (Shrawat et al., 2008).

Recently, DUR3, a urea transporter, has shown to contribute to grain filling and yield under limiting field conditions of nitrogen in rice by mutant study (Beier et al., 2019). Effective tillering in rice was promoted by overexpression of *Os*ATG8a leading to increased yield, grain N content, and NUpE (Yu et al., 2019). *Os*ATG8b overexpression has shown to be involved in increased N recycling efficiency leading to better grain quality in rice (Fan et al., 2020). Similarly, overexpression of *Md*ATG18a resulted in enhanced tolerance to N depletion in both *Arabidopsis* and apple (Raghuram and Sharma, 2019). In a recent study by mutant analysis, it has been found that *Os*LHT1 transporter in rice functions in both root uptake and root-to-shoot allocation of various amino acids in rice (Guo et al., 2020). This study gives the opportunity to explore its role in nitrogen translocation and NUE.

3.11 Manipulation of genes involved in carbon metabolism and its regulation

N and C metabolism are tightly linked at the cellular and whole-plant level regulated by glutamate, and maintaining C − N balance is crucial for plant growth and metabolism. Sugars are the source of carbon skeletons in the form of organic acids for amino acid synthesis in plants and they also influence various processes such as seed germination, embryogenesis, flowering, senescence, and hormone signaling. It has been reported that NRT2.1 acts a sensor of C/N ratio in plants. Since N metabolism is indirectly regulated by photosynthetic rates, one should also aim to improve photosynthesis when aiming for high NUE (Li et al., 2017).

Overexpression of a hexose transporter (STP13) and a protein kinase (SnRK1) has been shown to result in increased carbon metabolism and nitrogen uptake and use, reiterating the importance of C − N metabolism balance in plants for NUE and overall yield (Raghuram and Sharma, 2019). Recently, overexpression of maize C_4-PEPC phosphoenolpyruvate carboxylase in rice conferred low-nitrogen tolerance and resulted in higher grain yield per plant than wild-type by regulating photorespiration (Tang et al., 2018). Transgenic tobacco plants overexpressing isopentenyl transferase (IPT), an enzyme involved in cytokinin synthesis, resulted in high NUE in N-limiting conditions. GIF1 gene encoding cell wall invertase was shown to be required for carbon partitioning during early grain filling and its overexpression resulted in increased grain production in rice (Beatty and Good, 2018). The list of potential NUE genes involved in N metabolic pathways is summarized in Table 3.2.

3.12 Manipulation of genes involved in signaling

Nitrate is a key signaling molecule that modulates expression of thousands of genes that are involved in a wide range of diverse processes such as plant growth, RSA, leaf development, seed dormancy, and flowering time. Nitrate-induced gene expression is dependent on NPF6.3 sensor protein and mediated by calcium-dependent and calcium-independent pathways in roots (O'Brien et al., 2016). A comprehensive list of genomewide nitrate-responsive genes in *Arabidopsis* and rice has been prepared from analysis of microarray data in our lab (Pathak et al., 2011). Transcription factors are master regulators that coordinate the expression of an entire set of genes involved in a particular pathway making them attractive targets for NUE in plants. Several transcription factor gene families are implicated to be involved in regulating nitrate response including ANR1, NLP7, TGA1, Dof1, LBD37, SPL9, and TCP20. However, only few studies involving overexpression of these transcription factors have shown significant changes in NUE and/or N-response. These include Dof1, ANR1, TaNAC2−5A, SMOS1, NLP7, and PGL1.

Grain length is also one of the determinants of yield in rice; recent studies have reported the transgenic overexpression of PGL1 and Gnp4/LAX2 TFs in this context. Overexpression of PGL1 resulted in increased grain length and size in transgenic rice. Overexpression of Gnp4/LAX2 resulted in increased grain length and thousand-kernel weight in rice (Raghuram and Sharma, 2019). Recently, NRG5 (NITROGENMEDIATED TILLER GROWTH RESPONSE 5) has been reported to promote tillering in rice via chromatin modulation (Wu et al., 2020). This study has implicated epigenetic regulation as a new aspect of NUE that needs further validation. Many of these transgenic studies have been focused on *Arabidopsis*; investigations for similar mechanisms should also be done in other crops especially monocots, which account for the major part of cereal crops worldwide. The signaling targets and transcription factors for NUE are summarized in Tables 3.3 and 3.4 (McAllister et al., 2012). Other N-related genes are summarized in Table 3.5.

TABLE 3.3 A list of N-responsive genes (involved in growth and development processes) with potential roles in nitrogen use efficiency in plants.

Gene name	Gene family	Target plant
Signaling genes		
DEP1/ qNGR9	G-protein γ subunit, DENSE AND ERECT PANICLES 1 (Zhao et al., 2019)	Rice
SMG1	SMALL GRAIN 1 encodes Mitogen-activated protein kinase 4	Rice
SnRK1	Sucrose nonfermenting-1-related protein kinase 1	*Lycopersicon esculentum*
GLB1	P II N sensing and regulatory protein	*Arabidopsis*
CIPK8,CIPK23	Calcineurin B-like interacting protein kinases	
CEP	C-terminally Encoded Peptides, mobile signaling molecule	
CLE and CLV1 kinase	CLAVATA3/ESR-related peptides and CLAVATA 1 kinase	*Arabidopsis*
NRG2	Nitrate Regulatory Gene2	*Arabidopsis*
MEKK1	MAP3 kinase gene	*Arabidopsis*
iGluR, *At*GluR2	Ionotropic glutamate receptors	*Arabidopsis, Zea mays*
GCN2 kinase	General amino acid control nonderepressible 2 kinase	*Arabidopsis*
*Os*ENOD-93-1	Early nodulinlike protein (mitochondrial membrane protein)	Rice
Senescence-associated genes		
CKX2 mutation	Cytokinin oxidase/dehydrogenase	Rice
IPT	Isopentenyl transferase, cytokinin biosynthesis gene	*Nicotina tabacum, Z. mays, Arabidopsis*
Sgr-mutation	Stay-green protein	Rice

(*Continued*)

TABLE 3.3 (Continued)

*Os*ATG8a, *Os*ATG8b	Autophagy-related gene 8a (Yu et al., 2019), autophagy-related gene (Fan et al., 2020)	Rice
MdATG18a	Autophagy-related gene 18	*Arabidopsis*, apple

Source: Adapted from Raghuram, N. and Sharma, N. 2019. Improving crop nitrogen use efficiency. In: Comprehensive Biotechnology, vol. 4, Moo-Young, M., (Ed.), Elsevier:Pergamon; pp. 211−220; Lightfoot, D.A. 2009. Genes for use in improving nitrate use efficiency in crops, chapter 7. In: Genes for Plant Abiotic Stress, Wood, A.J., Jenks, M.A. (Eds.), Wiley Blackwell, New York, pp. 167-182; Mandal, V., Sharma, N., Raghuram, N. 2018. Molecular targets for improvement of crop nitrogen-use efficiency: current and emerging options. In: Engineering Nitrogen Utilization in Crop Plants, Shrawat, A., Zayed, A., Lightfoot, D.A. (Eds) Springer, ISBN: 978-3-319-92958-3 pp. 77-93; Xuan, W., Beeckman, T., Xu, G. 2017. Plant nitrogen nutrition: sensing and signaling. Curr. Opin. Plant Biol., 39, 57-65; Han, M., Wong, J., Su, T., Beatty, P.H., Good, A.G. 2016. Identification of nitrogen use efficiency genes in barley: searching for QTLs controlling complex physiological traits. Front. Plant Sci., 7, 1587; McAllister, C.H., Beatty, P.H., Good, A.G. 2012. Engineering nitrogen use efficient crop plants: the current status. Plant Biotechnol. J., 10(9), 1011-1025.

3.13 Conclusions

The list of potential targets for NUE manipulation is continuously expanding regarding genes, QTL, and miRNAs from diverse areas including genes involved in RSA, hormone signaling, C−N metabolic coordination, senescence, grain physiology, and epigenetic regulation apart from processes involved in N metabolic pathway. Precise spatial-temporal expression of the candidate gene should be taken into account in genetic manipulation studies of plants. Major agronomic improvements can be gained in the short term by incorporating legumes in every cropping system, replacing as much of the N fertilizer as possible with manure, compost, biochar, biofertilizers, and other organic sources, demand-based application of N-fertilizers using a leaf color chart, precision farming with drip irrigation coupled with fertigation, use of fertilizers incorporating nitrification inhibitors, etc. Introduction of prokaryotic N-fixing machinery into higher plants will take time and even if it achieves scientific success, its commercial success depends on the attitudes of farmers, consumers, and governments toward transgenic or genome-editing technologies. This leaves ample scope for continuing the quest for identification of NUE phenotype and the underlying genes/loci for crop improvement through marker-assisted selection and breeding. The advancements in phenomics and other measuring techniques for precise phenotyping of NUE-related traits and in other fields such as transcriptomics, proteomics, metabolomics, synthetic biology, and computational biology need to be integrated to fast-forward the process of shortlisting candidate genes or pathways

TABLE 3.4 A list of N-responsive transcription factor genes with potential roles in nitrogen use efficiency in plants.

Gene name	Gene family	Target plant
Transcription factor genes		
*Zm*Dof1,*Os*DOF18	DNA-binding with one finger 1 transcription factor, *Oryza sativa* DNA *binding with one finger* 18 gene (Wu et al., 2017)	Wheat, sorghum, *Arabidopsis*, rice
ANR1	*Arabidopsis* nitrate regulated 1 encodes MADS box family of TF	*Arabidopsis*
SAT1	Symbiotic Ammonium Transporter 1 encodes a bHLH transcription factor	Soybean
NF-Y, *At*NF-YB1, *Zm*NF-YB2,*Ta*NFYA-B1, *Cdt*-NF-YC1	B & C subunits of Plant Nuclear Factor Y family TF (CCAAT-box binding TF)	*Arabidopsis*, *Zea mays*, *Triticum aestivum*, rice
NLP7, NLP6	NIN LIKE PROTEIN 7, RWP-RK TF family	*Arabidopsis*
*Ta*NAC2–5A, *Hv*NAC005, *Os*Y37	NAM, ATAF, and CUC (NAC) transcription factor gene family, *Hv*NAC005 gene encodes a protein that belongs to subgroup NAC-a6 of NAC TF family, *Oryza sativa* Yellow37, ONAC011,NAC transcription factor gene (El Mannai et al., 2017)	*T. aestivum*, barley, rice
*Os*NAP	No apical meristem, *Arabidopsis* ATAF1/2, and cup-shaped cotyledon 2 (NAC) family TF	Rice
NGR5	NITROGEN-MEDIATED TILLER GROWTH RESPONSE 5 transcription factor gene (Wu et al., 2020)	Rice
TGA1, TGA4	TGACG MOTIF-BINDING FACTOR, a bZIP TF family protein	*Arabidopsis*
bZIP1	Basic-leucine zipper transcription factor 1 gene family	*Arabidopsis*
LBD37, LBD38, LBD39	Lateral organ boundary (LOB) Domain-LBD gene family of TF	*Arabidopsis*
TCP20	TEOSINTE BRANCHED1/ CYCLOIDEA/PROLIFERATING CELL FACTOR1–20, TCP family TF	*Arabidopsis*

(*Continued*)

TABLE 3.4 (Continued)

SPL9	SQUAMOSA PROMOTER BINDING PROTEIN-LIKE 9, SPB-box TF	*Arabidopsis*
GNC and CGA1/GNL	GATA family transcription factors, GNC (GATA, NITRATE-INDUCIBLE, CARBON-METABOLISM-INVOLVED) and CGA1/GNL (CYTOKININ-RESPONSIVE GATA 1/ GNC-LIKE)	*Arabidopsis*
HY5	ELONGATED HYPOCOTYL, light-responsive bZIP TF	*Arabidopsis*
AGL21	Agamouslike MADS-box protein	*Arabidopsis*
*Os*RDD1	Rice Dof daily fluctuations 1(RDD1) DoF family TF gene	Rice
Hap-2−3-5-Gln3 transcript reg.	Transcriptional activation complex, Haeme activator protein 3 (HAP3) TF family	*Saccharomyces cerevisiae*
HRS1	HYPERSENSITIVE TO LOW Pi-ELICITED PRIMARY ROOT SHORTENING1, GARP family TF	*Arabidopsis*
*Os*MADS25, *Os*MADS27, *Os*MADS57	MADS-box TF family protein	Rice
*Os*BT	Bric-a-Brac/Tramtrack/Broad (BTB) family TF	Rice
*Os*IDD10	INDETERMINATE DOMAIN 10 /C2H2 family TF	Rice
SMOS1	SMALL ORGAN SIZE1, AP2-type transcription factor	Rice
PGL1	POSITIVE REGULATOR OF GRAIN LENGTH1, an atypical bHLH TF	Rice
*Gm*NMHC5	Sucrose-regulatory MADS-box TF	*Glycine max*

Source: Adapted from Raghuram, N. and Sharma, N. 2019. Improving crop nitrogen use efficiency. In: Comprehensive Biotechnology, vol. 4, Moo-Young, M., (Ed.), Elsevier:Pergamon; pp. 211−220; Mandal, V., Sharma, N., Raghuram, N. 2018. Molecular targets for improvement of crop nitrogen-use efficiency: current and emerging options. In: Engineering Nitrogen Utilization in Crop Plants, Shrawat, A., Zayed, A., Lightfoot, D.A. (Eds) Springer, ISBN: 978-3-319-92958-3 pp. 77-93; Xuan, W., Beeckman, T., Xu, G. 2017. Plant nitrogen nutrition: sensing and signaling. Curr. Opin. Plant Biol., 39, 57-65; Beatty, P.H., & Good, A.G. 2018. Improving nitrogen use efficient in crop plants using biotechnology approaches. In: Engineering Nitrogen Utilization in Crop Plants (pp. 15-35). Springer, Cham; O'Brien, J.A., Vega, A., Bouguyon, E., Krouk, G., Gojon, A., Coruzzi, G., et al. 2016. Nitrate transport, sensing, and responses in plants. Mol. Plant, 9(6), 837-856; Li, H., Hu, B., Chu, C. 2017. Nitrogen use efficiency in crops: lessons from Arabidopsis and rice. J. Exp. Bot., 68(10), 2477-2488; Zhang, Z., Gao, S., Chu, C. 2020. Improvement of nutrient use efficiency in rice: current toolbox and future perspectives. Theor. Appl. Genet., 1-20. Kant, 2018.

TABLE 3.5 **A list of N-responsive genes (involved in miscellaneous processes related with nitrogen use efficiency) with potential roles in nitrogen use efficiency in plants.**

Gene name	Gene family	Target plant
Other genes		
CCA1	Central clock gene 1/ Master Clock Control gene 1	*Arabidopsis*
JR1	JASMONATE RESPONSIVE 1 gene	*Arabidopsis*
Calcium sensor receptor	Gene locus associated with root growth rate	*Arabidopsis*
PhcZ	Gene associated with root trait phenotype	*Arabidopsis*
RSA1	Root system architecture 1 gene	*Arabidopsis*
PHO1	PHOSPHATE 1 gene	*Arabidopsis*
KURZ AND KLEIN	F-box family gene, regulator of root development	*Arabidopsis*
14-3-3 proteins	Regulatory protein (regulates nitrate reductase)	*Arabidopsis*
HNI9/IWS1	High Nitrogen Insensitive 9 (HNI9)/Interact With STP6 (IWS1), epigenetic regulator	*Arabidopsis*
SAUR39	Small Auxin-Up RNA 39 gene	Rice
TAR2	Tryptophan Aminotransferase Related 2 gene (Auxin biosynthetic gene)	*Arabidopsis*
Gnp4/LAX2	NG-finger and *w*d40-*a*ssociated *u*biquitin-*l*ike (RAWUL) domain-containing protein	Rice
APO1	ABERRANT PANICLE ORGANIZATION 1, a gene at PBN6 QTL encodes F-Box protein	Rice
*Br*UGE	UDP-glucose 4-epimerase	*Brassica rapa*
Atl31	ATL31 UBI-ligase	*Arabidopsis*
TOND1	Tolerance Of Nitrogen Deficiency 1 encodes Thaumatin protein (important gene present in QTL for NUE)	Rice
qNGR2/GR4	Growth-regulating factor	Rice

(*Continued*)

TABLE 3.5 (Continued)

Gene name	Gene family	Target plant
DRO1	DEEPER ROOTING 1, a gene at QTL for root growth angle in rice	Rice

Source: Adapted from Raghuram, N. and Sharma, N. 2019. Improving crop nitrogen use efficiency. In: Comprehensive Biotechnology, vol. 4, Moo-Young, M., (Ed.), Elsevier:Pergamon; pp. 211–220; Mandal, V., Sharma, N., Raghuram, N. 2018. Molecular targets for improvement of crop nitrogen-use efficiency: current and emerging options. In: Engineering Nitrogen Utilization in Crop Plants, Shrawat, A., Zayed, A., Lightfoot, D.A. (Eds) Springer, ISBN: 978-3-319-92958-3 pp. 77-93; Xuan, W., Beeckman, T., Xu, G. 2017. Plant nitrogen nutrition: sensing and signaling. Curr. Opin. Plant Biol., 39, 57-65; Beatty, P.H., & Good, A.G. 2018. Improving nitrogen use efficient in crop plants using biotechnology approaches. In: Engineering Nitrogen Utilization in Crop Plants (pp. 15-35). Springer, Cham; Li, H., Hu, B., Chu, C. 2017. Nitrogen use efficiency in crops: lessons from Arabidopsis and rice. J. Exp. Bot., 68(10), 2477-2488; Zhang, Z., Gao, S., Chu, C. 2020. Improvement of nutrient use efficiency in rice: current toolbox and future perspectives. Theor. Appl. Genet., 1-20; Kant, 2018.

involved in NUE for crop improvement in the medium to long term for sustainable and climate-smart agriculture.

Acknowledgments

This work was supported in part by research Grant to Nandula Raghuram and fellowship to Aakansha Malik under the UKRI-GCRF project titled South Asian Nitrogen Hub (SANH) (NE/S009019/1). Bhumika Madan is a recipient of CSIR fellowship (09/806 (0038)/2019-EMR-I).

References

Abrol, Y.P., Raghuram, N., Sachdev, M.S. (Eds.), 2007. Agricultural Nitrogen Use and Its Environmental Implications. IK International Pvt Ltd., p. 552.

Abrol, Y.P., Raghuram, N., Hoysall, C., 2008. Special issue on reactive nitrogen in agriculture, environment and health (Guest Editors) Curr. Sci. 94 (11), 1343–1477.

Abrol, Y.P., Adhya, T.K., Aneja, V.P., Raghuram, N., Pathak, H., Kulshrestha, U., et al., 2017. The Indian Nitrogen Assessment: Sources of Reactive Nitrogen, Environmental and Climate Effects, Management Options, and Policies. Elsevier, ISBN: 978-0-12-811836-8, 538p.

Ali, A., Sivakami, S., Raghuram, N., 2007a. Effect of nitrate, nitrite, ammonium, glutamate, glutamine and 2-oxoglutarate on the RNA levels and enzyme activities of nitrate reductase and nitrite reductase in rice. Physiol. Mol. Biol. Plants 13 (1), 17.

Ali, A., Sivakami, S., Raghuram, N., 2007b. Regulation of activity and transcript levels of NR in rice (*Oryza sativa*): roles of protein kinase and G-proteins. Plant Sci. 172 (2), 406–413.

Ali, A., Jha, P., Sandhu, K.S., Raghuram, N., 2008. Spirulina nitrate-assimilating enzymes (NR, NiR, GS) have higher specific activities and are more stable than those of rice. Physiol. Mol. Biol. Plants 14 (3), 179–182.

Allen, R.S., Tilbrook, K., Warden, A.C., Campbell, P.C., Rolland, V., Singh, S.P., et al., 2017. Expression of 16 nitrogenase proteins within the plant mitochondrial matrix. Front. Plant Sci. 8, 287.

Ambrosio, R., Ortiz-Marquez, J.C.F., Curatti, L., 2017. Metabolic engineering of a diazotrophic bacterium improves ammonium release and biofertilization of plants and microalgae. Metabol. Eng. 40, 59–68.

Bargaz, A., Lyamlouli, K., Chtouki, M., Zeroual, Y., Dhiba, D., 2018. Soil microbial resources for improving fertilizers efficiency in an integrated plant nutrient management system. Front. Microbiol. 9, 1606.

Beatty, P.H., Good, A.G., 2018. Improving nitrogen use efficient in crop plants using biotechnology approaches. Engineering Nitrogen Utilization in Crop Plants. Springer, Cham, pp. 15–35.

Beier, M.P., Fujita, T., Sasaki, K., Kanno, K., Ohashi, M., Tamura, W., et al., 2019. The urea transporter DUR3 contributes to rice production under nitrogen-deficient and field conditions. Physiol. Plant 167 (1), 75–89.

Bennett, A.B., Pankievicz, V.C., Ané, J.M., 2020. A model for nitrogen fixation in cereal crops. Trends Plant Sci. 25 (3), 226–235.

Berry, J.C., Fahlgren, N., Pokorny, A.A., Bart, R.S., Veley, K.M., 2018. An automated, high-throughput method for standardizing image color profiles to improve image-based plant phenotyping. PeerJ 6, e5727.

Bhattacharya, S., Adhya, T.K., Pathak, H., Raghuram, N., Sharma, C., 2017. Issues and policies for reactive nitrogen management in the Indian region. In: Abrol, Y.P., Adhya, T.K., Aneja, V.P., Raghuram, N., Pathak, H., Kulshrestha, U., Sharma, C., Singh, B. (Eds.), The Indian Nitrogen Assessment: Sources of Reactive Nitrogen, Environmental and Climate Effects, Management Options, and Policies. Elsevier, pp. 491–513. , ISBN: 978-0-12-811836-8.

Bloch, S.E., Ryu, M.H., Ozaydin, B., Broglie, R., 2020. Harnessing atmospheric nitrogen for cereal crop production. Curr. Opin. Biotechnol. 62, 181–188.

Brauer, E.K., Shelp, B.J., 2010. Nitrogen use efficiency: re-consideration of the bioengineering approach. Botany 88 (2), 103–109.

Burén, S., Rubio, L.M., 2018. State of the art in eukaryotic nitrogenase engineering. FEMS Microbiol. Lett. 365 (2), fnx274.

Chakraborty, N., Raghuram, N., 2011. Nitrate sensing and signaling in genomewide plant N response. In: Jain, V., Anandakumar, P. (Eds.), Nitrogen Use Efficiency in Plants. New India Publishing Agency, New Delhi, pp. 45–62.

Chakraborty, N., Sharma, P., Kanyuka, K., Pathak, R.R., Choudhury, D., Hooley, R.A., et al., 2015a. Transcriptome analysis of *Arabidopsis* GCR1 mutant reveals its roles in stress, hormones, secondary metabolism and phosphate starvation. PLoS One 10 (2), e0117819.

Chakraborty, N., Sharma, P., Kanyuka, K., Pathak, R.R., Choudhury, D., Hooley, R., et al., 2015b. G-protein α-subunit (GPA1) regulates stress, nitrate and phosphate response, flavonoid biosynthesis, fruit/seed development and substantially shares GCR1 regulation in A. thaliana. Plant Mol. Biol. 89 (6), 559–576.

Chakraborty, N., Singh, N., Kaur, K., Raghuram, N., 2015c. G-protein signaling components GCR1 and GPA1 mediate responses to multiple abiotic stresses in *Arabidopsis*. Front. Plant Sci. 6, 1000.

Chakraborty, N., Kanyuka, K., Jaiswal, D.K., Kumar, A., Arora, V., Malik, A., et al., 2019. GCR1 and GPA1 coupling regulates nitrate, cell wall, immunity and light responses in *Arabidopsis*. Sci. Rep. 9 (1), 1–17.

Chietera, G., Chardon, F., 2014. Natural variation as a tool to investigate nutrient use efficiency in plants. Nutrient Use Efficiency in Plants. Springer, Cham, pp. 29–50.

Das, S.K., Pathak, R.R., Choudhury, D., Raghuram, N., 2007. Genomewide computational analysis of nitrate response elements in rice and *Arabidopsis*. Mol. Genet. Genomics 278 (5), 519–525.

Dent, D., Cocking, E., 2017. Establishing symbiotic nitrogen fixation in cereals and other non-legume crops: the greener nitrogen revolution. Agric. Food Secur. 6 (1), 1–9.

El Mannai, Y., Akabane, K., Hiratsu, K., Satoh-Nagasawa, N., Wabiko, H., 2017. The NAC transcription factor gene OsY37 (ONAC011) promotes leaf senescence and accelerates heading time in rice. Int. J. Mol. Sci. 18 (10), 2165.

Fagodiya, R.K., Pathak, H., Kumar, A., Bhatia, A., Jain, N., 2017. Global temperature change potential of nitrogen use in agriculture: a 50-year assessment. Sci. Rep. 7, 44928.

Fagodiya, R.K., Pathak, H., Bhatia, A., Jain, N., Kumar, A., Malyan, S.K., 2020. Global warming impacts of nitrogen use in agriculture: an assessment for India since 1960. Carbon Manage. 11 (3), 291–301.

Fan, X., Naz, M., Fan, X., Xuan, W., Miller, A.J., Xu, G., 2017. Plant nitrate transporters: from gene function to application. J. Exp. Bot. 68 (10), 2463–2475.

Fan, T., Yang, W., Zeng, X., Xu, X., Xu, Y., Fan, X., et al., 2020. A rice autophagy gene OsATG8b is involved in nitrogen remobilization and control of grain quality. Front. Plant Sci. 11.

Galloway, J., Raghuram, N., Abrol, Y.P., 2008. A perspective on reactive nitrogen in a global, Asian and Indian context. Curr. Sci. 94 (11).

Good, A.G., Shrawat, A.K., Muench, D.G., 2004. Can less yield more? Is reducing nutrient input into the environment compatible with maintaining crop production? Trends Plant Sci. 9 (12), 597–605.

Guo, N., Hu, J., Yan, M., Qu, H., Luo, L., Tegeder, M., et al., 2020. *Oryza sativa* Lysine-Histidine-type Transporter 1 functions in root uptake and root-to-shoot allocation of amino acids in rice. Plant J. 103 (1), 395–411.

Gupta, S.K., Gupta, A.B., Gupta, R., 2017. Pathophysiology of nitrate toxicity in humans in view of the changing trends of the global nitrogen cycle with special reference to India. In: Abrol, Y.P., Adhya, T.K., Aneja, V.P., Raghuram, N., Pathak, H., Kulshrestha, U., Sharma, C., Singh, B. (Eds.), The Indian Nitrogen Assessment: Sources of Reactive Nitrogen, Environmental and Climate Effects, Management Options, and Policies. Elsevier, pp. 459–468. , ISBN: 978-0-12-811836-8.

Hansen, N.J.S., Plett, D., Berger, B., Garnett, T., 2018. Tackling nitrogen use efficiency in cereal crops using high-throughput phenotyping. Engineering Nitrogen Utilization in Crop Plants. Springer, pp. 121–139.

Han, M., Okamoto, M., Beatty, P.H., Rothstein, S.J., Good, A.G., 2015. The genetics of nitrogen use efficiency in crop plants. Annu. Rev. Genet 49 (49), 269–289.

Han, M., Wong, J., Su, T., Beatty, P.H., Good, A.G., 2016. Identification of nitrogen use efficiency genes in barley: searching for QTLs controlling complex physiological traits. Front. Plant Sci. 7, 1587.

Hawkesford, M.J., Barraclough, P., 2011. An overview of nutrient use efficiency and strategies for crop improvement. The Molecular and Physiological Basis of Nutrient Use Efficiency in Crops. pp. 5–19.

Hirel, B., Tétu, T., Lea, P.J., Dubois, F., 2011. Improving nitrogen use efficiency in crops for sustainable agriculture. Sustainability 3 (9), 1452–1485.

Jaiswal, D.K., Raghuram, N., 2019. Nutrient perception and signaling in plants. Sensory Biology of Plants. Springer, Singapore, pp. 59–77.

Jangam, A.P., Raghuram, N., 2015. Nitrogen and stress. Elucidation of Abiotic Stress Signaling in Plants. Springer, New York, NY, pp. 323–339.

Jangam, A.P., Pathak, R.R., Raghuram, N., 2016. Microarray analysis of rice d1 (RGA1) mutant reveals the potential role of G-protein alpha subunit in regulating multiple abiotic stresses such as drought, salinity, heat, and cold. Front. Plant Sci. 7, 11.

Jat, R.A., Wani, S.P., Sahrawat, K.L., Singh, P., Dhaka, S.R., Dhaka, B.L., 2012. Recent approaches in nitrogen management for sustainable agricultural production and eco-safety. Arch. Agron. Soil Sci. 58 (9), 1033–1060.

Jha, P., Ali, A., Raghuram, N., 2007. Nitrate-induction of nitrate reductase and its inhibition by nitrite and ammonium ions in *Spirulina platensis*. Physiol. Mol. Biol. Plants 13 (2), 163.

Johnson, J.M.F., Franzluebbers, A.J., Weyers, S.L., Reicosky, D.C., 2007. Agricultural opportunities to mitigate greenhouse gas emissions. Environ. Pollut. 150 (1), 107–124.

Kanter, D.R., Winiwarter, W., Bodirsky, B.L., Bouwman, L., Boyer, E., Buckle, S., et al., 2020. A framework for nitrogen futures in the shared socioeconomic pathways. Global Environ. Change 61, 102029.

Kant, S., 2018. Understanding nitrate uptake, signaling and remobilisation for improving plant nitrogen use efficiency, Seminars in Cell & Developmental Biology, vol. 74. Academic Press, pp. 89–96.

Kumari, S., Raghuram, N., 2020. Protein phosphatases in N response and NUE in crops. In: Pandey, G.K. (Ed.), Protein Phosphatases and Stress Management in Plants: Functional Genomic Perspective. Springer (In press).

Lightfoot, D.A., 2009. Genes for use in improving nitrate use efficiency in crops, chapter 7. In: Wood, A.J., Jenks, M.A. (Eds.), Genes for Plant Abiotic Stress. Wiley Blackwell, New York, pp. 167–182.

Li, H., Hu, B., Chu, C., 2017. Nitrogen use efficiency in crops: lessons from *Arabidopsis* and rice. J. Exp. Bot. 68 (10), 2477–2488.

Lochab, S., Kumar, P.A., Raghuram, N., 2014. Molecular characterization of nitrate uptake and assimilatory pathway in *Arthrospira* platensis reveals nitrate induction and differential regulation. Arch. Microbiol. 196 (6), 385–394.

Lochab, S., Pathak, R.R., Raghuram, N., 2007. Molecular approaches for enhancement of nitrogen use efficiency in plants. Agricultural Nitrogen Use and Its Environmental Implications. IK International, New Delhi, pp. 327–350.

Lochab, S., Oberoi, H.S., Gothwal, M., Abbey, D., Raghuram, N., 2009. Nitrate assimilatory enzymes of *Spirulina* (*Arthospira*) platensis are more thermotolerant than those of rice. Physiol. Mol. Biol. Plants 15 (3), 277.

Malyan, S.K., Bhatia, A., Kumar, A., Gupta, D.K., Singh, R., Kumar, S.S., et al., 2016. Methane production, oxidation and mitigation: a mechanistic understanding and comprehensive evaluation of influencing factors. Sci. Total Environ. 572, 874–896.

Mandal, V., Sharma, N., Raghuram, N., 2018. Molecular targets for improvement of crop nitrogen-use efficiency: current and emerging options. In: Shrawat, A., Zayed, A., Lightfoot, D.A. (Eds.), Engineering Nitrogen Utilization in Crop Plants. Springer, pp. 77–93. , ISBN: 978-3-319-92958-3.

McAllister, C.H., Beatty, P.H., Good, A.G., 2012. Engineering nitrogen use efficient crop plants: the current status. Plant Biotechnol. J. 10 (9), 1011–1025.

Mus, F., Crook, M.B., Garcia, K., Costas, A.G., Geddes, B.A., Kouri, E.D., et al., 2016. Symbiotic nitrogen fixation and the challenges to its extension to nonlegumes. Appl. Environ. Microbiol. 82 (13), 3698–3710.

Nair, K.P., 2019. Intelligent Soil Management for Sustainable Agriculture. Springer, Cham.

Noor, M.A., 2017. Nitrogen management and regulation for optimum NUE in maize–a mini review. Cogent Food Agric. 3 (1), 1348214.

O'Brien, J.A., Vega, A., Bouguyon, E., Krouk, G., Gojon, A., Coruzzi, G., et al., 2016. Nitrate transport, sensing, and responses in plants. Mol. Plant 9 (6), 837–856.

Pathak, R.R., Ahmad, A., Lochab, S., Raghuram, N., 2008. Molecular physiology of plant nitrogen use efficiency and biotechnological options for its enhancement. Curr. Sci. 1394–1403.

Pathak, R.R., Das, S.K., Choudhury, D., Raghuram, N., 2009. Genomewide bioinformatic analysis negates any specific role for Dof, GATA and Ag/cTCA motifs in nitrate responsive gene expression in *Arabidopsis*. Physiol. Mol. Biol. Plants 15 (2), 145–150.

Pathak, R.R., Lochab, S., Raghuram, N., 2011. Improving nitrogen use efficiency. In: second ed. Moo-Young, M. (Ed.), Comprehensive Biotechnology, vol. 4. Elsevier, Oxford, pp. 209–218.

Pathak, H., Bhatia, A., 2017. Reactive nitrogen and its impacts on climate change: an Indian synthesis. In: Abrol, Y.P., Adhya, T.K., Aneja, V.P., Raghuram, N., Pathak, H., Kulshrestha, U., Sharma, C., Singh, B. (Eds.), The Indian Nitrogen Assessment: Sources of Reactive Nitrogen, Environmental and Climate Effects, Management Options, and Policies. Elsevier, pp. 491–513. , ISBN: 978-0-12–811836-8.

Pathak, R.R., Jangam, A.P., Malik, A., Sharma, N., Jaiswal, D.K., Raghuram, N., 2020. Transcriptomic and network analyses reveal distinct nitrate responses in light and dark in rice leaves (*Oryza sativa* Indica var. Panvell). Sci. Rep. (In press).

Perchlik, M., Tegeder, M., 2017. Improving plant nitrogen use efficiency through alteration of amino acid transport processes. Plant Physiol. 175 (1), 235–247.

Plett, D., Garnett, T., Okamoto, M., 2017. Molecular genetics to discover and improve nitrogen use efficiency in crop plants. Plant Macronutrient Use Efficiency. Academic Press, 122, p. 93.

Raghuram, N., Sopory, S.K., 1995a. Light regulation of nitrate reductase gene expression mechanism and signal response coupling. Physiol. Mol. Biol. Plants 1, 103–104.

Raghuram, N., Sopory, S.K., 1995b. Evidence for some common signal transduction events for opposite regulation of nitrate reductase and phytochrome-I gene expression by light. Plant Mol. Biol. 29 (1), 25–35.

Raghuram, N., Sopory, S.K., 1999. Roles of nitrate, nitrite and ammonium ion in phytochrome regulation of nitrate reductase gene expression in maize. IUBMB Life 47 (2), 239–249.

Raghuram, N., Chandok, M.R., Sopory, S.K., 1999. Light regulation of nitrate reductase gene expression in maize involves a G-protein. Mol. Cell Biol. Res. Commun. 2 (2), 86–90.

Raghuram, N., Pathak, R.R., Sharma, P., 2006. Signalling and the molecular aspects of N-use efficiency in higher plants. In: Singh, R.P., Jaiwal, P.K. (Eds.), Biotechnological Approaches to Improve Nitrogen Use Efficiency in Plants. Studium Press LLC, Houston, TX, pp. 19–40.

Raghuram, N., Sachdev, M.S., Abrol, Y.P., 2007. Towards an integrative understanding of reactive nitrogen. Agricultural Nitrogen Use and Its Environmental Implications. pp. 1–6.

Raghuram, N., Sharma, N., 2019. Improving crop nitrogen use efficiency. In: Moo-Young, M. (Ed.), Comprehensive Biotechnology, vol. 4. Elsevier, Pergamon, pp. 211–220.

Raghuram, N., Abrol, Y.P., Pathak, H.K., Adhya, T.K., Tiwari, M.M. 2020. South Asian nitrogen centre: capacity building for regional N assessment and management. In: Just Enough Nitrogen. In: Sutton, M., Mason, K., Bleeker, A., Hicks, K., Masso, C., Raghuram, N., Reis, S., Bekunda, M. (Eds.) Perspectives on How to Get There for Regions with too Much and too Little Nitrogen. Proceedings of the Sixth International Nitrogen Conference, Kampala, Uganda. Springer.

Ranjan, R., Yadav, R., 2019. Targeting nitrogen use efficiency for sustained production of cereal crops. J. Plant Nutr. 42 (9), 1086–1113.

Reddy, M.M., Ulaganathan, K., 2015. Nitrogen nutrition, its regulation and biotechnological approaches to improve crop productivity. Am. J. Plant Sci. 6 (18), 2745.

Rockström, J., Steffen, W., Noone, K., Persson, Å., Chapin, F.S., Lambin, E.F., et al., 2009. A safe operating space for humanity. Nature 461 (7263), 472–475.

Rosenblueth, M., Ormeño-Orrillo, E., López-López, A., Rogel, M.A., Reyes-Hernández, B.J., Martínez-Romero, J.C., et al., 2018. Nitrogen fixation in cereals. Front. Microbiol. 9, 1794.

Sharma, L.K., Bali, S.K., 2018. A review of methods to improve nitrogen use efficiency in agriculture. Sustainability 10 (1), 51.

Sharma, N., Kuchi, S., Singh, V., Raghuram, N., 2019. Method for preparation of nutrient-depleted soil for determination of plant nutrient requirements. Commun. Soil Sci. Plant Anal. 50 (15), 1878–1886.

Sharma, A.K., Raghuram, N., Chandok, M.R., Das, R., Sopory, S.K., 1994. Investigations on the nature of the phytochrome-induced transmitter for the regulation of nitrate reductase in etiolated leaves of maize. J. Exp. Bot. 45 (4), 485–490.

Sharma, N., Sinha, V.B., Prem Kumar, Subrahmanyam, D., Neeraja, C.N., Kuchi, S., Raghuram, N., 2021. Nitrogen use efficiency phenotype and associated genes: Roles of germination, flowering, root/shoot length and biomass. Frontiers in Plant Science 11, 2329.

Shrawat, A.K., Carroll, R.T., DePauw, M., Taylor, G.J., Good, A.G., 2008. Genetic engineering of improved nitrogen use efficiency in rice by the tissue-specific expression of alanine aminotransferase. Plant Biotechnol. J. 6 (7), 722–732.

Singh, A., Jaswal, A., Singh, M., 2019. Impact of neem coated urea on rice yield and nutrient use efficiency (NUE). Agric. Rev. 40 (1), 70–74.

Sinha, V.B., Jangam, A.P., Raghuram, N., 2020. Biological determinants of crop N use efficiency and biotechnological avenues for improvement. In: Sutton, M., Mason, K., Bleeker, A., Hicks, K., Masso, C., Raghuram, N., Reis, S., Bekunda, M. (Eds.), Just Enough Nitrogen. Perspectives on How to Get There for Regions with Too Much and Too Little Nitrogen. Proceedings of the 6th International Nitrogen Conference, Kampala, Uganda. Springer.

Steffen, W., Richardson, K., Rockström, J., Cornell, S.E., Fetzer, I., Bennett, E.M., et al., 2015. Planetary boundaries: guiding human development on a changing planet. Science 347 (6223).

Sutton, M.A., Howard, C.M., Erisman, J.W., Billen, G., Bleeker, A., Grennfelt, P., et al., (Eds.), 2011. The European Nitrogen Assessment: Sources, Effects and Policy Perspectives. Cambridge University Press, p. 664.

Sutton, M.A., Bleeker, A., Howard, C.M., Erisman, J.W., Abrol, Y.P., Bekunda, M., et al., 2013. Our Nutrient World. The Challenge to Produce More Food & Energy with Less Pollution. Centre for Ecology & Hydrologyfor GPNM and INI, Edinburgh, ISBN: 978-1-906698-40-9.

Sutton, M.A., Drewer, J., Moring, A., Adhya, T.K., Ahmed, A., Bhatia, A., et al., 2017. The Indian nitrogen challenge in a global perspective. The Indian Nitrogen Assessment. Elsevier, pp. 9–28, ISBN: 978-0-12-811836-8.

Sutton, M., Raghuram, N., Adhya, T.K., Baron, J., Cox, C., de Vries, W., et al., 2019. The nitrogen fix: from nitrogen cycle pollution to nitrogen circular economy-frontiers 2018/19: emerging issues of environmental concern, chapter 4. Frontiers 2018/19: Emerging Issues of Environmental Concern. United Nations Environment Programme, Nairobi, pp. 52–65.

Tang, Y., Li, X., Lu, W., Wei, X., Zhang, Q., Lv, C., et al., 2018. Enhanced photorespiration in transgenic rice over-expressing maize C4 phosphoenolpyruvate carboxylase gene contributes to alleviating low nitrogen stress. Plant Physiol. Biochem. 130, 577–588.

Tewatia, R.K., Chanda, T.K., 2017. Trends in fertilizer nitrogen production and consumption in India. The Indian Nitrogen Assessment. Elsevier, pp. 45–56.

UNEP, 2019. Sustainable Nitrogen Management. Resolution no. UNEP/EA.4/Res.14. United Nations Environment Assembly (15 March 2019). United Nations Environment Programme, Nairobi.

Van Deynze, A., Zamora, P., Delaux, P.M., Heitmann, C., Jayaraman, D., Rajasekar, S., et al., 2018. Nitrogen fixation in a landrace of maize is supported by a mucilage-associated diazotrophic microbiota. PLoS Biol. 16 (8), e2006352.

Winiwarter, W., Höglund-Isaksson, L., Klimont, Z., Schöpp, W., Amann, M., 2018. Technical opportunities to reduce global anthropogenic emissions of nitrous oxide. Environ. Res. Lett. 13 (1), 014011.

Wu, Y., Yang, W., Wei, J., Yoon, H., An, G., 2017. Transcription factor OsDOF18 controls ammonium uptake by inducing ammonium transporters in rice roots. Mol. Cells 40 (3), 178.

Wu, K., Wang, S., Song, W., Zhang, J., Wang, Y., Liu, Q., et al., 2020. Enhanced sustainable green revolution yield via nitrogen-responsive chromatin modulation in rice. Science 367, 6478.

Xuan, W., Beeckman, T., Xu, G., 2017. Plant nitrogen nutrition: sensing and signaling. Curr. Opin. Plant Biol. 39, 57–65.

Yu, J., Zhen, X., Li, X., Li, N., Xu, F., 2019. Increased autophagy of rice can increase yield and nitrogen use efficiency (NUE). Front. Plant Sci. 10, 584.

Zhang, Z., Gao, S., Chu, C., 2020. Improvement of nutrient use efficiency in rice: current toolbox and future perspectives. Theor. Appl. Genet. 133, 1–20.

Zhao, M., Zhao, M., Gu, S., Sun, J., Ma, Z., Wang, L., et al., 2019.). DEP1 is involved in regulating the carbon–nitrogen metabolic balance to affect grain yield and quality in rice (*Oryza sativa* L.). PloS One 14 (3), e0213504.

Zhou, W., Lyu, T.F., Yang, Z.P., Sun, H., Yang, L.J., Chen, Y., et al., 2016. Research advances on regulating soil nitrogen loss by the type of nitrogen fertilizer and its application strategy. J. Appl. Ecol. 27 (9), 3051–3058.

Chapter 4

Role of plant sulfur metabolism in human nutrition and food security

Anna Koprivova and Stanislav Kopriva
Institute for Plant Sciences, Cluster of Excellence on Plant Sciences, University of Cologne, Cologne, Germany

4.1 Introduction

Sulfur is present in nature in a number of inorganic and organic compounds of different redox states. It is this versatility of sulfur and the ability to undergo redox changes that makes sulfur a key active element in a large number of biochemical processes in all forms of life. Redox reactions of sulfur provide energy for sulfate-reducing or sulfur-oxidizing bacteria and archaea (Kelly 1979; Jobe and Kopriva, 2018). Sulfur is also assimilated to a number of organic compounds, many of which are essential building blocks of life (Fig. 4.1) (Takahashi et al., 2011b). The ubiquitous compounds include the amino acids cysteine and methionine and the cofactors, such as iron sulfur centers, thiamine, or lipoic acid, but many compounds, for example, glutathione, are essential for some organisms and not present in others. The source of sulfur for assimilation processes is sulfate but the ability to use sulfate is restricted to plants and algae, yeast and fungi, and some bacteria and archaea (Patron et al., 2008). Therefore animals and humans are dependent on these other organisms to provide dietary sulfur, for example, in the form of sulfur-containing amino acids (Parcell, 2002). For human nutrition, plant proteins are the major source of the essential amino acid methionine. However, plant proteins are usually low in Met; therefore strictly plant-based vegan and vegetarian diets come with a risk of Met deficiency (Kim et al., 2018).

To ensure food security for future generations presents several major challenges. The increasing human population poses increasing demands for agricultural production. Maintaining high crop yields and even increasing them is a challenge because of decreasing agricultural land area and the impact of changing climate (Myers et al., 2017). However, not only the

Plant Nutrition and Food Security in the Era of Climate Change. DOI: https://doi.org/10.1016/B978-0-12-822916-3.00005-6

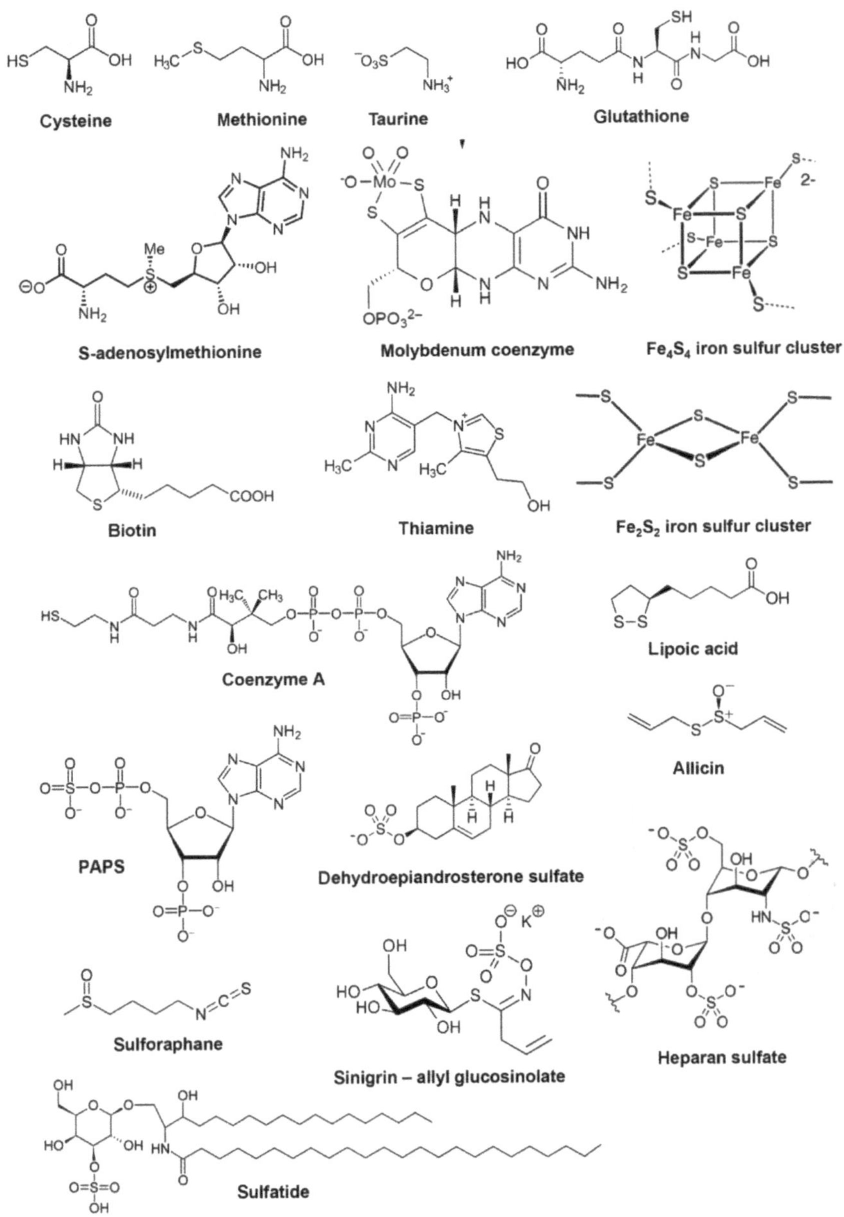

FIGURE 4.1 Structures of major sulfur-containing metabolites.

quantity of food is crucial for food security, also its quality, particularly a sufficient content of micronutrients. Deficiencies in micronutrients, such as iron, zinc, selenium, vitamins A and B12, or methionine, are common and already now affecting worldwide over 2 billion people (Amoroso, 2016).

Deficiencies in crop nutrient quality will be also exacerbated by the climate change. There is a large body of evidence that elevated atmospheric CO_2 concentration causes a reduction in contents of micronutrients (zinc, iron) and proteins in crops (Loladze, 2014; Zhu et al., 2018). Regarding sulfur, the reduction in protein content is particularly relevant, as less Met will be available for human nutrition. Thus to ensure sufficient sulfur for human consumption, better understanding of plant sulfur nutrition and how it is impacted by the changing climate is necessary. Here the metabolism of plant compounds important for human nutrition and health will be summarized, the impact of the changing environment on sulfur nutrition will be discussed, and research directions for improving plant sulfur homeostasis to ensure it does not negatively impact food security and human health will be proposed.

4.2 Sulfur compounds in human nutrition and health

Human contribution to the global sulfur cycle is mainly the oxidation of reduced organic sulfur compounds into sulfate, which can be again assimilated by bacteria, fungi, and plants (Parcell, 2002; Jobe and Kopriva, 2018). Thus, the main sources of sulfur intake are the amino acids methionine and cysteine, possessing sulfur in the reduced form. Biochemically, the primary source of sulfur in the human body is methionine, either directly or through cysteine. Cysteine can be synthesized from Met in a transsulfuration pathway, after Met is converted to homocysteine (Fig. 4.2). This conversion requires activation to S-adenosylmethionine (SAM), which is an essential donor of methyl groups for methylation of DNA, histones, and many metabolites. After transfer of the methyl group, the resulting S-adenosylhomocysteine is hydrolyzed by adenosylhomocysteinase. Homocysteine and serine are substrates for cystathionine β-synthase (CBS)-forming cystathionine, which is further metabolized by cystathionine γ-lyase (CSE) to cysteine, releasing α-ketobutyrate and ammonia (Stipanuk, 2004). Cysteine is directly incorporated into another essential metabolite, the tripeptide glutathione (GSH), in two ATP-dependent enzymatic steps. The γ-glutamyl synthetase joins cysteine with glutamate and glutathione synthetase adds to the resulting γ-glutamylcysteine a glycine to form GSH (Meister and Anderson, 1983). GSH has multiple functions as a redox buffer, as a coenzyme in enzymatic reactions, in defense against reactive oxygen species (ROS), or in detoxification, transport, and excretion of xenobiotics and some endogenous compounds (Meister and Anderson, 1983).

For synthesis of other metabolites, cysteine serves as a substrate or donor of reduced sulfur. The nonproteinogenic amino acid taurine accumulates highly in brain, retina, and other tissues and serves a number of functions in the central nervous system or as a substrate for synthesis of bile salts (Ripps and Shen, 2012). Due to its abundance and variety of functions, it is considered an essential metabolite. Taurine is synthesized from cysteine in an oxidative process initiated by cysteine deoxygenase, followed by cysteinesulfinate

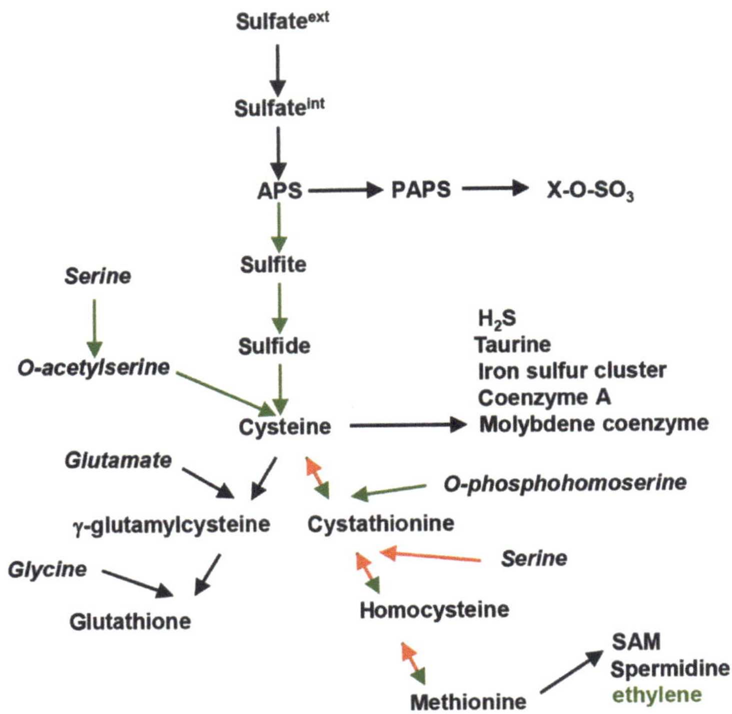

FIGURE 4.2 Scheme of the primary sulfur metabolism. *Green* and *red arrows* represent reactions occurring only in plants and animals, respectively.

decarboxylase and hypotaurine dehydrogenase (Fig. 4.1) (Ripps and Shen, 2012). Cysteine is also a donor of sulfur for iron-sulfur clusters, essential coenzymes of a large number of different enzymes involved in oxidation and reduction processes, electron transfer, radical generation, or DNA repair (Wachnowsky et al., 2018). While there are multiple cluster types and multiple assembly systems, in all cases the sulfur for FeS cluster synthesis is derived from cysteine by cysteine desulfurase (Land and Rouault, 1998). Interestingly, FeS clusters serve themselves as sulfur donors for the synthesis of lipoic acid in human mitochondria (Cronan, 2016). In addition, cysteine is a sulfur donor for two other essential cofactors, coenzyme A and molybdene cofactor (Daugherty et al., 2002; Schwarz, 2016). While cysteine is not an essential amino acid, as it can be synthesized from methionine, its presence in the diet reduces the demand for methionine (Parcell, 2002).

Another large group of compounds contains sulfur in the oxidized form of organic sulfate (Gunal et al., 2019). Different molecules can be sulfated, proteins, peptides, carbohydrates, or small metabolites. Sulfation was first recognized as important in metabolism and excretion of xenobiotics (Leung et al., 2016). However, sulfation is a reversible process, similar to phosphorylation, and the sulfate transferred onto the diverse acceptors by sulfotransferases can be removed again by sulfatases, allowing a rapid regulation of biological activity (Leung et al., 2016). This mechanism is particularly common in metabolism of steroid hormones (Foster and Mueller, 2018). Besides small molecules, a large number of proteins and peptides are sulfated, particularly secreted proteins (Monigatti et al., 2002). The sulfation occurs exclusively on tyrosine residues and seems to be particularly important for protein interactions with other biomolecules (Gunal et al., 2019). Another class of sulfated molecules are the glycosaminoglycans, anionic polysaccharides forming proteoglycans, key structural components of extracellular matrices and cell surfaces (Lindahl et al., 2017). These proteoglycans, such as heparin, chondroitin sulfate, keratan sulfate, or dermatan sulfate, have vital functions in cell adhesion, signaling, blood clotting, and other biological activities (Pomin et al., 2019). In addition, sulfated galactosylceramide, sulfatide, belonging to glycosphingolipids, is an abundant component in the myelin sheath of nerves (Boggs et al., 2010). This large variety of sulfated compounds is synthesized by the family of sulfotransferases with 52 members (Gunal et al., 2019). The sulfation of extracellular metabolites takes place in Golgi; the intracellular metabolites are sulfated by cytosolic sulfotransferases. All of these enzymes use the same donor of activated sulfate, the 3′-phosphoadenosine 5′-phosphosulfate (PAPS) (Gunal et al., 2019). PAPS is synthesized from sulfate in two ATP-dependent reactions: adenylation by ATP sulfurylase to adenosine 5′-phosphosulfate (APS) and phosphorylation of APS by APS kinase. In humans the two enzyme activities reside on a single protein, the PAPS synthetase (Gunal et al., 2019). Sulfate for synthesis of PAPS can be taken up directly from nutrition or derive from oxidation of the reduced sulfur compounds and thus ultimately from Met (Parcell, 2002). Therefore sulfate in the diet also results in sparing the need for dietary Met.

A special position among all sulfur-containing metabolites is occupied by hydrogen sulfide, whose role as an endogenous gaso-transmitter has been recognized only relatively recently (Abe and Kimura, 1996). Since the mid-1990s the number of reports on the function of H_2S in human physiology and a number of diseases exploded. H_2S has been shown to affect processes in the central nervous system, cardiovascular, renal, reproductive, respiratory, and digestive systems, as well as in oncology (Kimura, 2014; Talaei et al., 2013; Wallace and Wang, 2015; Cao et al., 2019). H_2S can easily react with metal cofactors, reduce disulfide bonds, or bind to thiol groups of proteins and modulate so their activity. H_2S is different from other compounds discussed so far, as it is not an organic compound but its release requires metabolism of S-containing metabolites. There are several pathways for H_2S synthesis with the main contribution to H_2S synthesis coming from enzymes of the transsulfuration pathway, cystathionine-β-synthase (CBS) and cystathionine γ-lyase (CSE) (Cao et al., 2019; Singh et al., 2009). CBS can produce H_2S directly from cysteine or much more efficiently in reaction of cysteine with homocysteine, to form cystathionine and H_2S, in analogy with the canonical reaction in cysteine synthesis (Singh et al., 2009). This reaction is prevalent in the brain and is regulated by GSH, SAM, or carbon monoxide (Niu et al., 2015). CSE can produce H_2S from homocysteine or cysteine, releasing ammonia and α-ketoglutarate or pyruvate, respectively (Chiku et al., 2009). Another enzyme producing H_2S is 3-mercaptopyruvate sulfurtransferase, which transfers the sulfur from 3-mercaptopyruvate onto an internal Cys residue forming a persulfide, from which H_2S is released by a reductant, such as thioredoxin (Shibuya et al., 2009). Mercaptopyruvate is synthesized from L-cysteine through cysteine aminotransferase or from D-cysteine through D-amino acid oxidase (Cao et al., 2019). However, H_2S can also be formed nonenzymatically by reduction of sulfane sulfur in thiosulfate, various persulfides, and polysulfides by GSH, NADH, and NADPH (Searcy and Lee, 1998). Chemical release of H_2S might underlie the health-promoting effects of organosulfur compounds from garlic and is exploited in a number of therapeutic treatments using artificial H_2S donors (Cao et al., 2019).

Besides the number of sulfur-containing compounds vital for life, which are synthesized in human cells, there are also several essential coenzymes that humans cannot produce and that need to be provided in food, namely biotin and thiamine (Parcell, 2002). However, many nonessential sulfur compounds taken up through food are also beneficial for human health. In particular, two classes of plant natural products contain sulfur and have proven positive influences on health. In cruciferous vegetables, such as broccoli, cabbage, or mustard, glucosinolates, or rather their degradation products, are responsible for the taste and smell but confer also a number of health benefits (Halkier and Gershenzon, 2006; Traka and Mithen, 2009). Upon tissue damage, glucosinolates come into contact with the enzyme myrosinase that hydrolyzes the thioglucosidic bond and the resulting products rearrange to create volatile isothiocyanates or nitriles (Halkier and Gershenzon, 2006). The most prominent molecule from this class of compounds is sulforaphane, the isothiocyanate produced from glucoraphanin (4-methylsulfinylbutyl glucosinolate) abundant in broccoli (Juge et al., 2007). Sulforaphane protects against various types of cancer, decreases risk of cardiovascular disease, and might help in autism and osteoporosis (Traka and Mithen, 2009). Sulforaphane is an antioxidant and reacts directly with ROS, but also triggers substantial transcriptional changes, particularly to the genes for xenobiotic defense (Juge et al., 2007). Allicin from garlic is similar to sulforaphane in that it is also produced chemically after enzymatic reaction by alliinase with an inactive precursor alliin (Borlinghaus et al., 2014). Allicin is instable and degrades rapidly in a number of sulfides, disulfides, or trisulfides. These compounds act as antioxidants, antimicrobials, and donors for nonenzymatical H_2S release (Borlinghaus et al., 2014). All the essential and beneficial sulfur compounds can be provided to humans by plants in their diet, as plants are capable of sulfate assimilation and synthesis of all amino acids, vitamins, and the secondary compounds.

4.3 Plant sulfate assimilation and methionine synthesis

Plants play such an important role in human sulfur nutrition as they are able to assimilate inorganic sulfate into all essential metabolites. Plants take sulfate up by high-affinity sulfate transporters SULTR1;1 and SULTR1;2 into the roots (Takahashi et al., 2011a). From there the sulfate can be translocated to the shoots with the help of sulfate transporters SULTR1;3, SULTR2;1, and SULTR2;2 (Takahashi et al., 2011a); stored in the vacuoles; or assimilated into organic sulfur-containing compounds. Plants possess two pathways of sulfate assimilation, one is reducing and forms cysteine and further large number of primary sulfur metabolites, the other is transferring sulfate in its oxidized form primarily forming secondary products, identical to human sulfate activation (Fig. 4.1) (Kopriva et al., 2012). Both pathways are initiated by sulfate activation by adenylation catalyzed by ATP sulfurylase yielding APS. For the primary reductive sulfate assimilation this reaction occurs in the plastids, after the sulfate was imported by sulfate transporters of the group 3 (Cao et al., 2013); however, ATP sulfurylase is also present in the cytosol (Bohrer et al., 2014). In the reductive pathway, APS is reduced by APS reductase to sulfite, which is subsequently reduced to sulfide by sulfite reductase, with both enzymes exclusively localized in the plastids (Suter et al., 2000; Khan et al., 2010). Sulfide is then incorporated into the amino acid skeleton of O-acetylserine (OAS), formed from serine and acetyl-coenzyme A by serine acetyltransferase (SAT), by OAS-(thiol)lyase (OASTL) (Wirtz and Hell, 2006). The product of this reaction, cysteine, is the end product of primary sulfate assimilation and a metabolite in which the metabolism of C, N, and S converge. SAT and OASTL form a multienzyme complex in plastids, cytosol, and mitochondria, and while Cys can be formed in all compartments, in normal conditions the major flux into Cys occurs in the cytosol, from OAS formed in mitochondria and sulfide transported from the plastids (Krueger et al., 2009).

Cysteine is the primary source of reduced sulfur for other compounds of both primary and secondary metabolism. The essential amino acid Met is synthesized from Cys in a transsulfurylation pathway (Hesse et al., 2004). First, cystathionine γ-synthase joins Cys with O-phosphohomoserine, which is derived from aspartate through homoserine, to form cystathionine. In the next step, cystathionine is broken up by cystathionine β-lyase to homocysteine, pyruvate, and ammonia. Finally homocysteine is methylated to Met by methionine synthase (Hesse et al., 2004). Free Met does not accumulate in large concentrations; it is either used for protein synthesis or metabolized to S-adenosyl-methionine (SAM) or S-methyl-methionine (SMM). SAM is not only an essential donor of methyl groups for methylation of DNA, histones, and many other metabolites but also a substrate for synthesis of ethylene and polyamines (Takahashi et al., 2011b). Another essential primary metabolite is GSH, which is formed from cysteine in the same way as in humans

(Noctor et al., 2012). GSH has multiple functions in plants as a redox buffer, in defense against reactive oxygen species, in detoxification of xenobiotics and heavy metals, and as a sulfur donor for synthesis of some secondary metabolites, such as glucosinolates and camalexin. As in human sulfur metabolism, cysteine provides sulfur for synthesis of iron sulfur centers, coenzyme A, molybdene cofactor, and in addition for thiamine, while biotin and lipoic acid synthesis require sulfur from an iron sulfur center.

In the oxidative branch of the pathway APS is phosphorylated by APS kinase to PAPS, which is the donor of activated sulfate for transfer onto diverse metabolites and peptides catalyzed by sulfotransferases, similar to humans. While the sulfotransferases are active in cytosol or in Golgi, PAPS is synthesized in the plastid or in the cytosol (Gunal et al., 2019). PAPS can be exported from plastids by PAPS transporters because the loss of cytosolic APS kinase has no impact on accumulation of sulfated compounds (Mugford et al., 2009). In contrast to animal systems with multiple sulfotransferases in both cytosol and Golgi, in plants the only confirmed Golgi localized enzyme is the tyrosylpeptide sulfotransferase (TPST) (Gunal et al., 2019). TPST is important for sulfation of a number of peptide hormones, such as phytosulfokines, root growth meristem factor (RGF), plant peptide containing sulfated tyrosine 1 (PSY1), or casparian strip integrity factor (CIF) (Kaufmann and Sauter, 2019). The cytosolic sulfotransferases modify a variety of diverse metabolites and form a large multienzyme family, for example, in *Arabidopsis* there are 21 isoforms (Hirschmann et al., 2014). The best characterized class of sulfated metabolites are the glucosinolates, important components of plant immunity in the *Brassicaceae* (Kopriva, 2016). Glucosinolates are derived from amino acids, which define the different types of glucosinolates, tryptophan for indolic glucosinolates, phenylalanine for aromatic ones and methionine, valine, leucine isoleucine, and alanine for aliphatic glucosinolates (Halkier and Gershenzon, 2006). There are more than 150 different glucosinolate structures, which are achieved by combinations of elongation of some initial amino acids, mainly Met and Phe, and by a number of modification reactions after the core molecules are synthesized (Sonderby et al., 2010). The core structure consists of a sulfated aldoxime bound to a glucose through a thioglucoside bond. Glucosinolates are phytoanticipines, that is they accumulate in vacuoles and become active after tissue disruption brings them into contact with myrosinase, which releases the toxic and deterrent volatiles (Halkier and Gershenzon, 2006). Glucosinolates possess two sulfur atoms in their core structure, a reduced one in the thioglucoside bond originating from GSH and one oxidized transferred from PAPS by sulfotransferases in the last step of the core synthesis pathway (Sonderby et al., 2010). In addition, many, but not all, Met-derived glucosinolates retain the sulfur from the donor amino acid. Besides glucosinolates, sulfation modifies other natural products, for example flavonoids, and a number of phytohormones, such as jasmonate, salicylate, and brassinosteroids (Hirschmann et al., 2014). The biological functions of these modifications are, however, not known.

The discovery of H_2S as a gaso-transmitter in regulation of human physiological processes triggered a rich field of research into the H_2S function in plants (Calderwood and Kopriva, 2014). Numerous papers reported protection against stress, abiotic and biotic, and roles in various physiological processes (Aroca et al., 2018). In particular, H_2S is part of the signaling processes controlling stomata opening and autophagy (Alvarez et al., 2012; Garcia-Mata and Lamattina, 2010). The function of H_2S in stress response is, however, less well characterized in plants, because unlike in humans, H_2S is rapidly incorporated into cysteine and then GSH, and increased GSH is associated with stress tolerance (Calderwood and Kopriva, 2014). In addition, H_2S can substitute sulfate as the sole sulfur source for plant growth (Ausma and De Kok, 2019).

4.4 Control of plant sulfate assimilation

The pathways of sulfur assimilation and metabolism are under strict control because of the importance of the S-containing compounds but also because of the high reactivity and toxicity of the reaction intermediates (Takahashi et al., 2011b). The pathway is regulated in a demand-driven manner; it is activated when sulfate is not available or when GSH concentration declines due to interrupted synthesis or increased use (Koprivova and Kopriva, 2014). On the other hand, excess reduced sulfur inhibits sulfate uptake and assimilation (Vauclare et al., 2002). Since cysteine synthesis is the merging point of sulfate assimilation with the assimilation of nitrogen and carbon, the pathway is also regulated by the availability of N and C compounds. Nitrate deficiency inhibits sulfate uptake and assimilation, whereas ammonium and amino acid acts stimulatory and, similarly, CO_2 deprivation leads to reduced sulfate metabolism, whereas sugars activate the pathway (Kopriva and Rennenberg, 2004; Takahashi et al., 2011b). This general concept of demand-driven regulation explains the upregulation of cysteine and GSH synthesis due to abiotic and biotic stresses, the diurnal changes in activity of the key enzymes with maximum during the day or the repression of glucosinolate synthesis during sulfate starvation (Kopriva et al., 1999; Hirai et al., 2005; Rausch and Wachter, 2005).

The physiological responses of sulfate uptake and assimilation to environmental triggers are well described and many compounds affecting the uptake, enzyme activity, or transcript levels of key genes have been identified (Takahashi et al., 2011b; Koprivova and Kopriva, 2014). Apart of the intermediates and products: sulfite, sulfide, cysteine, and GSH repressing the pathway and OAS inducing it, there are a number of metabolites that may possibly act as signals (Takahashi et al., 2011b). In particular, sulfate uptake and assimilation are extensively regulated by phytohormones (Koprivova and

Kopriva, 2016). The critical regulatory points of the pathway are the uptake of sulfate to the cells and the reduction of APS by APS reductase; however, other enzymes are also regulated and under certain conditions may form a bottleneck (Haas et al., 2008; Khan et al., 2010; Vauclare et al., 2002). This is apparent also from the fluid distribution of flux control over the pathway across multiple components (Vauclare et al., 2002; Feldman-Salit et al., 2019). A large part of the regulation is on the transcriptional level, as evident from coordinated regulation of enzyme activity, protein level, and transcript levels, for example for APS reductase or sulfate transporters (Koprivova et al., 2000; Yoshimoto et al., 2002). Nevertheless, some posttranscriptional control is also involved and the regulatory network might be more complex than originally thought (Yoshimoto et al., 2007; Koprivova et al., 2008). Such more complex regulation is found at the downstream parts of the pathway, for example, the stability of cysteine synthase complex, which is oppositely affected by sulfide and OAS and influences the activities of the constituting enzymes (Wirtz and Hell, 2006) or the allosteric activation of threonine synthase by SAM, which reduces availability of O-phosphohomoserine for Met synthesis (Hesse et al., 2004). In addition, several enzymes are regulated by redox, for example γ-glutamylcysteine synthetase, APS reductase, or APS kinase (Ravilious et al., 2012; Hothorn et al., 2006; Kopriva and Koprivova, 2004).

However, despite decades of research, the molecular mechanisms of the pathway regulation are largely unknown. It is not known how sulfate or its deficiency is sensed, how the signal is transduced, and how the transcriptional response is orchestrated (Koprivova and Kopriva, 2014). The only component of the signaling known so far is the key regulator of sulfate deficiency response, transcription factor SULFUR LIMITATION1 (SLIM1) (Maruyama-Nakashita et al., 2006), but even for this factor it is not known how the increased transcript levels of target genes are achieved. SLIM1 was identified in a genetic screen for loss of sulfate deficiency response, measured as green fluorescent protein fluorescence controlled by *SULTR1;2* promoter (Maruyama-Nakashita et al., 2006). While *slim1* mutants lose the induction of expression of sulfate deficiency marker genes, the transcript levels of *SLIM1* as well as its nuclear localization, two common mechanisms for activation of gene expression, are not affected (Maruyama-Nakashita et al., 2006). A different situation is with control of the biosynthesis of glucosinolates. Their synthesis is controlled by two groups of MYB transcription factors, MYB28, MYB29, and MYB76 for the aliphatic ones and MYB51, MYB34, and MYB122 for indolic glucosinolates (Gigolashvili et al., 2007a; Gigolashvili et al., 2007b; Hirai et al., 2007). Glucosinolates are strongly linked with sulfate assimilation, since reduced activity of APS kinase results in reduction of glucosinolate accumulation (Mugford et al., 2009) and, correspondingly, the MYB factors contribute also to regulation of ATP sulfurylase, APS kinase, and APS reductase (Yatusevich et al., 2010). The MYB factors cooperate in their function with bHLH factors forming larger regulatory complexes (Schweizer et al., 2013). In addition to these factors with large effect on the pathway, another 22 were shown to control accumulation of at least some glucosinolates (Li et al., 2014). Other approaches looking for regulatory elements of sulfur homeostasis using quantitative genetics pointed instead to the importance of metabolic processes. QTL analyses for sulfate and/or sulfur accumulation using *Arabidopsis* accessions revealed that low activity of APR2 isoform of APS reductase or ATPS1 isoform of ATP sulfurylase results in accumulation of sulfate and total sulfur in the leaves (Chao et al., 2014; Koprivova et al., 2013; Loudet et al., 2007). Also the variation in quantity and quality of glucosinolate within *Arabidopsis* accessions is controlled largely through polymorphisms in metabolic genes *MAM1* and *AOP2* (Kliebenstein et al., 2001).

4.5 Impact of changing environment on plant sulfur nutrition

Plant sulfur nutrition has been affected by environmental changes for centuries. Anthropogenic activities have been strongly influencing the concentration of sulfur gases in the atmosphere, resulting in changes in sulfur metabolism or in downstream consequences of modulated concentrations of sulfur-containing compounds. Industrial revolution and the burning of coal resulted in high increases of sulfur dioxide in the air. SO_2 has toxic effects on plants, leading to leaf injury, but at the same time plants can assimilate atmospheric sulfur into cysteine (Lee et al., 2017). From a metabolic perspective, SO_2 can be reduced by sulfite reductase and the sulfide can be directly incorporated to cysteine, or alternatively, sulfite can be oxidized to sulfate by sulfite oxidase (Brychkova et al., 2013). Exposure to SO_2 leads to reduction of sulfate uptake and activity of APS reductase, which has been observed in plants growing in areas with higher air pollution (Tschanz et al., 1986). Additional environmental damages caused by SO_2 include acid rain resulting in acidification of soils and surface water. From the beginning of the 1970s, a series of Environmental and Clean Air Acts resulted in significant reduction in SO_2 emissions in Europe and the United States (Crippa et al., 2016). Indeed, from 1970 to 2010 European SO_2 emissions were reduced by 80%; however, there was almost no change globally due to increased industrial production in other parts of the world (Crippa et al., 2016). However, besides the entirely positive influence on air quality and reduced forest and lake damage, the reduction in SO_2 emissions resulted in lower sulfur deposits on the earth's surface and triggered sulfur deficiency in many agricultural soils (Bloem et al., 2007).

Sulfur deficiency has a great impact on crop yields and quality. Obviously, it is detrimental to accumulation of sulfur-containing metabolites affecting proteins and natural products in the plants, leading to a lower nutritional value of crops. The composition of seed proteins is changed in favor of sulfur-poor proteins, for example, in wheat grains ω-gliadins and high-molecular-weight glutenin are produced at the expense of S-rich γ-gliadins and low-molecular-weight subunits of glutenin (Wrigley et al., 1984). The number of Cys residues in grain proteins and the ability to form disulfide bonds have a great impact on the baking quality of flour (Zorb et al., 2009). Similar decrease in protein quality is true for other crops, for example, oil-seed rape or soybean (D'Hooghe et al., 2014; Gayler and Sykes, 1985). The contents of health-promoting glucosinolates also decrease with sulfur limitation (Falk et al., 2007). Another big problem of sulfur limitation is connected to the physiological response to these conditions by increased accumulation of amino acids. The accumulated asparagine can react with reducing sugars in a Maillard reaction and form toxic acrylamide (Muttucumaru et al., 2006). The effect of sulfur deficiency on acrylamide content in baked products has been clearly demonstrated for wheat and in potatoes (Halford et al., 2012). In addition, through the regulatory interactions between sulfur and nitrogen nutrition, sulfur availability affects strongly the ability of plants to use nitrogen (Kopriva and Rennenberg, 2004).

However, there is also an indirect effect of sulfur deficiency on crop yields, through the higher susceptibility of such crops to diseases (Bloem et al., 2007) and abiotic stresses (Usmani et al., 2020). The phenomenon of sulfur-induced resistance, that is the strengthening of plants' ability to cope with pathogens through sulfur fertilization, has been initially observed in the field and has a number of possible mechanisms (Bloem et al., 2014; Rausch and Wachter, 2005). Several plant defense compounds contain sulfur: elemental S^0 used as fungicide since the 19th century, glucosinolates, sulfur-rich proteins like defensins, sulfur-containing indolic phytoalexins, such as camalexin or brassicin, or glutathione (Rausch and Wachter, 2005). Other possible mechanisms include formation of H_2S or other volatiles (Bloem et al., 2004). In the case of abiotic stress the role of sulfur-containing phytochelatins in protection against heavy metals is well established (Cobbett, 2000), as is the protection of maize from chilling and drought stress by glutathione (Kocsy et al., 2001; Usmani et al., 2020). Thus the reduced SO_2 emissions led to the need for sulfur fertilization in order to maintain the yields, health, and quality of crops.

The environment is, however, continuing to change, most prominently the content of CO_2 in the atmosphere adding further challenges for crop production. A number of studies with crop plants grown in elevated atmospheric CO_2 concentration, as predicted through different climatic models, showed increased yields but decreases in concentrations of mineral elements and proteins (Loladze, 2014; Zhu et al., 2018). This is particularly critical for zinc and iron, but also for proteins as sources of essential amino acids, nitrogen, and sulfur. This "hidden hunger" is a major obstacle in achieving food security and various models predict that it will lead to increased occurrence of malnutrition by mid 21st century (Ebi and Loladze, 2019). The reduction in proteins is particularly critical to providing sulfur for human nutrition as proteins are the source of essential Met (Parcell 2002). Indeed, several metaanalyses of FACE (free-air CO_2 enrichment) experiments showed 7%–15% decrease in grain protein in rice, barley, wheat, or potato but only an insignificant one for different legumes and C_4 crops, such as maize or sorghum (Myers et al., 2017; Zhu et al., 2018). The reduction in proteins was manifested in c. 7% reduction in total grain sulfur and an 8% decrease in content of Met and Cys (Hogy et al., 2009; Fernando et al., 2015). The increased availability of CO_2 in C_3 plants leads to its greater incorporation into carbohydrates at the expense of proteins, the so-called carbon penalty (Loladze, 2014; Zhu et al., 2018). In C_4 crops, however, the protein content is not affected by elevated CO_2, since C_4 plants do not increase carbon fixation rates in elevated CO_2 (von Caemmerer and Furbank, 2003).

4.6 Plant sulfur nutrition and food security—open questions

The importance of plant sulfur metabolism for food security is obvious. Although our knowledge about the control of sulfur homeostasis is increasing, there are still several gaps that need to be filled to ensure that in the future sulfur deficiency will not affect human health. One major challenge is to ensure sufficient Met supply in future crops. Even without the carbon penalty, Met content in plant proteins is low, particularly in legumes, so that synthetic methionine has to be added to animal feed (Ufaz and Galili, 2008). Decades of research attempted to increase Met in seeds, both through breeding and genetic engineering and using mainly two approaches, "push" and "pull" (Galili and Amir, 2013). In the "push" approach, synthesis of Met can be increased by manipulation of aspartate kinase or CGS (Kim et al., 2002; Hacham et al., 2008). However, it seems that regulation of Met synthesis and particularly CGS is species-specific and therefore no universal solutions are possible (Galili and Amir, 2013). In addition, this approach increases Met content in all organs, so it may be suitable for vegetative tissues but rather wasteful for increasing Met specifically in seeds. The alternative "pull" approach makes use of Met-rich proteins to increase the sink for Met in the seeds (Amir et al., 2012). Several such proteins have been tested, some synthetic ones, some naturally rich in Met, such as maize β-zein, or 2S albumins from Brazil nut, sesame, or sunflower (Galili and Amir, 2013). While expression of these proteins in seeds indeed increased Met content by 30%–60%, some of the albumins acted as allergens and therefore are not suitable for biotechnological purposes (Kelly and Hefle, 2000). In addition, expression of a heterologous Met-rich protein often led to change in pattern of endogenous storage proteins in favor of Met and Cys-poor isoforms (Hagan et al., 2003). Given the carbon penalty discussed above, it seems that particular effort should be put into improving Met content in C_4 crops such as maize. Indeed, several recent reports provided possible solutions to increase Met content in maize seeds. Two transgenic approaches, aimed at increasing the sulfate-reducing capacity of maize leaves by overexpressing either *Arabidopsis* SAT or *Escherichia coli* PAPS reductase, resulted in 40%–60% increase in Met in seeds (Planta et al., 2017; Xiang et al., 2018). In addition, a detailed analysis of zeins in different maize and teosinte populations revealed that domestication impacted the evolution and variation in the Met-rich zeins, and identified interesting haplotypes of the corresponding genes that might be used for breeding of high-Met maize (Li et al., 2018).

While improvement of Met content in maize is an important and promising way to balance its reduced availability in future crops, it cannot be the only strategy. It is important to understand the processes responsible for decrease in sulfur in C_3 crops grown at elevated CO_2 in order to attenuate their effects. However, very little is known about regulatory interactions of sulfur assimilation with carbon metabolism, in particular the effects of elevated CO_2 on sulfate assimilation and allocation. Whereas depletion of CO_2 in the atmosphere downregulates sulfate uptake and assimilation, feeding with sugars activates the pathway (Hesse et al., 2003; Kopriva et al., 1999; Kopriva et al., 2002). Both glucose and sucrose induced sulfate assimilation when fed to *Arabidopsis* in the dark, through transcriptional regulation of S assimilation genes, particularly APS reductase (Hesse et al., 2003; Kopriva et al., 1999). Interestingly, through measuring sulfur fluxes in sunflower plants grown at different O_2 and CO_2 ratios, Abadie and Tcherkez (2019) showed a stimulation of sulfur metabolism by photorespiration. Incorporation of ^{33}S into Cys and Met was positively correlated with a relative rate of photorespiration and negatively correlated with net CO_2 assimilation (Abadie and Tcherkez, 2019). The mechanism(s) behind the stimulation of sulfate assimilation by photorespiration are unclear, but the rapid response of the sulfur fluxes pointed to post-translational or metabolic regulation (Abadie and Tcherkez, 2019). It might be connected to photorespiratory Ser, which after acetylation to OAS is a substrate for Cys synthesis and a regulator of the cysteine synthase complex as well as the transcription of sulfur assimilation genes. Since elevated CO_2 is reducing photorespiratory flux, the corresponding decrease in sulfate assimilation might explain the reduced accumulation of S in grains of plants from the FACE experiments (Hogy et al., 2009).

On the other hand, elevated CO_2 stimulated synthesis of health-promoting glucosinolates in several crops (Schonhof et al., 2007; Klaiber et al., 2013). This has been accompanied by a better defense against generalist herbivores but increased damage by specialists (Klaiber et al., 2013). Similarly, organosulfur compounds from garlic seem to be stimulated by elevated CO_2. Naznin et al. (2015) showed that CO_2 levels used for cultivation of garlic correlated with gain of fresh weight per plant as well as with concentration of ajoene, an organosulfur compound derived from allicin (Block, 1985). The increased growth and production of ajoene synergistically caused an almost two-fold increase in ajoene content per plant at 900 μmol/mol CO_2

(Naznin et al., 2015). Thus potentially, elevated CO_2 may stimulate the health-promoting effects of crucifers, onion, and garlic, but this still needs to be tested. Both glucosinolates and allicin contain reduced sulfur and are derived from products of primary sulfate assimilation. As discussed before, however, sulfate assimilation is diminished in elevated CO_2, so how is synthesis of these compounds regulated? Why and how is it uncoupled from the primary assimilation? These are important question and answering them will help to mitigate the negative effects of elevated CO_2 on sulfur assimilation.

4.7 Conclusions

Plant sulfur metabolism produces numerous compounds essential for human nutrition and others beneficial for health. However, keeping adequate levels of these compounds in human diets seems to be under threat through the reduced availability because of lower sulfur deposits and because of the negative effect of elevated CO_2 on sulfate assimilation. To ensure sufficient supply of sulfur metabolites in food, more research into control of plant sulfur homeostasis is needed, particularly in regulation of Met content in seeds and in effects of elevated CO_2 on sulfate assimilation. The recent progress as discussed here promises that plant sulfur research is up to the challenge.

Acknowledgments

Research in Stanislav Kopriva's lab is funded by the Deutsche Forschungsgemeinschaft (DFG) under Germany's Excellence Strategy – EXC 2048/1 – project 390686111.

References

Abadie, C., Tcherkez, G., 2019. Plant sulphur metabolism is stimulated by photorespiration. Commun. Biol. 2, 379. Available from: https://doi.org/10.1038/s42003-019-0616-y.

Abe, K., Kimura, H., 1996. The possible role of hydrogen sulfide as an endogenous neuromodulator. J. Neurosci. 16 (3), 1066–1071.

Alvarez, C., Garcia, I., Moreno, I., Perez-Perez, M.E., Crespo, J.L., Romero, L.C., et al., 2012. Cysteine-generated sulfide in the cytosol negatively regulates autophagy and modulates the transcriptional profile in *Arabidopsis*. Plant Cell 24 (11), 4621–4634. Available from: https://doi.org/10.1105/tpc.112.105403.

Amir, R., Han, T., Ma, F., 2012. Bioengineering approaches to improve the nutritional values of seeds by increasing their methionine content. Mol. Breed. 29, 915–924.

Amoroso, L., 2016. The second international conference on nutrition: implications for hidden hunger. World Rev. Nutr. Diet 115, 142–152. Available from: https://doi.org/10.1159/000442100.

Aroca, A., Gotor, C., Romero, L.C., 2018. Hydrogen sulfide signaling in plants: emerging roles of protein persulfidation. Front Plant. Sci. 9:, 1369. Available from: https://doi.org/10.3389/fpls.2018.01369.

Ausma, T., De Kok, L.J., 2019. Atmospheric H2S: impact on plant functioning. Front Plant Sci. 10, 743. Available from: https://doi.org/10.3389/fpls.2019.00743.

Block, E., 1985. The chemistry of garlic and onions. Sci. Am. 252 (3), 114–119. Available from: https://doi.org/10.1038/scientificamerican0385-114.

Bloem, E., Riemenschneider, A., Volker, J., Papenbrock, J., Schmidt, A., Salac, I., et al., 2004. Sulphur supply and infection with Pyrenopeziza brassicae influence L-cysteine desulphydrase activity in *Brassica napus* L. J. Exp. Bot. 55 (406), 2305–2312. Available from: https://doi.org/10.1093/jxb/erh236.

Bloem, E., Haneklaus, S., Salac, I., Wickenhauser, P., Schnug, E., 2007. Facts and fiction about sulfur metabolism in relation to plant-pathogen interactions. Plant Biol. (Stuttg). 9 (5), 596–607. Available from: https://doi.org/10.1055/s-2007-965420.

Bloem, E., Haneklaus, S., Schnug, E., 2014. Milestones in plant sulfur research on sulfur-induced-resistance (SIR) in Europe. Front. Plant Sci. 5, 779. Available from: https://doi.org/10.3389/fpls.2014.00779.

Boggs, J.M., Gao, W., Zhao, J., Park, H.J., Liu, Y., Basu, A., 2010. Participation of galactosylceramide and sulfatide in glycosynapses between oligodendrocyte or myelin membranes. FEBS Lett. 584 (9), 1771–1778. Available from: https://doi.org/10.1016/j.febslet.2009.11.074.

Bohrer, A.S., Yoshimoto, N., Sekiguchi, A., Rykulski, N., Saito, K., Takahashi, H., 2014. Alternative translational initiation of ATP sulfurylase underlying dual localization of sulfate assimilation pathways in plastids and cytosol in *Arabidopsis thaliana*. Front. Plant Sci. 5, 750. Available from: https://doi.org/10.3389/fpls.2014.00750.

Borlinghaus, J., Albrecht, F., Gruhlke, M.C., Nwachukwu, I.D., Slusarenko, A.J., 2014. Allicin: chemistry and biological properties. Molecules 19 (8), 12591–12618. Available from: https://doi.org/10.3390/molecules190812591.

Brychkova, G., Grishkevich, V., Fluhr, R., Sagi, M., 2013. An essential role for tomato sulfite oxidase and enzymes of the sulfite network in maintaining leaf sulfite homeostasis. Plant Physiol. 161 (1), 148–164. Available from: https://doi.org/10.1104/pp.112.208660.

Calderwood, A., Kopriva, S., 2014. Hydrogen sulfide in plants: from dissipation of excess sulfur to signaling molecule. Nitric Oxide 41, 72–78. Available from: https://doi.org/10.1016/j.niox.2014.02.005.

Cao, M.J., Wang, Z., Wirtz, M., Hell, R., Oliver, D.J., Xiang, C.B., 2013. SULTR3;1 is a chloroplast-localized sulfate transporter in *Arabidopsis thaliana*. Plant J. 73 (4), 607–616. Available from: https://doi.org/10.1111/tpj.12059.

Cao, X., Ding, L., Xie, Z.Z., Yang, Y., Whiteman, M., Moore, P.K., et al., 2019. A review of hydrogen sulfide synthesis, metabolism, and measurement: is modulation of hydrogen sulfide a novel therapeutic for cancer? Antioxid. Redox Signal. 31 (1), 1-–38. Available from: https://doi.org/10.1089/ars.2017.7058.

Chao, D.Y., Baraniecka, P., Danku, J., Koprivova, A., Lahner, B., Luo, H., et al., 2014. Variation in sulfur and selenium accumulation is controlled by naturally occurring isoforms of the key sulfur assimilation enzyme ADENOSINE 5′-PHOSPHOSULFATE REDUCTASE2 across the *Arabidopsis* species range. Plant Physiol. 166 (3), 1593–1608. Available from: https://doi.org/10.1104/pp.114.247825.

Chiku, T., Padovani, D., Zhu, W., Singh, S., Vitvitsky, V., Banerjee, R., 2009. H_2S biogenesis by human cystathionine gamma-lyase leads to the novel sulfur metabolites lanthionine and homolanthionine and is responsive to the grade of hyperhomocysteinemia. J. Biol. Chem. 284 (17), 11601–11612. Available from: https://doi.org/10.1074/jbc.M808026200.

Cobbett, C.S., 2000. Phytochelatins and their roles in heavy metal detoxification. Plant Physiol. 123 (3), 825–832.

Crippa, M., Janssens-Maenhout, G., Dentener, F., Guizzardi, D., Sindelarova, K., Muntean, M., et al., 2016. Forty years of improvements in European air quality: regional policy-industry interactions with global impacts. Atmos. Chem. Phys. 16, 3825–3841.

Cronan, J.E., 2016. Assembly of lipoic acid on its cognate enzymes: an extraordinary and essential biosynthetic pathway. Microbiol. Mol. Biol. Rev. 80 (2), 429–450. Available from: https://doi.org/10.1128/MMBR.00073-15.

Daugherty, M., Polanuyer, B., Farrell, M., Scholle, M., Lykidis, A., de Crecy-Lagard, V., et al., 2002. Complete reconstitution of the human coenzyme A biosynthetic pathway via comparative genomics. J. Biol. Chem. 277 (24), 21431–21439. Available from: https://doi.org/10.1074/jbc.M201708200.

D'Hooghe, P., Dubousset, L., Gallardo, K., Kopriva, S., Avice, J.C., Trouverie, J., 2014. Evidence for proteomic and metabolic adaptations associated with alterations of seed yield and quality in sulfur-limited *Brassica napus* L. Mol. Cell. Proteomics 13 (5), 1165–1183. Available from: https://doi.org/10.1074/mcp.M113.034215.

Ebi, K.L., Loladze, I., 2019. Elevated atmospheric CO_2 concentrations and climate change will affect our food's quality and quantity. Lancet Planet Health 3 (7), e283–e284. Available from: https://doi.org/10.1016/S2542-5196(19)30108-1.

Falk, K.L., Tokuhisa, J.G., Gershenzon, J., 2007. The effect of sulfur nutrition on plant glucosinolate content: physiology and molecular mechanisms. Plant Biol. (Stuttg). 9 (5), 573–581. Available from: https://doi.org/10.1055/s-2007-965431.

Feldman-Salit, A., Veith, N., Wirtz, M., Hell, R., Kummer, U., 2019. Distribution of control in the sulfur assimilation in *Arabidopsis thaliana* depends on environmental conditions. New Phytol. Available from: https://doi.org/10.1111/nph.15704.

Fernando, N., Panozzo, J., Tausz, M., Norton, R., Fitzgerald, G., Khan, A., et al., 2015. Rising CO_2 concentration altered wheat grain proteome and flour rheological characteristics. Food Chem. 170, 448–454. Available from: https://doi.org/10.1016/j.foodchem.2014.07.044.

Foster, P.A., Mueller, J.W., 2018. SULFATION PATHWAYS: insights into steroid sulfation and desulfation pathways. J. Mol. Endocrinol. 61 (2), T271–T283. Available from: https://doi.org/10.1530/JME-18-0086.

Galili, G., Amir, R., 2013. Fortifying plants with the essential amino acids lysine and methionine to improve nutritional quality. Plant Biotechnol. J. 11 (2), 211–222. Available from: https://doi.org/10.1111/pbi.12025.

Garcia-Mata, C., Lamattina, L., 2010. Hydrogen sulphide, a novel gasotransmitter involved in guard cell signalling. New Phytol. 188 (4), 977–984. Available from: https://doi.org/10.1111/j.1469-8137.2010.03465.x.

Gayler, K.R., Sykes, G.E., 1985. Effects of nutritional stress on the storage proteins of soybeans. Plant Physiol. 78 (3), 582–585. Available from: https://doi.org/10.1104/pp.78.3.582.

Gigolashvili, T., Berger, B., Mock, H.P., Muller, C., Weisshaar, B., Flugge, U.I., 2007a. The transcription factor HIG1/MYB51 regulates indolic glucosinolate biosynthesis in *Arabidopsis thaliana*. Plant J. 50 (5), 886–901. Available from: https://doi.org/10.1111/j.1365-313X.2007.03099.x.

Gigolashvili, T., Yatusevich, R., Berger, B., Muller, C., Flugge, U.I., 2007b. The R2R3-MYB transcription factor HAG1/MYB28 is a regulator of methionine-derived glucosinolate biosynthesis in *Arabidopsis thaliana*. Plant J. 51 (2), 247–261. Available from: https://doi.org/10.1111/j.1365-313X.2007.03133.x.

Gunal, S., Hardman, R., Kopriva, S., Mueller, J.W., 2019. Sulfation pathways from red to green. J. Biol. Chem. 294 (33), 12293–12312. Available from: https://doi.org/10.1074/jbc.REV119.007422.

Haas, F.H., Heeg, C., Queiroz, R., Bauer, A., Wirtz, M., Hell, R., 2008. Mitochondrial serine acetyltransferase functions as a pacemaker of cysteine synthesis in plant cells. Plant Physiol. 148 (2), 1055–1067. Available from: https://doi.org/10.1104/pp.108.125237.

Hacham, Y., Matityahu, I., Schuster, G., Amir, R., 2008. Overexpression of mutated forms of aspartate kinase and cystathionine gamma-synthase in tobacco leaves resulted in the high accumulation of methionine and threonine. Plant J. 54 (2), 260–271. Available from: https://doi.org/10.1111/j.1365-313X.2008.03415.x.

Hagan, N.D., Upadhyaya, N., Tabe, L.M., Higgins, T.J., 2003. The redistribution of protein sulfur in transgenic rice expressing a gene for a foreign, sulfur-rich protein. Plant J. 34 (1), 1–11.

Halford, N.G., Curtis, T.Y., Muttucumaru, N., Postles, J., Elmore, J.S., Mottram, D.S., 2012. The acrylamide problem: a plant and agronomic science issue. J. Exp. Bot. 63 (8), 2841–2851. Available from: https://doi.org/10.1093/jxb/ers011.

Halkier, B.A., Gershenzon, J., 2006. Biology and biochemistry of glucosinolates. Ann. Rev. Plant Biol. 57, 303–333. Available from: https://doi.org/10.1146/annurev.arplant.57.032905.105228.

Hesse, H., Trachsel, N., Suter, M., Kopriva, S., von Ballmoos, P., Rennenberg, H., et al., 2003. Effect of glucose on assimilatory sulphate reduction in *Arabidopsis thaliana* roots. J. Exp. Bot. 54 (388), 1701–1709. Available from: https://doi.org/10.1093/Jxb/Erg177.

Hesse, H., Kreft, O., Maimann, S., Zeh, M., Hoefgen, R., 2004. Current understanding of the regulation of methionine biosynthesis in plants. J. Exp. Bot. 55 (404), 1799–1808. Available from: https://doi.org/10.1093/jxb/erh139.

Hirai, M.Y., Klein, M., Fujikawa, Y., Yano, M., Goodenowe, D.B., Yamazaki, Y., et al., 2005. Elucidation of gene-to-gene and metabolite-to-gene networks in *Arabidopsis* by integration of metabolomics and transcriptomics. J. Biol. Chem. 280 (27), 25590–25595. Available from: https://doi.org/10.1074/jbc.M502332200.

Hirai, M.Y., Sugiyama, K., Sawada, Y., Tohge, T., Obayashi, T., Suzuki, A., et al., 2007. Omics-based identification of *Arabidopsis* Myb transcription factors regulating aliphatic glucosinolate biosynthesis. Proc. Natl. Acad. Sci. USA 104 (15), 6478–6483. Available from: https://doi.org/10.1073/pnas.0611629104.

Hirschmann, F., Krause, F., Papenbrock, J., 2014. The multi-protein family of sulfotransferases in plants: composition, occurrence, substrate specificity, and functions. Front. Plant Sci. 5:, 556. Available from: https://doi.org/10.3389/fpls.2014.00556.

Hogy, P., Wieser, H., Kohler, P., Schwadorf, K., Breuer, J., Franzaring, J., et al., 2009. Effects of elevated CO_2 on grain yield and quality of wheat: results from a 3-year free-air CO_2 enrichment experiment. Plant Biol. (Stuttg). 11 (Suppl 1), 60–69. Available from: https://doi.org/10.1111/j.1438-8677.2009.00230.x.

Hothorn, M., Wachter, A., Gromes, R., Stuwe, T., Rausch, T., Scheffzek, K., 2006. Structural basis for the redox control of plant glutamate cysteine ligase. J. Biol. Chem. 281 (37), 27557–27565. Available from: https://doi.org/10.1074/jbc.M602770200.

Jobe, T.O., Kopriva, S., 2018. Sulfur metabolism in plants. eLS. John Wiley & Sons, Ltd., Chichester.

Juge, N., Mithen, R.F., Traka, M., 2007. Molecular basis for chemoprevention by sulforaphane: a comprehensive review. Cell. Mol. Life Sci. 64 (9), 1105–1127. Available from: https://doi.org/10.1007/s00018-007-6484-5.

Kaufmann, C., Sauter, M., 2019. Sulfated plant peptide hormones. J. Exp. Bot. 70 (16), 4267–4277. Available from: https://doi.org/10.1093/jxb/erz292.

Kelly D.P. 1979. The sulphur cycle: definitions, mechanisms and dynamics. In: Proceedings of the Ciba Foundation Symposium, 72, pp. 3–18. doi:10.1002/9780470720554.ch2

Kelly, J.D., Hefle, S.L., 2000. 2S methionine-rich protein (SSA) from sunflower seed is an IgE-binding protein. Allergy 55 (6), 556–560. Available from: https://doi.org/10.1034/j.1398-9995.2000.00498.x.

Khan, M.S., Haas, F.H., Samami, A.A., Gholami, A.M., Bauer, A., Fellenberg, K., et al., 2010. Sulfite reductase defines a newly discovered bottleneck for assimilatory sulfate reduction and is essential for growth and development in *Arabidopsis thaliana*. Plant Cell 22 (4), 1216–1231. Available from: https://doi.org/10.1105/tpc.110.074088.

Kimura, H., 2014. The physiological role of hydrogen sulfide and beyond. Nitric Oxide 41, 4–10. Available from: https://doi.org/10.1016/j.niox.2014.01.002.

Kim, S., Fenech, M.F., Kim, P.J., 2018. Nutritionally recommended food for semi- to strict vegetarian diets based on large-scale nutrient composition data. Sci. Rep. 8 (1), 4344. Available from: https://doi.org/10.1038/s41598-018-22691-1.

Kim, J., Lee, M., Chalam, R., Martin, M.N., Leustek, T., Boerjan, W., 2002. Constitutive overexpression of cystathionine gamma-synthase in *Arabidopsis* leads to accumulation of soluble methionine and S-methylmethionine. Plant Physiol. 128 (1), 95–107.

Klaiber, J., Dorn, S., Najar-Rodriguez, A.J., 2013. Acclimation to elevated CO_2 increases constitutive glucosinolate levels of *Brassica* plants and affects the performance of specialized herbivores from contrasting feeding guilds. J. Chem. Ecol. 39 (5), 653–665. Available from: https://doi.org/10.1007/s10886-013-0282-3.

Kliebenstein, D.J., Kroymann, J., Brown, P., Figuth, A., Pedersen, D., Gershenzon, J., et al., 2001. Genetic control of natural variation in *Arabidopsis* glucosinolate accumulation. Plant Physiol. 126 (2), 811–825.

Kocsy, G., Galiba, G., Brunold, C., 2001. Role of glutathione in adaptation and signalling during chilling and cold acclimation in plants. Physiol. Plant 113 (2), 158–164.

Kopriva, S., 2016. Glucosinolates, Advances in Botanical Research, vol. 80. Elsevier Ltd.

Kopriva, S., Koprivova, A., 2004. Plant adenosine 5′-phosphosulphate reductase: the past, the present, and the future. J. Exp. Bot. 55 (404), 1775–1783. Available from: https://doi.org/10.1093/Jxb/Erh185.

Koprivova, A., Kopriva, S., 2014. Molecular mechanisms of regulation of sulfate assimilation: first steps on a long road. Front. Plant Sci. 5:, 589. Available from: https://doi.org/10.3389/fpls.2014.00589.

Koprivova, A., Kopriva, S., 2016. Hormonal control of sulfate uptake and assimilation. Plant Mol. Biol. 91 (6), 617–627. Available from: https://doi.org/10.1007/s11103-016-0438-y.

Kopriva, S., Rennenberg, H., 2004. Control of sulphate assimilation and glutathione synthesis: interaction with N and C metabolism. J. Exp. Bot. 55 (404), 1831–1842. Available from: https://doi.org/10.1093/Jxb/Erh203.

Kopriva, S., Muheim, R., Koprivova, A., Trachsel, N., Catalano, C., Suter, M., et al., 1999. Light regulation of assimilatory sulphate reduction in *Arabidopsis thaliana*. Plant J. 20 (1), 37–44.

Koprivova, A., Suter, M., Op den Camp, R., Brunold, C., Kopriva, S., 2000. Regulation of sulfate assimilation by nitrogen in *Arabidopsis*. Plant Physiol. 122 (3), 737–746.

Kopriva, S., Suter, M., von Ballmoos, P., Hesse, H., Krahenbuhl, U., Rennenberg, H., et al., 2002. Interaction of sulfate assimilation with carbon and nitrogen metabolism in *Lemna minor*. Plant Physiol. 130 (3), 1406–1413. Available from: https://doi.org/10.1104/Pp.007773.

Koprivova, A., North, K.A., Kopriva, S., 2008. Complex signaling network in regulation of adenosine 5′-phosphosulfate reductase by salt stress in *Arabidopsis* roots. Plant Physiol. 146 (3), 1408–1420. Available from: https://doi.org/10.1104/pp.107.113175.

Kopriva, S., Mugford, S.G., Baraniecka, P., Lee, B.R., Matthewman, C.A., Koprivova, A., 2012. Control of sulfur partitioning between primary and secondary metabolism in *Arabidopsis*. Front. Plant Sci. 3, 163. Available from: https://doi.org/10.3389/fpls.2012.00163.

Koprivova, A., Giovannetti, M., Baraniecka, P., Lee, B.R., Grondin, C., Loudet, O., et al., 2013. Natural variation in the ATPS1 isoform of ATP sulfurylase contributes to the control of sulfate levels in *Arabidopsis*. Plant Physiol. 163 (3), 1133–1141. Available from: https://doi.org/10.1104/pp.113.225748.

Krueger, S., Niehl, A., Lopez Martin, M.C., Steinhauser, D., Donath, A., Hildebrandt, T., et al., 2009. Analysis of cytosolic and plastidic serine acetyltransferase mutants and subcellular metabolite distributions suggests interplay of the cellular compartments for cysteine biosynthesis in *Arabidopsis*. Plant Cell Environ. 32 (4), 349–367. Available from: https://doi.org/10.1111/j.1365-3040.2008.01928.x.

Land, T., Rouault, T.A., 1998. Targeting of a human iron-sulfur cluster assembly enzyme, nifs, to different subcellular compartments is regulated through alternative AUG utilization. Mol. Cell 2 (6), 807–815. Available from: https://doi.org/10.1016/s1097-2765(00)80295-6.

Lee, H.K., Khaine, I., Kwak, M.J., Jang, J.H., Lee, T.Y., Lee, J.K., et al., 2017. The relationship between SO_2 exposure and plant physiology: a mini review. Hortic. Environ. Biotechnol. 58, 523–529.

Leung, A.W., Backstrom, I., Bally, M.B., 2016. Sulfonation, an underexploited area: from skeletal development to infectious diseases and cancer. Oncotarget 7 (34), 55811–55827. Available from: https://doi.org/10.18632/oncotarget.10046.

Li, B., Gaudinier, A., Tang, M., Taylor-Teeples, M., Nham, N.T., Ghaffari, C., et al., 2014. Promoter-based integration in plant defense regulation. Plant Physiol. 166 (4), 1803–1820. Available from: https://doi.org/10.1104/pp.114.248716.

Li, X., Han, Y., Yan, Y., Messing, J., Xu, J.H., 2018. Genetic diversity and evolution of reduced sulfur storage during domestication of maize. Plant J. 94 (6), 943–955. Available from: https://doi.org/10.1111/tpj.13907.

Lindahl, U., Couchman, J., Kimata, K., Esko, J.D., 2017. Proteoglycans and Sulfated Glycosaminoglycans. Essentials of Glycobiology. Cold Spring Harbor Laboratory Press, Cold Spring Harbor, NY.

Loladze, I., 2014. Hidden shift of the ionome of plants exposed to elevated CO(2)depletes minerals at the base of human nutrition. eLife 3, e02245. Available from: https://doi.org/10.7554/eLife.02245.

Loudet, O., Saliba-Colombani, V., Camilleri, C., Calenge, F., Gaudon, V., Koprivova, A., et al., 2007. Natural variation for sulfate content in *Arabidopsis thaliana* is highly controlled by APR2. Nat. Genet. 39 (7), 896–900. Available from: https://doi.org/10.1038/Ng2050.

Maruyama-Nakashita, A., Nakamura, Y., Tohge, T., Saito, K., Takahashi, H., 2006. *Arabidopsis* SLIM1 is a central transcriptional regulator of plant sulfur response and metabolism. Plant Cell. 18 (11), 3235–3251. Available from: https://doi.org/10.1105/tpc.106.046458.

Meister, A., Anderson, M.E., 1983. Glutathione. Annu. Rev. Biochem. 52, 711–760. Available from: https://doi.org/10.1146/annurev.bi.52.070183.003431.

Monigatti, F., Gasteiger, E., Bairoch, A., Jung, E., 2002. The sulfinator: predicting tyrosine sulfation sites in protein sequences. Bioinformatics 18 (5), 769–770.

Mugford, S.G., Yoshimoto, N., Reichelt, M., Wirtz, M., Hill, L., Mugford, S.T., et al., 2009. Disruption of adenosine-5′-phosphosulfate kinase in *Arabidopsis* reduces levels of sulfated secondary metabolites. Plant Cell. 21 (3), 910–927. Available from: https://doi.org/10.1105/tpc.109.065581.

Muttucumaru, N., Halford, N.G., Elmore, J.S., Dodson, A.T., Parry, M., Shewry, P.R., et al., 2006. Formation of high levels of acrylamide during the processing of flour derived from sulfate-deprived wheat. J. Agric. Food Chem. 54 (23), 8951–8955. Available from: https://doi.org/10.1021/jf0623081.

Myers, S.S., Smith, M.R., Guth, S., Golden, C.D., Vaitla, B., Mueller, N.D., et al., 2017. Climate change and global food systems: potential impacts on food security and undernutrition. Annu. Rev. Public Health 38, 259–277. Available from: https://doi.org/10.1146/annurev-publhealth-031816-044356.

Naznin, M.T., Kitaya, Y., Shibuya, T., Endo, R., Hirai, H., Lefsrud, M.G., 2015. Ground based study on culturing garlic as a source of vegetable food and medicine in space - growth and ajoene accumulation in garlic plants cultured with different CO_2 regimes. Biol. Sci. Space 29, 1–7. 10.2187/bss.29.1.

Niu, W.N., Yadav, P.K., Adamec, J., Banerjee, R., 2015. S-glutathionylation enhances human cystathionine beta-synthase activity under oxidative stress conditions. Antioxid. Redox Signal. 22 (5), 350–361. Available from: https://doi.org/10.1089/ars.2014.5891.

Noctor, G., Mhamdi, A., Chaouch, S., Han, Y., Neukermans, J., Marquez-Garcia, B., et al., 2012. Glutathione in plants: an integrated overview. Plant Cell Environ. 35 (2), 454–484. Available from: https://doi.org/10.1111/j.1365-3040.2011.02400.x.

Parcell, S., 2002. Sulfur in human nutrition and applications in medicine. Altern. Med. Rev. 7 (1), 22–44.

Patron, N.J., Durnford, D.G., Kopriva, S., 2008. Sulfate assimilation in eukaryotes: fusions, relocations and lateral transfers. BMC Evol. Biol. 8, 10.1186/1471–2148-8–39 39.

Planta, J., Xiang, X., Leustek, T., Messing, J., 2017. Engineering sulfur storage in maize seed proteins without apparent yield loss. Proc. Natl. Acad. Sci. USA 114 (43), 11386–11391. Available from: https://doi.org/10.1073/pnas.1714805114.

Pomin, V.H., Vignovich, W.P., Gonzales, A.V., Vasconcelos, A.A., Mulloy, B., 2019. Galactosaminoglycans: medical applications and drawbacks. Molecules 24 (15). Available from: https://doi.org/10.3390/molecules24152803.

Rausch, T., Wachter, A., 2005. Sulfur metabolism: a versatile platform for launching defence operations. Trends Plant Sci. 10 (10), 503–509. Available from: https://doi.org/10.1016/j.tplants.2005.08.006.

Ravilious, G.E., Nguyen, A., Francois, J.A., Jez, J.M., 2012. Structural basis and evolution of redox regulation in plant adenosine-5′-phosphosulfate kinase. Proc. Natl. Acad. Sci. USA 109 (1), 309–314. Available from: https://doi.org/10.1073/pnas.1115772108.

Ripps, H., Shen, W., 2012. Review: taurine: a "very essential" amino acid. Mol. Vis. 18, 2673–2686.

Schonhof, I., Klaring, H.P., Krumbein, A., Schreiner, M., 2007. Interaction between atmospheric CO_2 and glucosinolates in broccoli. J. Chem. Ecol. 33 (1), 105–114. Available from: https://doi.org/10.1007/s10886-006-9202-0.

Schwarz, G., 2016. Molybdenum cofactor and human disease. Curr. Opin. Chem. Biol. 31, 179–187. Available from: https://doi.org/10.1016/j.cbpa.2016.03.016.

Schweizer, F., Fernandez-Calvo, P., Zander, M., Diez-Diaz, M., Fonseca, S., Glauser, G., et al., 2013. *Arabidopsis* basic helix-loop-helix transcription factors MYC2, MYC3, and MYC4 regulate glucosinolate biosynthesis, insect performance, and feeding behavior. Plant Cell 25 (8), 3117–3132. Available from: https://doi.org/10.1105/tpc.113.115139.

Searcy, D.G., Lee, S.H., 1998. Sulfur reduction by human erythrocytes. J. Exp. Zool. 282 (3), 310–322. 10.1002/(sici)1097-010x(19981015)282:3 < 310::aid-jez4 > 3.0.co;2-p.

Shibuya, N., Tanaka, M., Yoshida, M., Ogasawara, Y., Togawa, T., Ishii, K., et al., 2009. 3-Mercaptopyruvate sulfurtransferase produces hydrogen sulfide and bound sulfane sulfur in the brain. Antioxid. Redox Signal. 11 (4), 703–714. Available from: https://doi.org/10.1089/ARS.2008.2253.

Singh, S., Padovani, D., Leslie, R.A., Chiku, T., Banerjee, R., 2009. Relative contributions of cystathionine beta-synthase and gamma-cystathionase to H_2S biogenesis via alternative trans-sulfuration reactions. J. Biol. Chem. 284 (33), 22457–22466. Available from: https://doi.org/10.1074/jbc.M109.010868.

Sonderby, I.E., Geu-Flores, F., Halkier, B.A., 2010. Biosynthesis of glucosinolates--gene discovery and beyond. Trends Plant Sci. 15 (5), 283–290. Available from: https://doi.org/10.1016/j.tplants.2010.02.005.

Stipanuk, M.H., 2004. Sulfur amino acid metabolism: pathways for production and removal of homocysteine and cysteine. Annu. Rev. Nutr. 24, 539–577. Available from: https://doi.org/10.1146/annurev.nutr.24.012003.132418.

Suter, M., von Ballmoos, P., Kopriva, S., den Camp, R.O., Schaller, J., Kuhlemeier, C., et al., 2000. Adenosine 5′-phosphosulfate sulfotransferase and adenosine 5′-phosphosulfate reductase are identical enzymes. J. Biol. Chem. 275 (2), 930–936.

Takahashi, H., Buchner, P., Yoshimoto, N., Hawkesford, M.J., Shiu, S.H., 2011a. Evolutionary relationships and functional diversity of plant sulfate transporters. Front. Plant Sci. 2, 119. Available from: https://doi.org/10.3389/fpls.2011.00119.

Takahashi, H., Kopriva, S., Giordano, M., Saito, K., Hell, R., 2011b. Sulfur assimilation in photosynthetic organisms: molecular functions and regulations of transporters and assimilatory enzymes. Annu. Rev. Plant Biol. 62, 157–184. Available from: https://doi.org/10.1146/annurev-arplant-042110-103921.

Talaei, F., van Praag, V.M., Henning, R.H., 2013. Hydrogen sulfide restores a normal morphological phenotype in Werner syndrome fibroblasts, attenuates oxidative damage and modulates mTOR pathway. Pharmacol. Res. 74, 34–44. Available from: https://doi.org/10.1016/j.phrs.2013.04.011.

Traka, M., Mithen, R., 2009. Glucosinolates, isothiocyanates and human health. Phytochem. Rev. 8 (1), 269–282. Available from: https://doi.org/10.1007/s11101-008-9103-7.

Tschanz, A., Landoit, W., Bleuler, P., Brunold, C., 1986. Effect of SO_2, on the activity of adenosine 5′-phosphosulfate sulfotransferase from spruce trees (*Picea abies*) in fumigation chambers and under field conditions. Physiol. Plantarum. 67, 235–241.

Ufaz, S., Galili, G., 2008. Improving the content of essential amino acids in crop plants: goals and opportunities. Plant Physiol. 147 (3), 954–961. Available from: https://doi.org/10.1104/pp.108.118091.

Usmani, M.M., Nawaz, F., Majeed, S., Shehzad, M.A., Ahmad, K.S., Akhtar, G., et al., 2020. Sulfate-mediated drought tolerance in maize involves regulation at physiological and biochemical levels. Sci. Rep. 10 (1), 1147. Available from: https://doi.org/10.1038/s41598-020-58169-2.

Vauclare, P., Kopriva, S., Fell, D., Suter, M., Sticher, L., von Ballmoos, P., et al., 2002. Flux control of sulphate assimilation in *Arabidopsis thaliana*: adenosine 5′-phosphosulphate reductase is more susceptible than ATP sulphurylase to negative control by thiols. Plant J. 31 (6), 729–740.

von Caemmerer, S., Furbank, R.T., 2003. The C(4) pathway: an efficient CO(2) pump. Photosynth. Res. 77 (2-3), 191–207. Available from: https://doi.org/10.1023/A:1025830019591.

Wachnowsky, C., Fidai, I., Cowan, J.A., 2018. Iron-sulfur cluster biosynthesis and trafficking - impact on human disease conditions. Metallomics 10 (1), 9–29. Available from: https://doi.org/10.1039/c7mt00180k.

Wallace, J.L., Wang, R., 2015. Hydrogen sulfide-based therapeutics: exploiting a unique but ubiquitous gasotransmitter. Nat. Rev. Drug Discov. 14 (5), 329–345. Available from: https://doi.org/10.1038/nrd4433.

Wirtz, M., Hell, R., 2006. Functional analysis of the cysteine synthase protein complex from plants: structural, biochemical and regulatory properties. J. Plant Physiol. 163 (3), 273–286. Available from: https://doi.org/10.1016/j.jplph.2005.11.013.

Wrigley, C.W., Du Cros, D.L., Fullington, J.G., Kasarda, D.D., 1984. Changes in polypeptide composition and grain quality due to sulfur deficiency in wheat. J. Cereal Sci. 2, 15–24.

Xiang, X., Wu, Y., Planta, J., Messing, J., Leustek, T., 2018. Overexpression of serine acetyltransferase in maize leaves increases seed-specific methionine-rich zeins. Plant Biotechnol. J. 16 (5), 1057–1067. Available from: https://doi.org/10.1111/pbi.12851.

Yatusevich, R., Mugford, S.G., Matthewman, C., Gigolashvili, T., Frerigmann, H., Delaney, S., et al., 2010. Genes of primary sulfate assimilation are part of the glucosinolate biosynthetic network in *Arabidopsis thaliana*. Plant J. 62 (1), 1–11. Available from: https://doi.org/10.1111/j.1365-313X.2009.04118.x.

Yoshimoto, N., Takahashi, H., Smith, F.W., Yamaya, T., Saito, K., 2002. Two distinct high-affinity sulfate transporters with different inducibilities mediate uptake of sulfate in *Arabidopsis* roots. Plant J. 29 (4), 465–473. 1231 [pii].

Yoshimoto, N., Inoue, E., Watanabe-Takahashi, A., Saito, K., Takahashi, H., 2007. Posttranscriptional regulation of high-affinity sulfate transporters in *Arabidopsis* by sulfur nutrition. Plant Physiol. 145 (2), 378–388. Available from: https://doi.org/10.1104/pp.107.105742.

Zhu, C., Kobayashi, K., Loladze, I., Zhu, J., Jiang, Q., Xu, X., et al., 2018. Carbon dioxide (CO_2) levels this century will alter the protein, micronutrients, and vitamin content of rice grains with potential health consequences for the poorest rice-dependent countries. Sci. Adv. 4 (5), eaaq1012. Available from: https://doi.org/10.1126/sciadv.aaq1012.

Zorb, C., Steinfurth, D., Seling, S., Langenkamper, G., Koehler, P., Wieser, H., et al., 2009. Quantitative protein composition and baking quality of winter wheat as affected by late sulfur fertilization. J. Agric. Food Chem. 57 (9), 3877–3885. Available from: https://doi.org/10.1021/jf8038988.

Chapter 5

Potassium: an emerging signal mediator in plants?

Swati Mahiwal and Girdhar K. Pandey
Department of Plant Molecular Biology, University of Delhi South Campus, New Delhi, India

5.1 Introduction

5.1.1 K^+ as a signal in the biological world

Well-known for its tremendous importance as a nutrient in plants as well as in animals, K^+ needs no further introduction. Apart from being a nutrient, it plays crucial roles that are universally important in each living organism such as maintenance of osmotic balance (Wolfgang, 1986; Kay, 2017). Extensive research work in the field of K^+ biology has garnered important information about this ion. Its physiological importance in the context of nutrient transport systems and how its deficiency affects the cellular as well as physiological architecture have been well examined in past years (Dreyer et al., 2011; Sharma et al., 2013; Adams and Shin, 2014). This has been studied in single-cell bacteria to complex organisms, animals, and plants. In plants, the presence of K^+ has been proven to contribute to development and abiotic stress tolerance (Srivastava et al., 2019; Pandey and Mahiwal, 2020). However, plant biologists have pointed to several new areas where K^+ might be involved. One of these areas is plant signaling. At first, it may sound very unrealistic to think about K^+ as a signal, but this has been already established in bacteria as a cytoplasmic signaling molecule. In human T lymphocytes, upon perceiving mitogenic stimulations, voltage-gated K^+ channels contribute to transmembrane potential and cell proliferation (Leonard et al., 1992; Negulescu et al., 1994). Another support for this statement came from a study that demonstrated inhibition of cell proliferation and arrest of the G1 phase of the cell cycle upon suppression of channel activity of voltage-gated K^+ channels. Hence, it became evident that voltage-gated K^+ channels regulate G1/S transition in the cell cycle and contribute to control of cell growth as well as proliferation. This indicates the function of K^+ channels mediating signals involved in cell growth. However, the whole mechanism underlying

Plant Nutrition and Food Security in the Era of Climate Change. DOI: https://doi.org/10.1016/B978-0-12-822916-3.00012-3

this phenomenon has not been deciphered yet and requires further investigation (Wang et al., 2014).

Therefore taking a hint from these observations, one might think of K^+ as a signal or signal mediator in plants. Published reports have shown some clues that K^+ might act as a potential second messenger or signal mediator. This new concept has been discussed briefly in this chapter. However, a few properties of K^+ do not support this concept, and they are also discussed in this chapter.

5.2 K^+ as a signal in bacteria

Universally known for maintaining the essential cellular functions such as turgor maintenance, osmotic adaptations, and activation of few cytoplasmic enzymes, K^+ is maintained at relative high concentrations, that is 300–500 mM in bacteria (Bossemeyer et al., 1989; Epstein, 2003). In their lifetime, bacteria get exposed to various environments, including a wide range of external K^+ concentrations. In order to maintain intracellular K^+ levels, bacteria utilize numerous transporters and efflux pumps for transport of K^+ across the cell membrane. So far, in gram-negative bacteria, low-affinity K^+ transporter (Trk), K^+-dependent (Kdp), and K^+-uptake (Kup) are the best-characterized K^+ transport systems. According to the relative external concentrations of K^+, these transport systems get activated and regulate K^+ homeostasis (Su et al., 2009). At a neutral or alkaline pH, the low-affinity K^+ transporter (Trk) system acts as the major K^+ transport system, which is a rapid transport system (Bossemeyer, 1989; Trchounian and Kobayashi, 2000; Epstein, 2003). In acidic conditions, K^+-uptake (Kup), a high-affinity K^+ transporter, acts as a major transport system (Trchounian and Kobayashi, 1999; Zakharyan and Trchounian, 2001). Another high-affinity K^+ transport system, K^+ dependent (Kdp) is induced upon low external K^+ concentrations (less than 5 mM) (Frymier et al., 1997; Epstein, 2003). This Kdp system is a multiprotein complex, consisting of three main components KdpD, KdpE, and KdpFABC (Su et al., 2009). KdpD is a histidine kinase that acts as sensor and KdpE is a response regulator that regulates the Kdp system in a phosphorylation-dependent manner (Laermann et al., 2013; Ashraf et al., 2016; Schramke et al., 2017). Extensive research in the field has revealed the probable mechanism of this regulation. Under K^+-limiting conditions, KdpD perceives three chemical stimuli: extracellular/intracellular K^+ concentration, ionic strength, and ATP levels (Jung and Altendorf, 1998; Heermann et al., 2000; Jung et al., 2000; Laermann et al., 2013; Schramke et al., 2016). After perception of K^+-limiting conditions, KdpD gets autophosphorylated, which further phosphorylates KdpE. This results in dimerization of KdpE, which activates the high-affinity K^+ uptake system KdpFABC as shown in Fig. 5.1. Since KdpD also possesses a phosphatase activity, it can dephosphorylate KdpE. This function of KdpD enters

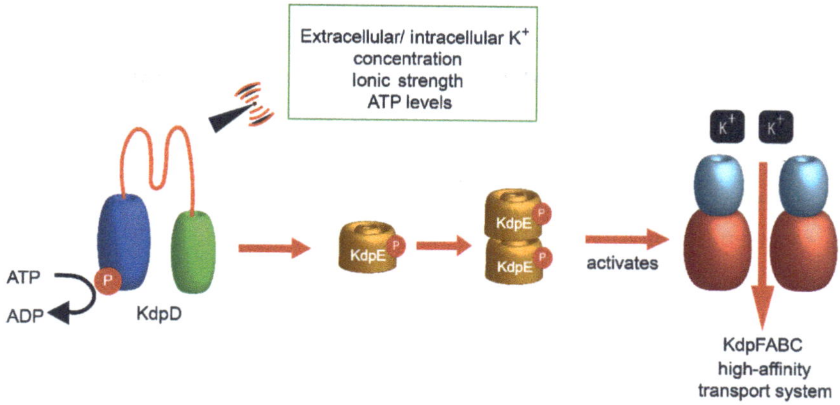

FIGURE 5.1 A schematic representation of cytoplasmic-K^+ signaling in bacteria. Under K^+-limiting conditions, KdpD (a histidine kinase) senses three different chemical stimuli resulting in its autophosphorylation. Further, it phosphorylates another component KdpE (a response regulator), which gets dimerized upon phosphorylation, finally activating high-affinity transport system KdpFABC.

into the picture under K^+-sufficient conditions. When sufficient K^+ is available, KdpD dephosphorylates KdpE, which further suppresses the activation of KdpFABC (Sugiura et al., 1992; Voelkner et al., 1993). Based on this, K^+ has been considered as a cytoplasmic signal in bacteria.

5.3 Does a similar mechanism exist in plants?

The past few years of research on K^+ have focused on studying the in-depth mechanism of K^+ deficiency response, as it is a major nutrient required by plants. The deficiency of K^+ causes many physiological aberrations in plants and disturbs several cellular processes. In order to cope with K^+ deficiency, activation of K^+ transporters and "switching on" of several signaling cascades are the primary responses of plants. An increasing number of reports on K^+ deficiency show these multidimensional arrays of events triggered by K^+ deficiency. Being sensed by the roots at first, as K^+ triggers all these responses (which is quite similar to signal amplification by a second messenger), can K^+ be considered as a signal or signal mediator in plants? To establish this concept in plants, a closer inspection of the properties of K^+ and mechanisms triggered by its deficiency is required (Pandey and Mahiwal, 2020).

5.4 How is potassium perceived and sensed in plants?

Plant cell roots acts as the primary site for absorption of water, nutrients, and minerals. Two main components of the root, that is epidermal cell and root hairs, are mainly responsible for absorption of K^+ from the soil. This suggests them as potential organs for K^+ sensing. Depending on K^+ availability in the soil, root surfaces come into contact with varying K^+ concentrations (ranges from 0.1 to 1 mM) while they are embedded in the soil. The typical concentration of K^+ around the root surface is comparatively lower than the concentration of K^+ present inside the plants (Schroeder et al., 1994; Maathuis, 2009). Therefore, passive diffusion is not possible and the only choice left for the root cells is to absorb K^+ against a concentration gradient. An impressive set of transport systems contributes to K^+ absorption against the concentration gradient in the root cell (Wang and Wu, 2013). This absorption takes place by dual-affinity mechanism, either a high-affinity uptake mechanism or a low-affinity uptake mechanism. Both of these mechanisms are governed by $[K^+]_{ext}$ concentration in the soil. High-affinity K^+ uptake mechanisms gets activated by a low $[K^+]_{ext}$ concentration below 0.2 mM, mediated majorly by K^+ transporters (Epstein et al., 1963; Maathuis and Sanders, 1994). At low $[K^+]_{ext}$, active transporters are required in order to trail K^+ inside the cells against the electrochemical gradient. This justifies the role of transporters in high-affinity uptake (Ragel et al., 2019). Low-affinity K^+ uptake gets activated at relatively high $[K^+]_{ext}$ above 0.3 mM, mediated by channels (Epstein et al., 1963; Schroeder et al., 1994). At high $[K^+]_{ext}$, K^+ simply moves down the electrochemical gradient through channels. This property of sensing $[K^+]_{ext}$ by the plasma membrane of the root cell points toward its possible function of being a potent K^+ sensor (Pandey and Mahiwal, 2020).

5.5 How does K^+ sensing take place?

The varying K^+ concentration around the root surface is perceived at the plasma membrane of epidermal cells and in a similar manner, the K^+ deficiency is also perceived. After perception, its subsequent cytosolic transduction results in generating either short- or long-term responses via a series of physiological and biochemical reactions (Wang and Wu, 2013). The short-term responses are important for the plants as they may provide an insight into molecular mechanisms involved in K^+ perception. Interestingly, the preference of the sensing mechanism is to generate a "quick response" rather than a delayed response (Wang and Wu, 2013). Since the concept of K^+ sensing came into existence, a constant quest for deciphering the exact mechanism has begun. Many hypotheses have been proposed so far in the literature and the more promising hypotheses are discussed below (Pandey and Mahiwal, 2020).

5.5.1 Membrane potential as a potent K^+ sensor

In animal systems, the two ions K^+ and Na^+ are classically known for their role in generating membrane potential, so it won't be a surprise if one says that external K^+ concentration influences the membrane potential. As far as the plasma membranes of root epidermal cells are concerned, the crucial job of membrane potential maintenance is primarily mediated by H^+-ATPase activity (Maathuis and Sanders, 1993; Palmgren, 2001). In a previous report, it was already established that the root cell membrane displays a linear relationship with external K^+ concentration (K^+_{ext}) as it exhibits high specificity and dependency on $[K^+]_{ext}$ (Maathuis and Sanders, 1993). The varying $[K]_{ext}$ concentration leads to changes in membrane potential. This fact has been previously examined by many reports and these reports point to a single observation that low $[K^+]_{ext}$ leads to hyperpolarization of the membrane whereas high $[K^+]_{ext}$ leads to depolarization of the membrane as shown in Fig. 5.2 (Schroeder and Fang, 1991; Maathuis and Sanders, 1993; Spalding et al., 1999; Nieves-Cordones et al., 2008). Interestingly, these changes in membrane also appear after a few minutes of K^+ starvation (which can be stated as a quick response). Hence the change in membrane potential is thought to be as one of the sensing mechanisms in response to K^+ deprivation in plant cells (Maathuis and Sanders, 1993; Nieves-Cordones et al., 2008).

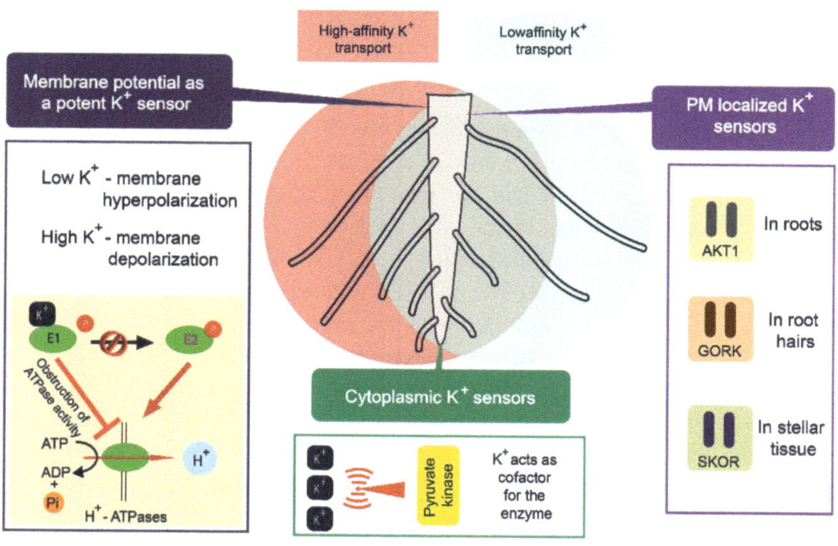

FIGURE 5.2 A schematic representation of potential mechanisms of K^+ sensing leading to activation of high-affinity and low-affinity K^+ transport systems. Three possible mechanisms have been described. The first mechanism suggests change in membrane potential as a K^+ sensor, which is majorly regulated by the H^+-ATPase upon switching between two forms E1 and E2. The second mechanism highlights the importance of K^+ transport systems that are localized to the plasma membrane in sensing as they come in contact with K^+ directly after its perception on the root surface. The third mechanism emphasizes cytosolic enzymes being a K^+ sensor as many enzymes require K^+ as a cofactor whose availability is dependent on its cytosolic concentration. *Adapted and modified from Pandey, G.K., Mahiwal, S., 2020. Role of Potassium in Plants. SpringerBriefs in Plant Science. Springer Nature, Switzerland AG. ISBN: 978-3-030-45953.*

Another clue for membrane potential acting as a K^+ sensor comes from a well-known H^+-ATPase that also exhibits sensitivity to external K^+. Depending on the phosphorylation status, this H^+-ATPase can switch between its two forms, E1 and E2, which are interconvertible into each other. The E1 form is an ATP-hydrolyzing form, which gets converted into E2 form upon phosphorylation. It is hypothesized that K^+ binding results in dephosphorylation of the E1 form of the ATPases (E1P) (ATP-hydrolyzing form). The key residue for this dephosphorylation event is an aspartate residue placed in the P-domain, finally preventing conversion of E1 form to the E2 form (E2P). This form of the H^+-ATPase creates an obstruction to H^+ extrusion resulting in uncoupling of ATP hydrolysis and H^+ pumping activity. Accumulation of the E1P form of ATPase negatively regulates the transmembrane electrochemical gradient as shown in Fig. 5.2. Hence, K^+ acts as a regulator of H^+/ATP coupling ratio and can serve as an intrinsic uncoupler of proton pump (Buch-Pedersen et al., 2006; Cherel et al., 2014). In order to establish this concept, more experiments are required with solid evidence as well as a clear understanding of the exact mechanism of membrane potential–triggered specific responses (Pandey and Mahiwal, 2020).

5.5.2 Plasma membrane localized K^+ sensors

For ion uptake, the only membrane that comes directly in contact with the ions is the plasma membrane, which also serves as a passage for ions via different means of transport. When K^+ ions come in contact with the root surface, the plasma membrane of the root epidermal cell is responsible for K^+ sensing. This indicates the possibility of K^+ sensors being located at the plasma membrane (Wang and Wu, 2013). Previous reports suggest outwardly rectifying K^+ channels might function as K^+ sensors when $[K^+]_{ext}$ is reduced. The potential candidates for this are guard cell outward rectifier (GORK) and stomatal K^+ outward rectifier (SKOR) as shown in Fig. 5.2. When the GORK expresses in the root hair, membrane depolarization affects its activity and gating largely depends on $[K^+]_{ext}$. Based on these observation GORK has been suggested to play a role in K^+ sensing (Sharma et al., 2013). Membrane potential along with the $[K^+]_{ext}$ modulate the activity of another outward-rectifying K^+ channel, SKOR. Depending on the apoplastic K^+ concentration, SKOR (shows specific expression in stellar tissue) controls the K^+ flux to the xylem. This observation points to the higher K^+ sensitivity of SKOR channel to $[K^+]_{ext}$ (Johansson et al., 2006; Sharma et al., 2013).

The typical dual-affinity character of major inward rectifier K^+ channel (AKT1) suggests AKT1 as a potential candidate for K^+ sensing at the plasma membrane as shown in Fig. 5.2 (Spalding et al., 1999; Li et al., 2017). Fluctuation in $[K^+]_{ext}$ leads to the change in between two states of AKT1, that is switch between low-affinity and high-affinity forms. This kinetic shift of AKT1 seems to be a phosphorylation-dependent phenomenon that is under the control of calcineurin B-like protein kinase 23 (CIPK23) (Hirsch et al., 1998; Li et al., 2006; Yuan et al., 2007). Interestingly, the properties of AKT1 show similarity to the nitrate transporter NRT1.1 (also known as CHL1, acts as nitrate sensor) in fact, in terms of their regulation as well. Both are regulated by the same set of calcineurin B- like proteins (CBL)-CIPK complexes, that is CBL1/9-CIPK23 (Ho et al., 2009). Based on these facts, assumptions have been made for AKT1 acting as a sensor, which gets activated via CBL-CIPK complexes triggering their mobilization (Pandey and Mahiwal, 2020).

5.5.3 Cytoplasmic K^+ sensors: an interplay of cytoplasmic enzymes

A common phenomenon in plant cells is to store compounds in the vacuole whenever they are present in excess amounts in the cytosol. This is also true for the K^+. In K^+ deficiency responses, one of the short-term responses is mobilization of K^+ from the vacuole to the cytosol. This mobilization from the vacuole compensates for the decreased K^+ absorption from external media. If K^+ deficiency occurs for a longer duration, many cytosolic activities get affected. In long-term K^+ deficiency, the whole physiology of the plant cell is affected (Walker et al., 1996). Now the question that arises is whether there is short-term or long-term response, a signal must exist to trigger these responses. How is the K^+ limitation sensed in the cytoplasm? Two hypotheses suggest that existence of K^+-sensitive cytosolic enzymes makes it possible to sense the cytosolic K^+ concentration. These cytosolic enzymes require K^+ as a cofactor; among all those enzymes, pyruvate kinase has been suggested as the most likely candidate for sensing K^+ concentration as shown in Fig. 5.2. It shows maximum sensitivity to cytosolic K^+ concentration (Ramirez-Silva et al., 2001; Amtmann and Armengaud, 2009; Armengaud et al., 2009). A previous study reports that long-term K^+ deficiency in plants leads to inhibition of pyruvate kinase activity in root cells (Armengaud et al., 2009). In crop plant wheat, an increased activity of pyruvate kinase was seen under K^+ deficiency (Sugiyama et al., 1968). Interestingly, these two separate experiments report contrasting observations probably because of experimental conditions. However, another possibility could be the variable K^+ requirement of different plant species, consisting of different mechanisms to deal with K^+ deficiency (Sugiyama et al., 1968; Armengaud et al., 2009).

Despite extensive research being performed, no canonical sensor has been identified so far in plants (Pandey and Mahiwal, 2020).

5.6 Is potassium deficiency a potential stress signal?

As mentioned earlier, since many enzymatic activities are dependent upon K^+, its fluxes in the plants alter functioning of metabolic processes, ROS homeostasis, and posttranslation modifications such as phosphorylation/dephosphorylation (Shin and Schachtman, 2004; Ma et al., 2012; Pandey et al., 2016; Sharma and Pandey, 2017; Pandey, 2018). Until now, no canonical K^+-dependent signaling cascades have been identified in plants as it had been identified in bacteria a signal that helps bacteria to survive under K^+-limiting conditions by switching on a high-affinity transport system. In plants, however, it has been presented as a different concept altogether. K^+ deficiency has been linked to stress signaling but there are speculations that this may act as a stress signal that turns on various defense responses in order to withstand K^+ deficiency (Ashley et al., 2005; Pandey and Mahiwal, 2020).

In this postgenomics and transcriptomic era, an outburst of information provided the first clue of K^+ being a stress signal. With a continuous rise in transcriptomics studies in plants, one study came up with exclusive details regarding K^+ deficiency. This study revealed that K^+ deprivation largely affects genes related to jasmonic acid (JA) signaling and metabolism (Armengaud et al., 2004; Sirhindi et al., 2017). The study pointed to the possibility that low-K condition somehow triggers the expression of these genes. The second clue was provided by an experimental demonstration, which showed involvement of root hair defective 2 (RHD2). As RHD2 encodes for NADPH oxidase generating H_2O_2 in response to K deprivation, this further suggests activation of a ROS-dependent signaling pathway (Foreman et al., 2003). The third clue came from another experiment upon measurement of direct ethylene levels released by K-deficient and K-sufficient plants. As more ethylene production in K-starved plants was seen, it was proposed that K starvation triggers ethylene-mediated signaling in response to K starvation (Shin and Schachtman, 2004). Another report highlights the importance of auxin in K^+ deficiency. Two Trp-dependent auxin biosynthetic genes Tryptophan N-monooxygenase1 (CYP79B2) and Tryptophan N-monooxygenase2 (CYP79B3) were downregulated in K-starved roots when resupplied with sufficient K. This points to the role of the auxin-dependent process in fine-tuning of K-deficiency responses (Armengaud et al., 2004). Another study highlights alterations in gene expression patterns of K^+-responsive genes, which play roles in stress responses, metabolic processes, and signaling pathways. Out of these genes, several show reversal upon K^+ resupply indicating K^+ deficiency as an activator of multiple gene networks (Shankar et al., 2013).

5.7 K^+ as a signal mediator in plants: connecting the dots between K^+ deprivation and Ca^{2+} signaling

One of the most important and pivotal signaling systems is Ca^{2+} signaling, which gets activated in response to various environmental cues such as biotic stress, abiotic stress, as well as nutrient deprivation (Pandey, 2008; Wang and Wu, 2010; Kanwar et al., 2014; Singh et al., 2014; Pandey et al., 2016). Reactive-oxygen species (ROS) became the first connecting dot between Ca^{2+} signaling and K^+ deprivation. Few earlier reports associate H_2O_2 production under low-K conditions with changes in intracellular Ca^{2+} fluxes (Foreman et al., 2003; Shin and Schachtman, 2004). Subsequent studies using ROS and Ca^{2+} sensors show that both ROS and Ca^{2+} signals play important roles in low-K^+ sensing and signaling (Xu et al., 2006). A series of reports described this phenomenon with experimental evidence. Elevation in Ca^{2+} concentration in cytoplasm stimulates NADPH oxidase-mediated ROS production (Yang and Poovaiah, 2002; Shankar et al., 2018), which leads to activation of Ca^{2+} permeable ion channels generating Ca^{2+} influx (Demidchik et al., 2007). Later on, a report spotlighted the possible mechanism of this phenomenon. It suggested that targeting of plasma membrane–localized NADPH oxidases by either open stomatal (OST1) or CBL1/CBL9-CIPK26 results in ROS production as shown in Fig. 5.3 (Drerup et al., 2013). Based on these reports, Ca^{2+} was also suggested as a second messenger in response to K^+ deficiency (Hafsi et al., 2014).

Recent reports have emerged with newer pieces of information that suggest the generation of two successively distinct Ca^{2+} signals in roots upon the onset

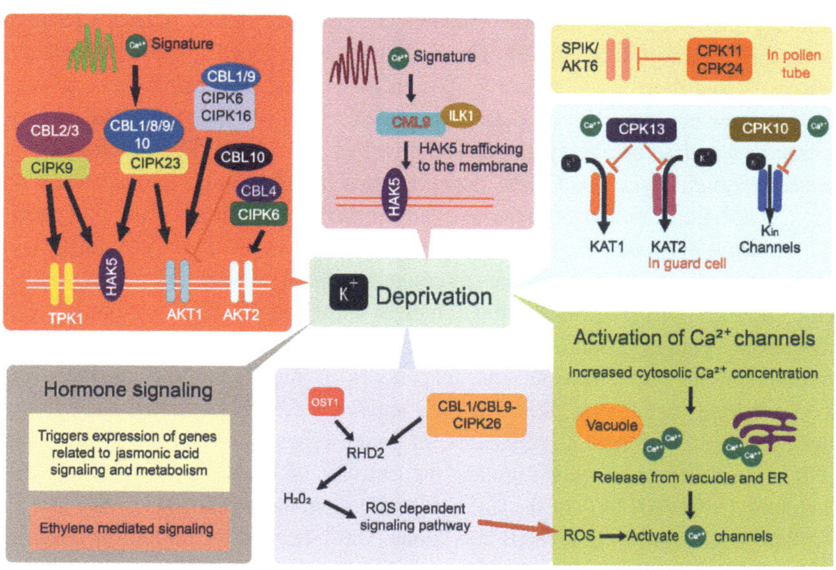

FIGURE 5.3 Schematic representation of responses triggered by K^+ deficiency in roots. Two types of major responses are generated, one of them is activation of Ca^{2+} signaling and its components such as Ca^{2+} channels. Another signaling pathway including JA signaling, ethylene signaling, and ROS-dependent signaling pathways gets activated by different mechanisms. However, the activation of the ROS-dependent signaling pathway occurs through the CBL-CIPK complex, which is a part of Ca^{2+} signaling. *ROS, reactive-oxygen species.*

of K^+ deficiency. Spatial and temporal specificity is the unique feature of these distinct signals that helps convey the response to K^+ deficiency (Behera et al., 2017). Activation of the root epidermis and root hair zone-specific K^+ channels via hyperpolarization of the plasma membrane induced by low K^+ conditions seems to be an obvious choice for the plant (Very and Davies, 2000). Earlier reports suggest several mechanisms for the activation of Ca^{2+} channels in response to K^+ deficiency. One mechanism highlights dependence on the increased cytosolic concentration, as it activates other Ca^{2+} channels present within inner membranes causing Ca^{2+} release from the Ca^{2+} pools such as vacuole and ER (Wang et al., 2018). Another mechanism suggests ROS-dependent activation of Ca^{2+} permeable ion channels further mediating Ca^{2+} influx (Pei et al., 2000; Demidchik et al., 2007). Taken together, it can be stated that both ROS and Ca^{2+} activate Ca^{2+} signaling as a response to K^+ deficiency. Interestingly, ROS along with Ca^{2+} is also proposed to be involved in long-distance signaling, probably for generating Ca^{2+} waves (Choi et al., 2017). However, a precise mechanism still needs to be established between transient increase in intracellular Ca^{2+}, that is $[Ca^{2+}]_i$ in response to low K^+ stress, and how plants distinguish this Ca^{2+} signal from other Ca^{2+} signatures generated in a cell (Pandey and Mahiwal, 2020).

The usual choice for the perception of Ca^{2+} signals in plant cells is Ca^{2+} sensors, such as calmodulin (CaM), calmodulinlike proteins (CML), calcineurin-B like proteins (CBL), and Ca^{2+}-dependent protein kinases (CDPK) (Sanyal et al., 2015; Sanyal et al., 2020). The basic reason for a cell to have many Ca^{2+} sensors is that these Ca^{2+} sensors differ in their mode of sensing Ca^{2+} as well as relay mechanisms. These sensors can be of two types in a plant cell, sensor responders and sensor relay proteins (Sanyal et al., 2019). The functioning of sensor responders such as CDPK depends upon intramolecular interactions with Ca^{2+}. Ca^{2+}-induced conformational changes modulate their structure and activity. On the other hand, the functioning of sensor relays such as CaM, CBL, and CML depends upon bimolecular interactions. First, sensor proteins combine with Ca^{2+} and bring conformational changes subsequently relaying signals to an interacting partner (Sanders et al., 2002; Hashimoto and Kudla, 2011). Involvement of these sensors and responders in regulating K^+ channels has been reported in the past. Some reports show the CDPK-mediated modulation mechanisms of different K^+ channels. In pollen tubes, two CDPKs, CPK11 and CPK24 together negatively regulate the activity of shaker pollen inward K^+ channels (SPIK/AKT6) in a Ca^{2+}-dependent manner (Zhao et al., 2013). Similarly, in guard cell CPK10 modulates Ca^{2+}-dependent inhibition of K^+_{in} channels (Zou et al., 2010). Another CDPK, that is CPK13, is responsible for specific inhibition of guard cell-expressed potassium channel in *Arabidopsis thaliana* (KAT2) and (KAT1) shaker K^+ channels (Ronzier et al., 2014).

Another protein module consisting of CBLs and CIPKs also regulates plant responses to K^+ deficiency by modulating the activity of two major K^+ transport systems, *Arabidopsis* (shaker-type) K^+ channel (AtAKT1)

(Rubio et al., 2008) and high-affinity K^+ transporters (AtHAK5) (Qi et al., 2008). Two simultaneously occurring pathways for low K^+ signaling have been identified in *Arabidopsis* regulating AtHAK5 activity. A complex of calcineurin B-like proteins (CBL)/calcineurin B-like protein kinases (CIPKs) encompassing CBL1/9 with CIPK23 and CBL2/3 with CIPK9 both regulate AtHAK5 activity (Pandey et al., 2007; Liu et al., 2013; Ragel et al., 2015; Singh et al., 2018; Yadav et al., 2018; Lara, 2020). Interestingly, another CIPK, named CIPK1, also regulates HAK5 activity in planta (Lara, 2020). Based on the complementation experiments performed in yeast, it has been suggested that other *Arabidopsis* CBLs (AtCBL8/9/10) interact with CIPK23 that results in the activation of AtHAK5 (Ragel et al., 2019). Based on this experimental evidence, it has been proposed that as plants are subjected to K deficiency, a specific, biphasic Ca^{2+} signature is generated. This signature when sensed by CBL, leads to the activation of AtHAK5 via CBL-CIPK complex (Pandey et al., 2014; Ragel et al., 2015).

Another important component of K response, AKT1 is known to be regulated by CBL1/9 and CIPK23 in a phosphorylation-dependent manner under low-K condition (Li et al., 2006; Xu et al., 2006; Luan et al., 2009). Increasing evidence suggests AKT1 regulation by multiple CBL-CIPK pairs. In these reports, besides CIPK23, CIPK6 and CIPK16 have been shown to regulate AKT1 activity in a CBL-dependent manner (Lan et al., 2011). AKT1 is also under direct regulation of CBL10, independent of any CIPK, in addition to its regulation by CBL1/9-CIPK23 pair. Here, CBL10 acts as a competitor of CIPK23 for binding with AKT1 in order to repress its activity and negatively regulates the process (Ren et al., 2013). However, CIPKs are also known to positively regulate AKT1 activity upon interaction with ankyrin repeat domains that are present within the C-terminal region of AKT1 (Lee et al., 2007). In a similar manner, another K^+ channel, AKT2 is regulated by a CBL4-CIPK6 pair resulting in its translocation from endoplasmic reticulum to plasma membrane (Held et al., 2011).

Another CIPK, that is CIPK9, has been identified as a crucial alternative regulator of low-K^+ signaling (Pandey et al., 2007). Two independent studies reported that at the tonoplast membrane CIPK9 interacts with CBL2/3, which in turn might regulate a vacuolar K^+ transport system. AP2C1 (a type 2C phosphatase) upon interaction with CIPK9 dephosphorylates it, which negatively regulates K^+ deficiency response (Liu et al., 2013; Singh et al., 2018; Yadav et al., 2018; Bheri et al., 2020; Sanyal et al., 2020). These findings suggested the possibility of an alternate signaling pathway consisting of CIPK9-AP2C1 module similar to CIPK23-AKT1-AIP module working under K^+ deficiency (Pandey et al., 2014; Sanyal et al., 2016; Srivastava et al., 2019).

In response to K^+ deficiency, another alternative pathway, a vacuolar Ca^{2+}-dependent CBL–CIPK pathway, has been reported recently. This modulates the activity of tonoplast two-pore K^+ channel (TPK)-type K^+ channels, with CBL3, CIPK9, and tonoplast two-pore K^+ channel 1(TPK1) being exogenous components that constitute a linear pathway (Tang, 2020). This pathway facilitates K^+ release from *Arabidopsis* vacuole establishing a molecular link between channel activity and cytosolic K^+. Based on the cellular K^+ status, channel activities of K^+ can be adjusted (Tang, 2020).

Sensor relay proteins such as ÇaM-related proteins (CML) are well known for their prime function in signaling crosstalk. In addition to this, they are involved in coordinating plant responses to abiotic and biotic stresses (Bender et al., 2013; Ranty et al., 2016; Wu et al., 2017). Few members of the CML family contribute to Ca^{2+}-mediated plant responses to K^+ deficiency. CML25 has been shown as an important transducer of Ca^{2+}-mediated regulation of K^+ influx (Wang et al., 2015). Another CML protein, AtCML9 together with Raf-like MAPKK kinase (AtILK1) directly interacts with AtHAK5, promoting AtHAK5 accumulation on the membrane (Brauer et al., 2016). Based on these studies, it can be stated that Ca^{2+} sensor proteins might act as connecting dots between Ca^{2+} signaling and downstream target proteins during plant responses to K^+ deficiency (Chae et al., 2010; Bender et al., 2013; Pandey and Mahiwal, 2020).

Activation of a Ca^{2+} signal in the roots upon K^+ starvation might result in different downstream responses in different parts of the plants (Lee et al., 2007; Pandey et al., 2010). These downstream signals probably arise from the distinct Ca^{2+} signatures that are sensed by different Ca^{2+} sensors. These Ca^{2+} sensors further convey these signals and regulate activities of the K^+ transport system throughout the plant (Li et al., 2006; Lee et al., 2007; Wang et al., 2018; Tang, 2020). Based on these, K^+ can be considered a signal mediator in the plants.

5.7.1 Can potassium act as a signaling molecule? What have we learned so far?

Retention of nutrients and other compounds during stress response is a common phenomenon in plants. A common feature for the stress tolerance in plants seems to be the ability of the plants to retain K^+ as seen in most cases. Under salinity and oxidative stress, high K^+ retention minimizes damage induced by programmed cell death, as K^+ inhibits the activity of caspaselike proteases and endonucleases to minimize the damage to cellular components (Demidchik et al., 2010). Recent reports emphasize the importance of stress-induced K^+ efflux in regulating growth and development under unfavorable conditions. This stress-induced K^+ efflux is mediated by K^+ efflux channels. These two contrasting observations in plants point to the existence of a unique mechanism that may function during stress response, which is yet unchartered. Looking at the phenomenon of K^+ retention in stress tolerance and the presence of K^+ efflux transport systems suggests that K^+ may play signaling roles during hostile conditions (Shabala, 2017). The only need for developing K^+ efflux channels in plants seems to be for performing a signaling function. Many hypotheses have been suggested for this. Some studies put a spotlight on the probable mechanism of this complex phenomenon in plants. Considering the fact that K^+ efflux contributes to a signaling phenomenon in plants, still it is questionable why a plant prefers to lose such an important ion like K^+. One study focusing on salinity stress provides a possible explanation for this. It suggests that under salinity stress, to counter with initial membrane depolarization caused by the salinity stress, K^+ efflux is thought of as a "safety valve" as the upregulated H^+-ATPases slowly and gradually compensate to restore the membrane potential. Once the membrane potential (MP) is restored, this is followed by K^+ retention by high-affinity transporters present in the plants (Wu and Seliskar, 1998; Chen et al., 2007; Alvarez-Pizarro et al., 2009). An idea called the "metabolic hypothesis" suggests K^+ as a metabolic switch. According to this hypothesis (Fig. 5.4), K^+ obstructs energy-consuming anabolic reactions and conserves

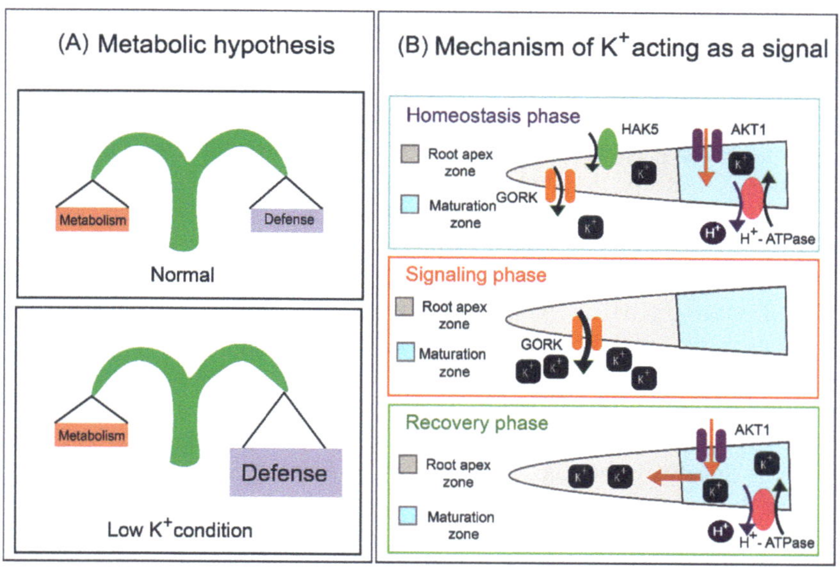

FIGURE 5.4 A schematic representation of K^+ as a potential signaling molecule. (A) Metabolic hypothesis depicting the normal conditions where metabolism and defense responses are in fine balance. In K^+ deficiency, disturbance in the balance occurs as a result of less metabolic activities, in order to divert major energy pool to defense responses. (B) A model for K^+ signaling depicting the homeostasis phase, signaling phase, and the recovery phase. In the homeostasis phase, three major transport systems in the root (AKT1, HAK5, and GORK) function to fine-tune K^+ levels in the root system. In the signaling phase, K^+ efflux occurs to a much greater extent via GORK as the membrane depolarizes upon perception of stress. This K^+ efflux acts as a switch for restoring the membrane potential. In the recovery phase, K^+ influx from the mature zone helps in restoring the membrane potential. *Adapted and modified from Shabala, S., 2017. Signalling by potassium: another second messenger to add to the list? J. Exp. Bot. 68, 4003–4007.*

energy for repair and adaptation (Demidchik, 2014). Why does a plant need to limit the anabolic reactions and spend on repair process? An opinionated review that came out in 2017 tried to answer this by explaining a complex process (Shabala, 2017). When plants encounter stress, they must maintain a sound balance between metabolism and defense response. Meanwhile focusing on the allocation of more ATP pools to the plant defense, which is accomplished by shutting down cellular metabolism to avoid competition between the two processes. Most of the metabolic reactions are inactivated as cytosolic K^+ decreases to subthreshold levels, thus redistributing ATP pools toward defense responses (Shabala, 2017). For cytosolic K^+ signaling, a hypothetical mechanism has been suggested consisting of three phases. These three phases include a homeostasis phase, signaling phase, and recovery phase. The K^+ concentration gradient between the root apex zone and root maturation zone is a characteristic signature of the homeostasis phase. Because of the high demand for K^+, the root apex zone exhibits slightly higher cytosolic K^+ concentration (Chen et al., 2005). In the maturation zone, maintenance of constant cytosolic K^+ levels is mediated by AKT1 channels, whereas H^+-ATPases provide electrical gradients for MP maintenance, hence keeping both at constant levels. However, in the apex zone less negative MP results in K^+ leak via GORK and HAK family transporters compensate for it (Chakraborty et al., 2016). Upon perception of stress, tremendous K^+ efflux through GORK channels is triggered because of the depolarization of MP to very low levels in the root apex zone. This is referred to as the signaling phase as this K^+ efflux acts as a switch and is responsible for restoring the MP as well as initiating defense operations of the cell. Once the MP is restored to normal, K^+ recovery in both the apex and maturation zones stresses induced activation of H^+-ATPases along with symplastic K^+ contribution from the mature zones to the apex. This is the recovery phase. Although the metabolic hypothesis (Fig. 5.4) seems to be a promising one, this hypothesis needs experimental validations to establish K^+ as a signaling molecule. It is still a big topic of debate whether or not K^+ can be considered as potential signaling molecule. Some key points are discussed in the next section of the chapter. Extensive research with appropriate experimental evidence is a prerequisite to support the hypothesis of adding K^+ as another second messenger to the existing list. Many questions need to be answered, and many mechanisms need to deciphered with experimental support to add K^+ as a second messenger in the near future (Shabala, 2017; Pandey and Mahiwal, 2020). (Fig. 5.4).

5.7.2 Why is K^+ not considered a second messenger? Does it have the potential qualities to qualify as a second messenger?

The messages carried by the second messengers are encoded in terms of their concentration. A very basic mechanism of conveying a signal is either rise or fall in the concentration from its source to its targets. In some cases, the fluctuation rate and fluctuation frequency in the concentration of the second messenger decides the intensity of the signal to be conveyed. Typically for second messengers, the local concentrations depend on the rate of production, diffusion from the production site, and the removal rate (Pollard et al., 2017). These are the general properties of the molecules that have been designated as second messenger. However, the first prerequisite for a molecule to act as second messenger is the need of its intracellular requirement at very low basal levels under normal conditions. The reason behind this is very simple: The cell must prevent prohibitive energy expenses while modulating the concentration of second messengers, as per the demand of the signaling function (Carafoli and Krebs, 2016). Because enzymes are responsible for the production of most second messengers, a rapid on-and-off switch allows modulation of their concentration on a millisecond time scale. Keeping all these points in consideration, comparing K^+ with all second messengers would not be a fair choice. Because K^+ is an ion, comparing its properties with another second messenger that is also an ion would be more relevant. A signaling molecule such as Ca^{2+} fits well into this criterion. Being an ion, Ca^{2+} is well known for its rapid and reversible change in concentration in the cytosol. Thus, the cell has developed mechanisms to sequester extra Ca^{2+} from the cytosol. Two famous storehouses, ER and vacuoles, help to store excess Ca^{2+} and release it only when required. Such fine-tuned mechanism is needed in order to generate either a Ca^{2+} signature or a Ca^{2+} response (Stael et al., 2012). In the case of K^+, its relative concentration in the cytosol is too high under normal conditions making it unfit to be called a second messenger (Pandey and Mahiwal, 2020).

The key reason for utilizing ions as second messengers is their ease and speed of generation without any requirement of enzymatic steps. Intracellular ion concentration and proximity of the ion to its targets are the deciding factors for the speed of response (Newton et al., 2016). Ca^{2+} has proven abilities to generate quick responses in milliseconds (Berridge, 2006). On the other hand, for K^+ it is unclear whether K^+ fluxes are quick enough to generate a millisecond response. So far, no such calculations for K^+ have been performed that can support the hypothesis of K^+ being a second messenger (Pandey and Mahiwal, 2020).

5.8 Conclusion and future perspectives

The availability of cutting-edge techniques has expanded the horizons of current K^+ biology. With the available technology, every aspect of K^+ has been studied extensively, with K^+ deficiency still being on the top. Because this is important from an agricultural point of view, it has become a central research topic in the past few years. Several new concepts have been established indicating K^+ as a versatile component in plant nutrition. A new spotlight on K^+ deprivation shows involvement of Ca^{2+} signaling and ROS in low-K^+ sensing and signaling. Since K^+ deprivation leads to the onset of various Ca^{2+} signaling pathways at the whole-plant level, it is convincing enough to be considered as signal mediator. Also, the pathways of K^+ deficiency response crosstalking to various signaling pathways further support this fact. Taking a hint from this, a few researchers hypothesize its role in plant signaling and hence point to its possible function as a second messenger. With the available evidence and limited information on this new concept, it is quite difficult to establish this phenomenon in plants. Also, a few fundamental properties of K^+ also contradict the concept of K^+ being a signal or signal mediator, as it doesn't fulfill all the criteria needed to call it a signaling molecule. Indeed more studies digging deeper into this phenomenon may provide an actual picture whether it can be stated as a signal or second messenger.

Acknowledgments

Research in Girdhar K. Pandey's lab is funded by Department of Biotechnology (DBT), Science and Engineering Research Board (SERB), University Grant Commission (UGC) and Delhi University (IoIoE/frp IoE/FRP grant) India. Swati Mahiwal acknowledges DBT for Junior Research Fellowship (DBT-JRF program).

References

Adams, E., Shin, R., 2014. Transport, signaling, and homeostasis of potassium and sodium in plants. J. Integr. Plant Biol. 56, 231–249.

Alvarez-Pizarro, J.C., Gomes-Filho, E., de Lacerda, C.F., Alencar, N.L.M., Prisco, J.T., 2009. Salt-induced changes on H^+-ATPase activity, sterol and phospholipid content and lipid peroxidation of root plasma membrane from dwarf-cashew (*Anacardium occidentale* L.) seedlings SpringerLink. Plant Growth Regul. 59, 125–135.

Amtmann, A., Armengaud, P., 2009. Effects of N, P, K and S on metabolism: new knowledge gained from multi-level analysis. Curr. Opin. Plant Biol. 12, 275–283.

Armengaud, P., Breitling, R., Amtmann, A., 2004. The potassium-dependent transcriptome of *Arabidopsis* reveals a prominent role of jasmonic acid in nutrient signaling. Plant Physiol. 136, 2556–2576.

Armengaud, P., Sulpice, R., Miller, A.J., Stitt, M., Amtmann, A., Gibon, Y., 2009. Multilevel analysis of primary metabolism provides new insights into the role of potassium nutrition for glycolysis and nitrogen assimilation in *Arabidopsis* roots. Plant Physiol. 150, 772–785.

Ashley, M.K., Grant, M., Grabov, A., 2005. Plant responses to potassium deficiencies: a role for potassium transport proteins. J. Exp. Bot. 57, 425–436.

Ashraf, K.U., Josts, I., Mosbahi, K., Kelly, S.M., Byron, O., Smith, B.O., et al., 2016. The potassium binding protein Kbp is a cytoplasmic potassium sensor. Structure 24, 741–749.

Behera, S., Long, Y., Schmitz-Thom, I., Wang, X.P., Zhang, C., Li, H., et al., 2017. Two spatially and temporally distinct Ca^{2+} signals convey *Arabidopsis thaliana* responses to K^+ deficiency. N. Phytol. 213, 739–750.

Bender, K.W., Rosenbaum, D.M., Vanderbeld, B., Ubaid, M., Snedden, W.A., 2013. The *Arabidopsis* calmodulin-like protein, CML39, functions during early seedling establishment. Plant J. 76, 634–647.

Berridge, M.J., 2006. Calcium microdomains: organization and functions. Cell Calcium 40, 405–412.

Bheri, M., Mahiwal, S., Sanyal, S.K., Pandey, G.K., 2020. Plant protein phosphatases: what do we know about their mechanism of action? FEBS J. 288 (3), 756–785.

Bossemeyer, D., Borchard, A., Dosch, D.C., Helmer, G.C., Epstein, W., Booth, I.R., et al., 1989. K^+-transport protein TrkA of *Escherichia coli* is a peripheral membrane protein that requires other *trk* gene products for attachment to the cytoplasmic membrane. J. Biol. Chem. 264, 16403–16410.

Bossemeyer, D., Schlösser, A., Bakker, E.P., 1989. Specific cesium transport via the *Escherichia coli* Kup (TrkD) K^+ uptake system. J. Bacteriol. 171, 2219–2221.

Brauer, E.K., Ahsan, N., Dale, R., Kato, N., Coluccio, A.E., Pineros, M.A., et al., 2016. The Raf-like kinase ILK1 and the high affinity K^+ transporter HAK5 are required for innate immunity and abiotic stress response. Plant Physiol. 171, 1470–1484.

Buch-Pedersen, M.J., Rudashevskaya, E.L., Berner, T.S., Venema, K., Palmgren, M.G., 2006. Potassium as an intrinsic uncoupler of the plasma membrane H^+-ATPase. J. Biol. Chem. 281, 38285–38292.

Carafoli, E., Krebs, J., 2016. Why calcium? How calcium became the best communicator. J. Biol. Chem. 291, 20849–20857.

Chae, L., Pandey, G.K., Cheong, Y.H., Kim, K.N., Luan, S., 2010. Protein kinases and phosphatases for stress signal transduction in plants. In: Pareekh, A., Sopory, S.K., Bohnart, H.J., Govindjee (Eds.), Abiotic Stress Adaptation in Plants: Physiological, Molecular and Genomic Foundation. Springer.

Chakraborty, K., Bose, J., Shabala, L., Eyles, A., Shabala, S., 2016. Evaluating relative contribution of osmotolerance and tissue tolerance mechanisms toward salinity stress tolerance in three *Brassica* species. Physiol. Plant 158, 135–151.

Chen, Z., Newman, I., Zhou, M., Mendham, N., Zhang, G., Shabala, S., 2005. Screening plants for salt tolerance by measuring K^+ flux: a case study for barley. Plant Cell Environ. 28, 1230–1246.

Chen, Z., Pottosin, I.I., Cuin, T.A., Fuglsang, A.T., Tester, M., Jha, D., et al., 2007. Root plasma membrane transporters controlling K^+/Na^+ homeostasis in salt-stressed barley. Plant Physiol. 145, 1714–1725.

Cherel, I., Lefoulon, C., Boeglin, M., Sentenac, H., 2014. Molecular mechanisms involved in plant adaptation to low K^+ availability. J. Exp. Bot. 65, 833–848.

Choi, W.G., Miller, G., Wallace, I., Harper, J., Mittler, R., Gilroy, S., 2017. Orchestrating rapid long-distance signaling in plants with Ca^{2+}, ROS and electrical signals. Plant J. 90, 698–707.

Demidchik, V., 2014. Mechanisms and physiological roles of K^+ efflux from root cells. J. Plant Physiol. 171, 696–707.

Demidchik, V., Cuin, T.A., Svistunenko, D., Smith, S.J., Miller, A.J., Shabala, S., et al., 2010. *Arabidopsis* root K^+-efflux conductance activated by hydroxyl radicals: single-channel properties, genetic basis and involvement in stress-induced cell death. J. Cell Sci. 123, 1468–1479.

Demidchik, V., Shabala, S.N., Davies, J.M., 2007. Spatial variation in H_2O_2 response of *Arabidopsis thaliana* root epidermal Ca^{2+} flux and plasma membrane Ca^{2+} channels. Plant J. 49, 377–386.

Drerup, M.M., Schlucking, K., Hashimoto, K., Manishankar, P., Steinhorst, L., Kuchitsu, K., et al., 2013. The Calcineurin B-like calcium sensors CBL1 and CBL9 together with their interacting protein kinase CIPK26 regulate the *Arabidopsis* NADPH oxidase RBOHF. Mol. Plant 6, 559–569.

Dreyer, I., 2011. Plant Biophysics and Heisenberg Group of Biophysics and Molecular Plant Biology, C.d.B.y.G.d.P., Universidad Politécnica de Madrid, Spain, Uozumi, N., and Department of Biomolecular Engineering, G.S.o.E., Tohoku University, Japan. Potassium channels in plant cells. FEBS J. 278, 4293–4303.

Epstein, E., Rains, D.W., Elzam, O.E., 1963. Resolution of dual mechanisms of potassium absorption by barley roots. Proc. Natl. Acad. Sci. USA 49, 684–692.

Epstein, W., 2003. The roles and regulation of potassium in bacteria. Prog. Nucleic Acid Res. Mol. Biol. 75, 293–320.

Foreman, J., Demidchik, V., Bothwell, J.H., Mylona, P., Miedema, H., Torres, M.A., et al., 2003. Reactive oxygen species produced by NADPH oxidase regulate plant cell growth. Nature 422, 442–446.

Frymier, J.S., Reed, T.D., Fletcher, S.A., Csonka, L.N., 1997. Characterization of transcriptional regulation of the KDP operon of Salmonella typhimurium. J. Bacteriol. 179, 3061–3063.

Hafsi, C., Debez, A., Abdelly, C., 2014. Potassium deficiency in plants: effects and signaling cascades | SpringerLink. Acta Physiol. Plant 36, 1055–1070.

Hashimoto, K., Kudla, J., 2011. Calcium decoding mechanisms in plants. Biochimie 93, 2054–2059.

Heermann, R., Altendorf, K., Jung, K., 2000. The hydrophilic N-terminal domain complements the membrane-anchored C-terminal domain of the sensor kinase KdpD of *Escherichia coli*. J. Biol. Chem. 275, 17080–17085.

Held, K., Pascaud, F., Eckert, C., Gajdanowicz, P., Hashimoto, K., Corratgé-Faillie, C., et al., 2011. Calcium-dependent modulation and plasma membrane targeting of the AKT2 potassium channel by the CBL4/CIPK6 calcium sensor/protein kinase complex. Cell Res. 21, 1116–1130.

Hirsch, R.E., Lewis, B.D., Spalding, E.P., Sussman, M.R., 1998. A role for the AKT1 potassium channel in plant nutrition. Science 280, 918–921.

Ho, C.H., Lin, S.H., Hu, H.C., Tsay, Y.F., 2009. CHL1 functions as a nitrate sensor in plants. Cell 138, 1184–1194.

Johansson, I., Wulfetange, K., Poree, F., Michard, E., Gajdanowicz, P., Lacombe, B., et al., 2006. External K^+ modulates the activity of the *Arabidopsis* potassium channel SKOR via an unusual mechanism. Plant J. 46, 269–281.

Jung, K., Altendorf, K., 1998. Truncation of amino acids 12-128 causes deregulation of the phosphatase activity of the sensor kinase KdpD of *Escherichia coli*. J. Biol. Chem. 273, 17406–17410.

Jung, K., Veen, M., Altendorf, K., 2000. K^+ and ionic strength directly influence the autophosphorylation activity of the putative turgor sensor KdpD of *Escherichia coli*. J. Biol. Chem. 275, 40142–40147.

Kanwar, P., Sanyal, S.K., Tokas, I., Yadav, A.K., Pandey, A., Kapoor, S., et al., 2014. Comprehensive structural, interaction and expression analysis of CBL and CIPK complement during abiotic stresses and development in rice. Cell Calcium 56, 81–95.

Kay, A.R., 2017. How cells can control their size by pumping ions. Front. Cell Dev. Biol. 5, 41.

Laermann, V., Cudic, E., Kipschull, K., Zimmann, P., Altendorf, K., 2013. The sensor kinase KdpD of *Escherichia coli* senses external K^+. Mol. Microbiol. 88, 1194–1204.

Lan, W.Z., Lee, S.C., Che, Y.F., Jiang, Y.Q., Luan, S., 2011. Mechanistic analysis of AKT1 regulation by the CBL-CIPK-PP2CA interactions. Mol. Plant 4, 527–536.

Lara, A., Ródenas, R., Andrés, Z., Martínez, V., Quintero, F.J., Nieves-Cordones, M., et al., 2020. AtHAK5-mediated root high-affinity K^+ uptake is regulated by the protein kinases AtCIPK1 and AtCIPK9. J. Exp. Bot. 71 (16), 5053–5060. Available from: https://doi.org/10.1093/jxb/eraa212.

Lee, S.C., Lan, W.Z., Kim, B.G., Li, L., Cheong, Y.H., Pandey, G.K., et al., 2007. A protein phosphorylation/dephosphorylation network regulates a plant potassium channel. Proc. Natl. Acad. Sci. USA 104, 15959–15964.

Leonard, R.J., Garcia, M.L., Slaughter, R.S., Reuben, J.P., 1992. Selective blockers of voltage-gated K^+ channels depolarize human T lymphocytes: mechanism of the antiproliferative effect of charybdotoxin. Proc. Natl. Acad. Sci. USA 89, 10094–10098.

Liu, L.L., Ren, H.M., Chen, L.Q., Wang, Y., Wu, W.H., 2013. A protein kinase, calcineurin B-like protein-interacting protein Kinase9, interacts with calcium sensor calcineurin B-like Protein3 and regulates potassium homeostasis under low-potassium stress in *Arabidopsis*. Plant Physiol. 161, 266–277.

Li, L., Kim, B.G., Cheong, Y.H., Pandey, G.K., Luan, S., 2006. A Ca^{2+} signaling pathway regulates a K^+ channel for low-K response in *Arabidopsis*. Proc. Natl. Acad. Sci. USA 103 (33), 12625–12630.

Li, J., Wu, W.H., Wang, Y., 2017. Potassium channel AKT1 is involved in the auxin-mediated root growth inhibition in *Arabidopsis* response to low K^+ stress. J. Integr. Plant Biol. 59, 895–909.

Luan, S., Lan, W., Chul Lee, S., 2009. Potassium nutrition, sodium toxicity, and calcium signaling: Connections through the CBL-CIPK network. Curr. Opin. Plant Biol. 12, 339–346.

Maathuis, F.J., 2009. Physiological functions of mineral macronutrients. Curr. Opin. Plant Biol. 12, 250–258.

Maathuis, F.J., Sanders, D., 1994. Mechanism of high-affinity potassium uptake in roots of *Arabidopsis thaliana*. Proc. Natl. Acad. Sci. USA 91, 9272–9276.

Maathuis, F.J.M., Sanders, D., 1993. Energization of potassium uptake in *Arabidopsis thaliana* | SpringerLink. Planta 191, 302–307.

Ma, T.L., Wu, W.H., Wang, Y., 2012. Transcriptome analysis of rice root responses to potassium deficiency. BMC Plant Biol. 12, 161.

Negulescu, P.A., Shastri, N., Cahalan, M.D., 1994. Intracellular calcium dependence of gene expression in single T lymphocytes. Proc. Natl. Acad. Sci. USA 91, 2873–2877.

Newton, A.C., Bootman, M.D., Scott, J.D., 2016. Second messengers. Cold Spring Harb. Perspect. Biol. 8, a005926.

Nieves-Cordones, M., Miller, A.J., Aleman, F., Martinez, V., Rubio, F., 2008. A putative role for the plasma membrane potential in the control of the expression of the gene encoding the tomato high-affinity potassium transporter HAK5. Plant Mol. Biol. 68, 521–532.

Palmgren, M.G., 2001. Plant plasma membrane H^+-ATPases: powerhouses for nutrient uptake. Annu. Rev. Plant Physiol. Plant Mol. Biol. 52, 817–845.

Pandey, G.K., 2008. Emergence of a novel calcium signaling pathway in plants: CBL-CIPK signaling network. Physiol. Mol. Biol. Plants 14, 51–68.

Pandey, G.K., Cheong, Y.H., Kim, B.G., Grant, J.J., Li, L., Luan, S., 2007. CIPK9: a calcium sensor-interacting protein kinase required for low-potassium tolerance in *Arabidopsis*. Cell Res. 17, 411–421.

Pandey, G.K., Kanwar, P., Pandey, A., 2014. Global comparative analysis of CBL-CIPK gene families in plants. SpringerBriefs in Plant Science. ISBN-13: 978-3319090771.

Pandey, G.K., Mahiwal, S., 2020. Role of potassium in plants. SpringerBriefs in Plant Science. Springer Nature, Switzerland AG., ISBN: 978-3-030-45953.

Pandey, G.K., Pandey, A., Prasad, M., Böhmer, M., 2016. Editorial: abiotic stress signaling in plants: functional genomic intervention. Front. Plant Sci. 7, 681.

Pandey, G.K., Yadav, A.K., Kanwar, P., Tuteja, N., 2010. Role of calcium in regulating potassium-sodium homeostasis and nutrition signaling under stress condition. Plant Stress 4, 94–103.

Pandey, G.K., 2018. Genomics and functional genomics of stress-mediated signaling in plants Vol I & Vol. II. Curr. Genomics 18 and 19, ISSN: 1389-2029.

Pei, Z.M., Murata, Y., Benning, G., Thomine, S., Klusener, B., Allen, G.J., et al., 2000. Calcium channels activated by hydrogen peroxide mediate abscisic acid signalling in guard cells. Nature 406, 731–734.

Pollard, D.T., Earnshaw, W.C., Schwartz, J.L., Johnson, G.T., 2017. Second messenger. In: Pollard, D.T., Earnshaw, W.C., Schwartz, J.L., Johnson, G.T. (Eds.), Cell Biology. Elsevier, Inc, Philadelphia, PA, pp. 443–462.

Qi, Z., Hampton, C.R., Shin, R., Barkla, B.J., White, P.J., Schachtman, D.P., 2008. The high affinity K^+ transporter AtHAK5 plays a physiological role in planta at very low K^+ concentrations and provides a caesium uptake pathway in *Arabidopsis*. J. Exp. Bot. 59, 595–607.

Ragel, P., Raddatz, N., Leidi, E.O., Quintero, F.J., Pardo, J.M., 2019. Regulation of K+ nutrition in plants. Front. Plant Sci. 10, 281.

Ragel, P., Rodenas, R., Garcia-Martin, E., Andres, Z., Villalta, I., Nieves-Cordones, M., et al., 2015. The CBL-interacting protein kinase CIPK23 regulates HAK5-mediated high-affinity K^+ uptake in *Arabidopsis* roots. Plant Physiol. 169, 2863–2873.

Ramirez-Silva, L., Ferreira, S.T., Nowak, T., Tuena de Gomez-Puyou, M., Gomez-Puyou, A., 2001. Dimethylsulfoxide promotes K^+-independent activity of pyruvate kinase and the acquisition of the active catalytic conformation. Eur. J. Biochem. 268, 3267–3274.

Ranty, B., Aldon, D., Cotelle, V., Galaud, J.P., Thuleau, P., Mazars, C., 2016. Calcium sensors as key hubs in plant responses to biotic and abiotic stresses. Front. Plant Sci. 7, 327.

Ren, X.L., Qi, G.N., Feng, H.Q., Zhao, S., Zhao, S.S., Wang, Y., Wu, W.H., 2013. Calcineurin B-like protein CBL10 directly interacts with AKT1 and modulates K+ homeostasis in Arabidopsis. The Plant Journ. 74, 258–266.

Ronzier, E., Corratge-Faillie, C., Sanchez, F., Prado, K., Briere, C., Leonhardt, N., et al., 2014. CPK13, a noncanonical Ca^{2+}-dependent protein kinase, specifically inhibits KAT2 and KAT1 shaker K^+ channels and reduces stomatal opening. Plant Physiol. 166, 314–326.

Rubio, F., Nieves-Cordones, M., Aleman, F., Martinez, V., 2008. Relative contribution of AtHAK5 and AtAKT1 to K^+ uptake in the high-affinity range of concentrations. Physiol. Plant 134, 598–608.

Sanders, D., Pelloux, J., Brownlee, C., Harper, J.F., 2002. Calcium at the crossroads of signaling. Plant Cell 14 (Suppl), S401–S417.

Sanyal, S.K., Mahiwal, S., Nambiar, D.M., Pandey, G.K., 2020. CBL–CIPK module-mediated phosphoregulation: facts and hypothesis. Biochem. J. 477, 853–871.

Sanyal, S.K., Mahiwal, S., Pandey, G.K., 2019. Calcium signaling: a communication network that regulates cellular processes. In: Sopory, S. (Ed.), Sensory Biology of Plants. Springer, Singapore, pp. 279–309.

Sanyal, S.K., Pandey, A., Pandey, G.K., 2015. The CBL–CIPK signaling module in plants: a mechanistic perspective. Physiol. Plant 155, 89–108.

Sanyal, S.K., Rao, S., Mishra, L.K., Sharma, M., Pandey, G.K., 2016. Plant stress responses mediated by CBL-CIPK phosphorylation network. Enzymes 40, 31–64.

Schramke, H., Laermann, V., Tegetmeyer, H.E., Brachmann, A., Jung, K., Altendorf, K., 2017. Revisiting regulation of potassium homeostasis in *Escherichia coli*: the connection to phosphate limitation. Microbiologyopen 6 (3), e00438.

Schramke, H., Tostevin, F., Heermann, R., Gerland, U., Jung, K., 2016. A dual-sensing receptor confers robust cellular homeostasis. Cell Rep. 16, 213–221.

Schroeder, J.I., Fang, H.H., 1991. Inward-rectifying K^+ channels in guard cells provide a mechanism for low-affinity K^+ uptake. Proc. Natl. Acad. Sci. USA 88, 11583–11587.

Schroeder, J.I., Ward, J.M., Gassmann, W., 1994. Perspectives on the physiology and structure of inward-rectifying K^+ channels in higher plants: biophysical implications for K^+ uptake. Annu. Rev. Biophys. Biomol. Struct. 23, 441–471.

Shabala, S., 2017. Signalling by potassium: another second messenger to add to the list? J. Exp. Bot. 68, 4003–4007.

Shankar, A., Fernandes, J.L., Kaur, K., Sharma, M., Kundu, S., Pandey, G.K., 2018. Rice phytoglobins regulate responses under low mineral nutrients and abiotic stresses in *Arabidopsis thaliana*. Plant Cell Environ. 41, 215–230.

Shankar, A., Singh, A., Kanwar, P., Srivastava, A.K., Pandey, A., Suprasanna, P., et al., 2013. Gene expression analysis of rice seedling under potassium deprivation reveals major changes in metabolism and signaling components. PLoS One 8, e70321.

Sharma, T., Dreyer, I., Riedelsberger, J., 2013. The role of K^+ channels in uptake and redistribution of potassium in the model plant *Arabidopsis thaliana*. Front. Plant Sci. 4.

Sharma, M., Pandey, G.K., 2017. Genomics and functional genomics of stress-mediated signaling in plants: volume I. Curr. Genomics 18 (6), 467–468.

Shin, R., Schachtman, D.P., 2004. Hydrogen peroxide mediates plant root cell response to nutrient deprivation. Proc. Natl. Acad. Sci. USA 101, 8827–8832.

Singh, A., Kanwar, P., Yadav, A.K., Mishra, M., Jha, S.K., Baranwal, V., et al., 2014. Genome-wide expressional and functional analysis of calcium transport elements during abiotic stress and development in rice. FEBS J. 281, 894–915.

Singh, A., Yadav, A.K., Kaur, K., Sanyal, S.K., Jha, S.K., Fernandes, J.L., et al., 2018. A protein phosphatase 2C, AP2C1, interacts with and negatively regulates the function of CIPK9 under potassium-deficient conditions in *Arabidopsis*. J. Exp. Bot. 69, 4003–4015.

Sirhindi, G., Mushtaq, R., Sharma, P., Kaur, H., Mir, M.A., 2017. Jasmonate signaling and stress management in plants. In: Pandey, G.K. (Ed.), Mechanism of Hormone Signaling Under Stress. Wiley-Blackwell, pp. 143–171.

Spalding, E.P., Hirsch, R.E., Lewis, D.R., Qi, Z., Sussman, M.R., Lewis, B.D., 1999. Potassium uptake supporting plant growth in the absence of AKT1 channel activity: Inhibition by ammonium and stimulation by sodium. J. Gen. Physiol. 113, 909–918.

Srivastava, A.K., Shankar, A., Chandran, A.K., Sharma, M., Jung, K.H., Suprasanna, P., et al., 2019. Emerging concepts of potassium homeostasis in plants. J. Exp. Bot. 71, 608–619.

Stael, S., Wurzinger, B., Mair, A., Mehlmer, N., Vothknecht, U.C., Teige, M., 2012. Plant organellar calcium signalling: an emerging field. J. Exp. Bot. 63, 1525–1542.

Sugiura, A., Nakashima, K., Tanaka, K., Mizuno, T., 1992. Clarification of the structural and functional features of the osmoregulated kdp operon of *Escherichia coli*. Mol. Microbiol. 6, 1769–1776.

Sugiyama, T., Goto, Y., Akazawa, T., 1968. Pyruvate kinase activity of wheat plants grown under potassium deficient conditions. Plant Physiol. 43, 730–734.

Su, J., Gong, H., Lai, J., Main, A., Lu, S., 2009. The potassium transporter Trk and external potassium modulate *Salmonella enterica* protein secretion and virulence. Infect. Immun. 77, 667–675.

Tang, R.J., Zhao, F.G., Yang, Y., Wang, C., Li, K., Kleist, T.J., et al., 2020. A calcium signalling network activates vacuolar K^+ remobilization to enable plant adaptation to low-K environments. Nat. Plants 6, 384–393.

Trchounian, A., Kobayashi, H., 1999. Kup is the major K^+ uptake system in *Escherichia coli* upon hyper-osmotic stress at a low pH. FEBS Lett. 447, 144–148.

Trchounian, A., Kobayashi, H., 2000. K^+ uptake by fermenting *Escherichia coli* cells: pH dependent mode of the TrkA system operating. Biosci. Rep. 20, 277–288.

Very, A.A., Davies, J.M., 2000. Hyperpolarization-activated calcium channels at the tip of *Arabidopsis* root hairs. Proc. Natl. Acad. Sci. USA 97, 9801–9806.

Voelkner, P., Puppe, W., Altendorf, K., 1993. Characterization of the KdpD protein, the sensor kinase of the K^+-translocating Kdp system of *Escherichia coli*. Eur. J. Biochem. 217, 1019–1026.

Walker, D.J., Leigh, R.A., Miller, A.J., 1996. Potassium homeostasis in vacuolate plant cells. Proc. Natl. Acad. Sci. USA 93, 10510–10514.

Wang, S.S., Diao, W.Z., Yang, X., Qiao, Z., Wang, M., Acharya, B.R., et al., 2015. *Arabidopsis thaliana* CML25 mediates the Ca^{2+} regulation of K^+ transmembrane trafficking during pollen germination and tube elongation. Plant Cell Environ. 38, 2372–2386.

Wang, W., Fan, Y., Wang, S., Wang, L., He, W., Zhang, Q., et al., 2014. Effects of voltage-gated K^+ channel on cell proliferation in multiple myeloma. Sci. World J. 2014, 785140.

Wang, X., Hao, L., Zhu, B., Jiang, Z., 2018. Plant calcium signaling in response to potassium deficiency. Int. J. Mol. Sci. 19, 3456. Available from: https://doi.org/10.3390/ijms19113456.

Wang, Y., Wu, W.H., 2010. Plant sensing and signaling in response to K^+-deficiency. Mol. Plant 3, 280–287.

Wang, Y., Wu, W.H., 2013. Potassium transport and signaling in higher plants. Annu. Rev. Plant Biol. 64, 451–476.

Wolfgang, E., 1986. Osmoregulation by potassium transport in *Escherichia coli*. FEMS Microbiol. Rev. 2, 73–78.

Wu, J., Seliskar, D.M., 1998. Salinity adaptation of plasma membrane H^+-ATPase in the salt marsh plant *Spartina patens*: ATP hydrolysis and enzyme kinetics. J. Exp. Bot. 49, 1005–1013.

Wu, X., Qiao, Z., Liu, H., Acharya, B.R., Li, C., Zhang, W., 2017. CML20, an *Arabidopsis* Calmodulin-like protein, negatively regulates guard cell ABA signaling and drought stress tolerance. Front. Plant Sci. 8, 824.

Xu, J., Li, H.D., Chen, L.Q., Wang, Y., Liu, L.L., He, L., et al., 2006. A protein kinase, interacting with two calcineurin B-like proteins, regulates K^+ transporter AKT1 in *Arabidopsis*. Cell 125, 1347–1360.

Yadav, A.K., Jha, S.K., Sanyal, S.K., Luan, S., Pandey, G.K., 2018. *Arabidopsis* calcineurin B-like proteins differentially regulate phosphorylation activity of CBL-interacting protein kinase 9. Biochem. J. 475, 2621–2636.

Yang, T., Poovaiah, B.W., 2002. Hydrogen peroxide homeostasis: activation of plant catalase by calcium/calmodulin. Proc. Natl. Acad. Sci. USA 99, 4097–4102.

Yuan, L., Loque, D., Kojima, S., Rauch, S., Ishiyama, K., Inoue, E., et al., 2007. The organization of high-affinity ammonium uptake in *Arabidopsis* roots depends on the spatial arrangement and biochemical properties of AMT1-type transporters. Plant Cell 19, 2636–2652.

Zakharyan, E., Trchounian, A., 2001. K^+ influx by Kup in *Escherichia coli* is accompanied by a decrease in H^+ efflux. FEMS Microbiol. Lett. 204, 61–64.

Zhao, L.N., Shen, L.K., Zhang, W.Z., Zhang, W., Wang, Y., Wu, W.H., 2013. Ca^{2+}-dependent protein kinase 11 and 24 modulate the activity of the inward rectifying K^+ channels in *Arabidopsis* pollen tubes. Plant Cell 25, 649–661.

Zou, J.J., Wei, F.J., Wang, C., Wu, J.J., Ratnasekera, D., Liu, W.X., et al., 2010. *Arabidopsis* calcium-dependent protein kinase CPK10 functions in abscisic acid- and Ca^{2+}-mediated stomatal regulation in response to drought stress. Plant Physiol. 154, 1232–1243.

Chapter 6

Exploring the relationship between plant secondary metabolites and macronutrient homeostasis

Kareem A. Mosa[1,2], Muna A. Ali[1], Kalidoss Ramamoorthy[1] and Ahmed Ismail[2,3]

[1]*Department of Applied Biology, College of Sciences, University of Sharjah, Sharjah, United Arab Emirates,* [2]*Department of Biotechnology, Faculty of Agriculture, Al-Azhar University, Cairo, Egypt,* [3]*Public Health Department, College of Applied Medical Science, Majmaah University, Al Majma'ah, Saudi Arabia*

6.1 Introduction

The major challenges of plant growth mainly depend on climate conditions as well as mineral content in soil. The process by which plants sense available nutrients in soil and regulate their intake or release to maintain relative concentrations within plant tissue is called nutritional homeostasis. In order to maintain regular physiology and metabolism, plants require essential nutrients in adequate quantities (Amtmann and Blatt, 2009; Clemens et al., 2002). Plant survival in specific environments mainly depends upon the availability of nutrients. It's crucial for plants to obtain enough nutrients from the soil to meet the demands of their basic cellular processes. Multiple factors can influence plant root absorption of nutrients from the soil, including soil pH, soil water content, nutrient availability, soil chemistry, and climate (Marschner and Rengel, 2012). Other issues may involve difficulty in the absorption of certain forms of nutrients, as well as the lack of their availability in particular forms of soils. Acquired plant nutrients may be disturbed under low pH through impairment of proton gradients across plasma membranes. For example, NO_3 uptake occurs either via nitrate transporters or proton symporters in plant roots. Hence transport of NO_3 largely influences rhizosphere pH, which in turn influences NO_3

Plant Nutrition and Food Security in the Era of Climate Change. DOI: https://doi.org/10.1016/B978-0-12-822916-3.00007-X

uptake as well (Bouguyon et al., 2015; Crawford and Forde, 2002; Fang et al., 2016; Khademi, 2004; Ortiz-Ramirez et al., 2011; Zhu et al., 2009).

In the event of nutrient starvation, genetic materials receive various signals to maintain nutrient homeostasis. Nutrients are absorbed by roots and transported into various parts of the plant body by specific transporters. Nutrient deficiency in soil triggers a series of signaling molecules that interact with specific sequences resulting in the expression of specific transporter genes to enhance nutrient uptake. Additionally, this situation may lead to modifications in root shape and structure (Jung and McCouch, 2013).

Plants also produce various small molecules known as secondary metabolites that are not directly involved in growth and development. These are mainly used for plant defense against various biotic and abiotic stresses (Mosa et al., 2017). Moreover, they are involved in attracting pollinators, enhancing plant recovery from injury, and providing protection from ultraviolet radiation. Recently, a link between plant secondary metabolites and plant nutrients has also been reported.

In this chapter, we discuss primary and secondary macronutrients in terms of their nutrient sensing, nutrient uptake, and regulation. We will also discuss the response of plant secondary metabolites to macronutrients.

6.2 Macronutrient cycling in soil

Plants require several nutrients that are classified into three main groups based on required amounts: primary macronutrients, secondary macronutrients, and micronutrients (Fig. 6.1). Primary macronutrients are utilized by plants in the largest amounts, followed by secondary macronutrients. On the other hand, only trace amounts of micronutrients are required (Kumar et al., 2017). Additionally, carbon (C), hydrogen (H), and oxygen (O) are essential primary macronutrients that are acquired from atmospheric carbon dioxide and water (Bhaduri et al., 2014). In order to obtain an optimum yield of crop plants, it is vital to know the rate, method, and time of nutrient application; the nutrient source; and the influence of soil and climatic conditions on nutritional elements (Bhaduri et al., 2014).

Different processes are involved in nutrient cycling through soil, plants, and animals. Similar to the water cycle, nutrients may be converted into plant-usable forms, retained in the soil, or even lost via air or water. Plants typically absorb inorganic essential nutrients from the soil. Prior to plant uptake, most nutrients should be converted into inorganic forms with the exception of certain metal elements that can be absorbed in the organic form (Guoqing et al., 2019). Nutrients must be in contact with the root surface for ion absorption to occur (Mussa et al., 2009). The three key mechanisms plants utilize to absorb nutrients from the soil include: mass flow, diffusion, and root interception (Karthika et al., 2018). The insoluble nutrients are converted into soluble forms in the soil. This is followed by their uptake by

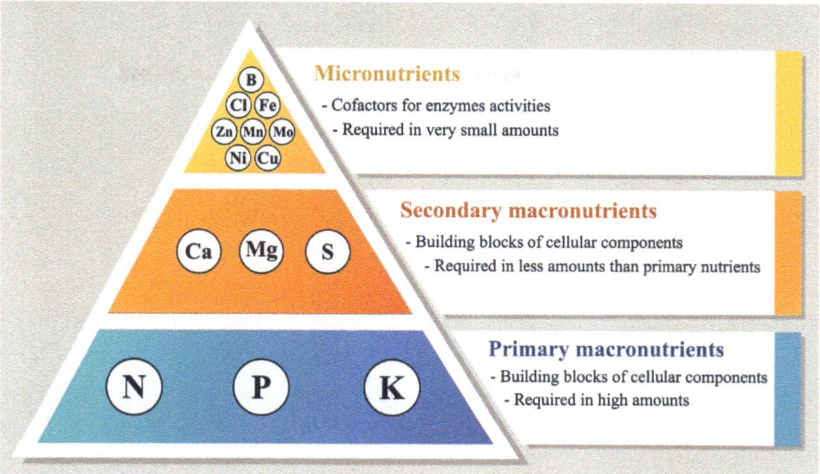

FIGURE 6.1 The classification of plant nutrients and their roles in plant growth. *Adapted from Kumar, S., Verma, S., Trivedi, P.K., 2017. Involvement of small RNAs in phosphorus and sulfur sensing, signaling and stress: current update. Front. Plant Sci. 8, 285.*

plant roots and translocation and utilization by plants. Then the nutrients are consumed by animals and excreted in different forms. Finally, after decomposition of animal excreta and plant residues, the nutrients are returned to the soil (Kayser and Isselstein, 2005). As examples of primary and secondary macronutrients, mechanisms involved in soil cycling, sensing, and uptake of nitrogen, phosphorus, potassium, and sulfur respectively will be highlighted in this section and the following section.

6.2.1 Nitrogen (N) cycling in soil

Nitrogen is the key organic element in soil and is required in significant amounts by plants. In prairie farming environments, it is also the first nutrient that is restrictive. N is generally the most deficient (e.g., the first restricting) mineral in most prairie soils, then P, K, and S (Jiao et al., 2019). The N cycles comprise seven types of N in the soil: nitrogen gas (N_2), organic N, nitrate (NO_3), ammonia (NH_3), ammonium (NH_4^+), nitrite (NO_2), and nitrogen oxide gases (NO, N_2O) (Xin et al., 2019). Each type of N occurs in a pool. For instance, organic N is part of the organic pool, NO_3 presents in the soil as a solution pool, and NH_4^+ presents in the soil as a solution or exchangeable pool.

Plants can rapidly utilize only inorganic N (NH_4^+ and NO_3) to meet their N requirements (Christodoulou et al., 2019). Compounds of nitrogen are part of organic soil matter, which must be decomposed before the N is usable. Soil species (for example, insects, small mammals, and microorganisms) are involved in a mechanism called mineralization, breaking complex

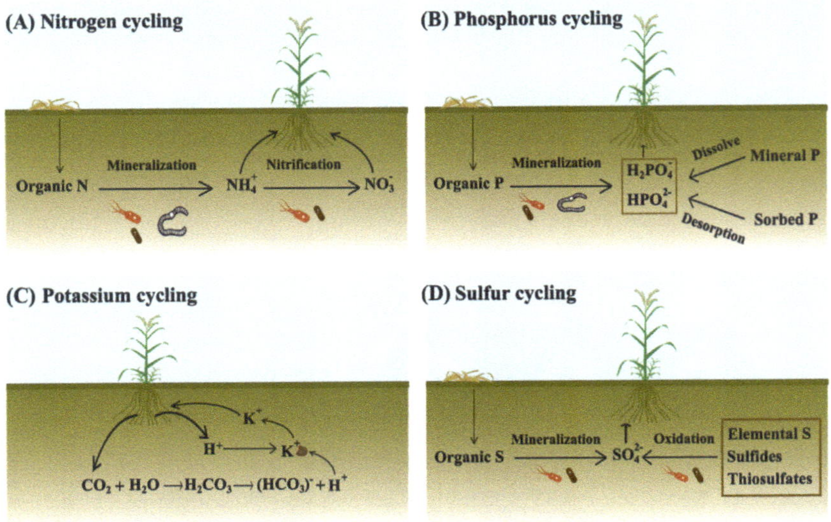

FIGURE 6.2 Macronutrient cycling in soil. (A) Nitrogen cycle in soil. Organic N compounds are mineralized by soil organisms to ammonium ion, which is further converted to nitrate by microorganisms through nitrification. The figure shows the reactions involved in making nitrogen bioavailable to plant roots. Adapted from Liu et al. (2020); Li et al. (2019). (B) Phosphorus cycle in soil. Orthophosphates are released in soil through mineralization of organic P, dissolve of mineral P, and desorption of P adsorbed to soil minerals. The figure shows the reactions involved in making phosphorus bioavailable to plant roots. Adapted from Mullins (2019). (C) Potassium cycling in the soil. Potassium is released into soil solution by cation exchange through proton ions released from plant roots. The figure shows the reactions involved in making potassium bioavailable to plant roots. Adapted from Urry et al. (2017). (D) Sulfur cycling in soil. Sulfate is released in soil upon mineralization of organic S and oxidation of elemental S, sulfides, and thiosulfates. The figure shows the reactions involved in making sulfur bioavailable to plant roots. *Adapted from Mullins, G.L., 2019. Phosphorus, agriculture & the environment. Virginia Cooperative Extension; Urry, L.A., Cain, M.L., Wasserman, S.A., Minorsky, P.V., Reece, J.B., 2017. Campbell Biology. Pearson; Kovar, J.L., Grant, C.A., 2011. Nutrient cycling in soils: sulfur. In: Hatfield, J.L., Sauer, T.J. (Eds.), Soil Management: Building a Stable Base for Agriculture. Madison, WI: American Society of Agronomy, Crop Science Society of America, Soil Science Society of America, pp. 103–115.*

N compounds into simpler forms (Liu et al., 2020). During N mineralization, organic N compounds are converted into NH_4^+, which can be absorbed from soil by plants. Soil NH_4^+ is normally rapidly transformed into NO_3 by a mechanism known as nitrification that involves bacteria (Li et al., 2019; Fig. 6.2A).

6.2.2 Phosphorus (P) cycling in soil

P is considered the second most limiting soil nutrient. The storage and transfer of energy are the key functions of P in plants. In molecules such as adenosine triphosphate (ATP) and adenosine diphosphate, high-energy phosphate bonds can generate virtually any biochemical reaction in plants (Dániel et al., 2020). Phosphorus is also essential to plant structure as a component of nucleic acids, phospholipids, and coenzymes, and membrane lipids.

Soil phosphorus is found in soil in two forms, inorganic P and organic P (Shen et al., 2011). Phosphorus forms in the soil in association with other elements such as O and H (Dai et al., 2020) and is mostly consumed by plant roots such as orthophosphate (H_2PO_4 or HPO_4^2). The primary form of orthophosphate (H_2PO_4) is dominant in soils below pH 7.2, while the secondary form (HPO_4^2) is present in soils above pH 7.2 (Ahmed et al., 2020). As a note, the primary form of orthophosphate can be absorbed faster than the secondary.

Phosphorus is released into soil solution in the forms of H_2PO_4 and HPO_4^2 upon the dissolve of phosphorus-bearing minerals, desorption (uncoupling) of phosphorus adsorbed to soil minerals, and mineralization (decomposition) of soil organic matter mediated by soil microorganisms (Shen et al., 2011; Mullins, 2019; Fig. 6.2B). The organic pool P is derived from residues of microbials, plants, and animals on or in the soil. Most organic pools are present in the top soil layers, with almost half of them being the phytic acid form. Every phytic acid molecule can release six orthophosphate molecules into the soil solution (H_2PO_4 or HPO_4^2) (Singh et al., 2018).

6.2.3 Potassium (K) cycling in soil

Potassium is an essential primary macronutrient required for plant growth and development (Clarkson and Hanson, 1980). About 10% of plant dry weight consists of potassium cations that are the most abundant in plant cells (Leigh and Wyn Jones, 1984). The highest amounts of available K are reported in brown and dark brown soils (Saha et al., 2018). Unlike other vital nutrients, K is not an important part of plant structural components. However, in many processes, it plays an important role in plant development (Pandey and Mahiwal, 2020). K function in plants includes enzyme activation, sugar transfer, water balance, and regulation of plant stomata. Potassium "activates" several plant growth enzymes. For instance, photosynthesis as well as the synthesis of starch and protein is focused on enzyme systems influenced by K (Rawat et al., 2016). Water balance within plants is regulated by potassium, along with sugars and other inorganic ions, which helps establish an inner concentration gradient between root and soil solution. Potassium also affects transpiration rate in response to internal and external environmental conditions by managing stomatal openings (Kooch & Noghre, 2020).

Most soil particles are negatively charged; thus cations (including potassium, calcium, and magnesium) adhere to these particles. Roots do not absorb cations directly; rather they absorb them from soil solutions by cation exchange, a process in which cations adsorbed to soil particles are displaced by other cations (Kayser and Isselstein, 2005). The process of cation exchange begins with releasing CO_2 and H^+ from roots to acidify the soil solution. Additionally, H_2CO_3 is formed upon the reaction of the released CO_2 and water. Upon the decomposition of H_2CO_3, H^+ is released. The H^+ ions will neutralize the negative charge of soil particles, leading to the release of cations that will be absorbed by plant roots (Urry et al., 2017; Fig. 6.2C).

6.2.4 Sulfur (S) cycling in soil

S concentrations can vary greatly by region and fields. The switch to continuous cropping, for example canola (a high-S user), has increased the incidence of S deficiencies (Kriel, 2020). S (sulfate-sulfur) can temporarily become available and subsequently lost via precipitation from the soil solution with magnesium, or calcium, or adsorption by clay particles containing aluminum or iron oxides. As a consequence, soluble and mobile deposits with sulfate-sulfur move to groundwater (Ma et al., 2020).

Most soil sulfur contents originate from organic wastes. Approximately less than 5% of soil sulfur is found as inorganic sulfate form, while 95% of soil sulfur is organically bound (organo-S) in the following forms: sulfur esters and sulfonates (Kovar and Grant, 2011; Gahan and Schmalenberger, 2014). Inorganic sulfate can be found in soil solution, adsorbed on mineral surfaces, or coprecipitated with Ca and Mg. The soil pH plays an important role in determining the form of inorganic sulfate. In alkaline soils, sulfate is mainly found as soluble salts of Ca and Mg (Kovar and Grant, 2011). In contrast, acidic soil sulfate is found adsorbed on the surfaces of hydrous oxides of Fe and Al and edges of aluminosilicate clay minerals (Bohn et al., 1986). Soil microorganisms are involved in the reduction of inorganic S, such as elemental sulfur, sulfides, and thiosulfates, to sulfates (Stevenson and Cole, 1999). Organo-S must be mineralized by microorganisms before being available for plant uptake. Bacteria and fungi in the soil and the rhizosphere mineralized organo-S into sulfate by different sulfatase enzymes (Scherer, 2009). Mineralization is the process that converts organo-S into sulfate form that is plant-available. In contrast, immobilization is a process that converts sulfate to organo-S by soil biota (Kovar and Grant, 2011; Fig. 6.2D).

6.3 Plant–soil interactions: macronutrient sensing, uptake, and regulation

Plants require a variety of nutritional elements for proper growth and metabolism and depend upon capturing a number of elements from the soil (Oldroyd and Leyser, 2020). It is important to study plant nutrient sensing as plant nutrition plays a major role in crop production. Roots are the first sites that sense nutrients in the soil. Plants uptake these nutrients from the soil as well as the associated microorganisms in the rhizosphere. Subsequently, the nutrients are transported to vascular cells where they will be further allocated to different plant tissues (Kumar et al., 2017; Fig. 6.3A). Any nutrient deficiency leads to reduced biomass and early plant senescence, which eventually results in less crop yield. Deficiencies in any nutrient reduce plant yield and/or fertility. The symptoms of nutrient deficiency include chlorosis due to low chlorophyll levels, death of plant tissue, and stunted growth (Morgan and Connolly, 2013). Nutrient deficiency initiates complex signaling cascades in plant shoots, which activates different biochemical compounds that are essential for nutrient uptake and transport. As a consequence, root reconstruction will occur increasing root-to-shoot ratios that will enhance nutrient sensing and uptake (Kumar et al., 2017; López-Bucio et al., 2003; Fig. 6.3B).

Nutritional limitations have negative impacts on plant growth and reduce crop optimal yields (Schachtman, 2012). To overcome the problem of nutrient deficiency, inorganic fertilizers are used to support global food

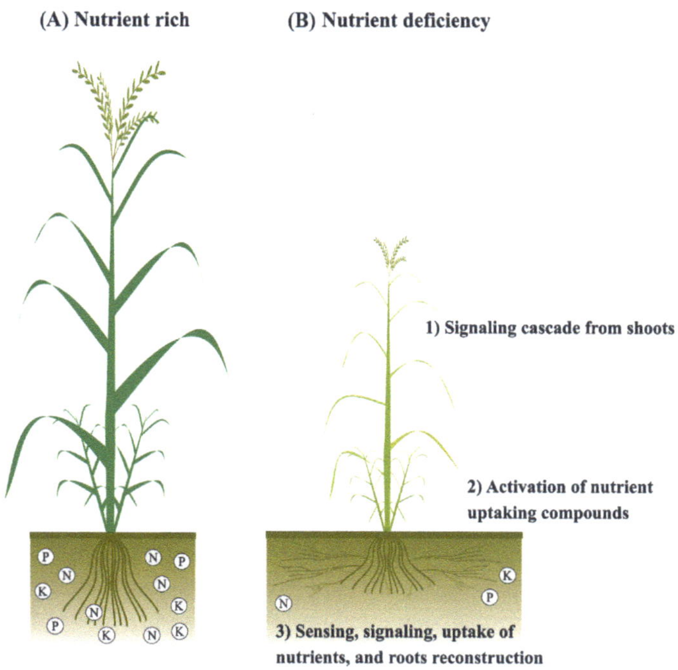

FIGURE 6.3 The effect of nutrient deficiency on plant growth. (A) In nutrient-rich soils, plants take nutrients and are allocated to plant tissues. (B) Upon nutrient deficiency, signaling cascades will activate nutrient uptake and induce root reconstruction. *Adapted from Morgan, J.B., Connolly, E.L., 2013. Plant-soil interactions: nutrient uptake. Nat. Educ. Knowl. 4, 2.*

production. The excess application of fertilizers can cause environmental issues due to the release of these nutritional elements that contribute to climate change by altering soil greenhouse gas emissions (Schachtman, 2012). This occurs by altering plant root exudates, affecting soil microbial communities (methanogens and methanotrophs), and altering soil physicochemical properties (Datta et al., 2013). Thus one of the main aims of studying plant nutrient sensing is to develop more sustainable agricultural methods that provide plants with optimum nutrition while being environmentally friendly. In this section, the sensing and uptake of the major macronutrients in the soil will be discussed.

6.3.1 Nitrogen sensing and uptake

Nitrogen sensing is the first step for N acquisition by plant roots that enables plants to control N uptake. This process is enabled by membrane proteins that are capable of activating downstream signaling pathways involved in N-dependent responses. A variety of nitrate and ammonium transporters have potential functions in N sensing and uptake (Xuan et al., 2017).

At least four nitrate (NO_3) transporter families have been identified. The plant nitrate transporter 1 (NRT1) gene family was renamed the NPF family, which stands for nitrate transporter 1/peptide transporter family (Léran et al., 2014). The NPF family not only transports NO_3, but it also transports nitrite (Sugiura et al. 2007), peptides (Komarova et al., 2008), amino acids (Zhou et al., 1998), and auxins (Krouk et al., 2010). The first identified NO_3 transporter was AtNRT1.1 (CHL1/AtNPF6.3), which was identified in *Arabidopsis* (Tsay, 1993). It acts as a transceptor (transporter and receptor) for NO_3. Further, it regulates lateral root formation under N deficiency (Krouk et al., 2010; Gojon et al., 2011). At low NO_3 concentration, NO_3 binds to high-affinity sites of NRT1.1, which will activate or recruit calcineurin B-like interacting protein kinase CIPK23 to phosphorylate NRT1.1 at threonine 101 (T101) in NRT1.1. This will cause NRT1.1 to switch to the high-affinity mode of NO_3 transport to maintain efficient NO_3 uptake. When NO_3 concentration is high (>1 mM), phosphorylation at T101 does not occur and NRT1.1 will remain in low-affinity NO_3 transport mode. CIPK8 is a positive regulator of the low-affinity mode of NRT1.1, although its target site remains to be identified (Ho et al., 2009; Vert and Chory, 2009; Fig. 6.4A).

Ammonium (NH_4^+) is taken mainly by transporter belonging to the ammonium transporter/methylammonium permeases (AMT/MEPs) gene family (Xuan et al., 2017). Four ammonium transporters AMT1.1, AMT1.2, AMT1.3, and AMT1.5 are involved in ammonium uptake. Expression of AtAMT1.1 in *Arabidopsis* roots is repressed by glutamine under high NH_4^+ levels. Under NH_4^+ deficiency the expression of AtAMT1.1 is derepressed (Yuan et al., 2007; Fig. 6.4B).

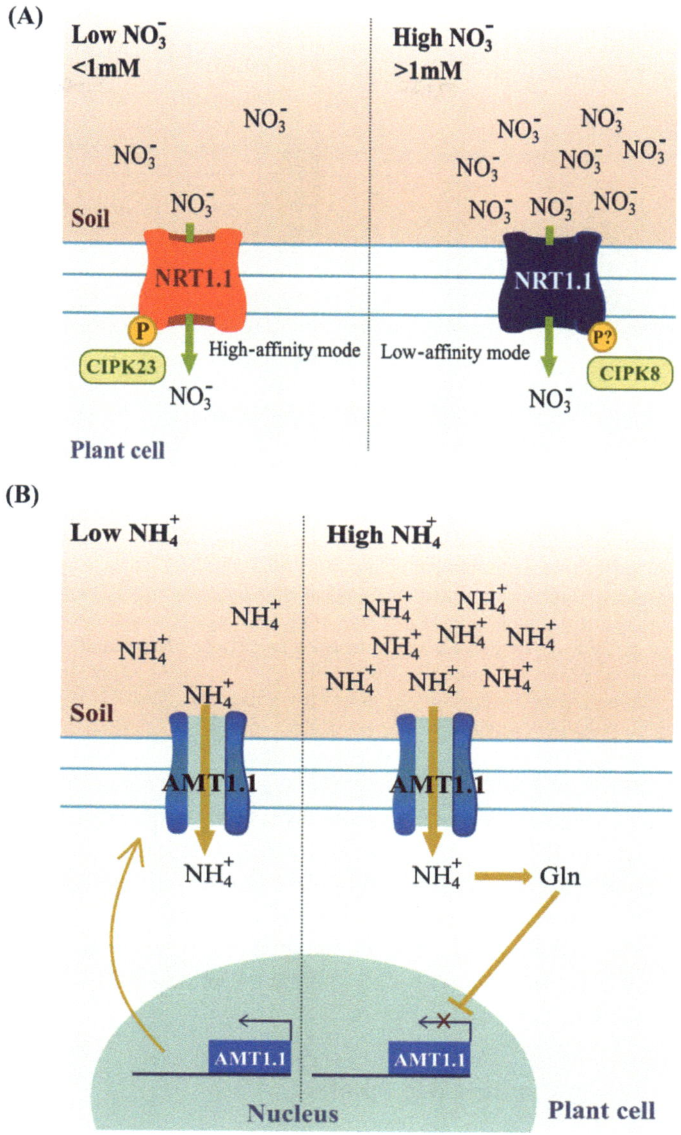

FIGURE 6.4 Nitrate and ammonium sensing. (A) Nitrate sensing by plant roots through NRT1.1. At low nitrate concentration, NrT1.1 is phosphorylated by CIPK23, which will render NRT1.1 at high-affinity mode. In contrast, under high nitrate, NRT1.1 is at low-affinity mode. Adapted from Vert and Chory (2009). (B) Ammonium sensing by AMT1.1. At high NH_4^+, the expression of AMT1.1 will be repressed by Gln. *Adapted from Vert, G., Chory, J., 2009. A toggle switch in plant nitrate uptake. Cell 138, 1064–1066; Yuan, L., Loqué, D., Kojima, S., Rauch, S., Ishiyama, K., Inoue, E., et al., 2007. The organization of high-affinity ammonium uptake in arabidopsis roots depends on the spatial arrangement and biochemical properties of AMT1-type transporters. Plant Cell, 19, 2636–2652.*

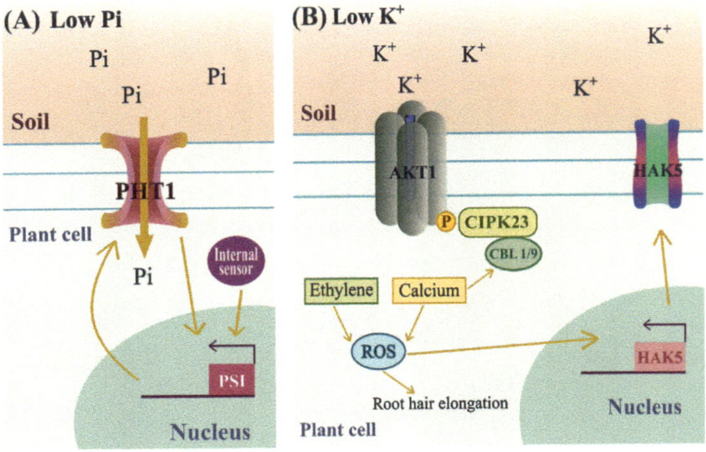

FIGURE 6.5 Phosphorus and potassium sensing. (A) Under low phosphorus, signals from an internal sensor and PHT1 will induce the transcription of PSI, which will lead to higher PHT1 in the plasma membrane. (B) Under low potassium concentration, calcium signals are induced to regulate ROS production. Calcium-binding proteins CBL1 and CBL9 will induce CIPK23 that will phosphorylate AKT1 and enhance its K uptake efficiency. Further, ROS induces root hair elongation and the transcription of HAK5, a high-affinity K transporter. *Adapted from Zhang, Z., Liao, H., Lucas, W.J., 2014. Molecular mechanisms underlying phosphate sensing, signaling, and adaptation in plants. J. Integr. Plant Biol. 56, 192–220; Ho, C., Tsay, Y., 2010. Nitrate, ammonium, and potassium sensing and signaling. Curr. Opin. Plant Biol. 13, 604–610.*

6.3.2 Phosphorus sensing and uptake

Phosphorus is taken from the soil in the form of orthophosphates (H_2PO_4 or HPO_4^2), which are found in the soil in the micromolar range; thus phosphorus sensing and uptake is mediated by high-affinity Pi/H^+ symporter that belongs to the PHT1 family (Shen et al., 2011). At Pi deficiency, signals from an internal sensor and PHT1 transporter upregulate the genes of Pi-starvation-inducible (PSI) genes. This will lead to higher levels of PHT1 in the plasma membrane (Zhang et al., 2014; Fig. 6.5A).

6.3.3 Potassium sensing and uptake

Potassium is acquired by plants with membrane channels and transporter proteins based on concentration gradients (Ashley et al., 2006; Véry and Sentenac, 2003; Ward et al., 2009). Various genes encoding for K^+ channels and transporters have been identified in different plant species (Chen et al., 2008; Gierth and Mäser, 2007; Lebaudy et al., 2007; Mäser et al., 2001; Véry and Sentenac, 2003).

Plant roots are the primary organs of K^+ sensors (Schachtman and Shin, 2007). Upon K^+ deficiency, hyperpolarization of the plant root cell membrane is triggered within a few minutes (Maathuis and Sanders, 1993;

Nieves-Cordones et al., 2008). Subsequently, the supply of potassium leads to depolarization of the membrane potential upon increase in external K^+ concentrations (Spalding et al., 1999). K^+ deficiency also induces H^+ extrusion and extracellular acidification (Behl and Raschke, 1987). Under K^+ deficiency, ethylene and calcium signals are activated to regulate reactive oxygen species (ROS) production (Ho and Tsay, 2010). Calcium-binding proteins CBL1 and CBL9 (calcineurin B-like protein) activate CIPK23, a CBL-interacting protein kinase, which in turn activates AKT1, an influx potassium channel, and enhance its potassium uptake activity under potassium deficiency (Gierth et al., 2005; Ho and Tsay, 2010). Furthermore, ROS production will induce the activation of HAK5, a high-affinity potassium (K) transporter (Kim et al., 2010). Additionally, ROS plays a role in root hair elongation, which also enhances potassium uptake by roots (Ho and Tsay, 2010; Fig. 6.5B).

6.3.4 Sulfur sensing and uptake

Translocation of sulfur into plants occurs via various transport proteins with different affinities (Kataoka et al., 2004; Shibagaki et al., 2002; Takahashi et al., 2000; Yoshimoto et al., 2002). Sulfate is the major form of sulfur taken by plants from the soil (Gigolashvili and Kopriva, 2014). The first cloned sulfate transporter plant gene was SULTR (Smith et al., 1995). According to their mechanism of action, SULTRs can be divided into several groups. In higher plants, the mechanism of SULTRs is H^+ sulfate symport (Takahashi et al., 2011). Furthermore, SULTR1;2 in plants has been proposed to act as a sulfur status sensor (Zhang et al., 2014).

Sulfate uptake and distribution largely depend upon the biochemical qualities of transport proteins as well as the cellular state of different membranes (Takahashi et al., 2011). Many diverse biochemical regulatory mechanisms and cellular responses are employed in plant sulfur sensing. Sulfur sensing and related downstream signaling pathways are also controlled by various biosynthetic and regulatory proteins that are involved in sulfur-assimilation processes.

Sulfur assimilation is the process of converting sulfate (SO_4^2) to sulfide (S^{2-}), which will be further incorporated in cysteine (Yi et al., 2010). ATP sulfurylase (ATPS) is the enzyme that catalyzes the conversion of SO_4^2 to adenosine-5′-phosphosulfate (APS). Next, APS reductase (APSR) will catalyze the reduction of APS to sulfite (SO_3^2). Eventually, sulfite reductase will catalyze the conversion of $SO_3^2$to S^{2-} (Kopriva and Koprivova, 2004; Saitoh et al., 2006). The incorporation of S^{2-} in cysteine involves two enzymes. Serine acetyltransferase (SAT) yields *O*-acetylserine by transferring acetate from acetyl coenzyme A into serine (Saito et al., 1995). Next, *O*-acetylserine sulfhydrylase (OASS) catalyzes the formation of cysteine from *O*-acetylserine and S^{2-} (Bonner et al., 2005; Fig. 6.6). Further,

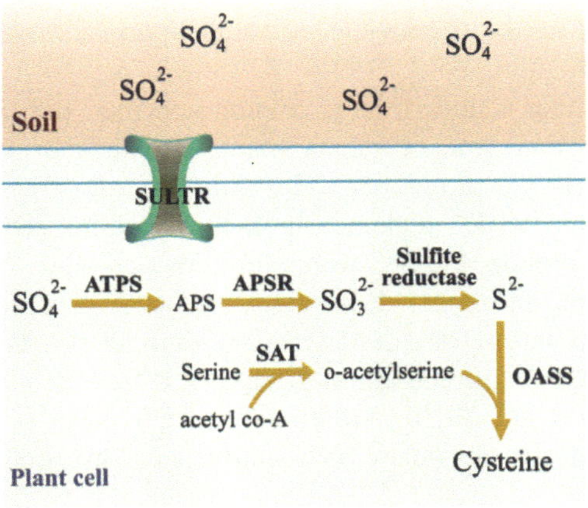

FIGURE 6.6 Pathways involved in sulfur assimilation. After the uptake of sulfate by SULTR, it is converted to APS by ATPS. Then, APSR will convert APS to sulfite that will be reduced by sulfite reductase to sulfide. Sulfide is incorporated in the synthesis of cysteine. *Adapted from Yi, H., Galant, A., Ravilious, G.E., Preuss, M.L., Jez, J.M., 2010. Sensing sulfur conditions: simple to complex protein regulatory mechanisms in plant thiol metabolism. Mol. Plant 3, 269–279.*

cysteine is incorporated into several sulfur-containing metabolites including glutathione.

6.3.5 Regulations of macronutrients within the circadian clock

Due to multiple factors, nutrient availability in soil can affect plant growth and metabolism (Hsu and Harmer, 2014). Circadian clocks, or molecular oscillators, are comprised of interlocking regulatory feedback loops. Depending upon environmental cues like light and temperature, the clock functions as a rhythmic physiological output. In general, the structure of circadian clocks is more likely to be highly conserved in higher plants (Song et al., 2010).

Transpiration initiates the key pathway for nutrient movement via the xylem. This process is controlled by the stomatal aperture and combined aquaporin channel proteins, which are regulated by the circadian clock (Caldeira et al., 2014; Dodd et al., 2005; Takase et al., 2011). Circadian clocks also regulate plant photosynthesis (Dodd et al., 2005; Harmer et al., 2000; Noordally et al., 2013), which utilizes nutrients and manages nutrient movement in the phloem. Evidence has also shown that the circadian clock regulates macronutrient homeostasis. For example, it has been reported that oscillations of cytosolic free calcium ($[Ca^{2+}]_{cyt}$) are dependent upon the rhythmic expression of the central clock gene CIRCADIAN CLOCK ASSOCIATED 1 (CCA1) (Xu et al., 2007). Additionally, CCA1 was identified as a master regulator in the circadian regulation of N assimilation in *Arabidopsis* (Gutiérrez et al., 2008). Interestingly, a recent study that evaluated the expression of sulfate transporters and analyzed *Brassica napus* sulfur-containing metabolites proposed that sulfur metabolism is regulated by a diurnal rhythm and not the circadian clock (Hornbacher et al., 2019).

6.4 Plant secondary metabolites and their response to soil fertility

Generally, plant phytochemicals are classified under two separate groups, namely the primary and secondary metabolites in accordance with their function in fundamental metabolic processes. The plant primary metabolites are participating in growth, development, and reproduction functions; the plant secondary metabolites constitute organic compounds that are not specifically involved in the natural development, growth, or reproduction of the plant. Therefore the lack of secondary metabolites does not lead to immediate mortality (Jain et al., 2019). The invention of analytical techniques such as chromatography permitted more and more of these molecules to be identified in the mid-20th century, which was the foundation for creating a metabolomics discipline (Mosa et al., 2017). Numerous reports have exhibited the significant benefits of plant secondary metabolites to both plants as well as humans. Plant secondary metabolites often play an important role in the protection of plants from a diverse spectrum of microorganisms and herbivores (Keswani et al., 2020). Humans often utilize numerous secondary metabolites of plants in drugs, pigments, fragrances, insecticides, dyes, and flavorings (Pagare et al., 2015; Marchev et al., 2020). In this section, we highlight the plant secondary metabolite classes as well as discuss the effect of soil macronutrients (nitrogen, phosphorus, potassium, and sulfur) on flavonoids, terpenoids, and alkaloids.

6.4.1 Classes of plant secondary metabolites

Plant secondary metabolites are generally classified into several major classes such as phenolics, alkaloids, saponins, and terpenes.

6.4.1.1 Phenolics

Phenolics are the largest groups of plant secondary metabolites and are present in most plants. These substances have common characteristics, such as one or more phenol groups, but can vary in structure at the level of single aromatic rings to more complex polymer compounds (Vuolo et al., 2019). It is possible to distinguish phenolics in accordance with their composition or biosynthetic source. Based on their structures, it is possible to classify phenolics into phenolic acids, simple phenols, flavonoids, lignans, tannins, stilbenes, coumarins, and curcuminoids (Basli et al., 2017).

6.4.1.2 Alkaloids

Alkaloids are organic compounds comprising a heterocyclic ring that contains at least one nitrogen atom. As such, no general description refers to all alkaloids except for the assumption that they all contain nitrogen. Alkaloids contain a proton-accepting atom of nitrogen and a donor of one or more protons of amine hydrogen atoms forming protein, enzyme, and receptor hydrogen bonds. In addition, they may comprise functional groups such as phenolic hydroxyl that are likely responsible for their robust bioactivity (Adejoke, Louis, Amusan, & Apebende, 2019). Alkaloids may be classified into various groups in accordance with their specific chemical structure: acridones, aromatics, oxindoles, pyrrolidines, carbolines, indoles, bisindoles, pyrroloindoles, ergots, imidazoles, purines, piperidines, phenylethylamines, phenylisoquinolines, quinolines, quinozolines, ephedras, indolizidines, manzamines, pyrrolizidines, pyridines, and simple tetrahydroisoquinolines (Aniszewski, 2015).

6.4.1.3 Saponins

Saponins are active surface compounds consisting of a steroid or triterpenoid aglycon and one or more sugar chains. They are found in substantial amounts in quinoa plants with highly bitter taste (Collar, 2016). The primary sources of saponins are legume plants, where they contain various saponins with different concentrations.

6.4.1.4 Terpenes

Terpenes are the largest and most complex groups of plant secondary metabolites and are chemically extracted from 5-carbon isoprene units assembled in various ways (Hoffmann, 2003). Terpenes are involved in defensive mechanisms such as the prevention of herbivores consuming or destroying plants, as well as tolerance to high temperatures. They also attract pollinators like bees (Abbas et al., 2017).

6.4.2 Response of plant secondary metabolites to soil macronutrient fertility

Soil nutrient types and quantities have major effects on both plant growth and secondary metabolite content. Several studies revealed that nutrient availability plays an important role in determining secondary metabolism in plants (Yang et al., 2018; Table 6.1). In this case, fertilizers may be applied to provide a number of substances that are not present in certain soils to enhance plant growth and development.

6.4.2.1 Response to N fertilization

Nitrogen is a crucial nutrient for plant growth. It controls the flow of carbon allocation between primary and secondary metabolism (Deng et al., 2019).

TABLE 6.1 Plant secondary metabolite response to fertilization or deficiency of macronutrients.

Metabolite class	Metabolite name	Plant species	Plant part/ component	Effect*	Reference
Nitrogen fertilization					
Flavonoids	Total flavonoids	*Labisia pumila* Benth.	Leaves, stems, roots	↓	Ibrahim et al. (2010)
	Quercitin, kaempferol, isorhamnetin	*Arabidopsis thaliana*	Seedlings	↓	Stewart et al. (2001)
	Quercitin, kaempferol, isorhamnetin	*Lycopersicon esculentum*		↓	Stewart et al. (2001)
Terpenes	Foliar volatile terpenes	*Pinus halpensis*	Foliar emissions	↓	Blanch et al. (2007)
Terpenes	Foliar volatile terpenes	*Quercus ilex*	Foliar emissions	(–)	Blanch et al. (2007)
Alkaloids	Total alkaloids	*Catharanthus roseus*	Leaves, roots	↑	Sreevalli et al. (2004)
Phosphorus fertilization					
Flavonoids	Quercitin, kaempferol, isorhamnetin	*A. thaliana*	Seedlings	↓	Stewart et al. (2001)
Flavonoids	Quercitin, kaempferol, isorhamnetin	*L. esculentum*	Seedlings	↓	Stewart et al. (2001)
Terpenes	Total foliar volatile terpenes	*P. halpensis*	Foliar emissions	↓	Blanch et al. (2007)

(*Continued*)

TABLE 6.1 (Continued)

Metabolite class	Metabolite name	Plant species	Plant part/ component	Effect*	Reference
Terpenes	Total foliar volatile terpenes	*Q. ilex*	Foliar emissions	(–)	Blanch et al. (2007)
Phosphorus deficiency					
Flavonoids	Kaempferol, daidzein, and their glycoside derivatives	*Stylothanthes*	Roots	↑	Luo et al. (2020)
Alkaloids	Lupanine, 13-hydroxylupanine, and angustifoline	*Lupinus angustifolius* L. (sweet varieties Yorrel, Gungurru, and Danja)	Seeds	↓	Gremigni et al. (2003)
Alkaloids	Lupanine, 13-hydroxylupanine, and angustifoline	*L. angustifolius* L. (bitter var. Fest)	Seeds	(–)	Gremigni et al. (2003)
Terpenes	Total terpenes	*Pinus pinaster*	Leaves	↑	Blanch et al. (2012)
Terpenes	Total foliar volatile terpenes	*Pinus pinaster*	Foliar emissions	↓	Blanch et al. (2012)
Nitrogen and phosphorus fertilization					
Terpenes	Monoterpenes and sesquiterpenes	*P. halepensis*	Leaves	↑	Ormeño et al. (2008)
Potassium fertilization					
Flavonoids	Total flavonoids	*Cicer arietinum*	Leaves	↑	Ahmad et al. (2016)
Flavonoids	Total flavonoids	*Labisia pumila*	Leaves, stems, roots	↑	Ibrahim et al. (2012)

Potassium deficiency					
Alkaloids	Total alkaloids	*Atropa acuminata*	Leaves, stems, roots	↑	Khana and Harborne (1991)
Sulfur deficiency					
Flavonoids	Total flavonoids	*Foeniculum vulgare*	Seeds	↓	Houhou et al. (2018)
Alkaloids	Total water-soluble alkaloids, total tertiary alkaloids	*Brassica campestris* L. var. *Sarson* Prain	Leaves	↓	Singh (1979)
Terpenes	α-Terpinene, p-cymene, myrcene, α-thujene, and α-pinene	*Satureja hortensis* L.	Essential oil	↑	Najafian and Zahedifar (2015)
Terpenes	γ-Terpinene	*Satureja hortensis* L.	Essential oil	↓	Najafian and Zahedifar (2015)
Sulfur fertilization					
Flavonoids	Total flavonoids	*Brassica oleracea* L. var. italica	Inflorescence	↑	Vallejo et al. (2003)

↑: Increase, ↓: decrease, (–): no significant change.

Based on the carbon nutrient balance hypothesis (CNBH), the level of carbon-based secondary metabolites (e.g., terpenes and flavonoids) is inversely correlated with nutrient availability (Heimler et al., 2017). In nitrogen-rich conditions, large amounts of carbohydrates are dedicated for primary metabolism and growth while limiting secondary metabolism. At limited nitrogen levels, carbohydrate accumulation can exceed growth requirements. This leads to the availability of carbohydrates for secondary metabolism (Deng et al., 2019).

Total flavonoid levels in *Labisia pumila* Benth. were significantly reduced at high nitrogen levels (Ibrahim et al., 2010). A similar trend was observed in the seedlings of *Arabidopsis thaliana* and *Lycopersicon esculentum* (Stewart et al., 2001). The response of terpenoids to nitrogen content showed inconsistent results that differed amongst various plant species, growth stages, and environmental conditions (Lee et al., 2005). Terpene concentrations decreased upon nitrogen fertilization in the seedlings of *Pinus halpensis*, while there was no significant effect in *Quercus ilex* terpene levels (Blanch et al., 2007). The effect of the growth stage on terpene levels in response to nitrogen was observed in *Thuja plicata*, demonstrating no significant effect of nitrogen fertilization on terpenoid production during May. However, during the active growth season in August, the monoterpenoid levels were lower at high nitrogen fertilization, indicating that the additional nutrient sources were allocated to plant growth rather than terpenoid production. In September, as plant growth declined, high nitrogen fertilization led to higher monoterpenoid concentrations, which indicated that the additional nutrients were shifted toward secondary metabolism rather than plant growth (Burney et al., 2012).

In contrast to flavonoids and terpenoids, alkaloids are nitrogen-based compounds. Thus they require nitrogen for their synthesis. Nitrogen fertilization has been found to elevate alkaloid content in plants. In *Catharanthus roseus*, nitrogen fertilization caused an increase in alkaloid content in leaves and roots, as well as increased leaf and root yields (Sreevalli et al., 2004). Similarly, alkaloid content was elevated by increasing levels of nitrogen application to *Datura stramonium* (Demeyer and Dejaegere, 1989).

6.4.2.2 Response to P fertilization

In the soil, phosphate ions can precipitate with cations (e.g., Ca, Al, and Fe) and form insoluble complexes. This reduces phosphate's bioavailability to plant roots. To overcome phosphate deficiency, plants will release root exudates rich in flavonoids, which acidifies the rhizosphere by promoting reduction-complexation processes that might facilitate phosphate remobilization and increase its availability (Valentinuzzi et al., 2019; Luo et al., 2020). Under P deficiency, apple tree roots were observed to release more flavonoids (Valentinuzzi et al., 2019). Similarly, upon subjecting *Stylothanthes* to low P stress, flavonoid metabolites, such as daidzein, kaempferol, and their glycoside derivatives, were increased and root growth was enhanced (Luo et al., 2020).

Phosphorus plays a crucial role in controlling the alkaloid levels of plants. Phosphorus deficiency leads to depression in the enzymatic activities of key enzymes involved in alkaloid biosynthesis (Marschner, 1995). Seed alkaloid concentrations in *Lupinus angustifolius* L. were observed to be reduced at low P levels (Gremigni et al., 2003). The response of terpenoids to P fertilization varies in different plant species. Phosphorus deficiency caused a 31% increase in foliar terpene concentrations in *Pinus pinaster* seedlings (Blanch et al., 2012). Likewise, at low P and N, most of the terpenes of *Pinus caribaea* needles increased (Felipe Barnola & Cedeño, 2000). This might be explained by the hypothesis of carbon/nutrient balance. However, this is not consistent for all plant species. For example, *Pinus halepensis* showed elevated terpene content upon nitrogen and phosphorus fertilization. This might be due to resin duct (specialized structures to store terpenes) formation that was favored at high nutrient availability (Ormeño et al., 2008).

6.4.2.3 Response to K fertilization

Potassium is important in activating enzymes that are involved in flavonoid synthesis (Chong et al., 2013). A significant increase in flavonoid content was observed in *Cicer arietinum* treated with potassium (Ahmad et al., 2019). Similarly, high potassium fertilization caused increased flavonoid content in *Labisia pumila*. At high potassium, the activity of the phenylalanine ammonia lyase enzyme was also increased (Ibrahim et al., 2012), which is a key enzyme in flavonoid synthesis (Borges et al., 2017).

Under potassium deficiency, plants accumulate large amounts of putrescine (Tachimoto et al., 1992), which plays a role in cation balance, mediating ROS signaling, and pH regulation (Cui et al., 2020). In addition, it is involved in the replacement of K as an organic cation in K-deficient plants (Tachimoto et al., 1992). Under 100-fold less potassium compared to controls, the total alkaloid levels in *Atropa acuminata* increased. Moreover, the activities of ornithine decarboxylase and arginine decarboxylase, which are accountable for putrescine production, were increased in accordance with putrescine levels. This supports the hypothesis that potassium ions affect the enzymes involved in *A. acuminata* alkaloid biosynthesis (Khana and Harborne, 1991).

6.4.2.4 Response to S fertilization

Sulfur is involved in the synthesis of cysteine, methionine, glutathione, coenzyme A, and phytochelatins. At poor sulfur levels, sulfur-containing metabolites like methionine and *S*-adenosylmethionine (SAM) are reduced (Houhou et al., 2018). SAM is a key substrate for the synthesis of many enzymes including coenzyme A. In addition, it donates methylene groups for the synthesis of flavonoids (Lewandowska and Sirko, 2008). As such, the total flavonoid content in *Foeniculum vulgare* seeds under sulfur deficiency was observed to be reduced significantly (Houhou et al., 2018). Upon sulfur fertilization, the total flavonoid content in broccoli inflorescences (*Brassica oleracea* L. var. italica) increased significantly (Vallejo et al., 2003). Several alkaloids contain S, such as benzothiazole alkaloids (Le Bozec and Moody, 2009). This indicates that under S deficiency, alkaloid levels are expected to decrease. Mustard plant leaves showed a significant reduction of tertiary as well as water-soluble alkaloids under sulfur deficiency (Singh, 1979). Sulfur is also involved in many forms of terpene synthesis (Pandey et al., 2009). The response of terpenoids to sulfur deficiency has shown to be variable. The application of 0.05 g S kg^{1} soil increased the levels of -terpinene, p-cymene, myrcene, -thujene, and -pinene in *Satureja hortensis* L. essential oil. However, under these conditions the levels of -terpinene were decreased (Najafian and Zahedifar, 2015).

6.5 Conclusion

Water and nutrients that are stored in soil are transported to plants via the root system. The amount and variety of nutrients determine the physiological functions of plants enabling proper plant growth. Macronutrients that are required in large quantities are essential for plant growth and development. As such, further investigations should be put forth toward better understanding the molecular mechanisms and regulation of macronutrients that are crucial for developing platforms that enhance crop productivity. On the other hand, recent reports demonstrated that nutrient availability is a critical factor in controlling secondary metabolism in plants. Thus the potential role of plant secondary metabolites in different applications with significant economic value necessitates future efforts to better elucidate the complex relationship between plant secondary metabolites and macronutrient homeostasis.

References

Abbas, F., Ke, Y., Yu, R., Yue, Y., Amanullah, S., Jahangir, M.M., Fan, Y., 2017. Volatile terpenoids: multiple functions, biosynthesis, modulation and manipulation by genetic engineering. Planta 246 (5), 803–816.

Adejoke, H.T., Louis, H., Amusan, O.O., Apebende, G., 2019. A review on classes, extraction, purification and pharmaceutical importance of plants alkaloid. Journal of Medicinal and Chemical Sciences 2 (4), 130–139. Available from: https://doi.org/10.26655/jmchemsci.2019.8.2.

Ahmad, P., Abdel Latef, A.A., Abd_Allah, E.F., Hashem, A., Sarwat, M., Anjum, N.A., et al., 2016. Calcium and potassium supplementation enhanced growth, osmolyte secondary metabolite production, and enzymatic antioxidant machinery in cadmium-exposed chickpea (*Cicer arietinum L.*). Front. Plant Sci. 7, 513.

Ahmed, A.A., Gypser, S., Freese, D., Leinweber, P., Kuehn, O., 2020. Molecular level picture of the interplay between pH and phosphate binding at the goethite–water interface. Physical Chemistry Chemical Physics.

Amtmann, A., Blatt, M.R., 2009. Regulation of macronutrient transport. New Phytol. 181, 35–52.

Aniszewski, T., 2015. Alkaloids: chemistry, biology, ecology, and applications. Elsevier.

Ashley, M.K., Grant, M., Grabov, A., 2006. Plant responses to potassium deficiencies: a role for potassium transport proteins. J. Exp. Bot. 57, 425–436.

Basli, A., Belkacem, N., Amrani, I., 2017. Health benefits of phenolic compounds against cancers. Phenolic compounds–Biological activity. IntechOpen, London, UK, pp. 193–210.

Behl, R., Raschke, K., 1987. Close coupling between extrusion of h^+ and uptake of k^+ by barley roots. Planta 172, 531–538.

Bhaduri, D., Rakshit, R., Chakraborty, K., 2014. Primary and secondary nutrients-a boon to defense system against plant diseases. Int. J. Bio-Resource Stress Manag. 5, 461–466.

Blanch, J., Sampedro, L., Llusià, J., Moreira, X., Zas, R., Peñuelas, J., 2012. Effects of phosphorus availability and genetic variation of leaf terpene content and emission rate in *Pinus pinaster* seedlings susceptible and resistant to the pine weevil, *Hylobius abietis*. Plant Biol. 14, 66–72.

Blanch, J., Peñuelas, J., Llusià, J., 2007. Sensitivity of terpene emissions to drought and fertilization in terpene-storing *Pinus halepensis* and non-storing *Quercus ilex*. Physiol. Plant. 131, 211–225.

Bohn, H.L., Barrow, N.J., Rajan, S., Parfitt, R.L., 1986. Reactions of inorganic sulfur in soils. In: Tabatabai, M.A. (Ed.), Sulfur in Agriculture. Wiley Online Library, Madison, WI, pp. 233–249.

Bonner, E.R., Cahoon, R.E., Knapke, S.M., Jez, J.M., 2005. Molecular basis of cysteine biosynthesis in plants: structural and functional analysis of O-acetylserine sulfhydrylase from *Arabidopsis thaliana*. J. Biol. Chem. 280, 38803–38813.

Borges, C.V., Minatel, I.O., Gomez-Gomez, H.A., Lima, G.P.P., 2017. Medicinal plants: Influence of environmental factors on the content of secondary metabolites. Medicinal Plants and Environmental Challenges. Springer, Cham, pp. 259–277.

Bouguyon, E., Brun, F., Meynard, D., Kubeš, M., Pervent, M., Leran, S., et al., 2015. Multiple mechanisms of nitrate sensing by arabidopsis nitrate transceptor NRT1.1. Nat. Plants 1, Article number: 15015.

Burney, O.T., Davis, A.S., Jacobs, D.F., 2012. Phenology of foliar and volatile terpenoid production for *Thuja plicata* families under differential nutrient availability. Environ. Exp. Bot. 77, 44–52.

Caldeira, C.F., Jeanguenin, L., Chaumont, F., Tardieu, F., 2014. Circadian rhythms of hydraulic conductance and growth are enhanced by drought and improve plant performance. Nat. Commun. 5, 5365.

Chen, Y., Wang, Y., Wu, W., 2008. Membrane transporters for nitrogen, phosphate and potassium uptake in plants. J. Integr. Plant Biol. 50, 835–848.

Chong, P.L., Jaafar, H., Ibrahim, M.H., 2013. Relationship of chlorophyll and flavonoids content in Labisia pumila benth. under nitrogen and potassium fertilization. In: Plant Physiology in Addressing Green Economy: Proceedings of the 23rd Malaysian Society of Plant Physiology Conference 2012 (MSPPC 2012), pp. 88–91.

Christodoulou, E., Agapiou, A., Anastopoulos, I., Omirou, M., Ioannides, I.M., 2019. The effects of different soil nutrient management schemes in nitrogen cycling. Journal of Environmental Management 243, 168–176.

Clarkson, D.T., Hanson, J.B., 1980. The mineral nutrition of higher plants. Annu. Rev. Plant Physiol. 31, 239–298.

Clemens, S., Palmgren, M.G., Krämer, U., 2002. A long way ahead: understanding and engineering plant metal accumulation. Trends Plant Sci. 7, 309–315.

Collar, C. (2016). Quinoa. In B. Caballero, P.M. Finglas & F. Toldrá (Eds), Encyclopedia of food and health (pp. 573-579). Oxford: Academic Press. https://doi.org/10.1016/B978-0-12-384947-2.00583-3.

Crawford, N.M., Forde, B.G., 2002. Molecular and developmental biology of inorganic nitrogen nutrition. Arabidopsis Book 1, e0011.

Cui, J., Pottosin, I., Lamade, E., Tcherkez, G., 2020. What is the role of putrescine accumulated under potassium deficiency? Plant Cell Environ. 43, 1331–1347.

Dai, Z., Liu, G., Chen, H., Chen, C., Wang, J., Ai, S., et al., 2020. Long-term nutrient inputs shift soil microbial functional profiles of phosphorus cycling in diverse agroecosystems. The ISME Journal 14 (3), 757–770.

Dániel, S., Chiba, P., Gergely, S., Stockner, T., 2020. Conversion of chemical to mechanical energy by the nucleotide binding domains of ABCB1. Scientific Reports (Nature Publisher Group) 10 (1).

Datta, A., Santra, S.C., Adhya, T.K., 2013. Effect of inorganic fertilizers (N, P, K) on methane emission from tropical rice field of India. Atmos. Environ. 66, 123–130.

Demeyer, K., Dejaegere, R., 1989. Influence of the ion-balance in the growth medium on the yield and alkaloid content of *Datura stramonium* L. Plant Soil. 114 (2), 289–294.

Deng, B., Li, Y., Xu, D., Ye, Q., Liu, G., 2019. Nitrogen availability alters flavonoid accumulation in *Cyclocarya paliurus* via the effects on the internal carbon/nitrogen balance. Sci. Rep. 9, 2370.

Dodd, A.N., 2005. Plant circadian clocks increase photosynthesis, growth, survival, and competitive advantage. Sci. (Am. Assoc. Adv. Sci.) 309, 630–633.

Fang, X.Z., Tian, W.H., Liu, X.X., Lin, X.Y., J-in, C.W., Zheng, S.J., 2016. Alleviation of proton toxicity by nitrate uptake specifically depends on nitrate transporter 1.1 in arabidopsis. New Phytol. 211, 149–158.

Felipe Barnola, L., Cedeño, A., 2000. Inter-population differences in the essential oils of *Pinus caribaea* needles. Biochem. Syst. Ecol. 28, 923–931.

Gahan, J., Schmalenberger, A., 2014. The role of bacteria and mycorrhiza in plant sulfur supply. Front. Plant Sci. 5, 723.

Gierth, M., Mäser, P., 2007. Potassium transporters in plants — involvement in K^+ acquisition, redistribution and homeostasis. FEBS Lett. 581, 2348–2356.

Gierth, M., Mäser, P., Schroeder, J.I., 2005. The potassium transporter AtHAK5 functions in K deprivation-induced high-affinity K uptake and AKT1 K channel contribution to K uptake kinetics in arabidopsis roots. Plant Physiol. 137, 1105–1114.

Gigolashvili, T., Kopriva, S., 2014. Transporters in plant sulfur metabolism. Front. Plant Sci. 5, 442.

Gojon, A., Krouk, G., Perrine-Walker, F., Laugier, E., 2011. Nitrate transceptor(s) in plants. J. Exp. Bot. 62, 2299–2308.

Gremigni, P., Hamblin, J., Harris, D., Cowling, W.A., 2003. The interaction of phosphorus and potassium with seed alkaloid concentrations, yield and mineral content in narrow-leafed lupin (*Lupinus angustifolius* L.). Plant Soil 253, 413–427.

Guoqing, X., Xiuqin, C., Liping, B., Hongtao, Q., Haibo, L., 2019. Absorption, accumulation and distribution of metals and nutrient elements in poplars planted in land amended with composted sewage sludge: A field trial. Ecotoxicology and Environmental Safety 182, 109360.

Gutiérrez, R.A., Stokes, T.L., Thum, K., Xu, X., Obertello, M., Katari, M.S., et al., 2008. Systems approach identifies an organic nitrogen-responsive gene network that is regulated by the master clock control gene CCA1. Proc. Natl Acad. Sci. U S A 105, 4939–4944.

Harmer, S.L., Hogenesch, J.B., Straume, M., Chang, H., Han, B., Zhu, T., et al., 2000. Orchestrated transcription of key pathways in arabidopsis by the circadian clock. Sci. (Am. Assoc. Adv. Sci.) 290, 2110–2113.

Heimler, D., Romani, A., Ieri, F., 2017. Plant polyphenol content, soil fertilization and agricultural management: a review. Eur. Food Res. Technol. 243, 1107–1115.

Ho, C., Tsay, Y., 2010. Nitrate, ammonium, and potassium sensing and signaling. Curr. Opin. Plant Biol. 13, 604–610.

Ho, C., Lin, S., Hu, H., Tsay, Y., 2009. CHL1 functions as a nitrate sensor in plants. Cell 138, 1184–1194.

Hoffmann, D. (2003). Medical herbalism: The science and practice of herbal medicine Simon and Schuster.

Hornbacher, J., Rumlow, A., Pallmann, P., Turcios, A.E., Riemenschneider, A., Papenbrock, J., 2019. The Levels of Sulfur-containing Metabolites in Brassica napus are Not Influenced by the Circadian Clock but Diurnally. Journal of Plant Biology 62 (5), 359–373.

Houhou, M., Joutei, K., Rais, C., Louahlia, S., 2018. Phenolic compounds, antioxidant and antibacterial activities are affected by sulfur deficiency in *Foeniculum vulgare* seeds. J. Crop Sci. Biotechnol. 21, 241–247.

Hsu, P.Y., Harmer, S.L., 2014. Wheels within wheels: the plant circadian system. Trends Plant Sci. 19, 240–249.

Ibrahim, M.H., Jaafar, H.Z.E., Rahmat, A., Rahman, Z.A., 2010. The relationship between phenolics and flavonoids production with total non structural carbohydrate and photosynthetic rate in *Labisia pumila* benth. under high CO_2 and nitrogen fertilization. Molecules 16, 162–174.

Ibrahim, M., Jaafar, H., Karimi, E., Ghasemzadeh, A., 2012. Primary, secondary metabolites, photosynthetic capacity and antioxidant activity of the Malaysian herb kacip fatimah (*Labisia pumila* benth) exposed to potassium fertilization under greenhouse conditions. Int. J. Mol. Sci. 13, 15321–15342.

Jain, C., Khatana, S., Vijayvergia, R., 2019. Bioactivity of secondary metabolites of various plants: A review. Int. J. Pharm. Sci. 10, 404–494.

Jiao, S., Xu, Y., Zhang, J., Hao, X., Lu, Y., 2019. Core microbiota in agricultural soils and their potential associations with nutrient cycling. Msystems 4 (2).

Jung, J.K.H., McCouch, S., 2013. Getting to the roots of it: genetic and hormonal control of root architecture. Front. Plant Sci. 4, 186.

Karthika, K.S., Rashmi, I., Parvathi, M.S., 2018. Biological functions, uptake and transport of essential nutrients in relation to plant growth. In Plant nutrients and abiotic stress tolerance. Springer, Singapore, pp. 1–49.

Kataoka, T., Watanabe-Takahashi, A., Hayashi, N., Ohnishi, M., Mimura, T., Buchner, P., et al., 2004. Vacuolar sulfate transporters are essential determinants controlling internal distribution of sulfate in arabidopsis. Plant Cell 16, 2693–2704.

Kayser, M., Isselstein, J., 2005. Potassium cycling and losses in grassland systems: a review. Grass Forage Sci. 60, 213–224.

Keswani, C., Singh, H.B., García-Estrada, C., Caradus, J., He, Y., Mezaache-Aichour, S., et al., 2020. Antimicrobial secondary metabolites from agriculturally important bacteria as next-generation pesticides. Applied Microbiology and Biotechnology 104 (3), 1013–1034.

Khademi, S., 2004. Mechanism of ammonia transport by amt/MEP/rh: structure of AmtB at 1.35 A. Sci. (Am. Assoc. Adv. Sci.) 305, 1587–1594.

Khana, M.B., Harborne, J.B., 1991. Potassium deficiency increases tropane alkaloid synthesis in *Atropa acuminata* via arginine and ornithine decarboxylase levels. Phytochemistry 30, 3559–3563.

Kim, M.J., Ciani, S., Schachtman, D.P., 2010. A peroxidase contributes to ROS production during arabidopsis root response to potassium deficiency. Mol. Plant 3, 420–427.

Komarova, N.Y., Thor, K., Gubler, A., Meier, S., Dietrich, D., Weichert, A., et al., 2008. AtPTR1 and AtPTR5 transport dipeptides in planta. Plant Physiol. 148, 856–869.

Kooch, Y., Noghre, N., 2020. Nutrient cycling and soil-related processes under different land covers of semi-arid rangeland ecosystems in northern iran. Catena 193, 104621.

Kopriva, S., Koprivova, A., 2004. Plant adenosine 5′-phosphosulphate reductase: the past, the present, and the future. J. Exp. Bot. 55, 1775–1783.

Kovar, J.L., Grant, C.A., 2011. Nutrient cycling in soils: sulfur. In: Hatfield, J.L., Sauer, T.J. (Eds.), Soil Management: Building a Stable Base for Agriculture. American Society of Agronomy, Crop Science Society of America, Soil Science Society of America, Madison, WI, pp. 103–115.

Kriel, G., 2020. Canola: The crop that requires perfect planning: Crops-canola. Farmer's Weekly 2020 (20023), 44–46.

Krouk, G., Lacombe, B., Bielach, A., Perrine-Walker, F., Malinska, K., Mounier, E., et al., 2010. Nitrate-regulated auxin transport by NRT1.1 defines a mechanism for nutrient sensing in plants. Dev. Cell 18, 927–937.

Kumar, S., Verma, S., Trivedi, P.K., 2017. Involvement of small RNAs in phosphorus and sulfur sensing, signaling and stress: current update. Front. Plant Sci. 8, 285.

Le Bozec, L., Moody, C.J., 2009. Naturally occurring nitrogen–sulfur compounds. the benzothiazole alkaloids. Aust. J. Chem. 62, 639–647.

Lebaudy, A., Véry, A., Sentenac, H., 2007. K^+ channel activity in plants: genes, regulations and functions. FEBS Lett. 581, 2357–2366.

Lee, K.D., Yang, M.S., Supanjani, Smith, D.L., 2005. Fertilizer effect on the yield and terpene components from the flowerheads of *Chrysanthemum boreale* M. (compositae). Agron. Sustain. Dev. 25, 205–211.

Leigh, R.A., Wyn Jones, R.G., 1984. A hypothesis relating critical potassium concentrations for growth to the distribution and functions of this ion in the plant cell. New Phytol. 97, 1–13.

Léran, S., Varala, K., Boyer, J., Chiurazzi, M., Crawford, N., Daniel-Vedele, F., et al., 2014. A unified nomenclature of nitrate transporter 1/peptide transporter family members in plants. Trends Plant Sci. 19, 5–9.

Lewandowska, M., Sirko, A., 2008. Recent advances in understanding plant response to sulfur-deficiency stress. Acta Biochim. Pol. 55, 457–471.

Li, M., Ren, L., Zhang, J., Luo, L., Qin, P., Zhou, Y., et al., 2019. Population characteristics and influential factors of nitrogen cycling functional genes in heavy metal contaminated soil remediated by biochar and compost. Science of the Total Environment 651, 2166–2174.

Liu, X., Guo, C., He, S., Zhu, H., Li, J., Yu, Z., et al., 2020. Divergent gross nitrogen transformation paths in the topsoil and subsoil between abandoned and agricultural cultivation land in irrigated areas. Science of the Total Environment 716, 137148.

López-Bucio, J., Cruz-Rammírez, A., Herrera-Estrella, L., 2003. The role of nutrient availability in regulating root architecture. Curr. Opin. Plant Biol. 6, 280–287.

Luo, J., Liu, Y., Zhang, H., Wang, J., Chen, Z., Luo, L., et al., 2020. Metabolic alterations provide insights into stylosanthes roots responding to phosphorus deficiency. BMC Plant Biol. 20, 85.

Ma, Q., Wen, Y., Pan, W., Macdonald, A., Hill, P.W., Chadwick, D.R., et al., 2020. Soil carbon, nitrogen, and sulphur status affects the metabolism of organic S but not its uptake by micro-organisms. Soil Biology and Biochemistry 149, 107943.

Maathuis, F.M., Sanders, D., 1993. Energization of potassium uptake in *Arabidopsis thaliana*. Planta 191, 302–307.

Marchev, A.S., Yordanova, Z.P., Georgiev, M.I., 2020. Green (cell) factories for advanced production of plant secondary metabolites. Critical Reviews in Biotechnology 40 (4), 443–458.

Marschner, H. (Ed.), 1995. Marschner's Mineral Nutrition of Higher Plants. Academic Press.

Marschner, P., Rengel, Z., 2012. Nutrient availability in soils. In: Marschner, P. (Ed.), Marschner's Mineral Nutrition of Higher Plants. Academic Press, pp. 315–330.

Mäser, P., Thomine, S., Schroeder, J.I., Ward, J.M., Hirschi, K., Sze, H., et al., 2001. Phylogenetic relationships within cation transporter families of arabidopsis. Plant Physiol. 126, 1646–1667.

Morgan, J.B., Connolly, E.L., 2013. Plant-soil interactions: nutrient uptake. Nat. Educ. Knowl. 4, 2.

Mosa, K.A., Ismail, A., Helmy, M., 2017. Plant Stress Tolerance. Springer International Publishing AG, Cham.

Mullins, G.L., 2019. Phosphorus, agriculture & the environment. Virginia Cooperative Extension.

Mussa, S.B., Elferjani, H.S., Haroun, F.A., Abdelnabi, F.F., 2009. Determination of available nitrate, phosphate and sulfate in soil samples. International Journal of PharmTech Research 1 (3), 598–604.

Najafian, S., Zahedifar, M., 2015. Antioxidant activity and essential oil composition of *Satureja hortensis* L. as influenced by sulfur fertilizer. J. Sci. Food Agric. 95, 2404–2408.

Nieves-Cordones, M., Miller, A.J., Alemán, F., Martínez, V., Rubio, F., 2008. A putative role for the plasma membrane potential in the control of the expression of the gene encoding the tomato high-affinity potassium transporter HAK5. Plant Mol. Biol. 68, 521–532.

Noordally, Z.B., Ishii, K., Atkins, K.A., Wetherill, S.J., Kusakina, J., Walton, E.J., et al., 2013. Circadian control of chloroplast transcription by a nuclear-encoded timing signal. Sci. (Am. Assoc. Adv. Sci.) 339, 1316–1319.

Oldroyd, G.E., Leyser, O., 2020. A plant's diet, surviving in a variable nutrient environment. Science 368.

Ormeño, E., Baldy, V., Ballini, C., Fernandez, C., 2008. Production and diversity of volatile terpenes from plants on calcareous and siliceous soils: effect of soil nutrients. J. Chem. Ecol. 34, 1219–1229.

Ortiz-Ramirez, C., Mora, S.I., Trejo, J., Pantoja, O., 2011. PvAMT1;1, a highly selective ammonium transporter that functions as H^+/NH^{4+}Symporter. J. Biol. Chem. 286, 31113–31122.

Pagare, S., Bhatia, M., Tripathi, N., Pagare, S., Bansal, Y.K., 2015. Secondary metabolites of plants and their role: Overview. Current Trends in Biotechnology and Pharmacy 9 (3), 293–304.

Pandey, G.K., Mahiwal, S., 2020. Potassium in Plant Growth and Development. Role of Potassium in Plants. Springer International Publishing, Cham, pp. 37–43.

Pandey, S.N., Naaz, S., Ansari, S.R., 2009. Growth, biomass and petroleum convertible hydrocarbons' yield of *Grindelia camporum* planted on an alluvial soil (entisol) of north India and its response to sulphur fertilization. Biomass Bioenerg. 33, 454–458.

Rawat, J., Sanwal, P., Saxena, J., 2016. Potassium and its role in sustainable agriculture. Potassium Solubilizing Microorganisms for Sustainable Agriculture. Springer, pp. 235–253.

Saha, S., Rajwar, G.S., Kumar, M., 2018. Soil properties along altitudinal gradient in himalayan temperate forest of garhwal region. Acta Ecol. Sin. 38, 1–8.

Saito, K., Yokoyama, H., Noji, M., Murakoshi, I., 1995. Molecular cloning and characterization of a plant serine acetyltransferase playing a regulatory role in cysteine biosynthesis from watermelon. J. Biol. Chem. 270, 16321–16326.

Saitoh, T., Ikegami, T., Nakayama, M., Teshima, K., Akutsu, H., Hase, T., 2006. NMR study of the electron transfer complex of plant ferredoxin and sulfite reductase: mapping the interaction sites of ferredoxin. J. Biol. Chem. 281, 10482–10488.

Schachtman, D.P., 2012. Recent advances in nutrient sensing and signaling. Mol. Plant 5, 1170–1172.

Schachtman, D.P., Shin, R., 2007. Nutrient sensing and signaling: NPKS. Annu. Rev. Plant Biol. 58, 47–69.

Scherer, H.W., 2009. Sulfur in soils. J. Plant Nutr. Soil. Sci. 172, 326–335.

Shen, J., Yuan, L., Zhang, J., Li, H., Bai, Z., Chen, X., et al., 2011. Phosphorus dynamics: from soil to plant. Plant Physiol. 156, 997–1005.

Shibagaki, N., Rose, A., McDermott, J.P., Fujiwara, T., Hayashi, H., Yoneyama, T., et al., 2002. Selenate-resistant mutants of *Arabidopsis thaliana* identify Sultr1;2, a sulfate transporter required for efficient transport of sulfate into roots. Plant J. Cell Mol. Biol. 29, 475–486.

Singh, O., 1979. Sulphur deficiency and alkaloid contents in mustard (*Brassica campestris* L. var. sarson, prain) plants. Indian J. Plant Physiol. 22, 78–80.

Singh, B., Sharma, K.K., Kumari, A., Kumar, A., Gakhar, S.K., 2018. Molecular modeling and docking of recombinant HAP-phytase of a thermophilic mould *Sporotrichum thermophile* reveals insights into molecular catalysis and biochemical properties. Int. J. Biol. Macromol. 115, 501–508.

Smith, F.W., Ealing, P.M., Hawkesford, M.J., Clarkson, D.T., 1995. Plant members of a family of sulfate transporters revel functional subtypes. Proc. Natl Acad. Sci. U S A 92, 9373.

Song, Y.H., Ito, S., Imaizumi, T., 2010. Similarities in the circadian clock and photoperiodism in plants. Curr. Opin. Plant Biol. 13, 594–603.

Spalding, E.P., Hirsch, R.E., Lewis, D.R., Qi, Z., Sussman, M.R., Lewis, B.D., 1999. Potassium uptake supporting plant growth in the absence of AKT1 channel activity inhibition by ammonium and stimulation by sodium. Journal of General Physiology 113 (6), 909–918.

Sreevalli, Y., Kulkarni, R.N., Baskaran, K., Chandrashekara, R.S., 2004. Increasing the content of leaf and root alkaloids of high alkaloid content mutants of periwinkle through nitrogen fertilization. Ind. Crop. Prod. 19, 191–195.

Stevenson, F.J., Cole, M.A. (Eds.), 1999. Cycles of Soil—Carbon, Nitrogen, Phosphorus, Sulfur, Micronutrients. John Wiley and Sons.

Stewart, A.J., Chapman, W., Jenkins, G.I., Graham, I., Martin, T., Crozier, A., 2001. The effect of nitrogen and phosphorus deficiency on flavonol accumulation in plant tissues. Plant Cell Environ. 24, 1189–1197.

Sugiura, M., Georgescu, M.N., Takahashi, M., 2007. A nitrite transporter associated with nitrite uptake by higher plant chloroplasts. Plant Cell Physiol. 48, 1022–1035.

Tachimoto, M., Fukutomi, M., Matsushiro, H., Kobayashi, M., Takahashi, E., 1992. Role of putrescine in lemna plants under potassium deficiency. Soil. Sci. Plant Nutr. 38, 307–313.

Takahashi, H., Watanabe-Takahashi, A., Smith, F.W., Blake-Kalff, M., Hawkesford, M.J., Saito, K., 2000. The roles of three functional sulphate transporters involved in uptake and translocation of sulphate in *Arabidopsis thaliana*. Plant J. Cell Mol. Biol. 23, 171–182.

Takahashi, H., Buchner, P., Yoshimoto, N., Hawkesford, M.J., Shiu, S., 2011. Evolutionary relationships and functional diversity of plant sulfate transporters. Front. Plant Sci. 2, 119.

Takahashi, H., Kopriva, S., Giordano, M., Saito, K., Hell, R., 2011. Sulfur assimilation in photosynthetic organisms: molecular functions and regulations of transporters and assimilatory enzymes. Annual review of plant biology 62, 157–184.

Takase, T., Ishikawa, H., Murakami, H., Kikuchi, J., Sato-Nara, K., Suzuki, H., 2011. The circadian clock modulates water dynamics and aquaporin expression in arabidopsis roots. Plant Cell Physiol. 52, 373–383.

Tsay, Y., 1993. The herbicide sensitivity gene CHL1 of arabidopsis encodes a nitrate-inducible nitrate transporter. Cell 72, 705–713.

Urry, L.A., Cain, M.L., Wasserman, S.A., Minorsky, P.V., Reece, J.B., 2017. Campbell Biology. Pearson.

Valentinuzzi, F., Venuti, S., Pii, Y., Marroni, F., Cesco, S., Hartmann, F., et al., 2019. Common and specific responses to iron and phosphorus deficiencies in roots of apple tree (malusdomestica). Plant Mol. Biol. 101, 129–148.

Vallejo, F., Tomás-Barberán, F.A., García-Viguera, C., 2003. Effect of climatic and sulphur fertilisation conditions, on phenolic compounds and vitamin C, in the inflorescences of eight broccoli cultivars. Eur. Food Res. Technol. 216, 395–401.

Vert, G., Chory, J., 2009. A toggle switch in plant nitrate uptake. Cell 138, 1064–1066.

Véry, A., Sentenac, H., 2003. Molecular mechanisms and regulation of K^+ transport in higher plants. Annu. Rev. Plant Biol. 54, 575–603.

Vuolo, M.M., Lima, V.S., Junior, M.R.M., 2019. Phenolic compounds: structure, classification, and antioxidant power. In: Campos, M.R.S. (Ed.), Bioactive Compounds: Health Benefits and Potential Applications. Woodhead Publishing, pp. 33–50.

Ward, J.M., Mäser, P., Schroeder, J.I., 2009. Plant ion channels: gene families, physiology, and functional genomics analyses. Annu. Rev. Physiol. 71, 59–82.

Xin, J., Liu, Y., Chen, F., Duan, Y., Wei, G., Zheng, X., et al., 2019. The missing nitrogen pieces: a critical review on the distribution, transformation, and budget of nitrogen in the vadose zone-groundwater system. Water Res. 165, 114977.

Xu, X., Hotta, C.T., Dodd, A.N., Love, J., Sharrock, R., Lee, Y.W., Webb, A.A., 2007. Distinct light and clock modulation of cytosolic free Ca2+ oscillations and rhythmic CHLOROPHYLL A/B BINDING PROTEIN2 promoter activity in Arabidopsis. The Plant Cell 19 (11), 3474–3490.

Xuan, W., Beeckman, T., Xu, G., 2017. Plant nitrogen nutrition: sensing and signaling. Curr. Opin. Plant Biol. 39, 57–65.

Yang, L., Wen, K., Ruan, X., Zhao, Y., Wei, F., Wang, Q., 2018. Response of plant secondary metabolites to environmental factors. Molecules 23, 762.

Yi, H., Galant, A., Ravilious, G.E., Preuss, M.L., Jez, J.M., 2010. Sensing sulfur conditions: simple to complex protein regulatory mechanisms in plant thiol metabolism. Mol. Plant 3, 269–279.

Yoshimoto, N., Takahashi, H., Smith, F.W., Yamaya, T., Saito, K., 2002. Two distinct high-affinity sulfate transporters with different inducibilities mediate uptake of sulfate in arabidopsis roots. Plant J. Cell Mol. Biol. 29, 465–473.

Yuan, L., Loqué, D., Kojima, S., Rauch, S., Ishiyama, K., Inoue, E., et al., 2007. The organization of high-affinity ammonium uptake in arabidopsis roots depends on the spatial arrangement and biochemical properties of AMT1-type transporters. Plant Cell 19, 2636–2652.

Zhang, B., Pasini, R., Dan, H., Joshi, N., Zhao, Y., Leustek, T., et al., 2014. Aberrant gene expression in the arabidopsis SULTR1;2 mutants suggests a possible regulatory role for this sulfate transporter in response to sulfur nutrient status. Plant J. Cell Mol. Biol. 77, 185–197.

Zhang, Z., Liao, H., Lucas, W.J., 2014. Molecular mechanisms underlying phosphate sensing, signaling, and adaptation in plants. J. Integr. Plant Biol. 56, 192–220.

Zhou, J., Theodoulou, F.L., Muldin, I., Ingemarsson, B., Miller, A.J., 1998. Cloning and functional characterization of a brassica napus transporter that is able to transport nitrate and histidine. J. Biol. Chem. 273, 12017–12023.

Zhu, Y., Di, T., Xu, G., Chen, X., Zeng, H., Yan, F., et al., 2009. Adaptation of plasma membrane H^+-ATPase of rice roots to low pH as related to ammonium nutrition. Plant Cell Environ. 32, 1428–1440.

Chapter 7

Water and nitrogen fertilization management in light of climate change: impacts on food security and product quality

A.P.G. Fernandes[1,*], J. Machado[1,2,3,*], T.R. Fernandes[1], M.W. Vasconcelos[2] and S.M.P. Carvalho[1]

[1]*GreenUPorto — Sustainable Agrifood Production Research Centre / Inov4Agro & DGAOT, Faculty of Sciences, University of Porto, Vairão, Portugal,* [2]*Centro de Biotecnologia e Química Fina (CBQF) – Laboratório Associado, Escola Superior de Biotecnologia, Universidade Católica Portuguesa, Porto, Portugal,* [3]*Horticulture and Product Physiology Group, Department of Plant Sciences, Wageningen University, Wageningen, The Netherlands*

7.1 Introduction

The human population is exponentially growing and may double in the next 100 years, so a substantial increase in agricultural productivity worldwide will be needed (United Nations, 2019). By 2050, productivity must be increased by around 70% in order to feed the 9.6 billion people that are going to inhabit the planet by that time (FAO, 2015). However, most of the world's croplands have yields far below their potential, and the current global trajectory of agricultural expansion has serious long-term implications for the environment (Tilman et al., 2011). Moreover, efforts to grow and maximize yields can be threatened by climatic change (Bita and Gerats, 2013). As previously described by the Intergovernmental Panel on Climate Change (IPCC), climate change can be defined as any change in climate over time, either due to natural variability or as a result of human activity (IPCC, 2007). An analysis on climate change impact shows that there is a general reduction of potential crop yield and a decrease in water availability and quality for agriculture and populations in many parts of the developing world, whereas nearly 50% of the population rely entirely on agriculture

* These authors contributed equally to this study.

Plant Nutrition and Food Security in the Era of Climate Change. DOI: https://doi.org/10.1016/B978-0-12-822916-3.00013-5

(Bita and Gerats, 2013; IPCC, 2014). As established by the 2030 Agenda SDG 2 "Zero Hunger" (United Nations, 2015), the agriculture sector must "ensure sustainable food production systems and implement resilient agricultural practices that increase productivity and production, that help maintain ecosystems, that strengthen capacity for adaptation to climate change, extreme weather, drought, flooding, and other disasters and that progressively improve land and soil quality".

Currently, irrigated agriculture uses 80% of the world's freshwater resources and it is considered as a major sector responsible for water misuse, excessive water waste, and water quality degradation. It has been reported that approximately 2.4 billion people, or 36% of the global population, are already subjected to water scarcity. Forecasts predict that global water demand will increase by 55% by 2050 with the greatest increases in emerging economies and developing countries that are already under water deficit (Wang et al., 2014). For these reasons, water is now located at the core of the United Nations Sustainable Development Goals (UN-SDGs) of the 2030 Agenda (ex. SDGs 6 "Clean Water and Sanitation") (United Nations, 2015). Further, to respond to these and other challenges, FAO has developed the Global Framework for Action to Cope with Water Scarcity in Agriculture in the Context of Climate Change, which is addressing the linkages between water scarcity, agricultural production, food security, and climate change (FAO, 2012).

On the other hand, the intensive use of agrochemicals by modern agriculture (fertilizers and phytopharmaceutical products), combined with the increased water temperature caused by climate change, has increased the risks of eutrophication and pollution of freshwater ecosystems by excess nutrients, pesticides, salts, heavy metals, sediments, and organic matter (Lu et al., 2015; Machado et al., 2017a; Evans et al., 2019). More specifically, N fertilization plays a key role in agriculture because it is the main crop yield-determining nutrient. Since the early 1960s, N fertilization has grown approximately sevenfold (Goulding et al., 2008), and it is estimated to reach around 240 million tons by 2050, almost doubling the 2020 amount (Liang et al., 2018). N fertilizers have promoted economic development, spared forest areas that would likely have been destroyed for agriculture fields to meet food demand, and enabled food security. In fact, more than half of the world's population is nourished by crops grown with synthetic N fertilizers (Zhang et al., 2015; Zhang, 2017). Thus large amounts are applied to crops to increase their yield, often resulting in fertilizer overuse that exceeds crops' N requirements by 30%–80% (Goulding et al., 2008, Liang et al., 2018). The N excess induces unintended environmental consequences, such as leaching of nitrate (NO_3^-) to the underground water and rivers and emission of nitrous oxide (N_2O) and ammonia (NH_3). This became a major concern for scientists, environmental groups, farmers, and policymakers worldwide due to its potential for increasing the eutrophication of freshwater ecosystems, the acidification of agricultural soils, as well as its impact in greenhouse gas (GHG) emissions

(Liang et al., 2018). Further, N fertilization over-usage has detrimental effects on human health, mainly linked to toxic levels of NO_3^- either in the water or in the edible tissue of crop plants (Mosier et al., 2004; Ahmed et al., 2017). For that reason, the World Health Organization determined a maximum standard of 10 ppm NO_3^-–N for safe drinking water and the European Commission set maximum levels for NO_3^- in specific food products, such as vegetables, that vary depending on the season (Townsend et al., 2003; Anjana and Iqbal, 2007). Recently, the Green Deal has established that fertilizers application should be reduced by at least 20% by 2030 and it also proposes that all EU farmers would contribute in reducing ammonia and N_2O emissions from the agricultural sector and improve water quality in Europe (European Union, 2020). Therefore, it has never been more critical to increase the N use efficiency (NUE) of crops—the fraction of N input harvested as product—in order to assure environmental sustainability without compromising food security and human health (Gu et al., 2017; Kanter and Searchinger, 2018; Houlton et al., 2019).

In order to achieve the challenge of a balanced use of resources in agricultural production, for ensuring global food security and food safety while protecting the ecosystems, it is of utmost importance to understand how water and N fertilizers are related with climate change queues (Fig. 7.1). Therefore, in this chapter we will analyse (1) the impact of low water availability and quality and (2) the impact of wide-ranging N fertilization, on both crop yield and product quality. Finally, an overview on current and novel environmentally friendly strategies able to enhance water and NUE will be provided.

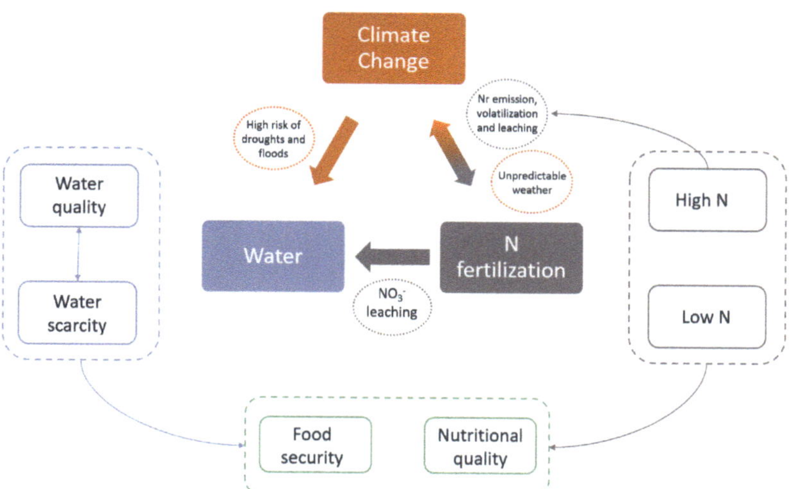

FIGURE 7.1 Schematic overview of water, N fertilization, and climate change interactions and their connection with food security and product quality.

7.2 Impact of climate change on water resources

According to the IPCC, the impact of climate change on water resources is mainly related to the intensity and frequency of the rainfall, as well as its seasonal and interannual variability and geographical distribution (Bates et al., 2008; IPCC, 2014; Wang et al., 2013). As a consequence, climate change is likely to alter the runoff of rivers, dramatically shifting between high flows due to changes in precipitation patterns and low flows due to more frequent droughts (Bates et al., 2008). Therefore, extreme events, such as heat waves, droughts, or floods, are likely to be more frequent and violent and to affect water resources both qualitatively and quantitatively (IPCC, 2007; IPCC, 2014). Additionally, the impact of climate change on sea level may also affect groundwater levels and quality, strongly influencing the availability of water resources (Bates et al., 2008).

Importantly, the impacts of climate change on water demand will be particularly relevant on agriculture, as intensification of evapotranspiration and/or reduction of soil moisture will negatively affect agronomical yields and product quality (Singh et al., 2014).

7.2.1 The effect of water scarcity on agricultural yield and product quality

Almost all regions of the world are expected to experience a net negative impact of climate change on water resources and freshwater ecosystems (IPCC, 2007). The IPCC anticipates that during this century atmospheric air temperature will increase from 2.2°C to 5.1°C and rainfall will decrease between 4% and 27% (IPCC, 2007). Further, according to FAO, for each 1°C of global warming, 7% of the global population will perceive a decrease of 20% or even more in renewable water resources (FAO, 2012). Nowadays, the world needs to deal with the biggest challenge of having more food to feed the increasing world population (Bogardi et al., 2012; FAO, 2015). Although irrigated agriculture has been central for food security, through the promotion of higher yields, an expansion of the irrigated area is at risk due to climate change–triggered water scarcity (Booker and Trees, 2020).

Drought is one of the most severe abiotic stresses limiting plant growth and development, consequently reducing crop yield, with tremendous economic and societal impacts (Farooq et al., 2009; Salehi-Lisar and Bakhshayeshan-Agdam, 2016). Drought impairs plant growth due to altered photosynthesis, respiration, nutrient uptake, nutrients and photoassimilates translocation, and carbohydrates and hormones regulation. Briefly, drought reduces plant water potential and turgor thus affecting leaf size, stem extension, and root proliferation. It causes the accumulation of abscisic acid and of compatible osmolytes such as proline (Sun et al., 2020). It also limits CO_2 assimilation (photosynthesis) rates and reduces leaf transpiration, which are mainly due to lower stomatal conductance, membrane damage, and disturbed activity of various enzymes (Osakabe et al., 2014). Enhanced metabolite flux through the photorespiratory pathway increases the oxidative load on plant tissues, which triggers growth inhibition (Farooq et al., 2009; Wallace et al., 2016) and finally programmed cell death (Gill and Tuteja, 2010).

Considering the magnitude of morphological and physiological processes affected by drought, this stress triggers dramatic crop yield losses worldwide (Siracusa et al., 2018). Between 2000 and 2007, approximately 3% of the global cereal production was lost to extreme drought disasters (Lesk et al., 2016). In barley, Samarah (2005) found that drought reduced the grain yield due to smaller numbers of tillers, spikes, and number of grains per plant.

Although several crops have been affected by drought, its effect on yield highly depends on numerous factors including crop development stage, vulnerability of each crop specie/cultivar to water stress, and even on the water availability to compensate for this stress (Pereira, 2007; Knox et al., 2016; Rey et al., 2017). Further, it also depends on drought severity and duration (Pereira, 2007).

In general, the impact of drought stress on plant growth is more pronounced at the early developmental stages and also very critical at the flowering stage (Gonzalez-Dugo et al., 2010). Accordingly, the International Maize and Wheat Improvement Center (CIMMYT) estimates that around 50% of yield losses in cereal crops worldwide are due to drought stress before flowering. Indeed, Gupta et al. (2001) observed 38% − 43% less yield when water stress was applied at boot stage and 39% − 51% when it was applied at the anthesis stage. Additionally, drought during grain filling, especially if accompanied by high temperatures, as is common, hastens leaf senescence, reduces the duration of grain filling, and reduces the mean weight of the grains (Austin, 1989). For barley, it was also revealed that water deficit prior to anthesis negatively impacts grain yield (Samarah, 2005). Regarding legumes, the impact of drought on yield is extremely complex, and associated with reduced germination and declined photosynthetic activity, decreased assimilates translocation and carbon fixation, and repressed flowering time and function of reproductive organs. This leads to pollen grain sterility, fewer pods and lower grain set, and declined sink activity, comprising thus a high number of processes such as fertilization, gametogenesis, embryogenesis, and grain formation (Farooq et al., 2017; Nadeem et al., 2019). For legumes, the impact of water stress on yield is also significantly dependent on the development stage of the plant life cycle (Samarah, 2005; Fang et al., 2010). For lentil, water deficit during early pod development led to 70% yield loss (Shrestha et al., 2006), while during the reproductive phase it decreased only 24% of the yield (Allahmoradi et al., 2013). For horticulture crops such as tomato, the flowering and fruit-setting stages are the most sensitive to water deficits (Chen et al., 2013; Khapte et al., 2019). These crops generally consist of more than 90% water, making drought highly impactful on their yields (Rao et al., 2016). Water deficit may lead to problems like abscission of floral organs, and consequently lower number of fruits, and/or lower fruit size and less marketable yield (Giardini et al., 1988; Rao et al., 2016).

Similarly to yield, the effect of drought on crop quality is highly dependent on the stress severity and plant development stage. Ripoll et al. (2014) and Stagnari et al. (2016) have reviewed the main effects of drought stress on several quality parameters of different crops. Drought stress can induce physiological disorders, namely fruit cracking (Srinivasa Rao et al., 2001) and blossom end rot (a physiological disorder caused by the limited capacity to internally translocate calcium towards the fruits, which is enhanced at lower transpiration rates particularly in horticultural crops such as tomato and sweet pepper) (Tabatabaie et al., 2004; Djangsou et al., 2019; Millones-Chanamé et al., 2019). However, several studies have demonstrated that water stress can actually be used to enhance product quality (Nora et al., 2012; Stagnari et al., 2016; Khapte et al., 2019) as shown in Fig. 7.2. Generally, a later and moderated stress is more effective on improving nutritional quality and can result in the amelioration of the water stress impact on yield, keeping the balance between yield and quality (Ozturk and Aydin, 2004; Wang and Frei, 2011; Hao et al., 2019). Thus, it is important to evaluate the timing and degree of water deficit application in order to compensate for the loss of yield with a higher quality product (Nora et al., 2012). Enhancement of nutritional quality and taste by moderate water stress results from the biosynthesis of several metabolites, such as amino acids, sugars, indoles, phenolics, and glucosinolates that are induced under water deficiency conditions, or by early leaf senescence that encourages remobilization

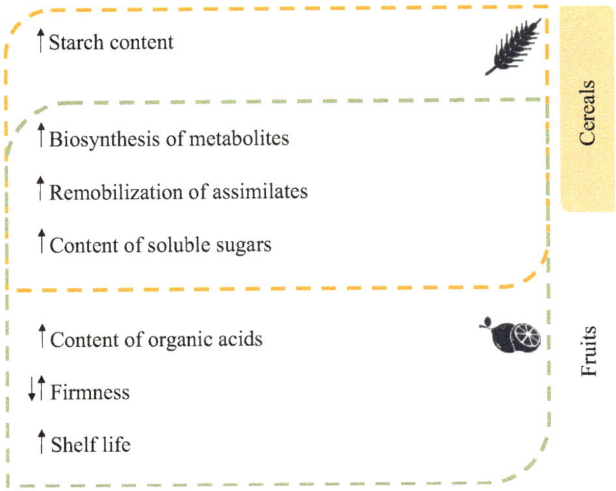

FIGURE 7.2 General impacts of water deficit on crop quality.

of assimilates to the grain, thus promoting the grain-filling (Yang et al., 2000; Stagnari et al., 2016). In grapes (Du et al., 2008), citrus fruits (Navarro et al., 2010), and peaches (Faci et al., 2014), the content of soluble sugars and other substances contributing to turgor maintenance was increased by water deficits. On apples (Mpelasoka et al., 2001) and tomatoes (Patanè et al., 2011) a more general increase of Total Soluble Solids (TSS; measured as degrees Brix) was reported. In many circumstances, higher levels of fruit TSS can be explained by the lower water content in fruits, although an increased active biosynthesis is often well defined (Stagnari et al., 2016). Soluble sugars and organic acids (primarily malic and citric acids) have an important role on osmotic potential and adjustment, particularly sucrose cleavage into two molecules that allows to double the osmotic contribution by facilitating water flux into fruit cells (Stagnari et al., 2016; Hou et al., 2020). The conversion of sucrose to starch lowers the amount of osmotically active solutes, but an induction of such process in tomatoes was observed upon water stress (Mitchell et al., 1991; Ripoll et al., 2016). One possible explanation for this occurrence is that it can help to maintain a sucrose concentration gradient between the source and sink facilitating the entrance of this sugar to the fruit (Mitchell et al., 1991; Hou et al., 2020). In cereals, starch content is an important grain quality factor greatly influenced by drought (Gous et al., 2013; Stagnari et al., 2016). Similarly to tomato fruits, in corn a higher sugar content and a lower protein content were observed upon water stress (Aydinsakir et al., 2013). On the contrary, in cereals like wheat (Ozturk and Aydin, 2004; Singh et al., 2008) and barley (Thitisaksakul et al., 2012; Gous et al., 2013), water stress tends to lower the starch amount, in part due to the degradation of the photosynthetic machinery. This in turn reduces carbon assimilation and increases protein content by a concentration effect and by a reduction in synthesis and storage of carbohydrates, allowing more N concentration per unit of starch to be accumulated in the grain (Ozturk and Aydin, 2004; Gous et al., 2013; Silva et al., 2020). Furthermore, although the negative effect on the photosynthetic machinery translates into a decrease in carbohydrate supply to the fruits or grains, it creates conditions that are conducive to photooxidative stress (Grassmann et al., 2002). Nonetheless, this often leads to an increase in the accumulation of antioxidant compounds that could also be seen as a beneficial effect of water deficit on the nutritional value of the fruits (Nora et al., 2012; Ripoll et al., 2014). A review from Wang and Frei (2011) highlighted some typical patterns in crop-quality response to several abiotic stressors and indicated a tendency toward an increase on crop nutritional value in response to water deficit.

Besides yield, chemical, and nutritional quality, the physical proprieties are also important quality features. Studies on cotton have demonstrated the impact of water stress on fiber quality traits (Pettigrew, 2004a,b). It was found that while irrigation promotes the vegetative growth and delays maturity in cotton, drought leads to the production of shorter and weaker fibers with low micronair value (Mert, 2005). Several descriptive studies have reported an increase of the firmness in pears (Lopez et al., 2011), tomatoes (Nangare et al., 2016; Cui et al., 2020), and peaches (Alcobendas et al., 2013). Ripoll et al. (2014) proposed that water deficit impacts firmness through its effect on cell size, cell turgor, solute transport, and through the accumulation of osmotically active solutes at the cell level. However, an absence of effect on firmness was also reported for apricots (Pérez-Pastor et al., 2007), kiwi fruits (Miller et al., 1998), strawberries (Weber et al., 2016), and apples (Hooijdonk et al., 2007), suggesting that differences may arise from different methods of texture evaluation (compression or puncture test), different stress intensities, and genotype interactions with the environment. Additionally, water stress seems to influence shelf life of crops, with several studies reporting an improved shelf-life quality in water deficit treatments (Zaicovski et al., 2008; Peña-Estévez et al., 2014; Hachmann et al., 2019). This improvement seems to be dose-dependent with severe water stress resulting in longer shelf lives compared to moderate water deficit (Zaicovski et al., 2008; Hachmann et al., 2019). These observations were explained by an increase in cytokinin content, decrease of stem-end rot incidence, and by prevention of the growth of the majority of spoilage bacteria (Gould, 1996; Zaicovski et al., 2008).

Overall, several studies have demonstrated that a moderate water deficit produced the best yield-quality relation. However, further studies are still required to understand how the timing and magnitude of water stress can enhance crop yield and quality, supporting farmers in their goals of more targeted water management.

7.2.2 Effect of poor water quality on yield and product quality

Besides water availability, climate change also directly affects the quality of water resources (Wilk and Wittgren, 2009). Indeed, the increase in water temperature and in precipitation intensity in short periods of time leading to floods, as well as more frequent and longer droughts, are strongly believed to exacerbate many forms of water pollution (IPCC, 2007). The effects of climate change on water quality can include decreased dissolved oxygen levels, enhanced contaminant load to water bodies, and reduced stream and river flows, which can result in higher mineralization of organic N in soil, heightened harmful algal blooms (HABs), and increased likelihood of saltwater intrusion near coastal regions (Singh et al., 2014). All these sources of contamination can harm aquatic ecosystems and make water unsuitable for agronomical use.

Excessive nutrients can impair crop growth and development through induction of osmotic stress, ion toxicity, and nutrients imbalance and can reduce glucose, fructose, starch, total carbohydrates, and proteins content, directly affecting crop yields and quality. Furthermore, high NO_3^- concentration in leafy vegetables can result in human health hazards (Fallovo et al., 2009). Associated with eutrophication, outbreaks of HABs have become a major environmental threat worldwide as they may contain strains that are able to produce secondary metabolites with a demonstrated toxic activity, thus limiting the use of available water resources and representing a risk to human health (Machado et al., 2017a,b). Cyanobacterial blooms are one of the principal HABs affecting freshwaters, and they may benefit from temperature increase and more severe drought scenarios; this is of great concern due to their ability to produce toxins (cyanotoxins) that can accumulate in the food chain and contaminate drinking water (Machado et al., 2018). In addition, water contaminated with these toxins is often used in agriculture and may represent a risk to product yield and safety. Indeed, recent studies have suggested that cyanotoxins can cause toxic effects on terrestrial plants by inducing morphological and physiological alterations, and consequently by causing a putative loss of productivity due to the inhibition of germination, growth, and development as well as a loss of quality of crop food products. In addition, it may pose concerns for food safety if the possible absorption of toxins by plants lead to its bioaccumulation on edible tissues, as already demonstrated (Corbel et al., 2014; Machado et al., 2017a,b).

Pesticides became an important component of worldwide agriculture systems during the last century as farmers use them to increase crop yield (Sulaiman et al., 2019). However, their use can lead to long-term hazards for the environment, biodiversity, and food safety because of their inherent toxicity. In developed countries pesticide legislation and proper maximum residue limits are well established, but in developing countries highly toxic pesticides are still used because legislation is inadequate or poorly enforced (Handford et al., 2015). In general, it has been estimated that only about 0.1% of pesticides reach the target organisms and the remaining bulk contaminates the surrounding environment (Carriger et al., 2006). Although soil is the main reservoir of pesticides, water from excessive rainfall and irrigation cannot always be held within the soil structure meaning that pesticides and residues can be quickly transported, thus contaminating groundwater and freshwater supplies (Carvalho, 2017). Similarly, heavy metal pollution of surface and underground water sources results in considerable soil pollution and is increasing at an alarming rate.

Pollution of water by heavy metals is of great concern due to the potential risk to human health, through direct intake or bioaccumulation through the food chains, and its impact on ecological systems (Srivastava et al., 2017). Although plants require certain metals for their growth and upkeep, excessive amounts of heavy metals can become toxic (Nagajyoti et al., 2010). Additionally, the ability of plants to accumulate essential metals equally enables them to acquire other nonessential metals (Djingova and Kuleff, 2000). As metals cannot be broken down, when concentrations within the plant exceed optimal levels they adversely affect the plant and may pose concerns for food safety. The routes and mechanisms by which heavy metals pose a risk to human health through the consumption of grains and vegetables were previously reviewed (Peralta-Videa et al., 2009; Rai et al., 2019).

The increased salinization of water is also of great concern for the agricultural sector, since this is an important abiotic stress for most crops, resulting in an estimated economic impact surpassing 22€ billion per year concerning salinization of irrigated land (Zörb et al., 2019). Besides a direct effect of salinity on crop growth and development, salinization also affects the soil structure, reducing water infiltration, soil aeration, and consequently oxygen supply to the roots, as well as increasing osmotic potential that restricts water absorption from the roots (Levy and Veilleux, 2007). Further, the direct effect of salinity in plant growth emerges from the disruption of ionic equilibrium and elicitation of sodium toxicity that results in deleterious effects on the function of important enzymes (Agarwal et al., 2012). Interestingly, salt tolerance appears to be highly dependent on a crop's species, cultivar, developmental stage, and growth conditions and is based on the ability of the crop to extract water and nutrients from saline soils as well as to avert high tissue accumulation of salt ions (Katerji et al., 2000; del Amor et al., 2001; Zörb et al., 2019). Several crops such as rice, sweet potato, durum wheat, and maize have shown high sensitivity to salinity stress (Munns and Tester, 2008; Zörb et al., 2019), whereas other crops like alfalfa, barley, cotton, and sugarbeet are known to be more tolerant to saline environments (Skaggs et al., 1999; Ashraf and Harris, 2004; Munns and Tester, 2008; Hossain et al., 2017; Zhu et al., 2020). Further, high salinity levels in the early stages of plant development (e.g., in plantlets) seems to strongly reduce plant growth and yield (Zörb et al., 2019).

As access to noncontaminated water for proper agricultural practices is now extremely difficult (Cross et al., 2009), efforts to find new and effective water quality management strategies are of utmost urgency. A large amount of freshwater is required in agriculture, but in turn, agriculture is also a massive polluter of water resources (FAO, 2015). Thus, the implementation of novel, sustainable, and cost-effective technologies is eminently necessary in order to decrease the leakage of contaminants to water resources.

7.3 Nitrogen fertilization management in light of climate change

N is the main nutrient required for plant growth and its exogenous application in modern agriculture has allowed, in the past century, a great increase in crop yield. This has ensured the farmers an adequate economic return and has helped to guaranteed the fulfillment of an increased demand for plant food products (Lawlor et al., 2001; Ahmed et al., 2017).

Reactive N (Nr) is essential for all life forms, being involved in the foundation of essential molecules, like enzymes and DNA, and acting as a key biological signaling molecule (UNEP, 2019). However, the massive scale-up of N fertilizer use in agriculture resulted in great losses of Nr to the environment, mainly through NO_3^- leaching, N oxides (NOx) and NH_3 volatilization, and N_2O emission, altering severely the global N cycle. This has generated serious threats to ecosystem health and biodiversity, at local and global scales, and implications in all aspects of climate change (Suddick et al., 2013; Basso et al., 2016; Zhang, 2017; UNEP, 2019). From all Nr losses, N_2O emissions are considered the most direct way of aggravating climate change problems through this agricultural practice (Suddick et al., 2013; Kanter, 2018; Houser and Stuart, 2019). N fertilization alone accounts for 70% of the N_2O emissions from anthropogenic activities (Reay et al., 2012; Skiba and Rees, 2014; Pauleta et al., 2019). N_2O is an important GHG and is considered to have 300 times the global warming impact of CO_2 because of its capacity to absorb infrared radiation and long half-life (more than 100 years) in the atmosphere (Choudhury and Kennedy, 2005; Suddick et al., 2013; Pauleta et al., 2019). Also, N_2O is an ozone-depleting substance when it reaches the stratosphere (Choudhury and Kennedy, 2005; Pauleta et al., 2019). Besides N_2O, N cycling also alters, in a complex way, the atmospheric concentration of two other highly important GHGs, methane and CO_2. On the one hand, climate change variables will alter the rate and timing of N losses to the environment, weakening the ability of farmers to meet crop needs with appropriate N fertilizer applications (Suddick et al., 2013). Furthermore, high N levels can ultimately have a negative impact on plant growth, yield, and quality (Goyal and Huffaker, 1984; Sánchez et al., 2002).

Considering the relevance of proper N fertilization management, improving NUE in agricultural production, by diminishing Nr losses to the environment and optimizing plant yield and quality, is fundamental to face the challenges of guaranteeing food security, avoiding environmental degradation, and reducing climate change (Zhang et al., 2015). For that, is of great significance to understand the current negative impacts of excessive N fertilization as well as the disadvantages of low N supply on crop yield and product quality, as summarized in this sub-chapter.

7.3.1 Effects of low nitrogen levels on yield and nutritional quality

N supply is crucial for plant development and growth mainly because N is required for protein synthesis, affecting direct or indirectly almost all processes of the cell, such as the production of metabolites, carbohydrates, fats, pigments, and many others (Lawlor et al., 2001). Therefore, N-assimilates are transported to the developing plant tissues and are responsible for the synthesis of new vegetative cells and tissues that give rise to reproductive organs, seeds and yield components (Lawlor, 2002).

As a consequence of the essential role of N for cellular function, N-starved plants often show stunted growth mainly related to photosynthesis impairment (Lemaire et al., 2008; Jangam and Raghuram, 2015). Indeed, changes on photosynthetic capacity upon N deficiency have been reported for several crops, such as corn, maize, rice, tomato, sugarcane, sorghum, mangrove species, and many others (Kao et al., 2001; Tei et al., 2002; Zhao et al., 2003; Zhao et al., 2005; Wei et al., 2016; Bassi et al., 2018; Xiong et al., 2018). N deficiency can also indirectly limit cell expansion and elongation by the deposition of unutilized carbohydrates on cell walls, as fewer proteins are produced (Okunlola et al., 2015). Phytohormones (auxins, cytokinin, ethylene, gibberellic acid, and abscisic acid) are also known to control plant growth and development from embryogenesis to reproduction, playing a major role on crop yield (Zaman et al., 2014; Chandran et al., 2019). All these phytohormones are related to N availability, especially abscisic acid, auxins, and cytokinins, which are closely linked to N signaling (Pavlíková et al., 2012; Zaman et al., 2014). N deficiency or starvation often leads to an increase in the growth-inhibitor hormones (abscisic acid and ethylene) and a decrease in the growth-promoting hormones (auxins, cytokinins, and gibberellic acid), resulting in impaired development of the reproduction system (Gawronska et al., 2003; Zaman et al., 2014; Khan et al., 2015; Xiong et al., 2018; Chandran et al., 2019).

The impact of N deficiency on yield is highly dependent on the crop and its developmental stage (Lawlor et al., 2001; Fageria, 2003; Xiong et al., 2018). Studies in hybrid rice demonstrated that absence of N fertilization throughout the cultivation cycle led to more than 50% yield losses, when compared with plants with constant supply of N during the different stages of growth (Xiong et al., 2018). Interestingly, authors have also found that N deficiency had little impact on yield when occurring during the seedling and heading stages, while at the tillering and young panicle differentiation stages, the yield per plant was significantly reduced (Xiong et al., 2018). In *Capsicum chinense*, no N supplementation led to lower numbers of flowers (less than 50%), and number of fruits (less than 78%) and consequently lower yield (Medina-Lara et al., 2008). Similarly, in cotton a reduction on N supplementation from 200 to 80 kg/ha resulted in a reduction of about 1000 kg/ha on seed yield (Singh et al., 2010). In spinach, the absence of N fertilization greatly reduced shoot fresh and dry weights (by 85% and 79%, respectively), specific leaf area, chlorophyll content, and photochemical efficiency and yield (Xu and Mou, 2016). Further, in three different potato cultivars (Yanshu 4, Xiabodi, and Chunshu), N depletion triggered a reduction of about 16% in the fresh tuber weight (Zhang et al., 2020a).

N availability can also directly or indirectly affect almost all crop quality parameters, including physical and chemical properties, such as external appearance (size and color of the product), texture, flavor, and nutritional composition (including carbohydrates, amino acids, proteins, lipids, vitamins, and minerals). Not surprisingly, in cereals like rice, wheat, and corn, an increase in N supply is usually accompanied by an increase in grain protein and kernel quality and strength (Blumenthal et al., 2008; Maheswari et al., 2017). Reduced N levels usually result in lower yield and size of deciduous fruits and vegetables but during fruit formation can also be associated with better fruit color (Goyal and Huffaker, 1984). Further, fruit quality of tomato was improved due to lower acid (10% − 16%) and increased soluble sugar content (5% − 17%), and the content of phenolic compounds and total vitamin C also tended to be higher in fruits with the lowest N supply (Bénard et al., 2009). However, a review by Maheswari et al. (2017) reported a general increase in vegetable quality with N fertilization for plants like tomato and zucchini, with higher fruit firmness and high TSS for tomato and higher dry matter for zucchini, while for cabbage and lettuce, the crispness of heads was reduced at high N levels. In cucumber fruit, ascorbate might be accumulated under N deficiency, but the content of TSS was reduced by increasing glucose consumption and reducing photosynthesis (Hao et al., 2021).

7.3.2 Effects of high nitrogen levels on the environment, yield, and nutritional quality

Agriculture is the main sector responsible for the increase of Nr global rate input to terrestrial ecosystems through the production of synthetic N fertilizers and their inefficient management (Howarth, 2008). Importantly, a report by Zhang (2017) noticed that only about 50% of the N supplemented to croplands is harvested as crop product, with a huge portion lost to the environment. Concerning greenhouse vegetable production specifically, an alarming meta-analysis by Zhao et al. (2019) reported that adding excessive N dramatically affects the environment without significantly improving the crop yield. In particular, for a yield gain of 36%, there is an aggravation of 217% in NO_3^- leaching, 202% in N_2O emission, 176% in NH_3 emission, and 543% in nitric oxide (NO) emission, when compared with a control group where no fertilizer was applied. Relatedly, a reduction of 20%−40% of N fertilization in greenhouse vegetable production decreases NO_3^- leaching (32.4%), total N leaching (37.3%), NH_3 emission (28.4%), N_2O emission (38.6%), and NO emission (8.0%), while having no significant effect on yield.

High levels of N supply are also known for extending the vegetative phase and delaying the reproductive development, affecting the yield (Goyal and Huffaker, 1984). In French bean, increasing the N supply from 5.8 mM to 23.2 mM resulted in a drastic decrease in pod and seed production (92% and 90%, respectively) (Sánchez et al., 2002). Further, in wheat, fertilizations above 240 kg/ha significantly reduced grain yield (Wang et al., 2011). In cereals, high N rates may also cause the plants to grow unusually tall, increasing the danger of grain loss through lodging (Goyal and Huffaker, 1984).

Although it usually increases the protein content in cereals, excessive N can result in lower protein biological quality by a decrease of the relative proportion of lysine and threonine (Blumenthal et al., 2008). Also, a lower content in sugars with higher N supplies was observed in sugarbeet (Goyal and Huffaker, 1984). In oil seeds, an increase in N supply, and consequentially in protein content, resulted in decreased oil content (Maheswari et al., 2017). Further, excessive N supply often leads to lower vitamin C values and shorter shelf lives (Goyal and Huffaker, 1984; Maheswari et al., 2017). In fruits, studies on apple demonstrated that high N-fertilizer levels resulted in lower TSS and negatively affected fruit color, texture, and firmness (Nava et al., 2008; Sharma, 2016). The increase in fruit color in N-deficiency treatments may be due to the accumulation of unutilized carbohydrates that can be used for anthocyanin synthesis (Nava et al., 2008). On potato, low sucrose content is relatable to tuber maturity. In a study by Zhang et al. (2020b), sucrose content increased with excessive N (315 kg N/ha) demonstrating that high N dosages can delay potato maturity.

Altogether, an excess of N fertilization in agriculture is producing a cocktail of Nr that endangers ecosystems and climate, which turns N pollution into a major environmental issue, often not even justified by higher crop productivity and quality. Notwithstanding, the true scale of this problem is still largely unknown, and the solution will definitely require an integral, multidisciplinary, and complex approach (Erisman et al., 2011; UNEP, 2019).

7.4 Overview on strategies to enhance water use efficiency and N use efficiency and future prospects

Improving WUE and NUE in agronomical crops is highly important to guarantee environmental, economic, and social sustainability. Climate change has prompted many studies to seek strategies to overcome this major challenge. These strategies include both biotechnological and agronomic approaches, namely: (1) plant breeding for increased water and N stress resilience, (2) use of novel and sustainable products and techniques (e.g. application of biostimulants, elicitors, and seed priming nanofertilizers), and (3) agronomic practices such as plant grafting, water management strategies (deficit irrigation), and recycling of nutrient solutions in soilless cultivation.

Concerning plant breeding, as revised by Singhal et al. (2016), conventional plant breeding methods (involving interspecific or intergeneric hybridization and *in vitro*–induced variation) have been applied to improve the abiotic stress tolerance of various crops, but their success has been limited. The bottlenecks that have been identified are related to the complexity of the quantitative traits, low genetic variation of yield components under stress conditions, and lack of efficient selection criteria (Singhal et al., 2016). On the other hand, the combination of several high-throughput molecular techniques, such as transcriptomics, proteomics, and metabolomics, are being widely used as they have potential to provide a system-based approach capable of elucidating the complex biological processes (spanning from gene expression to metabolite accumulation and phenotype translation) involved in plant responses to abiotic stress (Jain et al., 2019). Traits identified with these techniques can be used as investigation tools to monitor plant status and to identify candidate genes for WUE and NUE improvement.

Novel studies have increasingly focused on biostimulants applications (growth-promoting bacteria, seaweed extracts, humic and fulvic acids, and protein hydrolysates) as they are able to enhance nutrient uptake, stimulate growth, and enhance stress tolerance or crop quality, thus allowing a more rational use of mineral fertilizers, which has important implications on environmental protection (Le Mire et al., 2016; Van Oosten et al., 2017; Rouphael and Colla, 2020; Del Buono, 2020). Although the physiological effects of some biostimulants have been partly documented (Conrath, 2011; Savvides et al., 2016), the information on the molecular basis regarding their mode of action at the plant level is still very limited (Fleming et al., 2019; De Saeger et al., 2020; Nephali et al., 2020). Moreover, plant elicitors (e.g., abscisic acid, salicylic acid, methyl jasmonate, stragloctones, silicon, and NO) are hazard-free compounds capable of inducing defense responses in plants against various abiotic stresses, which play a vital role in helping the plant to mitigate the stress-induced changes by modulating an array of processes at the cellular and molecular levels (Vasconcelos, 2014; Ahmad et al., 2019; Chakraborty et al., 2019; Venegas-Molina et al., 2020; Bokor et al., 2021). The exogenous application of elicitors before stress events has been shown to result in the modulation of gene expression to reduce stress impacts on plant development and growth and several phytohormones can be used as a specific metabolic target (Samota et al., 2017).

Seed priming is another environmentally friendly technique recommended by many researchers for better crop stand establishment and growth under adverse conditions. Seed priming (e.g., hydropriming, osmopriming, chemical priming, hormonal priming, biological priming, and redox priming) is defined as a physiological process by which a plant prepares to respond to imminent abiotic stress more quickly or aggressively and is induced by the treatment with natural or synthetic compounds before seed germination (Jisha et al., 2013; Bhanuprakash and Yogeesha, 2016; Rhaman et al., 2020). Plants raised from primed seeds showed sturdy and quick cellular defense response against abiotic stresses (Bhanuprakash and Yogeesha, 2016; Choudhary et al., 2019; Marthandan et al., 2020).

Nanotechnology appeared as a great promise of an innovative tool for delivering agrochemicals in a more efficient manner (Fraceto et al., 2016; Shang et al., 2019; Jain and Das, 2021). Nanofertilizers' use offers benefits in nutrition management through its strong potential to increase NUE. Furthermore, nanofertilizers may also be used for enhancing abiotic stress tolerance and used in combination with microorganisms (the so-called nanobiofertilizers) to provide great additional benefits (Raliya et al., 2017; Shang et al., 2019; Zulfiqar et al., 2019; Pirzadah et al., 2020). The use of these products minimizes nutrient leaching into groundwater, but the risks of nanofertilizer application should be carefully examined before being generalized. Therefore, further biotechnological advances are required for a precise and safe application of nanomaterials in agriculture (Kah, 2015; Fraceto et al., 2016; Zulfiqar et al., 2019; Zhang et al., 2020c; Jain and Das, 2021).

Another promising strategy for water and N stress mitigation is the targeted grafting, which provides opportunities to exploit natural genetic variability that exists for interesting root-specific traits (Kumar et al., 2017; Gaion et al., 2018; Mozafarian Meimandi and Kappel, 2020; Zhang et al., 2020c). In this sense, grafting could be used as an additional tool to counteract environmental stresses and increase tolerance by selecting the appropriate rootstock that will manipulate scion morphology through regulation of phytohormones (Kumar et al., 2017; Nanda and Melnyk, 2018). Thus, grafting might offer a unique alternative to conventional and transgenic breeding for compiling tolerance traits, improving yield potential and stability, and even the product quality (Rouphael et al., 2018).

With respect to the use of water in agriculture, higher WUE has been obtained either by increasing irrigation precision (e.g., use of drip irrigation), thus allowing to diminish water losses through evaporation, or by improving plant physiological responses through applying, for example, deficit irrigation. This technique, which consists of supplying water below the evapotranspiration requirements of plants (e.g., regulated deficit irrigation or partial root drying), is an example of a well-recognized water-saving strategy to maximize water productivity and to stabilize, rather than maximize, yields while conserving irrigation water (Costa et al., 2007; Ćosić et al., 2015). However, there is little information and poor understanding about the level of deficit irrigation that could be used without compromising yield, since the overall effect of deficit irrigation highly depends on the crop species and certain specifications concerning the irrigation strategy adopted (Khapte et al., 2019). Nonetheless, the use of this technique has often demonstrated the potential for fruit quality improvement (mostly due to higher TSS content), reduction of production costs, and minimization of groundwater pollution owing to reduced leaching of agrochemicals (Kannan and Anandhi, 2020).

Concerning soilless cultivation systems, a highly relevant aspect that would strongly contribute to enhanced WUE and NUE is the generalized use of closed production systems (i.e., recycling of drained nutrient solutions, often following desinfection). Whereas these recirculating systems are widely implemented in several central and northern European countries (van der Salm et al., 2020), they have been poorly implemented in regions with lesser availability of good-quality irrigation water and/or in countries with lower economic availability for the required investment (Massa et al., 2010; Gianquinto et al., 2013; Santos et al., 2017). This is the case of the Mediterranean region, where prevail production systems with free-drainage or those in which drainages are applied to soil-grown crops (semi-open systems) (Massa et al., 2020). These nonrecirculating fertigation strategies are currently seen as obsolete, for their low use efficiency of resources and environmental degradation resulting from soil salinization and emission of contaminants (particularly N, phosphorus, and plant protection product residues) to natural compartments (Kwon et al., 2021). Hence, a large margin of progression exists, regarding WUE and NUE in soilless cultivation, related to the potential of a generalized implementation of closed production systems, which would strongly decrease the water footprint and the soil and water contamination in horticultural production. To this end the implementation of strict regulations for soil and water preservation at national level would be highly relevant.

Although a considerable amount of literature concerning strategies that seek to achieve high WUE and NUE exists, some knowledge gaps are clear. Research on most of these strategies has been focused on plant performance improvement under a single stress, however, under the ligh of the climate change it urges to understand the benefits of their application when a combination of stresses occurs. Several studies revealed that plants respond very differently when a combination of stresses (biotic and/or abiotic) occurs, in comparison with each one alone, by activating an entirely new program of gene expression. The same is valid for the effect of simultaneous N and water stresses on crop performance, particularly in terms of yield and quality. A comprehensive and integrated analysis will be absolutely essential to understand the complexity of this phenomenon and will be crucial to create molecular target breeding programs, resulting in more tolerant cultivars with increased WUE and NUE. A better understanding of the metabolic events and signaling crosstalk underlying the application of BSs, elicitors, and seed priming will also require further investigation. Many aspects related to rootstock/scion interactions are poorly understood, making it more difficult to overcome yield losses and decreased product quality resulting from unsuccessful grafting. Thus, further research is required before grafting can be recommended. Additionaly, the literature suggests that it is possible to manage water stress by applying water-saving irrigation strategies, continuing to explore and improve the use of deficit irrigation techniques, namely the timing and magnitude of water stress, especially when integrated with other novel sustainable agronomical tools (e.g., grafting, elicitors, BSs, and other promoting growth substances). These efforts will support farmers in the quest for more targeted water management, which integrated with N-saving strategies, will be of utmost importance.

7.5 Conclusions

Reducing the use of water and N without affecting agricultural production remains a challenging task. However, the overuse of these resources is aggravating climate change, which is strongly compromising food security, safety, and product quality. The effects of climate change on water resources are becoming more difficult to predict, with higher frequency of extreme rainfall events (droughts and floods) threatening water availability, quality, and, consequently, food security and human health. Further, climate change queues are being strongly affected by the excess of N that is not absorbed by crops (more than 50%), being often leached to deep soil layers, or volatilized to the atmosphere. Therefore, loss of N not only increases the cost of agricultural production, but it also impacts climate change, the environment, biodiversity, and human health. Overall, the studies presented in this chapter strongly reinforce that water (quantity and quality) and N inputs have huge impacts on crop yield and product quality, but this is highly dependent on many factors, such as the crop needs, the development stage, and the severity and duration of the stresses. Therefore, to define reliable and environmental-friendly strategies able to enhance the efficient use of these limiting resources is of utmost importance. In recent years a huge effort has been made in that sense, which includes both biotechnological and agronomic approaches. However, so far, the crosstalk between both abiotic stresses has not been deeply explored. Understanding the complex interaction between water and N stresses will definitely be a step forward to adapt and combine strategies aiming to enhance environmental and economic sustainability in agricultural production.

Acknowledgments

The authors would like to thank Fundação para a Ciência e a Tecnologia (FCT) for JM and APGF the PhD scholarships (SFRH/BD/116147/2016 and UI/BD/151039/2021, respectively). This research was also supported by national funds through FCT within the scope of UIDB/05748/2020, UIDP/05748/2020, UIDB/50016/2020.

References

Agarwal, P.K., Shukla, P.S., Gupta, K., Jha, B., 2012. Bioengineering for salinity tolerance in plants: state of the art. Mol. Biotechnol. 54 (1), 102–123. Available from: https://doi.org/10.1007/s12033-012-9538-3.

Ahmad, B., Zaid, A., Sadiq, Y., Bashir, S., Wani, S.H., 2019. Role of selective exogenous elicitors in plant responses to abiotic stress tolerance. In: Hasanuzzaman, M., Hakeem, K., Nahar, K., Alharby, H. (Eds.), Plant Abiotic Stress Tolerance. Springer Cham. Available from: https://doi.org/10.1007/978-3-030-06118-0_12.

Ahmed, M., Rauf, M., Mukhtar, Z., Saeed, N.A., 2017. Excessive use of nitrogenous fertilizers: an unawareness causing serious threats to environment and human health. Environ. Sci. Pollut. Res. 24 (35), 26983–26987. Available from: https://doi.org/10.1007/s11356-017-0589-7.

Alcobendas, R., Mirás-Avalos, J.M., Alarcón, J.J., Nicolás, E., 2013. Effects of irrigation and fruit position on size, colour, firmness and sugar contents of fruits in a mid-late maturing peach cultivar. Sci. Hortic. 164, 340–347. Available from: https://doi.org/10.1016/j.scienta.2013.09.048.

Allahmoradi, P., Mansourifar, C., Saiedi, M., Jalali Honarmand, S., 2013. Effect of different water deficiency levels on some antioxidants at different growth stages of lentil (*Lens culinaris* L.). Adv. Environ. Biol. 7, 535–543.

Anjana, S.U., Iqbal, M., 2007. Nitrate accumulation in plants, factors affecting the process, and human health implications. A review. Agron. Sustainable Dev. 27 (1), 45–57. Available from: https://doi.org/10.1051/agro:2006021.

Ashraf, M., Harris, P.J.C., 2004. Potential biochemical indicators of salinity tolerance in plants. Plant Sci. 166 (1), 3–16. Available from: https://doi.org/10.1016/j.plantsci.2003.10.024.

Austin, R.B., 1989. Maximizing crop production in water-limited environments. In: Baker, F.W. G. (Ed.), Drought Resistance in Cereals. ICSU Press, pp. 13–25.

Aydinsakir, K., Erdal, S., Buyuktas, D., Bastug, R., Toker, R., 2013. The influence of regular deficit irrigation applications on water use, yield, and quality components of two corn (*Zea mays* L.) genotypes. Agric. Water Manage. 128, 65–71. Available from: https://doi.org/10.1016/j.agwat.2013.06.013.

Bassi, D., Menossi, M., Mattiello, L., 2018. Nitrogen supply influences photosynthesis establishment along the sugarcane leaf. Sci. Rep. 8, 2327. Available from: https://doi.org/10.1038/s41598-018-20653-1.

Basso, B., Dumont, B., Cammarano, D., Pezzuolo, A., Marinello, F., Sartori, L., 2016. Environmental and economic benefits of variable rate nitrogen fertilization in a nitrate vulnerable zone. Sci. Total Environ. 545-546, 227–235. Available from: https://doi.org/10.1016/j.scitotenv.2015.12.104.

Bates, B.C., Kundzewicz, Z.W., Wu, S., Palutikof, J.P. (Eds.), (2008). Climate change and water. Technical Paper, International Panel on Climate Change (IPCC) Secretariat, Geneva, p. 210.

Bénard, C., Gautier, H., Bourgaud, F., Grasselly, D., Navez, B., Caris-Veyrat, C., et al., 2009. Effects of low nitrogen supply on tomato (*Solanum lycopersicum*) fruit yield and quality with special emphasis on sugars, acids, ascorbate, carotenoids, and phenolic compounds. J. Agric. Food Chem. 57 (10), 4112–4123. Available from: https://doi.org/10.1021/jf8036374.

Bhanuprakash, K., Yogeesha, H.S., 2016. Seed priming for abiotic stress tolerance: an overview. In: Rao, N., Shivashankara, K., Laxman, R. (Eds.), Abiotic Stress Physiology of Horticultural Crops. Springer, New Delhi. Available from: https://doi.org/10.1007/978-81-322-2725-0_6.

Bita, C.E., Gerats, T., 2013. Plant tolerance to high temperature in a changing environment: scientific fundamentals and production of heat stress-tolerant crops. Front. Plant Sci. 4, 273. Available from: https://doi.org/10.3389/fpls.2013.00273.

Blumenthal, J.M., Baltensperger, D.D., Cassman, K.G., Mason, S.C., Pavlista, A.D., 2008. Importance and effect of nitrogen on crop quality and health. Nitrogen Environ. 51–70. Available from: https://doi.org/10.1016/b978-0-12-374347-3.00003-2.

Bogardi, J.J., Dudgeon, D., Lawford, R., Flinkerbusch, E., Meyn, A., Pahl-Wostl, C., 2012. Water security for a planet under pressure: interconnected challenges of a changing world call for sustainable solutions. Curr. Opin. Environ. Sustainability 4 (1), 35–43. Available from: https://doi.org/10.1016/j.cosust.2011.12.002.

Bokor, B., Santos, C.S., Kostoláni, D., Machado, J., Nunes da Silva, M., Carvalho, S.M.P., et al., 2021. Mitigation of climate change and environmental hazards in plants: potential role of beneficial metalloid silicon. Journal of Hazardous Materials 416, 126193. Available from: https://doi.org/10.1016/j.jhazmat.2021.126193.

Booker, J.F., Trees, W.S., 2020. Implications of water scarcity for water productivity and farm labor. Water 12, 308. Available from: https://doi.org/10.3390/w12010308.

Carriger, J.F., Rand, G.M., Gardinali, P.R., Perry, W.B., Tompkins, M.S., Fernandez, A.M., 2006. Pesticides of potential ecological concern in sediment from south Florida canals: an ecological risk prioritization for aquatic arthropods. Soil Sediment Contam. 15, 21–45. Available from: https://doi.org/10.1080/15320380500363095.

Carvalho, F.P., 2017. Pesticides, environment, and food safety. Food Energy Secur. 6 (2), 48–60. Available from: https://doi.org/10.1002/fes3.108.

Chakraborty, N., Sarkar, A., Acharya, K., 2019. Elicitor-mediated amelioration of abiotic stress. Plants Molecular Plant Abiotic Stress: Biology and Biotechnology. Wiley, pp. 105–122. Available from: https://onlinelibrary.wiley.com/doi/abs/10.1002/9781119463665.ch6.

Chandran, V., Shahena, S., Rajan, M., Mathew, L., 2019. Controlled release of plant hormones for modifying crop yield. Controlled Release of Pesticides for Sustainable Agriculture. Springer, pp. 253–266. Available from: https://doi.org/10.1007/978-3-030-23396-9_11.

Chen, J., Kang, S., Du, T., Qiu, R., Guo, P., Chen, R., 2013. Quantitative response of greenhouse tomato yield and quality to water deficit at different growth stages. Agric. Water Manage. 129, 152–162. Available from: https://doi.org/10.1016/j.agwat.2013.07.011.

Choudhury, A.T.M.A., Kennedy, I.R., 2005. Nitrogen fertilizer losses from rice soils and control of environmental pollution problems. Commun. Soil Sci. Plant Anal. 36 (11-12), 1625–1639. Available from: https://doi.org/10.1081/CSS-200059104.

Choudhary, V.K., Chander, S., Chethan, C.R., Kumar, B., 2019. Effect of seed priming on abiotic stress tolerance in plants. In: Hasanuzzaman, M., Fujita, M., Oku, H., Islam, M.T. (Eds.), Plant Tolerance to Environmental Stress: Role of Phytoprotectants. CRC Press, Florida, pp. 29–40.

Conrath, U., 2011. Molecular aspects of defence priming. Trends Plant Sci. 16 (10), 524–531. Available from: https://doi.org/10.1016/j.tplants.2011.06.004.

Corbel, S., Mougin, C., Bouaïcha, N., 2014. Cyanobacterial toxins: modes of actions, fate in aquatic and soil ecosystems, phytotoxicity and bioaccumulation in agricultural crops. Chemosphere 96, 1–15. Available from: https://doi.org/10.1016/j.chemosphere.2013.07.056.

Ćosić, M., Djurović, N., Todorović, M., Maletić, R., Zečević, B., Stričević, R., 2015. Effect of irrigation regime and application of kaolin on yield, quality and water use efficiency of sweet pepper. Agric. Water Manage. 159, 139–147. Available from: https://doi.org/10.1016/j.agwat.2015.05.014.

Costa, J.M., Ortuño, M.F., Chaves, M.M., 2007. Deficit irrigation as a strategy to save water: physiology and potential application to horticulture. J. Integr. Plant Biol. 49 (10), 1421–1434. Available from: https://doi.org/10.1111/j.1672-9072.2007.00556.x.

Cross, K.M., Lu, Y., Zheng, T., Zhan, J., McPherson, G., John, V., 2009. Water decontamination using iron and iron oxide nanoparticles. Nanotechnology Applications for Clean Water. Elsevier, pp. 347–364. Available from: https://doi.org/10.1016/b978-0-8155-1578-4.50033-0.

Cui, J., Shao, G., Lu, J., Keabetswe, L., Hoogenboom, G., 2020. Yield, quality and drought sensitivity of tomato to water deficit during different growth stages. Sci. Agric. 77 (2). Available from: https://doi.org/10.1590/1678-992x-2018-0390.

De Saeger, J., Van Praet, S., Vereecke, D., Park, J., Jacques, S., Han, T., et al., 2020. Toward the molecular understanding of the action mechanism of *Ascophyllum nodosum* extracts on plants. J. Appl. Phycol. 32, 573–597. Available from: https://doi.org/10.1007/s10811-019-01903-9.

del Amor, F.M., Martinez, V., Cerda, A., 2001. Salt tolerance of tomato plants as affected by stage of plant development. HortScience 36 (7), 1260–1263. Available from: https://doi.org/10.21273/HORTSCI.36.7.1260.

Del Buono, D., 2020. Can biostimulants be used to mitigate the effect of anthropogenic climate change on agriculture? It is time to respond. Sci. Total Environ. 141763. Available from: https://doi.org/10.1016/j.scitotenv.2020.141763.

Djangsou, H., Enrico, F., Domenico, R., Matteo, B., 2019. Blossom end-rot in tomato (*Solanum lycopersicum* L.): a multi-disciplinary overview of inducing factors and control strategies. Sci. Hortic. 249, 49–58. Available from: https://doi.org/10.1016/j.scienta.2019.01.042.

Djingova, R., Kuleff, I., 2000. Instrumental techniques for trace analysis. In: Vernet, J.P. (Ed.), Trace Elements: Their Distribution and Effects in the Environment, 4. Elsevier, London, pp. 137–185.

Du, T., Kang, S., Zhang, J., Li, F., Yan, B., 2008. Water use efficiency and fruit quality of table grape under alternate partial root-zone drip irrigation. Agric. Water Manage. 95, 659–668. Available from: https://doi.org/10.1016/j.agwat.2008.01.017.

Erisman, J.W., Galloway, J., Seitzinger, S., Bleeker, A., Butterbach-Bahl, K., 2011. Reactive nitrogen in the environment and its effect on climate change. Curr. Opin. Environ. Sustainability 3 (5), 281–290. Available from: https://doi.org/10.1016/j.cosust.2011.08.012.

European Union, 2020. Farm to fork strategy. For a fair, healthy and environmentally-friendly food system. Available from: <https://ec.europa.eu/food/system/files/2020-05/f2f_action-plan_2020_strategy-info_en.pdf> (accessed 10.06.2021).

Evans, A.E., Mateo-Sagasta, J., Qadir, M., Boelee, E., Ippolito, A., 2019. Agricultural water pollution: key knowledge gaps and research needs. Curr. Opin. Environ. Sustainability 36, 20–27. Available from: https://doi.org/10.1016/j.cosust.2018.10.003.

Faci, J.M., Medina, E.T., Martínez-Cob, A., Alonso, J.M., 2014. Fruit yield and quality response of a late season peach orchard to different irrigation regimes in a semi-arid environment. Agric. Water Manage. 143, 102–112. Available from: https://doi.org/10.1016/j.agwat.2014.07.004.

Fageria, N.K., 2003. Plant tissue test for determination of optimum concentration and uptake of nitrogen at different growth stages in lowland rice. Commun. Soil Sci. Plant Anal. 34 (1-2), 259–270. Available from: https://doi.org/10.1081/css-120017430.

Fallovo, C., Rouphael, Y., Rea, E., Battistelli, A., Colla, G., 2009. Nutrient solution concentration and growing season affect yield and quality of *Lactuca sativa* L. var. acephalain floating raft culture. J. Sci. Food Agric. 89 (10), 1682–1689. Available from: https://doi.org/10.1002/jsfa.3641.

Fang, X., Turner, N.C., Yan, G., Li, F., Siddique, K.H.M., 2010. Flower numbers, pod production, pollen viability,and pistil function are reduced and flower and pod abortion increased in chickpea (*Cicer arietinum* l.) under terminal drought. J. Exp. Bot. 61, 335–345. Available from: https://doi.org/10.1093/jxb/erp307.

FAO (2012). Coping with water scarcity in agriculture: a global framework for action in a changing climate. Available from: <http://www.fao.org/3/bq656e/bq656e.pdf> (accessed 20.12.2020.).

FAO. (2015). Climate change and food systems: global assessments and implications for food security and trade. Available from: <http://www.fao.org/3/a-i4332e.pdf> (accessed 20.12.2020.).

Farooq, M., Gogoi, N., Barthakur, S., Baroowa, B., Bharadwaj, N., Alghamdi, S.S., et al., 2017. Drought stress in grain legumes during reproduction and grain filling. J. Agron. Crop Sci. 203 (2), 81–102. Available from: https://doi.org/10.1111/jac.12169.

Farooq, M., Wahid, A., Kobayashi, N., Fujita, D., Basra, S.M.A., 2009. Plant drought stress: effects, mechanisms and management. Agron. Sustainable Dev. 29 (1), 185–212. Available from: https://doi.org/10.1051/agro:2008021.

Fleming, T.R., Fleming, C.C., Levy, C.C.B., Repiso, C., Hennequart, F., Nolasco, J.B., et al., 2019. Biostimulants enhance growth and drought tolerance in *Arabidopsis thaliana* and exhibit chemical priming action. Ann. Appl. Biol. 174 (2), 153–165. Available from: https://doi.org/10.1111/aab.12482.

Fraceto, L.F., Grillo, R., de Medeiros, G.A., Scognamiglio, V., Rea, G., Bartolucci, C., 2016. Nanotechnology in agriculture: which innovation potential does it have? Front. Environ. Sci. 4, 20. Available from: https://doi.org/10.3389/fenvs.2016.00020.

Gaion, L.A., Braz, L.T., Carvalho, R.F., 2018. Grafting in vegetable crops: a great technique for agriculture. Int. J. Veg. Sci. 24 (1), 85–102. Available from: https://doi.org/10.1080/19315260.2017.1357062.

Gawronska, H., Deji, A., Sakakibara, H., Sugiyama, T., 2003. Hormone-mediated nitrogen signaling in plants: implication of participation of abscissic acid in negative regulation of cytokinin-inducible expression of maize response regulator. Plant Physiol. Biochem. 41 (6-7), 605–610. Available from: https://doi.org/10.1016/s0981-9428(03)00059-7.

Gianquinto, G., Muñoz, P., Pardossi, A., et al., 2013. Soil fertility and plant nutrition. In: FAO (Ed.), Good Agricultural Practices for Greenhouse Vegetable Crops: Principles for Mediterranean Climate Areas, 205–. Food and Agriculture Organization of the United Nations, p. 269.

Giardini, L., Giovanardi, R., Borin, M., 1988. Water consumption and yield response of tomato in relation to water availability at different soil depth. Acta Hortic. 228, 119–126. Available from: https://doi.org/10.17660/ActaHortic.1988.228.12.

Gill, S.S., Tuteja, N., 2010. Reactive oxygen species and antioxidant machinery in abiotic stress tolerance in crop plants. Plant Physiol. Biochem. 48, 909–930. Available from: https://doi.org/10.1016/j.plaphy.2010.08.016.

Gonzalez-Dugo, V., Durand, J.L., Gastal, F., 2010. Water deficit and nitrogen nutrition of crops. A review. Agron. Sustainable Dev. 30, 529–544. Available from: https://doi.org/10.1051/agro/2009059.

Gould, G.W., 1996. Methods for preservation and extension of shelf life. Int. J. Food Microbiol. 33 (1), 51–64. Available from: https://doi.org/10.1016/0168-1605(96)01133-6.

Goulding, K., Jarvis, S., Whitmore, A., 2008. Optimizing nutrient management for farm systems. Philos. Trans. R. Soc. B Biol. Sci. 363 (1491), 667–680. Available from: https://doi.org/10.1098/rstb.2007.2177.

Gous, P.W., Hasjim, J., Franckowiak, J., Fox, G.P., Gilbert, R.G., 2013. Barley genotype expressing "stay-green"-like characteristics maintains starch quality of the grain during water stress condition. J. Cereal Sci. 58 (3), 414–419. Available from: https://doi.org/10.1016/j.jcs.2013.08.002.

Goyal, S.S., Huffaker, R.C., 1984. Nitrogen toxicity in plants. In: Hauck, R.D. (Ed.), Nitrogen in Crop Production. American Society of Agronomy. Available from: https://doi.org/10.2134/1990.nitrogenincropproduction.c6.

Grassmann, J., Hippeli, S., Elstner, E.F., 2002. Plant's defence and its benefits for animals and medicine: role of phenolics and terpenoids in avoiding oxygen stress. Plant Physiol. Biochem. 40, 471–478. Available from: https://doi.org/10.1016/S0981-9428(02)01395-5.

Gu, B., Ju, X., Chang, S.X., Ge, Y., Chang, J., 2017. Nitrogen use efficiencies in chinese agricultural systems and implications for food security and environmental protection. Reg. Environ. Change 17 (4), 1217–1227. Available from: https://doi.org/10.1007/s10113-016-1101-5.

Gupta, N.K., Gupta, S., Kumar, A., 2001. Effect of water stress on physiological attributes and their relationship with growth and yield in wheat cultivars at different growth stages. J. Agron. 86, 1437–1439. Available from: https://doi.org/10.1046/j.1439-037x.2001.00457.x.

Hachmann, T.L., Rezende, R., Matumoto-Pintro, P.T., Saath, R., Anjo, F.A., Menezes, C.S.L., 2019. Yield, antioxidant activity and shelf-life of cauliflower inflorescences under drought stress and foliar spraying of selenium. Ciênc. Agrotecnol. 43, e017819. Available from: https://doi.org/10.1590/1413-7054201943017819.

Handford, C.E., Elliott, C.T., Campbell, K., 2015. A review of the global pesticide legislation and the scale of challenge in reaching the global harmonization of food safety standards. Integr. Environ. Assess. Manage. 11, 525–536. Available from: https://doi.org/10.1002/ieam.1635.

Hao, S.X., Cao, H.X., Wang, H.B., Pan, X.Y., 2019. Effects of water stress at different growth stages on comprehensive fruit quality and yield in different bunches of tomatoes in greenhouses. Int. J. Agric. Biol. Eng. 12 (3), 67–76.

Hao, J., Li, Q., Yu, H., Wang, H., Chai, L., Miao, T., et al., 2021. Comparative proteomic analysis of cucumber fruits under nitrogen deficiency at the fruiting stage. Hortic. Plant J. 7 (1), 59–72. Available from: https://doi.org/10.1016/j.hpj.2020.12.002.

Hooijdonk, B.M.V., Dorji, K., Behboudian, M.H., 2007. Fruit quality of 'Pacific Rose' TM apple grown under partial rootzone drying and deficit irrigation. J. Food Agric. Environ. 5, 173–178.

Hossain, M.S., El Sayed, A.I., Moore, M., Dietz, K.J., 2017. Redox and reactive oxygen species network in acclimation for salinity tolerance in sugar beet. J. Exp. Bot. 68, 1283–1298. Available from: https://doi.org/10.1093/jxb/erx019.

Hou, X., Zhang, W., Du, T., Kang, S., Davies, W.J., 2020. Responses of water accumulation and solute metabolism in tomato fruit to water scarcity and implications for main fruit quality variables. J. Exp. Bot. 71 (4), 1249–1264. Available from: https://doi.org/10.1093/jxb/erz526.

Houlton, B.Z., Almaraz, M., Aneja, V., Austin, A.T., Bai, E., Cassman, K.G., et al., 2019. A world of co-benefits: solving the global nitrogen challenge. Earth's Future 7, 1–8. Available from: https://doi.org/10.1029/2019EF001222.

Houser, M., Stuart, D., 2019. An accelerating treadmill and overlooked contradiction in industrial agriculture: climate change and nitrogen fertilizer. J. Agrar. Change 20 (2), 215–237. Available from: https://doi.org/10.1111/joac.12341.

Howarth, R.W., 2008. Coastal nitrogen pollution: a review of sources and trends globally and regionally. Harmful Algae 8 (1), 14–20. Available from: https://doi.org/10.1016/j.hal.2008.08.015.

IPCC. (2007). Climate change 2007: impacts, adaptation, and vulnerability. Available from: <https://www.ipcc.ch/report/ar4/wg2/> (accessed 20.12.2020.).

IPCC. (2014). Climate change 2014: synthesis report of the IPCC fifth assessment report: working group I, II and III. Available from: <http://www.ipcc.ch/glossary/> (accessed 20.12.2020.).

Jain, D., Ashraf, N., Khurana, J., Kameshwari, M.S., 2019. The 'Omics' approach for crop improvement against drought stress. In: Rajpal, V., Sehgal, D., Kumar, A., Raina, S. (Eds.), Genetic Enhancement of Crops for Tolerance to Abiotic Stress: Mechanisms and Approaches, Vol. I. Sustainable Development and Biodiversity, 20. Springer, pp. 183–204. Available from: https://doi.org/10.1007/978-3-319-91956-0_8.

Jain, A., Das, S., 2021. Regulatory requirements for nanopesticides and nanofertilizers. In: Jogaiah, S., Sing, H.B., Fraceto, L.F., Lima, R. (Eds.), Woodhead Publishing Series in Food Science, Technology and Nutrition, Advances in Nano-Fertilizers and Nano-Pesticides in Agriculture. Woodhead Publishing, pp. 145–152. Available from: https://doi.org/10.1016/B978-0-12-820092-6.00006-9.

Jangam, A.P., Raghuram, N., 2015. Nitrogen and stress. In: Pandey, G.K. (Ed.), Elucidation of Abiotic Stress Signaling in Plants, 2. Springer, New York, pp. 323–339. Available from: https://doi.org/10.1007/978-1-4939-2540-7_12.

Jisha, K.C., Vijayakumari, K., Puthur, J.T., 2013. Seed priming for abiotic stress tolerance: an overview. Acta Physiol. Plant. 35, 1381–1396. Available from: https://doi.org/10.1007/s11738-012-1186-5.

Kah, M., 2015. Nanopesticides and nanofertilizers: emerging contaminants or opportunities for risk mitigation? Front. Chem. 3, 64. Available from: https://doi.org/10.3389/fchem.2015.00064.

Kanter, D.R., 2018. Nitrogen pollution: a key building block for addressing climate change. Clim. Change 147 (1-2), 11–21. Available from: https://doi.org/10.1007/s10584-017-2126-6.

Kanter, D.R., Searchinger, T.D., 2018. A technology-forcing approach to reduce nitrogen pollution. Nat. Sustainability 1, 544–552. Available from: https://doi.org/10.1038/s41893-018-0143-8.

Kao, W.-Y., Tsai, H.-C., Tsai, T.-T., 2001. Effect of NaCl and nitrogen availability on growth and photosynthesis of seedlings of a mangrove species, *Kandelia candel* (L.) druce. J. Plant Physiol. 158 (7), 841–846. Available from: https://doi.org/10.1078/0176-1617-00248.

Kannan, N., Anandhi, A., 2020. Water management for sustainable food production. Water 12, 778. Available from: https://doi.org/10.3390/w12030778.

Katerji, N., van Hoorn, J., Hamdy, A., Mastrorilli, M., 2000. Salt tolerance classification of crops according to soil salinity and to water stress day index. Agric. Water Manage. 43 (1), 99–109. Available from: https://doi.org/10.1016/s0378-3774(99)00048-7.

Khan, M.I.R., Trivellini, A., Fatma, M., Masood, A., Francini, A., Iqbal, N., et al., 2015. Role of ethylene in responses of plants to nitrogen availability. Front. Plant Sci. 6, 927. Available from: https://doi.org/10.3389/fpls.2015.00927.

Khapte, P.S., Kumar, P., Burman, U., Kumar, P., 2019. Deficit irrigation in tomato: agronomical and physio-biochemical implications. Sci. Hortic. 248, 256–264. Available from: https://doi.org/10.1016/j.scienta.2019.01.006.

Kwon, M.J., Hwang, Y., Lee, J., Ham, B., Rahman, A., Azam, H., et al., 2021. Waste nutrient solutions from full-scale open hydroponic cultivation: dynamics of effluent quality and removal of nitrogen and phosphorus using a pilot-scale sequencing batch reactor. J. Environ. Manage. 281, 111893. Available from: https://doi.org/10.1016/j.jenvman.2020.111893.

Knox, J.W., Daccache, A., Hess, T.M., Haro, D., 2016. Meta-analysis of climate impacts and uncertainty on crop yields in Europe. Environ. Res. Lett. 11, 113004. Available from: https://doi.org/10.1088/1748-9326/11/11/113004.

Kumar, P., Rouphael, Y., Cardarelli, M., Colla, G., 2017. Vegetable grafting as a tool to improve drought resistance and water use efficiency. Front. Plant Sci. 8, 1130. Available from: https://doi.org/10.3389/fpls.2017.01130.

Lawlor, D.W., Lemaire, G., Gastal, F., 2001. Nitrogen, plant growth and crop yield. In: Lea, P. J., Morot-Gaudry, J.F. (Eds.), Plant Nitrogen. Springer, Berlin, Heidelberg, pp. 343–367. Available from: https://doi.org/10.1007/978-3-662-04064-5_13.

Lawlor, D.W., 2002. Carbon and nitrogen assimilation in relation to yield: mechanisms are the key to understanding production systems. J. Exp. Bot. 53 (370), 773–787. Available from: https://doi.org/10.1093/jexbot/53.370.773.

Le Mire, G., Nguyen, M.L., Fassott, B., du Jardin, P., Verheggen, F., Delaplace, P., et al., 2016. Review: implementing plant biostimulants and biocontrol strategies in the agroecological management of cultivated ecosystems. Biotechnol. Agron. Soc. Environ. 20, 299–313. Available from: https://doi.org/10.25518/1780-4507.12717.

Lemaire, G., van Oosterom, E., Jeuffroy, M.-H., Gastal, F., Massignam, A., 2008. Crop species present different qualitative types of response to N deficiency during their vegetative growth. Field Crops Res. 105 (3), 253–265. Available from: https://doi.org/10.1016/j.fcr.2007.10.009.

Lesk, C., Rowhani, P., Ramankutty, N., 2016. Influence of extreme weather disasters on global crop production. Nature 529, 84–87. Available from: https://doi.org/10.1038/nature16467.

Levy, D., Veilleux, R.E., 2007. Adaptation of potato to high temperatures and salinity-a review. Am. J. Potato Res. 84 (6), 487–506. Available from: https://doi.org/10.1007/bf02987885.

Liang, S., Li, Y., Zhang, X., Sun, Z., Sun, N., Duan, Y., et al., 2018. Response of crop yield and nitrogen use efficiency for wheat-maize cropping system to future climate change in northern China. Agric. For. Meteorol. 262, 310–321. Available from: https://doi.org/10.1016/j.agrformet.2018.07.019.

Lopez, A., Zon, K., Dussi, M.C., Reeb, P., Giardina, G., Leskovar, M., et al. (2011). Economic evaluation between chemical thinning vs. hand thinning in 'Williams' pear. In: XI International Pear Symposium. Acta Horticulturae. https://doi.org/29–37.10.17660/ActaHortic.2011.909.1

Lu, Y., Song, S., Wang, R., Liu, Z., Meng, J., Sweetman, A.J., et al., 2015. Impacts of soil and water pollution on food safety and health risks in China. Environ. Int. 77, 5–15. Available from: https://doi.org/10.1016/j.envint.2014.12.010.

Machado, J., Campos, A., Vasconcelos, V., Freitas, M., 2017a. Effects of microcystin-LR and cylindrospermopsin on plant-soil systems: a review of their relevance for agricultural plant quality and public health. Environ. Res. 153, 191–204. Available from: https://doi.org/10.1016/j.envres.2016.09.015.

Machado, J., Azevedo, J., Freitas, M., Pinto, E., Almeida, A., Vasconcelos, V., et al., 2017b. Analysis of the use of microcystin-contaminated water in the growth and nutritional quality of the root-vegetable, *Daucus carota*. Environ. Sci. Pollut. Res. 24 (1), 752–764. Available from: https://doi.org/10.1007/s11356-016-7822-7.

Machado, J., Azevedo, J., Vasconcelos, V., Campos, A., 2018. Mode of action and toxicity of major cyanobacterial toxins and corresponding chemical variants. In: Gopalakrishnakone, P., Stiles, B., Alape-Girón, A., Dubreuil, J., Mandal, M. (Eds.), Microbial Toxins. Toxinology. Springer, Dordrecht. Available from: https://doi.org/10.1007/978-94-007-6449-1_30.

Maheswari, M., Murthy, A.N.G., Shanker, A.K., 2017. Nitrogen nutrition in crops and its importance in crop quality. the Indian nitrogen assessment. In: Abrol, Y.P., Adhya, T.K., Aneja, V.P., Raghuram, N., Pathak, H., Kulshrestha, U., Sharma, C., Singh, B. (Eds.), The Indian Nitrogen Assessment. Springer, pp. 175–186. Available from: https://doi.org/10.1016/b978-0-12-811836-8.00012-4.

Marthandan, V., Geetha, R., Kumutha, K., Renganathan, V.G., Karthikeyan, A., Ramalingam, J., 2020. Seed priming: a feasible strategy to enhance drought tolerance in crop plants. Inter. J. Mol. Sci 21 (21), 8258. Available from: https://doi.org/10.3390/ijms21218258.

Massa, D., Incrocci, L., Maggini, R., Carmassi, G., Campiotti, C.A., Pardossi, A., 2010. Strategies to decrease water drainage and nitrate emission from soilless cultures of greenhouse tomato. Agric. Water Manage. 97, 971–980. Available from: https://doi.org/10.1016/j.agwat.2010.01.029.

Massa, D., Magán, J.J., Montesano, F.F., Tzortzakis, N., 2020. Minimizing water and nutrient losses from soilless cropping in southern Europe. Agric. Water Manage. 241, 106395. Available from: https://doi.org/10.1016/j.agwat.2020.106395.

Medina-Lara, F., Echevarría-Machado, I., Pacheco-Arjona, R., Ruiz-Lau, N., Guzmán-Antonio, A., Martinez-Estevez, M., 2008. Influence of nitrogen and potassium fertilization on fruiting and capsaicin content in habanero pepper (*Capsicum chinense* Jacq.). HortScience 43 (5), 1549–1554. Available from: https://doi.org/10.21273/HORTSCI.43.5.1549.

Mert, M., 2005. Irrigation of cotton cultivars improves seed cotton yield, yield components and fibre properties in the Hatay Region, Turkey. Acta Agric. Scand., Sec. B. Soil Plant Sci. 55 (1), 44–50. Available from: https://doi.org/10.1080/09064710510008658.

Miller, S.A., Smith, G.S., Boldingh, H.L., Johansson, A., 1998. Effects of water stress on fruit quality attributes of kiwifruit. Ann. Bot. 81, 73–81. Available from: https://doi.org/10.1006/anbo.1997.0537.

Millones-Chanamé, C.E., Souza de Oliveiras, A.M., Castro, E.M., Maluf, W.R., 2019. Inheritance of blossom end rot resistance induced by drought stress and of associated stomatal densities in tomatoes. Euphytica 215, 120. Available from: https://doi.org/10.1007/s10681-019-2444-z.

Mitchell, J.P., Shennan, C., Grattan, S.R., 1991. Developmental changes in tomato fruit composition in response to water deficit and salinity. Physiol. Plant. 83, 177–185. Available from: https://doi.org/10.1111/j.1399-3054.1991.tb01299.x.

Mosier, A.R., Syers, J.K., Freney, J.R., 2004. Nitrogen fertilizer: an essential component of increasing food, feed and fiber production. In: Mosier, A.R., Syers, J.K., Freney, J.R. (Eds.), Agriculture and the Nitrogen Cycle: Assessing the Impacts of Fertilizer Use on Food Production and the Environment. Island Press, Washington, Covelo, London, pp. 3–15.

Mozafarian Meimandi, M., Kappel, N., 2020. Grafting plants to improve abiotic stress tolerance. In: Hasanuzzaman, M. (Ed.), Plant Ecophysiology and Adaptation under Climate Change: Mechanisms and Perspectives II. Springer, Singapore, pp. 477–490. Available from: https://doi.org/10.1007/978-981-15-2172-0_16.

Mpelasoka, B., Behboudian, M., Mills, T., 2001. Effects of deficit irrigation on fruit maturity and quality of "Braeburn" apple. Sci. Hortic. 90 (3-4), 279–290. Available from: https://doi.org/10.1016/s0304-4238(01)00231-x.

Munns, R., Tester, M., 2008. Mechanisms of salinity tolerance. Ann. Rev. Plant Biol. 59 (1), 651–681. Available from: https://doi.org/10.1146/annurev.arplant.59.032607.092911.

Nadeem, M., Li, J., Yahya, M., Sher, A., Ma, C., Wang, X., et al., 2019. Research progress and perspective on drought stress in legumes: a review. Int. J. Mol. Sci. 20 (10), 2541. Available from: https://doi.org/10.3390/ijms20102541.

Nagajyoti, P.C., Lee, K.D., Sreekanth, T.V.M., 2010. Heavy metals, occurrence and toxicity for plants: a review. Environ. Chem. Lett. 8, 199–216. Available from: https://doi.org/10.1007/s10311-010-0297-8.

Nanda, A.K., Melnyk, C.W., 2018. The role of plant hormones during grafting. J. Plant Res. 131, 49–58. Available from: https://doi.org/10.1007/s10265-017-0994-5.

Nangare, D.D., Singh, Y., Kumar, P.S., Minhas, P.S., 2016. Growth, fruit yield and quality of tomato (*Lycopersicon esculentum* Mill.) as affected by deficit irrigation regulated on phenological basis. Agric. Water Manage. 171, 73–79. Available from: https://doi.org/10.1016/j.agwat.2016.03.016.

Nava, G., Dechen, A.R., Nachtiga, G.R., 2008. Nitrogen and potassium fertilization affect apple fruit quality in southern Brazil. Commun. Soil Sci. Plant Anal. 39 (1-2), 96–107. Available from: https://doi.org/10.1080/00103620701759038.

Navarro, J.M., Pérez-Pérez, J.G., Romero, P., Botía, P., 2010. Analysis of the changes in quality in mandarin fruit, produced by deficit irrigation treatments. Food Chem. 119, 1591–1596. Available from: https://doi.org/10.1016/j.foodchem.2009.09.048.

Nephali, L., Piater, L.A., Dubery, I.A., Patterson, V., Huyser, J., Burgess, K., et al., 2020. Biostimulants for plant growth and mitigation of abiotic stresses: a metabolomics perspective. Metabolites 10 (12), 505. Available from: https://doi.org/10.3390/metabo10120505.

Nora, L., Dalmazo, G.O., Nora, F.R., Rombaldi, C.V., 2012. Controlled water stress to improve fruit and vegetable postharvest quality. In: Ismail, M., Mofizur, R., Hiroshi, H. (Eds.), Water Stress. IntechOpen, pp. 59–72. Available from: http://doi.org/10.5772/30182.

Okunlola, G.O., Olatunji, O.A., Afolabi, A.M., Gbadegesin, K.K., 2015. Effects of nitrogen nutritional stress on the morphological and yield parameters of tomato (*Solanum lycopersicum* L.). Sci. Cold Arid Reg. 7 (2), 0137–0145. Available from: https://doi.org/10.3724/SP.J.1226.2015.00137.

Osakabe, Y., Osakabe, K., Shinozaki, K., Tran, L.S., 2014. Response of plants to water stress. Front. Plant Sci. 5 (86), 1–8. Available from: https://doi.org/10.3389/fpls.2014.00086.

Ozturk, A., Aydin, F., 2004. Effect of water stress at various growth stages on some quality characteristics of winter wheat. J. Agron. Crop Sci. 190 (2), 93–99. Available from: https://doi.org/10.1046/j.1439-037x.2003.00080.x.

Patanè, C., Tringali, S., Sortino, O., 2011. Effects of deficit irrigation on biomass, yield, water productivity and fruit quality of processing tomato under semi-arid Mediterranean climate conditions. Sci. Hortic. 129, 590–596. Available from: https://doi.org/10.1016/j.scienta.2011.04.030.

Pavlíková, D., Neuberg, M., Žižková, E., Motyka, V., Pavlík, M., 2012. Interactions between nitrogen nutrition and phytohormone levels in *Festulolium* plants. Plant Soil Environ. 58 (8), 367–372. Available from: https://agris.fao.org/agris-search/search.do?recordID = CZ2012000628.

Pauleta, S.R., Carepo, M.S.P., Moura, I., 2019. Source and reduction of nitrous oxide. Coord. Chem. Rev. 387, 436–449. Available from: https://doi.org/10.1016/j.ccr.2019.02.005.

Peña-Estévez, M.E., Gómez, P.A., Artés, F., Aguayo, E., Martínez-Hernández, G.B., Otón, M., et al., 2014. Quality changes of fresh-cut pomegranate arils during shelf life as affected by deficit irrigation and postharvest vapour treatments. J. Sci. Food Agric. 95 (11), 2325–2336. Available from: https://doi.org/10.1002/jsfa.6954.

Peralta-Videa, J.R., Lopez, M.L., Narayan, M., Saupe, G., Gardea-Torresdey, J., 2009. The biochemistry of environmental heavy metal uptake by plants: implications for the food chain. Int. J. Biochem. Cell Biol. 41 (8–9), 1665–1677. Available from: https://doi.org/10.1016/j.biocel.2009.03.005.

Pereira, L.S., 2007. Drought impacts in agriculture: water conservation and water saving practices and management. In: Rossi, G., Vega, T., Bonaccorso, B. (Eds.), Methods and Tools for Drought Analysis and Management. Water Science and Technology Library, vol. 62. Springer. Available from: https://doi.org/10.1007/978-1-4020-5924-7_17.

Pérez-Pastor, A., Ruiz-Sánchez, M.C., Martínez, J.A., Nortes, P.A., Artés, F., Domingo, R., 2007. Effect of deficit irrigation on apricot fruit quality at harvest and during storage. J. Sci. Food Agric. 87 (13), 2409–2415. http://doig.org/10.1002/jsfa.2905.

Pettigrew, W.T., 2004a. Moisture deficit effects on cotton lint yield, yield components, and boll distribution. Agron. J. 96, 377–383. Available from: https://doi.org/10.2134/agronj2004.0377.

Pettigrew, W.T., 2004b. Crop physiology and metabolism: physiological consequences of moisture deficit stress in cotton. Crop Sci. 44 (4), 1265–1272. Available from: https://doi.org/10.2135/cropsci2004.1265.

Pirzadah, B., Pirzadah, T.B., Jan, A., Hakeem, K.R., 2020. Nanofertilizers: a way forward for green economy. In: Hakeem, K., Pirzadah, T. (Eds.), Nanobiotechnology in Agriculture. Nanotechnology in the Life Sciences. Springer. Available from: https://doi.org/10.1007/978-3-030-39978-8_5.

Rai, P.K., Lee, S.S., Zhang, M., Tsang, Y.F., Kim, K.-H., 2019. Heavy metals in food crops: health risks, fate, mechanisms, and management. Environ. Int. 125, 365–385. Available from: https://doi.org/10.1016/j.envint.2019.01.067.

Raliya, R., Saharan, V., Dimkpa, C., Biswas, P., 2017. Nanofertilizer for precision and sustainable agriculture: current state and future perspectives. J. Agric. Food Chem. 66 (26), 6487–6503. Available from: https://doi.org/10.1021/acs.jafc.7b02178.

Rao, N.K.S., Laxman, R.H., Shivashankara, K.S., 2016. Physiological and morphological responses of horticultural crops to abiotic stresses. In: Rao, N., Shivashankara, K., Laxman, R. (Eds.), Abiotic Stress Physiology of Horticultural Crops. Springer. Available from: https://doi.org/10.1007/978-81-322-2725-0_1.

Reay, D.S., Davidson, E.A., Smith, K.A., Smith, P., Melillo, J.M., Dentener, F., et al., 2012. Global agriculture and nitrous oxide emissions. Nat. Clim. Change 2 (6), 410–416. Available from: https://doi.org/10.1038/nclimate1458.

Rey, D., Holman, I.P., Knox, J.W., 2017. Developing drought resilience in irrigated agriculture in the face of increasing water scarcity. Reg. Environ.menatl Change 17, 1527–1540. Available from: https://doi.org/10.1007/s10113-017-1116-6.

Rhaman, M.S., Imran, S., Rauf, F., Khatun, M., Baskin, C.C., Murata, Y., et al., 2020. Seed priming with phytohormones: an effective approach for the mitigation of abiotic stress. Plants (Basel) 25 (10(1)), 37. Available from: https://doi.org/10.3390/plants10010037.

Ripoll, J., Urban, L., Staudt, M., Lopez-Lauri, F., Bidel, L.P.R., Bertin, N., 2014. Water shortage and quality of fleshy fruits—making the most of the unavoidable. J. Exp. Bot. 65 (15), 4097–4117. Available from: https://doi.org/10.1093/jxb/eru197.

Ripoll, J., Urban, L., Brunel, B., Bertin, N., 2016. Water deficit effects on tomato quality depend on fruit developmental stage and genotype. J. Plant Physiol. 190, 26–35. Available from: https://doi.org/10.1016/j.jplph.2015.10.006.

Rouphael, Y., Kyriacou, M.C., Colla, G., 2018. Vegetable grafting: a toolbox for securing yield stability under multiple stress conditions. Front. Plant Sci. 8. Available from: https://doi.org/10.3389/fpls.2017.02255.

Rouphael, Y., Colla, G., 2020. Editorial: biostimulants in agriculture. Front. Plant Sci. 11. Available from: https://doi.org/10.3389/fpls.2020.00040.

Salehi-Lisar, S.Y., Bakhshayeshan-Agdam, H., 2016. Drought stress in plants: causes, consequences, and tolerance. In: Hossain, M., Wani, S., Bhattacharjee, S., Burritt, D., Tran, L.S. (Eds.), Drought Stress Tolerance in Plants, vol. 1. Springer. Available from: https://doi.org/10.1007/978-3-319-28899-4_1.

Samarah, N.H., 2005. Effects of drought stress on growth and yield of barley. Agron. Sustainable Dev. 25, 145–149. Available from: https://doi.org/10.1051/agro:2004064.

Samota, M.K., Sasi, M., Awana, M., Yadav, O.P., Amitha Mithra, S.V., Tyagi, A., et al., 2017. Elicitor-induced biochemical and molecular manifestations to improve drought tolerance in rice (*Oryza sativa* L.) through seed-priming. Front. Plant Sci. 8. Available from: https://doi.org/10.3389/fpls.2017.00934.

Sánchez, E., Soto, J.M., García, P.C., López-Lefebre, L.R., Rivero, R.M., Ruiz, J.M., et al., 2000. Phenolic compounds and oxidative metabolism in green bean plants under nitrogen toxicity. Funct. Plant Biol. 27 (10), 973. Available from: https://doi.org/10.1071/pp00008.

Sánchez, E., Ruiz, J.M., Romero, L., 2002. Proline metabolism in response to nitrogen toxicity in fruit of French bean plants (*Phaseolus Vulgaris* L. Cv Strike). Sci. Hortic. 93 (3–4), http://doi.org/10.1016/S0304–4238(01)00342-9.

Santos, M.G., Roncon, I., Pereira, R., Carvalho, S.M.P., 2017. Characterization of the Portuguese management practices of fertigation and phytochemicals' application in soilless cultivation. Actas Portuguesas de Horticultura 29, 611–619.

Savvides, A., Ali, S., Tester, M., Fotopoulos, V., 2016. Chemical priming of plants against multiple abiotic stresses: mission possible? Trends Plant Sci. 21 (4), 329–340. Available from: https://doi.org/10.1016/j.tplants.2015.11.003.

Shang, Y., Hasan, M.K., Ahammed, G.J., Li, M., Yin, H., Zhou, J., 2019. Applications of nanotechnology in plant growth and crop protection: a review. Molecules 24 (14), 2558. Available from: https://doi.org/10.3390/molecules24142558.

Sharma, L.K., 2016. Effect of nutrients sprays on growth, yield and fruit quality of apple under cold desert condition of Himachal Pradesh. J. Appl. Nat. Sci. 8, 207e300. Available from: https://doi.org/10.31018/jans.v8i1.789.

Shrestha, R.A., Turner, N.C.A., Siddique, K.H.M.A., Turner, D.W.B., Speijers, J.C., 2006. A water deficit during pod development in lentils reduces flower and pod numbers but not seed size. Aust. J. Agric. Res. 57, 427–438. Available from: https://doi.org/10.1071/AR05225.

Silva, A.N., Ramos, M.L.G., Ribeiro, W.Q., Alencar, E.R., Silva, P.C., Lima, C.A., et al., 2020. Water stress alters physical and chemical quality in grains of common bean, triticale and wheat. Agric. Water Manage. 231, 106023. Available from: https://doi.org/10.1016/j.agwat.2020.106023.

Singh, S., Singh, G., Singh, P., Singh, N., 2008. Effect of water stress at different stages of grain development on the characteristics of starch and protein of different wheat cultivars. Food Chem. 108, 130–139. Available from: https://doi.org/10.1016/j.foodchem.2007.10.054.

Singh, Y., Rao, S.S., Regar, P.L., 2010. Deficit irrigation and nitrogen effects on seed cotton yield, water productivity and yield response factor in shallow soils and semi-arid environment. Agric. Water Manage. 97, 965–970. Available from: https://doi.org/10.1016/j.agwat.2010.01.028.

Singh, V.P., Mishra, A.K., Chowdhary, H., Khedun, C.P., 2014. Climate change and its impact on water resources. In: Wang, L.K., Yang, C.T. (Eds.), Modern Water Resources Engineering, vol. 15. Humana Press. Available from: http://doi.org/10.1007/978-1-62703-595-8_11.

Singhal, P., Jan, A.T., Azam, M., Haq, Q., 2016. Plant abiotic stress: a prospective strategy of exploiting promoters as alternative to overcome the escalating burden. Front. Life Sci. 9 (1), 52–63. Available from: https://doi.org/10.1080/21553769.2015.1077478.

Siracusa, L., Patanè, C., Rizzo, V., Cosentino, S.L., Ruberto, G., 2018. Targeted secondary metabolic and physico-chemical traits analysis to assess genetic variability within a germplasm collection of “long storage” tomatoes. Food Chem. 244, 275–283. Available from: https://doi.org/10.1016/j.foodchem.2017.10.043.

Skaggs, R.W., van Schilfgaarde, J., Maas, E.V., Grattan, S.R., 1999. Crop yields as affected by salinity. In: Skaggs, R.W., van Schilfgaard, J. (Eds.), Book Series: Agronomy Monographs. American Society of Agronomy. Available from: http://doi.org/10.2134/agronmonogr38.c3.

Skiba, U.M., Rees, R.M., 2014. Nitrous oxide, climate change and agriculture. CAB Rev. 10. Available from: https://doi.org/10.1079/PAVSNNR20149010.

Srinivasa Rao, N.K., Bhatt, R.M., Sadasivan, A.T., 2001. Tolerance to water stress in tomato cultivars. Photosynthetica 38 (3), 465–467. Available from: https://doi.org/10.1023/A:1010902427231.

Srivastava, V., Sarkar, A., Singh, S., Singh, P., de Araújo, A.S.F., Singh, R.P., 2017. Agroecological responses of heavy metal pollution with special emphasis on soil health and plant performances. Front. Environ. Sci. 5. Available from: https://doi.org/10.3389/fenvs.2017.00064.

Stagnari, F., Galieni, A., Pisante, M., 2016. Drought stress effects on crop quality. In: Ahmad, P. (Ed.), Water Stress And Crop Plants. Wiley Online Library, pp. 375–392. Available from: http://doi.org/10.1002/9781119054450.ch23.

Suddick, E.C., Whitney, P., Townsend, A.R., Davidson, E.A., 2013. The role of nitrogen in climate change and the impacts of nitrogen–climate interactions in the United States: foreword to thematic issue. Biogeochemistry 114 (1-3), 1–10. Available from: https://doi.org/10.1007/s10533-012-9795-z.

Sulaiman, N.S., Rovina, K., Joseph, V.M., 2019. Classification, extraction and current analytical approaches for detection of pesticides in various food products. J. Consum. Prot. Food Saf. 14, 209–221. Available from: https://doi.org/10.1007/s00003-019-01242-4.

Sun, Y., Wang, C., Chen, H.Y.H., Ruan, H., 2020. Response of plants to water stress: a meta-analysis. Front. Plant Sci. 11, 978. Available from: https://doi.org/10.3389/fpls.2020.00978.

Tabatabaie, S.J., Gregory, P.J., Hadley, P., 2004. Uneven distribution of nutrients in the root zone affects the incidence of blossom end rot and concentration of calcium and potassium in fruits of tomato. Plant Soil 258 (2), 169–178. Available from: https://doi.org/10.1023/B:PLSO.0000016548.84566.62.

Tei, F., Benincasa, P., Guiducci, M., 2002. Effect of N availability on growth, N uptake, light interception and photosynthetic activity in processing tomato. Acta Hortic. 571, 209–216. Available from: https://doi.org/10.17660/actahortic.2002.571.25.

Thitisaksakul, M., Jimenez, R.C., Arias, M.C., Beckles, D.M., 2012. Effects of environmental factors on cereal starch biosynthesis and composition. J. Cereal Sci. 56, 67–80. Available from: https://doi.org/10.1016/j.jcs.2012.04.002.

Tilman, D., Balzer, C., Hill, J., Befort, B.L., 2011. Global food demand and the sustainable intensification of agriculture. Proc. Natl. Acad. Sci. 108 (50), 20260–20264. Available from: https://doi.org/10.1073/pnas.1116437108.

Townsend, A.R., Howarth, R.W., Bazzaz, F.A., Booth, M.S., Cleveland, C.C., Collinge, S.K., et al., 2003. Human health effects of a changing global nitrogen cycle. Front. Ecol. Environ. 1 (5), 240–246. Available from: https://doi.org/10.1890/1540-9295(2003)001[0240:hheoac]2.0.co;2.

UNEP. (2019). Frontiers 2018/19 emerging issues of environmental concern. United Nations Environment Programme, Nairobi.

United Nation. (2015). Sustainable development goals. Available from: <https://www.un.org/sustainabledevelopment/> (accessed 20.12.2020.).

United Nation. (2019). Population facts. Volume 6. Available from: <https://www.un.org/en/development/desa/population/publications/pdf/popfacts/PopFacts_2019-6.pdf> (accessed 20.12.2020.).

Van der Salm, C., Voogt, W., Beerling, E., van Ruijven, J., Van Os, E., 2020. Minimising emissions to water bodies 791 from NW European greenhouses with focus on Dutch vegetable cultivation. Agric. Water Manage. 242, 106398. Available from: https://doi.org/10.1016/j.agwat.2020.106398.

Van Oosten, M.J., Pepe, O., De Pascale, S., Silleti, S., Maggio, A., 2017. The role of biostimulants and bioeffectors as alleviators of abiotic stress in crop plants. Chem. Biol. Technol. Agric. 4, 5. Available from: https://doi.org/10.1186/s40538-017-0089-5.

Vasconcelos, M.W., 2014. Chitosan and chitooligosaccharide utilization in phytoremediation and biofortification programs: current knowledge and future perspectives. Front. Plant Sci. 5, 616. Available from: https://doi.org/10.3389/fpls.2014.00616.

Venegas-Molina, J., Proietti, S., Pollier, J., Orozco-Freire, W., Ramirez-Villacis, D., Leon-Reyes, A., 2020. Induced tolerance to abiotic and biotic stresses of broccoli and arabidopsis after treatment with elicitor molecules. Sci. Rep. 10 (1). Available from: https://doi.org/10.1038/s41598-020-67074-7.

Wallace, J.G., Zhang, X.C., Beyene, Y., Semagn, K., Olsen, M., Prasanna, B.M., et al., 2016. Genome-wide association for plant height and flowering time across 15 tropical maize populations under managed drought stress and well-watered conditions in Sub-Saharan Africa. Crop Sci. 56, 2365–2378. Available from: https://doi.org/10.2135/cropsci2015.10.0632.

Wang, Y., Frei, M., 2011. Stressed food – the impact of abiotic environmental stresses on crop quality. Agric. Ecosyst. Environ. 141 (3-4), 271–286. Available from: https://doi.org/10.1016/j.agee.2011.03.017.

Wang, D., Xu, Z., Zhao, J., Wang, Y., Yu, Z., 2011. Excessive nitrogen application decreases grain yield and increases nitrogen loss in a wheat–soil system. Acta Agric. Scand. Sec. B Soil Plant Sci. 61 (8), 681–692. Available from: https://doi.org/10.1080/09064710.2010.534108.

Wang, X., Zhang, J., Shahid, S., Guan, E., Wu, Y., Gao, J., et al., 2014. Adaptation to climate change impacts on water demand. Mitigation and Adaptation Strategies for Global Change 21 (1), 81–99. Available from: http://doi.org/10.1007/s11027-014-9571-6.

Wang, X., Zhang, J., Yang, Z., Shahid, S., He, R., Xia, X., et al., 2013. Historic water consumptions and future management strategies for Haihe River Basin of Northern China. Mitigation Adapt. Strategies Global Change 20 (3), 371–387. Available from: https://doi.org/10.1007/s11027-013-9496-5.

Weber, N., Zupanc, V., Jakopic, J., Veberic, R., Mikulic-Petkovsek, M., Stampar, F., 2016. Influence of deficit irrigation on strawberry (*Fragaria* × *ananassa* duch.) fruit quality. J. Sci. Food Agric. 97 (3), 849–857. Available from: https://doi.org/10.1002/jsfa.7806.

Wei, S., Wang, X., Shi, D., Li, Y., Zhang, J., Liu, P., et al., 2016. The mechanisms of low nitrogen induced weakened photosynthesis in summer maize (*Zea mays* L.) under field conditions. Plant Physiol. Biochem. 105, 118–128. Available from: https://doi.org/10.1016/j.plaphy.2016.04.007.

Wilk, J., and Wittgren, H.B. (2009). Adapting Water Management To Climate Change. Swedish Water House Policy Brief No. 7. Stockholm International Water Institute, Stockholm. Available from: <https://www.siwi.org/wp-content/uploads/2015/09/SWHWaterClimate.pdf> (accessed 20.12.2020).

Xiong, Q., Tang, G., Zhong, L., He, H., Chen, X., 2018. Response to nitrogen deficiency and compensation on physiological characteristics, yield formation, and nitrogen utilization of rice. Front. Plant Sci. 9, 1075. Available from: https://doi.org/10.3389/fpls.2018.01075.

Xu, C., Mou, B., 2016. Responses of spinach to salinity and nutrient deficiency in growth, physiology, and nutritional value. J. Am. Soc. Hortic. Sci. 141 (1), 12–21. Available from: https://doi.org/10.21273/JASHS.141.1.12.

Yang, J., Zhang, J., Huang, Z., Zhu, Q., Wang, L., 2000. Remobilization of carbon reserves is improved by controlled soil-drying during grain filling of wheat. Crop Sci. 40 (6), 1645. Available from: https://doi.org/10.2135/cropsci2000.4061645x.

Zaicovski, C.B., Zimmerman, T., Nora, L., Nora, F.R., Silva, J.A., Rombaldi, C.V., 2008. Water stress increases cytokinin biosynthesis and delays postharvest yellowing of broccoli florets. Postharvest Biol. Technol. 49 (3), 436–439. Available from: https://doi.org/10.1016/j.postharvbio.2008.02.001.

Zaman, M., Kurepin, L.V., Catto, W., Pharis, R.P., 2014. Enhancing crop yield with the use of N-based fertilizers co-applied with plant hormones or growth regulators. J. Sci. Food Agric. 95 (9), 1777–1785. Available from: https://doi.org/10.1002/jsfa.6938.

Zhang, X., Davidson, E.A., Mauzerall, D.L., Searchinger, T.D., Dumas, P., Shen, Y., 2015. Managing nitrogen for sustainable development. Nature 528, 51–59. Available from: https://doi.org/10.1038/nature15743.

Zhang, X., 2017. A plan for efficient use of nitrogen fertilizers. Nature 543, 322–323. Available from: https://doi.org/10.1038/543322a.

Zhang, J., Wang, Y., Zhao, Y., Zhang, Y., Zhang, J., Ma, H., et al., 2020a. Transcriptome analysis reveals nitrogen deficiency induced alterations in leaf and root of three cultivars of potato (*Solanum tuberosum* L.). PLoS ONE 15 (10), e0240662. Available from: https://doi.org/10.1371/journal.pone.0240662.

Zhang, H., Liu, X., Song, B., Nie, B., Wei zhang, Zhao, Z., 2020b. Effect of excessive nitrogen on levels of amino acids and sugars, and differential response to post-harvest cold storage in potato (*Solanum tuberosum* L.) tubers. Plant Physiol. Biochem. 157, 38–46. Available from: https://doi.org/10.1016/j.plaphy.2020.09.040.

Zhang, Z., Liu, Y., Cao, B., Chen, Z., Xu, K., 2020c. The effectiveness of grafting to improve drought tolerance in tomato. Plant Growth Regul. 91, 157–167. Available from: https://doi.org/10.1007/s10725-020-00596-2.

Zhao, D., Reddy, K.R., Kakani, V.G., Read, J.J., Carter, G.A., 2003. Corn (*Zea mays* L.) growth, leaf pigment concentration, photosynthesis and leaf hyperspectral reflectance properties as affected by nitrogen supply. Plant Soil 257, 205–217. Available from: https://doi.org/10.1023/A:1026233732507.

Zhao, D., Reddy, K.R., Kakani, V.G., Reddy, V.R., 2005. Nitrogen deficiency effects on plant growth, leaf photosynthesis, and hyperspectral reflectance properties of sorghum. Eur. J. Agron. 22 (4), 391–403. Available from: https://doi.org/10.1016/j.eja.2004.06.005.

Zhao, H., Li, X., Jiang, Y., 2019. Response of nitrogen losses to excessive nitrogen fertilizer application in intensive greenhouse vegetable production. Sustainability 11 (6), 1513. Available from: https://doi.org/10.3390/su11061513.

Zhu, J., Fan, Y., Shabala, S., Li, C., Lv, C., Guo, B., et al., 2020. Understanding mechanisms of salinity tolerance in barley by proteomic and biochemical analysis of near-isogenic lines. Int. J. Mol. Sci. 21 (4), 1516. Available from: https://doi.org/10.3390/ijms21041516.

Zörb, C., Geilfus, C.-M., Dietz, K.-J., 2019. Salinity and crop yield. Plant Biol. 21 (1), 31–38. Available from: https://doi.org/10.1111/plb.12884.

Zulfiqar, F., Navarro, M., Ashraf, M., Akram, N.A., Munné-Bosch, S., 2019. Nanofertilizer use for sustainable agriculture: advantages and limitations. Plant Sci. 110270. Available from: https://doi.org/10.1016/j.plantsci.2019.110270.

Chapter 8

Soilless indoor smart agriculture as an emerging enabler technology for food and nutrition security amidst climate change

Arun Banerjee, Kenny Paul, Alok Varshney, Rajesh Nandru, Rahul Badhwar, Ajit Sapre and Santanu Dasgupta
Reliance Research and Development Centre, Reliance Industries Ltd., Navi Mumbai, India

8.1 Climate change and its impact on food and nutrition security

A symbol of progress and necessity, agriculture is at the heart of human civilization. The rapid acceleration of agriculture-mediated food production and nutrition security has always been tempered by the constraints of nature, especially due to the process of climate change. Climate change is a multifaceted, most pressing challenge of the 21st century and for the forthcoming decades toward achieving sustainable agriculture, food security, and nutrition security. Climate change is associated with changes in greenhouse gasses in the atmosphere (e.g., CO_2, methane, and N_2O), increasing temperature, changes in precipitation pattern, and extreme weather events. It alters relationships among crops, pests, pathogens, and weeds; and it aggravates several trends including decline of pollinating insects, increasing water scarcity, and increasing ground-level ozone concentrations (Myers et al., 2017). Climate change will severely impact global food security and nutritional status, both of which are likely to deteriorate further due to the health and socioeconomic conditions of developing nations. If emerging enabling technologies are not adopted in a timely manner (Fig. 8.1), climate change will likely bring significant disproportionality and inequalities between the climates in which we reside and accessibility to food and nutrition security.

Plant Nutrition and Food Security in the Era of Climate Change. DOI: https://doi.org/10.1016/B978-0-12-822916-3.00004-4

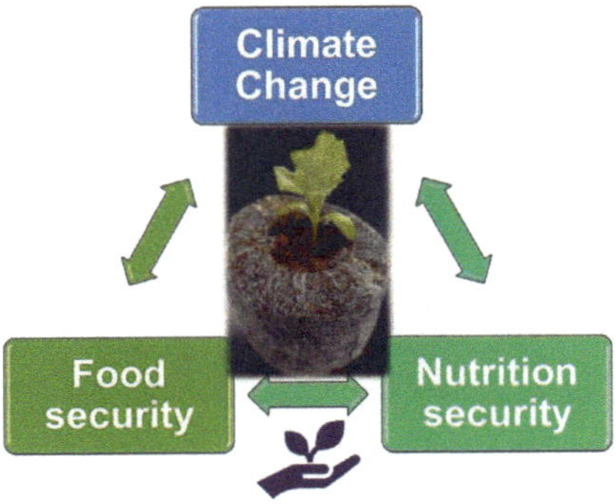

FIGURE 8.1 **The dichotomy between climate change, and food and nutrition security can be improved through soilless cultivation systems as an emerging enabler technology.**

TABLE 8.1 Global food and nutrition insecurity key metrics.

Parameter or index	Total number globally
Nutrition insecurity, 2019	3.0 billion
Food insecurity, 2019—lack of regular access to safe, nutritious, and enough food	2.0 billion
Undernourished/hungry, 2019	690.0 million
Moderate food insecurity, 2019	1.25 billion
Severe food insecurity, 2019	750.0 million
Stunted/chronic undernutrition children under 5 years of age, 2019	144.0 million
Children with micronutrient deficiencies, 2019	340.0 million
Number of people affected by hunger by 2030 (estimated)	840.0 million
Undernourished due to COVID-19 pandemic by 2020	83.0–132.0 million

Source: Adapted from FAO, IFAD, UNICEF, WFP, WHO, 2020. The State of Food Security and Nutrition in the World 2020. Building Climate Resilience for Food Security and Nutrition. FAO, Rome. https://doi.org/10.4060/ca9692en.

Recent data from the Food and Agriculture Organization of the United Nations (FAO et al., 2020; Table 8.1) described that almost 3 billion people are under nutrition insecurity, 2 billion people are facing food insecurity, and among them 690 million people, that is, 8.9% of the global population, is hungry (FAO et al., 2020). FAO projections further suggest that the world is

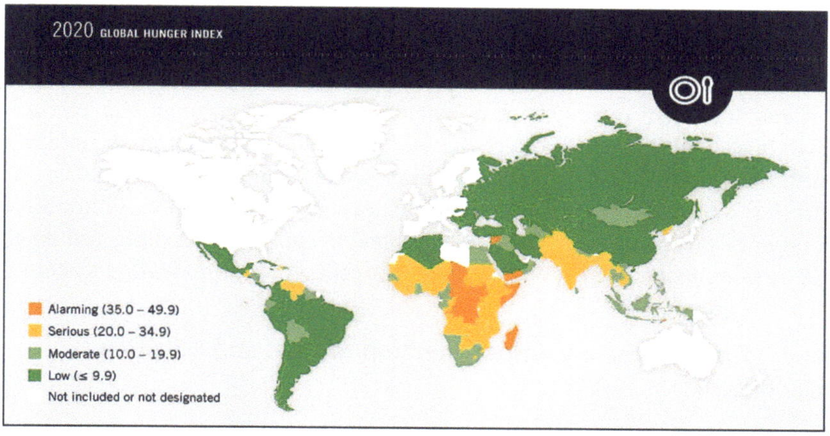

FIGURE 8.2 Global Hunger Index by Severity 2020 (Welthungerhilfe, Germany, and Concern Worldwide). *From Grebmer, v.K., Bernstein, J., Alders, R., Dar, O., Kock, R., Rampa, F., et al., 2020. Figure 1.4: 2020 Global Hunger Index by Severity. In: Map in 2020 Global Hunger Index: One Decade to Zero Hunger: Linking Health and Sustainable Food Systems. Welthungerhilfe, Bonn; Concern Worldwide, Dublin. https://www.globalhungerindex.org/results.html. For the 2020 global hunger index (GHI) data on the proportion of undernourished are for 2017–2019; data on child stunting and wasting are for the latest year in the period 2015–2019 for which data are available; and data on child mortality are for 2018. The 2020 GHI assessed 132 countries and assigned GHI scores to 107 of those countries based on GHI indicators. Based on other known data, it assigned provisional hunger severity categories to another 18 countries. These provisional severity categories do not reflect exact GHI scores for those 18 countries; rather, they show the range within which those countries' GHI scores are highly likely to fall and thus the severity of hunger there. For the remaining 7 countries, data were insufficient to allow for either calculating GHI scores or assigning provisional categories. GHI scores were not calculated for certain high-income countries, countries with small populations, and non-independent territories. The boundaries shown and the designations used on this map do not imply official endorsement or acceptance by Welthungerhilfe (WHH) or Concern Worldwide.*

not on the track to achieve zero hunger by 2030 and, despite some progress, most indicators are also not on track to meet the global nutrition targets. If such a scenario continues, then by 2030 the number of people who will remain hungry might exceed 840 million (FAO et al., 2020). This is very alarming, and such global problems are amplified further through extreme weather events such as El Nino, global conflicts, damaging pests (such as locusts and fall armyworms), and a pandemic such as COVID-19 severely impacting food production, disrupting food supply chains, and increasing the inequalities between food and nutrition security (The World Bank, 2020). Fig. 8.2 displays the global hunger index by a severity map that depicts the severity level in selected countries as alarming, serious, moderate, or low (Grebmer et al., 2020). This analysis excludes countries with high income, countries with small populations, and countries with non-independent territories.

This chapter aims to provide a better understanding of present food and nutrition insecurity due to climate change mostly mediated by elevated CO_2 and rise in land temperature. We discuss probable mitigation strategies towards the food insecurity and nutrition insecurity through the climate-smart agriculture practices such as soilless indoor cultivation technology. Furthermore, different types of soilless technologies and their emerging enabling roles toward food and nutrition security are explained. In addition, the role of precision agriculture and automation platforms is discussed as an empowering technology towards a climate-resilient food production system.

8.1.1 Climate change and connection to food and nutrition insecurity

Climate change is a multifactorial stress. Overall, 1.0°C increase in global warming since the preindustrial era has been observed and is expected to reach 1.5°C by 2050 if increases at the current rate are maintained (IPCC, 2019; Soares et al., 2019). Anthropogenic increases in CO_2 concentration and other greenhouse gases have already increased land surface temperature by ~1.0°C since 1990, and it is projected that global mean surface temperatures will likely to increase by 2.6°C–4.8°C by the end of the century (Leisner, 2020). Food security is one of the pivotal themes of the new millennium and, reasonably, the most urgent challenge for the agricultural sector. Climate change disproportionately affects poorer nations and poorer communities and widens the inequalities in food and nutrition security. This even brings agroclimatic zone disparities like rice yields; for example, a 1.0oC temperature rise predicted an approximate 20% yield decrease in India, but an approximate 10% yield increase in Russia (Agnolucci et al., 2020). Hence, it can be surmised that temperature rise due to the climate change might have more pronounced effects in tropical and temperate countries.

8.1.2 Food security scenario and climate change

According to the 1996 World Food Summit, "Food security exists when all people, at all times, have physical and economic access to sufficient, safe, and nutritious food that meets their dietary needs and food preferences for an active and healthy life." This definition is built upon four pillars of food security: (1) food availability, (2) physical and economic access to food, (3) utilization to reach a state of nutritional well-being, and (4) the stability of these dimensions (FAO et al., 2020). Out of the 2 billion people suffering from food insecurity, Asia represents 1.03 billion, Africa represents 675 million, Latin America and the Caribbean corresponds to 205 million, Northern America and Europe represent 88 and 5.9 million in Oceania (FAO et al., 2020). With this alarming 2 billion people struggling with food security, nearly 690 million people are hungry, which is almost 8.9% of the total world population. Reports indicate that if a similar trend continues more than 840 million people will be affected by hunger by 2030. Adding to this catastrophe, a preliminary analysis suggests that the COVID-19 pandemic could add around additional 83–132 million people to the category of undernourished in the world in 2020 (FAO et al., 2020).

Studies revealed that each degree-Celsius increase in global mean temperature could reduce global yields of wheat, rice, maize, and soybean by 6.0%, 3.2%, 7.4%, and 3.1%, respectively (Zhao et al., 2017). Climate change–mediated shortened chilling durations have been reported to have negative impacts on vegetable quality, which directly impacts access to good food quality (Leisner, 2020). Lack of chilling hours may lead to delayed bloom or leaf emergence, but delays can be compensated for by quicker heat accrual for plants in the spring. Several studies have demonstrated that in temperate and subtropical regions winter chill is declining, and this decline is likely to continue in the future. Breaking dormancy is assumed to require both a chilling requirement and a heating recruitment in perennial species, which makes them particularly sensitive to changes in seasonal temperatures. Hence, the shifting of seasons or delayed appearance due to climate change is of paramount concern. Scientific investigations on the effect of elevated CO_2 (eCO_2) concentration on vegetable quality revealed a decrease in macronutrients and micronutrients in lettuce and spinach, and decreased nitrate content in lettuce, celery, and Chinese cabbage. Similarly, extended exposure to eCO_2 could affect the photosynthetic acclimation process and lead to an imbalanced C:N ratio, accelerated leaf senescence, and limited growth rate (Soares et al., 2019).

8.1.3 Nutrition security scenario and climate change

Along with food security where caloric need or access to food at all levels of society at all the time is made available, meeting nutrition security is a major concern worldwide. It can be achieved through the secure access to an appropriately nutritious diet that provides not only acceptable calorie but also balanced macronutrients and micronutrients in combination with a sanitary environment and adequate health services (FAO et al., 2020). The present global scenario is alarming in terms of nutrition security with an estimated 662 million people deficient in protein, 1.5 billion deficient in zinc (Zn), and around 2 billion people encountering iron (Fe) deficiency (Smith and Myers, 2018). Global assessment by FAO estimated that by 2050 more than 175 million people worldwide (1.9% of global population) would be deficient in Zn, around 122 million additional people would be deprived of protein, and approximately 1.4 billion women and children under the age of 5 years might lose an additional 4% of their dietary Fe intake and face concurrent increased rates of anemia (FAO et al., 2019; 2020).

Climate change influences the buildup of minerals and protein in crop plants, with eCO_2 being the underlying factor of most of the reported changes (Soares et al., 2019). Recent work in C3 crops has described a substantial decrease in several minerals (~8%) in foliar and edible tissues (Loladze, 2014) and around 3%–17% reduction in protein, zinc, and iron concentrations under eCO_2 conditions (Leisner, 2020). The effect of eCO_2 is described to possess a role in decreasing nitrogen (N) and protein concentrations in many crops (e.g., potato and soybean) (Leisner, 2020). It is noteworthy to mention that at eCO_2 an alarming reduction in protein concentration is reported in wheat (6.3%) and rice grains (7.8%) (Myers et al., 2014; Soares et al., 2019).

Medek et al. (2017) established that under eCO_2 conditions (500–700 ppm) a significant decrease in protein concentration is reported in wheat (7.8%), rice (7.6%), barley (14.1%), potato (6.4%), vegetables (17.3%), and fruit (23.0%). With the constant increasing atmospheric CO_2, it is predicted that 1.4 billion people worldwide would be at risk of protein deficiency by 2050. It is anticipated that an additional 148.4 million people will face protein deficiency, if projected CO_2 levels go above 500 ppm by 2050 (Medek et al., 2017; Smith and Myers, 2018). Increasing CO_2 levels have been reported to decrease the Zn levels in many food crops such as wheat (9.1%), rice (3.1%), barley (13.6%), field peas (6.8%), and soybean (5.0%) and is a significant global health problem (Myers et al., 2017). Similarly, eCO_2 decreased the Zn concentration by 9.4% in vegetables as described by Dong et al. (2018).

8.1.4 Climate-smart agriculture and soilless cultivation as an "emerging enabler" technology

In the current context of massive deforestation, water scarcities, soil depletion, and high levels of greenhouse gas emissions, high-input and resource-intensive farming systems cannot deliver sustainable food and agricultural production. Advanced, alternative systems that protect and augment the use of natural resources while at the same time enhancing crop productivity are the need of the hour (FAO, OECD, 2018). To overcome bottlenecks and challenges associated with the food and nutrition insecurity created through the nuisance of climate change, many climate-smart agriculture practices or technologies have been adopted globally. Some of the climate-smart technologies and practices are crop rotation, crop diversification, use of renewable sources of energy like solar and wind energy for driving farm machinery, reduced tiller, novel breeding crop variety development, and genetic intervention including CRISPR technology. Although global reduction of arable land, insufficient fresh water supply, population outburst, rampant use of fertilizers, toxic chemicals, and urbanization create a demand for additional alternative technologies to emerge.

Furthermore, outdoor soil-based agriculture poses a multitude of global problems including soil erosion, changes in soil organic carbon (SOC), soil health in terms of nutrient imbalance, geography-specific soil salinization, soil alkalinity enhancement, soil acidification, loss of microbiome, soil contamination, and water logging (FAO, OECD, 2018). Hence, soilless crop cultivation and allied controlled environment agronomic practice symbolize a promising technology toward environment friendly agriculture and a sustainable vision towards mitigating general challenges in terms of food and nutrition security.

8.2 Soilless indoor cultivation as climate-smart agriculture

Soilless cultivation in controlled environment agriculture (CEA) cannot completely eradicate food and nutrition security. Although, it cannot provide complete sustainability, soilless cultivation does provide an "emerging enabling technology" for contributing significantly toward improving food availability, food quality, and nutritional qualities. Especially with regards to urban and periurban populations, soilless CEA is smart and pivotal technology. In addition, combining other emerging technologies such as the use of nanoparticles, beneficial microorganisms, microbiota, and precision and automated agriculture practices might improve the management of soilless cropping cultivation ecosystems.

In the following Sections (8.2.1 to 8.5.3), the theme is to describe soilless cultivation system, types of soilless cultivation methods, and to what extent this technology can enable food and nutrition security toward meeting the UN sustainable development goals 3 and 13. The role of precision agriculture and automation platforms is opined as empowering for the management of soilless smart CEA.

8.2.1 Controlled environment agriculture and soilless cultivation technology

Traditional agriculture system for the production of food crops encounters several challenges, including abiotic stresses, food safety issues. Hence, many progressive farmers growing these crops hydroponically in a controlled environment. Greenhouse and indoor-grown crops provide many advantages over field-grown crops, including a more stable climate, more consistent time to harvest, increased food safety, and the ability to control lighting, shading, and CO_2. Low ambient light levels in the winter months and on cloudy days make supplemental light essential for producing high-quality, uniform greens year-round.

CEA is defined as an integrated science and engineering-based approach of farming technology that allows a grower to control and optimize a crop's growth environment including microclimatic conditions to improve plant productivity and optimize the use of resources. CEA technologies include greenhouse, hydroponics, aquaculture, and aquaponics. Variables controlled include temperature, humidity, pH, and nutrient analysis. Growers can create optimal growing conditions throughout the year from anywhere in the world (CEAC Arizona, 2020).

Growing in a controlled environment also vastly improves a farm's ability to predict crop timing, grow high-quality plants, and maintain high food safety standards (Benke and Tomkins, 2017). Many CEA farms practice integrated pest management to prevent and deal with pest problems thereby avoiding use of pesticides. CEA provides simple protected solutions such as tunnels and walk-in tunnels, and advanced solutions such as high-tech greenhouses, naturally ventilated poly-houses, anti-insect net houses, and computerized irrigation and fertigation technologies including utilization of green energy; with these techniques, a significant increase of 5–10 times in crop yields was observed. This also allows reduced use of pesticides and prolongs the harvesting period from 3 to 9 months (IIAP CoE, 2018).

Soilless farming can be good for the environment because it significantly reduces the amount of fossil fuels needed for farming equipment, which is not required to sow, fertilize, weed, or harvest crops. It allows water to be recycled and reused, reducing water wastage and requirements. The growth of soilless farming is limited by factors that include initial and operational costs, insurance options, and business model development (El-Kazzaz and El-Kazzaz, 2017). In soilless farming agriculture, macronutrients and micronutrients provided for plant growth are premixed into a water reservoir, along with high levels of oxygen, efficiently supplied to the plant roots. Soilless cultivation systems can produce 100-fold increases in yield per square meter output if grown in vertically stacked trays (Beacham et al., 2019; Touliatos et al., 2016). According to Plenty farms (United States), water use is dramatically reduced by up to 90% through circulating soilless cultivation practices. This is a significant advantage when compared with the traditional outdoor soil-based agriculture where 70% of the world's fresh water is used, creating uncertainty in terms of water security globally.

8.2.2 Different types of soilless cultivation

Soilless crop cultivation systems employ either circulating or non-circulating modes of nutrient supply dissolved in water through smart irrigation technology. Depending upon the growing substrate medium (liquid or solid), soilless cultivation technology is majorly divided into circulating and non-circulating types. In a closed soilless circulating system, the fertigated solutions are recycled and the nutrient concentrations are monitored continuously and optimized. This system maintains regular mixing of growth solution by repeatedly pumping the solution into the growing channel at regular intervals and allowing it to drain. Continuous circulation of solution during the pumping cycles facilitates delivery of oxygen to the roots. This type of system maintains aeration and does not hamper plant growth. Continuous monitoring and adjustment of optimum pH and electrical conductivity (EC) as per crop type and growth stages are key to the nutrient uptake process by plants in a circulating system. A closed soilless system incorporates both basic and advanced soilless cultivation frameworks. To ensure food-grade quality, polycarbonate plastic fragments with characteristic chemical inertness (no changes in the composition of available nutrients provided by the growth solution) and surface wetness (facilitates the continued supply of moisture and nutrients during draining) are being used. An open soilless system works with constant aeration and less complexity. Commercially available aquarium pumps, plastic tubing, and aeration stones are used to achieve continuous aeration. In open soilless cultivation, the roots of very young seedlings are directly submerged in aerated nutrient solution and the shoots are supported to grow above the solution. Newly dissolved nutrient dosage is involved for every irrigation cycle of an open soilless cultivation system. The dissolved nutrients can also be supplied to the plants by using a drip irrigation system (El-Kazzaz and El-Kazzaz, 2017; Shavrukov et al., 2012).

8.2.2.1 Circulating soilless cultivation systems

Circulating soilless cultivation systems are further divided into aeroponics, aquaponics, and hydroponics cultivation systems.

Aeroponics is the process of growing plants in an air or mist environment without the use of soil or misting the roots with hydroponic solutions, which are suspended in the air. It does not use soil or aggregate medium. Aeroponics is a subgroup of hydroponics where plant roots are suspended in a dark enclosure, while a nutrient-dense solution is sprayed on the roots at certain intervals (Alshrouf, 2017). Nutrient delivery in this system is very precise and faster growth is possible as roots get ample oxygen. As very tiny holes are used for spraying, this system can get clogging issues (El-Kazzaz and El-Kazzaz, 2017).

Aquaponics is a symbiotic system of cultivation environment in which aquatic animals like snails, fishes, etc. are grown in reservoirs with a combination of hydroponics in which plants are grown in water. While the fish provide fertilizer for the plants, the plant roots filter the water for the fish (El-Kazzaz and El-Kazzaz, 2017).

Using solely water as a substrate, ***hydroponic*** system types include nutrient film technique (NFT), ebb-and-flow system, deep water culture (DWC), drip system, wicking system, and aquaponics (El-Kazzaz and El-Kazzaz, 2017). Some systems involve an aggregate as a substrate, for example bag culture, container culture, trough culture, thin layer systems, and various alternative systems. Within hydroponics, there are several techniques to get the best customized operation. Depending on the space and crop type, growers can choose one of these techniques.

Ebb-and-flow systems are similar in operation as that of NFT, but the grow tray is temporarily flooded with solution every few hours submerging the roots before returning to the reservoir. Ebb-and-flow systems are low maintenance and produce high yields unlike NFT systems wherein pump failure can quickly become catastrophic for the plants.

In the ***DWC*** method, nutrient-rich oxygenated water is the medium wherein plant roots are suspended in about 6–18 inches until harvest. DWC systems hold a large volume of water slowing any swings.

A ***drip system*** is another technique wherein a pump delivers a slow feed of the solution to the base of each plant individually based on a timer, and the remaining solution will be either returned to the reservoir or not collected.

A ***wicking system*** is a technique by which solution is delivered to a tray through the wicks and then to the roots via the plant's capillary action.

NFT is a popular and versatile technique in which the system uses a pump to deliver water and the unused nutrient solution is recycled through a drainpipe. In NFT, continuous flowing over the roots is happening with the nutrient solution through gravity.

By closing the loop and recycling the water, an NFT hydroponics system uses only 10% the amount of water used in conventional farming methods. As only a small portion of water and nutrients is utilized by plant roots on traditional outdoor soil farms, the rest is unused and wasted. In vertical hydroponics stack systems, the yield per square foot increases highly (Touliatos et al., 2016). This is a critical component as we need to grow enough food for growing populations without needing to expand farmland into wetlands, forests, or other important natural ecosystems. Crop lifecycle in a typical NFT hydroponics soilless cultivation system is described in Fig. 8.3. Physical and hydraulic properties are better in soilless substrate compared to those of soil medium. Economic losses caused by soil-borne pathogens could be avoided in soilless media for commercial greenhouse production purposes (Asaduzzaman et al., 2015).

8.2.2.2 Non-circulating systems

In non-circulating soilless cultivation systems, growing media function as a reservoir for holding water and nutrients including a zone for gaseous exchange for the plant root system and anchorage for plant roots. Growing media components when mixed appropriately can provide appropriate water, air, nutrients, and support to plants. Healthy growing media will be free from soil pests, pathogenic organisms, soil insects, nematodes, and weed seeds. They are porous, well drained, and relatively low in soluble salts. The pH will be retained between 5.2–6.8 and EC around 2.0–3.5 mS/cm.

FIGURE 8.3 Schematic representation of complete cycle of leafy vegetable production in soilless indoor hydroponics cultivation system (© Reliance Industries Ltd; 2020). *Different steps of crop growth cycle have been depicted here starting with seed sowing in cocopeat plugs, followed by germination in optimum conditions of temperature, relative humidity and other microclimate conditions followed by appearance of true leaves in plantlets. After appearance of true leaves plantlets are transferred in food-grade NFT (nutrient film technology) channels in food-grade net-pot systems without disturbing the root system and without detaching from substrate and this is called transplantation. Under optimum conditions of growth plant depending of variety and genus completes its early vegetative growth, mid-vegetative growth and matures within a month a crop cycle to generate fresh soilless produce without the use of any pesticides or toxic chemicals in a climate friendly manner. Soluble nutrients are supplied along with circulating water system in an optimized concentration using a submersible pump system. The same is recirculated in its entire growth cycle and thereby conserving total freshwater requirement in hydroponics cultivation system and thus providing climate smart agriculture via water security (© Reliance Industries Ltd; 2020).*

Understanding the composition, functions, and intended use can make the selection of growing media formulations challenging. Most commercial growing media are pH adjusted with either calcific or dolomitic limestone and contain a balanced starter fertilizer to help plants acclimate after planting (Gruda et al., 2013). It is generally recommended to begin fertilization once new plant leaves begin to emerge and new roots develop. Based on the crop type, stage of development and container size, the volume and frequency of applications will vary (Gruda et al., 2016).

Commercially used growing media components in soilless cultivation are either organic or inorganic. Organic components include cocopeat, peat moss, bark, tree, wood fiber, etc., and inorganic components include perlite, pumice, vermiculite, sand, and hydrogel (Gruda et al., 2013). Some of these components hold water on their surface, while some can hold water on their surface as well as within their structure, and few perlite-like hold very little water. Depending on the material's origin and how a specific type of

TABLE 8.2 Evaluation of different types of commercially used substrate grow media in soilless farming system (Gruda et al., 2013, 2016; Popsimonova et al., 2017; Alshrouf, 2017).

Substrate	Physiochemical properties	Advantages	Limitations
Cocopeat	Physical stability; natural antifungal property; air porosity 12.2 (V/V) %; EC ≤ 0.5–0.8 mS/cm; density 0.05–0.16 g/cm^3; pH 5.5–6.8; WHC 79.8 (V/V) %; CEC 138.7 (cmol/kg)	Low pH, low density; excellent cation exchange and wet ability; retains nutrients; good porosity water-holding capacity	Potentially high salt level
Rockwool	High pH and capillary action; air porosity 10.3 (V/V) %; WHC 83.8 (V/V) %; CEC 0.0 (cmol/kg); bulk density 0.06 g/cm^3	Easy to handle; weightless; totally inert media; nutrition control possible	Disposal problems; lot of energy required during manufacture
Perlite	pH 7–7.5; density 1.1 gm/cm^3 (raw), 0.03–0.15 gm/cm^3; air porosity 41.1 (V/V) %; WHC 25.2(V/V) %; CEC 0.0 (cmol/kg); bulk density 0.13 g/cm^3	Lightness and uniformity; sterility; low density; good aeration and drainage	Very dusty when dry; float at the top during irrigation, toxic levels of fluorine
Vermiculite	Hydrous mineral heating at 745°C; ability to supply potassium and magnesium; pH 7–7.5 and density 172 kg/m^3	High water-holding capacity, helps in aeration and drainage; excellent cation exchange and buffering capacities	Less durable than sand and perlite
Peat and peat-like material	pH 3.0–4.5; bulk density 0.207 kg/m^3; EC 0.02–0.29 dS/m; air porosity 13.2 (V/V) %; WHC 76.8 (V/V) %; CEC110.5 (cmol/kg); bulk density 0.11 g/cm^3	Low pH, easy to adjust; spongy, fibrous, high water-holding capacity; low microbial activity; lightweight	Strong acidity possible; expensive

(*Continued*)

TABLE 8.2 (Continued)

Substrate	Physiochemical properties	Advantages	Limitations
Gravel	Density 1601.84 kg/m^3; hardness 6.0–7.0	Inexpensive; easy to keep clean; drains well and will not become waterlogged	Heavy; dries fast if continuous water supply is not provided

FIGURE 8.4 Schematic representation of (A) NFT circulating soilless cultivation system, and (B) grow bag noncirculating cocopeat cultivation system (© Reliance Industries Ltd., 2020).
Representation of hydroponics based NFT circulating cultivation system in vertical stacks for round the year leafy vegetable production system under artificial light conditions in controlled environment system (A). Representation of non-circulating grow bag substrate based non-circulating soilless cultivation system in poly-house based indoor cultivation system (B).

ingredient is processed, it can vary in its physical structure and water-holding capacity (Gruda et al., 2016; Popsimonova et al., 2017). Physiochemical properties, advantages, and disadvantages of these different growing substrates are described in Table 8.2. Growing substrate-based cultivation is carried out in different types of grow bags such as vertistack system, Dutch bucket type, trough system, single planter, and multiplanter. In grow bag soilless cultivation systems, a multitude of crops are grown including vine crops, tomato, pepper, strawberry, cucumber, and muskmelon. Fig. 8.4 shows a circulating NFT system and non-circulating grow bag system. Taken together, both kinds of systems can provide prudent climate-friendly cultivation systems for various vegetable and fruits crops including leafy vegetables.

8.2.2.3 Comparisons between indoor soilless versus soil-based conventional agriculture systems

To overcome certain limitations of soil-based cultivation systems, soilless indoor cultivation provides enabling solutions such as year-round cultivation, prevention of soil-borne or soil-mediated diseases, no propensity of weed contamination, higher yield per unit area, less required horizontal cultivation area, a significant reduction in fresh water use, reduced use of toxic chemicals, minimization of preharvest and postharvest food loss, and opportunities for precision nutrition management. Table 8.3 compares the major differences between soilless indoor cultivation systems and soil-based conventional cultivation systems.

8.3 Soilless systems as an enabling technology for food security

Due to intensive farming practices and climate change, the evitable outcomes are land degradation, soil erosion, flooding, drought, and decline in soil fertility (Tajudeen and Taiwo, 2018). Hence, there is an urgent need to deploy sustainable alternatives to soil-based farming systems. Soilless farming is a method of growing plants by which nutrients absorbed by plants roots are supplied through irrigation water without the use of soil as a rooting medium (Orsini et al., 2013). To achieve sustainable food security in different parts of the world, soilless farming could be a viable agriculture practice that can provide all the credentials of soil while overcoming the challenges faced in conventional soil-based farming (Tajudeen and Taiwo, 2018). The success of soilless systems relies much on the fertigation solution supplied to the plants: its availability, affordability, and reliability in developing countries for small-scale growers in a daily or weekly quantity. The production of horticultural goods shows the main benefits of climate-smart agriculture. With more efficient use of agricultural inputs and high-value additions, fruit and vegetable crops give significantly high yields on soilless farming, up to 50 $kg/m^2/year$, that can easily fulfill the rural production for the local market (Orsini et al., 2013).

8.3.1 Year-round productivity and yield improvement in soilless cultivation systems

By the virtue of efficient use of water and fertilizer, better physiochemical and hydraulic properties, lack of soil-borne pathogens, lack of weeds, and amidst controlled environment conditions, a soilless cultivation system facilitates year-round crop production with increased yield compared with soil-based cultivation (Asaduzzaman et al., 2015). As per the UN estimate, by 2050 global population will surpass 9 billion with 80% of the population

TABLE 8.3 Comparison of soilless cultivation system and conventional farming system.

Parameter	Soilless cultivation system	Soil-based conventional agriculture	Reference(s)
Climate resource efficiency and land size	Cultivation year-round; control over the environment such as temperature, humidity, and photoperiod; vertical farming minimizes land size	Year-round cultivation not possible; highly fluctuating environment; size of arable and fertile land is decreasing	Noren et al. (2004); Lee and Lee (2015)
Crop quality and yield	Higher yields (e.g., tomato yield 12–72 kg/m^2); superior quality of production; fresh with better color, taste, and aroma	Lower yields (e.g., tomato yield 1.0–2.5 kg/m^2); low-quality produce and food loss is a major problem	Lee and Lee (2015); Resh et al. (2013); Touliatos et al. (2016)
Water use efficiency	30%–70% water can be saved	Poor water use efficiency and climate change or extreme precipitation decreases water use efficiency	Resh et al. (2013); Lee and Lee (2015)
Micronutrient uptake	Superior and can be controlled through precision agriculture	Highly dependent on soil health, quality, and microenvironment	Savvas et al. (2013)
Diseases and pests	Prevent soil-borne diseases; easy to control insects and animals; reduced amount of pesticide usage; no weeds	Soil-borne diseases; hard to control insects and pests; removing weeds is laborious; rampant use of toxic chemicals for pest, disease control	Jones et al. (1997)
Sanitation	Easy to maintain sanitary conditions; ease of handling materials and media	Difficult to maintain sanitized conditions and sanitize soil; soil nematodes and plant pathogens are bigger problem	Knutson et al. (2000)

(Continued)

TABLE 8.3 (Continued)

Parameter	Soilless cultivation system	Soil-based conventional agriculture	Reference(s)
Carbon footprint and GHG emission	Reduce GHG emissions; stored carbon compounds improve fertility and helps retaining moisture	Higher emission of greenhouse gases	Romeo et al. (2018)

TABLE 8.4 Comparison of vegetable crop yield between soilless controlled environment agriculture and open-field soil-based agriculture (IIAP CoE, 2018).

Crop name	Soilless system (t/acre)	Open-field soil-based system (t/acre)
Cucumber	18	1.4
Capsicum	29	4.8
Tomato	38	6.5
Cherry tomato	29	Cannot be cultivated in open field

residing in urban areas with demand of food supply increasing by 70%. Hence, a soilless system is an alternative system for food production amidst farmland shortage, reducing "food miles," producing healthy and good-quality food, and thereby reducing food wastage and food loss (Al-Kodmany, 2018).

A very recent study by Asseng et al. (2020) demonstrated that in a land surface area of 1 hectare using a 10-layer soilless indoor vertical system 700 t/ha of wheat production is achieved. This wheat production capacity is around 220 times higher than the current world average annual wheat yield of 3.2 t/ha (Asseng et al., 2020). Under the Indo-Israel Center of Excellence initiatives using CEA along with soilless cultivation, automated irrigation and fertigation technologies significantly enhancing vegetable crop yield has been reported. This yield value is almost 5–10 times higher than the soil-based cultivation system (Table 8.4). In addition, such technologies utilize reduced pesticides and toxic chemicals, extend the harvesting period from 3 to 9 months, and produce fresh-quality harvest (IIAP CoE, 2018).

TABLE 8.5 Comparison of crop yield between soilless controlled environment agriculture and soil-based outdoor agriculture (Singh and Singh, 2012).

Crop name	Soilless CEA hydroponics yield (t/acre)	Soil-based outdoor Ag yield (t/acre)	Increased fold
Lettuce	9.50	4.00	2.30
Cucumber	12.70	3.10	4.10
Tomato	16.30	4.50	3.62
Peas	6.30	0.90	7.00
Soybean	0.67	0.27	2.48
Okra	8.60	2.20	3.90
Beet root	9.00	4.00	2.25
Cabbage	8.10	5.80	1.39
Oats	1.35	0.38	3.55
Maize	3.60	0.68	5.29
Cauliflower	13.50	4.50	3.00
Rice	5.40	0.34	15.88
Potato	63.00	7.25	8.68
Wheat	2.20	0.27	8.10

Singh and Singh (2012) reported comparative yield data between soilless hydroponics cultivation and a soil-based outdoor cultivation system for 14 crops including vegetables such as tomato, cucumber, cabbage, and cauliflower, and cereals such as rice, wheat, and maize. Table 8.5 depicts yield comparison data that clearly touts the potential of soilless cultivation systems as evident by 2-fold to 10-fold enhancement in productivity compared with soil-based farming systems. Michelon et al. (2020) demonstrated adoption of a simplified soilless cultivation system over a traditional soil-based system resulted in lettuce crop yield enhancement by 35% and 72% in northeast Brazil and the central dry zone of Myanmar, respectively. This study has demonstrated water use reduction by 76% and 59% in Brazil and Myanmar respectively in the soilless system compared with the soil-based cultivation system. Tomato growers in Slovenia reported an astounding annualized yield with use of a soilless glass greenhouse-based intensified cultivation system. Data analysis revealed annualized tomato yield of 220 t/acre for soilless cultivation, whereas the soil-based open cultivation system produced 86 t/acre (FAO, 2018). Such results suggest that soilless cultivation systems could become one of the efficient strategies for contributing to sustainably feeding the world's growing population with certain technological advancements for a variety of crops.

Agriculture production in countries like Qatar is facing a number of challenges such as harsh environment, water scarcity, groundwater depletion, and limited availability of suitable soil. In 2010, a technical assessment report by the International Center for Agricultural Research in the Dry Areas (ICARDA, 2020) carried out on Qatar national food security with special emphasis on protected agriculture and hydroponics for high-value cash crops. This technical report analyzed comparative productivity studies between soilless cultivation systems and open field soil-based cultivation systems along with productivity per unit of water and land for different vegetable crops and fruits. The reported yield data for soilless cultivation systems are 56 t/acre for tomato, 40 t/acre for cucumber, 31 t/acre for sweet melon, 20 t/acre for sweet pepper, and 16 t/acre for green bean. The open field soil-based cultivation system yield data revealed 12 t/acre for tomato, 6 t/acre for cucumber, 3.6 t/acre for sweet melon, 2.4 t/acre for sweet pepper, and 3 t/acre for green bean (Moustafa, 2010). Similarly, agronomic assessment of a greenhouse soilless production system in the Arabian Peninsula described doubling of production per unit area for strawberry cultivation when compared with a greenhouse soil-bed-based cultivation system. In this soilless production system, 40 plant/m^2 of strawberry plant density has been achieved with a land surface productivity of 16 kg/m^2 with superior crop quality and 50%–60% reduced production inputs. Comparatively, in the case of the greenhouse soil-bed system, plant density is found to be 5 plants/m^2 and land surface productivity is 2.9 kg/m^2 (Moustafa and Ouraifan, 2004). Thus, increasing evidence suggests that soilless cultivation along with a CEA production system is an evolving technology and an alternate strategy for enabling food security through increased crop yield and an extended cultivation and harvesting period.

8.3.2 Improving resource use efficiency—water use efficiency and land surface use efficiency

Depleting groundwater, decline of available fresh water for agriculture, reduced arable land surface area, along with desertification due to climate change demand sustainable (Pastor et al., 2019; Sambo et al., 2019; Verones et al., 2017) and enabling technologies that can improve the resource use efficiency, such as minimizing water use and producing more in less land surface area. By 2050, the per capita availability of agricultural land is expected to be reduced by half as a result of increasing population and economic use of land for many other purposes (Fischer and Shah, 2010). This in

TABLE 8.6 Comparison of water use efficiency: crop water requirement (liters per kilogram) in soilless systems and conventional soil-based cultivation systems (Sambo et al., 2019).

Crops	Soilless cultivation systems	Soil-based conventional Ag
Tomato	35	78
Hot pepper	58	110
Sweet pepper	17	121
Zucchini squash	39	97
Muskmelon	42	170
Lettuce	1.6	76
Strawberry	136	544
Brassica	5.0	129
Spinach	8.3	106

turn is pressuring the agricultural community for expanding the land under cultivation. Water use efficiency (WUE) has been proven to be significantly much higher in soilless cultivation systems compared with the soil-based cultivation systems.

Table 8.6 describes the comparison of crop WUE for different crops between soilless systems and conventional soil-based cultivation systems and clearly demonstrates that soilless cultivation systems can minimize water consumption by 50%–90% (Barbosa et al., 2015; Dukes et al., 2012; Sambo et al., 2019; Van Ginkel et al., 2017). In an ICARDA technical assessment report, crop productivity per unit of water for selected vegetable and fruit crops has been compared between soilless cultivation systems and soil-based cultivation systems around Qatar, Oman, and United Arab Emirates (UAE). In Table 8.7, it is evident that crop productivity per unit of water is 2-fold to 15-fold higher in case of soilless cultivation systems when compared with the soil-based cultivation systems (Moustafa, 2010). Furthermore, detailed analysis revealed that to produce 1 kg of tomato total water requirement was estimated to be around 110 L in soil-based cultivation systems, whereas a similar amount of water produced 2–3 kg of tomato in soilless systems. To produce 1 t of strawberry in soil-based systems, water utilization was found to be around 507 m^3, while the same amount of water could produce 7.9 t of strawberry in hydroponics soilless cultivation systems. In the same technical report, it has been observed that 200 m^3 of water is consumed to produce 1 t of muskmelon in a soil-based open cultivation field, whereas with the same water input a soilless hydroponics system could produce 15 t of muskmelon

TABLE 8.7 Comparison of crop productivity per unit of water use in soilless systems and conventional soil-based cultivation systems in the Arabian Peninsula (Moustafa, 2010).

Crops	Soilless cultivation systems productivity (kg/m^3)	Soil-based conventional agriculture productivity (kg/m^3)
Tomato	38	19
Cucumber	56	11
Lettuce	13	1
Green bean	14	10
Strawberry	33	4
Muskmelon	76	5

(Moustafa, 2010). Ouled Belgacem et al. (2017) reported 90% water savings for strawberry production through hydroponics when compared with conventional soil-based systems in Kuwait along with 20% reduction in fertilizer consumption and strawberry yield improvement by 4.6-fold in hydroponics cultivation. An ICARDA (2020) report of agroeconomic case studies shows examples of improved WUE in Yemen, with soilless cultivation medium resulting in productivity of 73.9 kg/m^3 for cucumber and conventional soil system producing 20 kg/m^3 in a greenhouse-protected environment (ICARDA-APRP Report, 2020). It is noteworthy to mention that cost-benefit analysis for the same case study has indicated the soilless system helped to improve the net income by 65%, which is a positive impact toward reducing the gap between food security and income disparity. Data from Agrinatura (2020), suggest that there has been an increase in controlled environment cultivation systems around Saudi Arabia, United Arab Emirates, and it has been reported that such agronomic practices save up to 120 m^3 water per ton of tomato production. In Kuwait and Oman vertical soilless production systems resulted in 50% less water consumption while at the same time increasing strawberry yield by 5-fold when compared with conventional greenhouse systems (Agrinatura, 2020).

Such vertical soilless systems are equally contributing toward producing more in less land surface area and also helping to improve "land surface use efficiency". With rapid urbanization, shrinkage of arable land, and increased demand for nutritionally rich food, improving the land surface use efficiency is of paramount importance (Orsini et al., 2020). In this scenario, a soilless vertical hydroponics system allows a greater number of plants per unit area, which can potentially increase yield per unit area of land. In such a system plants grow rapidly and with a shorter lifecycle from the germinating to the harvesting steps maximizing productivity and profitability (Beacham et al., 2019). Hydroponics lettuce cultivation in Arizona in conjunction with modeling studies estimated annualized yield of 41 ± 6 kg/m^2/year with an annual water demand of 20 ± 3.8 L/kg/year. Using a conventional soil-based system, annualized lettuce yield has been reported to be 3.9 ± 0.21 kg/m^2/year with an annual water demand of 250 ± 25 L/kg/year (Barbosa et al., 2015). Hence, a soilless cultivation system provides an opportunity to increase land surface use efficiency by 10-fold, which is a promising step toward enabling food security to some extent. However, innovation and superior technologies are needed in order to minimize energy expenditures and initial development costs.

8.3.3 Role of microbiome and biostimulants in improving crop performances in soilless systems

Pesticides, inorganic fertilizers, and other chemicals have increased yield for intensive agriculture practices but also have contributed to increased susceptibility of crops to pathogens/pest, loss of biodiversity, and increased soil degradation. All these consequences significantly impact food security and human health. Sustainable crop production demands additional emerging technologies. Manipulating the plant holobiont by microbiome engineering is one such biotechnological strategy to improve crop yields and resilience (Tilman et al., 2002).

The microbial symbiotic association with a plant is called a "plant microbiome" (Vorholt, 2012). Plant-associated microbiomes whether in core or satellite ecosystems can boost plant growth, enhance plant development, improve nutrient use efficiency, and mitigate against pathogens as well as abiotic stresses. Microorganisms like bacteria, fungi, and archaea were recognized consistently in association with plants in different spheres. Microbial colonies can exist inside plants (endosphere), externally on roots (rhizosphere), and on leaves (phyllosphere) (Arif et al., 2020). Beneficial microbes result in improved crop traits and performance via a plethora of mechanisms. Some of the mechanistic aspects are modulating plant endogenous hormone levels by producing phytohormones and secreting 1-aminocyclopropane-1-carboxylic acid deaminase leading to reduced stress levels, improved nitrogen fixation, phosphate solubilization, improved nutrient uptake, triggering microbe-associated molecular patterns for defense against pests and diseases, reducing oxidative stress, and decreasing metal toxicity (Arif et al., 2020).

Promising results using microbiome applications from large-scale field trials have demonstrated an increase of 10%–20% in crop production on economically important crop plants advocating the strength of technology (Perez-Jaramillo et al., 2016). Recently, industrial demonstration of microbiome engineering for agriculture application was carried out by a commercial organization, "Indigo Ag." Wheat, cotton, and corn seed coated with microbiome have shown improvement of crop yield by 12.7%, 14%, and 40%, respectively (Indigo Ag, 2019).

In soilless systems, use of plant microbiome is a sustainable and emerging approach for global food production enhancement and food quality improvement (Dere et al., 2019). In the hydroponic solution *Bacillus sphaericus* UPMB10 and *Azospirillum brasilense* Sp7 carries biological nitrogen fixation and reduces the N2 utilization (Baset Mia et al., 2010). *A. brasilense* enhanced the nutraceutical qualities and sweetness in strawberries (Pii et al., 2018). However, plant microbial association in soilless cultivation is different from soil plant microbe association. Major advantages are efficiency on plant nutrition, which can decrease the cost of fertilization, improve the growth and yield, and enhance the plant resistance to biotic and abiotic stress. Commercial *Bacillus amyloliquefaciens* (FZB24 and FZB42) improved 8%–9% of tomato yield (Gül et al., 2008).

In hydroponics, water is a substrate and it plays a key role in the microbial community; water storage condition also plays a role in microbial diversity and its population. In the hydroponic system *Mycorrhiza* improves the melon leaves, nutritional aspects, and fruit quality (Dere et al., 2019). Hydroponic cultivation of tomatoes showed 160 bacterial strains identified from nutrient solutions and they diversified into five taxonomic groups: Bacteroidetes, γ-Proteobacteria, Actinobacteria, α- and β-Proteobacteria (Berkelmann et al., 1994). Microbial changes in the hydroponics system are particularly vulnerable to external influences (Xiao et al., 2010). Beneficial microbes control the plant pathogens by producing the siderophore or antipathogenic substances and enhancing the bioavailability of nutrients (Pii et al., 2015). When lettuce was grown in a hydroponic system treated with *Pseudomonas chlororaphis*, transcriptome analysis of treated test sampling proved that *P. chlororaphis* treatment significantly promoted the growth of lettuce by upregulating the genes that were related to stress response, defense response, and metabolic pathways (Lee and Lee, 2015). *P. chlororaphis* treatment even influences the signaling activity, stimulates transport activity, regulates phosphorylation activity, and promotes resistance to ecological and osmotic stresses (Pii et al., 2012). Table 8.8 describes different examples of applications of microbiome for crop trait and performance improvement, which may contribute toward food and nutrition security.

8.3.3.1 Root zone microbiome application in soilless systems

Plant and microbe association is highly dynamic over a plant's developmental stage and time. In a hydroponic system, beneficial microbe contribution makes the host plant abiotic stress-tolerant and improves productivity. However, the proper substrate (soilless)-plant–microbe system and the

TABLE 8.8 Microbiome application examples for crop trait improvement.

Microbe	Crop	Growing media	Trait	Reference
Azospirullum brasilense	Cucumber	Hydroponic	Trigger iron uptake gene machinery	Pii et al. (2016)
Pseudomonas chlororaphis	Lettuce	Hydroponic system	Improved quality and quantity; reduced crop damage caused by *Pythium ultimum*	Lee and Lee (2015)
Pseudomonas sp. strain *WCS417r*	Tomato, potato, cucumber, lettuce	Hydroponic	Increased fresh weights of shoots and roots	Peer and Schippers (1989)
Pseudomonas putida, Erratia marcescens, Bacillus amyloliquefaciens, Pseudomonas fluorescens	Cucumber	Perlite	Increased fruit yield; increased plant tolerance to Fusarium wilt	Gül et al. (2013)
Pseudomonas fluorescens	Tomato	Perlite	Crown and root rot tolerance; increased yield	Gül et al. (2012)
Pseudomonas chlororaphis	Sweet pepper	Hydroponics	Strongly antagonized *Pythium aphanidermatum* and controlled root rot	Khan et al. (2003)
Pseudomonas chlororaphis	Cucumber	Hydroponics	Increased yield; mucilage management and root rot control	Zheng et al. (2000)
Consortia microbiota (Sphingomonadaceae, Lactobacilliaceae and Pseudomonadaceae family)	Lettuce	Aquaponics	Suppressed root rot (*Pythium aphanidermatum*)	Stouvenakers et al. (2020)

(*Continued*)

TABLE 8.8 (Continued)

Microbe	Crop	Growing media	Trait	Reference
Bacillus spp. *66/3, Pseudomonas* sp.	Tomato	Hydroponics	Increased yield by 36%	Kıdoğlu et al. (2009)
Pantoea agglomerans	Cucumber	Greenhouse conditions—soil	Increased dry matter, plant length, fruit number, and mineral content of fruits	Dursun et al. (2010)
Pseudomonas fluorescens strain MTCC103	Broccoli	Greenhouse conditions in the pot	Enhanced nutrient uptake and yield	Tanwar et al. (2014)
Arbuscular mycorrhiza	Melon	Hydroponics	Enhanced plant growth, total fruit yield, and nutrient contents of leaves	Dere et al. (2019)
Bacillus sp. *FS-3*	Strawberry	Greenhouseconditions in the pots	Increased fruit and leaf nutrient concentrations (N, P, K, Ca, and Fe)	Günes et al. (2009)
Pseudomonas fluorescens strain	Pepper	Greenhouse conditions in the pots	Increased root area, number and length of tips	Paul and Sharma (2006)
Bacillus licheniformis	Pepper	Greenhouse conditions in the pots	Increase shoot, root length, and dry weight	Lim and Kim (2013)

microclimatic conditions need to be optimized for successful colonization (Sheridan et al., 2017). Root microbiota varied for species and variety depend on root exudate (Fitzpatrick et al., 2018). Microbe–plant interaction in the rhizosphere takes place in three steps: (1) signaling or secretion of metabolites from host and microbe, (2) microbe movement toward the root, and (3) colonizing the root (Vacheron et al., 2013). *Pseudoalteromonas* sp. is efficient in colonizing the plant root zone in bread, wheat, soybean, and potato, which has promoted plant growth in the stress environment (Dimitrieva et al., 2006) and protected the plant system from bacteria and fungi infections (Franks et al., 2006). Mixed species microbiome provides more advantageous to the host plant system (Rao et al., 2005). *Arbuscular mycorrhiza* increases photosynthesis, growth-regulating substances, and tolerance to osmotic stress and pest (Al-Karaki, 2006). The effective increase in the yield of tomato plants by *Bacillus* sp. strain 66/3 when grown in soilless culture, and also by adding *Pseudomonas flourescens* strain 70 and *Pseudomonas putida* strain 18/1 K, showed favorable results. *Bacillus* and *Pseudomonas* species are primarily from plant-beneficial microbiome and these strains are mostly used in commercial products (Kloepper et al., 2004). *Bacillus* sp. 66/3 native strain specifically may be developed as a commercial product in the coming years. Multiple mechanisms of action were shown by most of the plant growth promoting rhizobium (PGPR). Nevertheless, according to the environmental conditions, there will be a change in the effect of PGPR. In tomato plants, response to *Bacillus amyloliquefaciens* in soilless culture changed according to the nutrition level. After root inoculation with *Bacillus* sp. strain 66/3, the marketable yield increased by 37% and 18% compared to the control in spring and fall seasons respectively (Gül et al., 2008).

8.3.3.2 Phyllosphere applications of microbiome

Filamentous green algae, archaea, yeast, and fungi are known to be harbored in leaves. However, the most prevailing microbial colonizers of the phyllosphere are bacteria, which is found to be on an average of 106–107 cells cm^{-2} of the leaf surface (Lindow and Brandl, 2003). Multiple factors that influence the microbial composition in the phyllosphere and endosphere include salinity, substrate type, pH, moisture, organic matter, and exudates (Fierer, 2017). In pepper and tomato plants *Bacillus licheniformis* significantly enhanced the height of the plant, leaf area, and diameter of tomato fruits (García et al., 2004). The microbial load identified in the phyllosphere was phyla *Firmicutes* (40.4%), *Proteobacteria* (22.5%), *Actinobacteria* (15.7%), *Deinococcus* (10.4%), *Arthrobacter* (4.6%), *Sphingomonas* (4.4%), *Bacillus* (4.2%), and one unidentified genus from the family Bacillaceae (30.1%) (Ortega et al., 2016). Pepper leaves infected with *Xanthomonas campestris* reduced the herbivory of armyworm (Cardoza and Tumlinson, 2006).

Phyllosphere bacteria *Sphingomonas Erwinia* and *Pseudomonas* provide the plant protection from pathogens (Zarraonaindia et al., 2015).

8.3.3.3 Endophytic bacteria

Bacteria found inside plant tissue is classified as endophytic bacteria. Some of the most common genera are *Pseudomonas*, *Bacillus*, *Klebsiella*, *Enterobacter*, *Streptomyces*, and *Serratia* (Micallef et al., 2009). Endophytic bacteria can be transmitted through a direct approach like directly entering into the roots and later reaching and colonizing other organs of the plant. Endophytic bacteria transmitted from plant tissue to seed, served as inoculum of bacteria and subsequently passed later stages of the plant in a process called *vertical microbial transmission* (Micallef et al., 2009). Endophytes can even occur through the passive process or emerging points of lateral roots or root cracks by active mechanism (Compant et al., 2005).

8.3.3.4 Biostimulant application in soilless systems

Biostimulants are compounds or substances (organic/inorganic) that when applied in plants in small quantities facilitate nutrient utilization efficiency, tolerance to abiotic stress, improved crop quality and/or performance, and defense against pest and diseases. They are neither biofertilizer or biopesticides nor herbicides nor plant growth-promoting rhizobacteria (Calvo et al., 2014; Povero et al., 2016). Biological molecules such as amino acids, phytohormones, polyphenols, exopolysaccharides, osmoprotectants, vitamins, and fatty acids are known biostimulants (Paul et al., 2019). Cultivation of indoor crops in hydroponic systems with biostimulants is a promising and environmentally friendly strategy for sustainable farming. The biostimulant compound seaweed extract "Actiwave®, (Valagro S.p.a.)" added with a nutrient solution in hydroponics systems increased the growth and stimulated nutrient uptake and nitrogen metabolism of a leafy vegetable arugula (*Eruca sativa* Mill.) (Vernieri et al., 2005, 2006). Humic substances have auxin-like activity and promote the root, leaf, and shoot growth of the plant (Trevisan et al., 2010). The Viva commercial biostimulant under NPK-deficient conditions compensated nutrient deficiency and enhanced tomato growth (Koleška et al., 2017). In a recent trend, algae-based biostimulants have a vital role in the sustainable increase of productivity in indoor cultivation. In lettuce and pepper cultivation, algal extracts stimulated seed germination, seedling growth, root biomass, and shoot (Arroussi et al., 2018; Faheed et al., 2008). Cocultivation of microalgae *Scenedesmus quadricauda* and tomato plants grown in a hydroponic system with Hoagland's nutrient solution resulted in the mutual growth effect of plants and microalgae biomass (Barone et al., 2018a). Further studies revealed that the usage of microalgae as a biostimulant modulates the plant growth-promoting microbes in the root and phyllosphere of the host plant system (Ranjan et al., 2016). Kelpak seaweed-based product helped to overcome the nutrient deficiency in okra production (Papenfus et al., 2013). A summary of examples for biostimulant-mediated crop trait improvement is described in Table 8.9. Biostimulants represent an environment-friendly, soil-friendly technology and could play an important role in climate-smart agriculture processes.

TABLE 8.9 **Biostimulant application examples for crop trait improvement.**

Microbe/substance	Plant	Effect	Reference
Chlorella vulgaris—dry powder	Lettuce	Enhances the physiological reactions and improves growth	Faheed et al. (2008)
Acutodesmus dimorphus—algae extracts	Roma tomato	Increases crop yield, seed germination, and lateral root number and its development	Garcia-Gonzalez and Sommerfeld (2016)
Dunaliella salina exopolysaccharides	Tomato	Salt stress decreases the plant length and root; activation/deactivation of jasmonic acid-dependent pathways	Arroussi et al. (2018)
Chlorella vulgaris/ Scenedesmus quadricauda microalgae extracts	Sugar beet	Higher root number, root tips, root length and surface	Barone et al. (2018b)
Arthospira platensis, Dunaliella salina and *Porphorydium* sp. crude polysaccharides extract	Tomato	Significantly increases the carotenoid, shoot length, nodes number, and shoot dry weight	Farid et al. (2020)
Peptone	Lettuce	Tolerant to cold stress	Polo et al. (2006)
Moringa leaf extract	Pumpkin	Tolerant to drought stress; enhances growth	Abd El-Mageed et al. (2017)
Ascorbic acid	Mung bean	Enhances overall growth and tolerance to heat stress	Kumar et al. (2011)
Humic acid	Common bean	Improved nutrient uptake and tolerance to salt stress	Aydin et al. (2012)

8.4 Soilless systems as an enabling technology for nutrition security

Conventional agricultural practices often focus on commercial gain and usually pay less attention to environmental degradation and the nutritional quality of food. Today globally, along with food security, nutrition security is of paramount importance. Nutrition security is achieved through secure access to an appropriately nutritious diet (adequate calories with balanced macronutrients and micronutrients) in conjunction with a sanitary environment and good health services (FAO et al., 2020). Indoor soilless farming can overcome some of the alarming problems associated with insufficient nutritious food produced through outdoor soil-based agriculture, including low sanitary environments and inadequacy for human health. It can be done in less area, using infertile land year-round under controlled conditions, with precise use of macronutrients and micronutrients, 50%–80% less water consumption, and negligible use of insecticide, pesticides, and weedicides.

8.4.1 Healthy and fresh produce through soilless systems

Soilless cultivation systems could help face the challenges due to climatic change and mitigating the problem of malnutrition by using naturally available resources appropriately (Butler and Oebker, 2006). The produce coming from soilless culture can be of high quality and needs less washing (El-Kazzaz and El-Kazzaz, 2017) inputs, which saves a lot of water and human resources. A significant benefit of this technology is to create a farm-to-fork food chain and enable the concept of hyperlocal food supply. In such farming, contactless harvesting, packaging, and transporting could be carried out, reducing transmission of contagious diseases or pandemics, which generally spread by touch or contact. Such farming provides more healthy, fresh (Hussain et al., 2014) and crispy (in terms of leafy vegetables), nutrient-rich food to customers. In regular agricultural practices, the chances of nutrient degradation of the produce are higher due to the time required for packaging and supply to the consumers. People in urban areas are more cautious about their health, which is directly dependent on their food intake, especially on minerals and vitamin sources provided by consuming vegetables and fruits in their daily consumption. One can grow various kinds of leafy veggies like lettuces, Swiss chard, spinach, purslane, coriander, mint, basil, cherry tomatoes, strawberries, blueberries, or other types of berries (Hussain et al., 2014).

8.4.2 Improving nutrient use efficiency via biofortification of micronutrients

Biofortification in various agronomic crop species is successful by increasing the accumulation of target nutrients in edible plant tissues in staple food or vegetables through fertilization or other eliciting factors. The soilless cultivation system also provides uptake of such micronutrients through the growing nutrient solution, which can address the desired element's micronutrient deficiencies (Sambo et al., 2019). However, various research activities are carried out to significantly refine the concentrations of multiple nutrients to restrain the nitrate content in edible tissues (Fageria, 2010) or improve nutritional quality (Putra and Yuliando, 2015) in soilless technology to get higher, safe, and healthy yields. Several reports are available for the biofortification of various vegetable crops by using these elements in the nutrient solution of the growing plants in hydroponics or aquaponics (Croft et al., 2017; Gioia et al., 2019).

Hickman (2016) stated that about 3.5% of the total world agriculture area had adopted cultivation of vegetables under tunnels and greenhouses with soilless agriculture techniques such as hydroponics, aquaponics, or aeroponics. This transformation of the technology from soil to soilless cultivation could help the efficient use of nutritional resources including water (Rodriguez-Ortega et al., 2017) and reduce the effect of various greenhouse gases released to the environment. A hydroponic system can be used efficiently in a diverse biofortification program with oligo elements like iodine, selenium, silicon, and calcium (Sambo et al., 2019) in *Brassicas* (Wiesner-Reinhold et al., 2017), chili (Li et al., 2017), chicory (D'Imperio et al., 2016a,b), strawberry (Mimmo et al., 2017), lettuce (Smoleń et al., 2014), potato (Nancy and Arulselvi, 2014), spinach (D'Imperio et al., 2016a), and common cornsalad (Gottardi et al., 2012). Such a soilless system can also improve the quality of produce and shelf life demanded by the end user (Giro and Ferrante, 2017; Islam et al., 2018).

Soilless cultivation systems could be one of the useful techniques that could help to increase the beneficial elements in produce by manipulating the nutrient solution of the system. Silicon is one of the recommended components in food for its beneficial effect on bone health. Montesano et al. (2016) reported the elevated concentration of silicon (Si) in green bean pods by fortifying a Si-enriched nutrient solution in a soilless system. Interestingly the Si concentrations became almost tripled after biofortification while the overall crop performance was not negatively influenced by the quality of the product. Selenium (Se) is one of the essential elements in the human diet and a key component of selenoproteins, which are responsible for fundamental biological and nutraceutical activities. Gioia et al. (2019) examined Se biofortification in two different Salanova (green and red) lettuce in an open gas-exchange growth chamber using closed soilless cultivation for delivering Se-rich food. In the given treatments, the green lettuce yield reduced to 9% and decreased red lettuce at 32–40 μM Se (11% and 21% respectively). But the leaf Se content increased 57% more in red lettuce in comparison to green lettuce. The application of selenium slightly reduced the carotenoid contents of green Salanova, whereas in red Salanova treated with 32 μM Se violaxanthin, neoxanthin, lutein, and β-cryptoxanthin raised by 38.6%, 27.4%, and 23.1%, respectively.

Barrameda-Medinaa et al. (2017) investigated the effect of biofortification of Zn in *Brassica oleracea* cv. *Bronco* and found the increase in Zn supply could accumulate the total amino acids and enhanced the enzymatic activity to accumulated glucosinolates and phenolic compounds. The uptake of such leafy greens could be beneficial further for human health. White et al. (2018) tried to explain that zinc concentration biofortification could be achieved more profoundly in leafy vegetables compared to the intake of fruits, seeds, or tubers, and *Brassicas* such as cabbage and broccoli. Hence, providing an increased dietary Zn intake through the biofortification of leafy vegetables seems to be a prudent approach towards nutritional security.

8.4.3 Enriching antioxidants, vitamins, and essential nutrients through soilless systems

It is imperative to have appropriate mineral composition of the nutrient solution for crop productivity and quality of the hydroponics grown crop. Nazarideljou et al. (2018) studied the evaluation of effects of K^+:Ca^{++} ratio on growth, yield, the content of vitamin C in fruits, and total carbohydrate and phenolic compounds of the hydroponically grown strawberry. The highest phenolic compounds were shown at a 1.4 rate (K > Ca). Photosynthetic capacity including leaf area, photosynthetic pigments, and stomatal conductance and K^+:Ca^{++} ratio also affected the enzymatic (peroxidase, superoxide dismutase, and catalase) and nonenzymatic (total phenol, phenolic compounds, and anthocyanin) antioxidant capacities. So, it was highly recommended to maintain the K^+:Ca^{++} ratio for growth, productivity, and strawberry produce quality. They also recorded a slight increase in polyphenolic acids at 16 μM Se while anthocyanin content also increased by 184% in red Salanova. A soilless production system has been reported to produce crop with superior taste, color, uniformity, higher nutrition value in fruits, richer in vitamin C level in tomato, and better firmness than in the soil-based cultivation system. It even helps in synthesizing more sugar, higher citric acid percentage, and sodium level resulting in better taste (Asaduzzaman et al., 2015). Soilless-produced fruit and vegetable crops have better shelf life, organoleptic and hygienic-sanitary properties, phenolic compounds, higher ratio of reducing sugar:organic acid, higher total flavonoids, higher 3-quercetin rutinoside, quercetin-3-*O*-glucoside, myricetin, chlorogenic acid, and kaempferol content (Asaduzzaman et al., 2015; Islam et al., 2018).

Iron (Fe) and zinc (Zn) are essential micronutrient elements in human dietary uptake. The suboptimal presence of these components creates a latent health issue in many global populations, especially in children and women living in impoverished areas at risk of undernutrition. Fe and Zn biofortification has proven successful in the soilless cultivation system in Brassicaceae microgreens (Gioia et al., 2019). They examined the effect of nutrient solutions supplemented with zinc sulfate (Exp-1; 0, 5, 10, 20 mg/L) and iron sulfate (Exp-2; 0, 10, 20, 40 mg/L) on the growth, yield, and mineral concentration of various Brassicaceae plants (arugula, red cabbage, and red mustard microgreens) and the Fe and Zn accumulation were increased as a quadratic model. The response of Fe and Zn enrichment was found to be genotype-specific. Ropokis et al. (2018) also reported the genotype specificity with pepper plants' growth in the same growing conditions with the uniform nutrient enrichment. It was observed that the plants treated with 5–10 mg/L Zn resulted in 75%–281% enhanced concentrations in comparison to untreated controls. Similar results were also found in the case of 10–20 mg/L Fe supplemented solution with a 64% increase in arugula and 278% in red cabbage. The biofortification in the soilless cultivation system was proven successful by simple manipulation of the nutrient solution composition. It is still not very clear for the biofortification performance at the commercial level due to risks related to even nutrient uptake to the variable interactions between different genotypes and environment and the interactions among various growth conditions during uptake (Gioia et al., 2019). In contrast, Giordano et al. (2019) reported in experiments on lettuce that the supplementation of 2 mM iron in the nutrient solution resulted in increased Fe concentration overall in the leaves of lettuce, genotype independently. Soilless-protected cultivation technology provides a plethora of advantages. However, specific challenges include precise nutrition management, crop growth performance tracking, accurate irrigation management, and precise climate control. This could be circumvented through precision agriculture and automation technologies.

8.5 Role of precision agriculture and automation for enabling food and nutrition security

At the cusp of fourth industrial revolution, the amalgamation of the physical, biological, and digital world brings us leapfrogging into the management of agriculture production systems including CEA. With the advent of big data, the availability of various types of smart sensors, artificial intelligence, machine learning, and data analytics opened many avenues in modern agriculture. Soilless farming utilizes smart sensors, and monitoring and control systems to improve yield and nutrition content of the crop in the least amount of time and at reduced prices. Automation reduces labor costs and human involvement in adverse weather conditions and can even contribute to protect the field throughout the day from different biotic and abiotic conditions. In this section, we will highlight the major automation components available, their uses, increases in productivity, the possibility of contactless farming, and the standing limitations of this utility.

8.5.1 Precision agriculture and automation in context with soilless cultivation system

Precision agriculture or site-specific crop management requires use of precise machinery that is tailor-made for the crop and environment. Prevailing technologies must be developed further to solve problems such as variability in environment and products, continuously changing geological conditions (like location, soil type and properties, size and shape of the plant, weather, etc.), produce handling, and unfriendly environmental conditions (extreme temperatures, humidity, dust, dirt, etc.) (Edan et al., 2009). Thus, intelligent control systems are pertinent for dynamic, real-time monitoring and interpretation of the environment and the plant performance. Keeping plant performance in mind, precision agriculture provides precise measures for efficient farming. Furthermore, precision agriculture and smart farming offer an environmentally friendly agriculture practice that is sustainable by improving resource use efficiency (Gebbers and Adamchuk, 2010). This is not only helpful in food security but also provides an appropriate balance of nutrition requirements, by providing the ability to grow crops year-round in any weather condition. This can be achieved by a combination of human and machine-based intelligence, which plays a pivotal role. Soilless farming, specifically hydroponics-based cultivation, demands real-time monitoring (García et al., 2020). Sophisticated sensors and controllers are available to delineate data at specific time points providing valuable information. This insightful information along with interpretation algorithms based on machine-learned logic offers specific decisions to enhance the food and nutrition security. With this plethora of information and use of big data analytics a virtual field can be created. The sheer volume, velocity, variety, and veracity of the data achieved from a single field can be used for generating an in silico farm (Zamora-Izquierdo et al., 2019).

Simulations over this digital field offer optimized parameters in given conditions. IoT (Internet of things) devices consist of sensors to measure soil properties (e.g., moisture, electrical conductivity), environmental conditions (e.g., temperature, humidity, and rainfall), and ambient light conditions (e.g., irradiation, light levels). All these parameters relate closely to the growth of plants which if managed using tools of precision agriculture can provide savings in energy and water consumption, as well as reduction of fertilizer and chemicals used to support plant growth (Georgiadis et al., 2019). This supports the future goals of reduced resources, maximizing productivity, and producing more crop per drop of water.

Precision agriculture starts with acquiring crop knowledge from the field and converting it into a digital ecosystem. Crop data is collected through intelligent sensors followed by curation and processing of same. This curated data is used to develop artificial intelligence and machine learning (AI-ML) based models that have decision-making abilities. These statistics-based

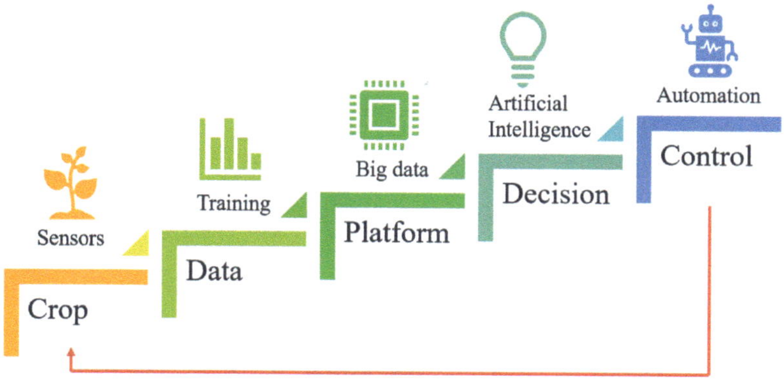

FIGURE 8.5 Schematic representation of site-specific crop precision management system.

decisions are translated through actuation technologies like precision harvesting and precision chemical spraying to bring about changes in the field (Fig. 8.5). The field is constantly monitored for sudden changes and thus precise decisions are made on the spot.

8.5.2 Automation and IoT tools to manage soilless cultivation systems

Field automation in farms, greenhouses, and soilless farming plays an important role in precisely regulating the content of water, nutrition, and continuous biotic and abiotic monitoring. The production of crops is affected by many factors like unusual weather changes, lack of water, and the lack of enough arable lands available to grow the crops (Resh, 2016). Many different methods are used to reduce space and water consumption; one of the most famous methods in soilless farming is a vertical hydroponic system. Current hydroponic systems are developed to control parameters automatically, which affect plant growth. People have devised computer-based vertical hydroponics systems that analyze and determine the optimal nutrient that can be wired or wireless for better reliability (Ibayashi et al., 2016; Rius-Ruiz et al., 2014). Automatic aeroponic, NFT systems also came into the picture, which use Arduino platforms and microcontrollers for measuring parameters such as temperature, humidity, pH, total dissolved solids (TDS), and proximity (Eridani et al., 2017; Montoya et al., 2017). Complete automation involves IoT platforms, machine learning, artificial intelligence, and robotics providing LED lighting control, water spray, water pump control, etc. through smart phone applications (Marques et al., 2019; Palande et al., 2018; Ruengittinun et al., 2017; Ruscio et al., 2019). A basic framework of an autonomous system is depicted in Fig. 8.6. In many agricultural operations, an operator is required to perform a couple of basic functions concurrently,

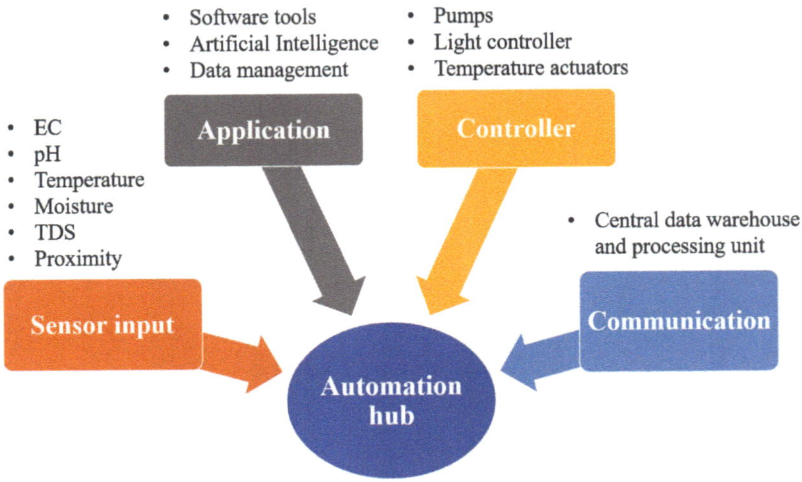

FIGURE 8.6 **Basic framework of autonomous device with its important layers.**

for example, navigating the vehicle and working the equipment. The need to relieve the operator for making navigation adjustments continuously has been the main reason for the development of automatic guidance systems, unmanned vehicles, and AI-based bots (Jha et al., 2019). Autoguidance will continue to be the focus of future development.

Several sensors have been developed and applied for internal quality determination, including sugar content, acidity, rind puffing, rotten core, and other internal defects at early stages. This can lead to early intervention and saving a large amount of crop at the farm level. For example, it has been estimated that robotic scouting of crop saves around $4.5 per hectare per year, and overall increment of $88 per hectare per year was estimated by use of continuous monitoring and automation (Lowenberg-DeBoer et al., 2019).

Architectures of distributed sensor networks for site-specific irrigation automation combining smart soil moisture sensors and sprinkler valve controllers have been developed and are commercially available. This can be expanded to closed-loop control for automated irrigation based on in-field sensing feedback of plant and soil conditions. Further developments include spot-spraying of herbicides based on real-time weed detection using optical sensors (Yan et al., 2018).

Newer methods include use of hyperspectral imagery to understand the growth stage differences, plant pathology, and nutrition-related information in real time, which can be mounted over AI bots (Bodner et al., 2018). Real-time monitoring using computer vision and AI-based detection is rapidly expanding from detection of nutrition deficiency to pests and weeds (An et al., 2009). Illumination is one of the most important components for the machine-vision system to monitor and inspect products, because it determines the quality of images acquired, especially for glossy products whose cuticular layers are thick. Countering this problem requires light polarizers in front of camera lenses or illumination sources to reduce halation on the acquired images. Soilless cultivation can easily incorporate all these requirements thus acting as an enabler in the future of sustainable agriculture.

8.5.3 Reducing labor costs and contactless farming using modern tools

In the current situation of pandemic and other widespread diseases, contactless farming has shown a promising way of sustaining agriculture. In a contactless farm the only involvement of the farm owner would be to refill the reservoirs and monitor the indoor farm for any leakage or maintenance. All the updates and notifications collected from the sensors installed within the field are available over the Internet and any special instructions can be set for the AI bots for the type variety of crop (Jha et al., 2019). This reduces the labor to bare minimum and requires zero human contact with the crop during growing seasons, thus providing not only crop protection measures but also measures to continue agriculture during pandemic lockdowns or other health problems of the farm owner.

The exhaustive nature of precision agriculture practices requires a great amount of automated machines and equipment. Automation provides effective on-farm use of resources like water, fertilizers, pesticides, etc. and tools for all measures by enabling rate, flexible frequency, and duration of water supply with regulation of the irrigation at the required spot (García et al., 2020). Because of cheap labor availability and government policies, the adaptation of automation and indoor farming is poor; initial setup is expensive and continuous maintenance is required. Aside from these limitations, with incremental demands and lesser availability of land vertical soilless farms are gaining interest in terms of being economical over long periods of time. With emerging technology in automation and mass production, the setup costs are coming down, which enables farmers to explore this area of newer farming methods that will lead to year-round high-quality product.

8.6 Challenges and future perspectives

Technological improvement and soilless cultivation systems provide a great opportunity to mitigate climate change–related agriculture bottlenecks especially in terms of food and nutrition security. However, there are many unexpected and uncertain factors which directly or indirectly become rate-limiting steps and impede mitigation strategies. These factors are income inequality, dichotomy between technology and ecology, pandemic, trade policies, capital investments, government policies, regulatory environment, socioeconomic contexts, geography, and divided demographics.

Furthermore, detailed research is needed regarding the molecular mechanisms responsible for reduction in crop yield, crop quality, nutrient quality, and quantity due to the elevated CO_2 and temperature and their combinatorial effects. In terms of availability of comprehensive and big data on the effect of elevated CO_2, and elevated temperatures on different crops, in different geographies in different seasons, will create new opportunities for data science intervention, AI, and deep learning methodologies to develop intelligent predictive models. Such predictive models will help to determine strategies and policies for mitigation of climate change effects. The energetic cost in controlled environment soilless cultivation systems is one of the critical area of improvement and warrants immediate innovation and affordable technologies. Solar photovoltaic-driven greenhouse technologies might answer such energy minimization and cost optimization aspects.

8.7 Conclusion

Considering the multitude of climate-smart agriculture technologies and practices, soilless protected cultivation technology brings a completely new dimension to hyperlocal food production, and thereby reducing food loss and food waste. In today's volatile, uncertain, complex, and ambiguous world of agriculture, we need to have ancillary and modern technologies readily available and scalable. Deployment of soilless indoor cultivation technology could be one such sustainable and scalable strategy. Adoption of soilless technology provides a great opportunity to produce more with less cultivation area, more crop per drop of water, reduce toxic chemical load, and improve the ecological footprint. Soilless cultivation and vertical farming coupled with precision and automation technologies in CEA might close the gap between climate change and assurance for food and nutrition security by 2050. This technology is a pivotal solution towards providing stability in terms of nutritious fresh food supply amidst the rapid urbanization process. At the same time, preventing both preharvest and postharvest food loss with reduced food miles. Furthermore, it will serve as an alternative opportunity for soil-based agriculture, due to the diminishing arable land, deteriorating soil health, and anticipated freshwater insecurity. Urban and periurban soilless food production systems could bring us one step closer to affordable nutritious food. Although, soilless protected cultivation technology cannot provide a complete solution for food and nutrition insecurity or food sustainability, but it can be considered as an "emerging enabling technology". This in conjunction with other climate-smart agriculture practices would disrupt the conundrum of food and nutrition security.

Acknowledgments

The research and development work in Reliance Industries Ltd. is supported through internal funding. The authors acknowledge the research and development facility at Reliance Industries Ltd. AB acknowledges Puja Pai, Mahadev G, Namrata G, Siddhi J, Sushant K, and Venkatesh Prasad for development of experimental setup and infrastructure establishment at Reliance Industries Ltd. R&D laboratory.

References

Abd El-Mageed, T.A., Semida, W.M., Rady, M.M., 2017. Moringa leaf extract as biostimulant improves water use efficiency, physio-biochemical attributes of squash plants under deficit irrigation. Agric. Water Manag. 193, 46–54. Available from: https://doi.org/10.1016/j.agwat.2017.08.004.

Agnolucci, P., Rapti, C., Alexander, P., Vincenzo De Lipsis, V., Robert, A., Holland, R.A., et al., 2020. Impacts of rising temperatures and farm management practices on global yields of 18 crops. Nat. Food 1, 562–571. Available from: https://doi.org/10.1038/s43016-020-00148-x.

Agrinatura, January 12, 2020. Protection technologies for family agriculture in desert conditions. Available online: https://agrinatura-eu.eu/news/protection-technologies-for-family-agriculture-in-desert-conditions/ (accessed 20.11. 2020.).

Al-Karaki, G.N., 2006. Nursery inoculation of tomato with arbuscular mycorrhizal fungi and subsequent performance under irrigation with saline water. Sci. Hortic. 109, 1–7. Available from: https://doi.org/10.1016/j.scienta.2006.02.019.

Al-Kodmany, K., 2018. The vertical farm: a review of developments and implications for the vertical city. Buildings 8, 24. Available from: https://doi.org/10.3390/buildings8020024.

Alshrouf, A., 2017. Hydroponics, aeroponic and aquaponic as compared with conventional farming. Am. Sci. Res. J. Eng. Technol. Sci. 27 (1), 247–255.

An, Q., Li, Z., Ji, C., Zhou, J., 2009. Agricultural robot vision navigation algorithm based on illumination invariant image. Nongye Gongcheng Xuebao/Trans. Chin. Soc. Agric. Eng. 25, 208–212.

Arif, N., Batool, M., Schenk, P.M., 2020. Plant microbiome engineering: expected benefits for improved crop growth and resilience. Trends Biotechnol. 38, 1385–1396. ISSN 01677799.

Arroussi El, H., Benhima, R., Elbaouchi, A., Sijilmassi, B., El Mernissi, N., Aafsar, A., et al., 2018. *Dunaliella salina* exopolysaccharides: a promising biostimulant for salt stress tolerance in tomato (*Solanum lycopersicum*). J. Appl. Phycol. 30, 2929–2941. Available from: https://doi.org/10.1007/s10811-017-1382-1.

Asaduzzaman, M., Saifullah, M., Mollick, A.S.R., Hossain, M.M., Halim, G.M.A., Asao, T., 2015. Influence of soilless culture substrate on improvement of yield and produce quality of horticultural crops. *Intech Open*, Chapter 1, ISBN: 978-953-51-1739-1. https://doi.org/10.5772/59708.

Asseng, S., Guarin, J.R., Raman, M., Monje, O., Kiss, G., Dickson, D., et al., 2020. Wheat yield potential in controlled-environment vertical farms. Proc. Natl. Acad. Sci. U S A 117 (32), 19131–19135. Available from: http://www.pnas.org/cgi/doi/10.1073/pnas.2002655117.

Aydin, A., Kant, C., Turan, M., 2012. Humic acid application alleviate salinity stress of bean (*Phaseolus vulgaris* L.) plants decreasing membrane leakage. Afr. J. Agric. Res. 7, 1073–1086. Available from: https://doi.org/10.5897/AJAR10.274.

Barbosa, G.L., Gadelha, F.D.A., Kublik, N., Proctor, A., Reichelm, L., Weissinger, E., et al., 2015. Comparison of land, water, and energy requirements of lettuce grown using hydroponic vs. conventional agricultural methods. Int. J. Env. Res. Public Health 12, 6879–6891. Available from: https://doi.org/10.3390/ijerph120606879.

Barone, V., Puglisi, I., Fragalà, F., Lo Piero, A.R., Giuffrida, F., Baglieri, A., 2018a. Novel bioprocess for the cultivation of microalgae in hydroponic growing system of tomato plants. J. Appl. Phycol. Available from: https://doi.org/10.1007/s10811-018-1518-y.

Barone, V., Baglieri, A., Stevanato, P., Broccanello, C., Bertoldo, G., Bertaggia, M., et al., 2018b. Root morphological and molecular responses induced by microalgae extracts in sugar beet (*Beta vulgaris* L.). J. Appl. Phycol. 30, 1061–1072. Available from: https://doi.org/10.1007/s10811-017-1283-3.

Barrameda-Medinaa, Y., Blascoa, B., Lentinib, M., Espositob, S., Baenasc, N., Morenoc, D.A., et al., 2017. Zinc biofortification improves phytochemicals and amino-acidicprofile in *Brassica oleracea* cv. Bronco. Plant Sci. 58, 45.

Baset Mia, M.A., Shamsuddin, Z.H., Wahab, Z., Marziah, M., 2010. Effect of plant growth promoting rhizobacterial (PGPR) inoculation on growth and nitrogen incorporation of tissue-cultured musa plantlets under nitrogen-free hydroponics condition. Aust. J. Crop. Sci. 4, 85–90. ISSN:1835-2707.

Beacham, A.M., Vickers, L.H., Monaghan, J.M., 2019. Vertical farming: a summary of approaches to growing skywards. Hortic. Sci. Biotechnol. 94, 277–283.

Benke, K., Tomkins, B., 2017. Future food-production systems: vertical farming and controlled-environment agriculture. Sustain. Sci., Pract. Policy 13 (1), 13–26. Available from: https://doi.org/10.1080/15487733.2017.1394054.

Berkelmann, B., Wohanka, W., Wolf, G.A., 1994. Characterization of the bacterial flora in circulating nutrient solutions of a hydroponic system with rockwool. Acta Hortic. 361, 372–381. Available from: https://doi.org/10.17660/ActaHortic.1994.361.37.

Bodner, G., Nakhforoosh, A., Arnold, T., Leitner, D., 2018. Hyperspectral imaging: a novel approach for plant root phenotyping. Plant Methods 14.

Butler, J.D., Oebker, N.F., 2006. Hydroponics as a hobby—growing plants without soil. Circular 844. Information Office, College of Agriculture, University of Illinois, Urbana, IL.

Calvo, P., Nelson, L., Kloepper, J.W., 2014. Agricultural uses of plant biostimulants. Plant Soil 383, 3–41. Available from: https://doi.org/10.1007/s11104-014-2131-8.

Cardoza, Y.J., Tumlinson, J.H., 2006. Compatible and incompatible *Xanthomonas* infections differentially affect herbivore-induced volatile emission by pepper plants. J. Chem. Ecol. 32, 1755–1768. Available from: https://doi.org/10.1007/s10886-006-9107-y.

CEAC Arizona, 2020. The University of Arizona controlled environment agriculture center. https://ceac.arizona.edu/sites/default/files/Ch1.pdf (accessed 10.11.2020.).

Compant, S., Duffy, B., Nowak, J., Clément, C., Barka, E.A., 2005. Use of plant growth-promoting bacteria for biocontrol of plant diseases: principles, mechanisms of action, and future prospects. Appl. Env. Microbiol. 71, 4951–4959. Available from: https://doi.org/10.1128/aem.71.9.4951-4959.

Croft, M.M., Hallett, S.G., Marshall, M.I., 2017. Hydroponic production of vegetable Amaranth (Amaranthus cruentus) for improving nutritional security and economic viability in Kenya. Renew. Agri. Food Sys. 1–10. Available from: https://doi.org/10.1017/S1742170516000478. Available from: http://doi:10.1080/15487733.2017.1394054.

Dere, S., Coban, A., Akhoundnejad, Y., Ozsoy, S., Dasgan, H.Y., 2019. Use of *Mycorrhiza* to reduce mineral fertilizers in soilless melon. cultivation. Not. Botanicae Horti Agrobotanici Cluj-Napoca 47, 1331–1336. Available from: https://doi.org/10.15835/nbha47411738.

Dimitrieva, G.Y., Crawford, R.L., Yüksel, G.Ü., 2006. The nature of plant growth-promoting effects of a *pseudoalteromonas* associated with the marine algae *Laminaria japonica* and linked to catalase excretion. J. Appl. Microbiol. 100, 1159–1169. Available from: https://doi.org/10.1111/j.1365-2672.2006.02831.x.

D'Imperio, M., Renna, M., Cardinali, A., Buttaro, D., Santamaria, P., Serio, F., 2016a. Silicon biofortification of leafy vegetables and its bioaccessibility in the edible parts. J. Sci. Food Agric. 96, 751–756. Available from: https://doi.org/10.1002/jsfa.7142.

D'Imperio, M., Renna, M., Cardinali, A., Buttaro, D., Serio, F., Santamaria, P., 2016b. Calcium biofortification and bioaccessibility in soilless "baby leaf" vegetable production. Food Chem. 213, 149–156. Available from: https://doi.org/10.1016/j.foodchem.2016.06.071.

Dong, J., Gruda, N., Lam, S.K., Li, X., Duan, Z., 2018. Effects of elevated CO_2 on nutritional quality of vegetables: a review. Front. Plant Sci. 9, 924. Available from: https://doi.org/10.3389/fpls.2018.00924.

Dukes, M.D., Zotarelli, L., Liu, G.D., Simonne, E.H., 2012. Principles and practices of irrigation management for vegetables.

Dursun, A., Ekinci, M., Donmez, M.F., 2010. Effects of foliar application of plant growth promoting bacterium on chemical contents, yield and growth of tomato (*Lycopersicon esculentum L.*) and cucumber (*Cucumis sativus L.*). Pak. J. Bot. 42, 3349–3356.

Edan Y., Han S., Kondo N. (2009) Automation in Agriculture. In: Nof S. (eds) Springer Handbook of Automation. Springer Handbooks. Springer, Berlin, Heidelberg. Available from: https://doi.org/10.1007/978-3-540-78831-7_63.

El-Kazzaz, K.A., El-Kazzaz, A.A., 2017. Soilless agriculture a new and advanced method for agriculture development: an introduction. Agric. Res. Tech 3, 555610. Available from: https://doi.org/10.19080/ARTOAJ.2017.03.555610. Open Access Journal.

Eridani, D., Wardhani, O., Widianto, E.D., 2017. Designing and implementing the arduino-based nutrition feeding automation system of a prototype scaled nutrient film technique, NFT hydroponics using total dissolved solids, TDS sensor. In: Proceedings of the Fourth International Conference on Information Technology, Computer, and Electrical Engineering, ICITACEE.

Fageria, N.K., 2010. The Use of Nutrients in Crop Plants. CRC Press, Boca Raton, FL.

Faheed, F.A., Abd-El Fattah, Z., 2008. Effect of *Chlorella vulgaris* as bio-fertilizer on growth parameters and metabolomic aspects of lettuce plant. J. Agric. Soc. Sci. 4, 165–169.

FAO, 2018. Good agricultural practices for greenhouse vegetable production in the South East European countries. In: Principles for Sustainable Intensification of Smallholder Farms, vol. 230, FAO Plant Production and Protection Papers, ISSN 0259-2517. http://www.fao.org/3/a-i6787e.pdf.

FAO, OECD, 2018. Food Security and Nutrition: Challenges for Agriculture and the Hidden Potential of Soil. FAO, Rome and OECD Publishing, Paris. Available from: https://doi.org/10.1787/9789264183780-en.

FAO, IFAD, UNICEF, WFP, WHO, 2019. The State of Food Security and Nutrition in the World 2019. Building Climate Resilience for Food Security and Nutrition. FAO, Rome. Available from: http://www.fao.org/3/ca5162en/ca5162en.pdf.

FAO, IFAD, UNICEF, WFP, WHO, 2020. The State of Food Security and Nutrition in the World 2020. Building Climate Resilience for Food Security and Nutrition. FAO, Rome. Available from: https://doi.org/10.4060/ca9692en.

Farid, R., Benhima, R., Sbabou, L., El Arroussi, H., 2020. Microalgae polysaccharides biostimulating effect on tomato plants: growth and metabolic distribution. Biotech. Rep. 25, e00426. Available from: https://doi.org/10.1016/j.btre.2020.e00426. ISSN 2215-017X.

Fierer, N., 2017. Embracing the unknown: disentangling the complexities of the soil microbiome. Nat. Rev. Microbiol. 15, 579–590. Available from: https://doi.org/10.1038/nrmicro.2017.87.

Fischer, G., Shah, M., 2010. Farmland investments and food security: statistical annex. In: Report prepared under a World Bank and International Institute for Applied Systems Analysis Contract. Luxembourg, World Bank, Washington, DC.

Fitzpatrick, C.R., Copeland, J., Wang, P.W., Guttman, D.S., Kotanen, P.M., Johnson, M.T.J., 2018. Assembly and ecological function of the root microbiome across angiosperm plant species. Proc. Natl. Acad. Sci. U S A 115, E1157–E1165. Available from: https://doi.org/10.1016/j.jare.2019.03.004.

Franks, A., Egan, S., Holmström, C., James, S., Lappin-Scott, H., Kjelleberg, S., 2006. Inhibition of fungal colonization by *Pseudoalteromonas tunicata* provides a competitive advantage during surface colonization. Appl. Environ. Microbiol. 72, 6079–6087. Available from: https://doi.org/10.1128/aem.00559-06.

García, L., Parra, L., Jimenez, J.M., Lloret, J., Lorenz, P., 2020. IoT-based smart irrigation systems: an overview on the recent trends on sensors and IoT systems for irrigation in precision agriculture. Sensors 20, 1042.

García L.A.J., Probanza A., Ramos B., Palomino M., Mañero G.J.F., 2004. Effect of inoculation of Bacillus licheniformis on tomato and pepper, Agronomie. 24, 2004, 169–176. Available from: https://hal.archives-ouvertes.fr/hal-00886016/document.

Garcia-Gonzalez, J., Sommerfeld, M., 2016. Biofertilizer and biostimulant properties of the microalgae *Acutodesmus dimorphus*. J. Appl. Phycol. 28, 1051–1061. Available from: https://doi.org/10.1007/s10811-015-0625-2.

Gebbers, R., Adamchuk, V.I., 2010. Precision agriculture and food security. Science 327, 828–831.

Georgiadis, G., Komninos, A., Koskeris, A., Garofalakis, J., Cuomo, F., Lorenzo, P.D., et al., 2019. New publication: improving Hydroponic Agriculture Through IoT-enabled Collaborative Machine Learning.

Gioia, F.D., Petropoulos, S.A., Ozores-Hampton, M., Morgan, K., Rosskopf, E.N., 2019. Zinc and iron agronomic biofortification of *Brassicaceae* microgreens. Agronomy 9, 677. Available from: https://doi.org/10.3390/agronomy9110677.

Giordano, M., El-Nakhel, C., Pannico, A., Kyriacou, M.C., Stazi, S.R., De Pascale, S., et al., 2019. Iron biofortification of red and green pigmented lettuce in closed soilless cultivation impacts crop performance and modulates mineral and bioactive composition. Agronomy 9, 290.

Giro, A., Ferrante, A., 2017. Postharvest physiology of *Corchorus olitorius* baby leaf growing with different nutrient solutions. J. Hortic. Sci. Biotechnol. 93, 1–9. Available from: https://doi.org/10.1080/14620316.2017.1382313.

Global Hunger Index, 2020. Global Hunger Index by Severity. Concern Worldwide and Welthungerhilfe, Germany https://www.globalhungerindex.org/results.html (accessed 15.11.2020.).

Gottardi, S., Iacuzzo, F., Tomasi, N., Cortella, G., Manzocco, L., Pinton, R., 2012. Beneficial effects of silicon on hydroponically grown corn salad (*Valerianella locusta* (L.) Laterr) plants. Plant. Physiol. Biochem. 56, 14–23. Available from: https://doi.org/10.1016/j.plaphy.2012.04.002.

Grebmer, v.K., Bernstein, J., Alders, R., Dar, O., Kock, R., Rampa, F., et al., 2020. Figure 1.4: 2020 global hunger index by severity. In: Map in 2020 Global Hunger Index: One Decade to Zero Hunger: Linking Health and Sustainable Food Systems. Welthungerhilfe, Bonn; Concern Worldwide, Dublin. https://www.globalhungerindex.org/results.html

Gruda, N., Qaryouti, M.M., Leonardi, C., 2013. Growing media. Good Agricultural Practices for Greenhouse Vegetable Crops. Principles for Mediterranean Climate Areas. FAO, Rome, pp. 271–302, Plant Production and Protection Paper 217.

Gruda, N., Caron, J., Prasad, M., Maher, M., 2016. Growing media. In: Lal, R. (Ed.), Encyclopedia of Soil Science, third ed. Taylor & Francis Group, pp. 1053–1058.

Gül, A., Kıdoglu, F., Tuzel, Y., Tüzel, H., 2008. Effects of nutrition and *Bacillus amyloliquefaciens* on tomato (*Solanum lycopersicum*) growing in perlite. Spanish J. Agric. Res. 6, 422–429.

Gül, A., Özaktan, H., Yolageldi, L., Cakir, B., Sahin, M., Akat, S., 2012. Effect of rhizobacteria on yield of hydroponically grown tomato plants. Acta Hortic. 952, 777–784. https://www.actahort.org/books/952/952_98.htm.

Gül, A., Özaktan, H., Kıdoğlu, F., Tüzel, Y., 2013. Rhizobacteria promoted yield of cucumber plants grown in perlite under *Fusarium* wilt stress. Sci. Hortic. (Amst.) 153, 22–25. Available from: https://doi.org/10.1016/j.scienta.2013.01.004.

Günes, A., Ataoglu, N., Turan, M., Es Itken, A., Ketterings, Q.M., 2009. Effects of phosphate-solubilizing microorganisms on strawberry yield and nutrient concentrations. J. Plant. Nutr. Soil. Sci. 172, 385–392. Available from: https://doi.org/10.1002/jpln.200800121.

Hickman, G., 2016. International Greenhouse Vegetable Production – Statistics. Cuesta Roble Greenhouse Vegetable Consulting, Mariposa, CA.

Hussain, A., Iqbal, K., Aziem, S., Mahato, P., Negi, A.K., 2014. Review on the science of growing crops without soil (soilless culture) – a novel alternative for growing crops. Int. J. Agric. Crop Sci. 7, 833–842.

Ibayashi, H., Kaneda, Y., Imahara, J., Oishi, N., Kuroda, M., Mineno, H., 2016. A reliable wireless control system for tomato hydroponics. Sensors 16, 644.

ICARDA–APRP Report, 2020. Improving Food Security and Sustainable Natural Resources Management Through Enhancing Integrated Agricultural Production Systems in the Arabian Peninsula 2014-2019. International Centre for Agricultural Research in the Dry Areas (ICARDA), Beirut, Lebanon. vi + 81 pages. ISBN: 92-9127-534-4. https://www.icarda.org/research/innovations/protected-agriculture (accessed 10.11.2020.).

IIAP CoE, 2018. Indo-Israel Agriculture Project Center of Excellence Karnal, Haryana-Vegetables Success Story. https://embassies.gov.il/delhi/Relations/Mashav/Documents/IIAP%202018.pdf (accessed 20.11.2020.).

Indigo Ag, 2019, Available online: https://non-gmoreport.com/articles/is-the-plant-microbiome-a-key-to-higher-yielding-healthier-food-crops/, and https://www.agriculture.com/technology/crop-management/startup-spotlight-indigo-ag. (accessed 15.11.2020.).

Intergovernmental Panel on Climate Change (IPCC), 2019. Summary for policymakers. In: Climate Change and Land: An IPCC Special Report on Climate Change, Desertification, Land Degradation, Sustainable Land Management, Food Security, and Greenhouse Gas Fluxes in Terrestrial Ecosystems. P.R. Shukla, J. Skea, E. Calvo Buendia, V. Masson-Delmotte, H.-O. Pörtner, D.C. Roberts, et al. (Eds.). In press. https://www.ipcc.ch/srccl/chapter/summary-for-policymakers

Islam, M.Z., Mele, M.A., Baek, J.P., Kang, H., 2018. Iron, iodine and selenium effects on quality, shelf life and microbial activity of cherry tomatoes. Not. Botanicae Horti Agrobotanici Cluj-Napoca 46, 388–392. Available from: https://doi.org/10.1186/s41182-018-0085-x.

Jha, K., Doshi, A., Patel, P., Shah, M., 2019. A comprehensive review on automation in agriculture using artificial intelligence. Artific. Intel. Agric. 2, 1–12.

Jones, J.B., 1997. Hydroponics: A Practical Guide for the Soilless Grower. St. Lucie Press, Boca Raton, FL.

Khan, A., Sutton, J.C., Grodzinski, B., 2003. Effects of *Pseudomonas chlororaphis* on *Pythium aphanidermatum* and root rot in peppers grown in small-scale hydroponic troughs. Biocont. Sci. Technol. 13, 615–630. Available from: https://doi.org/10.1080/0958315031000151783.

Kıdoğlu, F., Gül, A., Tüzel, Y., Özaktan, H., 2009. Yield enhancement of hydroponically grown tomatoes by Rhizobacteria. Acta Hortic. 807, 475–480. Available from: https://doi.org/10.17660/ActaHortic.2009.807.6.

Kloepper, J.W., Ryu, C.M., Zhang, S.A., 2004. Induced systemic resistance and promotion of plant growth by *Bacillus* spp. Phytopathol 94, 1259–1266. Available from: https://doi.org/10.1094/PHYTO.2004.94.11.1259.

Knutson, A., 2000. The Best of the Growing Edge 2: Popular Hydroponics and Gardening for Small-commercial Growers and Hobbyists. New Moon Publishing, Corvallis, MA.

Koleška, I., Hasanagíc, D., Todorovíc, V., Murtíc, S., Klokíc, I., Paradikovíc, N., et al., 2017. Biostimulant prevents yield loss and reduces oxidative damage in tomato plants grown on reduced NPK nutrition. J. Plant Interact. 12, 209–218.

Kumar, S., Kaur, R., Kaur, N., Bhandhari, K., Kaushal, N., Gupta, K., et al., 2011. Heat-stress induced inhibition in growth and chlorosis in mungbean (*Phaseolus aureus* Roxb.) is partly mitigated by ascorbic acid application and is related to reduction in oxidative stress. Acta Physiol. Plant 33, 2091. Available from: https://doi.org/10.1007/s11738-011-0748-2.

Lee, S., Lee, J., 2015. Beneficial bacteria and fungi in hydroponic systems: types and characteristics of hydroponic food production methods. Sci. Hortic. 195, 206–215. Available from: https://doi.org/10.1016/j.scienta.2015.09.011.

Leisner, C.P., 2020. Climate change impacts on food security-focus on perennial cropping systems and nutritional value. Plant Sci. 293, 110412. Available from: https://doi.org/10.1016/j.plantsci.2020.110412. ISSN 0168-9452.

Li, R., Li, D.W., Yan, A.L., Hong, C.L., Liu, H.P., Pan, L.H., 2017. The bioaccessibility of iodine in the biofortified vegetables throughout cooking and simulated digestion. J. Food Sci. Technol. 55, 366–375. Available from: https://doi.org/10.1007/s13197-017-2946-4.

Lim, J.-H., Kim, S.-D., 2013. Induction of drought stress resistance by multi-functional PGPR *Bacillus licheniformis* K11 in pepper plant. Pathol. J. 29, 201–208. Available from: https://doi.org/10.5423/PPJ.SI.02.2013.0021.

Lindow, S.E., Brandl, M.T., 2003. Microbiology of the phyllosphere. Appl. Environ. Microbiol. Available from: https://doi.org/10.1128/AEM.69.4.1875-1883.2003.

Loladze, I., 2014. Hidden shift of the ionome of plants exposed to elevated CO_2 depletes minerals at the base of human nutrition. eLife 3, e02245. Available from: https://doi.org/10.7554/eLife.02245.001.

Lowenberg-DeBoer, J., Huang, I.Y., Grigoriadis, V., Blackmore, S., 2019. Economics of robots and automation in field crop production. Precis. Agric. 21, 278–299.

Marques, G., Aleixo, D., Pitarma, R., 2019. Enhanced hydroponic agriculture environmental monitoring: an internet of things approach. In: Rodrigues J. et al. (eds) Computational Science – ICCS 2019. ICCS 2019. Lect. Notes Comput. Sci. 11538, 658–669. Springer, Cham. Available from: https://doi.org/10.1007/978-3-030-22744-9_51.

Medek, D.E., Schwartz, J., Myers, S.S., 2017. Estimated effects of future atmospheric CO_2 concentrations on protein intake and the risk of protein deficiency by country and region. Environ. Health Perspect. 125, 87002. Available from: https://doi.org/10.1289/EHP41.

Micallef, S.A., Shiaris, M.P., Colón-Carmona, A., 2009. Influence of Arabidopsis thaliana accessions on rhizobacterial communities and natural variation in root exudates. J. Exp. Bot. 60, 1729–1742. Available from: https://doi.org/10.1093/jxb/erp053.

Michelon, N., Pennisi, G., Myint, N.O., Dall'Olio, G., Batista, L.P., Salviano, A.A.C., et al., 2020. Strategies for improved yield and water use efficiency of lettuce (*Lactuca sativa* L.) through simplified soilless cultivation under semi arid climate. Agronomy 10, 1379. Available from: https://doi.org/10.3390/agronomy10091379.

Mimmo, T., Tiziani, R., Valentinuzzi, F., Lucini, L., Nicoletto, C., Sambo, P., 2017. Selenium biofortification in Fragaria × ananassa: implications on strawberry fruits quality, content of bioactive health beneficial compounds and metabolomic profile. Front. Plant. Sci. 8, 1887. Available from:. Available from: https://www.frontiersin.org/article/10.3389/fpls.2017.01887.

Montesano, F.F., D'Imperio, M., Parente, A., Cardinali, A., Renna, M., Serio, F., 2016. Green bean biofortification for Si through soilless cultivation: plant response and Si bioaccessibility in pods. Sci. Rep. 6, 31662. Available from: https://doi.org/10.1038/srep31662.

Montoya, A.P., Obando, F.A., Morales, J.G., Vargas, G., 2017. Automatic aeroponic irrigation system based on Arduino's platform. J. Phys. Conf. Ser. 850, 012003.

Moustafa, A.T., 2010. Potential of protected agriculture and hydroponics for improving the productivity and quality of high value cash crops in Qatar. Technical Report Arabian Peninsula Regional Program. ICARDA. Available from: https://doi.org/10.13140/RG.2.2.21300.55686.

Moustafa, A., Ouraifan, S., 2004. Agro-Economic Assessment of Hydroponic System in the Arabian Peninsula. Sustainable Production of Cash Crops Using Less Water. ICARDA. Available from: https://doi.org/10.13140/RG.2.2.23590.78408.

Myers, S., Zanobetti, A., Kloog, I., Huybers, P., Leakey, A.D.B., Bloom, A.J., et al., 2014. Risen CO_2 threatens human nutrition. Nature 510, 139–142. Available from: https://doi.org/10.1038/nature13179.

Myers, S.S., Smith, M.R., Sarah, G., Christopher, D., Golden Vaitala, B., et al., 2017. Climate change and global food systems: potential impacts on food security and undernutrition. Annu. Rev. Public Health 38 (1), 259–277. Available from: https://doi/10.1146/annurev-publhealth-031816-044356.

Nancy, D., Arulselvi, P.I., 2014. Effect of selenium fortification on biochemical activities of tomato (*Solanum lycopersicum*) plants. Indo Am. J. Pharm. Res. 4, 3997–4005. Available from:. Available from: http://www.iajpr.com.

Nazarideljou, M.J., Haghshenasa, M., Hamedanb, H.J., Ferrante, A., 2018. Growth, yield and antioxidant capacity of strawberry under various K^+: Ca^{++} ratios in hydroponic culture. Acta Agric. Scand. Sect. B Soil Plant Sci. 69 (2), 1–9. Available from: https://doi.org/10.1080/09064710.2018.1506500.

Noren, H., Svensson, P., Andersson, B., 2004. A convenient and versatile hydroponic cultivation system for *Arabidopsis thaliana*. Physiol. Plant 121, 343–348. Available from: https://doi.org/10.1111/j.0031-9317.2004.00350.x.

Ouled Belgacem, A., Nejatian, A., Ben Salah, M., Moustafa, A., 2017. Water and food security in the Arabian Peninsula: struggling for more actions. J. Exp. Biol. Agric. Sci. 5, S50–S62, ISSN No. 2320–8694. Available from: http://doi.org/10.18006/2017.5(Spl-1-SAFSAW).S50.S62.

Orsini, F., Kahane, R., Nono-Womdim, R., Gianquinto, G., 2013. Urban agriculture in the developing world: a review. Agronom. Sustain. Dev. 33, 695–720. Available from: http://doi:10.1007/s13593-013-0143-z.

Orsini, F., Pennisi, G., Zulfiqar, F., Gianquinto, G., 2020. Sustainable use of resources in plant factories with artificial lighting (PFALs). Euro. J. Hortic. Sci. 85, 297–309. Available from: https://doi.org/10.17660/eJHS.2020/85.5.1.

Ortega A.R., Mahnert A., Berg C., Müller H., and Berg G., 2016. The plant is crucial: specific composition and function of the phyllosphere microbiome of indoor ornamentals, FEMS Micro. Eco. 92, 1–12. Available from: https://doi.org/10.1093/femsec/fiw173

Palande, V., Zaheer, A., George, K., 2018. Fully automated hydroponic system for indoor plant growth. Proc. Comput. Sci. 129, 482–488.

Papenfus, H.B., Kulkarni, M.G., Stirk, W.A., Finnie, J.F., Van Staden, J., 2013. Effect of a commercial seaweed extract (Kelpak®) and polyamines on nutrient-deprived (N, P and K) okra seedlings. Sci. Hortic. (Amst.) 151, 142–146.

Pastor, A.V., Palazzo, A., Havlik, P., Biemans, H., Wada, Y., Obersteiner, M., et al., 2019. The global nexus of food–trade–water sustaining environmental flows by 2050. Nat. Sustain. 2, 499. Available from: https://doi.org/10.1038/s41893-019-0287-1.0.

Paul, D., Sharma, Y., 2006. Antagonistic effects of metabolites of *Pseudomonas fluorescens* strains on the different growth phases of *Phytophthora capsici*, foot rot pathogen of black pepper (*Piper nigrum* L.). Arch. Phytopathol. Plant Protect. 39, 311–314. Available from: https://doi.org/10.1080/03235400500301182.

Paul, K., Sorrentino, M., Lucini, L., Rouphael, Y., Cardarelli, M., Bonini, P., et al., 2019. Understanding the biostimulant action of vegetal-derived protein hydrolysates by high-throughput plant phenotyping and metabolomics: a case study on tomato. Front. Plant. Sci. 10, 47. Available from: https://doi.org/10.3389/fpls.2019.00047.

Peer, R., Schippers, B., 1989. Plant growth responses to bacterization with selected *Pseudomonas spp.* strains and rhizosphere microbial development in hydroponic cultures. Can. J. Microbiol. 35, 456–463. Available from: https://doi.org/10.1139/m89-070.

Perez-Jaramillo J.E, Mendes R., Raaijmakers J.M., 2016. Impact of plant domestication on rhizosphere microbiome assembly and functions, Plant Mol Biol. 90, 635–644. Available from: https://link.springer.com/article/10.1007/s11103-015-0337-7.

Pii, Y., Graf, H., Valentinuzzi, F., Cesco, S., Mimmo, T., 2018. The effects of plant growth-promoting rhizobacteria (PGPR) on the growth and quality of strawberries. Acta Hortic. 1217, 231–238. Available from: https://doi.org/10.17660/ActaHortic.2018.1217.29.

Pii, Y., Molesini, B., Masiero, S., Pandolfini, T., 2012. The non-specific lipid transfer protein N5 of *Medicago truncatula* is implicated in epidermal stages of rhizobium-host interaction. BMC Plant. Biol. 12, 233. Available from: https://doi.org/10.1186/1471-2229-12-233.

Pii, Y., Mimmo, T., Tomasi, N., Terzano, R., Cesco, S., Crecchio, C., 2015. Microbial interactions in the rhizosphere: beneficial influences of plant growth-promoting rhizobacteria on nutrient acquisition process. A review. Biol. Fertil. Soils 51, 403–415. Available from: https://doi.org/10.1007/s00374-015-0996-1.

Pii, Y., Marastoni, L., Springeth, C., Fontanella, M.C., Beone, G.M., Cesco, S., et al., 2016. Modulation of Fe acquisition process by *Azospirillum brasilense* in cucumber plants. Environ. Exp. Bot. 130, 216–225. Available from: https://doi.org/10.1016/j.envexpbot.2016.06.011.

Polo, J., Barroso, R., Ródenas, J., Azcón-Bieto, J., Cáceres, R., Marfà, O., 2006. Porcine hemoglobin hydrolysate as a biostimulant for lettuce plants subjected to conditions of thermal stress. Hortic. Technol. 16, 483–487.

Popsimonova, G., Benko, B., Karic, L., Gruda, N., 2017. Production systems: integrated and organic production, and soilless culture. FAO, Rome, pp. 207–226, Plant Production and Protection Paper 230.

Povero, G., Mejia, J.F., Di Tommaso, D., Piaggesi, A., Warrior, P., 2016. A systematic approach to discover and characterize natural plant biostimulants. Front. Plant. Sci. 7, 435. Available from: https://doi.org/10.3389/fpls.2016.00435.

Putra, P.A., Yuliando, H., 2015. Soilless culture system to support water use efficiency and product quality: a review. Agric. Agric. Sci. Procedia 3, 283–288. Available from: https://doi.org/10.1016/j.aaspro.2015.01.054.

Ranjan, K., Priya, H., Ramakrishnan, B., Prasanna, R., Venkatachalam, S., Thapa, S., et al., 2016. Cyanobacterial inoculation modifies the rhizosphere microbiome of rice planted to a tropical alluvial soil. Appl. Soil. Ecol. 108, 195–203. Available from: https://doi.org/10.1016/j.apsoil.2016.08.010.

Rao, D., Webb, J.S., Kjelleberg, S., 2005. Competitive interactions in mixed-species biofilms containing the marine bacterium *Pseudoalteromonas tunicata*. Appl. Environ. Microbiol. 71, 1729–1736. Available from: https://doi.org/10.1128/aem.71.4.1729-1736.2005.

Resh, H.M., 2013. Hydroponic Food Production: A Definitive Guidebook for the Advanced Home Gardener and the Commercial Hydroponic Grower. CRC Press, Boca Raton, FL.

Resh, H.M., 2016. Hydroponic Food Production A Definitive Guidebook for the Advanced Home Gardener and the Commercial Hydroponic Grower, seventh ed. CRC Press, Baton Raton, FL.

Rius-Ruiz, F.X., Andrade, F.J., Riu, J., Rius, F.X., 2014. Computer-operated analytical platform for the determination of nutrients in hydroponic systems. Food Chem. 147, 92–97.

Rodriguez-Ortega, W.M., Martinez, V., Rivero, R.M., Camara-Zapata, J.M., Mestre, T., Garcia-Sanchez, F., 2017. Use of a smart irrigation system to study the effects of irrigation management on the agronomic and physiological responses of tomato plants grown under different temperatures regimes. Agric. Water Manag. 183, 158–168. Available from: https://doi.org/10.1016/j.agwat.2016.07.014.

Romeo, D., Vea, E.B., Thomsen, M., 2018. Environmental impacts of urban hydroponics in Europe: a case study in Lyon. Procedia CIRP 69, 540–545. Available from: https://doi.org/10.1016/j.procir.2017.11.048.

Ropokis, A., Ntatsi, G., Kittas, C., Katsoulas, N., 2018. Impact of cultivar and grafting on nutrient and water uptake by sweet pepper (*Capsicum annuum* L.) grown hydroponically under mediterranean climatic conditions. Front. Plant. Sci. 24, 1244. Available from: https://doi.org/10.3389/fpls.2018.01244.

Ruengittinun, S., Phongsamsuan, S., Sureeratanakorn, P., 2017. Applied internet of thing for smart hydroponic farming ecosystem, HFE. In: Proceedings of the 10th International Conference on Ubi-media Computing and Workshops, Ubi-Media.

Ruscio, F., Paoletti, P., Thomas, J., Myers, P., Fichera, S., 2019. Low-cost monitoring system for hydroponic urban vertical farms. World Acad. Sci. Eng. Technol. Int. J. Agricult. Biosys. Eng. 13, 267–271. Open Science Index 154.

Sambo, P., Nicoletto, C., Giro, A., Pii, Y., Valentinuzzi, F., Mimmo, T., et al., 2019. Hydroponic solutions for soilless production systems: issues and opportunities in a smart agriculture perspective. Front. Plant. Sci. 10, 923. Available from: https://doi.org/10.3389/fpls.2019.00923.

Savvas, D., Gianquinto, G., Tuzel, Y., Gruda, N., 2013. Soilless Culture. FAO, Rome, pp. 303–354, Plant Production And Protection Paper 217.

Shavrukov, Y., Genc, Y., Hayes, J., 2012. The use of hydroponics in abiotic stress tolerance research. In: Asao, T. (Ed.), Hydroponics - A Standard Methodology for Plant Biological Researches. In Tech, pp. 39–66. , ISBN: 978-953-51-0386-8.

Sheridan, C., Depuydt, P., De Ro, M., Petit, C., Van Gysegem, E., Delaere, P., et al., 2017. Microbial community dynamics and response to plant growth-promoting microorganisms in the rhizosphere of four common food crops cultivated in hydroponics. Microb. Ecol. 73 (2), 378–393. Available from: https://doi.org/10.1007/s00248-016-0855-0.

Singh, S., Singh, B.S., 2012. Hydroponics – A technique for cultivation of vegetables and medicinal plants. In: Proceedings of Fourth Global conference on Horticulture for Food, Nutrition and Livelihood Options Bhubaneshwar, Odisha, India, p. 220.

Smith, M.R., Myers, S.S., 2018. Impact of anthropogenic CO_2 emissions on global human nutrition. Nat. Clim. Change 8, 834–839. Available from: https://doi.org/10.1038/s41558-018-0253-3.

Smoleń, S., Kowalska, I., Sady, W., 2014. Assessment of biofortification with iodine and selenium of lettuce cultivated in the NFT hydroponic system. Sci. Hortic. 166, 9–16. Available from: https://doi.org/10.1016/j.scienta.2013.11.011.

Soares, J.C., Santos, C.S., Carvalho, S.M.P., Pintado, M.M., Vasconcelos, M.W., 2019. Preserving the nutritional quality of crop plants under a changing climate: importance and strategies. Plant Soil. 443, 1–26. Available from: https://doi.org/10.1007/s11104-019-04229-0.

Stouvenakers, G., Massart, S., Depireux, P., Jijakli, M.H., 2020. Microbial origin of aquaponic water suppressiveness against *Pythium aphanidermatum* lettuce root rot disease. Microorganisms 8, E1683. Available from: https://doi.org/10.3390/microorganisms8111683.

Tajudeen, A.L., Taiwo, O.S., 2018. Soilless farming – a key player in the realisation of "zero hunger" of the sustainable development goals in Nigeria. Int. J. Ecol. Sci. Environ. Eng. 5, 1–7. ISSN: 2375-3854.

Tanwar, A., Aggarwal, A., Parkash, V., 2014. Effect of bioinoculants and superphosphate fertilizer on the growth and yield of broccoli (*Brassica oleracea* L. var. italica Plenck). N Z J. Crop. Hortic. Sci. 42, 288–302. Available from: https://doi.org/10.1080/01140671.2014.924537.

The World Bank. Oct 2020. Understanding Poverty: Topic-Agriculture and Food: Focus Area Food Security. https://www.worldbank.org/en/topic/food-security (accessed 15.11.2020.).

Tilman, D., Cassman, K., Matson, P., Naylore, R., Polasky, S., 2002. Agricultural sustainability and intensive production practices. Nature 418, 671–677. Available from: https://doi.org/10.1038/nature01014.

Touliatos, D., Dodd, I.C., McAinsh, M., 2016. Vertical farming increases lettuce yield per unit area compared to conventional horizontal hydroponics. Food Energy Secur 5, 184–191. Available from: https://doi.org/10.1002/fes3.83.

Trevisan, S., Francioso, O., Quaggiotti, S., Nardi, S., 2010. Humic substances biological activity at the plant-soil interface: from environmental aspects to molecular factors. Plant Signal. Behav. 5, 635–643. Available from: https://doi.org/10.4161/psb.5.6.11211.

Vacheron, J., Desbrosses, G., Bouffaud, M.L., Touraine, B., Moënne-Loccoz, Y., Muller, D., et al., 2013. Plant growth-promoting rhizobacteria and root system functioning. Front. Plant Sci. 4, 356. Available from: https://doi.org/10.3389/fpls.2013.00356.

Van Ginkel, S.W., Igou, T., Chen, Y., 2017. Energy, water and nutrient impacts of California-grown vegetables compared to controlled environmental agriculture systems in Atlanta, GA. Resour. Conserv. Recycl. 122, 319–325. Available from: https://doi.org/10.1016/j.resconrec.2017.03.003.

Vernieri, P., Borghesi, E., Ferrante, A., Magnani, G., 2005. Application of biostimulants in floating system for improving rocket quality. J. Food Agricult. Env. 3, 86–88.

Vernieri, P., Borghesi, E., Tognoni, F., Serra, G., Ferrante, A., Piagessi, A., 2006. Use of biostimulants for reducing nutrient solution concentration in floating system. Acta Hortic. 718, 477–484. Available from: https://doi.org/10.17660/ActaHortic.2006.718.55.

Verones, F., Pfister, S., van Zelm, R., Hellweg, S., 2017. Biodiversity impacts from water consumption on a global scale for use in life cycle assessment. Int. J. Life Cycle Assess. 22, 1247–1256. Available from: https://doi.org/10.1007/s11367-016-1236-0.

Vorholt, J.A., 2012. Microbial life in the phyllosphere. Nat. Rev. Microb. 10, 828–840. Available from: https://doi.org/10.1038/nrmicro2910.

White, P.J., Pongrac, P., Sneddon, C.C., Thompson, J.A., Wright, G., 2018. Limits to the biofortification of leafy brassicas with zinc. Agriculture 8, 32. Available from: https://doi.org/10.3390/agriculture8030032.

Wiesner-Reinhold, M., Schreiner, M., Baldermann, S., Schwarz, D., Hanschen, F.S., Kipp, A.P., 2017. Mechanisms of selenium enrichment and measurement in brassicaceous vegetables, and their application to human health. Front. Plant Sci. 8, 1365. Available from: https://doi.org/10.3389/fpls.2017.01365.

Xiao, H., Cheng, S., Wu, Z., 2010. Microbial community variation in phytoremediation of triazophos by *Canna indica Linn.* in a hydroponic system. J. Environ. Sci. (China) 22, 1225–1231. Available from: https://doi.org/10.1016/S1001-0742(09)60242-4.

Yan, X., Li, H., Su, X., 2018. Review of optical sensors for pesticides. Trends Anal. Chem. 103, 1–20.

Zamora-Izquierdo, M.A., Santa, J., Martínez, J.A., Martínez, V., Skarmeta, A.F., 2019. Smart farming IoT platform based on edge and cloud computing. Biosyst. Eng. 177, 4–17.

Zarraonaindia, I., Owens, S.M., Weisenhorn, P., et al., 2015. The soil microbiome influences grapevine-associated microbiota. mBio 6, e02527–e02614. Available from: https://doi.org/10.1128/mBio.02527-14.

Zhao, C., Liu, B., Piao, S., Wang, X., Lobell, D.B., Huang, Y., et al., 2017. Temperature increase reduces global yields of major crops in four independent estimates. Proc. Natl. Acad. Sci. U S A 114, 9326–9331. Available from: https://doi.org/10.1073/pnas.1701762114.

Zheng, J., Sutton, J.C., Yu, H., 2000. Interactions among Pythium aphanidermatum, roots, root mucilage, and microbial agents in hydroponic cucumbers. Can. J. Plant Pathol. 22, 368–379. Available from: https://doi.org/10.1080/07060660009500455.

Chapter 9

Plant ionomics: toward high-throughput nutrient profiling

Anshuman Singh[1], Ayushi Jaiswal[1], Ashutosh Singh[2], Ram Sewak Tomar[3] and Anil Kumar[4]
[1]*Genetics and Plant Breeding, Rani Lakshmi Bai Central Agricultural University, Jhansi, India,* [2]*Biotechnology and Crop Improvement, Rani Lakshmi Bai Central Agricultural University, Jhansi, India,* [3]*Biotechnology, Rani Lakshmi Bai Central Agricultural University, Jhansi, India,* [4]*Rani Lakshmi Bai Central Agricultural University, Jhansi, India*

9.1 Introduction

Ionomics represents a wholesome approach for upgrading the particle substance within plants. Micronutrient deficits (e.g., selenium, zinc, and calcium) affect nearly 2 billion individuals in developing as well as developed nations (Tulchinsky, 2010). Most of those tormented are lacking staple crops such as wheat, rice, and maize for their food. Those tormented cannot manage the fortified foods to meet their mineral necessities. Hence mineral enhancement, that is, biofortification (genetic improvement), of staple food crops has been considered a feasible technique to handle the issue of mineral deficiencies. Be that as it may, the primary and preeminent prerequisite to developing biofortified crops for improved mineral substances is understanding the genetic control for mineral accumulation inside the edible parts of the plants (e.g., seed, fruit). The sum of mineral substances in these tissues depends on various forms such as mobilization from the soil, uptake by the roots, transport and redistribution inside the plant, import and accumulation within the seeds, etc. To dissect such a complex characteristic, examination of the molecular apparatus and understanding its working through high-throughput technologies are essential tasks. In this regard, plant ionomics may well be a valuable innovation to explore the relation of gene(s) to the transport and accumulation of ions. However, compared with other "-omics," ionomics is generally a new field, since the majority of genes and gene networks in ionome regulation are still obscure. An ionome is the composition of whole mineral nutrients and trace elements found in an organism and represents the inorganic component of cellular system of an organism.

Plant Nutrition and Food Security in the Era of Climate Change. DOI: https://doi.org/10.1016/B978-0-12-822916-3.00015-9

(Salt et al., 2008). It is the dynamic network of elements, controlled by physiology and organic chemistry of the plant, which is eventually controlled by genetic and environmental factors (Baxter, 2010). Hence, the issue of plant ionomes and their regulation is very germane with regard to development of newer strategies for the advancement of nutritional quality through biofortification of crops. In this chapter, attempts will be made to explain the intricacy of and possible strategies to explore their regulation.

9.2 Concept of ionomics

An ionome is a powerful approach to functional gene and gene networks. Living organisms are organized and sustained by the genome through the activity of the transcriptome, proteome, metabolome, and ionome, the four fundamental biochemical pillars of functional genomics. These pillars represent the sum of all the expressed genes, proteins, metabolites, and elements inside an organism. The dynamic response and interaction of those biochemical "-omes" define how a living system functions, and the study of these biological systems is presently among the most important challenges within the life sciences. Studies on the functional connections between the genome and the transcriptome (Martzivanou and Hampp, 2003; Leonhardt et al., 2004), proteome (Koller et al., 2002), and metabolome (Fiehn et al., 2000) are well underway. In contrast, the study of the ionome is still in its earliest stages (Lahner et al., 2003; Hirschi, 2003; Rea, 2003), with the majority of genes and gene networks involved in its regulation still unknown. In addition, because the ionome is involved in such a wide range of important biological phenomenon, including electrophysiology, signaling, enzymology, osmoregulation, and transport, its study promises to yield new and significant biological insight. An understanding of the ionome and the way it interacts with other cellular systems such as the genome, proteome, metabolome, and environment are integral to our full understanding of how plants coordinate their organic and inorganic metabolisms. Lahner et al. first described the ionome to incorporate all the metals, metalloids, and nonmetals present in a living organism (Lahner et al., 2003), extending the term metallome (Outten and O'Halloran, 2001; Williams, 2001; Szpunar, 2004) to incorporate biologically significant nonmetals such as nitrogen, phosphorus, sulfur, selenium, chlorine, and iodine. Compounds containing the nonmetals phosphorus, sulfur, or nitrogen, for example, would fall inside both the ionome and metabolome, and metals such as zinc, copper, manganese, and iron in metalloproteins would drop inside the proteome or metalloproteome because it has been depicted (Szpunar, 2004).

9.3 Important events in ionomics

The thought of plant ionomics starts with the mixing of metabolomics and mineral nutrition. It was begun with the conclusion given by Robinson and

Pauling in the late 1960s and early 1970s that the metabolite profile of a life form contains a rich source of data that is a reflection of the physiological status of that living being. Since then, remarkable progress has been made in describing and understanding the basic biology of nutrition homeostasis in plants, since its foundation as a scientific discipline within the 19th century (Marschner, 2011). Lahner et al. (2003) for the first time included all the metals, metalloids, and nonmetals present in an organism within the ionome. In 2008, Salt and colleagues used the term *plant ionome* for the full concentration (i.e., all shapes) of chosen components (elements) in a test of plant tissue. Studies on the functional connections between the genome, transcriptome (Leonhardt et al., 2004), metabolome (Fiehn et al., 2000), and proteome (Koller et al., 2002) are underway (Fleet et al., 2011).

9.4 Spectrum of mineral elements

Plants require a variety of components for their survival, generally referred to as mineral elements. They are ingested by the roots of plants in ionic form or in the form of salts. Deficiency of these elements in plants leads to specific types of indications. Approximately 112 components have been found but this time 17 component element that stimulate growth and may be essential to species are defined as beneficial elements are for physiological development, reproduction and other availability of plant genome sequences and the development of biometrical techniques have accelerated the identification of mineral element. (Ohkama-Ohtsu and Wasaki, 2010).

9.5 Mineral acquisition, distribution, and storage in plants

The total process of mineral acquisition, distribution, and storage in plants is exceptionally complex. It includes multiple molecular components including transporters, channels, chelators, and a few specific genes that encode and regulate them. Before learning the complete ionome, we have to grasp how minerals are retained from the soil and after that conveyed all through the plant. Acquisition of ions by roots is more pronounced in the root hair zone than within the meristem and elongation zone (Gilroy and Jones, 2000). In *Arabidopsis*, out of 25,500 proteins, only 1800 (7%) perform transport functions (Schwacke et al., 2003). Channels are transmembrane proteins that work as particular pores for a specific type of element. Transport through channels is always passive, and the specificity of transport primarily depends on pore size and electric charge more than on selective binding (Alberts et al., 2002). Our understanding with respect to phloem mineral stacking into functional tissue is very restricted, but it is obvious that a pool of mineral particles in source tissues, a solid transporter framework, and sequestering particles in sink must be shown for their conveyance and capacity.

9.6 Mineral interaction in plants

Mineral interactions can influence absorption and bioavailability of other nutrients by a number of mechanisms. Minerals with chemical similitude can compete for transport proteins or other uptake instruments, encouraging or preventing assimilations. The concentration and proportions of different ions can impact the uptake and transport of a specific nutrient. Interactions between nutrients occur both within the rhizosphere as well as inside plant tissues, which may cause insufficiency or toxicity of that specific nutrient and may lead to changed plant growth and yield. Even though the genetic foundation of a plant is the sole determinant of mineral assimilation, transport, and functions, these activities are briefly affected by other variables too (Clárk, 1983).

9.7 Element–element interactions

Minerals that are chemically similar can compete for transport proteins or other uptake mechanism by facilitating or hindering their absorption. Transport proteins offer assistance in uptake and transport of elements within the cell. In spite of the fact that a specific transport protein is more often than not profoundly particular for the kind of substances it transports, its specificity is regularly not outright. For example, CAX2 and CAX3 are vacuolar Ca^{2+}/H^{+1} antiporters, but they also transport cadmium (Cd) and manganese (Mn^{2+}) instead of calcium (Ca^{2+}) (Hirschi et al., 2000; Punshon et al., 2012).

9.8 Element–gene interactions

There are numerous genes whose expressions, specifically or by implication, are regulated by availability and unavailability of that element whose transport, storage, and numerous other functions are done by protein products of that gene. A few atomic instruments viz, arrangement of specialized auxiliary structure of mRNA, customized utilize of halt codon and specialized particles as metal-activated translation components offer assistance to direct the interpretation and soundness of those qualities. Making plant mutants for particular transporters for their ionome consider is the finest technique to know the distinction in concentration of one or more components.

9.9 Element–environment interactions

The studies on interaction among genes, elements, and the environment may be progressed by performing genetics and modeling that determine the mechanisms by which plants adopt to distinctive situations and clarify the whole elemental and ionic composition of the plants. Studies of ionomes in genetically diverse population provide insight about plant–environment interactions that control elemental accumulation and identification proof of the underlying genes (Baxter and Dilkes, 2012). Long-term exposure of plants to saline soil results in mineral ion imbalance and thus, alters metabolism (Diego et al. 2008). Hence, the interaction of genotype and environment is clearly responsible for alterations in the ionome of plant cell or tissue.

9.10 Interaction with mineral chelating or sequestering molecules

The movement of mineral elements from source to sink involves their mobilization from soil to roots, translocation to the shoot, redistribution within the plant, and deposition in seeds (Grusak and Della Penna, 1999; White and Broadley, 2009). In general, plants absorb minerals through diffusion from the soil, but in response to low metal availability, plants activate uptake systems. For example, in the case of iron, a few plants secrete phytosiderophores from roots that bind Fe^{+3} for subsequent acquisition of the Fe^{+3} chelate, and some plants acquire Fe^{+2} after reductions by ferric reductases. Similarly, in the case of metal toxicity, various chelating and sequestering molecules in plants actively bind with the elements. The binding of these molecules with these elements reduce their toxic effects in plant. Chelating molecules contribute to metal detoxification by buffering cytosolic metal concentration. The principal class of metal chelators includes phytochelatins, metallothioneins, amino acids, and organic acids (Clemens, 2001). Citrate, malate, and oxalate are members of organic acid chelators that have been implicated in a range of important processes, including differential metal tolerance, metal transport through the xylem, and vacuolar metal sequestration (Rauser, 1999). The presence of minerals bound with chelating or sequestering provides an added sink, so that more of the mineral ions could have moved to seeds by diffusional routes.

9.11 Bioinformatics involved in ionomics

The Purdue Ionomics Information Management System (PiiMS) described here is founded on our National Science Foundation Plant Genome and Arabidopsis 2010 funded ionomics project. The ionome is a new conceptual framework for thinking about plant mineral nutrition and was described by Salt et al. (for review, see Salt, 2004) to include all the metals, metalloids, and nonmetals present in an organism (Lahner et al., 2003), extending the term metallome (Outten and O'Halloran, 2001; Williams, 2001; Szpunar, 2004) to include biologically significant nonmetals, such as N, P, S, Se, Cl, and I. PiiMS couples flexible data storage and search tools with our high-throughput inductively coupled plasma (ICP)–mass spectrometry (MS) phenotyping platform and allows Web-based open access to our large ionomics dataset. PiiMS can be accessed at www.purdue.edu/dp/ionomics. In PiiMS, we have developed workflow controls and data storage systems to facilitate the high-throughput data acquisition required by our ionomics projects. PiiMS models the physical workflow in the laboratory and divides it into stages based on the physical activities and the information and data generated throughout the experiment life cycle. In this way, it can provide workflow management support as well as the capability to capture contextual information (metadata) necessary to fully describe the experiment. This system allows us to control and capture, through a series of input portals, the flow of information in our high-throughput ionomics system. Information mapping the position of the plants in the experimental unit to the sample analysis vial is captured, helping to reduce errors in sample identification after analysis. Information relating to the plant's genotype and genealogy, date planted and harvested, environmental, and ICP-MS analytical conditions are also captured. After ICP-MS analysis is completed, all analytical data are uploaded into PiiMS via a Web-based upload portal. All experimental metadata and ICP-MS analytical data are associated in the database. Tools are also implemented in the system to manage various aspects of the PiiMS environment, including the creation of new sample maps, the addition of new genetic and genealogy information, and the management of users and instruments. PiiMS also provides on-demand data reanalysis and open-access search and retrieval capability across the complete ionomics dataset. Similar ideas for the development of supportive data collection tools for plant metabolomics have recently been proposed, but to our knowledge, PiiMS is the first true implementation of these ideas integrated into an existing high-throughput analytical phenotyping platform. Ionomics consists of an experimental design, which includes sample preparation, digestion, and analysis ICP-MS/ICP-Optical Emission Spectrometry (OES) followed by data processing (Fig. 9.1), storage, and retrieval. Because of these commonalities, the overall architecture of PiiMS is adaptable across various ionomics platforms, providing a good model for the management of functional genomics approaches in general.

9.12 Different technology used in plant element profiling

The estimation of composition of ions and elements of the whole plant, tissue, and indeed a single cell is required for plant ionomics. From sample preparation protocols to specialized instruments, different advanced technologies and equipment are needed. These may change according to the elements to be measured, availability of sample size, sample throughput, dynamic quantification range, sensitivity, reliability quality, and accuracy.

9.13 Techniques based on electronic properties of elements

In this section, we are progressing to examine a few methods that are based on the electronic properties of atoms. In these strategies energy is being

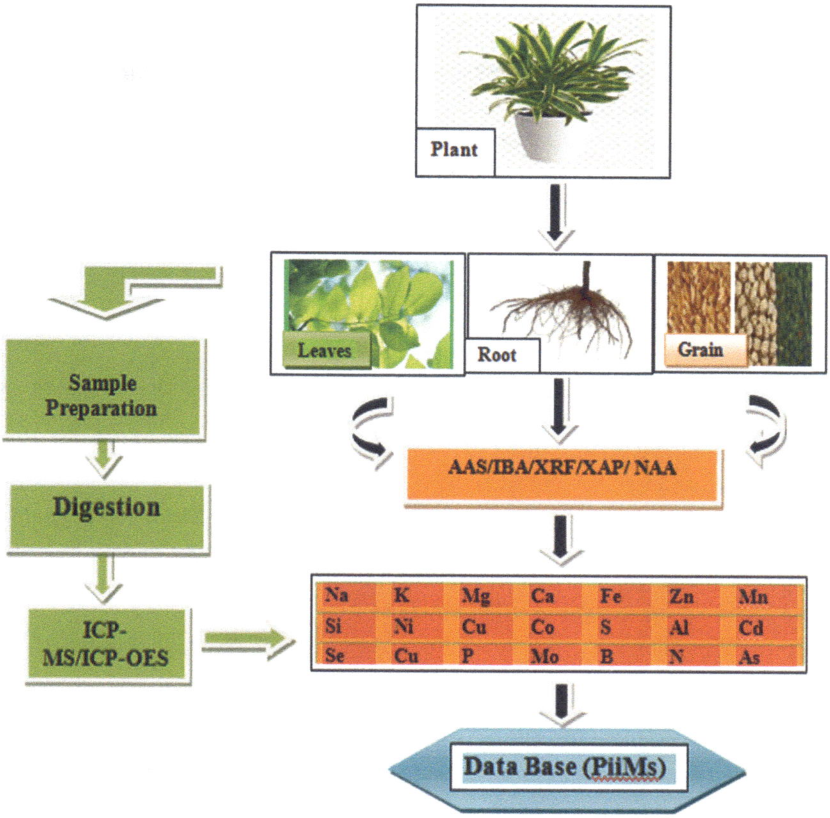

FIGURE 9.1 Plant Ionomics Study Platform; ICP-MS/ICP-OES-Inductively coupled plasma-mass spectroscopy/Optical Emission spectroscopy, AAS-Atomic absorption spectrometry, IBA-Ion beam analysis, XRF-X-ray fluorescence spectroscopy, XAF-X-ray Absorption spectroscopy, NAA-Neutron activation analysis, PiiMs: Purdue Ionomics Information Management System.

provided during sample estimation. It collides with the mixture of ions and electrons are moved from inward shells to external shells of atoms. During this process, certain energy is released and therefore the value of this energy is element-specific. A brief summary of these strategies are mentioned below:

9.13.1 Atomic absorption spectrometry

In atomic absorption spectrometry (AAS), free atoms lie within the gaseous state and absorb light within the form of optical radiation for quantitative detection of the chemical components display in a test (Lvov, 2005). To arrange to analyze a test for nuclear constituents by AAS, it must be atomized by any of the two atomizers, that is, fire or electrothermal (graphite tube) atomizers. A fire atomizer is utilized in fire nuclear retention spectrometry

(FAAS). In electrothermal nuclear retention spectrometry (ETAAS), a graphite tube is utilized as the atomizer. In this, the test is presented into the graphite tube and subjected to a temperature program. The graphite tubes are warmed with ohmic resistance employing a low-voltage, high-current control supply. In ETAAS a temporal flag is produced, the range of which specifically corresponds to the mass of analyte (not its concentration) presented into the graphite tube (Török and Žemberyová, 2012).

9.13.2 Optical emission spectroscopy

When ionized analyte atoms in the ICP plasma fall back to ground state, they emit photons at a wavelength characteristic of a given element. For quantification of these analytes the light from the plasma, representing a summation of emitted light from all the atoms introduced into the plasma from the sample, can be focused and passed through optical slits into a spectrometer. Within the spectrophotometer an optical filter is used to separate the collected photons by wavelength, and a charge injection device (CID) detector simultaneously measures the intensities of photons at multiple wavelengths. By comparing these energy intensities to reference standards, a quantitative measurement of every element within the sample is often obtained.

9.13.3 Ion beam analysis

The ion beam analysis (IBA) is a group of advanced and powerful techniques for quantitative determination of elements within the sample. In IBA, a beam of accelerated charged particles moving at exceptionally high speed strikes on the sample material, which leads to the further emission of ions or secondary radiation as c- and X-rays from the target material. IBA is an exceedingly delicate method to identify the composition of the target sample without hurting or aggravating the sample material.

9.13.4 X-ray fluorescence spectroscopy

X-ray fluorescence spectroscopy (XRF) is another method widely used for plant ionomic study. It is a dependable technique for assurance of chemical elements and their concentrations in samples shown in liquid or powdered (solid) states. XRF is generally a nondestructive procedure. In XRF analysis, a source of X-rays or c-rays is utilized to illuminate a sample, in which the source radiations may undergo either scatterings or absorption by the sample molecules. When an atom assimilates the source X-rays, the occurrence radiation can unstick electrons from the deepest shell of the ions, making openings. Electrons from external shells fill the internal shell opening and radiate X-ray photons. Ionomic analysis can be performed by XRF on living specimens without compromising the potential viable form. X-ray fluorescence spectroscopy (XRF) X-ray fluorescence spectroscopy (XRF) is another method widely used for plant ionomic study. It is a dependable technique for assurance of chemical elements and their concentrations in samples present in liquid or powdered (solid) states. XRF is generally a nondestructive procedure. In XRF analysis, a source of X-rays or c-rays is utilized to illuminate a sample, in which the source radiations may undergo either scatterings or absorption by the sample molecules. When an atom absorbs the source X-rays, the occurrence radiation can dislodge electrons from the innermost shell of the ions, creating vacancies. Electrons from external shells fill the internal shell vacancy and emit X-ray photons. The energy of the emitted X-ray depends on the difference in energy of the shell with the initial vacancy and the energy of the electron that fills the vacancy. Ionomic analysis can be performed by XRF on living specimens without compromising the potential viable form. The XRF method can be used for multi element screening seedlings to identify FRD3 mutants with altered ionomes. A second novel application of XRF to plants as a rapid screening tool is the recent use of synchrotron-based micro-XRF for the possible identification of A. thaliana seeds with mutant ionomic phenotypes (Delhaize et al.,1993). XRF is capable of measureing elements from beryllium (Be) to uranium (U) and beyondat trace levels often below one parts per million and up to 100% (Beckhoff et al. 2006). Recently confocal imaging and XRF microtomography were used to create in various plant samples three-dimensional distribution of multiple elements. Three-dimensional distribution of various elements was quantitatively mapped in intact A. thaliana seeds for V1T1 transporter by XRF microtomography. Principle excitation of the specific element of interest is being used by X-ray absorption spectroscopy (XAP). XRF analyzers are for the most part classified as being either energy dispersive (EDXRF) or wavelength dispersive (WDXRF), but there are some additional subcategories inside these two frameworks.

9.13.5 Inductively coupled plasma–mass spectrometry

The most recent technique for ionome analysis utilization is inductively coupled plasma (ICP) spectroscopy (Li et al., 2013). ICP may be a gaseous source and is typically being generated for ionizing the sample elements. ICP may be a vaporous source and is regularly being created for ionizing the test components. In inductively coupled plasma–mass spectroscopy (ICP-MS), ionized atoms are to first carry out to sampling and skimmer cones that are made up of nickel (Ni). They are at that point exchanged to the quadrupole mass analyzer in which they pass through a series of ion lenses according to their mass-to-charge ratio. Here, an electron multiplier identifies these ions. At long last, these are measured by comparing with reference measures for each element (Bertoldi et al., 2011; Cizdziel et al., 2011). For ICP-MS analyses of shoots and roots of axenically grown plants, metals were desorbed from samples for 10 min on ice with cold 5 mM $CaSO_4$, 1 mM MES, pH 5.7, and then washed twice with cold ultrapure water (Haydon et al., 2012).

9.14 Techniques based on nuclear properties of atoms

The nuclear properties based on protons and neutrons are also utilized for estimation of elements from sample mixtures. The nuclear spectrum is particular for each element and is able to estimate the amounts of element present in sample. The method discussed in the following section is based on the nuclear property.

9.14.1 Neutron activation analysis

Neutron activation analysis (NAA) is a vital strategy for elemental analysis of local environmental studies of different materials. It was applied for the first time in 1936 by Hevesy and Levi utilizing the [Ra/Be] isotopic neutron source for irradiation of the sample. The NAA procedure accomplishes a subjective and quantitative examination of obscure tests by illuminating the test with neutrons [cold, warm, epi-warm, and quick], which are based on the estimation of the characteristic radiation from radio nuclides shaped straightforwardly or in a roundabout way by neutron light of such tests. In comparison with other strategies, such as ICP-(MS and OES) and XRF, NAA has not however been utilized as a high-throughput elemental analysis tool for the identification of ionomic mutants (Galinha et al., 2011). Neutron activation is the process whereby free neutrons are captured by atomic nuclei. This neutron capture results in the formation of new nuclei that are frequently radioactive. These radioactive nuclei will decay with time, and a proportion of the energy is released as gamma radiation. This emitted gamma radiation can be detected, and gamma rays at a particular energy are indicative of the presence of a specific radionuclide. Processing of the gamma-ray spectra released from a neutron-irradiated sample can therefore yield the concentration of various elements in the sample. This type of technique can allow the nondestructive quantification of multiple major and trace elements in a sample. Because neutrons have no charge they can penetrate and pass through most samples, activating the whole sample. Gamma rays released from the activated sample are also very penetrating and can escape the sample for efficient detection. Therefore elemental analysis by NAA in general does not suffer from errors introduced due to the sample matrix. NAA can be performed on a sample directly, without sample preparation. In this configuration the analysis is termed instrument neutron activation analysis (INAA), and it is this approach that is most useful for the large sample numbers required for ionomic analysis. Unfortunately, NAA does require access to a nuclear reactor as a source of neutrons to perform the initial activation step, which could limit the usefulness of NAA for a long-term ionomics project. Elemental analysis of plant samples by NAA dates back to at least the mid-1960s. Analysis was initially focused on single elements, but by the early 1970s researchers were performing simultaneous analysis of macro, micro, and trace elements in plants by NAA (54). The power of elemental profiling using NAA was rapidly applied to environmental toxicology, where the transport of potentially toxic trace elements from the environment, through crop plants and livestock to humans, has been monitored using NAA (15,23,33,83). Multielement analysis of plant samples by NAA was also used to identify and monitor areas of heavy-metal pollution across broad geographical areas and multiple years (8,25). Interestingly, NAA was also used to perform multielement quantification of nonplant samples collected within and across broad phylogenetic groupings, for the identification of trends in mineral nutrient and trace element accumulation in plants across taxa (62,87). However, even though the use of NAA for plant samples goes back more than 40 years, to our knowledge NAA has not yet been used as a high-throughput elemental analysis tool for ionomic analyses, as defined here. For example, unlike ICP-(OES and MS) and XRF, NAA has not been used as a screening tool for the identification of ionomic mutants. This is surprising given the high-throughput and minimal sample preparation required for INAA. We do note that NAA has been applied to perform ionomics in the study of breast cancer (24,57), colorectal cancer (3,81), and brain cancer (2); in these studies the ionome was shown to be present in the diseased tissues or organisms.

9.15 Recent advances in plant ionomic techniques

The prior element estimation methods are based on either electronic or atomic properties and these are not able to study the impact of any change within the plants and their surroundings at the cellular or subcellular level. Indeed, a few methods that are being utilized these days for sample preparation to study the changes in elemental composition are so interfering that they disturb the elemental profile inside the cell. Be that as it may, further investigations are required to get the component control of particle homeostasis, which incorporates transport and capacity of a specific particle at cellular and subcellular levels. In this context, synchrotron-based X-ray fluorescence microscopy (SXRF) could be a very valuable spatial method for collecting resolved metal abundance data and will be ideal for answering fundamental questions about metal homeostasis in plants such as how these ions are taken, transported, and stored in a cell, which sort of films and cellular compartments are included in these processes, etc.

9.16 Applications of plant ionomics

Ionomics is used to understand the mechanism of mineral transport in plant by identifying the candidate transporter genes and their further functional validation. It requires the use of high-throughput basic investigation innovations and their combination with both bioinformatics and hereditary instruments. The physiological status of plants is evaluated and biomarkers (ionome profile) for each status can be recognized through this strategy (Baxter, 2009). Individuals also utilize ionomic information for phylogenetic examination of plant species (White et al., 2012). Ionomics has yielded progress in our understanding of the loci that control common ionomic varieties. Be that as it may, the biological capacities of allelic variations of ionomic loci, in any case, are still generally obscure. Ionomics has developed over the final decade as a capable apparatus for effectively distinguishing qualities and quality systems that control mineral supplements and follow component homeostasis. Such victory has required the opportune integration of the explanatory strategies of high-throughput basic profiling with present-day atomic hereditary qualities. The molecular, cellular, and physiological functions of genes discovered using this approach are starting to be unravelled, although the road from gene to function (and publication) can still be a long one.

9.17 Paradigm shift from ionome to gene regulating network

In plant system biology, several studies now involve the observation of one or more different nutrient parameters such as nitrogen, carbon, sulfate, and light treatments. For example, transcriptome studies have been used to explore the regulatory interaction between manifold inputs such as carbon and nitrogen (Price et al., 2004; Gutierrez et al., 2007), carbon and light (Thum et al., 2004), as well as circadian rhythms and carbon (Blasing et al., 2005), or more recently carbon-light-nitrogen in two different organs (Krouk et al., 2009). Interestingly, in the latter dataset this complete combinatorial experiment involving three inputs (carbon, light, and nitrogen), which enabled the measurement of mRNA accumulation with a potential high degree of freedom, highlighted the fact that multiple signals coordinate gene expression precisely. Indeed, gene expression can be explained by a relatively limited number of behaviors/models of expression compared to the potential freedom given by the combinatorial experiment (Krouk et al., 2009). This kind of analysis has some potential to cover the relationships between signal interactions in the control of gene expression.

9.18 Ionomics in Identifying of QTLs/genes

Ionomics study is one of the powerful methods to identify novel genes regulating ionome profiling within the plant system. Hence ionomics is called the fourth pillar of functional genomics. The ionomic profile study helps describe the interaction of the metal ion with the function of a particular gene for its function and also the interaction of the ion with the protein for transportation within the plant system. For example, Cu ion is involved in the interaction with protein transporters such as RAN1 and HMA5. By the mutational inactivation method in *A. thaliana* the function of the gene FRD3 regulating Fe level in the plant has been identified (Rogers and Guerinot, 2002). The function of gene AtHKT1 regulating the accumulation of the Na^{+} ion in the shoot and root has been identified using a reverse genetic approach. Gene expression profiling has been done by the DNA microarray technique using the BSA in *A. thaliana* (Rus et al., 2006). Nullifying the action of the gene regulating Mo accumulation by transposon tagging has been identified in *A. thaliana* and the function analyzed by the GUS expression found that the gene is expressed in almost the entire plant (Tomastu et al., 2007). The same study has been reported for the identification of the gene accumulating Cd in rice plants. A gene (OsHMA3) was reported found to be responsible for low Cd accumulation in rice. The findings indicated that OsHMA3 limits translocation of Cd from the roots to the aboveground tissues by selectively sequestrating Cd into the root vacuoles from the low Cd-accumulating cultivar (Norton et al., 2010). Using the RNAi method of gene silencing, the function of MTP3 has been identified, which regulates Fe and Zn homeostasis in the plant system. The ectopic expression of the gene and the vacuolar localization have been identified using GFP as a selectable marker (Arrivault et al., 2006). For the increased micronutrients like Fe and Zn in maize plants, in vitro mutagenesis has been done and found two genes present on the chromosome 1 s (Raboy, 2010). In case of maize using the map-based cloning approach, the MIK gene has been identified and denoted as the lpa3 mutant. This gene helps in lowering the phytate level in the plant and thereby increasing the level of divalent cations like Fe and Zn. The study of genetic variation in microelements and their stability in different environmental conditions suggests constitutive expression of the gene (Sobhana et al., 2010). In *Arabidopsis*, the iron-regulated 1 and iron-regulated 2 gene regulates the ferroportin metal efflux proteins FPN1 and FPN2 that function in iron and cobalt homeostasis (Joe et al., 2009). A study using real-time RT-PCR gene expression profiling for Cu transporters in *A. thaliana* revealed the function of some novel transporter and the chaperons in the regulation of Cu^{++} ion homeostasis in the plant. The results indicated a common transcriptional regulation pattern of transporters and chaperone components, in particular transcriptional changes of COPT2, ZIP2, and CCH showed an inverse relation with Cu content suggesting that these proteins are required to avoid excesses and deficits of Cu (Pozo et al., 2010). In recent years some other Fe-regulator protein has been identified like IRT1, which comes under ZIP family protein (Yang et al., 2010). Identification and mapping the location of the gene helps in studying the gene interaction and gene pyramiding in the segregating population for the introgression of the trait into offspring. In dicot plants like *A. thaliana*, the complete ionomic profile has been done with the help of the ICP-MS method (Elizabeth et al., 2010). Researchers conducted a complete genomic study of the plant with all the quantitative characteristic loci (QTLs) regulating the function of the gene associated with the ionome. QTL mapping for the ions was done after selecting widely variant populations collected from different environmental conditions and growing them in a controlled environment. This study resulted in the identification of nearly 280 QTLs for nearly 19 elements (Elizabeth et al., 2010). Genetic mapping for rice has been done for the identification of the QTLs associated with 17 essential elements both in the leaves and in grains. In grains there are 41 QTLs and in leaves there are 36 QTLs associated with the elements. These QTLs also show epistatic effect among themselves. Only Se was found to have correlation between leaf and grain [26]. QTL analysis has been done in wheat for the identification and location of the gene regulating Fe and Zn in the plant. QTLs governing Fe regulation in the plant are present on two chromosomes, one in chromosome 2 and another in chromosome 7. However, for the regulation of Zn there is only one QTL present on chromosome 2 (Tiwari et al., 2009). A related study has been done for the QTL study in *Brassica napus* for the ions like B, Ca, Fe, Cu, Mg, P, and Zn using DH population. A total of 35 QTLs have been identified (Liu et al., 2009). In *B. napus* the QTL study was done for B ion (Zhao et al., 2008). QTL associated with the low phytic acid in maize plant has been mapped to be located on chromosome 1.

Understanding ionomic information of molecularly characterized differing germplasms of a given species has the capacity to connect hereditary information to worldwide mineral nourishment (Ghandilyan et al., 2009; Norton et al., 2010). Connecting hereditary information with ionome is a newly developed and precise approach for identification of QTLs related with characteristic of interest. As of late, QTL mapping has been broadly utilized for partner hereditary variety with the phenotypic variation and gives a dependable apparatus for quality revelations. The essential advantage of QTL mapping is that examiners require no earlier information of the physiology or science fundamentals of the pathway that is to be examined. Various QTLs have been distinguished utilizing recombinant innate lines (RILs) for various imperative characteristics related to the ionome; these incorporate phosphate aggregation in seeds and shoots (Bentsink et al., 2003), nitrogen (N) utilization proficiency (Rauh et al., 2002; Loudet et al., 2003), aluminum (Al) resistance (Hoekenga et al., 2006), and shoot cesium (Cs) aggregation (Payne et al., 2004). The related locale or connected locale (i.e., hereditary markers) distinguished through QTL mapping can at that point be separated or cloned for identification and investigation of mindful qualities.

9.19 Functional validation of gene(s)

Functional validation of genes to reveal a gene's precise role in nutrient transport has been a big challenge in plant science. The establishment of a new branch of "-omics," that is, ionomics, is contributing a lot in understanding the role of specific genes in ion transport by creating alterations in their nucleotide sequence. A common strategy for making alterations in genomes is induced either through various mutagens or to generate populations of DNA insertion mutants and recognize their impacts by screening and characterizing the mutants. As of late, Borghi et al. (2011) practically approved the part of quick neutron mutagenized *A. thaliana* allele CPR5 in moo leaf K^{+1} concentration. Considering the significance of such data in functional genomics in rice, a Rice Mutant Database (RMD, http://rmd.ncpgr.cn) is accessible (Hirochika et al., 2004). Along with other insertional rice mutant collections, this database might be exceptionally informative for quickening and productivity and also ionomics of rice and other monocots (Hirochika et al., 2004; Sallaud et al., 2004).

9.20 Ionomics for coping with abiotic stresses

Recently, several approaches have focused on investigating the regulation system adapted by plant ions in response to stresses. The results revealed that this network system is composed of various components including growth-inducing stress-responsive genes, and signaling and regulatory expressive elements. Also, it has been revealed that plant traits regulating stress responses are QTLs to a large extent. These traits are difficult to be genetically selected. In addition, various proteins are needed during stress response. For example, metabolic proteins inducing osmoprotectant production and regulatory protein kinases or transcriptional factors are required for signaling transduction. Ionomics is the better approach for coping with abiotic stress. Abiotic stresses are the outcome of numerous environmental disturbances as a result of the continuous encroachment of industrialization, urbanization, and some human interference with the natural ecosystem that affects the quality and quantity of agriculture production per year (Mantri et al., 2012). Abiotic stresses encompass heat, cold, freezing, drought, salinity, flooding agents, UV, and heavy-metal stresses that have significant effects on a plant life cycle (Macedo, 2012; Mantri et al., 2012; Tripathi et al., 2016c). These stresses not only affect plant biodiversity but can also interfere with the food web and the ecosystem. Abiotic stresses cause worldwide problems as they reduce crop production (Cramer et al., 2011). Moreover, the occurrence of heavy metals in food crops causes some severe diseases like cancer and asthma (Brigham et al., 2015). Plants, being the most precious gift of nature that fulfill various important basic requirements of human beings, are severely influenced by various abiotic stress factors. Significant reduction in growth and yield of plants happens due to abiotic stresses via inducing oxidative stress through enhanced reactive oxygen species (ROS) production and by lowering the antioxidant activities, level of nutrients, and modification of anatomical structures (Nagajyoti et al., 2010;

Nazar et al., 2012; Vaculík et al., 2012; Tripathi et al., 2016b, 2017a). Hoagland and Arnon (1950) pioneered the most popular commercial technique for growing plants with their roots in solutions containing mineral nutrients required for the growth of plants.

In order to protect plants from the adverse effects of abiotic stresses, various techniques are used, including exogenous supplementations of silicon, nitric oxide, growth-promoting hormones, enzymes, and nutrient management (Nagajyoti et al., 2010; Nazar et al., 2012; Tripathi et al., 2012, 2016a, 2017a,b,c,d; Vaculík et al., 2012; Saxena and Shekhawat, 2013). Among the remedies for abiotic stress, nutrient regulations or management are considered as cost-effective and ecofriendly techniques (Tripathi et al., 2015; Yadav et al., 2016). It has been reported that plant nutrients (micro and macro) play important roles in growth and development of plants (White and Brown, 2010; Waraich et al., 2011; Tripathi et al., 2014, 2015). Therefore, an adequate and balanced supply of nutrients at the correct time is required for the proper growth and development of plants, maintenance of better soil fertility, and preservation of an intensive cropping system to maintain the global food production, particularly in developing countries. (Dordas, 2009; Hansch and Mendel, 2009; Sarwar et al., 2010; Moharana et al., 2012; Waraich et al., 2012). In addition, studies also showed that an exogenous supply of nutrients plays a crucial role in the enhancement of plant tolerance against various abiotic stresses (Pankovic et al., 2000; Hassan et al., 2005; Tlustos et al., 2006; Dheri et al., 2007; Sarwar et al., 2010). While examining under salinity, drought, and heavy-metal stresses, some nutrients like calcium (Ca), magnesium (Mg), sulfur (S), zinc (Zn), and iron (Fe) have shown significant results (Sarwar et al., 2010; Singh et al., 2011; Nazar et al., 2012). These days, the application of Fe as a nutrient enhancement and its role in providing tolerance to plants against abiotic stresses are gaining interest as an area of research. As evident from recent studies, application of Fe reflects considerable and/or potential impacts in lowering the stress imposed by metal ions (Liu et al., 2007; Yadav et al., 2007; Garnier et al., 2010; Emamverdian et al., 2015).

Iron—the fourth most abundant and essential microelement on the earth's crust—is unavailable to higher plants, largely due to neutral and alkaline soils because of its existence in an insoluble form (Shao et al., 2007). In the plant system, Fe is regarded as an essential element in regulating life-sustaining processes like respiration, photosynthesis, chloroplast development, and chlorophyll biosynthesis where it takes part in electron transportation (Kim and Guerinot, 2007). Inadequate Fe supply in plants causes iron deficiency, which results in decline in yield of plant, interveinal chlorosis in plant leaves, etc. (Kim and Guerinot, 2007). In addition, excess levels of Fe are often fatal for plant health and productivity (Anjum et al., 2015). Therefore an appropriate supply of Fe is required for sustaining plant productivity under stress and nonstress conditions. It has been reported that out of the total cultivated areas of the world, one-third has Fe deficiency (Kim and Guerinot, 2007), which is causing significant decline in yield of crops every year. Iron is a chief component of the cell redox systems and also acts as a cofactor of various antioxidant enzymes like catalase (CAT), peroxidase (POD), and ascorbate peroxidase (APX) (Marschner, 1995; Sharma et al., 2004; Kumar et al., 2010). Therefore Fe homeostasis plays an important role in the life cycle of plants under stressed conditions (Fig. 9.1) (Gratao et al., 2005; Meda et al., 2007; Sarwar et al., 2010). In addition to providing an overview of major Fe acquisition strategies in plants, this review briefly appraises the literature available on the role of Fe in plants exposed to major abiotic stresses like salinity, drought, and heavy metals.

9.21 Ionomics for biofortification

Biofortification means to upgrade the mineral focuses and additionally bioavailability in plants through genetic improvement to tackle the issues of worldwide mineral deficits for healthy sustenance (for a survey, Bouis and Welch, 2010). The identification of qualities that impact biofortification characteristics in staple food, especially in oat grain, could assist with expanding the convergences of bioavailable mineral components in crop cultivars. Bioavailability can be characterized as the extent of the aggregate sum of mineral component that is conceivably absorbable in a metabolically dynamic structure (House, 1999). The fundamental substance known to repress the ingestion of iron (Fe), zinc (Zn), and magnesium (Mg) from oat grain is phytate (myo-inositol hexaphosphate) (for audit, White and Broadley, 2009). The bioavailability of Fe is decreased at dietary phytate/Fe molar remainders more noteworthy than one, and the bioavailability of Zn is diminished when the phytate/Zn molar remainder surpasses around six (Lönnerdal, 2002). Specifically, high phytate/zinc molar proportions are seen in the eating regimens of children from Malawi, Kenya, Mexico, and Guatemala, who devour just unfermented results of maize (*Zea mays*) (Gibson, 2006). There are low-phytate strains in maize (Raboy et al., 2000), yet they express some less-than-ideal agronomic highlights on the grounds that phytate is significant for seed germination and seedling development. Moreover, phytate has some recipient impacts, for example, hindrance of various sorts of tumors (Somasundar et al., 2005). Thus biofortification programs generally do exclude diminishing phytate beyond normally happening fluctuation as a goal (Ortiz-Monasterio et al., 2007).

9.22 Conclusion and future prospects

The changes in ionome build in any life form happen by the impact of a few variables that are biochemical, physiological, hereditary, or natural depending upon their condition. The onset of present-day high-throughput ionome analysis gives us the information about the effect of these changes on the complete ionome and helps in understanding their regulation at the gene level. The examination at quality level requires changes in their groupings. Because most forms are connected to each other, a change in any quality sequence may influence the direction of other mineral particles. Such a modification within the control of mineral particles in any single quality may cause serious issues in doling out the part of that specific quality and its related qualities. In order to grasp the precise impact of that specific modification within the genome at cellular and subcellular levels, a quick, high-throughput and in situ ionome analysis system is absolutely required. Studies of mineral dynamics at the cellular and subcellular level are conceivable by the utilization of more sophisticated and specialized strategies [like laser capture microdissection (LCM) that can isolate even a single cell from a sample] in combination with online ionome analysis methods like SXRF, which can evaluate minerals at the cellular and subcellular levels in their native state.

Other than this, ionomics has moreover been an imperative approach for giving data almost more extended gene network-controlling developmental and physiological processes that indirectly influence the ionome.

With the appearance of these unused frameworks, potential candidate qualities included in mineral securing, transport, distribution, and storage can be recognized. These gene(s) may well be utilized for improvement of designer crops having higher mineral context as a part of a biofortification program either through genetic engineering or through molecular breeding approaches. Candidate genes involved in different processes of mineral accumulation could be transferred with tissue-specific promoters in any organism. However, transfer of complex characteristic, that is, QTLs through marker-assisted selection, is presently the more satisfactory approach to fulfill the objective of nutritional security.

Another approach is based on the understanding of mineral storage in edible parts of plants and understanding the method of making them bioavailable. This in turn will assist in planning methodologies for the advancement of functional food and nutraceuticals.

To encourage such efforts, an online database (http://hort.agriculture.purdue.edu/Ionomics/database.asp) containing ionomic data of about 8300 fast neutron mutagenized *A. thaliana* plants has been developed by Salt and his colleagues and several other studies are in the pipeline universally. This database depicts the importance of mineral supplement and trace element profiling as a modern tool to determine the biological significance of associations between a plant's genome and its elemental profile. This complicated information will empower us in developing plants for biofortification of vital minerals and also for making procedures for the improvement of nutraceuticals for viable bioavailability of nutrients.

References

Alberts, B., Johnson, A., Lewis, J., Raff, M., Roberts, K., Walter, P., 2002. Molecular Biology of the Cell. Garland Science, New York.

Anjum, N.A., Singh, H.P., Khan, M.I.R., Mason, A., Per, T.S., Negi, A., et al., 2015. Too much is bad—an appraisal of phytotoxicity of elevated plant-beneficial heavy metal ions.

Arrivault, S., Senger, T., Kramer, U., 2006. The *Arabidopsis* metal tolerance protein AtMTP3 maintains metal Ho-meostasis by mediating Zn exclusion from the shoot under Fe deficiency and Zn oversupply. Plant J. 46 (5), 861–879. Available from: https://doi.org/10.1111/j.1365-313X.2006.02746.x.

Baxter, I., 2009. Ionomics: studying the social network of mineral nutrients. Curr. Opin. Plant. Biol. 12, 381–386.

Baxter, I., 2010. Ionomics: the functional genomics of elements. Brief Funct. Genomics 9, 149–156.

Baxter, I., Dilkes, B., 2012. Elemental profiles reflect plant adaptations to the environment. Science 336, 1661–1663.

Beckhoff, B., Kanngießer, B., Langhoff, N., Wedell, R., Wolff, H., 2006. Handbook of Practical X-ray Fluorescence Analysis. Springer, New York.

Bentsink, L., Yuan, K., Koorneef, M., Vreugdenhil, D., 2003. The genetics of phytate and phosphate accumulation in seeds and leaves of *Arabidopsis thaliana*, using natural variation. Theor. Appl. Genet. 106, 1234–1243.

Bertoldi, D., Larcher, R., Bertamini, M., Otto, S., Concheri, G., Nicolini, G., 2011. Accumulation and distribution pattern of macro, micro and trace elements in *Vitis vinifera* L. cv. Chardonnay berries. J. Agric. Food Chem. 59 (13), 7224–7236.

Blasing, O.E., Gibon, Y., Gunther, M., Hohne, M., Morcuende, R., Osuna, D., et al., 2005. Sugars and circadian regulation make major contributions to the global regulation of diurnal gene expression in Arabidopsis. Plant Cell 17, 3257–3281.

Borghi, M., Rus, A., Salt, D.E., 2011. Loss-of-function of constitutive expresser of pathogenesis related Genes5 affects potassium homeostasis in *Arabidopsis thaliana*. PLoS One 6 (10), e26360.

Bouis, H.E., Hotz, C., McClafferty, B., Meenakshi, J.V., Pfeiffer, W., 2009. Biofortifi cation: a new tool to reduce micro-nutrient malnutrition. In: International Congress of Nutrition, 19th, Bangkok, 4–9 October.

Bouis, H.E., Welch, R.M., 2010. Biofortification—a sustainable agricultural strategy for reducing micronutrient malnutrition in the global south. Crop Sci. 50, S-20–S-32.

Brigham, E.P., McCormack, M.C., Takemoto, C.M., Matsui, E.C., 2015. Iron status is associated with asthma and lung function in US women. PLoS ONE 10, e0117545. Available from: https://doi.org/10.1371/journal.pone.0117545.

Cizdziel, J., Bu, K., Nowinski, P., 2011. Determination of elements in situ in green leaves by laser ablation ICP-MS using pressed reference materials for calibration. Anal. Methods 4, 564–569.

Clárk, R.B., 1983. Plant genotype differences in the uptake, translocation, accumulation, and use of mineral elements required for plant growth. Plant Soil 72 (2–3), 175–196.

Clemens, S., 2001. Molecular mechanisms of plant metal tolerance and homeostasis. Planta 212 (4), 475–486.

Cramer, G.R., Urano, K., Delrot, S., Pezzotti, M., Shinozaki, K., 2011. Effects of abiotic stress on plants: a systems biology perspective. BMC Plant Biol 11, 163. Available from: https://doi.org/10.1186/1471-2229-11-163.

Delhaize, E., Randall, P.J., Wallace, P.A., Pinkerton, A., 1993. Screening Arabidopsis for mutants in mineral nutrition. Plant Soil 155 (1), 131–134.

Dheri, G.S., Brar, M.S., Malhi, S.S., Xie, Y., 2007. Influence of phosphorus application on growth and cadmium uptake of spinach in two Cd contaminated soils. J. Plant Nutr. Soil Sci. 170, 495–499. Available from: https://doi.org/10.1002/jpln.200625051.

Diego, H.S., Henning, R., Ute, K., Michael, K.U., Joachim, K., 2008. Metabolome–ionome–biomass interactions: what can we learn about salt stress by multiparallel phenotyping. Plant Signal. Behav. 3 (8), 598–600.

Dordas, C., 2009. Role of nutrients in controlling plant diseases in sustainable agriculture: a review. In: Lichtfouse, E., Navarrete, M., Debaeke, P., Souchere, V., Alberola, C. (Eds.), Sustainable Agriculture. Springer, pp. 443–460. Available from: http://doi.org/10.1007/978-90-481-2666-8_28.

Emamverdian, A., Ding, Y., Mokhberdoran, F., Xie, Y., 2015. Heavy metal stress and some mechanisms of plant defense response. Sci. World J. 2015, 756120. Available from: https://doi.org/10.1155/2015/756120.

Elizabeth, B., Tilman, A., Anthony, I., Cherie, O., Ana, R., Lahner, B., et al., 2010. Natural genetic variation in selected populations of *Arabidopsis thaliana* is associated with ionomic differences. PLoS One 5 (6), 11081–11088. Available from: https://doi.org/10.1371/journal.pone.0011081.

Fiehn, O., Kopka, J., Dörmann, P., Altmann, T., Trethewey, R.N., Willmitzer, L., 2000. Metabolite profiling for plant functional genomics. Nat. Biotechnol. 8, 1157–1161.

Fleet, J.C., Replogle, R., Salt, D.E., 2011. Systems genetics of mineral metabolism. J. Nutr. 141, 520–525.

Galinha, C., Anawar, H.M., Freitas, M.C., et al., 2011. Neutron activation analysis of wheat samples. Appl. Radiat. Isot. 69 (11), 1596–1604.

Garnier, J.M., Travassac, F., Lenoble, V., Rose, J., Zheng, Y., Hossain, M.S., et al., 2010. Temporal variations in arsenic uptake by rice plants in Bangladesh: the role of iron plaque in paddy fields irrigated with ground water. Sci. Total Environ. 408, 4185–4193. Available from: https://doi.org/10.1016/j.scitotenv.2010.05.019.

Ghandilyan, A., Ilk, N., Hanhart, C., et al., 2009. A strong effect of growth medium and organ type on the identification of QTLs for phytate and mineral concentrations in three *Arabidopsis thaliana* RIL populations. J. Exp. Bot. 60 (5), 1409–1425.

Gibson, R.S., 2006. Zinc: the missing link in combating micronutrient malnutrition in developing countries. Proc. Nutr. Soc. 65, 51–60.

Gilroy, S., Jones, D.L., 2000. From form to function: development and nutrient uptake in root hairs. Trends Plant Sci. 5, 56–60.

Gratao, P.L., Polle, A., Lea, P.J., Azevedo, R.A., 2005. Making the life of heavy metal stressed plants a little easier. Funct. Plant Biol. 32, 32481–32494. Available from: https://doi.org/10.1071/FP05016.

Grusak, M.A., Della Penna, D., 1999. Improving the nutrient composition of plants to enhance human nutrition and health. Annu. Rev. Plant Physiol. Plant Mol. Biol. 50, 133–161.

Guerinot, M.L., 2000. The ZIP family of metal transporters. Biochim. Biophys. Acta 1465, 190–198. Available from: https://doi.org/10.1016/S0005-2736(00)00138-3.

Gutierrez, R.A., Lejay, L., Dean, A., Chiaromonte, F., Shasha, D.E., Coruzzi, G.M., 2007. Qualitative network models and genome-wide expression datadefine carbon/nitrogen-responsive molecular machines in Arabidopsis. Genome. Biol 8, R7.

Hansch, R., Mendel, R.R., 2009. Physiological functions of mineral micronutrients (Cu, Zn, Mn, Fe, Ni, Mo, B, Cl). Curr. Opin. Plant Biol. 12, 259–266. Available from: https://doi.org/10.1016/j.pbi.2009.05.006.

Hassan, M.J., Zhang, G., Wu, F., Wei, K., Chen, Z., 2005. Zinc alleviates growth inhibition and oxidative stress caused by cadmium in rice. J. Plant Nutr. Soil Sci. 168, 255–261. Available from: https://doi.org/10.1002/jpln.200420403.

Haydon, M.J., Kawachi, M., Wirtz, M., Hillmer, S., Hell, R., Krämer, U., 2012. Vacuolar nicotianamine has critical and distinct roles under iron deficiency and for zinc sequestration in *Arabidopsis*. Plant Cell 24, 724–737. Available from: https://doi.org/10.1105/tpc.111.095042.

Hirochika, H., Guiderdoni, E., An, G., et al., 2004. Rice mutant resources for gene discovery. Plant Mol. Biol 54, 325–334.

Hirschi, K.D., 2003. Striking while the ionome id hot: making the most of plant genomic advances. Trends Biotechnol 21, 520–521.

Hirschi, K.D., Korenkov, V.D., Wilganowski, N.L., Wagner, G.J., 2000. Expression of *Arabidopsis* CAX2 in tobacco. Altered metal accumulation and increased manganese tolerance. Plant Physiol. 124, 125–133.

Hoagland, D.R., Arnon, D.I., 1950. The Water-Culture Method for Growing Plants Without Soil, second ed. Circular. California Agricultural Experiment Station, p. 347.

Hoekenga, O.A., Maron, L.G., Piñeros, M.A., et al., 2006. AtALMT1, which encodes a malate transporter, is identified as one of several genes critical for aluminum tolerance in *Arabidopsis*. Proc. Natl. Acad. Sci. U S A 103, 9738–9743.

House, W.A., 1999. Trace element bioavailability as exemplified by iron and zinc. Field Crops Res 60, 115–141.

Joe, M., Baxter, I.R., Lee, J., Li, L., Lahner, B., Natasha Kaplan, G., et al., 2009. The ferroportin metal efflux proteins function in iron and cobalt Ho-meostasis in *Arabidopsis*. Plant Cell 21, 3326–3338. Available from: https://doi.org/10.1105/tpc.109.069401.

Kim, S.A., Guerinot, M.L., 2007. Mining iron: iron uptake and transport in plants. FEBS Lett 581, 2273–2280. Available from: https://doi.org/10.1016/j.febslet.2007.04.043.

Koller, A., Washburn, M.P., Lange, B.M., Andon, N.L., Deciu, C., Haynes, P.A., 2002. Proteomic survey of metabolic pathways in rice. Proc. Natl. Acad. Sci. U S A 99, 11969–11974.

Krouk, G., Tranchina, D., Lejay, L., Cruikshank, A.A., Shasha, D., Coruzzi, G.M., et al., 2009. A systems approach uncovers restrictions forsignal interactions regulating genome-wide responses to nutritional cues in Arabidopsis. PLoS Comput. Biol. 5, e1000326.

Kumar, P., Tewari, R.K., Sharma, P.N., 2010. Sodium nitroprusside-mediated alleviation of iron deficiency and modulation of antioxidant responses in maize plants. AoB Plants 2, 1–11. Available from: https://doi.org/10.1093/aobpla/plq002.

Lahner, B., Gong, J., Mahmoudian, M., et al., 2003. Genomic scale profiling of nutrient and trace elements in *Arabidopsis thaliana*. Nat. Biotechnol. 21 (10), 1215–1221.

Leonhardt, N., Kwak, J.M., Robert, N., Waner, D., Leonhardt, G., Schroeder, J.I., 2004. Microarray expression analysis of *Arabidopsis* guard cells and isolation of a recessive abscisic acid hypersensitive protein phosphatase 2C mutant. Plant Cell 16, 596–615.

Li, G., Nunes, L., Wang, Y., Williams, P.N., Zheng, M., Zhang, Q., et al., 2013. Profiling the ionome of rice and its use in discriminating geographical origins at the regional scale, China. J. Env. Sci. 25 (1), 144–154.

Li, Y.P., Ye, W., Wang, M., Yan, X.D., 2009. Climate change and drought: a risk assessment of crop yield impacts. Clim. Res. 39, 31–46. Available from: https://doi.org/10.3354/cr00797.

Liu, H.J., Zhang, J.L., Zhang, F.S., 2007. Role of iron plaque in Cd uptake by and translocation within rice (*Oryza sativa* L.) seedlings grown in solution culture. Environ. Exp. Bot. 59 (3), 314–320.

Liu, J., Yang, J., Li, R., Lei, S., Zhang, C., Yan, L., et al., 2009. Analysis of genetic factors that control shoot mineral concentrations in rapeseed (*Brassica napus*) in different boron environments. Plant Soil 320 (1-2), 255–266. Available from: https://doi.org/10.1007/s11104-009-9891-6.

Lönnerdal, B., 2002. Phytic acid–trace element (Zn, Cu, Mn) interactions. Int. J. Food Sci. Technol. 37, 749–758.

Loudet, O., Chaillou, S., Merigout, P., Talbotec, J., Daniel-Vedele, F., 2003. Quantitative trait loci analysis of nitrogen use efficiency in *Arabidopsis*. Plant Physiol. 131, 345–358.

Lvov, B., 2005. Fifty years of atomic absorption spectroscopy. J. Anal. Chem. 60, 382–392.

Macedo, A.F., 2012. Abiotic stress responses in plants: metabolism to productivity. In: Ahmad, P., Prasad, M. (Eds.), Abiotic Stress Responses in Plants. Springer, New York, NY. Available from: http://doi.org/10.1007/978-1-4614-0634-1_3.

Mantri, N., Patade, V., Penna, S., Ford, R., Pang, E., 2012. Abiotic stress responses in plants: present and future. In: Ahmad, P., Prasad, M. (Eds.), Abiotic Stress Responses in Plants. Springer, New York, NY. Available from: http://doi.org/10.1007/978-1-4614-0634-1_1.

Marschner, H., 1995. Mineral Nutrition of Higher Plants. Academic Press, London.

Marschner, P., 2011. Marschner's Mineral Nutrition of Higher Plants, third ed. Academic Press, London.

Martzivanou, M., Hampp, R., 2003. Hyper-gravity effects on the Arabidopsis transcriptome. Physiol. Plant 118, 221–231.

Meda, A.R., Scheuermann, E.B., Prechsl, U.E., Erenoglu, B., Schaaf, G., Hayen, H., et al., 2007. Iron acquisition by phytosiderophores contributes to cadmium tolerance. Plant Physiol 143, 1761–1773. Available from: https://doi.org/10.1104/pp.106.094474.

Moharana, P.C., Sharma, B.M., Biswas, D.R., Dwivedi, B.S., Singh, R.V., 2012. Long-term effect of nutrient management on soil fertility and soil organic carbon pools under a 6-year-old pearl millet-wheat cropping system in an Inceptisol of subtropical India. Field Crop Res 136, 32–41. Available from: https://doi.org/10.1016/j.fcr.2012.07.002.

Nagajyoti, P.C., Lee, K.D., Sreekanth, T.V.M., 2010. Heavy metals, occurrence and toxicity for plants: a review. Environ. Chem. Lett. 8, 199–216. Available from: https://doi.org/10.1007/s10311-010-0297-8.

Nazar, R., Iqbal, N., Masood, A., Khan, M.I.R., Syeed, S., Khan, N.A., 2012. Cadmium toxicity in plants and role of mineral nutrients in its alleviation. Am. J. Plant Sci. 3, 1476–1489. Available from: https://doi.org/10.4236/ajps.2012.310178.

Norton, G.J., Deacon, C.M., Xiong, L., Huang, S., Meharg, A.A., Price, A.H., 2010. Genetic mapping of the rice ionome in leaves and grain: identification of QTLs for 17 elements including arsenic, cadmium, iron and selenium. Plant Soil 329, 139–153.

Ohkama-Ohtsu, N., Wasaki, J., 2010. Recent progress in plant nutrition research: cross-talk between nutrients, plant physiology and soil microorganisms. Plant Cell Physiol. 51 (8), 1255–1264.

Ortiz-Monasterio, J.I., Palacios-Rojas, N., Meng, E., Pixley, K., Trethowan, R., Peña, R.J., 2007. Enhancing the mineral and vitamin content of wheat and maize through plant breeding. J. Cereal Sci. 46, 293–307.

Outten, C.E., O'Halloran, T.V., 2001. Femtomolar sensitivity of metalloregulatory proteins controlling zinc homeostasis. Science 292, 2488–2492.

Pankovic, D., Plesnicar, M., Maksimoviic, I.A., Petrovic, N., Sakac, Z., Kastori, R., 2000. Effect of nitrogen nutrition on photosynthesis in Cd treated sunflower plants. Ann. Bot. 86, 841–847. Available from: https://doi.org/10.1006/anbo.2000.1250.

Payne, K.A., Bowen, H.C., Hammond, J.P., Hampton, C.R., Lynn, J.R., Mead, A., et al., 2004. Natural genetic variation in caesium (Cs) accumulation by *Arabidopsis thaliana*. New Phytol. 162 (2), 535–548.

Pozo, T., Cambiazo, V., González, M., 2010. Gene expression profiling analysis of copper homeostasis in *Arabidopsis thaliana*. Biochem. Biophys. Res. Commun. 393 (2), 248–252. Available from: https://doi.org/10.1016/j.bbrc.2010.01.111.

Price, J., Laxmi, A., St Martin, K.S., Jang, J.C., 2004. Global transcription profiling reveals multiple sugar signal transduction mechanisms in Arabidopsis. Plant Cell 16, 2128–2150.

Punshon, T., Hirschi, K., Yang, J., Lanzirotti, A., Lai, B., Mary, L.G., 2012. The role of CAX1 and CAX3 in elemental distribution and abundance in *Arabidopsis* seed. Plant Physiol. 158, 352–362.

Quadir, Q.F., Watanabe, T., Chen, Z., Osaki, M., Shinano, T., 2011. Ionomic response of *Lotus japonicus* to different root-zone temperatures. Soil Sci. Plant Nutr 57, 221–232.

Raboy, V., 2010. Origin and seed phenotype of maize low phytic acid 1-1 and low phytic acid 2-1. Plant Physiol 124 (1), 355–368. Available from: https://doi.org/10.1104/pp.124.1.355.

Raboy, V., Gerbasi, P.F., Young, K.A., Stoneberg, S.D., Pickett, S.G., Bauman, A.T., et al., 2000. Origin and seed phenotype of maize low phytic acid 1-1 and low phytic acid 2-1. Plant Physiol. 124 (1), 355–368. Available from: https://doi.org/10.1104/pp.124.1.355.

Rauh, L., Basten, C., Buckler, S., 2002. Quantitative trait loci analysis of growth response to varying nitrogen sources in *Arabidopsis thaliana*. Theor. Appl. Genet. 104, 743–750.

Rauser, W.E., 1999. Structure and function of metal chelators produced by plants. Cell Biochem. Biophys. 31, 19–48.

Rea, P.A., 2003. Ion genomics. Nat. Biotechnol. 21, 1149–1151.

Rogers, E.E., Guerinot, M.L., 2002. FRD3, A member of the multidrug and toxin efflux family, controls iron deficiency responses in *Arabidopsis*. Plant Cell 14 (8), 1787–1799. Available from: https://doi.org/10.1105/tpc.001495.

Rus, A., Baxter, I., Muthukumar, B., Gustin, J., Lahner, B., Yakubova, E., et al., 2006. Natural variants of AtHKT1 enhance Na^+ accumulation in two wild populations of *Arabidopsis*. PLoS Genet. 2 (12), e210. Available from: https://doi.org/10.1371/journal.pgen.0020210.

Sairam, R.K., Tyagi, A., 2004. Physiology and molecular biology of salinity stress tolerance in plants. Curr. Sci. 86, 407–421.

Sallaud, C., Gay, C., Larmande, P., et al., 2004. High throughput T-DNA insertion mutagenesis in rice: a first step towards in silico reverse genetics. Plant J. 39, 450–464.

Salt, D.E., 2004. Update on ionomics. Plant Physiol. 136, 2451–2456.

Salt, D.E., Baxter, I., Lahner, B., 2008. Ionomics and the study of the plant ionome. Annu. Rev. Plant Biol. 59, 709–733.

Samarah, N., Mullen, R., Cianzio, S., 2004. Size distribution and mineral nutrients of soybean seeds in response to drought stress. J. Plant Nutr. 27, 815–835. Available from: https://doi.org/10.1081/PLN-120030673.

Sarwar, N., Malhi, S.S., Zia, M.H., Naeem, A., Bibi, S., Farid, G., 2010. Role of mineral nutrition in minimizing cadmium accumulation by plants. J. Sci. Food Agric. 90, 925–937. Available from: https://doi.org/10.1002/jsfa.3916.

Saxena, I., Shekhawat, G.S., 2013. Nitric oxide (NO) in alleviation of heavy metal induced phytotoxicity and its role in protein nitration. Nitric Oxide 32, 13–20. Available from: https://doi.org/10.1016/j.niox.2013.03.004.

Schwacke, R., Schneider, A., Van Der Graaff, E., Fischer, K., Catoni, E., Desimone, M., 2003. Aramemnon, a novel database for *Arabidopsis* integral membrane proteins. Plant Physiol. 131, 16–26.

Shao, G., Chen, M., Wang, W., Mou, R., Zhang, G., 2007. Iron nutrition affects cadmium accumulation and toxicity in rice plants. Plant Growth Regul. 53, 33–42. Available from: https://doi.org/10.1007/s10725-007-9201-3.

Sharma, S.S., Kaul, S., Metwally, A., Goyal, K.C., Finkemeier, I., Dietz, K.J., 2004. Cadmium toxicity to barley (Hordeumvulgare) as affected by varying Fe nutritional status. Plant Sci. 166, 1287–1295. Available from: https://doi.org/10.1016/j.plantsci.2004.01.006.

Singh, R., Gautam, N., Mishra, A., Gupta, R., 2011. Heavy metal and living systems: an overview. Ind. J. Pharmacol. 43, 246–253. Available from: https://doi.org/10.4103/0253-7613.81505.

Sobhana V.G., Senthil N., Raveendran M., Satismruti K., Nagarajan P., Sangeetha (2010) Genetic variation and stability analysis of grain iron and zinc concentrations in maize (*Zea mays* L.) under three environments, Plant Soil. Environ. (Submitted Article).

Somasundar, P., Riggs, D., Jackson, B., Cunnigham, C., Vona-Davis, L., McFadden, D.W., 2005. Inositol hexaphosphate (IP6) a novel treatment for pancreatic cancer. J. Surg. Res. 126, 199–203.

Szpunar, J., 2004. Metallomics: a new frontier in analytical chemistry. Anal. Bioanal. Chem. 378, 54–56.

Thum, K., Shin, M., Palenchar, P., Kouranov, A., Coruzzi, G., 2004. Genome-wide investigation of light and carbon signaling interactions in Arabidopsis. Genome Biol. 5, R10.

Tiwari, V.K., Nidhi, R., Paeveen, C., Neelam, K., Renuka, A., Randhawa, G.S., et al., 2009. Mapping of quantitative trait loci for grain iron and zinc concentration in diploid a genome wheat. J. Heredity 100 (6), 771–776. Available from: https://doi.org/10.1093/jhered/esp030.

Tlustos, P., Szakova, J., Korinek, K., Pavlikova, D., Hanc, A., Balik, J., 2006. The effect of liming on cadmium, lead and zinc uptake reduction by spring wheat grown in contaminated soil. Plant Soil Environ 52, 16–24.

Török, P., Žemberyová, M., 2012. Direct solid sampling electrothermal atomic absorption spectrometric determination of toxic and potentially toxic elements in certified reference materials of brown coal fly ash. Spectrochim. Acta B 71–72, 80–85.

Tomastu, H., Takano, J., Takahasi, H., Akiko, W.T., Shibagaki, N., Fujiwara, T., 2007. An *Arabidopsis thaliana* high-affinity molybdate transporter required for efficient uptake of molybdate from Soil. Proc. Natl. Acad. Sci. U S A 104, 18807–18812. Available from: https://doi.org/10.1073/pnas.0706373104.

Tripathi, D.K., Singh, V.P., Kumar, D., Chauhan, D.K., 2012. Impact of exogenous silicon addition on chromium uptake, growth, mineral elements, oxidative stress, antioxidant capacity, and leaf and root structures in rice seedlings exposed to hexavalent chromium. Acta Physiol. Plantar. 34, 279–289. Available from: https://doi.org/10.1007/s11738-011-0826-5.

Tripathi, D.K., Singh, V.P., Chauhan, D.K., Prasad, S.M., Dubey, N.K., 2014. Role of macronutrients in plant growth and acclimation: recent advances and future prospective. In: Ahmad, P., Wani, M.R., Azooz, M.M., Tran, L.-S.P. (Eds.), Improvement of Crops in the Era of Climatic Changes. Springer, New York, NY, pp. 197–216.

Tripathi, D.K., Singh, S., Singh, S., Mishra, S., Chauhan, D.K., Dubey, N.K., 2015. Micronutrients and their diverse role in agricultural crops: advances and future prospective. Acta Physiol. Plantar. 37, 1–14. Available from: https://doi.org/10.1007/s11738-015-1870-3.

Tripathi, D.K., Singh, S., Singh, S., Chauhan, D.K., Dubey, N.K., Prasad, R., 2016a. Silicon as a beneficial element to combat the adverse effect of drought in agricultural crops. In: Ahmad, P. (Ed.), Water Stress and Crop Plants: A Sustainable Approach. John Wiley & Sons, Chichester, pp. 682–694.

Tripathi, D.K., Singh, S., Singh, V.P., Prasad, S.M., Chauhan, D.K., Dubey, N.K., 2016b. Silicon nanoparticles more efficiently alleviate arsenate toxicity than silicon in maize cultiver and hybrid differing in arsenate tolerance. Front. Environ. Sci. 4, 46. Available from: https://doi.org/10.3389/fenvs.2016.00046.

Tripathi, D.K., Singh, V.P., Prasad, S.M., Dubey, N.K., Chauhan, D.K., Rai, A.K., 2016c. LIB spectroscopic and biochemical analysis to characterize lead toxicity alleviative nature of silicon in wheat (*Triticum aestivum* L.) seedlings. J. Photochem. Photobiol. B: Biol. 154, 89–98. Available from: https://doi.org/10.1016/j.jphotobiol.2015.11.008.

Tripathi, D.K., Bashri, G., Singh, S., Ahmad, P., Singh, V.P., Prasad, S.M., et al., 2017a. Efficacy of silicon against aluminum toxicity in plants. In: Tripathi, D.K., Singh, V.P., Ahmad, P., Chauhan, D.K., Prasad, S.M. (Eds.), Silicon in Plants: Present and Future Prospects. CRC Press; Taylor and Francis, pp. 355–366.

Tripathi, D.K., Shewata, G., Singh, S., Yadav, V., Arif, N., Singh, S., et al., 2017b. Silicon: a potential element to combat adverse impact of UV-B in plants. In: Singh, V.P., Singh, S., Prasad, S.M., Parihar, P. (Eds.), UV-B Radiation: From Environmental Stressor to Regulator of Plant Growth. John Wiley & Sons, Hoboken, NJ, pp. 175–195.

Tripathi, D.K., Singh, S., Singh, V.P., Prasad, S.M., Dubey, N.K., Chauhan, D.K., 2017c. Silicon nanoparticles more effectively alleviated UV-B stress than silicon in wheat (*Triticum aestivum*) seedlings. Plant Physiol. Biochem. 110, 70–81. Available from: https://doi.org/10.1016/j.plaphy.2016.06.026.

Tripathi, D.K., Tripathi, A., Gaur, S., Singh, S., Singh, Y., Vishwakarma, K., et al., 2017d. Uptake, accumulation and toxicity of silver nanoparticle in autotrophic plants, and heterotrophic microbes: a concentric review. Front. Microbiol. 8, 7. Available from: https://doi.org/10.3389/fmicb.2017.00007.

Tulchinsky, T.H., 2010. Micronutrient deficiency conditions: global health issues. Public Health Rev. 32 (1), 243–255.

Vaculík, M., Landberg, T., Greger, M., Luxová, M., Stoláriková, M., Lux, A., 2012. Silicon modifies root anatomy, and uptake and subcellular distribution of cadmium in young maize plants. Ann. Bot. 110, 433–443. Available from: https://doi.org/10.1093/aob/mcs039.

Waraich, E.A., Ahmad, R., Ashraf, M.Y., 2011. Role of mineral nutrition in alleviation of drought stress in plants. Aus. J. Crop Sci. 5, 764–777.

Waraich, E.A., Ahmad, R., Halim, A., Aziz, T., 2012. Alleviation of temperature stress by nutrient management in crop plants: a review. J. Soil Sci. Plant Nutr. 12, 221–244. Available from: https://doi.org/10.4067/S0718-95162012000200003.

White, P.J., Broadley, M.R., 2009. Biofortification of crops with seven mineral elements often lacking in human diets-iron, zinc, copper, calcium, magnesium, selenium and iodine. New Phytol. 182, 49–84.

White, P.J., Brown, P.H., 2010. Plant nutrition for sustainable development and global health. Ann. Bot. 105 (7), 1073–1080.

White, P.J., Broadley, M.R., Thompson, J.A., McNicol, J.W., Crawley, M.J., Poulton, P.R., et al., 2012. Testing the distinctness of shoot ionomes of angiosperm families using the rothamsted park grass continuous hay experiment. New Phytol. 196 (1), 101–109.

Williams, R.J.P., 2001. Chemical selection of elements by cells. Coordin. Chem. Rev. 216–217, 583–595.

Yadav, V., Arif, N., Singh, S., Srivastava, P.K., Sharma, S., Tripathi, D.K., et al., 2016. Exogenous mineral regulation under heavy metal stress: advances and prospects. Biochem. Pharmacol. 5, 220. Available from: https://doi.org/10.4172/2167-0501.1000220.

Yadav, S.S., Verma, A., Sani, S., Sharma, Y.K., 2007. Likely amelioration of chromium toxicity by Fe and Zn in maize. J. Ecophysiol. Occup. Health 7, 111–117.

Yang, T., Lin, W.D., Schmidt, W., 2010. Transcriptional profiling of the *Arabidopsis* iron deficiency response reveals conserved transition metal homeostasis networks. Plant Physiol. 152, 2130–2141. Available from: https://doi.org/10.1104/pp.109.152728.

Zhao, H., Lei, S., Duan, X.L., Xu, F.S., Wang, Y.H., Meng, J.L., 2008. Mapping and validation of chromosome regions conferring a new boron-efficient locus in *Brassica napus*. Mol. Breed. 22 (3), 495–506. Available from: https://doi.org/10.1007/s11032-008-9193-3.

Further reading

Adebooye, O.C., Schmitz-Eiberger, M., Lankes, C., Noga, G.J., 2010. Inhibitory effects of sub-optimal root zone temperature on leaf bioactive components, photosystem II (PS II) and minerals uptake in *Trichosanthes cucumerina* L. Cucurbitaceae. Acta Physiol. Plant. 32, 67–73.

Akbaba, U., Sahin, Y., Türkez, H., 2012. Comparison of element contents in haricot beans grown under organic and conventional farming regimes for human nutrition and health. Acta Sci. Pol. Hortorum Cultus 11 (2), 117–125.

Apaydın, G., Aylıkc, V., Cengiz, E., Saydam, M., Küp, N., Tıraşoğlu, E., 2010. Analysis of metal contents of seaweed (Ulva lactuca) from Istanbul, Turkey by EDXRF. Turkish J Fish Aquat Sci 10, 215–220.

Azpiroz-Leehan, R., Feldmann, K.A., 1997. T-DNA insertion muta-genesis in *Arabidopsis*: going back and forth. Trends Genet. 13, 152–156.

Baxter, I., Ouzzani, M., Orcun, S., Jandhyala, S.S., Salt, D.E., 2007. Purdue Ionomics Information Management System (PiiMS): An integrated functional genomics platform. Plant Physiol. 143, 600–601.

Borevitz, J.O., Nordborg, M., 2003. The impact of genomics on the study of natural variation in *Arabidopsis*. Plant Physiol. 132, 718–725.

Burnay, M., Crambert, G., Kharoubi-Hess, S., Geering, K., Horisberger, J.D., 2003. Electrogenicity of Na, K- and H, K-ATPase activity and presence of a positively charged amino acid in the fifth transmembrane segment. J. Biol. Chem. 278, 19237–19244.

Chen, Z., Watanabe, T., Shinano, T., Okazaki, K., Osaki, M., 2008. Ionomic study of *Lotus japonica*. New Phytol. 181 (4), 795–801.

Chen, Z., Shinano, T., Ezawa, T., Wasaki, J., Kimura, K., Osaki, M., et al., 2009. Elemental interconnections in *Lotus japonicus*: a systematic study of the affects of elements additions on different natural variants. Soil. Sci. Plant. Nutr. 55 (1), 91–101.

Connolly, E.L., Fett, J.P., Guerinot, M.L., 2002. Expression of the IRT1 metal transporter is controlled by metals at the levels of transcript and protein accumulation. Plant Cell 14, 1347–1357.

Eide, D., Broderius, M., Fett, J., Guerinot, M.L., 1996. A novel iron-regulated metal transporter from plants identified by functional expression in yeast. Proc. Natl Acad. Sci. U S A 93, 5624–5628.

Fahrni, C.J., 2007. Biological applications of X-ray fluorescence microscopy: exploring the subcellular topography and speciation of transition metals. Curr. Opin. Chem. Biol. 11, 121–127.

Grattan, S.R., Grieve, C.M., 1998. Salinity–mineral nutrient relations in horticultural crops. Sci. Hortic. 78, 127–157.

Hacker, S.D., Mark, D.B., 1995. Morphological and physiological consequences of a positive plant interaction. Ecology 76, 2165–2175.

Hemphill, D.D., 1972. Availability of trace elements to plants with respect to soil–plant interaction. Ann. N. Y. Acad. Sci. 199, 46–61.

Hevesy, G., Levi, H., 1936. Action of slow neutrons on rare earth elements. Nature 137, 185.

Induri, B.R., Ellis, D.R., Slavov, G.T., Yin, T., Zhang, X., Muchero, W., et al., 2012. Identification of quantitative trait loci and candidate genes for cadmium tolerance in Populus. Tree Physiol. 32 (5), 626–638.

Korshunova, Y.O., Eide, D., Clark, W.G., Guerinot, M.L., Pakrasi, H.B., 1999. The IRT1 protein from *Arabidopsis thaliana* is a metal transporter with a broad substrate range. Plant. Mol. Biol. 40, 37–44.

Kramer, P.J., Boyer, J.S., 1995. Water Relations of Plants and Soils. Academic Press, San Diego.

Lee, S.C., Lan, W.Z., Kim, B.G., Li, L., Cheong, Y.H., Pandey, G.K., 2007. A protein phosphorylation/dephosphorylation network regulates a plant potassium channel. Proc. Natl. Acad. Sci. U S A 104 (40), 15959–15964.

Martienssen, R.A., 1998. Functional genomics: probing plant gene function and expression with transposons. Proc. Natl. Acad. Sci. U S A 95, 2021–2026.

Matzke, M.A., Matzke, A.J.M., 1995. How and why do plants inactivate homologous (trans) genes. Plant Physiol. 107, 679–685.

Medved', J., Streško, V., Kubová, J., Chmielewská, E., 2003. Evaluation of atomic spectrometry methods for determination of some heavy metals in soils, soil extracts, plants, and biota. Chem. Pap. 57 (3), 169–171.

Nadkarni, R.A., Morrison, G.H., 1973. Multi-element instrumental neutron activation analysis of biological materials. Anal. Chem. 45, 1957–1960.

Nelson, M.T., 1986. Interactions of divalent cations with single calcium channels from brain synaptosomes. J. Gen. Physiol. 87, 201–222.

Prasad, M.N., Strzalka, K., 2002. Physiology and biochemistry of metal toxicity and tolerance in plants. New Phytol. 162, 535–548.

Qin, Z., Caruso, J.A., Lai, B., Matusch, A., Becker, J.S., 2011. Trace metal imaging with high spatial resolution: applications in biomedicine. Metallomics 3, 28–37.

Quadir, Q.F., Watanabe, T., Chen, Z., Osaki, M., Shinano, T., 2011. Ionomic response of *Lotus japonicus* to different root-zone temperatures. Soil. Sci. Plant Nutr. 57, 221–232.

Remeteiová, D., Ružicková, S., Matherny, M., Dirner, V., 2011. FAAS method and the extraction process applied for fractionation analysis of gravitation dust sediments and their evaluation. J. Chem. Metrol. 5 (1), 1–10.

Rivetta, A., Negrini, N., Cocucci, M., 1997. Involvement of Ca^{2+}-calmodulin in Cd^{2+} toxicity during the early phases of radish (*Raphanus sativus* L.) seed germination. Plant. Cell Env. 20, 600–608.

Sha, Z., Oka, N., Watanabe, T., Tampubolon, B.D., Okazaki, K., Osaki, M., et al., 2012. Ionome of soybean seed affected by previous cropping with mycorrhizal plant and manure application. J. Agric. Food Chem. 60, 9543–9552.

Šmit, Ž., 2005. Recent developments of material analysis with PIXE. Nucl. Instrum. Methods Phys. Res. B 240, 258–264.

Soetan, K.O., Olaiya, C.O., Oyewole, O.E., 2010. The importance of mineral elements for humans, domestic animals and plants: a review. Afr. J. Food Sci. 4 (5), 200–222.

Sundaresan, V., Springer, P., Volpe, T., Haward, S., Jones, J.D.G., Dean, C., et al., 1995. Patterns of gene action in plant development revealed by enhancer trap and gene trap transposable elements. Genes. Dev. 9, 1797–1810.

Vallapragada, V.V., Inti, G., Ramulu, J.S., 2011. A validated inductively coupled plasma–optical emission spectrometry (ICP–OES) method to estimate free calcium and phosphorus in in vitro phosphate binding study of eliphos tablets. Am. J. Anal. Chem. 2, 718–725.

Wheal, M.S., Fowles, T.O., Palmer, L.T., 2011. A cost-effective acid digestion method using closed polypropylene tubes for inductively coupled plasma optical emission spectrometry (ICP–OES) analysis of plant essential elements. Anal. Methods 3 (12), 2854–2863.

White, P.J., Biskup, B., Elzenga, J.T.M., Homann, U., Thiel, G., Wissing, F., et al., 1999. Advanced patch-clamp techniques and single-channel analysis. J. Exp. Bot. 50, 1037–1054.

Williams, L., Salt, D.E., 2009. The plant ionome coming into focus. Curr. Opin. Plant. Biol. 12, 247.

Wu, C., Li, X., Yuan, W., et al., 2003. Development of enhancer trap lines for functional analysis of the rice genome. Plant J. 35 (3), 418–427.

Yang, G., Livingston, K., Jones, R., Klein, F., 2012. High resolution X-ray diffraction study of single crystal diamond radiators. Phys. Status Solidi A 209, 1786–1791.

Chapter 10

Cobalt and molybdenum: deficiency, toxicity, and nutritional role in plant growth and development

Aryadeep Roychoudhury and Swarnavo Chakraborty
Department of Biotechnology, St. Xavier's College (Autonomous), Kolkata, India

10.1 Introduction

Cobalt (Co) is a transition element and is one of the essential micronutrients in plants. Co is an essential component of a number of plant enzymes and can also serve as a coenzyme. Co also induces drought tolerance in seeds. This element plays an important role in the fixation of nitrogen in legumes and regulates ethylene metabolism in plants (Wenzel et al., 1995). The concentration of Co and the properties of the complexes that Co tends to form with other elements in the rhizospheric soil, governs the overall growth and functioning of the plants. Co has two oxidation states in nature—II and III. Plants have the potential of utilizing both the oxidation states of the element from soil, but the rate of uptake is the highest for Co(II). Usually, 8 mg of Co is found per kg of soil, varying from 0.1 to 70 mg/kg of soil throughout the earth (Bowen et al., 1979). On average, 1−2 mg of Co per kg of rhizospheric soil is required for the optimum growth of plants. Plants start to develop deficiency symptoms below the abovementioned concentration. However, Co toxicity in plants can crop up at concentrations $>$100 mg/mL of capillary water (Williams et al., 2000). The rate of absorption and distribution of Co in plants is governed by the type of plant species. Moreover, this can also be modulated by various conditions like temperature, pH, water potential of the soil, and the concentration of other metals and salts in the soil. Co tends to remain associated with other trace metals like arsenic, copper, nickel, and manganese. It is usually released into the environment from industries like breweries, by the combustion of oil and coal, mining of Co compounds, and from exhausts released from vehicles and airplanes. Cobalt

Plant Nutrition and Food Security in the Era of Climate Change. DOI: https://doi.org/10.1016/B978-0-12-822916-3.00021-4

60, a radioisotope of Co used for treatment of cancer patients in hospitals and research laboratories, acts as a potential biomedical waste (Khan et al., 2010). Co can persist in air for a short period of time ranging up to few days but tends to remain in water and soil for several years. In plants, Co tends to demonstrate both detrimental as well as beneficial effects. The toxic effects include prevention of leaf greening, premature falling of leaves, reduction in shoot weight and size, and discoloration of leaf veins (Palit and Sharma, 1994). On the other hand, the beneficial effects include delaying of leaf senescence by downregulation of ethylene synthesis, enhancement of tolerance of seeds to various abiotic stresses like drought, and modulation of alkaloid biosynthesis pathways in various plants, having potential medicinal values (Palit and Sharma, 1994).

Molybdenum (Mo) is also a transition element, found in trace amounts in soil (found at an average concentration of 1.2 mg/kg of soil) and is essential for both plants and animals. However, on a dry matter basis, Mo leaf concentrations of 0.2–2.0 mg/kg are optimally required for most plants. Irrespective of its minute requirements by the plant system, Mo serves as a potent component of a range of plant enzymes. These enzymes primarily include the enzymes associated with nitrogen assimilation like nitrate reductase (NR) and nitrogenase, the nitrogen-fixing enzyme found in bacteroids associated with the root nodules of leguminous plants (Kaiser et al., 2005). Apart from its role in plant growth and metabolism, Mo forms an active part of some plant proteins, including enzymes that participate in oxidation–reduction pathways. Mo is rendered biologically active only when it forms an active component of the molybdenum cofactor (Moco), a pterin compound that acts as the key interacting partner of Mo-requiring enzymes in plants, animals, as well as many bacteria (Williams and Frausto da Silva, 2002). Basically, Mo alone cannot activate plant enzymes; it can do so when in conjunction with this organic complex (Moco). Mo generally exists in the soil in a bound state with other compounds (Smith et al., 1995, Kaiser et al., 2005). Mo can exist either in a dissolved or water-soluble form in the capillary water; in association with other metal oxides like manganese, iron, and aluminum oxides; or in conjunction with different organic compounds (like Moco) in soil (Reddy et al., 1997; Williams and Frausto da Silva, 2002) as well as in the form of powellite ($CaMoO_4$), molybdenite (MoS_2), ferrimolybdite [$Fe_2(MoO_4)_3$], and wulfenite ($PbMoO_4$) in solid-state in the soil. Mo can exist in nature in a number of oxidation states ranging from zero to VI; however, VI is the most commonly obtained form and zero is the rarest form found in soil. A number of factors govern the availability of Mo for plant uptake and growth. These include the amount of other metal oxides and organic compounds that tend to complex with Mo, pH of the soil, and water content of the soil. The most probable reason behind the presence of Mo in soil is recycling of Mo-containing proteins and from various fertilizers, but it is almost absent in acidic soil (Gupta, 1997). Mo deficiency in plants was first reported by Arnon and Stout (1939) in tomatoes grown hydroponically. Mo-deficiency symptoms in plants are similar to nitrogen-deficiency symptoms (Behboudian et al., 2016), probably due to the downregulation of Mo-dependent NR-catalyzed reactions. The symptoms associated with Mo deficiency include stunted growth, reduction in leaf size, leaf chlorosis and necrosis, cupping state in aging leaves, decrease in germination potential, loss of seed vigor, progressive fall in seed weight, and development of immature seeds (Chatterjee and Nautiyal, 2001). Moreover, Mo toxicity in plants is quite rare and is usually misdiagnosed, due to marked similarities with the symptoms associated with the altered expression of many enzymes involved in oxidation–reduction reactions in plants.

10.2 Toxicity and deficiency of cobalt and molybdenum in plants

10.2.1 Cobalt toxicity and deficiency

Due to excessive mining of Co-containing ore, combustion of coal and fuel, contamination of wastewater into agricultural fields, and uncontrolled application of fertilizers, the levels of Co in agricultural lands are alarmingly increasing day by day (Saaltink et al., 2014). Irrespective of its beneficial role, higher levels of Co above its optimum can lead to development of toxic symptoms in plants associated with altered physiology and morphology (Fig. 10.1;

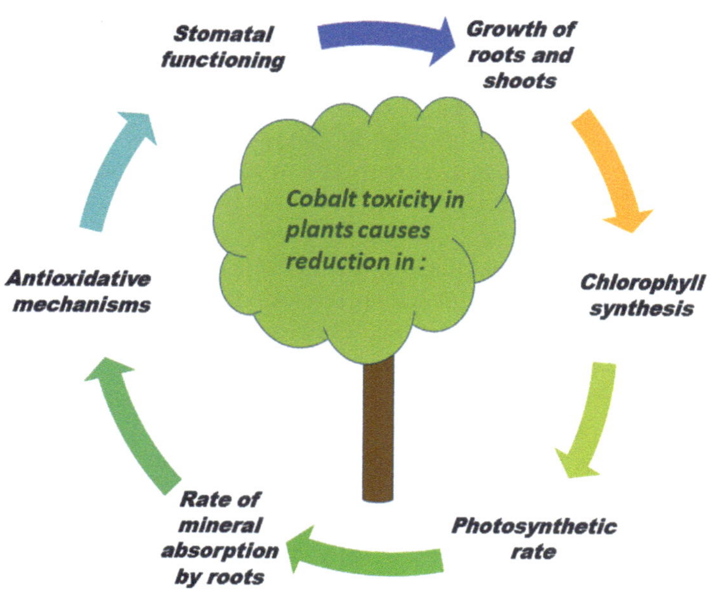

FIGURE 10.1 Cobalt toxicity symptoms in plants.

El-Sheekh et al., 2003; Jayakumar et al., 2007). Co demonstrates a higher value of redox potential in acidic medium, due to the variable valency of Co. This property of Co has not made this element an essential constituent of any of the enzymes involved in the respiratory chain. Fig. 10.1 depicts the various symptoms associated with cobalt toxicity in plants.

It has been demonstrated that Co can uncouple the mitochondrial respiration, along with the inhibition of the tricarboxylic acid (TCA) cycle in yeast, a number of fungi, and many plants (Tuppy and Sieghart, 1973; Banik, 1976). Excess Co in soil rather showed a disproportionate effect on the activities of ATPases. The activities of membrane-bound ATPases of the tonoplast and plasma membrane were upregulated, while those of cytosolic ATPases were downregulated in response to excessive elevation of Co levels (Kim and Weber, 1980). Moreover, overaccumulation of Co in leaves of some plants like sugarbeet has a rather negative effect on the rate of respiration and also reduces the rate of transpiration and water potential in cauliflower leaves. Marked alterations in the external morphology, structure, and number of plastids, followed by the change in the amount of chlorophyll were observed in plants (Zeid, 2001) treated with a progressively increasing concentration of Co. In some leguminous plants like pea, subsequent decline in photosystem II activity (along with considerable negative effects on photosystem I activities) and the rate of Hill reaction in the chloroplasts are usually associated with plants grown in Co-contaminated soil (Mohanty et al., 1989). On the other hand, monocots like maize when exposed to toxic concentrations of Co, demonstrated an inhibition of carbon dioxide fixation in the dark reaction. Therefore a fall in photosynthetic rate and photosynthetic yield are commonly associated with plants experiencing Co toxicity.

In higher plants as well as some algal species, Co is rendered as a preprophase poison, due to inhibition of both karyokinesis and cytokinesis. Thus, Co is responsible for the decline in mitotic index in affected plants. Moreover, Co can also cause a sharp fall in total nucleic acid content of plant cells, by downregulation of the biosynthesis of DNA and RNA. In a nutshell, Co toxicity adversely affects the overall plant growth and development by modulating several plant physiological and metabolic pathways, as well as downregulating the overall process of mitosis (Palit and Sharma, 1994). Interestingly, it was found that Co toxicity in plants led to iron deficiency in the shoot apices (Chatterjee and Chatterjee, 2003). Therefore excess Co can lead to fall in activities of enzymes and proteins containing iron at their active sites, like the iron–sulfur proteins, nitrogenase, and others. Co inhibits the translocation of other elements like phosphorus, sulfur, and copper in plants (Nagajyoti et al., 2010). The usually observed symptoms associated with plants growing in soil having excess of Co include chlorosis (Liu et al., 2000) and necrosis (Caselles et al., 1997) of young leaves and in extreme cases, leaf withering due to inhibition of root growth and thereby reducing the transport of nutrients and water in these plants

(Jayakumar et al., 2008). Toxicity symptoms also include an overall fall in plant growth and development due to shortening of root and shoot length and slowing of germination in seeds. Moreover, Co stress in plants led to huge increases in the expression of stress-responsive genes that encode different antioxidative enzymes and proteins, rise in phenolic content (Mittal et al., 2017), and reactive oxygen species-mediated mechanisms, in the form of a response to relieve the stress symptoms due to Co toxicity. However, some plants when exposed to very high levels of this metal, display rather lowering of levels of some antioxidative enzymes like peroxidase, catalase, acid phosphatase, and ribonuclease (Gopal et al., 2003).

Very little information is available to date about deficiency symptoms of Co in plants. In legumes, it is found that Co deficiency is usually associated with symptoms of nitrogen deficiency. Therefore low Co can lead to hampering of association between the nitrogen-fixing bacteria and root nodules. Plants grown in Co-depleted media, when supplied with fixed forms of nitrogen (in the form of nitrates), did not demonstrate symptoms of Co-induced nitrogen deficiency. Hence it can be inferred that Co deficiency in legumes probably had an adverse effect on the step associated with the fixation of atmospheric nitrogen. Huwait et al. (2015) explained the decline in the food value of crops grown in Co-deficient soils, due to marked impairment in the synthesis of vitamin B_{12} in these crops. Moreover, many plants when depleted of Co showed lowering of resistance to many diseases and susceptibility to different types of abiotic stresses like drought.

10.2.2 Molybdenum toxicity and deficiency

Deficiency symptoms of Mo are more commonly observed in the members of the family *Brassicaceae*. The members belonging to this family display a range of phenotypes associated with Mo deficiency and are also reproducible amongst the other plant species under *Brassicaceae*. The common manifestations include stunted growth, cupping, twisting, curling and mottling of leaves, and sometimes chlorosis, gray tinting and yellowing of leaves, with dwarf and dying seedlings (Hewitt and Bolle-Jones, 1952a). Leaves resembling whip-tail are one of the earliest deficiency symptoms in Mo-deprived higher plants. Due to improper development of leaf lamina and chloroplasts exhibiting a spherical and bulbous appearance, depicting enlarged spheres bound by membranes of tonoplast and chloroplast, leaves tend to look like a whip-tail (Fido et al., 1977). Some plants having whip-tail leaves also develop leaves with leathery appearance with necrosis of meristems (Hewitt and Bolle-Jones, 1952b; Kaiser et al., 2005). Fig. 10.2 depicts the various symptoms associated with molybdenum deficiency in plants.

Mo deficiency has a negative effect on the activity of nitrogen assimilatory enzymes like NR in plants. Therefore many crops including legumes fail to utilize nitrate-containing fertilizers, if grown in Mo-deprived soils.

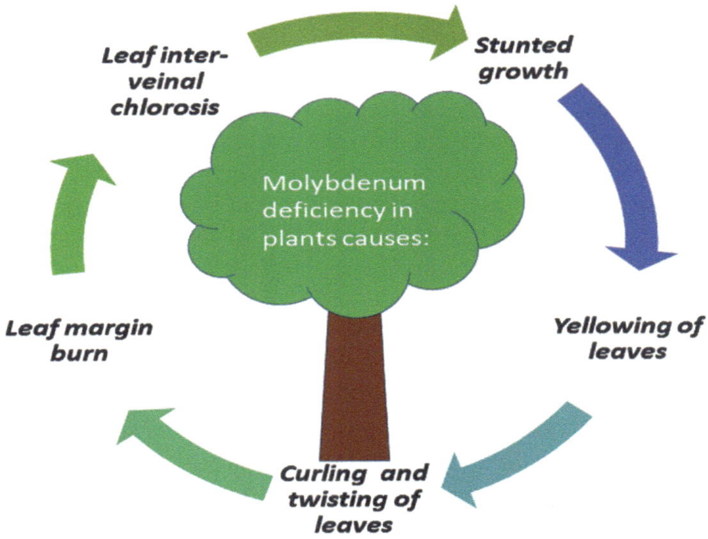

FIGURE 10.2 Molybdenum deficiency symptoms in plants.

These plants develop many visible symptoms marked by stunted and slow growth, such as yellow to pale green color of growing leaves and necrosis of leaves (Agarwala et al., 1978; Chatterjee et al., 1985; Chatterjee and Nautiyal, 2001). Monocots like wheat and oat also develop necrosis of the leaf blades and seeds are underdeveloped (Anderson, 1956; Chatterjee and Nautiyal, 2001). On the other hand, another monocot, maize, rarely develops necrotic lesions due to low Mo but develops chlorosis of leaves with an overall decrease in plant growth associated with smaller leaves with lesser surface area, short internodes, and poorly developed flowers having anthers and pollens with reduced germination potential (Agarwala et al., 1978; Agarwala et al., 1979; Kaiser et al., 2005). Millerandage is a commonly observed Mo deficiency symptom in *Vitis vinifera* 'Merlot' variety. Millerandage is characterized by a disorder in the development of bunches in grapevines, where already matured berries are present in a bunch having immature berries and unfertilized swollen ovaries that are yet to develop (Mullins et al., 2000; Williams et al., 2004). Therefore millerandage shows an uneven pattern of berry development. Apart from symptoms of millerandage, these grape plants also display other deficiency symptoms of Mo including necrosis of leaf margins, cupped and pale leaves, and short internodes with overall slowing of growth (Fig. 10.2).

Being an essential component of many important plant enzymes, Mo governs the proper functioning of a number of metabolic pathways in plants. Plants have been reported to contain a range of Mo-containing enzymes, commonly referred to as molybdoenzymes (Kaiser et al., 2005). These molybdoenzymes are required for many plant metabolic processes like reduction and assimilation of nitrogen by Mo-dependent NR; nitrogen fixation by nitrogenase, which is a Fe-Mo protein; biosynthesis of a number of phytohormones like auxins (indole-3-acetic acid) and abscisic acid (ABA); purine catabolism by Mo-requiring enzyme, xanthine dehydrogenase/oxidase (XDH); and metabolism of sulfur by another molybdoenzyme sulfite oxidase (SO). Therefore plants display a range of molybdoenzymes performing their functions during key steps of plant development and metabolism. Deficiency of Mo will have an adverse effect on each of these metabolic pathways, due to disruption of activities of these molybdoenzymes, thereby declining plant growth, yield, and development with higher susceptibility to diseases and pests (Graham and Stangoulis, 2007). Moreover, a number of these enzymes interact with the Mo cofactor, which upon interaction causes activation of the molybdoenzyme. For instance, in enzymes like aldehyde oxidase (AO) and XDH, after the generation of the enzyme-Mo cofactor complex, a sulfuration of the active Mo center causes the enzymes to function optimally (Mendel and Haensch, 2002). Therefore, a deficiency of a single element can lead to a complex series of manifestations, eventually dilapidating the entire metabolic and biosynthetic machinery of the plant system.

Mo toxicity in plants is very rare and shows similarities with the symptoms associated with the deficiency of iron and changed expression of a few redox enzymes in plants. Mo toxicity in plants is not characterized by any explicit symptoms and thus is often misdiagnosed. Thus, Mo deficiency in plants is usually difficult to address, due to lack of proper characteristic symptoms. However, McGrath et al. (2010) have shown reduction in overall yield of a selected number of plants like tomato, rye, oilseed rape, and red clover when subjected to almost 66- to 609-fold higher Mo levels than the optimum. Moreover, the fold decrease in yield of these plants changed with the alteration in the soil properties that usually affect the solubility of Mo in the growth medium.

10.3 Role in plant growth and development

10.3.1 Role of cobalt

Co is one of the beneficial elements for plants and an essential constituent of cobalamin, which serves as a key constituent of many plant enzymes and coenzymes. A series of these enzymes play an important role in the synthesis of leghaemoglobin, associated with nitrogen fixation in the legumes of leguminous plants, thereby also modulating the size and the number of nodules formed in the roots of these plants (Yadav and Khanna, 1988). Like most other metals, Co also exerts both positive and negative roles in plant growth and development. Exogenous application of Co in an optimum concentration to seeds improved their rate of germination and elevated the overall growth, quality, and yield of the crops. However, excessive application can lead to lowering of the number of rhizospheric bacteria, thereby disrupting the nodulation process and lowering crop yield and quality. Therefore, Co plays a critical role in the fixation of atmospheric nitrogen by bacteria in the nodules of leguminous plants (Witte et al., 2002).

The role of Co in different stages of plant growth and development is indispensable, yet highly variable depending on the amount and status of Co in soil and properties of the rhizospheric soil (Palit and Sharma, 1994). Depending on these properties, Co has the following positive impacts on the plant system when applied in optimum concentrations: elongation of coleoptiles, growth and increase in the length of stem, expansion of leaf lamina, delaying of senescence due to downregulation of biosynthesis of ethylene, reduction of peroxidase expression to prevent degradation of auxins, upregulation of alkaloid concentration, and mediation of drought tolerance in seeds (Palit and Sharma, 1994). Extensive experiments have been carried out on leguminous crops like soybean to reveal the role of Co in leguminous plants. The maxima of all forms of growth parameters including shoot and root length, weight in terms of fresh and dry weight, number of leaves and branches, and number of root nodules per plant were attained when a concentration of 12 mg/kg of Co was supplied to the plants. However, on further increase above these concentrations, the growth parameters rather worsened, thereby proving that 12 mg/kg is the optimum concentration essential for proper growth and development of plants (Kandil et al., 2013). Similar experiments on groundnut plants suggested that a concentration of 8 ppm of Co was optimally required for efficient nitrogen fixation and nodule formation. The highest number, fresh and dry weight of nodules was achieved at this concentration. This overall effect was due to the increased activity of nitrogen-fixing bacteria like *Rhizobium* in the rhizospheric soil with this optimal Co concentration (Gad, 2012a). This proves that in comparison with untreated plants, groundnut when grown under an optimal Co concentration has an enhanced expression of the enzyme nitrogenase, thereby increasing the efficiency of nitrogen fixation (Gad, 2012b). Inoculation of soil with nitrogen-fixing bacteria like *Rhizobium* and *Phosphobacterium* along with controlled amounts of Co, enhanced the uptake of some other elements like nitrogen, potassium, phosphorus, etc. (Basu and Bhadoria, 2008). Moreover, these treated plants displayed an overall improvement of other growth parameters and food value due to an increase in dry weight of stem, roots, pods, and nodules with respect to the untreated plants. These plants could fix higher amounts of atmospheric nitrogen with a greater efficiency (Fig. 10.3).

Co can also modulate the uptake of other elements by the plant system, for instance, nitrogen, potassium, and phosphorus (Manal et al., 2016) and in some cases copper, manganese, and zinc. Exogenous application of Co in an optimum concentration to plants led to a rise in the endogenous levels of these elements within the plant system (Kandil et al., 2013). Interestingly, it

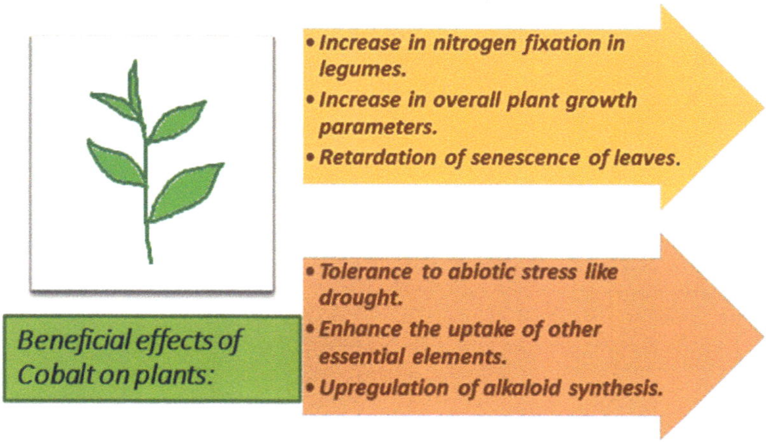

FIGURE 10.3 Beneficial effects of Co on plants.

was found that the application of phosphorus in the form of potassium dihydrogen phosphate along with requisite amounts of Co (II) nitrate to leguminous plants, has a rather appreciable effect on the number of root nodules and the rate of nitrogen fixation. Therefore, it can be concluded that the positive effects of Co were enhanced several-fold when Co was applied in conjunction with phosphorus (Yadav and Khanna, 1983; Yadav et al., 1984; Yadav and Khanna, 1988). Co at lower concentrations has been shown to increase the rate of photosynthesis in some plants, but a special role of Co revolves around the fixation of nitrogen in leguminous plants. Fig. 10.3 depicts the various beneficial effects of cobalt on plants.

10.3.2 Role of molybdenum

Mo displays quite a dynamic redox chemistry, by virtue of which Mo-dependent enzymes have the property to catalyze a range of redox reactions in plants (Mendel, 2013). However, it was found that the metal was not directly associated with the enzyme active site; rather, Mo complexed with a special scaffold, interacting with the enzyme catalytic site for its physiological functioning and optimal enzyme activity (Mendel and Schwarz, 2011). This unique scaffold is a ubiquitous organic molecule containing a pterin moiety called the Moco. The Moco constitutes the active sites of almost all Mo-dependent enzymes, with an exception to nitrogenase, which contains an iron–Mo cofactor instead of the Moco (Hu and Ribbe, 2013). Therefore, this Moco is essential for the activation of these molybdoenzymes and without the scaffold, the metal alone is not sufficient for the optimal activation of these enzymes.

The chief function of Mo lies in the process of nitrogen metabolism in plants. Being an essential component of the bacterial nitrogen-fixing enzyme

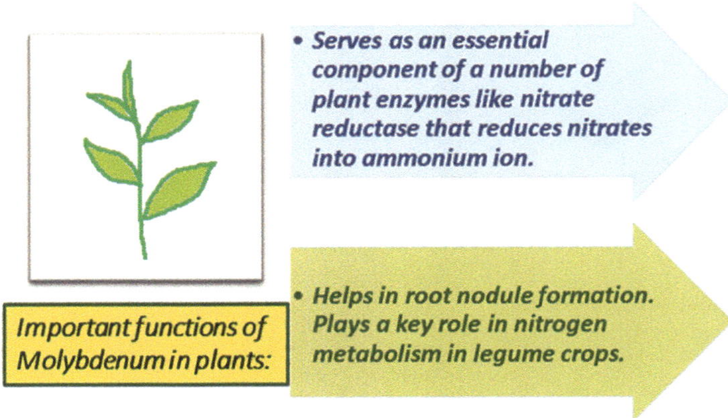

FIGURE 10.4 Functions of molybdenum in plants.

nitrogenase, Mo plays a pivotal role in symbiotic nitrogen fixation in legume crops. The nitrogenase enzyme consists of two subunits, of which one subunit is the iron–Mo center that catalyzes the reduction of atmospheric nitrogen into ammonia. Thus, the availability of nitrogen to leguminous plants is necessary for the development of root nodules (Anderson and Spencer, 1950; Anderson, 1956). Legumes have to maintain an optimal concentration of Mo in order to support nodule and seed development. Plants tend to accumulate higher levels of Mo in root nodules as compared to other parts of the plant such as shoots. Therefore, it is evident that nodules act as efficient sinks for the storage of Mo as compared to other storage sites in plants. Exogenous application of Mo in the form of fertilizers leads to more efficient symbiotic nitrogen fixation and development of larger nodules due to enhancement in the expression of the enzyme nitrogenase in legume crops like soybean (Adams, 1997; Vieira et al., 1998). Fig. 10.4 depicts the various beneficial effects of molybdenum on plants.

After the fixation of atmospheric nitrogen, it is necessary for the fixed nitrogen to be mobilized and exported out of the nodules. This process is controlled by another Mo-dependent enzyme called xanthine dehydrogenase/oxidase (XDH). This fixed form of nitrogen can either be mobilized as amides like asparagine and glutamine, or in the form of ureides like allantoic acid and allantoin, derived from purine catabolism. This enzyme mediates the conversion of hypoxanthine to uric acid, via a xanthine intermediate (Mendel and Haensch, 2002). The XDH enzyme activity is associated with the generation of hydrogen peroxide and superoxide radicals in the form of an after-effect of biotic (pathogenic fungal attack) or abiotic stress exposure. In legumes like pea, this process is directly correlated with the expression of superoxide dismutase, thereby modulating oxidative defenses in affected plants (Pastori and Rio, 1997; Hesberg et al., 2004). Therefore, Mo can improve plant health due to improved disease resistance, for instance, application of exogenous Mo to tomato plants generates development of resistance to verticillium wilt (Graham and Stangoulis, 2007). However, detailed mechanisms are less understood.

Apart from its role in nitrogen metabolism, Mo can activate other enzymes involved in some biological pathways in plants. The molybdoenzymes not involved in nitrogen fixation include AO, which is involved in the pathways contributing to production of phytohormones like indole-3-acetaldehyde (IAA) and ABA in plants. AO gets activated on interaction with Mo cofactor and mediates the conversion of abscisic aldehyde and IAA into ABA and indole-3-acetic acid, respectively. A key sulfuration step is essential for the proper functioning of Moco, and any disruption in this step or biosynthetic step will severely hamper biosynthesis of these phytohormones in plants (Schwartz et al., 1997; Marin and Marion-Poll, 1997; Sagi et al., 2002; Hesberg et al., 2004). It is commonly observed that plants with any disruption in the activation of Moco, develop symptoms associated with lower ABA expression, commonly observed in *flacca* and *aba3* mutants (Bittner et al., 2001; Sagi et al., 2002). The symptoms include huge rates of transpiration and unregulated stomatal movement leading to wilty phenotype, with lesser resistance to different stress and seeds having alteration in dormancy status (Mendel and Haensch, 2002). Moreover, the phenotypes of *flacca* and *aba3* mutants often tend to merge with those of Merlot varieties (Mo-deficient varieties), like flaccidity and cupping phenotypes of the leaves (Robinson and Burne, 2000). However, the molecular mechanisms and role of AO in Merlot varieties in response to Mo deficiency need to be addressed more meticulously.

NR is another molybdoenzyme that catalyzes important steps associated with nitrogen acquisition in algae, plants, and even in fungi. The cytosolic NR catalyzes the rate-limiting step in the nitrogen assimilation pathway (Campbell, 2001) and converts nitrates to nitrite by the process of reduction. After reduction, the nitrite is transported to the chloroplast via specific transporters. In *Chlamydomonas*, a type of formate/nitrite transporter called NAR1.1 carries out this task (Fernandez and Galvan, 2007). However, the presence of NAR1.1 is still not confirmed in higher plants, and the system carrying out this function is yet to be determined. This equilibrium that exists between the cytosol and the chloroplast for nitrite transport and the expression of different forms of NR (cytosolic and chloroplast NR) is essential for optimal nitrate utilization by the plants (Fernandez and Galvan, 2007; Sanz-Luque et al., 2015). Posttransport into the chloroplast, the nitrite is again reduced into ammonium ions by the plastid localized NR. This ammonium is subjected to incorporation into carbon skeletons and leads to generation of glutamate via the GS/GOGAT pathway (glutamine synthetase/glutamine oxoglutarate aminotransferase cycle) or via glutamate synthase (Sanz-Luque et al., 2015). Just because nitrate and ammonium are the most easily available and readily utilizable sources of inorganic nitrogen for the plants, NR plays an indispensable role in plant physiological, molecular, and metabolic pathways.

Sulfur dioxide is a notorious atmospheric pollutant that is potentially detrimental to plants. This gas is rendered as a phytotoxin (Follmann, 2000), which in the presence of water gets converted to a strong nucleophile that can damage many biological macromolecules inside the cells. Due to the nucleophilic activity of sulfites, many disulfide bridges in proteins like thioredoxins (Würfel et al., 1990) can get disrupted leading to inactivation of these proteins and decline in plant overall growth. So it is necessary to detoxify the sulfites generated within the plant cells. SO, a Mo-dependent plant enzyme associated with sulfur metabolism in plants, catalyzes the detoxification of excess sulfites formed inside cells, thereby promoting plant survival and growth even in sulfur dioxide–polluted environments.

10.4 Conclusion and future perspectives

Due to anthropogenic as well as natural effects, there has been an overaccumulation of different metals in the environment. Irrespective of their indispensable roles in plant growth and development, these metals can lead to toxicity in plants when their concentrations rise above optimal levels. It is quite conclusive that these elements when applied within a requisite range are effective for the plant system, but overaccumulation will be disastrous. However, toxicity due to Mo in plants has been rarely reported in the literature, while Co toxicity is quite commonly observed in plants. The morphological and phenotypic changes that occur in plants due to deficiency and toxicity of an essential element in plants have probably been enlisted properly. However, the molecular changes taking place within the plants due to toxicity or deficiency have to be understood in more detail. Plant metabolic pathways are quite diverse and each of them is linked at some point or another, thereby generating a huge and complex circuit governed by a series of different enzymes. Hence, disruption of a single step in this complex circuit has ripple effects throughout the entire system. Effects of excess or lower metal concentration on some enzymes or biological reactions have been determined, but their proper connection with this robust plant metabolic network need to be elucidated. For this reason, alteration in the concentration of a single micronutrient can have multidimensional effects. Once the disrupted step is repaired, the entire network gets properly established. Therefore, it is necessary to start from the roots, which are the key organs responsible for the uptake of these metals. A vivid study of the way these metals are taken up by the plant system needs to be done. More research is needed regarding the structure and working of most metal transporters in higher plants. Scientists often suggest administration techniques such as phytoremediation, rhizoremediation, bioremediation, and other processes to detoxify contaminated soil. Even then, it is necessary to look into the entire plant transcriptome and metabolome, for the better understanding of the molecular cause behind each of the altered phenotypes in plants due to alteration in the levels of a particular metal. Much research is needed on the ways that plants gain access to these elements and also on the alterations that occur in the levels of the plant molecular mediators due to changes in the levels of the essential elements.

Acknowledgments

Financial assistance from Science and Engineering Research Board, Government of India through the grant (EMR/2016/004799) and Department of Higher Education, Science and Technology and Biotechnology, Government of West Bengal, through the grant [264(Sanc.)/ST/P/S&T/1G-80/2017] to Dr. Aryadeep Roychoudhury is gratefully acknowledged.

References

Adams, J.F., 1997. Yield response to molybdenum by field and horticultural crops. In: Gupta, U.C. (Ed.), Molybdenum in Agriculture. Cambridge University Press, Cambridge, pp. 182–201.

Agarwala, S.C., Chatterjee, C., Sharma, P.N., Sharma, C.P., Nautiyal, N., 1979. Pollen development in maize plants subjected to molybdenum deficiency. Can. J. Bot. 57, 1946–1950.

Agarwala, S.C., Sharma, C.P., Farooq, S., Chatterjee, C., 1978. Effect of molybdenum deficiency on the growth and metabolism of corn plants raised in sand culture. Can. J. Bot. 56, 1905–1909.

Anderson, A.J., 1956. Molybdenum deficiencies in legumes in Australia. Soil. Sci. 81, 173–192.

Anderson, A.J., Spencer, D., 1950. Molybdenum in nitrogen metabolism oflegumes and non-legumes. Aust. J. Sci. Res. 3, 414–430.

Arnon, D.L., Stout, P.P., 1939. The essentiality of certain elements in minute quantity for plants with special reference to copper. Plant Physiol. 14, 1460–1470.

Banik, A.K., 1976. Mineral nutrition of *Aspergillus niger* for citric acid production. Folia Microbiol. 21, 139–143.

Basu, M., Bhadoria, P.B.S., 2008. Performance of groundnut under nitrogen fixing and phosphorous microbial inoculates with different levels of cobalt in alluvial soils of Eastern India. Agron. Res. 6, 15–25.

Behboudian, M.H., Pickering, A.H., Dayan, E., Hanegev, I., 2016. Deficiency diseases, principles. Encycl. Appl. Plant Sci. 1, 219–224.

Bittner, F., Oreb, M., Mendel, R.R., 2001. ABA3 is a molybdenum cofactor sulfurase required for activation of aldehyde oxidase and xanthine dehydrogenase in *Arabidopsis thaliana*. J. Biol. Chem. 276, 40381–40384.

Bowen, H.J.M., 1979. Environmental Chemistry of the Elements. Academic Press, London, pp. 1–333.

Campbell, W.H., 2001. Structure and function of eukaryotic NAD(P)H:nitrate reductase. Cell Mol. Life Sci. 58, 194–204.

Caselles, J.M., Pérez-Espinosa, A., Pérez-Murcia, M.D., Moral, R., Gómez, I., 1997. Effect of increased cobalt treatments on cobalt concentration and growth of tomato plants. J. Plant. Nutr. 20, 805–811.

Chatterjee, J., Chatterjee, C., 2003. Management of phytotoxicity of cobalt in tomato by chemical measures. Plant Sci. 164, 793–801.

Chatterjee, C., Nautiyal, N., 2001. Molybdenum stress affects viability and vigour of wheat seeds. J. Plant Nutr. 24, 1377–1386.

Chatterjee, C., Nautiyal, N., Agarwala, S.C., 1985. Metabolic changes in mustard plants associated with molybdenum deficiency. New Phytol. 100, 511–518.

El-Sheekh, M.M., El-Naggar, A.H., Osman, M.E.H., El-Mazaly, E., 2003. Effect of cobalt on growth, pigments and the photosynthetic electron transport in *Monoraphidium minutum* and *Nitzchiaper minuta*. Braz. J. Plant Physiol. 15, 159–166.

Fernandez, E., Galvan, A., 2007. Inorganic nitrogen assimilation in *Chlamydomonas*. J. Exp. Bot. 58, 2279–2287.

Fido, R.J., Gundry, C.S., Hewitt, E.J., Notton, B.A., 1977. Ultrastructural features of molybdenum deficiency and whiptail of cauliflower leaves: effects of nitrogen source and tungsten substitution for molybdenum. Aust. J. Plant Physiol. 4, 675–689.

Follmann, H., 2000. Light–dark and thioredoxin-mediated metabolic redox control. In: Vanden, T. (Ed.), Plant Cells. The Redox State and Circadian Rhythms. Driessche Kluwer Academic Publishers, Dordrecht, pp. 59–83.

Gad, N., 2012a. Physiological and chemical response of groundnut (*Arachis hypogaea*) to cobalt nutrition. World Appl. Sci. J. 20, 327–335.

Gad, N., 2012b. Role and importance of cobalt nutrition on groundnut (*Arachis hypogaea*) production. World Appl. Sci. J. 20, 359–367.

Gopal, R., Dube, B.K., Sinha, P., Chatterjee, C., 2003. Cobalt toxicityeffects on growth and metabolism of tomato. Commun. Soil. Sci. Plant Anal. 34, 619–628.

Graham, R.D., Stangoulis, J.R.C., 2007. Molybdenum and plant disease. In: Nutrition and Plant Disease, APS Press, Wisconsin, pp. 203–205.

Gupta, U., 1997. Introduction to Molybdenum in Agriculture. Molybdenum in Agriculture. Cambridge University Press, Cambridge.

Hesberg, C., Haensch, R., Mendel, R.R., Bittner, F., 2004. Tandem orientation of duplicated xanthine dehydrogenase genes from *Arabidopsis thaliana*. J. Biol. Chem. 279, 13547–13554.

Hewitt, E.J., Bolle-Jones, E.W., 1952a. Molybdenum as a plant nutrient. II. The effects of molybdenum deficiency on some horticultural and agricultural crop plants in sand culture. J. Hortic. Sci. 27, 257–265.

Hewitt, E.J., Bolle-Jones, E.W., 1952b. Molybdenum as a plant nutrient. I. The influence of molybdenum on the growth of some Brassica crops in sandculture. J. Hortic. Sci. 27, 245–256.

Huwait, E.A., Kumosani, T.A., Moselhy, S.S., Mosaoa, R.M., Yaghmoor, S.S., 2015. Relationship between soil cobalt and vitamin B12 levels in the liver of livestock in Saudi Arabia: role of competing elements in soils. Afr. Health Sci. 15, 993–998.

Hu, Y., Ribbe, M.W., 2013. Biosynthesis of the iron-molybdenum cofactor of nitrogenase. J. Biol. Chem. 288, 13173–13177.

Jayakumar, K., Jaleel, C.A., Vijayarengan, P., 2007. Changes ingrowth, biochemical constituents, and antioxidant potentialsin radish (*Raphanus sativus* L.) under cobalt stress. Turkish J. Biol. 31, 127–136.

Jayakumar, K., Jaleel, C.A., Azooz, M.M., 2008. Impact of cobalt on germination and seedling growth of *Eleusine coracana* L. and *Oryza sativa* L. under hydroponic culture. Glob. J. Mol. Sci. 3, 18–20.

Kaiser, B.N., Gridley, K.L., Brady, J.N., Phillips, T., Tyerman, S.D., 2005. The role of molybdenum in agricultural plant production. Ann. Bot. 96, 745–754.

Kandil, H., Ihab, M.F., Amal, E.M., 2013. Effect of cobalt level and nitrogen source on quantity and quality of soybean plant. J. Basic Appl. Sci. Res. 3, 185–192.

Khan, S., Syed, A.T., Ahmad, R., Rather, T.A., Ajaz, M., Jan, F.A., 2010. Radioactive waste management in hospital. Int. J. Health Sci. 4, 39–46.

Kim, C., Weber, D.J., 1980. Isolation and characterisation of ATPase from *Salicornia pacifica* var. *utahensis*. Plant Cell Physiol. 21, 755–763.

Liu, J., Reid, R.J., Smith, F.A., 2000. The mechanism of cobalt toxicity in mung beans. Physiol. Plant. 110, 104–110.

Manal, A.A., AbdAlla, M.A., Allam, S.M.M., 2016. Effect of zinc and cobalt applied with different methods and rates on the yield components of *Vicia faba* L. World Wide J. Multidiscip. Res. Dev. 2, 52–58.

Marin, A., Marion-Poll, A., 1997. Tomato *flacca* mutant is impaired in ABA aldehyde oxidase and xanthine dehydrogenase activities. Plant Physiol. Biochem. 35, 369–372.

McGrath, S.P., Mico, C., Zhao, F.J., Stroud, H., Zhang, H., Fozard, S., 2010. Predicting molybdenum toxicity to higher plants: estimation of toxicity threshold levels. Environ. Pollut. 158, 3085–3094.

Mendel, R.R., 2013. The molybdenum cofactor. J. Biol. Chem. 288, 12172–13165.

Mendel, R.R., Haensch, R., 2002. Molybdoenzymes and molybdenum cofactor in plants. J. Exp. Bot. 53, 1689–1698.

Mendel, R.R., Schwarz, G., 2011. Molybdenum cofactor biosynthesis in plants and humans. Coord. Chem. Rev. 255, 1145–1158.

Mittal, N., Sharma, K., Oshin Kaur, A., 2017. Toxicity of metals such as iron, nickel and cobalt on seedlings of broad beans (*Vicia faba* L.). Biotechnology 3, 809–810.

Mohanty, N., Vass, I., Demeter, S., 1989. Impairment of photosystem II activity at the level of secondary quinone electron acceptor in chloroplasts treated with cobalt, nickel, zinc ions. Physiol. Plant. 76, 386–390.

Mullins, M.G., Bouquet, A., Williams, L.E., 2000. Biology of the Grapevine. Cambridge University Press, Cambridge.

Nagajyoti, P.C., Lee, K.D., Sreekanth, T.V.M., 2010. Heavy metals, occurrence and toxicity for plants: a review. Environ. Chem. Lett. 8, 199–216.

Palit, S., Sharma, A., 1994. Effects of cobalt on plants. Bot. Rev. 60, 150–161.

Pastori, G.M., Rio, L.A., 1997. Natural senescence of pea leaves: an activated oxygen-mediated function for peroxisomes. Plant Physiol. 113, 411–418.

Reddy, K., Munn, L., Wang, L., 1997. Chemistry and Mineralogy of Molybdenum in Soils. Molybdenum in Agriculture. Cambridge University Press, Cambridge.

Robinson, B., Burne, P., 2000. Another look at the Merlot problem—could it be molybdenum deficiency? The Australian Grapegrower and Winemaker. In: Proceedings of the 28th Annual Technical Issue, 427a, pp. 21–22.

Saaltink, R., Griffioen, J., Mol, G., Birke, M., 2014. Geogenic and agricultural controls on the geochemical composition of European agricultural soils. J. Soils Sediment. 14, 121–137.

Sagi, M., Scazzocchio, C., Fluhr, R., 2002. The absence of molybdenum cofactor sulfuration is the primary cause of the *flacca* phenotype in tomato plants. Plant J. 31, 305–317.

Sanz-Luque, E., Chamizo-Ampudia, A., Llamas, A., Galvan, A., Fernandez, E., 2015. Understanding nitrate assimilation and its regulation in microalgae. Front. Plant Sci. 6, 899.

Schwartz, S.S., Leon-Kloosterziel, K.M., Koorneef, M., Zeevart, J.A.D., 1997. Biochemical characterisation of the *ab2* and *ab3* mutants in *Arabidopsis thaliana*. Plant Physiol. 117, 161–166.

Smith, F.W., Ealing, P.M., Hawkesford, M.J., Clarkson, D.T., 1995. Plant members of sulphate transporters reveal functional subtypes. Proc. Natl Acad. Sci. U S A 92, 9373–9377.

Tuppy, H., Sieghart, W., 1973. Effect of Co^{+2} on yeast mitochondria. Monatshefte für Chem. 104, 1433–1443.

Vieira, R.F., Cardoso, E.J.B.N., Vieira, C., Cassini, S.T.A., 1998. Foliar application of molybdenum in common beans. I. Nitrogenase and reductase activities in a soil of high fertility. J. Plant Nutr. 21, 169–180.

Wenzel, A.A., Schlautmann, H., Jones, C.A., Kuppers, K., Mehlhorn, H., 1995. Aminoethyoxyvinylglycine, cobalt and ascorbic acid all reduce ozone toxicity in mung beans by inhibition of ethylene biosynthesis. Physiol. Plant. 93, 286–290.

Williams, R.J.P., Frausto da Silva, J.J.R., 2002. The involvement of molybdenum in life. Biochem. Biophys. Res. Commun. 292, 293–299.

Williams, C.M.J., Maier, N.A., Bartlett, L., 2004. Effect of molybdenum foliar sprays on yield, berry size, seed formation, and petiolar nutrient composition of 'Merlot' grapevines. J. Plant Nutr. 27, 1891–1916.

Williams, L.E., Pittman, J.K., Hall, J.L., 2000. Emerging mechanisms for heavy metal transport in plants. Biochem. Biophys. Acta 1456, 104–126.

Witte, C.P., Tiller, S.A., Taylor, M.A., Davies, H.V., 2002. Addition of nickel to Murashige and Skoog medium in plant tissue culture activates urease and may reduce metabolic stress. Plant Cell Tissue Organ. Cult. 86, 103–104.

Würfel, M., Häberlein, I., Follmann, H., 1990. Inactivation of thioredoxin by sulfite ions. FEBS Lett. 268, 146–148.

Yadav, D.V., Khanna, S.S., 1983. Effect of cobalt and phosphorous on their availability and nitrogen fixation by legumes. Haryana Agric. Univ. J. Res. 13, 396–401.

Yadav, D.V., Khanna, S.S., 1988. Role of cobalt in nitrogen fixation: a review. Agric. Rev. 9, 180–182.

Yadav, R.P., Yadav, D.V., Khanna, S.S., 1984. Improved nodulation and nitrogen fixation in berseem with the application of cobalt and phosphorous. Indian J. Agric. Sci. 54, 506–508.

Zeid, I.M., 2001. Responses of *Phaseolus vulgaris* to chromium and cobalt treatments. Biol. Plant. 44, 111–115.

Chapter 11

Interplay between sodium and chloride decides the plant's fate under salt and drought stress conditions

Koushik Chakraborty[1], Subhankar Mondal[1], Debarati Bhaduri[1], Ankita Mohanty[1] and Alivia Paul[2]
[1]*ICAR-National Rice Research Institute, Cuttack, India,* [2]*Ballygunge Science College, University of Calcutta, Kolkata, India*

11.1 Introduction

Plants have lived in an uncertain and changing environment ever since they came into existence. Many abiotic stresses like high or low temperature, salinity, excess or scarce water, and heavy metal toxicity are major threats to crop growth and production (Wani et al., 2016). Reports show that globally 1125 million hectares of land are salt-affected, and among those, 76 million hectares of land are affected by anthropogenic factors resulting in soil salinization and sodification (Hossain, 2019). Among them, 6.73 million hectares of land in India are affected by salinity stress (Sharma and Singh, 2015). Reports also show that nearly 42% of the land in India is affected by drought stress (Gogoi et al., 2019). Both salinity and drought are detrimental for plant survival and are reported to affect the productivity of almost every crop. Additionally, both these stresses limit water availability in plants, which alters various physiological and metabolic processes. Together both salinity and drought restrict plant growth and production of photosynthates, which ultimately reduces the energy supply and disturbs cellular homeostasis (Zhu et al., 2010). At the initial stage, the physiological manifestations of drought and salinity stresses are the same, as in both cases osmotic stress is exerted in the plants, which lowers the water potential and causes net water loss from the plant cell (Uddin et al., 2016). But at a later stage, drought stress results in extreme water scarcity in plants; in salinity stress, excessive accumulation of sodium (Na^+) and chloride (Cl^-) ions leads to ion toxicity as

Plant Nutrition and Food Security in the Era of Climate Change. DOI: https://doi.org/10.1016/B978-0-12-822916-3.00020-2

the major limiting factor for plant growth. An influx of excess Na^+ hampers the uptake of other ions from the soil as well, leading to an overall ionic imbalance at the whole-plant level (Munns and Tester, 2008). A high concentration of Na^+ ions in soil hinders the uptake of essential macroelements like potassium (K^+), by competitive inhibition due to the physiochemical similarity of sodium (Na^+) and potassium (K^+) ions. Similarly, the activity of some cation transporters (CNGC and GLRs) was hampered under high saline conditions, and the transporters lost their specificity and promoted transport of Na^+ ion instead of other cations like K^+, Ca^{2+}, Li^+, and Ba^{2+} due to the concentration effect (Chakraborty et al., 2018). These factors altogether change the overall ionomics or ionic status of the plant. Apart from Na^+, chloride (Cl^-) is another important anion for plants and essential for maintaining osmotic potential under normal environmental conditions. Previously its contribution and effects under high salinity were addressed adequately by scientists (Wu and Li, 2019), but recent studies show that an excessive accumulation of chloride (Cl^-) in the cell could be equally toxic for the plant and can act as a potent inhibitor of plant growth and development. Chloride (Cl^-) is reported to interact with the photosynthetic machinery (chloroplast) of the plant to hamper the photosynthesis process under salt stress (Tavakkoli et al., 2010).

To survive under this adversity and maintain a convenient level of water potential in the cell, plants generally produce large amounts of osmoprotectants (proline, glycine betaine, and trehalose) to annihilate the damage caused by drought and salinity stress (Xu et al., 2010). Most interestingly, instead of synthesizing energy-requiring osmolytes, some glycophytic, xerophytic, and halophytic members utilize sodium (Na^+) and chloride (Cl^-) ions as an osmoticum (Chakraborty et al., 2020). Along with this, plants generally adopt several tolerance mechanisms to maintain cellular homeostasis and nullify the toxic effects of sodium (Na^+) and chloride (Cl^-) ions. Plants may employ an ion-exclusion process, where an excessive amount of sodium is thrown out of the cell by the action of different plasma membrane-localized ion transporters/pumps with active energy utilization. Similarly, plants may employ an ion-compartmentalization process, where an excess amount of sodium (Na^+) and chloride (Cl^-) ion is sequestrated into the vacuoles or in the older leaves to maintain favorable ionic homeostasis in the active tissues/cells to prevent damage to the more sensitive plant parts. The molecular mechanism identified for Na^+-exclusion showed that enhanced activity of the SOS and HKT family members and plasma membrane located H^+-pumps belong to the AHA family plays a key role in the Na^+-exclusion process. NHX family transporters and vacuolar H^+-pumps majorly contribute to sequester the excess amount of sodium (Na^+) ions into the vacuoles to maintain an ionic homeostasis in the cytoplasm (Chakraborty et al., 2018). Further studies identified NAC, MYB, bZIP, WRKY, and GATA family transcription factors as important for regulating the function of genes and orchestrating the signal transduction process that is directly involved in the abiotic stress tolerance process (Golldack et al., 2014).

Apart from the adaptive responses to abiotic stresses like salinity and drought, several management strategies were also worked out by the different researchers like modified agronomic and nutrient management practices, seed priming, mulching, and other soil amendment techniques, etc. Among these, seed priming with sodium chloride (NaCl) or other salt compounds like KCl, $CaCl_2$, KNO_3, etc. and additional doses of K^+-supplementation were found to have a positive influence in ameliorating the ill effects of salt and drought stresses (Chakraborty et al., 2016a; Theerakulpisut et al., 2017). In general, it is reported that seed priming improved the overall stress tolerance ability of the plant. Priming often raises the stress-responsive signals or physiochemical acclimation in the seeds that provide an elevated level of tolerance when they encounter actual stress conditions (Farooq et al., 2006; Sarkar et al. 2019). Reports also showed that seed priming effectively improved the growth and photosynthetic ability of a plant by regulating antioxidant enzymes such as catalase and peroxidase and also effectively minimizing the salt (NaCl)-induced DNA damage in the cell (Savvides et al., 2016; Iqbal et al., 2006).

The existing literature suggests diverse biological functions are associated with sodium (Na^+) and chloride (Cl^-) ions under different abiotic stresses like salinity and drought. In this compilation, we covered the diverse responses of different genotypes belonging to glcophytic, halophytic, and xerophytic families, under the influence of salinity and drought stresses. We further emphasized exploring the dominant role of sodium (Na^+) and chloride (Cl^-) ions that play a crucial role in determining the fate of plants under salinity and drought stresses.

11.2 Relative impact of dominant ions on plants under ionic stress

All soils essentially contain a mixture of soluble salts, some of which are considered essential for plant growth. Plant growth becomes stunted once the total concentration of salts becomes excessive, and eventually plants may also die. The rate of reduction in plant growth and the threshold of tolerance vary widely among different crop species (Francois and Maas, 1999). Besides osmotic effects, nutritional imbalances or ion toxicity may also exist due to specific ion effects. Excess influx of Na^+ causes ionic imbalance in plants under salinity stress (Munns and Tester, 2008). Plants growing in saline conditions accumulate either high quantities of inorganic ions or synthesize organic compounds with significant negative charge to neutralize the excess cation taken up by the plants in order to maintain stable intracellular pH (Yang et al., 2007, 2008; Chakraborty et al., 2016b). In the majority of cases, inorganic anions such as Cl^- play a dominant role in maintaining the ionic balance, whereas the contribution of organic acids (total negative charge) was reported to be very low in rice (Wang et al., 2011).

However, the relative contribution of salinity-induced nutrient imbalances or ion toxicity and the osmotic effects toward their impact on overall crop yield is hard to quantify. As summarized from many reports, yield losses from osmotic stress can be significant for most crops, including tree species. The role of ion toxicities is generally additive while considering its effects on yield, with prominent growth-suppressive effects caused by osmotic stress. Horticultural crops, such as woody fruit and nut crops, tend to accumulate Cl^- or Na^+ up to toxic levels then show symptoms like leaf burn, necrosis, and defoliation (Francois and Maas, 1999). The salt-tolerance ability of rice cultivar Panvel 3 was attributed to its ability to maintain low shoot Na^+/K^+ ratios (<1.0) and to exclude both Na^+ and Cl^- ions from the shoots. As reported in the study, the magnitude of ion toxicity showed a trend of NaCl$>$ $Na^+>$ Cl^- (Kumar and Khare, 2015). Further, it was also reported that the damage incurred by Na^+-induced hypertoxicity is greater than the toxicity imposed by Cl^{--} or other anions like NO_2^-, NO_3^-, etc.; however, additive effects of all these ions were found to be greater than individual ionic stresses in rice (Khare et al., 2020).

The high Cl^- and Na^+ concentrations of saline soils are a major limiting factor in reducing plant growth. Previous studies reported that NaCl treatment caused a greater reduction in growth and development of barley plants as compared to separate treatments of Na^+ and Cl^- (Tavakkoli et al., 2011). It suggests high concentrations of Na^+ or Cl^- possess additive and/or synergistic effects on stress imposition. They further emphasized that Na^+ and Cl^- exclusion are two independent mechanisms having different consequences on overall salt tolerance. High Na^+ hinders K^+ and Ca^{2+} uptake and stomatal regulation in plants, while the accumulation of high Cl^- disturbs photosynthetic capacity due to chlorophyll degradation in barley (Tavakkoli et al., 2011).

Under salinity stress, external K^+ application (30 kg K_2O/ha) resulted in better ionic balance (K^+/Na^+) in different plant parts of peanut crop, which imparted better salinity tolerance along with associated factors like higher plant water status, photosynthetic efficiency, biomass produced, and osmotic adjustment under stress (Chakraborty et al., 2016a). Further, accumulation of phosphorus ions in leaves and roots and their ratio were diverse among the groundnut cultivars under influence of salinity, and the phosphorus uptake by plants was found to be affected at a higher level of salinity, that is 6.0 EC_{iw} (Bhaduri et al., 2016). Nitrogen and micronutrient contents of different plant parts in *Brassica* cultivars were reduced due to salinity. Relatively tolerant species *Brassica juncea* showed less reduction in micronutrient uptake as compared with *Bactris campestris*, a relatively sensitive species to salt stress (Chakraborty et al., 2016c). In Tables 11.1 and 11.2, we compiled the major impact of ionic imbalances in plant parts under salinity and drought stress studied over the last few decades.

TABLE 11.1 Effect of salt stress on the availability of essential ions and their implications on physiological and metabolic functions in plants.

Nutrients	Available form	Effect on different physiological function	References
Nitrogen	NO_3^- and NH_4^+	• In vine plants nitrate content of leaves reduced • High Cl^- concentration in soil affects the NO_3^- content in the leaves resulting in yellowing and chlorosis	Hu and Schmidhalter (2005); Hussin et al. (2013)
Phosphorous	$H_2PO_4^-$, HPO_4^{2-}	Excess Cl^- and SO_4^{2-} severely affects P uptake and causes dead patches in leaves of sunflower	Zhukovskaya (1973)
Potassium	K^+	Excess Na^+ present in soil hinders the K^+ uptake and influences the enzymatic activity and photosynthesis in citrus and avocado	(Faber, 2015)
Calcium	Ca^{2+}	• An influx of excess Na^+ inhibits the calcium in durum wheat • High NaCl content in soil causes curling of younger leaves in tomato	Allen et al. (1995); Tuna et al. (2007)
Magnesium	Mg^{2+}	Mg^{2+} content in both root and shoot decreases due to antagonism with Na^+	Grattan and Grieve (1999)
Manganese and zinc	Mn^{2+} and Zn^2	• In barley and rice, concentration of Mn^{2+} and Zn^{2+} showed an increment in the shoot region • In corn shoot, Mn^{2+} and Zn^{2+} content reduced under saline stress	Hassan et al. (1970); Verma and Neue (1984)
Iron	Fe^{2+} or Fe^{3+}	• In pea, tomato, and rice, increase in soil salinity increases the shoot iron content and improves the dry matter content of plant • In tomato, salinity causes a decrement in the Fe^{2+} content in leaves as well as in roots	Dahiya and Singh (1976); Verma and Neue (1984); El-Fouly et al. (2002)
Copper	Cu^{2+}	In roots and leaves of tomato, Cu^{2+} content reduced	El-Fouly et al. (2002)

TABLE 11.2 Effect of drought stress on the availability of essential ions and their implications on physiological and metabolic functions in plants.

Nutrients	Available form	Effect on different physiological function	References
Nitrogen	NO_3^- and NH_4^+	• Decrease in root, shoot dry weight and nitrogen accumulation in lettuce under drought • Decrease in shoot biomass and nitrate reductase activity in the roots of *Vigna unguiculata* • Decrease in N_2 fixation activity in legumes	Ruiz-Lozano and Azcon (1996); Silveira et al. (2001); Streeter (2003)
Phosphorous	$H_2PO_4^-$ and HPO_4^-	Reduction in phosphorous uptake and translocation in the shoot is severely affected in some plants	Pinkerton and Simpson (1986)
Potassium	K^+	Under drought stress, K^+ content and chlorophyll content in flag leaves of barley decreased	(Hosseini et al., 2016)
Calcium	Ca^{2+}	• In *Cynodon dactylon*, the Ca^{2+} content increased during water stress • In *Dalbergia sissoo* under drought, Ca^{2+} content severely decreased in leaves and roots	Utrillas et al. (1995); Singh and Singh (2004)
Magnesium	Mg^{2+}	Reduction in shoot and root Mg^{2+} content and accomplished with leaf chlorosis in *Spartina alterniflora*	Brown et al. (2006)
Sulphur	SO_4^{2-}	S concentration in leaves, roots, and shoot of *Fagus sylvatica* L. remain unchanged after 3 weeks of drought	Peuke and Rennenberg (2004)
Chloride	Cl^-	Drought increases Cl^- as well as K^+ and Na^+ concentrations in both leaves and roots of papaya	Mahouachi et al. (2006)

11.3 Impact of sodium and other associated cations

11.3.1 Sodium: its role as a nutrient and an osmoticum in plants under stress

Sodium (Na^+), one of the significant cations prevailing in soils, is found to be influenced by salt stress. Normally, it is evenly distributed in the biosphere and comprises up to 3% of the earth's crust. It is regarded as a nonessential or functional nutrient except in some C_4 plants and some halophytes (Pessarakli 2001; Subbarao et al., 2003). Under normal conditions, the cytosolic concentration of Na^+ is 1–10 mM in the plant cell, but under salinity stress its hyperaccumulation is observed in the cytosol or organelles, which inhibits normal metabolic processes (Bassil et al., 2012). Higher Na^+ content inhibits the movement of guard cells of stomata in photosynthetic tissues and causes oxidative stress through increased reactive oxygen species (ROS) production (Chakraborty et al. 2016d). High salt concentration causes osmotic imbalance, membrane disintegration, and ceased cell division that leads to growth inhibition (Tuteja, 2007). Under these circumstances, plants need to maintain the positive turgor pressure by the process of osmotic adjustment by maintaining high solute concentrations inside the cell as compared with the external solution. For that, most plants synthesize different organic solutes (sucrose, sugar, proline, glycine betaine, etc.) to contribute to osmotic adjustment inside the vacuoles (Allen et al., 1995). From an energy utilization point of view, synthesis of organic solutes for osmotic adjustment and ion-exclusion requires a greater amount of ATP than sodium sequestration in some cases. Thus most of the dicotyledonous halophytes use NaCl as osmoticum and accumulate a high concentration of Na^+ and Cl^- inside the tissue through intracellular compartmentalization to maintain their osmotic potential (Flowers and Colmer, 2008). Tonoplast-located proton pumps create an electrochemical gradient across the membrane that helps in sequestering Na^+ and Cl^- inside the vacuole through a tonoplast ion exchanger (Hasegawa, 2013) (Shabala, 2013). Apart from transport, effective retention and restriction in leakage of ions to the cytoplasm are essential to avoid the toxic effect of salt stress.

11.3.2 Na^+ uptake and transport under stress

The large amount of Na^+ from the root-soil interface is transported passively through the root hair epidermal cells of the plant and then moves through the xylem to the upper parts of the plants especially leaf blades. Sodium ion moves via symplastic or apoplastic pathway once it gets inside the root epidermis. Apoplastic flow contributes to less than 1% of the total transpirational flow until and unless the transpiration pool demands more. Some portion of the Na^+ is transported via an apoplastic pathway and the rest moves completely until it reached the casparian band of the root endodermis, which acts as an obstacle in Na^+ movement, but sometimes it can pass the casparian band through "bypass flow" (Craig-Plett and Moller, 2010). A significant amount of Na^+ moves through bypass flow into the shoots of rice. Net uptake employing the symplastic pathway of Na^+ is catalyzed by specific transporters. Many candidates like HKT, HAK/KUP (high-affinity potassium uptake transporter), and nonselective cation channels (NSCCs) are the major groups of transporters responsible for the uptake of Na^+ from the soil. Besides that, glutamate receptors (GLRs) and LTCs (low-affinity cation transporters) are also believed to uptake Na^+ efficiently. HKT/KUP/HAKs are known to work as low affinity for Na^+ transporters (class I HKTs) but regarded to possess high specificity for K^+ uptake (Kobayashi et al., 2010; Haro et al., 2013). In rice, the HKT family transporters OsHKT2;1 and OsHKT2;2 have been known to be involved in Na^+ influx into the roots (Horie et al., 2007). In *Arabidopsis*, AtHKT1;1 was involved in the uptake of Na^+ ions. Also, it was reported to cause rapid accumulation of Na^+ ions in the upper parts of the plant rather than the lower part, that is root (Rus et al., 2004; Pardo, 2010; Kronzucker and Britto, 2011). Members of NSCCs, such as cyclic-nucleotide gated channels (CNGCs) and GLRs are two families that account for Na^+ influx in plants (Apse and Blumwald, 2007). The CNGC group of transporters was reported to be nonselective. In general, CNGC1, CNGC3, and CNGC4 were observed as efficient in the transport of both Na^+ and K^+ (Balague et al., 2003; Hua et al., 2003; Gobert et al., 2006). It was also reported that in *Arabidopsis*, AtCNGC1 and AtCNGC4, members of the AtCNGC family, were involved in the transport of monovalent cations like K^+ and Na^+ (Balague et al., 2003), whereas CNGC10 was studied for selective transport Na^+ over K^+ (Guo et al., 2008). GLRs, that is GLR1.1 and GLR1.4, were responsible for Na^+, K^+, Ba^+, and Ca^+ transport (Roy et al., 2008a; Tapken and Hollmann, 2008), while LCT1 transporter was shown to transport Na^+, Ca^{2+}, and Li^+. Numerous K^+/Na^+ transporters were found to be responsible for either upregulation or downregulation as a reaction to salt stress at the translational level (Chakraborty et al., 2018).

11.3.3 Impact of Na^+ on K^+ and its regulation for maintaining cellular homeostasis

Potassium (K^+) is one of the most important macronutrients for plants, comprising up to 1%–3% of plant dry weight (White and Karley, 2010) and essentially is required for enzyme activation, protein biosynthesis, osmoregulation, solute transport, and photosynthesis in plants. It is responsible for most of the fundamental physiological processes that have direct consequences in plant growth (Very and Sentenac, 2003). Under drought stress, the role of K^+ ion in the maintenance of turgor and regulation of stomata is well known. K^+ inside the cell under water stress balances the metabolic activity and produces compatible solutes like sugars, sugar-alcohols, and proline for regulating cell turgor. Due to their chemical and ionic similarity, Na^+ replaces K^+ in some physiological activities, particularly in maintaining turgor pressure under saline conditions. The high competence between Na^+ and K^+ for the binding sites in different biophysical processes reduces the potential binding of K^+ under salinity. An increase in Na^+ concentration leads to membrane depolarization and a reduction in K^+ influx (Adams and Shin, 2014; Chakraborty et al., 2016e). The dominance of Na^+ ions in soil under salinity hinders the uptake of K^+ by binding itself to HKT transporters located in the plasma membrane of roots (Rus et al., 2004; (Kronzucker and Britto, 2011);(Kronzucker and Britto, 2011) (Alemán et al., 2011)) and NSCCs and passively enters into the plant. In *Arabidopsis*, high-affinity K^+ transporters AKT1 and AtHAK5 were involved in Na^+ transport. Uptake and transport of K^+ are suppressed by Na^+ and AKT1 is downregulated (Qi and Spalding, 2004) and thus contributes to Na^+ uptake.

High cytosolic Na^+/K^+ ratio inhibits the synthesis of protein, enzyme activities, and vital physiological functions in plants under stress. Thus, it is essential to maintain a minimal cytosolic Na^+/K^+ level to overcome the ill effects of salt stress. In response to that, plants should selectively uptake ions for maintaining the ionic balance and efflux excessive Na^+ out from the apoplastic space to compartmentalize it into the vacuole (Cuin et al., 2011). Two mechanisms play vital roles to control the immoderate amount of Na^+ buildup (Cabot et al., 2014), while intracellular NHX transporter, a Na^+/H^+ antiporter, is identified as the most efficient in the compartmentalization of excess Na^+ into the vacuoles (Blumwald 2000). The function of NHX depends on the proton gradient generated by the activity of V-ATPase (Bassil et al., 2012). It is reported that cation selectivity of AtNHX1 decreased under stress and the transport of Na^+ was compromised, keeping the K^+ transport unchanged (Yamaguchi et al., 2005). Recently, four antiporter genes (OsNHX1–4) have been identified in rice, which were found to be efficient in controlling the excess Na^+ buildup (Fukuda et al., 2004) Kumar et al., 2013). Several others reported a plasma membrane located SOS1, a Na^+/H^+ antiporter that attributed to Na^+ exclusion across the plasma membrane against the electrochemical potential driven by H^+ in *Arabidopsis* (Shi et al., 2002). According to some reports, in several crops including rice (Martinez-Antienza et al., 2007), wheat (Xu et al., 2008), and tomato, the SOS1 activity regulated the transport and redistribution of Na^+ between the root and shoot xylem tissues and also found arresting Na^+ to reach leaf cells (Shi et al., 2002) (Olías et al., 2009). Once Na^+ is loaded into the transpiration pool through the xylem, the members of SOS and HKTs (high-affinity potassium transporters) are involved in long-distance transport and redistribution of Na^+ in the xylem of root cells to avoid excess Na^+ accumulation in the above-ground part (Rus et al., 2004). Upregulation of HKT family genes in *Arabidopsis* root and shoot, AtHKT 1;1 (Horie et al., 2006), OsHKT1;5, and OsHKT1;4 (Ren et al., 2005; Kronzucker and Britto, 2011) in the rice leaf sheath and root respectively have been shown to be effective in reducing the Na^+ transport. Thus it is noteworthy that shoot Na^+ exclusion by HKT1 from the plant cell is considered as an effective mechanism in salt tolerance in different cereals such as rice (Ren et al., 2005; Chakraborty et al., 2019) and durum wheat (James et al., 2006). Several QTLs have provided evidence of the HKT1 gene conferring accumulation of Na^+ in the leaves upon salt stress. Out of the two members of the HKT family, HKT2 family transporters exhibit more selectivity for K^+ over Na^+. It is believed that shoot Na^+ exclusion is one of the major mechanisms contributing toward salt stress, but both shoot and root Na^+ accumulation hamper plant growth and some studies also suggested that accumulation of Na^+ at the expense of K^+ has a more detrimental effect on plant health as compared to Na^+ accumulation alone.

11.3.4 Interaction between Na^+ and Ca^{2+}

Calcium is an essential macronutrient and a structural element that helps in maintaining the rigidity of the cell wall. In addition to that, it appeared as a sensing and signaling molecule in plants' responses to various abiotic and biotic stress (Lynch and Läuchli, 1988). Under the presence of excess Na^+ and Cl^-, uptake of calcium and nitrate is reduced and in some cases, Na^+ may replace Ca^{2+} from its binding site and disrupt its acquisition (Hadi and Karimi, 2012). Calcium deficiency in tomato (Tuna et al., 2007) and *Vigna unguiculata* (Murillo-Amado et al., 2006) was reported under high NaCl concentration. (Hadi and Karimi, 2009)indicated that high Na^+ at the root zone reduces Ca^{2+} uptake, which results in a low Ca^{2+}/Na^+ ratio in plants. In *Aloe vera* under salinity, Ca^{2+} deficiency is observed due to restricted uptake by root under the influence of Na^+ (Jin et al., 2007). So the interplay between Ca^{2+} and Na^+ might have some importance in salinity tolerance (Buschmann et al., 2000; Chakraborty et al. 2012). As there was evidence of low Na^+/Ca^{2+} ratio, higher cytosolic Ca^{2+} maintenance in the roots of *Aloe vera* showed less membrane damage (Jin et al., 2007). A rapid buildup of Ca^{2+}concentration inside the cell could hinder the uptake of Na^+ increasing K^+ selectivity for absorption to nullify the harmful effects of salinity (Hasegawa et al., 2000). Further to nullify the adverse effect of Na^+, Ca^{2+} competes at the membrane-binding site and protects the cell membrane (Shabala et al., 2007). It has been seen that most plants responded favorably to calcium supplementation during salinity stress (Shabala et al., 2006). It was also noticed that a rapid increase in cytoplasmic calcium level occurs in plants within a fraction of seconds under NaCl stress (Knight et al., 1997). A three-component system SOS pathway consisting of a Ca^{2+}-dependent protein kinase and calmodulin like Ca^{2+}-binding domains was identified in maintaining the ionic homeostasis in plant cells through CBL-CIPK interactions (Qiu et al., 2002) by extrusion of Na^+ from the cytosol to apoplastic space (Kumar et al., 2013). The SOS3 protein is typically a Ca^{2+}-binding protein that is responsible for decoding Ca^{2+} signals in response to salt stress (Liu and Zhu, 1997). Under salt stress, SOS3 physically interacts with SOS2, which ultimately phosphorylates SOS1 to stimulate the process of Na^+ exclusion by exchanging Na^+ with H^+. The process might also trigger osmotically responsive or voltage-gated channels and Ca^{2+} influx during the signaling cascade (Chakraborty et al., 2018). The ortholog of VCX1P, CAX1, and CAX2 transporters may influence the salt stress-induced transient as well as steady-state Ca^{2+} ($[Ca^{2+}]_{ext}$) signaling cascade (Sze et al., 2000).

11.4 Impact of chloride and other associated anions

Unlike Na^+, Cl^- is an essential micronutrient as it acts as a cofactor for enzymes involved in photosynthesis and also plays role in stomatal movements. It helps to regulate pH and stabilizes the membrane potential by regulating the turgor pressure (Xu et al., 1999; White and Broadley, 2001; Franco-Navarro et al., 2016). Cl^- ion is an essential cofactor for PSII during photosynthesis (Kawakami et al., 2009). Besides that, Cl^- has a major activity in regulating asparagine synthetase enzyme and tonoplast V-ATPase (Churchill and Sze, 1984). In glycophytes, the tissue chloride content is around 100–200 μg/g of shoot dry weight and does not exceed dry weight of 4–7 mg/g in Cl^--sensitive species and 15–50 mg/g for Cl^--tolerant species (Xu et al., 1999). Some plants like citrus and soybean are effective in excluding Na^+ but are sensitive to chloride toxicity and could not prevent excess Cl^- accumulation in leaves (Munns and Tester, 2008). In tobacco, Cl^- concentration in plants less than 5 mM has been effective in an increase in leaf area and plant biomass (Franco-Navarro et al., 2016), but in barley a concentration exceeding 120 mM resulted in a decrease in plant biomass (Tavakkoli et al., 2011). The decrease in plant biomass and chlorophyll content, along with an increase in malondialdehyde and proline content, and electrolyte leakage under Cl^- stress, were observed in *Chrysanthemum* (Guan et al., 2012).

11.4.1 Cl^- uptake and regulation in response to salt stress

Accumulation of excess Cl^- at the root zone during salt stress reduces the water and nutrient uptake due to the reduction of water potential at the root zone (Munns, 2002). Chloride ions do not form compounds rapidly inside the soil and possess specificity in adsorption to soil particles. Under high soil salinity Cl^- enters into the plant through both active and passive transport. About 50% of Cl^- translocated inside the rice shoots apoplastically (Shi et al., 2013). Passive influx or efflux of Cl^- occurs through vacuolar H^+/Cl^-

symporter present in the plasma membrane of root hair cells of green algae (*Chara coralline*) and root hairs of wild mustard (*Sinapis alba*) (White Broadley, 2001). To date, many anion channels are reported and characterized. Among them, the most common are slow anion and cation channels (SLAC channels) and their homologs SLAH, which are responsible for anion (nitrate, sulfate, and chloride) loading in the root xylem cells. Another is aluminum-activated malate transporters (ALMTs), which got activated by malate under high chloride concentration and involved in the transport of chloride across the plasma membrane (Gruber et al., 2010). De Angeli et al. (2016) reported that Cl^- channels (CLCs) are located in the tonoplast of *Arabidopsis*, which are of seven types. These are mainly found in many intracellular membranes including vacuolar membrane, Golgi membrane, and thylakoid membranes (AtCLCa to AtCLCg) and are possibly involved in Cl^- fluxes across the tonoplast (Barbier-Brygoo et al., 2011). Cation chloride cotransporters (CCCs) reported by Colmenero-Flores et al. (2007) have been seen contributing to long-distance chloride transport through xylems under salt stress (Flower and Colmer, 2008). DTX/MATE (detoxification efflux carrier/multidrug and toxic compound extrusion transporters) present in endomembrane (Zhang et al., 2017) and AtNPF2 (Li et al., 2017a) located in tonoplast were also reported to play roles in chloride ion transport in the plant. As compared to Na^+, much less information is available regarding the transport and regulation of Cl^- under stress. A minimum accumulation of chloride ions in root and shoot is one of the important features to control Cl^- toxicity. Like Na^+, the exclusion of Cl^- from root tissues was reported in wild soybean plants (Chen et al., 2013), but in barley, less accumulation of chloride in the shoot was reported due to the regulation of Cl^- transport, rather than root chloride exclusion (Tavakkoli et al., 2011). SLAC1 responsible for Cl^- and malate transport across the plasma membrane has been supported by mutation of SLAC1, which showed an increase in the concentration of chloride in the guard cell (Negi et al., 2008). In *Arabidopsis*, both SLAC1 and its homologous SLAH1 conferring salt tolerance by accumulation Cl^- in the shoot was reported (Cubero-Font et al., 2016).

11.4.2 Differential accumulation of chloride and its compartmentalization

Chloride sequestration inside the vacuole is possible due to the difference in membrane potential between the cytosol and the vacuole. The difference in the potential present across the vacuolar membrane perhaps facilitates the movement of Cl^- through the membrane transporters, if the vacuolar Cl^- concentration is higher than the cytosol. In barley leaf mesophyll cells, a more negative tonoplast potential was found under saline conditions (Cuin et al., 2003), which allowed more cytosolic Cl^- movement into the vacuole.

To resolve the harmful effects of salinity some plants can partition the excess Cl^- between the root and shoot tissues. It was seen in barley, salt-tolerant cultivars had a capacity greater than the sensitive ones to exclude chloride from the mesophyll cells (Huang and Van Steveninck, 1989). Members of AtCLC and AtALMT9 are some CLCs and transporters present in the tonoplast, which might be found to contribute toward vacuolar sequestration of Cl^- (Wei et al., 2015). Tonoplast-locating AtCLCs over expression lines of wild *Arabidopsis* showed higher Cl^- accumulation inside vacuoles (Nguyen et al., 2016; Hu et al., 2017). The evidence supported less transportation of Cl^- into the shoot because of the sequestration in roots due to overexpression of GmCLC1 in soybean (Wei et al., 2016). AtALMT9 is a tonoplast Cl^- channel that is transcriptionally upregulated under salt stress and recorded in less shoot accumulation of Cl^- in ALMT9 knockout mutant (Baetz et al., 2016).

11.4.3 Interaction between Cl^- and NO_3^-

Nitrate (NO_3^-) is one of the most ubiquitous anions present in the soil under normal conditions. High salt concentration results in a lower nitrogen level inside the plant due to the enmity between Cl^- and NO_3^- (Li et al., 2017a), which hinders its uptake and storage. The limited supply of nitrogen may cause Cl^- ion toxicity. It is assumed that a high concentration of Cl^- inadvertently leaks into the cell through NO_3^- transporters. Under the saline condition, a high amount of uptake and accumulation of chloride ion in shoot vacuole hinders the uptake of nitrate by competing with nitrate transporter (Cubero-Font et al., 2016) as CLCs reported to show higher permeability for Cl^- over nitrate (Accardi et al., 2015). There are various anion channels, which are mostly selective toward NO_3^- rather than Cl^- (Roberts, 2006). SLAC1 is reported to be 10 times more selective for NO_3^- over Cl^- (Geiger et al., 2009). It was also reported that culture media with high NO_3^- decrease the concentration of Cl^-, while the NO_3^-/Cl^- balance might have some relationship with salinity as of Na^+/K^+ as well as with Cl^- efflux mechanism (Li et al., 2017a).

11.5 Regulation of uptake and in-planta movement of Na^+ and Cl^- ions

11.5.1 Regulation of uptake and in-planta movement of sodium (Na^+) ions

Sodium, one of the major cations present in soil, is not considered essential for maintaining growth, development, and reproduction. Though it is nonessential for terrestrial plants, reports showed that a low level of sodium (Na^+) in the cell is beneficial for plants, especially when under conditions of low potassium (K^{+}). Studies showed that the hydrated form of Na^{+} is structurally very similar to K^{+}, and therefore it might be used as a substitution for K^{+} ion in some cases. Low levels of sodium can be beneficial but a moderate to high level of sodium in the cell is phytotoxic. Excessive accumulation of salt to the upper part of the cell is one of the prime reasons behind the cellular destruction process (Pradhan et al., 2019). In general, under normal conditions, glycophytes can maintain a lower level of sodium in cells, but the "control" is shattered under high saline conditions for salt-sensitive genotypes. For salt-tolerant genotypes, maintaining a nontoxic level of sodium in the cell is essential for maintaining cellular functions. Therefore proper regulation of sodium (influx/efflux) is necessary under salt stress. To check the excessive accumulation of sodium in the cell, plants generally use several mechanisms. The balanced level of sodium in the cell can be maintained by (1) reducing the net uptake of sodium, (2) reducing net translocation in the shoot, (3) enhancing sodium retrieval or selective transportation of K^{+} over Na^{+}, (4) applying sodium exclusion, and (5) preferential use of Na^{+} and partitioning sodium into vacuoles (Assaha et al., 2017). Among these, the reduction of net uptake of sodium in the cell needed downregulation of the genes encoding the Na^{+} influx channels. Senadheera et al. (2009) showed that the downregulation of OsCNGC1 imparts better tolerance to salt-stressed FL478 genotype of rice. Similarly, Mekawy et al. (2015) reported Egyptian jasmine showed improved tolerance to salt stress due to the downregulation of the OsCNGC1 gene. Apart from the reduction in uptake, excess accumulation of sodium in the cell might be controlled by the radial transport of sodium into the xylem. Oh et al. (2009) reported radial transportation of sodium in the xylem can be controlled by the enhanced expression of the SOS1 gene. Plett and Møller (2010) reported AtHKT1.1 transporter is important in *Arabidopsis*, as it mediates sodium retrieval of Na^{+} from the xylem cells. Similarly, in rice and wheat OsHKT1.5 and TaHKT1;5-D/TmHKT1;5D localized in the xylem parenchyma cells retrieve sodium from xylem cells and are therefore important for maintaining a balanced level of sodium to the upper plant parts (Ren et al., 2005; Munns et al., 2012; Byrt et al., 2014). Parallel to this study, two loci were identified in wheat Nax1 and Nax2, which are involved in the sodium retrieval process (James et al., 2006). Also, to maintain cellular homeostasis, some salt-tolerant genotypes of rice like FL478, Nona Bokara directly exclude sodium from the cell. A molecular study showed SOS (salt overly sensitive) and HKT gene families are involved in facilitating sodium exclusion. Shi et al. (2000) identified AtSOS1 as a proton-mediated sodium excluder (Na^{+}/H^{+}), which facilitates sodium out of the cell under high saline conditions. Similarly, in rice OsSOS1 is involved in the transport of sodium out of the cell under salt stress. Furthermore, studies identified the activity of SOS1 regulates the SOS2-SOS3 protein complex. Fukuda et al. (2004) reported a Na^{+}/H^{+} antiporter (OsNHX1) located in the vacuolar membrane facilitates the transport of Na^+ into the vacuoles and has played a major role in maintaining cellular homeostasis in rice.

11.5.1.1 Uptake of sodium from soil

In plant cells, uptake/influx of sodium ion is mediated by the activity of several NSCCs, including CNGCs and GLRs (Keisham et al., 2018). Previous reports showed that few members of high-affinity potassium transporters belonging to the HKT and HAK family had been involved in the uptake of sodium from the soil. An HKT member involved in the transportation of sodium from soil was identified and validated by the work of Horie et al. (2007). From the study, an HKT family transporter identified in rice, that is OsHKT2.1, was found to mediate sodium (Na^+) influx under potassium (K^+)-starved conditions. Along with this, Schachtman et al. (1997) previously identified and reported a transporter protein in wheat LCT1 that was involved in the influx of sodium ions. Recently, Byrt et al. (2017) identified an aquaporin (AtPIP2.1) in *Arabidopsis* that has been involved in the influx of Na^+ ions. A different group of transporters is involved in the transport and uptake of sodium, but amongst them, voltage-insensitive NSCC members are considered to be an important candidate for Na^+ influx (Keisham et al., 2018). Apart from this, in *Arabidopsis*, two subgroups of NSCC are present, which are reported to be involved in sodium influx. Among different members of CNGCs, AtCNGC3 is involved in the influx of sodium (Gobert et al., 2006). In rice, OsCNGC1 is identified as a protein that has been involved in the influx of sodium ion (Senadheera et al., 2009). Similar to these findings, Roy et al. (2008b) and (Tapken and Hollmann, 2008) identified two GLRs from *Arabidopsis*, AtGLR1.1 and AtGLR1.4, that are reported to be involved in the sodium influx process.

11.5.1.2 Long-distance transport of sodium

Once sodium enters into the root cells from the soil, it is radially transported in the xylem by apoplastic or symplastic pathways. Little information is available in the literature for the transporters that radially transport sodium into the xylem. In *Arabidopsis*, members of CHX family transporters (cation/H^+ mediated channels) might be involved in the transport of sodium ions into the xylem (Hall et al., 2006). Molecular characterization studies indicated that the protein localized and expressed in the root epidermal cells and mutants of CHX21 under high salinity accumulates less amounts of sodium than functional CHX21, indicating that CHX21 might be involved in the transport of Na^+ ions from endodermis to stele. Once the sodium is loaded into the xylem and enters the transportation stream, it is transported to the shoot and increases the concentration of sodium in the upper plant parts.

11.5.2 Regulation of uptake and in-planta movement of chloride (Cl^-) ions

Growth, development, and other important processes of higher plants are associated with the availability of micronutrients and macronutrients. Among them, chloride (Cl^-) is considered an essential micronutrient for all plants (Raven, 2017). Though it is an essential micronutrient, a high concentration of chloride in the cell or accumulation of high chloride under saline conditions can be toxic for plants and limit the overall growth and development of the plant. Responses of plants toward chloride ions are not static and might differ by their requirement and utilization of chloride ions in the cell. Reports showed that damage associated with excessive accumulation of sodium and chloride differs in plants. In crop species, the sodium ion is considered more lethal than the chloride ion. However, in soybeans, citrus, and woody plant species, the toxicity and ill effects of chloride are more significant than those of sodium ions. In general, plants take up chloride from the soil solution and utilize it for photosynthetic enhancement, maintaining the osmotic potential, stomatal movements, signal transduction process, and maintaining cellular pH. Different kinds of anion channels are involved in the transport and movement of chloride ions from soil to the upper parts of the plants. There are two different ways to transport chloride ions in plants: (1) symplastic (cytoplasmic), and (2) apoplastic (extracellular). The symplastic movement of anions across the plasma membrane is mediated by the activity of some channel proteins or by carrier proteins. Among them, channel proteins are the integral protein that facilitates the transportation of specific ions across the membranes. Carrier proteins are the integral proteins moving larger molecules across the plasma membrane and can transport molecules against a concentration gradient by utilizing ATP or energy. Until recently, very little information was available regarding the uptake of chloride ions in the plant. Li et al. (2016) identified a stelar-localized chloride transporter in *Arbidopsis* (AtNPF2.4), which facilitates the transportation of chloride ions from root to shoot. Based on the transcriptome, mutant, and functional analysis of plants under normal and stressed conditions, different families of anion transporters were identified, which directly or indirectly facilitate the transport of chloride ions and are involved in maintaining cellular homeostasis under saline environments. Myriad pieces of evidence were present, where members of these families—bacterial mechanosensitive channels of small conductance (MscS)-like (MSL), CLCs, nitrate transporter family (NRT), ABA transporter, CCC, slow anion channels (SLAC), and ALMTs—were reported to contribute to transport chloride ions or specifically exclude anions from cells (Teakle and Tyreman, 2010; Wu and Li, 2019).

11.5.2.1 Chloride channel

CLC is a group of anion transporters that transports chloride and other anions across the plasma membrane. Candidates of this family transporter are generally made up of 10–12 transmembrane helices that form homodimers, where each monomer is involved in the anion transport process. The CLC family of transporters are proton-mediated chloride pumps that shuttle Cl^-/H^+ across the plasma membrane. In plants, seven different types of the CLC group of transporters were identified so far from *Arabidopsis thaliana* (AtCLCa, AtCLCb, AtCLCc, AtCLCd, AtCLCe, AtCLCf, and AtCLCg). Among them, AtCLCa, AtCLCb, AtCLCc, and AtCLCg are localized in the membranes of tonoplast and other intercellular membranes. Reports showed that AtCLCe is present in the thylakoid membrane and other homologs such as AtCLCd and AtCLCf are present in the Golgi membrane. Similarly, in rice plants, seven different types of CLC members are involved in chloride transport and expressed in the different endomembranes to maintain cellular homeostasis. In soybean, eight different types of CLC transporters are present that contribute to maintaining cellular homeostasis under high saline conditions. Wei et al. (2016) functionally proved that the constitutive overexpression of GmCLC1 enhanced the overall salt-tolerance ability of transgenic *Arabidopsis* plants and reduced chloride ion accumulation in the upper plant parts. Later, Wei et al. (2019) again identified and reported another transporter from soybean. The GmCLC-c2 gene, localized in tonoplast membrane, contributes to chloride and NO^{3-} homeostasis and imparts salt tolerance to wild soybean. Parallel to this work, Sun et al. (2013) showed that GmCLC1 is a chloride transporter located in the vacuole and helps in transportation and accumulates chloride ions from the cytosol to the vacuole. Results showed that transgenic *Populus* plants overexpressing GmCLC1 were able to maintain better photosynthetic pigments by enhanced activity of antioxidant enzymes like SOD and POD. Overexpression of GmCLC1 increased the overall salt-tolerance ability of transgenic *Populus* plants. Expression of chloride channels in plant cells has played pivotal roles in maintaining the cellular homeostasis under saline stress also examined by Diedhiou and Golldack (2006). In rice, based on the expressional analysis data it was observed that OsCLC1 expression was induced in leaves and roots of salt-tolerant *Pokkali* rice genotype around xylem parenchyma and phloem cells, while it had repressed expression in salt-sensitive IR29 rice genotype.

11.5.2.2 Cation chloride cotransporter

CCC is the family of membrane-integral solute carrier proteins 12 (SLC12) that cotransports chloride (Cl^-) ions along with potassium (K^+) and sodium (Na^+) and might be involved in the retrieval of chloride ions around the xylem vessels. The whole process is electroneutral as two negatively charged chloride (Cl^-) ions are translocated with one positively charged K^+ and one positively charged Na^+ ion. Previous studies by the researcher showed that in xylem vessels transportation of chloride ion was an electroneutral process and later, it was considered that CCC groups of transporters located in the xylem vessels are a possible candidate for the xylem retrieval process. To date, several CCC groups of transporters have been identified from different dicots and monocots like *Arabidopsis, Brassica, Glycine*, rice, sorghum, and corn. Phylogenetic analysis among different transporters showed that different members of plant CCC belong to two distinct clades (CCC1 and CCC2). In rice, more than one CCC was present in the genome (OsCCC1.1 and OsCCC1.2) that are members of the CCC1 clade. The importance of CCC transporters for maintaining cellular homeostasis as well as for maintaining ionic homeostasis was functionally proved in *Arabidopsis* and rice plants (Colmenero-Flores et al., 2007; Chen et al., 2016) and loss of function of CCC genes altered the growth and development process, shunted growth of root and shoot, and reduced the fitness of the plant in general. Henderson et al. (2015) showed that constitutive expression of grapevine CCC gene in tobacco and *Arabidopsis* improved the overall plant salt-tolerance ability. The roles of different CCC transporters in maintaining Cl^- homeostasis was also investigated by scientists. Wei et al. (2017) reported constitutive expression of PtrCCC1 gene in *Nicotiana* plants able to maintain a low level of chloride in the leaves and roots of transgenic tobacco lines when compared with control plants, thus improving the overall plant salt tolerance.

11.5.2.3 Slow anion channels (SLAC/SLAH family)

SLAC1/SLAH is the member of S (slow)-type anion channel proteins located in the stomatal cells (guard cells) and facilitates the transportation of chloride, nitrate, and other kinds of anions. It consists of 10 transmembrane helices with 556–557 amino acids, showing homology with tellurite-resistance multidrug transporter TehA from *Haemophilus influenza* (HiTehA), C4 dicarboxylate transporter family, and malate transporter family (Mae1) from fission yeast (*Schizosaccharomyces pombe*). The activity or function of SLAC proteins majorly facilitates the phosphorylation process. Recently, it was identified that the activity of SLAC1 was blocked by the phenylalanine residue present at the 450th position. Generally, phosphorylation of SLAC1 removed the Phe450 residue and activated the protein. This group of channels was first identified from the two independent mutant lines of *Arabidopsis*—one was insensitive for ozone-mediated stomatal closure and another was deficient for CO_2-mediated leaf temperature change. In *Arabidopsis*, four types of SLAC1 homologs (SLAH) are present in the genome. Plentiful evidence shows that expression of SLAC1/SLAH is important in maintaining cellular homeostasis under saline conditions or specifically maintaining chloride ion homeostasis in the guard cells. Negi et al. (2008) showed that mutant slac1 plants had accumulated more chloride ions in the guard cells compared to wild plants. Similarly, Kurusu et al. (2013) showed that overexpression of AtSLAC1 in tobacco cells induced the ROS-generation process and increased the chloride ion exclusion. The importance of SLAC1 in facilitating stomatal movement under salt and drought stress was identified by scientists. Mustilli et al. (2002) and Geiger et al. (2009) reported the process of phosphorylation was mediated by the activity of a kinase (OST1 kinase). Open stomata 1 (OST1) is activated by the hormone ABA (abscisic acid), and phosphorylate serine120 residues of SLAC1 activate the protein. Later, a protein belonging to the group LRR-RLK (leucine-rich repeat kinase) named GHR1 (guard cell hydrogen peroxide resistant 1) was also identified that facilitates the activity of SLAC1 and both these enzymes worked in parallel to control the stomatal closure, which directly or indirectly improved the plant response to salt/drought stress.

11.5.2.4 Aluminum-activated malate transporters

ALMTs are the members of R (rapid)-type anion channels with six transmembrane helices, located in the stomatal guard cells and facilitating transportation of chloride, malate, and nitrate. In general, members of this family were stimulated by aluminum (Al^{3+}) and they help in transportation of anions across the plasma membrane. Sasaki et al. (2004) isolated a member of this family transporter TaALMT1 from wheat (*Triticum aestivum*). Meyer et al. (2010) identified *A. thaliana* AtALMT12, which is essential for maintaining cellular homeostasis and influenced substantial chloride efflux from cells. In *Arabidopsis*, different types of ALMT transporters are located in the tonoplast—ALMT4, 6, and 9. Among them, the AtALMT9 protein acts as a chloride transporter and efflux chloride ions to maintain cellular function. Parallel to this, Ligaba et al. (2012) reported in *Zea mays* ZmALMT2 a plasma membrane-bound anion transporter, which might be able to transport chloride and malate across the membrane. Pineros et al. (2008) identified *T. aestivum* TaALMT1 involved in the transportation of malate and chloride. At a lower concentration of chloride, TaALMT1 prefer transporting malate over chloride ions. The selectivity of chloride and malate ion transport is almost similar under a high concentration of chloride.

Regulation of gene expression is an important phenomenon that controls the expression of a particular gene. Based on the need of the plant, the particular gene expression is increased or decreased in the cell. Activation of a gene (transporter) is not simple and needs sophisticated downstream mechanism/signaling that unwinds the DNA and allows transcription and translation. Therefore the activity of a transporter located in the plasma membrane or any parts of the plant cell or others needs a signal or interaction to become active. Previous reports showed that calcium and ABA are important for controlling SLAC1/SLAH family proteins. Later, (Lee et al., 2008) reported SRK2E/ OST1/SnRK2.6 are involved in SLAC1/SLAH gene activation. Further, Geiger et al. (2009) identified calcium-dependent protein kinases (CDPK/CPK) are involved in the regulation of SLAC1/SLAH types of transporter under drought stress. They identified CPK21 and CPK23 as important for the functioning of SLAC1/SLAH and phosphorylate SLAC1 protein at the N-terminal domain. Brandt et al. (2012) identified another two members of the CDPK family; CPK3 and CPK6 are also involved in the phosphorylation of SLAC1/SLAH types of proteins. Similar to this, Imes et al. (2013) identified an R (rapid)-type anion channel that is locating in the tonoplast membrane also regulated by the activity of ABA signaling factors. According to the reports of Imes et al. (2013) an R-type anion channel ALMT12 might be activated by the SRK2E/OST1 protein.

11.5.2.5 Nitrate transporter family

The NRT is a member of the peptide transporter family (PTR) that actively transports nitrate across the plasma membrane. Though it transports nitrate, some members of this family also transport chloride in the cells to maintain cellular or ionic homeostasis. Phylogenetic analysis with *A. thaliana* showed that members of these families can be subdivided into three families—NRT1, NRT2, and NRT3. Among them, NRT1-mediated chloride efflux or exclusion was proved by Brumos et al. (2009), where different citrus genotypes respond depending on the nature of chloride exclusion or accumulation. Also, Li et al. (2017b) reported in *Arabidopsis* a PTR family member AtNPF2.5 (NRT1/PTR family protein) expression was important for maintaining the chloride in the root cortical cells. Overexpression of this gene had improved the chloride exclusion process in the root cortical cells, therefore increasing tolerance to salt.

11.5.2.6 Bacterial mechanosensitive channels of small conductance (MscS)-like

Plant MSL (MscS-like) is a member of the bacterial mechanosensitive channel of small conductance proteins (MscS) that functions to maintain cellular homeostasis under stressed conditions. Haswell (2007) and Hamilton et al. (2015) identified 10 different types of MSL proteins that are present in *A. thaliana*. Haswell and Meyerowitz (2006) reported that the different MSL members in *Arabidopsis* MSL2 and MSL3 that were located in the plastids have played an important role in maintaining the osmotic balance in plastids. In contrast, mutants of this gene exhibit altered plastid shape. Additionally, Haswell et al. (2008) reported that two candidate genes of that family, MSL9 and MSL10, located in the plasma membrane of root cells might be important for maintaining chloride transport across the membrane. But under high saline conditions, quintuple mutants of MSL4, MSL5, MSL6, MSL9, and MSL10 showed no significant differences with wild-type plants. This suggests that it might be important for maintaining osmotic balance and chloride transport but not directly contribute to providing tolerance under high salinity.

11.5.2.7 Regulation of chloride ion uptake in the plant

Plants uptake chloride ions from the soil solution and utilize it for further maintaining various metabolic processes like photosynthesis, stomatal movement, and activity of some key enzymes. In the plant cell, the absolute concentration of chloride ion has to be maintained by some regulatory proteins that are involved in either influx or efflux of chloride based on the need of the plant parts or cell. As per the reports of White and Broadley (2001) the net uptake by root was almost the same in the control and stressed plant. Under normal environmental conditions, chloride ions are distributed equally in root and shoot cells. Xylem and phloem cells maintain a 1:1 ratio for chloride ions. Whereas the scenario was quite complex in cases of high salinity and the ratio was 5:1 for xylem and phloem cells, accumulation of chloride ions was more dominant in shoots under the high saline condition that rapidly increased the chloride concentration to the upper parts of the cells. To maintain a nontoxic level of chloride in the upper parts of the cell, plants generally used some adaptive strategies that include exclusion of chloride by the transporters located in the root/shoot cells, reduction of net uptake of chloride ions, reduction of net xylem loading, applying intracellular compartmentation, and phloem recirculation. The exclusion of ions under stress is considered an important phenomenon for maintaining cellular homeostasis that imparts tolerance to salt stress. Tavakkoli et al. (2010) reported under saline conditions, a salt-sensitive faba bean cultivar Nura accumulates a higher concentration of chloride than the salt-tolerant cultivar 1487/7. Similarly, Tavakkoli et al. (2011) reported a significantly lower level of chloride was present on the upper parts of salt-tolerant barley varieties than salt-sensitive, suggesting that salt-tolerant genotypes of bean and barley invested a lot of concern to check the upper transport of chloride and maintain a lower level of chloride toward the upper parts of the cells. Root chloride exclusion is important to provide an improved level of salt tolerance (Chen et al. 2013). Wild-type soybean (BB52) had shown higher root exclusion ability compared with a salt-sensitive cultivar ZH13. Similarly, Liu et al. (2014) reported shoot chloride exclusion is also important for providing salt stress tolerance in barley. In barley leaves, HvSLAC1 and HvSLAH1 played important roles to exclude chloride ions. Apart from the exclusion of chloride ions from root or shoot cells, some genotypes minimize the toxic effects of chloride by reducing the net uptake of chloride ions in the xylem. In grapevine, citrus, and lotus net uptake of chloride in xylem was found significantly lower in the salt-tolerant genotype as compared to the salt-sensitive genotype. Tolerant genotypes were able to maintain a higher level of chloride in the roots and a lower level of chloride in shoots, which indicates better compartmentation of chloride in the root vacuoles, which might be involved in providing salt stress tolerance in tolerant genotypes. To avoid chloride-mediated ill effects, some plants developed a highly sophisticated structure like a gland and stored chloride ions in the salt glands or bladders that compartmentalize nearly 20% leaf chloride as found in *Leptochloa fusca* (Jeschke et al., 1995).

11.6 Differential impact of sodium and chloride on glycophytes, halophytes, and xerophytes

11.6.1 Impact of sodium (Na^+) and chloride (Cl^-) on glycophytes

Glycophytes are the plants that maintain a low level of sodium in the upper parts of tissues. In the plant kingdom, the majority of plants are glycophytes (98%–99%), nontolerant to salinity and incapable of maintaining overall plant growth and development under saline conditions (Maathuis, 2014). Generally, in salt-sensitive genotypes, salt-mediated damages are primarily associated with osmotic imbalance/dehydration due to high salt concentration (Pardo, 2010; Roy et al., 2014; Mondal and Chakraborty, 2020). In later stages, excessive accumulation of sodium toward the upper parts of tissues played a major role in degrading the pigment chlorophyll and hampering photosynthesis, respiration, and other important processes (Roy et al., 2014). A prolonged period of salt exposure makes the plant weak as overall transportation of other essential micronutrients and macronutrients is also compromised under stress, creating an imbalance in the different parts of the cell and disrupting the cellular homeostasis, which is important for maintaining normal growth and development. Though an excessive amount of sodium and chloride accumulation is toxic for plants, some salt-tolerant glycophytic members have adapted mechanisms to maintain a nontoxic level of sodium in the cells or are capable of maintaining cellular homeostasis. Reports showed, in general, glycophytic plants have used two independent strategies to cope with the saline environment, which includes ionic exclusion and sodium compartmentalization (Munns and Tester, 2008; Roy et al., 2014). Between them, ionic exclusion mostly utilizes different transporters belonging to the SOS, HKT, AHA families to exclude sodium from cells (Reddy et al., 2017) or preferentially loading K^+ over Na^+ in the xylem by the activity of HKT1.5 (Ren et al., 2005; Wang et al., 2005; Chakraborty et al., 2019). In rice, FL478 is a popular genotype that exclusively uses ion exclusion as a salt-adaptive mechanism. Further studies indicate that ion exclusion has been mediated by the induced activity of SOS1, SOS2, SOS3, AHA1, AHA7, HKT1.5 genes, and others (Walia et al., 2005; Platten et al., 2013; Prusty et al., 2018). In the sodium compartmentalization strategy, instead of exclusively focusing on Na^+ exclusion, plants preferentially stored sodium and other ions in the vacuole or other organelles and utilized sodium and chloride ions as an osmoticum for maintaining the overall cellular homeostasis. Several gene

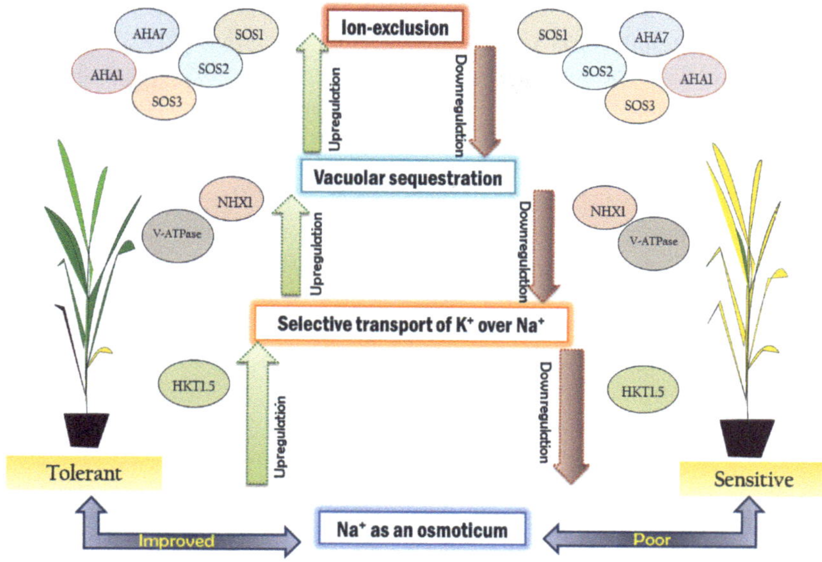

FIGURE 11.1 Comparative mechanistic differences of salt-tolerant strategies employed by tolerant and sensitive genotypes in glycophytes.

families belonging to NHX and V-ATPase families localized in the vacuolar membrane play a key role in this process. Recently, Prusty et al. (2018) showed a wild relative of cultivated rice (*Oryza nivara*) uses sodium compartmentalization and tissue tolerance as a primary salt-adaptive mechanism. Similarly, Chakraborty et al. (2020) identified a landrace of *Oryza sativa indica*, Kamini, which can potentially use both these ion-exclusion and sequestration mechanisms to withstand salinity stress (Fig. 11.1).

11.6.2 Impact of sodium and chloride on xerophytes

Previous studies on salinity and drought showed that Na^+ (in the form of NaCl) has played a positive role in multiple abiotic stresses. Sometimes, it has improved the growth and development process of the plants. Among the different members of plants, xerophytes are those that are adapted to grow in arid or semiarid conditions and have the efficiency of utilizing sodium (Na^+) as a beneficial product to survive under drought stress. To understand the amelioration effect or impact of Na^+ ion in xerophytes, (Ma et al., 2011), working with a succulent xerophytic plant *Zygophyllum xanthoxylum*, identified sodium-supplemented plants of *Z. xanthoxylum* with 50 mM of NaCl can enhance the ability to survive under drought conditions. Also, sodium-treated plants of *Z. xanthoxylum* were able to maintain a high level of chlorophyll, the relative water content in leaves, and showed an improved photosynthesis rate as compared to nontreated plants. Similarly, sodium-mediated stimulation of growth under drought stress in xerophyte was also studied by Yue et al. (2012), where an induced accumulation of sodium (Na^+) increases the overall Na^+ ion status in leaves that acts as an osmoregulator; furthermore, it helps to mitigate the deleterious impact of drought stress. According to Xi et al. (2018), some sodium-mediated adaptations are present in the xerophytic *Z. xanthophylum* plant. Sodium (Na^+)-treated plants showed improved survivability by maintaining a low frequency/density of stomata under drought stress that minimizes the transpiration process. Also, it decreases the size of guard cells to put a limit on the transpiration process. Apart from the sodium-mediated growth and development, the nutrient availability was also improved in xerophytes (*Z. xanthophylum*) as a result of NaCl treatment under drought stress. Wang et al. (2019) reported NaCl helps in the transport of other essential micronutrients [iron (Fe) and silicon (Si)] and macronutrients [nitrogen (N), phosphate (P), and calcium (Ca^{2+})] that are necessary for plants. Recently, He et al. (2019) in their study with a salt-secreting xerophyte *Reaumuria soongorica* showed that along with an improved level of chlorophyll concentration and photosynthesis rate, the convenience of NaCl or specifically Na^+ ion in leaves was important for maintaining the leaf morphology and helps to improve the overall tolerance under drought stress (Fig. 11.2).

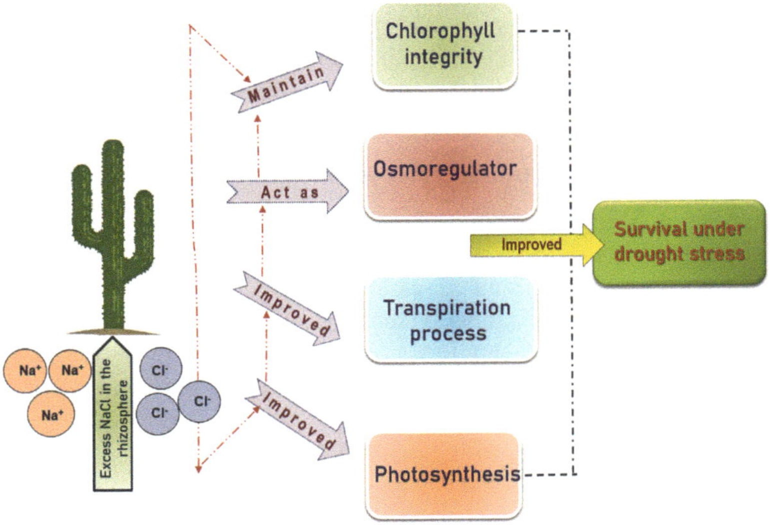

FIGURE 11.2 Sodium chloride plays a deciding role in survival and adaptation of xerophytic plants under drought stress.

11.6.3 Molecular mechanism of hypertolerance of xerophytes to salt and osmotic stress

The *Z. xanthophylum* plant is well known due to its excellent adaptation ability to salt and osmotic stress. Genes that confer tolerance to multiple abiotic stresses can further be divided into two classes—functional and regulatory. Among these, regulatory genes are the members of the signaling pathway that includes transcription factors and mediators for proteolysis degradation mechanism, etc. Recently, Yin et al. (2019) found some regulatory networks or transcription factors that secured tolerance under extreme salt and osmotic stress. In this study based on the transcriptome analysis data, they identified two transcription factor genes belonging to NAC transcription factor family (ZxNAC083 and ZxNAC035) that were involved in providing tolerance to salt and osmotic stresses. To further reinvestigate and validate the effectiveness of these candidate genes under salt and osmotic stress, they transformed the identified genes into a model plant system *A. thaliana* (Col-0). For this, PCR amplified fragments of DNA were cloned into a binary vector pBIB-BASTA-35S-GWR-GFP. The complete construct was transformed into the *Agrobacterium tumefaciens* strain GV3101 and finally transformed into the Col-0 plants by the floral dip method. Comparative analysis between control and transgenic plants showed that transgenic plants constitutively harbored ZxNAC083 and ZxNAC035 genes and had improved tolerance to salt and osmotic stress.

11.6.4 Impact of sodium (Na^+) and chloride (Cl^-) on halophytes

An excessive amount of sodium and chloride in the cell is considered a major impediment to growth and development for the plant. Indeed, it has ill effects on plant growth but is not inimical to all plants. According to Schimper, halophytes are plants that are capable of maintaining normal growth and development under saline habitats and are also able to grow on normal soil. In halophytes, better management and preferential distribution of sodium and chloride ions have made the plant strong and able to successfully survive under high salinity. Myriad evidence is present where distribution, preferential compartmentalization, and secretion of sodium and chloride ions play important roles in maintaining the cellular homeostasis in halophytes (Munns and Tester, 2008; Carillo et al., 2011). Hasanuzzaman et al. (2014) reported salt uptake by halophytes does not directly interfere with the growth or essential metabolism of the plant, such as photosynthesis and respiration. Scientists reported in halophytes, three different mechanisms are involved with tolerance—reduction in net Na^+ influx, Na^+ compartmentalization, and secretion of excess Na^+ through salt glands. Reports show vacuolar sequestration of sodium and chloride is much more dominant in halophytes compared to glycophytes. Expressions of the tonoplast-localized

Na^+/H^+ transporters (NHX) and proton pumps belonging to the H^+-ATPase and H^+-PPase family are constitutive, suggesting much more effective adaptive strategies and vacuolar sequestration are present in halophyte to maintain a balanced ratio of ions. In addition, Glenn et al. (1999) reported a thick layer of lipid present outside the vacuole to check and prevent the leakage of sodium into the cytoplasm. In *Suaeda maritime*, the size of the vacuole occupied 77% of the total cell area and was capable of storing more sodium ions than normal glycophytes (Hajibagheri et al., 1984). Reports showed that in a halophyte (*S. maritime*) nearly 500 mM concentration of sodium ions could be stored in the vacuole, while the sodium-storing capacity of the vacuoles of glycophytes was nearly 100 mM. Apart from vacuolar sequestration, preferential compartmentalization into older parts of cells or leaves also beneficial in halophytes was reported by Carillo et al. (2011) that maintains homeostasis in the transpiring, photosynthetic young leaf cells. Also, vacuolar sequestration, secretion of salts into salt hairs, or salt glands are also beneficial in halophytes. Previous reports showed that some halophytic members excrete an excess amount of sodium out of the cell in the form of liquid droplets that have made crystals when coming in contact with air.

11.6.5 Gene family members that are involved in regulating drought and salinity stress tolerance

A gene that controls the expression of several other functional and regulatory genes is called a regulatory gene or regulator gene. In plants, several gene families are involved in controlling and regulating the activity of other genes. Among these, stress-responsive NAC, MYB, bZIP, WRKY, and GATA families' transcription factors are important to securing tolerance under drought and salt stress. In general, NAC family proteins are well known and important for maintaining the growth and development process under stress. Recently, Guan et al. (2019) identified a NAC family transcription factor (OoNAC72), which has improved the performance of a legume (*Oxytropis ochrocephala*) under drought and salt stress and specifically under multiple abiotic stresses. Based on the study, OoNAC72 protein is localized in the nuclear membrane and might be induced by drought, salinity, and exogenous application of abscisic acid. To test the effectiveness of this transcription factor, when it overexpressed constitutively in tobacco, it showed interactions with several different stress-responsive genes like RD29A, RD29B, RD26, LEA14, ANACOR19, ZAT10, PP2CA, and NCED3. Similar to this, MYB family transcription factors are also important for survival and better performance under salinity and drought stress. Tang et al. (2019) identified that OsMYB6 is related to better growth and development processes in rice. For functional validation, transgenic lines overexpressing OsMYB6 showed better performance compared to the wild-type rice cultivars. Biochemical studies showed that transgenic lines overexpressing

OsMYB6 were able to maintain a lower level of MDA in the cell, showed elevated production of proline and maintained higher antioxidant activities. Recently, Du et al. (2018) identified another member of this family from soybean GmMYB118 also important for providing drought and salt stress tolerance. Also, (Nutan et al., 2020) reported the GATA family transcription factor in rice (OsGATA8) localized in the nucleus played an important role under multiple abiotic stresses. OsGATA8 imparts tolerance to multiple abiotic stresses by targeting the activity of SOD, CAT, APX, OsDREB1A, OsbZIP23, OsNAC6, POR A, POR B, and POR C. Activity of these genes positively regulated and improved the leaf greening process, photosynthetic efficiency, stress tolerance ability, ion homeostasis, antioxidant enzyme activity, and chloroplast integrity, while negatively regulating and stunting the membrane leakage process, senescence, and ROS buildup process. Apart from this, Orellana et al. (2010) reported a bZIP family transcription factor, which is important for providing multiple abiotic stress tolerance. According to this, in *Solanum lycopersicon*, abscisic acid-responsive element-binding proteins (AREBs)—AREB1 and AREB2—have been induced in root and leaf tissues under drought and salinity stress. When assessed functionally, it was observed that overexpressing lines of SlAREB1 showed improved tolerance to salt stress and were able to maintain better tissue integrity than wild-type tomato. Parallel to this study, Zong et al. (2018) reported in maize a member of bZIP family transcription member ABP2 also able to provide an improved level of tolerance against salinity and drought stress when overexpressed in *Arabidopsis*. Besides these transcription families, the WRKY family of transcriptional activators is also important for providing multiple stress tolerance. Qin et al. (2015) reported in wheat (*T. aestivum*) salinity-induced transcription factor belonging to the WRKY transcriptional family played a major role in providing multiple abiotic stress tolerance. A class II member of WRKY transcription factor in wheat TaWRKY93 induced several folds under high salt and ABA conditions. When overexpressed in *Arabidopsis*, transgenic plants showed an improved level of interaction with other stress-responsive genes including SR3 (important for root development) and provided an improved tolerance against salinity by regulating with proteins that impart membrane stability against multiple abiotic stresses.

11.7 Changes in overall ionomics in plants under Na^+ and Cl^- stress

The composition of mineral nutrients and trace elements of an organism or tissue holistically represents the inorganic component of cellular and organismal systems known as the ionome. Ionomics is an in-depth study of the ionome that encompasses the quantitative and simultaneous measurement of the elemental composition of a particular organism or its tissues. It accounts for the changes in this composition owing to physiological processes and thus needs the application of high-throughput elemental analysis technologies in combination with both bioinformatics and genetic tools (Salt et al., 2008).

The idea of plant ionomics started with the amalgamation of two branches: metabolomics and mineral nutrition. The concept was given by Robinson and Pauling during the late 1960s and early 1970s that the metabolite profile is an expression of the physiological status of that organism. In due course, substantial progress has been made in the depiction of the basic biology of nutrient ion homeostasis in plants, since its inception as a separate branch from the 19th century (Marschner, 2011). The four techniques mostly used in ionomics studies are (1) inductively coupled plasma-mass spectrometry (ICP-MS), (2) inductively coupled plasma-atom/optical emission spectrometry (ICP-AES/OES), (3) X-ray fluorescence (XRF), and (4) neutron activation analysis (NAA).

Studying G × E interaction along with elements was an advanced step through ionomics by performing genetics and modeling that determines the mechanism by which plants adapt to diverse environments while providing a detailing of whole elemental and ionic composition of the plant under specific circumstances. Further, such studies of ionomes deliver a deep insight into genetically diverse populations that control elemental accumulation under impact-specific plant-environment interactions and thus are able to identify the underlying gene (BaxterDilkes, 2012). A poor association was found between element-to-element and element-to-environment when plants were grown in diverse environmental conditions due to highly variable and unstable features (Ghandilyan et al., 2009). Baxter et al. (2012) reported that the ionome of 96 wild accessions of *A. thaliana* grown in hydroponic culture and soil and concluded that ionome of a plant tissue showed high plasticity but was governed by genes and G × E interactions. Nevertheless, any environmental aberrations modify the process of the source to sink transport of inorganic ions and thus change the whole plant ionome. Even the root zone temperature impacts the transport and storage of mineral ions in plants (Adebooye et al., 2010; Quadir et al., 2011). Long-term exposure to saline soils affects ionic imbalance in plant tissues holistically and thus affects plant metabolism (Diego et al., 2008).

Of late, the field of ionomics has also been suitably applied to learn more about the mechanisms of salt tolerance. (Sanchez et al., 2011) applied ICPAES for studying ionomics in integration with metabolite profiling to compare the extremophile species *Lotus creticus* (growing in highly saline coastal regions), and two other cultivated glycophytic grassland forage species, *Lotus corniculatus* and *Lotus tenuis*. This study demonstrated a distinctive readjustment of nutrient levels in the shoots of extremophiles under saline environments. Currently, ionomics are useful in understanding the occurrence of multiple physiological processes in plants, in combination with other new-generation scientific disciplines like transcriptomics, proteomics, and metabolomics (Salt et al., 2004, Salt et al., 2008). A metabolomic method was applied to disclose the physiological and molecular differences in salt-tolerance mechanism, which was functional between wild barley (*Hordeum spontaneum*) and cultivated barley (*Hordeum vulgare*) by observing the changes in the contents of as high as 82 metabolites, both in roots and leaves due to salinity stress (Wu et al., 2013). Plant ionomics studies were also found useful in seven walnut cultivars of varied stress tolerance backgrounds for identifying the ion-partitioning capacity of roots and shoots after exposure of NaCl stress of 50–250 mM (Lotfi et al., 2009) and they further concluded that magnitude of K^+ leakage from tissues may be considered as a useful strategy to screen salt-tolerant walnut cultivars.

Further, in a combined approach of ionomics and metabolomics on the three barley genotypes differing in their salt tolerance behavior, (Wu et al., 2013) proposed that restriction of Na^+ accumulation and relocation of other nutrient element levels with an effect of that along with the metabolite production in plant tissues under saline environment are attributable factors determining enhanced salt tolerance. As per the findings, the possible ionomics-governed mechanisms at the germination stage of plant tissues are (1) hindrance in Na^+ accumulation, (2) protecting the Cu, P, S, Mg, and Zn contents in roots, and (3) readjustment of K^+ and S contents, and maintenance of Cu, P, Mn, Zn, and Fe contents in leaves. However, these aforementioned functions operated under salinity and further contributed to salt tolerance at the seedling stage.

11.8 Ameliorative role of sodium and chloride on stress tolerance

Seeds provide a platform for conserved transmission of the genetic particulars, connecting generations. Maintaining seed viability and vigor is therefore a primary concern to support crop production and yield. Halopriming is a pre-germination process to fortify seeds, improving their viability, maintaining seed vigor, and enhancing resistance against several biotic or abiotic stresses. The germination course of seeds comprises several physiological and biochemical changes that may be broadly distributed in three phases (Lutts et al., 2016). Seed invigoration techniques primarily target the first two reversible phases of germination. Priming involves hydration of seeds with solutions having low water potentials, permitted to soak for a fixed period, followed by immediate dehydration, thus preventing seeds to enter the final phase of seed establishment (Manonmani et al., 2014). To set the seeds off for germination, passive imbibition of water driven by matric force of the cell wall polysaccharides and hydrophilic proteins present on the seed coat is mandatorily the first phase. After a 60% rise in water content, the level of soluble sugars and other osmotica rises, pulling in around 74% water-promoting seed to enter the second phase of germination (Obroucheva et al., 2017). The second phase appears to be a "lag phase" since water uptake ceases until the third phase of radical protrusion, but a lot of biochemical activations and molecular changes take place at this time (Lutts et al., 2016). Such a line of events, most importantly the activation of the second phase of germination, induces activation of several hydrolytic enzymes to mobilize storage molecules, hydrolysis of various inhibitory compounds, parallel to the biogenesis of several subcellular organelles, induction of enzymes driving glycolytic, pentose phosphate, and tricarboxylic acid pathways and association of several enzyme subunits with the cristae triggers the initiation of mitochondrial respiration and amino acid metabolism. Several genes including those related to stress tolerance are activated during this phase. DNA repair mechanisms are turned on to repair the damaged DNA of seeds (Ali and Elozeiri, 2017). On completion of this phase, seeds are dehydrated back to their initial moisture content but these biochemical and physiological activations remain irreversible. Thus priming not only brings all the seeds to a stage where several biochemical processes are uniformly activated but also sets a synchronization laid before they are allowed to regerminate, this time letting the third phase of cell elongation and radical formation occur (Kubala et al., 2015).

11.8.1 Priming as an ameliorative technique for drought and salt stress

Depending on the purpose of priming, which may include simple increase in seed vigor or to make seeds more tolerant to particular abiotic stress, one may choose from a variety of mediums in which to soak seeds. On this basis, priming may be further divided (Lutts et al., 2016) as discussed in the following sections.

11.8.1.1 Hydropriming

In hydropriming, water is used as the priming agent without the use of any additional chemical substances. Because hydropriming is governed by the affinity of tissue toward water, it leads to non-uniform water absorption and therefore non-synchronized activation of metabolic processes.

11.8.1.2 Halopriming or osmopriming

The use of chemicals like PEG, NaCl, KCl, KNO_3, KH_2PO_4, etc. reduces the water potential enabling slow and controlled absorption, thus overcoming the disadvantage of hydropriming. Salt, temperature, and water stress can largely be overcome by such treatments.

11.8.1.3 Hormopriming

Growth regulators like gibberellins, abscisic acid, ethylenes, and auxins are used for seed imbibitions.

11.8.1.4 Biopriming

Used to fight various biotic stresses, where seeds are soaked in medium containing antagonistic bacterial or other pathogen inoculums, these beneficial microorganisms or bioactive compounds help plants to stand strong against these pathogens. In some cases, the endophytic association of plant-specific bacteria/fungus may help in establishing a strong resistance to abiotic stresses.

11.8.2 Halopriming and growth enhancement of seedlings

Halopriming improves the growth and vigor of various cereals like rice, wheat, etc., as well as other economically important crops like pulses, tomatoes, linola, etc. Halopriming was found to be more advantageous than hydropriming in *T. aestivum* (Arif et al., 2019) and *Oryza sativa* (Jisha and Puthur, 2014), evident right from the seedling stage. Enhancement of early seedling growth is measured in terms of time taken for radicle emergence, germination rate, germination percentage, root and shoot length, seedling dry and fresh weight, etc. Halopriming with 1.25 MPa $CaCl_2$ improved germination rate, narrowed down the mean germination time, and increased the rate of seedling emergence by around 1.5-fold in linola compared to its non-haloprimed control seedlings (Rehman et al., 2014). Similarly, halopriming of *Solanum lycopersicum* cv. GT-2 seeds with 2% KNO_3, increased its germination index to great extents; final germination percentage increased by 89% along with fresh weight and dry weight (approximately by 1.5 g) of the seedlings over unprimed control seedlings (Vaktabhai and Kumar, 2017). Priming seeds with a predetermined concentration of salts like NaCl, KCl, KNO_3, etc. imparted tolerance to a range of abiotic stresses. Halopriming of seeds brings about positive changes in metabolism, which alleviates salt (NaCl)-induced growth hindrances compared to their respective non-primed control seedlings. Along with seedling length, fresh weight, and dry weight, seedling moisture content increased considerably in wheat when grown in the presence of excess sodium chloride (Gholami et al., 2015). In two NaCl-stressed aromatic cultivars of *O. sativa*, Shaheen Basmati and Basmati 2000, the mean germination time reduced, final seedling emergence percentage improved, and root and shoot length and seedling dry mass increased on halopriming seeds with $CaCl_2$ (2.2%) and KCl (2.2%) for 36 h, proving the pre-sowing treatment to be beneficial in alleviating salt-induced growth impairment. A similar ameliorative effect was observed when different concentrations of NaCl for 12 h primed rice cultivars like Neeraja (salt-sensitive), Vytilla 6 (salt-tolerant), and Vaisakh (drought-tolerant) were exposed to different concentrations of NaCl and PEG (Jisha and Puthur, 2014). Additionally, IR 64 (salt-sensitive) and Banyuasin (salt-tolerant) seeds, primed with 150 mM NaCl for 36 and 39 h respectively, showed similar improvement in seedling growth on salt exposure (Panga et al., 2019). It may be mentioned here that priming induces several genes for activating starch mobilization enzymes in the seeds, which increases sugar content thereby increasing the seeds' emergence ability, reducing the mean germination time (Rehman et al., 2014). Seeds exposed to drought stress or salt stress are subjected to low water potential; priming seeds increased water uptake ability and enhanced turgidity of cells, enhancing seedling emergence by promoting cell enlargement followed by cell division (Jisha and Puthur, 2014).

11.9 Conclusions

Sodium chloride (NaCl) is reported to play diverse roles in different genotypes for deciding the fate of the plant under saline and other challenging environments. Generally, in glycophytes, excessive accumulation of sodium (Na^+) and chloride (Cl^-) is toxic for the plant and a myriad of evidence is present where it results in disaster to plant growth. In glycophytes, sodium chloride is supposed to be a notorious compound that is involved in cellular damage and metabolic impairments in plants, interacting with other essential cations and anions, and creating an ionic imbalance in the cell. It holistically dampens the accumulation of other essential ions (K^+ and Ca^{2+}), which are necessary for maintaining cellular homeostasis and function. Tolerant members of glycophytes adopted to different mechanisms to withstand multiple abiotic stresses. Conversely, in halophytes and xerophytes, the Na^+ and Cl^- ions play a beneficial role and help to maintain cellular homeostasis, which ultimately supports in sustaining growth and survival under extreme environments. Usually, in halophytes and xerophytes, NaCl acts as an osmoticum and further helps to improve the overall performance of the plant under salinity and drought stress. Therefore based on the previous knowledge and reports, it is evident that a unique contribution of these two ions present under salinity and drought stress, which differs in glycophytes and halophytes. In this chapter, we tried to summarize the cumulative roles and contribution of sodium (Na^+) and chloride (Cl^-) ions for deciding the fate of the plant. We also showed the nature of damage or amelioration by a neutralized salt (NaCl) is not universal for all and plants' fate to salt and drought stress is decided by the overall genetic constitution and adaptability of the plant to sodium (Na^+) and chloride (Cl^-) ions.

References

Accardi, A., Kolmakova-Partensky, L., Williams, C., 2015. Structure and gating of CLC channels and exchangers. J. Physiol. 593, 4129–4138.

Adams, E., Shin, R., 2014. Transport, signaling, and homeostasis of potassium and sodium inplants. J. Integr. Plant Biol 56, 231–249.

Adebooye, O.C., Schmitz-Eiberger, M., Lankes, C., Noga, G.J., 2010. Inhibitory effects of suboptimal root zone temperature on leafbioactive components, photosystem II (PS II) and minerals uptake in *Trichosanthes cucumerina* L. Cucurbitaceae. Acta Physiol. Plant 32, 67–73.

Alemán, F., Nieves-Cordones, M., Martínez, V., Rubio, F., 2011. Root K^+ acquisition in plants: the *Arabidopsis thaliana* model. Plant Cell Physiol. 52, 1603–1612.

Ali, S.A., Elozeiri, A.A. 2017. Metabolic Processes During Seed Germination. Intechopen. doi: 10.5772/intechopen.70653.

Allen, G.J., Wyn-Jones, R.G., Leigh, R.A., 1995. Sodium transport in plasma membrane vesicles isolated from wheat genotypes with differing K^+/Na^+ discrimination traits. Plant Cell Environ. 18 (1), 105–115.

Apse, M.P., Blumwald, E., 2007. Na^+ transport in plants. FEBS Lett. 581, 2247–2254.

Arif, Z., Ashger, A., Zia, A., Riaz, M., 2019. Effect of halopriming along with hydropriming on germination of wheat (*Triticum aestivum* L.) seeds. Asian J. Res. Crop Sci. 3 (1), 1–7.

Assaha, D.V.M., Ueda, A., Saneoka, H., Al-Yahyai, R., Yaish, M.W., 2017. The role of Na^+ and K^+ transporters in salt stress adaptation in glycophytes. Front. Physiol. 8, 509.

Baetz, U., Eisenach, C., Tohge, T., Martinoia, E., De Angeli, A., 2016. Vacuolar chloride fluxes impact ion content and distribution during early salinity stress. Plant Physiol. 172, 1167–1181.

Balague, C., Lin, B., Alcon, C., Flottes, G., Malmstrom, S., Kohler, C., et al., 2003. HLM1, an essential signalling component in the hypersensitive response, is a member of the cyclic nucleotide gated channel ion channel family. Plant Cell 15, 365–379.

Barbier-Brygoo, H., De Angeli, A., Filleur, S., Frachisse, J.M., Gambale, F., Thomine, S., 2011. Anion channels/transporters in plants: from molecular bases to regulatory networks. Annu. Rev. Plant Biol. 62, 25–51.

Bassil, E., Coku, A., Blumwald, E., 2012. Cellular ion homeostasis: emerging roles of intracellular NHX Na^+/H^+ antiporters in plant growth and development. J. Exp. Bot. 63, 5727–5740.

Baxter, I., Hermans, C., Lahner, B., Yakubova, E., Tikhonova, M., Verbruggen, N., et al., 2012. Biodiversity of mineral nutrient and trace element accumulation in *Arabidopsis thaliana*. PLoS One 7, e35121.

BaxterDilkes, I.B.P., 2012. Elemental Profiles Reflect Plant Adaptations to the Environment. Science 336 (6089), 1661–1663.

Bhaduri, D., Meena, H.N., Chakraborty, K., 2016. Variation in phosphorus accumulation in groundnut cultivars as influenced by water salinity. Legume Res. Int. J. 39 (2), 215–220.

Blumwald, E., 2000. Sodium transport and salt tolerance in plants. Curr. Opin. Cell Biol. 12, 431–434.

Brandt, B., Brodsky, D.E., Xue, S., Negi, J., Iba, K., Kangasjarvi, J., et al., 2012. Reconstitution of abscisic acid activation of SLAC1 anion channel by CPK6 and OST1 kinases and branched ABI1 PP2C phosphatase action. Proc. Natl. Acad. Sci. U S A. 109, 10593–10598.

Brown, C.E., Pezheski, S.R., Delaune, R.D., 2006. He effects of salinity and soil drying on nutrient uptake and growth of *Spartina alterniflora* in a simulated tidal system. Environ. Exp. Bot. 58, 140–148.

Brumos, J., Colmenero-Flores, J.M., Conesa, A., Izquierdo, P., Sanchez, G., Iglesias, D.J., et al., 2009. Membrane transporters and carbon metabolism implicated in chloride homeostasis differentiate salt stress responses in tolerant and sensitive Citrus rootstocks. Funct. Integr. Genomics 9, 293–309.

Buschmann, P.H., Vaidyanathan, R., Gassmann, W., Schroeder, J.I., 2000. Enhancement of Na^+ uptake currents, time-dependent inward-rectifying K+ channel currents, and K^+ channel transcripts by K^+starvation in wheat root cells. Plant Physiol 122, 1387–1398.

Byrt, C.S., Xu, B., Krishnan, M., Lightfoot, D.J., Athman, A., Jacobs, A.K., et al., 2014. The Na^+ transporter, TaHKT 1;5-D, limits shoot Na^+ accumulation in bread wheat. Plant J. 80 (3), 516–526.

Byrt, C.S., Zhao, M., Kourghi, M., Bose, J., Henderson, S.W., Qiu, J., et al., 2017. Non-selective cation channel activity of aquaporin AtPIP2;1 regulated by Ca^{2+} and pH. Plant Cell Environ. 40, 802–815.

Cabot, C., Sibole, J.V., Barceló, J., Poschenrieder, C., 2014. Lessons from crop plants struggling with salinity. Plant Sci. 226, 2–13.

Carillo, P., Grazia, M.A., Pontecorvo, G., et al., 2011. Salinity stress and salt tolerance. In: Shanker, A.K., Venkateswarlu, B. (Eds.), Abiotic Stress in Plants—Mechanisms and Adaptations. InTech, Rijeka, pp. 21–38.

Chakraborty, K., Sairam, R.K., Bhattacharya, R.C., 2012. Differential expression of salt overly sensitive pathway genes determines salinity stress tolerance in Brassica genotypes. Plant Physiol. Biochem. 51, 90–101.

Chakraborty, K., Bhaduri, D., Meena, H.N., Kalariya, K., 2016a. External potassium (K^+) application improves salinity tolerance by promoting Na^+-exclusion, K^+-accumulation and osmotic adjustment in contrasting peanut cultivars. Plant Physiol. Biochem. 103, 143–153.

Chakraborty, K., Bose, J., Shabala, L., Eylis, A., Shabala, S., 2016b. Evaluating contribution of osmo- and tissue- tolerance mechanism towards salinity stress tolerance in three Brassica species. Physiol. Plant 158, 135–151.

Chakraborty, K., Sairam, R.K., Bhaduri, D., 2016c. Effects of different levels of soil salinity on yield attributes, accumulation of nitrogen, and micronutrients in *Brassica* spp. J. Plant Nutr. 39 (7), 1026–1037.

Chakraborty, K., Bishi, S.K., Goswami, N., Singh, A.L., Zala, P.V., 2016d. Differential fine-regulation of enzyme driven ROS detoxification network imparts salt tolerance in contrasting peanut genotypes. Environ. Exp. Bot. 128, 79–90.

Chakraborty, K., Bose, J., Shabala, L., Shabala, S., 2016e. Difference in root K^+ retention ability and reduced sensitivity of K^+-permeable channels to reactive oxygen species confer differential salt tolerance in three *Brassica* species. J. Exp. Bot. 67, 4611–4625.

Chakraborty, K., Basak, N., Bhaduri, D., Ray, S., Vijayan, J., Chattopadhyay, K., et al., 2018. Ionic basis of salt tolerance in plants. In: Hasanuzzaman, M. (Ed.), Plant Nutrients and Abiotic Stress Tolerance. Springer Nature Singapore, pp. 325–362.

Chakraborty, K., Chattaopadhyay, K., Nayak, L., Ray, S., Yeasmin, L., Jena, P., et al., 2019. Ionic selectivity and coordinated transport of Na^+ and K^+ in flag leaves render differential salt tolerance in rice at the reproductive stage. Planta 250 (5), 1637–1653.

Chakraborty, K., Mondal, S., Ray, S., Samal, P., Pradhan, B., Chattopadhyay, K., et al., 2020. Tissue tolerance coupled with ionic discrimination can potentially minimize the energy cost of salinity tolerance in rice. Front. Plant Sci. 11, 265.

Chen, P., Yan, K., Shao, H., Zhao, S., 2013. Physiological mechanisms for high salt tolerance in wild soybean (*Glycine soja*) from Yellow River Delta, China: photosynthesis, osmotic regulation, ion flux and antioxidant capacity. PloS One 8, 1–12.

Chen, Z.C., Yamaji, N., Kashino-Fujii, M., Ma, J.F., 2016. A cation-chloride cotransporter gene is required for cell elongation and osmoregulation in rice. Plant Physiol 171, 494–507.

Churchill, K.A., Sze, H., 1984. Anion-sensitive, H^+ pumping ATPase of oat roots: direct effects of Cl, NO(3), and a disulfonic stilbene. Plant Physiol. 76, 490–497.

Colmenero-Flores, J.M., Martinez, G., Gamba, G., Vazquez, N., Iglesias, D.J., Brumos, J., et al., 2007. Identification and functional characterization of cation-chloride cotransporters in plants. Plant J. 50, 278–292.

Craig Plett, D., Moller, I.S., 2010. Na^+ transport in glycophytic plants: what we know and would like to know. Plant Cell Environ. 33, 612–626.

Cuin, T., Miller, A., Anthony, J., Laurie Sophie, A., Leigh Roger, A., 2003. Potassium activities in cell compartments of salt-grown barley leaves. J. Exp. Bot. 54, 657–661.

Cubero-Font, P., Maierhofer, T., Jaslan, J., 2016. Silent S-type anion channel subunit SLAH1 gates SLAH3 open for chloride root-to-shoot translocation. Current Biology 26, 2213–2220.

Cuin, T.A., Bose, J., Stefano, G., Jha, D., Tester, M., Mancuso, S., et al., 2011. Assessing the role of root plasma membrane and tonoplast Na^+/H^+ exchangers in salinity tolerance in wheat: in planta quantification methods. Plant Cell Environ. 34, 947–961.

Dahiya, S.S., Singh, M., 1976. Effect of salinity, alkalinity and iron application on the availability of iron, manganese, phosphorus and sodium in pea (*Pisum sativum* L.) crop. Plant Soil 44, 697–702.

De Angeli, A., Thomine, S., Franchisse, J., 2016. Anion channel blockage by ATP as a mean for membrane to perceive the energy status of the cell. Mol. Plant 9, 320–322.

Diedhiou, C., Golldack, D., 2006. Salt dependent regulation of chloride channel transcripts in rice. Plant Sci. 170, 793–800.

Diego, H.S., Henning, R., Ute, K., Michael, K.U., Joachim, K., 2008. Metabolome–ionome–biomass interactions: what can we learnabout salt stress by multiparallel phenotyping. Plant Signal Behav. 3 (8), 598–600.

Du, Y., Zhao, M., Wang, C., et al., 2018. Identification and characterization of *GmMYB118* responses to drought and salt stress. BMC Plant Biol. 18, 320.

El-Fouly, M.M., Mobarak, Z.M., Salama, Z.A., 2002. Micronutrient foliar application increases salt tolerance of tomato seedlings. Acta Hortic. 573, 467–474.

Faber, B., 2015. Salt, sodium and potassium and drought. Topics in subtropics.

Farooq, M., Basra, S.M.A., Tabassum, R., Afzal, I., 2006. Enhancing the performance of direct seeded fine rice by seed priming. Plant Prod. Sci. 9 (4), 446–456.

Flower, T.J., Colmer, T.D., 2008. Salinity tolerance in halophytes. New Phytol. 179, 945–963.

Francois, L.E., Maas, E.V., 1999. Crop response and management of salt-affected soils. In: Pessarakli, M. (Ed.), Handbook of Plant and Crop Sciences. Marcel-Dekker Inc, New York, pp. 169–202.

Franco-Navarro, J.D., Brumós, J., Rosales, M.A., Cubero-Font, P., Talón, M., Colmenero-Flores, J.M., 2016. Chloride regulates leaf cell size and water relations in tobacco plants. J. Exp. Bot. 67, 873–891.

Fukuda, A., Nakamura, A., Tagiri, A., et al., 2004. Function, intracellular localization and the importance in salt tolerance of a vacuolar $Na(^+)/H(^+)$ antiporter from rice. Plant Cell Physiol. 45 (2), 146–159.

Geiger, D., Scherzer, S., Mumm, P., Stange, A., Marten, I., Bauer, H., et al., 2009. Activity of guard cell anion channel SLAC1 is controlled by drought-stress signaling kinase-phosphatase pair. Proc. Natl. Acad. Sci. U S A. 106, 21425–21430.

Ghandilyan, A., Ilk, N., Hanhart, C., et al., 2009. A strong effect of growthmedium and organ type on the identification of QTLs for phytateand mineral concentrations in three *Arabidopsis thaliana* RIL populations. J. Exp. Bot. 60 (5), 1409–1425.

Gholami, M., Mokhtarian, F., Baninasab, B., 2015. Seed halopriming improves the germination performance od black seeds (*Nigella sativa*) under salinity stress conditions. J. Crop Sci. Biotechnol. 18 (1), 21–26.

Glenn, E.P., Brown, J.J., Blumwald, E., 1999. Salt tolerance and crop potential of halophytes. Crit. Rev. Plant Sci. 18 (2), 227–255.

Gobert, A., Park, G., Amtmann, A., Sanders, D., Maathuis, F.J., 2006. *Arabidopsis thaliana* cyclic nucleotide gated channel 3 forms a non-selective ion transporter involved in germination and cation transport. J. Exp. Bot. 57 (4), 791–800.

Gogoi, P.P., Vinoj, V., Swain, D., et al., 2019. Land use and land cover change effect on surface temperature over Eastern India. Sci Rep 9, 8859.

Golldack, D., Li, C., Mohan, H., Probst, N., 2014. Tolerance to drought and salt stress in plants: unraveling the signaling networks. Front. Plant Sci. 5, 151.

Grattan, S.R., Grieve, C.M., 1999. Salinity-mineral nutrient relations in horticultural crops. Sci. Hortic 78, 127–157.

Gruber, B.D., Ryan, P.R., Richardson, A.E., 2010. HvALMT1 from barley is involved in the transport of organic anions. J. Expt. Bot 61, 1455–1467.

Guan, Z., Chen, S., Chen, F. et al. 2012. Comparison of stress effect of NaCl, Na^+ and Cl^-.

Guan, H., Liu, X., Niu, F., Zhao, Q., Fan, N., Cao, D., et al. 2019. OoNAC72, a NAC-Type Oxytropis ochrocephala transcription factor, conferring enhanced drought and salt stress tolerance in Arabidopsis. Front. Plant Sci. 10, salt

Guo, K.M., Babourina, O., Christopher, D.A., Borsics, T., Rengel, Z., 2008. The cyclic nucleotide gated channel, AtCNGC10, influences salt tolerance in Arabidopsis. Physiol. Plant 134, 499–507.

Hadi, M.R., Karimi, N., 2012. The role of calcium in salt tolerance. J. Plant Nutr. 35 (13), 2037–2054.

Hadi, M.R., Karimi, N., 2009. THE ROLE OF CALCIUM IN PLANTS. SALT TOLERANCE. journal of plant nutrition. 35 (13), 2037–2054.

Hajibagheri, M.A., Hall, J.L., Flowers, T.J., 1984. Stereological analysis of leaf cells of the halophyte *Suaeda maritima* (L.) dum,. J. Exp. Bot. 35 (10), 1547–1557.

Hall, D., Evans, A.R., Newbury, H.J., Pritchard, J., 2006. Functional analysis of CHX21: a putative sodium transporter in Arabidopsis. J. Exp. Bot. 57 (5), 1201–1210.

Hamilton, E.S., Jensen, G.S., Maksaev, G., Katims, A., Sherp, A.M., Haswell, E.S., 2015. Mechanosensitive channel MSL8 regulates osmotic forces during pollen hydration and germination. Science 350 (80), 438–441.

Haro, R., Fraile Escanciano, A., Gonzalez Melendi, P., Rodriguez Navarro, A., 2013. The potassium transporters HAK2 and HAK3 localize to endomembranes in *Physcomitrella patens*. HAK2 is required in some stress conditions. Plant Cell Physiol. 54, 1441–1454.

Hasanuzzaman, M., Nahar, K., Alam, M., Bhowmik, P.C., Hossain, M.M., et al., 2014. Potential use of halophytes to remediate saline soils. Biomed. Res. Int. Available from: https://doi.org/10.1155/2014/589341.

Hasegawa, M., Bressan, R., Pardo, J.M., 2000. The dawn of plant salt tolerance genetics. Trends Plant Sci. 5, 317–319.

Hasegawa, P.M., 2013. Sodium (Na^+) homeostasis and salt tolerance of plants. Environ Exp Bot 92, 19–31.

Hassan, N.A., Drew, J.V., Knudsen, D., Olsen, R.A., 1970. Influence of soil salinity on production of dry matter and uptake and distribution of nutrients in barley and corn: I. barley (*Hordeum vulgare* L.). Agron J. 62, 43–45.

Haswell, E.S., 2007. MscS-like proteins in plants. In: Hamil, D.P. (Ed.), Mechanosensitive Ion Channels, Part A. Elsevier Academic Press Inc, San Diego, CA, pp. 329–359.

Haswell, E.S., Meyerowitz, E.M., 2006. MscS-like proteins control plastid size and shape in *Arabidopsis thaliana*. Curr. Biol. 16, 1–11.

Haswell, E.S., Peyronnet, R., Barbier-Brygoo, H., Meyerowitz, E.M., Frachisse, J.M., 2008. Two MscS homologs provide mechanosensitive channel activities in the Arabidopsis root. Curr. Biol. 18, 730–734.

Henderson, S.W., Wege, S., Qiu, J., Blackmore, D.H., Walker, A.R., Tyerman, S.D., et al., 2015. Grapevine and Arabidopsis cation-chloride cotransporters localize to the Golgi and trans-Golgi network and indirectly influence long-distance ion transport and plant salt tolerance. Plant Physiol. 169, 2215–2229.

He, F.L., Bao, A.K., Wang, S.M., XiJin, H., 2019. NaCl stimulates growth and alleviates drought stress in the salt-secreting xerophyte *Reaumuria soongorica*. Environ. Exp. Bot. 162, 433–443.

Horie, T., Costa, A., Kim, T.H., Han, M.J., Horie, R., Leung, H.Y., et al., 2007. Rice *OsHKT2;1* transporter mediates large Na^+ influx component into K^+-starved roots for growth. EMBO J. 26, 3003–3014.

Horie, T., Horie, R., Chan, W-Y., Leung, H-Y., Schroder, J.I., 2006. Calcium Regulation of Sodium Hypersensitivities of sos3 and athktl Mutants. Plant and cell physiology 47 (5), 622–633.

Hossain, S., 2019. Present scenario of global salt affected soils, its management and importance in salinity research. Int. Res. J. Biol. Sci. 1 (1), 13.

Hosseini, S.A., Hajirezaei, M.R., Seiler, C., Sreenivasulu, N., von Wirén, N., 2016. A Potential Role of Flag Leaf Potassium in Conferring Tolerance to Drought-Induced Leaf Senescence in Barley. Frontiers in plant science 7, 206. Available from: https://doi.org/10.3389/fpls.2016.00206.

Hu, Y., Schmidhalter, U., 2005. Drought and salinity: a comparison of their effects on mineral nutrition of plants. J. Plant Nutr. Soil Sci. 168, 541–549.

Hu, R., Zhu, Y., Wei, J., Huazhong, S., Guoxin, S., Hong, J., 2017. Overexpression of *PP2A-C5* that encodes the catalytic subunit 5 of protein phosphatase 2A in *Arabidopsis* confers better root and shoot development under salt conditions. Plant Cell Environ. 40 (1), 150–164.

Hua, B.G., Mercier, R.W., Leng, Q., Berkowitz, G.A., 2003. A new basis for potassium/sodium selectivity in the pore of an ion channel. Plant Physiol. 132, 1353–1361.

Huang, C.X., Van Steveninck, M.E., 1989. Maintenance of low Cl^- concentrations in mesophyll cells of leaf blades of barley seedlings exposed to salt stress. Plant Physiol. 90, 1440–1443.

Hussin, S., Geissler, N., Koyro, H.W., 2013. Effect of NaCl salinity on *Atriplex nummularia* (L.) with special emphasis on carbon and nitrogen metabolism. Acta Physiol. Plant 35, 1025–1038.

Imes, D., Mumm, P., Böhm, J., Al-Rasheid, K.A.S., Marten, I., Geiger, D., et al., 2013. Open stomata 1 (OST1) kinase controls R-type anion channel QUAC1 in Arabidopsis guard cells. Plant J. 74, 372–382.

Iqbal, M, Ashraf, A., Jamil, A., Rehman, S.U., 2006. Does Seed priming induce changes in the levels of some endogenous plant hormones in hexaploid wheat plants under salt stress. J. Integr. Plant Biol. 48, 181–189.

James, R.A., Davenport, R.J., Munns, R., 2006. Physiological characterization of two genes for Na + exclusion in durum wheat, Nax1 and Nax2. Plant physiology 142 (4), 1537–1547. Available from: https://doi.org/10.1104/pp.106.086538.

Jeschke, W., Klagges, S., Hilpert, A., Bhatti, A., Sarwar, G., 1995. Partitioning and flows of ions and nutrients in salt-treated plants of *Leptochloa fusca* L. Kunth. I. Cations and Chloride. New Phytol. 130, 23–35.

Jin, Z.M., Wang, C.H., Liu, Z.P., Gong, W.J., 2007. Physiological and ecological characters studies on *Aloe vera* under soil salinity and seawater irrigation. Process Biochem. 42, 710–714.

Jisha, K.C., Puthur, J.T., 2014. Seed halopriming outdo hydropriming in enhancing seedling vgor and osmotic stress tolerance potential of rice varieties. J. Crop Sci. Biotech. 17 (4), 209–219.

Kawakami, K., Umena, Y., Kamiya, N., Shen, J.R., 2009. Location of chloride and its possible functions in oxygen-evolving photosystem II revealed by X-ray crystallography. Proc. Natl. Acad. Sci. USA 106, 8567–8572.

Keisham, M., Mukherjee, S., Bhatla, S.C., 2018. Mechanisms of sodium transport in plants-progresses and challenges. Int. J. Mol. Sci. 19 (3), 647.

Khare, T., Srivastava, A.K., Suprasanna, P., Kumar, V., 2020. Individual and additive stress Impacts of Na + and Cl^- on proline metabolism and nitrosative responses in rice. Plant Physiol. Biochem. 152, 44–52.

Knight, H., Trewavas, A.J., Knight, M.R., 1997. Calcium signalling in *Arabidopsis thaliana* responding to drought and salinity. Plant J. 12, 1067–1078.

Kobayashi, D., Uozumi, N., Hisamatsu, S., Yamagami, M., 2010. AtKUP/HAK/ KT9, a K^+ transporter from *Arabidopsis thaliana*, mediates Cs^+ uptake in *Escherichia coli*. Biosci. Biotech. Biochem. 74, 203–205.

Kronzucker, H.,J., Britto, D.T., 2011. Sodium transport in plants: a critical review. New Phytol. 189, 54–81.

Kubala, S., Garnczarska, M., Wojtyla, L., Clippe, A., Kosmala, A., Zmieńko, A., et al., 2015. Deciphering priming-induced improvement of rapeseed (*Brassica napus* L.) germination through an integrated transcriptomic and proteomic approach. Plant Sci. 231, 94–113.

Kumar, V., Khare, T., 2015. Individual and additive effects of Na^+ and Cl^- ions on rice under salinity stress. Arch. Agron. Soil Sci. 61 (3), 381–395.

Kumar, K., Kumar, M., Kim, S., 2013. Insights into genomics of salt stress response in rice. Rice 6, 27.

Kurusu, T., Saito, K., Horikoshi, S., Hanamata, S., Negi, J., Yagi, C., et al., 2013. An S-type anion channel SLAC1 is involved in cryptogein-induced ion fluxes and modulates hypersensitive responses in tobacco BY-2 cells. PloS One 8, e70623.

Lee, M., Choi, Y., Burla, B., Kim, Y.Y., Jeon, B., Maeshima, M., et al., 2008. The ABC transporter AtABCB14 is a malate importer and modulates stomatal response to CO_2. Nat. Cell Biol. 10, 1217–1223.

Li, B., Byrt, C., Qiu, J., Baumann, U., et al., 2016. Identification of a stelar-localized transport protein that facilitates root-to-shoot transfer of chloride in arabidopsis. Plant Physiol. 170 (2), 1014–1029.

Li, B., Qiu, J., Jayakannan, M., Xu, B., Li, Y., Mayo, G.M., et al., 2017a. *AtNPF2.5* modulates chloride (Cl^-) efflux from roots of *Arabidopsis thaliana*. Front. Plant Sci. 7, 2013.

Li, B., Tester, M., Gilliham, M., 2017b. Chloride on the move. Trends Plant Sci. 22, 236–248.

Liu, X., Mak, M., Babla, M., Wang, F., Chen, G., Veljanoski, F., et al., 2014. Linking stomatal traits and expression of slow anion channel genes *HvSLAH1* and *HvSLAC1* with grain yield for increasing salinity tolerance in barley. Front. Plant Sci. 5, 1–12.

Liu, J., Zhu, J.K., 1997. An *Arabidopsis* mutant that requires increased calcium for potassium nutrition and salt tolerance. Proc. Natl. Acad. Sci.USA. 94, 14960–14964.

Lotfi, N., Vahdati, K., Kholdebarin, B., Hassani, D., Amiri, R., 2009. Mineral composition of some walnut cultivars (*Juglans regia* L.) for evaluation of ionome and ionomics under salt stress condition. Acta Hortic. 839, 293–300.

Lutts, S., Benincasa, P., Wojtyla, L., Kubala, S.S., Pace, R., Lechowska, K., et al. 2016. Seed Priming: New Comprehensive Approaches for an Old Empirical Technique. IntechOpen. doi: 10.5772/64420.

Lynch, J., Läuchli, A., 1988. Salinity affects intracellular calcium in root protoplasts. Plant Physiol. 87, 351–356.

Maathuis, F.J.M., 2014. Sodium in plants: perception, signalling, and regulation of sodium fluxes. J. Exp. Bot. 65 (3), 849–858.

Mahouachi, J., Socorro, A.R., Talon, M., 2006. Responses of papaya seedlings (*Caricapapaya* L.) to water stress and re-hydration: growth, photosynthesis and mineral nutrient imbalance. Plant Soil 281, 137–146.

Manonmani, V., Ameer Junaithal Begum, M., Jayanthi, M., 2014. Halo priming of seeds. Res. J. Seed Sci. 7 (1), 1–13.

Marschner, P., 2011. Marschner's Mineral Nutrition of Higher Plants, third ed. Academic Press, London.

Martinez-Antienza, J., Jhiang, X., Garciadebles, B., Mendoza, I., Zhu, J.-K.Pardo, 2007. Conservation of salt overly sensitive pathway in rice. Plant Physiol. 143, 1001–1012.

Mekawy, A.M.M., Assaha, D.V., Yahagi, H., Tada, Y., Ueda, A., Saneoka, H., 2015. Growth, physiological adaptation, and gene expression analysis of two Egyptian rice cultivars under salt stress. Plant Physiol. Biochem. 87, 17–25.

Meyer, S., Mumm, P., Imes, D., Endler, A., Weder, B., Al-Rasheid, K.A.S., et al., 2010. AtALMT12 represents an R-type anion channel required for stomatal movement in Arabidopsis guard cells. Plant J. 63, 1054–1062.

Mondal, S., Chakraborty, K., 2020. Brassicaceae plant response and tolerance to salinity. In: Hasanuzzaman, M. (Ed.), The Plant Family Brassicaceae. Spiringer Nature Singapore, pp. 203–228.

Munns, R., 2002. Comparative physiology of salt and water stress. Plant cell and. Environment 25 (2), 239–250.

Munns, R., James, R.A., Xu, B., Athman, A., Conn, S.J., Jordans, C., et al., 2012. Wheat grain yield on saline soils is improved by an ancestral Na^+ transporter gene. Nat. Biotechnol. 30, 360–364.

Munns, R., Tester, M., 2008. Mechanisms of salinity tolerance. Annu. Rev. Plant Biol. 59, 651–681.

Murillo-Amado, B., Jones, H.G., Kaya, C., Aguilar, R.L., Garcia-Hernandez, J.L., Troyo-Di̇́eguez, E., et al., 2006. Effects of foliar application of calcium nitrate on growth and physiological attributes of cowpea (*Vigna unguiculata* L. Walp.) grown under salt stress. Environ. Exp. Bot. 58, 188–196.

Mustilli, A.C., Merlot, S., Vavasseur, A., Fenzi, F., Giraudat, J., 2002. Arabidopsis OST1 protein kinase mediates the regulation of stomatal aperture by abscisic acid and acts upstream of reactive oxygen species production. Plant Cell 14, 3089–3099.

Negi, J., Matusda, O., Nagasawa, T., Oba, Y., Takahashi, H., Kawai-Yamada, M., et al., 2008. CO_2 regulator SLAC1 and its homologues are essential for anion homeostasis in plant cells. Nature 452, 483–488.

Nguyen, C.T., Agorio, A., Jossier, M., et al., 2016. Characterization of the chloride channel-like, AtCLCg, involved in chloride tolerance in *Arabidopsis thaliana*. Plant Cell Physiol. 57, 764–775.

Nutan, K.K., Singla-Pareek, S.L., Pareek, A., 2020. The Saltol QTL-localized transcription factor *OsGATA8* plays an important role in stress tolerance and seed development in Arabidopsis and rice. J. Exp. Bot. 71 (2), 684–698.

Obroucheva, N.V., Sinkevich, I.A., Lityagina, S.V., Novikova, T.G.V., 2017. Water relations in germinative seeds. Russ. J. Plant Physiol. 64, 625–633.

Oh, D.H., Leidi, E., Zhang, Q., Hwang, S.M., Li, Y., Quintero, F.J., et al., 2009. Loss of halophytism by interference with SOS1 expression. Plant Physiol. 151, 210–222. on two Chrysanthemum species. Acta Hortic. 937, 369–375.

Olías, R., Eljakaoui, Z., Pardo, J.M., Belver, A., 2009. The Na(+)/H(+) exchanger SOS1 controls extrusion and distribution of Na(+) in tomato plants under salinity conditions. Plant signaling & behavior 4 (10), 973–976. Available from: https://doi.org/10.4161/psb.4.10.9679.

Orellana, S., Yanez, M., Espinoza, A., Verdugo, I., et al., 2010. The transcription factor SlAREB1 confers drought, salt stress tolerance and regulates biotic and abiotic stress-related genes in tomato. Plant Cell Environ. 33, 2191–2206.

Panga, N.J., Sjahril, R., Kaimuddin, 2019. Seed priming effect on Banyuasin and IR rice varieties growth at the seedling stage under salinity stress. IOP Conf. Ser. Earth Environ. Sci. 343, 012171.

Pardo, J.M., 2010. Biotechnology of water and salinity stress tolerance. Curr. Opin. Biotechnol. 21, 185–196.

Pessarakli, M., 2001. Handbook of plant and crop stress. Biol. Plant 44, 301–304.

Peuke, A.D., Rennenberg, H., 2004. Carbon, nitrogen, phosphorus, and sulphur concentration and partitioning in beech ecotypes (*Fagus sylvatica* L.): phosphorus most affected by drought. Trees 18, 639–648.

Pineros, M.A., Cancado, G.M.A., Maron, L.G., Lyi, S.M., Menossi, M., Kochian, L.V., 2008. Not all ALMT1 type transports mediate aluminium activated organic acid responses: that case of ZmALMT1- an anion selective transporter. Plant J. 53, 352–377.

Pinkerton, A., Simpson, J.R., 1986. Interactions of surface drying and subsurface nutrients affecting plant-growth on acidic soil profiles from an old pasture. Aust. J. Exp. Agric. 26, 681–689.

Platten, J.D., Egdane, J.A., Ismail, A.M., 2013. Salinity tolerance, Na^+ exclusion and allele mining of HKT1;5 in *Oryza sativa* and *O. glaberrima*: many sources, many genes, one mechanism? BMC Plant Biol. 13, 32.

Plett, D.C., Møller, I.S., 2010. Na^+ transport in glycophytic plants: what we know and would like to know. Plant Cell Environ. 33, 612–626.

Pradhan, B., Chakraborty, K., Prusty, N., Mukherjee, A.K., Chattopadhyay, K., Sarkar, R.K., 2019. Distinction and characterisation of rice genotype tolerance to combined stress of salinity and partial submergence, proved by a high resolution chlorophyll flouroscence imaging system. Funct. Plant Biol. 46 (3), 248–261.

Prusty, M.R., Kim, S.R., Vinarao, R., Entila, F., Egdare, J., Diaz, M.G.Q., et al., 2018. Newly identified wild rice accessions conferring high salt tolerance might use a tissue tolerance mechanism in leaf. Front. Plant Sci. 9, 417.

Qi, Z., Spalding, E.P., 2004. Protection of plasma membrane K^+ transport by the salt overly sensitive Na^+/H^+ antiporter during salinity stress. Plant Physiol. 136, 2548–2555.

Qin, Y., Tian, Y., Liu, X., 2015. A wheat salinity-induced WRKY transcription factor TaWRKY93 confers multiple abiotic stress tolerance in *Arabidopsis thaliana*. Biochem. Biophys. Res. Commun. 464 (2), 428–433.

Qing Ma, Li-Jun Yue, Jin-Lin Zhang, Guo-Qiang Wu, Ai-Ke Bao, Suo-Min Wang, 2011 Sodium chloride improves photosynthesis and water status in the succulent xerophyte Zygophyllum xanthoxylum, Tree Physiology, Volume 32, Issue 1, January 2012, Pages 4–13.

Qiu, Q.S., Guo, Y., Dietrich, M.A., Schumaker, K.S., Zhu, J.K., 2002. Regulation of SOS1, a plasma membrane Na^+/H^+ exchanger in *Arabidopsis thaliana*, by SOS2 and SOS3. Proc. Natl. Acad. Sci. USA 99, 8436–8441.

Quadir, Q.F., Watanabe, T., Chen, Z., Osaki, M., Shinano, T., 2011. Ionomic response of *Lotus japonicus* to different root-zone temperatures. Soil Sci. Plant Nutr. 57, 221–232.

Raven, J.A., 2017. Chloride: essential micronutrient and multifunctional beneficial ion. J. Exp. Bot. 68 (3), 359–367.

Reddy, I.N.B.L., Kim, B., Yoon, I., Kim, K., Kwon, T., 2017. Salt tolerance in rice: focus on mechanisms and approaches. Rice Sci. 24, 123–144.

Rehman, H., Nawaz, M.Q., Basra, S.M.A., Afzal, I., Yasmeen, A., Hassan, F., 2014. Seed priming influence on early crop growth, phonological development and yield performance of linola (*Linum usitatissimum* L.). J. Integr. Agric. 13 (5), 990–996.

Ren, Z.H., Gao, J.P., Li, L.G., Cai, X.L., Huang, W., Chao, D.Y., et al., 2005. A rice quantitative trait locus for salt tolerance encodes a sodium transporter. Nat. Genet. 37, 1141–1146.

Roberts, S.K., 2006. Plasma membrane anion channels in higher plants and their putative functions in roots. New Phytol. 169, 647–666.

Roy, S.J., Gilliham, M., Berger, B., Essah, P.A., Cheffings, C., Miller, A.J., et al., 2008a. Investigating glutamate receptor like gene co expression in *Arabidopsis thaliana*. Plant Cell Environ. 31, 861–871.

Roy, S.J., Gilliham, M., Berger, B., Essah, C., et al., 2008b. Investigating glutamate receptor-like gene co-expression in *Arabidopsis thaliana*. Plant Cell Environ. 31 (6), 861–871.

Roy, S.J., Negrao, S., Tester, M., 2014. Salt resistant crop plants. Curr. Opin. Biotechnol. 26, 115–124.

Ruiz-Lozano, J.M., Azcon, R., 1996. Mycorrhizal colonization and drought stress as factors affecting nitrate reductase activity in lettuce plants. Agric. Ecosyst. Environ. 60, 175–181.

Rus, A., Lee, B.-h, Muñoz-Mayor, A., Sharkhuu, A., Miura, K., Zhu, J.-K., et al., 2004. AtHKT1 facilitates Na^+ homeostasis and K^+ nutrition in plant. Plant Physiol. 136, 2500–2511.

Salt, D.E., 2004. Update on plant ionomics. Plant Physiol. 136, 2451–2456.

Salt, D.E., Baxter, I., Lahner, B., 2008. Ionomics and the study ofthe plant ionome. Annu. Rev. Plant Biol. 59, 709–733.

Sanchez, D.H., Pieckenstain, F.L., Escaray, F., Erban, A., Kraemer, U., Udvardi, M.K., et al., 2011. Comparative ionomics and metabolomics in extremophile and glycophytic Lotus species under salt stress challenge the metabolic pre-adaptation hypothesis. Plant Cell Environ. 34, 605–617.

Sarkar, R.K., Mukherjee, A.K., Chakraborty, K., 2019. Seed priming alleviates salt tolerance in rice. In: Hasanuzzaman, M., Fotopoulos, V. (Eds.), Priming and Pre-treatment of Seeds and Seedlings. Springer Nature Singapore, pp. 181–204.

Sasaki, T., Yamamoto, Y., Ezaki, B., Katsuhar, M., Ahn, S.J., Ryan, P.R., et al., 2004. A wheat gene encoding an aluminium activated malate transporter. Plant J. 37, 645–653.

Savvides, A., Ali, S., Tester, M., Fotopoulos, V., 2016. Chemical priming of plants against multiple abiotic stresses: Mission Possible? Trends in Plant Science 21 (4), 329–340.

Schachtman, D.P., Kumar, R., Schroeder, J.I., Marsh, E.L., 1997. Molecular and functional characterization of a novel low-affinity cation transporter (LCTI) in higher plants. Proc. Natl. Acad. Sci. USA 94, 11079–11084.

Senadheera, P., Singh, R.K., Maathuis, F.J., 2009. Differentially expressed membrane transporters in rice roots may contribute to cultivar dependent salt tolerance. J. Exp. Bot. 60, 2553–2563.

Shabala, S., Cuin, T.A., 2007. Potassium transport and plant salt tolerance. Physiol. Plant 133, 651–669.

Shabala, S., Demidchik, V., Shabala, L., Cuin, T.A., Smith, S.J., Miller, A.J., et al., 2006. Extracellular Ca^{2+} ameliorates NaCl-induced K^+ loss from *Arabidopsis* root and leaf cells by controlling plasma membrane K^+-permeable channels. Plant Physiol. 141, 1653–1665.

Shabala, S., 2013. Learning from halophytes: physiological basis and strategies to improve abiotic stress tolerance in crops. Annals of botany 112 (7), 1209–1221. Available from: https://doi.org/10.1093/aob/mct205.

Sharma, D.K., Singh, A., 2015. Salinity research in India, achievements, challenges and future prospects. Water Energy Int. 58 (6).

Shi, H., Ishitani, M., Kim, C., Zhu, J.K., 2000. The Arabidopsis thaliana salt tolerance gene SOS1 encodes a putative Na + /H + antiporter. Proc Natl Acad Sci U S A 97 (12), 6896–6901.

Shi, H.Z., Quintero, F.J., Pardo, J.M., Zhu, J.K., 2002. The putative plasma membrane $Na(^{+})$/ $H(^{+})$ antiporter SOS1 controls long-distance $Na(^{+})$ transport in plants. Plant Cell 14 (2), 465–477.

Shi, Y., Wang, Y., Flowers, T.J., Gong, H., 2013. Silicon decreases chloride transport in rice (Oryza sativa L.) in saline conditions. J Plant Physiol 170 (9), 847–853.

Silveira, J., Costa, R., Oliveira, J., 2001. Drought-induced effects and recovery of nitrate assimilation and nodule activity in cowpea plants inoculated with *Bradyrhizobium* spp. Under moderate nitrate level. Braz. J. Microbiol. 32, 187–194.

Singh, B., Singh, G., 2004. Influence of soil water regime on nutrient mobility and uptake by *Dalbergia sissoo* seedlings. Trop. Ecol. 45, 337–340.

Streeter, J.G., 2003. Effects of drought on nitrogen fixation in soybean root nodules. Plant Cell Environ. 26, 1199–1204.

Subbarao, G.V., Ito, O., Berry, W.L., Wheeler, R.M., 2003. Sodium-a functional plant nutrient. Crit. Rev. Plant Sci. 22, 391–416.

Sun, W., Deng, D., Yang, L., Zheng, X., Yu, J., Pan, H., et al., 2013. Overexpression of the chloride channel gene (GmCLC1) from soybean increases salt tolerance in transgenic *Populus deltoides* × *P. euramericana* 'Nanlin895'. Plant Omics J. 6 (5), 347–354.

Sze, H., Liang, F., Hwang, I., Curran, A.C., Harper, J.F., 2000. Diversity and regulation of plant Ca^{2+} pumps: insights from expression in yeast. Annu. Rev. Plant Physiol. Plant Mol. Biol. 51, 433–462.

Tang, Y., Bao, X., Zhi, Y., et al., 2019. Overexpression of a MYB family gene, *OsMYB6*, increases drought and salinity stress tolerance in transgenic rice. Front Plant Sci. 10, 168.

Tapken, D., Hollmann, M., 2008. *Arabidopsis thaliana* glutamate receptor ion channel function demonstrated by ion pore transplantation. J. Mol. Biol. 383 (1), 36–48.

Tavakkoli, E., Fatehi, F., Coventry, S., Rengasamy, P., McDonald, G.K., 2011. Additive effects of Na^{+} and Cl^{-} ions on barley growth under salinity stress. J. Exp. Bot. 62 (6), 2189–2203.

Tavakkoli, E., Rengasamy, P., McDonald, G.K., 2010. High concentrations of Na^{+} and Cl^{-} ions in soil solution have simultaneous detrimental effects on growth of faba bean under salinity stress. J. Exp. Bot. 61, 4449–4459.

Tavakkoli, E., Fatehi, F., Coventry, S., Rengasamy, P., McDonald, G.K., 2011. Additive effects of Na + and Cl − ions on barley growth under salinity stress. J. Exp. Bot. 62 (6), 2189–2203.

Teakle, N.L., Tyreman, S.D., 2010. Mechanisms of Cl − transport contributing to salt tolerance. Plant Cell Environ. 33 (4), 566–589.

Theerakulpisut, P., Kanawapee, N., Panwong, B., 2017. Seed Priming Alleviated Salt Stress Effects on Rice Seedlings by Improving Na + /K + and Maintaining Membrane Integrity. International journal of plant biology. Available from: https://doi.org/10.4081/pb.2016.6402.

Tuna, A.L., Kaya, C., Ashraf, M., Altunlu, H., Yokas, I., Yagmur, B., 2007. The effects of calcium sulfate on growth, membrane stability and nutrient uptake of tomato plants grown under salt stress. Environ. Exp. Bot. 59, 173–178.

Tuteja, N., 2007. Mechanisms of high salinity tolerance in plants. Methods Enzymol. 428, 419–438.

Uddin, Md. N., Hossain, Burritt, D.J., 2016. Salinity and drought stress Similarities and differences in oxidative responses and cellular redox regulation. In: book: Water stress and Crop plants: A sustainable approach, Ahmad P.

Utrillas, M.J., Alegre, L., Simon, E., 1995. Seasonal changes in production and nutrient content of *Cynodon dactylon* (L.) Pers. subjected to water deficits. Plant Soil 175, 153–157.

Vaktabhai, C.K., Kumar, S., 2017. Seedling invigouration by halo priming in tomato against salt stress. J. Pharmacogn. Phytochem. 6 (6), 716–722.

Verma, T.S., Neue, H.U., 1984. Effect of soil salinity level and zinc application on growth, yield and nutrient composition of rice. Plant Soil 82, 3–24.

Very, A.A., Sentenac, H., 2003. Molecular mechanism and regulation of K^+ transport in higher plants. Annu. Rev. Plant Biol. 54, 560–575.

Walia, H., Wilson, C., Condamine, P., et al., 2005. Comparative transcriptional profiling of two contrasting rice genotypes under salinity stress during the vegetative growth stage. Plant Physiol. 139 (2), 822–835.

Wang, S.M., Zhao, G.Q., Gao, Y.S., Tang, Z.C., Zhang, C.L., 2005. *Puccinellia tenuiflora* exhibits stronger selectivity for K^+ over Na^+ than wheat. J. Plant Nutr. 27, 1841–1857.

Wang, H., Wu, Z., Chen, Y., Yang, C., Shi, D., 2011. Effects of salt and alkali stresses on growth and ion balance in rice (*Oryza sativa* L.). Plant Soil Environ. 57 (6), 286–294.

Wani, S.H., Kumar, V., Shriram, V., Sah, S.K., 2016. Phytohormones and their metabolic engineering for abiotic stress tolerance in crop plants. Crop J. 4, 162–176.

Wang, W.Y., Chai, W.W., Zhao, C.Y., Rowland, O., Wang, B.S., Song, X., et al., 2019. Under drought conditions NaCl improves the nutritional status of the xerophyte *Zygophyllum xanthoxylum* but not of the glycophyte *Arabidopsis thaliana*. Plant Nutr. Soil Sci. 182 (5), 857.

Wei, P., Wang, L., Liu, A., Yu, B., Lam, H.M., 2016. *GmCLC1* confers enhanced salt tolerance through regulating chloride accumulation in soybean. Front. Plant Sci. 7, 1–11.

Wei, Q.J., Gu, Q.Q., Wang, N.N., Yang, C.Q., Peng, S.A., 2015. Molecular cloning and characterization of the chloride channel gene family in trifoliate orange. Biol. Plant 59, 645–653. Available from: https://doi.org/10.1007/s10535-015-0532-z.

Wei, Q., Ma, Q., Ning, S., Su, S., Gu, Q., 2017. Molecular characterization and functional analysis of a cation-chloride cotransporter gene from trifoliate orange (*Poncirus trifoliata* L.). Trees. Available from: https://doi.org/10.1007/s00468-017-1621-8.

Wei, P., Che, B., Shen, L., et al., 2019. Identification and functional characterization of the chloride channel gene, *GsCLC-c2* from wild soybean. BMC Plant Biol. 19, 121.

White, P.J., Broadley, M.R., 2001. Chloride in soils and its uptake and movement within the plant: a review. Ann. Bot. 88, 967–988.

White, P., Karley, A., 2010. Potassium. In: Hell, R., Mendel, R.-R. (Eds.), Cell Biology of Metals and Nutrients. Springer, Berlin/ Heidelberg, pp. 199–224.

Wu, H., Li, Z., 2019. The importance of Cl^- exclusion and vacuolar Cl^- sequestration: revisiting the role of Cl^- transport in plant salt tolerance. Front. Plant Sci. 10, 1418.

Wu, D., Shen, Q., Cai, S., Chen, Z-C., Dai, F., Zhang, G., 2013. Ionomic Responses and Correlations Between Elements and Metabolites Under Salt Stress in Wild and Cultivated Barley. Plant and Cell Physiology 5412, 1976–1988.

Xi, J.J., Chen, H.Y., Bai, W.P., et al., 2018. Sodium-related adaptations to drought: new insights from the xerophyte plant *Zygophyllum xanthoxylum*. Front. Plant Sci. 9, 1678.

Xu, G., Magen, H., Tarchitzky, J., Kafkafi, U., 1999. Advances in chloride nutrition of plants. Adv. Agron 68, 97–150.

Xu, H., Jiang, X., Zhan, K., Cheng, X., Chen, X., Pardo, J.M., Cui, D., 2008. Functional characterization of a wheat plasma membrane Na^+/H^+ antiporter in yeast. Arch Biochem Biophys 473, 8–15.

Xu, Z., Zhou, G., Shimizu, H., 2010. Plant responses to drought and rewatering. Plant Signal. Behav. 5, 649–654. Available from: https://doi.org/10.4161/psb.5.6.11398.

Yamaguchi, T., Blumwald, E., 2005. Developing salt tolerant crop plants: challenges and opportunities. Trends Plant Sci. 10, 615–620.

Yang, C., Chong, J., Li, C., Kim, C., Shi, D., Wang, D., 2007. Osmotic adjustment and ion balance traits of an alkali resistant halophyte *Kochia sieversiana* during adaptation to salt and alkali conditions. Plant Soil 294 (1-2), 263–276.

Yang, C., Shi, D., Wang, D., 2008. Comparative effects of salt and alkali stresses on growth, osmotic adjustment and ionic balance of an alkali-resistant halophyte *Suaeda glauca* (Bge.). Plant Growth Regul. 56 (2), 179.

Yin, H., Li, M., Li, D., et al., 2019. Transcriptome analysis reveals regulatory framework for salt and osmotic tolerance in a succulent xerophyte. BMC Plant Biol. 19, 88.

Yue, L.J., Li, S.X., Ma, Q., Zhou, X.R., et al., 2012. NaCl stimulates growth and alleviates water stress in the xerophyte *Zygophyllum xanthoxylum*. Agris 87, 153–160.

Zhang, H., Zhao, F.-G., Tang, R.-J., Yu, Y., Song, J., Wang, Y., 2017. Two tonoplast MATE proteins function as turgor-regulating chloride channels in *Arabidopsis*. Proc. Natl. Acad. Sci.USA 114, 2036–2045.

Zhu, J., Lee, B.H., Dellinger, M., Cui, X., Zhang, C., Wu, S., et al., 2010. A cellulose synthase-like protein is required for osmotic stress tolerance in *Arabidopsis*. Plant J. 63, 128–140.

Zhukovskaya, N.V., 1973. Absorption and accumulation of phosphateby plants under conditions of soil salinization. Soviet Plant Physiol. 20, 55–61.

Zong, N., Li, X., Wang, L., et al., 2018. Maize *ABP2* enhances tolerance to drought and salt stress in transgenic *Arabidopsis*. J. Integr. Agric. 17 (11), 2379–2393.

Chapter 12

Drought and nitrogen stress effects and tolerance mechanisms in tomato: a review

J. Machado[1,2,3,*], A.P.G. Fernandes[1,*], T.R. Fernandes[1], E. Heuvelink[3], M.W. Vasconcelos[2] and S.M.P. Carvalho[1]

[1]GreenUPorto – Sustainable Agrifood Production Research Centre / Inov4Agro & DGAOT, Faculty of Sciences, University of Porto, Vairão, Portugal, [2]Centro de Biotecnologia e Química Fina (CBQF) – Laboratório Associado, Escola Superior de Biotecnologia, Universidade Católica Portuguesa, Porto, Portugal, [3]Horticulture and Product Physiology Group, Department of Plant Sciences, Wageningen University, Wageningen, The Netherlands

12.1 Introduction

In less than 30 years, humanity will face the challenge of having to feed 9.8 billion inhabitants, which requires a dramatic increase in global agriculture production. The excessive use of water and nitrogen (N) fertilizers is unquestionably the most negatively impactful practice in modern agriculture. Irrigated agriculture accounts for 80% of the total available fresh water, which is making this natural resource increasingly scarce and of poor quality (FAO, 2015). N fertization is a major limiting factor determining crop yield (Liang et al., 2019) . However, increased N fertilization has serious impacts on the human health and on the environment due to the eutrophication of freshwater ecosystems, acidification of agricultural soils, and emission of greenhouse gases (Liang et al., 2019). Therefore, new efforts allowing sustainable management of water and N fertilizers usage will be imperative for feeding the world's growing population.

Despite the need for using less resources, it is important to take into account that water and N deficits will significantly affect the crop physiology and yield. Therefore, it is highly important to understand how plants perceive, are affecred, and respond to both stresses. Drought stress induces a wide range of physiological and biochemical processes in plants such as

* These authors contributed equally to this study.

Plant Nutrition and Food Security in the Era of Climate Change. DOI: https://doi.org/10.1016/B978-0-12-822916-3.00014-7

stomatal regulation, photosynthetic rapid reaction, and activation of the antioxidant defense system (Osakabe et al., 2014; Laxa et al., 2019) and can reduce crop yields up to 70% (Siracusa et al., 2018). N is one of the major components of proteins, nucleic acids, lipids, and some hormones in plants thus having a direct effect on cell division and growth (Tian et al., 2011). Hence, N stress symptoms result from impaired metabolism causing premature senescence and decreased growth, and consequently reduced yields (Fageria and Baligar, 2005).

Tomato (*Solanum lycopersicum* Mill.) is the most important vegetable crop in the world, both for fresh and processing markets, with approximately 182.2 million tons produced on more than 5.7 million hectares (Heuvelink et al., 2020; FAO, 2020). In Europe, the total value of annual tomato production exceeds EUR 7.3 billion, accounting for 21.1% of total fresh vegetables produced with the highest planted area (10.8% of the European Union's total fresh vegetables area) (Cicco, 2019). Moreover, tomato is emerging as a model system because of its (1) economical importance, (2) large genotypic variation, and (3) genome that has been recently completely sequenced (Heuvelink et al., 2020). Tomato production requires high amounts of water and N fertilizers, due to its long growing seasons and high greenhouse temperatures, tomato is quite sensitive to water and N deficits (Du et al., 2018). An adequate water supply is essential for vegetative and reproductive growth being particularly important in flowering and fruit enlargement stages (Srinivasa Rao et al., 2001). Similarly, a proper amount of N fertilizer during plant growth allows maximum yield and economic benefits, because N is an essential macronutrient that can affect several aspects of plant growth and metabolic pathways, including water relations and photosynthesis (Du et al., 2018; Fageria and Baligar, 2005). To halt the excessive use of water and N fertilizers worldwide for economic gains, it is imperative to improve drought and N stress tolerance in tomato cultivars for a sustainable and environmental-friendly production. Knowledge of the morphophysiological, biochemical, and molecular responses to drought and N stress is essential for a holistic perception of plant tolerance mechanisms to water and N-limited conditions in higher plants.

In this chapter, we will review the effect of both drought and N stresses on the morphological, physiological, biochemical, and molecular mechanisms in tomato plants and identify potential knowledge gaps. The identification of the machinery related to water use efficiency and N use efficiency (WUE/NUE) will be crucial to identify new key targets for breeding interventions to make agriculture more productive and sustainable.

12.2 Tomato plant responses to drought stress

The mechanisms involved in response to drought stress and its effects on plants have been reviewed often in the last decades (e.g., Chaves et al., 2003;

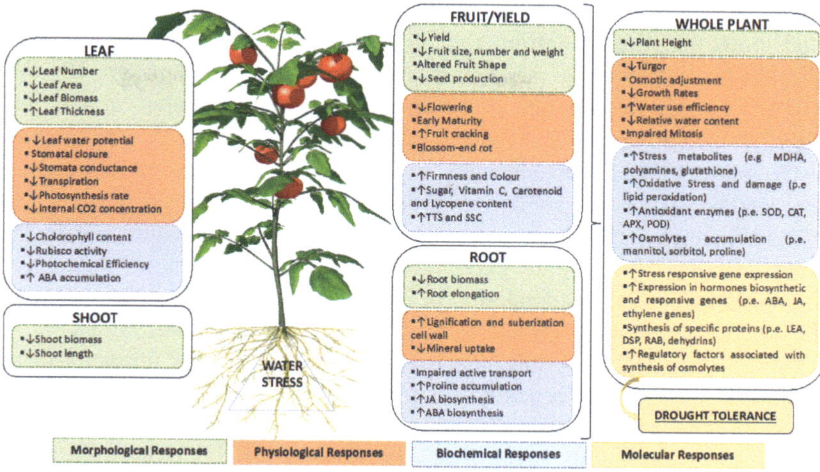

FIGURE 12.1 Schematic overview of the morphological, physiological, biochemical, and molecular processes occuring at specific tomato plant organs or at the whole-plant level in response to drought stress.

Nadeem et al, 2019; Kapoor et al., 2020; Sun et al., 2020) and are synthetized in Fig. 12.1. Drought is one of the most relevant environmental stresses affecting plant metabolism at different levels. Approximately one-third of agricultural land experiences an inadequate water supply (Klunklin and Savage, 2017). Plants respond and adapt to drought through important biochemical and physiological changes at the cellular and molecular levels (Osakabe et al., 2014). Drought alters water relations that lead to the formation of reactive oxygen species (ROS), reduced seed germination, impaired plant growth and photosynthesis, and biomass and yield losses (Shao et al., 2008).

In the section below we review the impact of drought on the tomato crop at several integration levels and we thoroughly explore the reported tolerance mechanisms to drought.

12.2.1 Impact on plant growth, yield, and fruit quality

12.2.1.1 Plant growth and development

Drought has effects on plant growth from the onset of seed germination. A successful seed germination is crucial for seedling establishment and crop yield. However, one of the initial effects of drought stress is poor or slow germination, and reduced seedling vigor and establishment (Bhatt and Srinivasa Rao, 1987; Foolad et al., 2003; Pervez et al., 2009; Esan et al., 2018). Studies have demonstrated that the extension growth of the radicles is affected by drought stress suggesting that the effect on germination occurs prior to this process (Bhatt and Srinivasa Rao; 1987). Accordingly, it was found that a total recovery of germination (100%) can occur if nongerminated seeds are displaced from this stressful condition (Esan et al., 2018). Drought stress can have major impacts on subsequent plant growth and development due to reduction of water potential, relative water content, and turgor, which consequently increases the concentration of solutes in the cytosol (Nonami, 1998). These changes impair both cell elongation and expansion, leading to growth inhibition (Jaleel et al., 2009; Salehi-Lisar et al., 2012). When plants are exposed to drought, the water pressure inside the leaves decreases, leading to a reduced carbon assimilation, imbalance of mineral nutrition, and accumulation of abscisic acid (ABA), causing wilt and reduction of plant growth (Jaleel et al. 2009; Salehi-Lisar et al., 2012).

Development of an optimal leaf area (LA) is of extreme importance for photosynthesis and, consequently, for plant growth. Leaf expansion depends on the turgor pressure and supply of assimilates. Drought stress limits leaf expansion by disrupting water flow from xylem to the elongating cells, reducing turgor pressure, and impairing photosynthesis (Nonami, 1998; Hsiao and Xu, 2000). Accordingly, a reduction of LA is one of the main traits affected by drought stress which limits crop productivity (Sinclair and Muchow, 2001). Moreover, data suggest that reduction of specific leaf area (SLA) is likely a strategy to cope with drought stress. Thicker leaves commonly have a higher content of chlorophyll and proteins per LA unit, thus having a greater photosynthetic capability, and therefore a higher biomass production (Chatterjee and Solankey, 2015). Pokluda et al. (2010) have shown that tomato plants respond to drought stress by significantly decreasing SLA and leaf weight content (LWC). Other parameters related to leaf growth, namely number of leaves per plant, number of branches per plant, total LA, SLA, LA ratio, leaf weight ratio, and LA index are significantly affected by drought stress in tomato (Selim and El-Nady, 2011). In particular, studies in three different tomato cultivars ('Arvento', 'LA1994', and 'LA2093') have demonstrated that drought stress significantly decreases the LA, leaf number, and leaf angle, and increases the leaf temperature, leading to plant wilting (Zhou et al., 2017). Al Hassan et al. (2015) also showed that for cherry tomato, drought stress impairs stem growth and leaf number. Results from Kamanga also demonstrated that drought stress resulted in significant reductions in shoot dry weight, LA, and shoot height of 14 different commercial tomato cultivars (Kamanga, 2020).

In many agronomically important crops, drought stress is initially perceived by the root system, which continues to grow underneath the soil even though shoot growth is inhibited under these conditions (Spollen and Sharp, 1991). Although the growth of the primary root is usually not severely affected by drought stress, the lateral roots growth is significantly reduced mainly by suppression of the activation of the lateral root meristems (Deak and Malamy, 2005; Franco et al., 2011; Koevoets et al., 2016; Du and Scheres, 2018). Experiments in tomato plants have demonstrated that drought stress triggers a decrease in lateral root number and density, and that elongation of the primary root axis is favored over that of the lateral ones (Hooker and Thorpe, 1997). Further, Nuruddin (2001) showed that tomato plants tend to develop a deeper root system at lower soil water potentials. Interestingly, data on sandy soils have also demonstrated that drought stress strongly induces the development of deeper roots (Alaoui et al., 2014). Besides root architecture, root weight, length, and volume are also affected and are important indicators of WUE (Senthilkumar et al., 2017). Water deficit leads to a decreased root hydraulic conductivity and long-lasting root anatomical changes, such as xylem elements with smaller diameters and lower total root xylem area (Hernandez-Espinoza and Barrios-Masias, 2020).

Although drought stress may decrease root elongation and LA expansion, these two processes are not equally affected, as roots are less sensitive than shoots to growth inhibition by low water potentials (Wu and Cosgrove, 2000). Thus increased root/shoot ratio is a frequent trait observed in some tolerant tomato genotypes in response to different drought conditions (Mingo et al., 2004; Manoj and Uday, 2007; Moles et al., 2018; Du et al., 2018) as shown by Nahar and Gretzmache (2011) in seven tomato cultivars. However, an opposite pattern was observed on the economically relevant cultivar 'Moneymaker' (Moles et al., 2018), suggesting that a reduced root/shoot ratio can be part of an alternative strategy activated by certain cultivars to cope with drought in which losses in biomass are compensated by the maintenance of the water status in the tissues (Liu and Li, 2005).

As a consequence of the effect on root and shoot development, the plant growth rate has been shown to be significantly impacted by drought stress. Different studies have reported a negative effect on fresh and dry weight of leaves, stems, and roots, and consequently in the total plant fresh and dry weight when tomato plants were exposed to drought conditions (Al Hassan et al., 2015; Khan et al., 2015; Selim and El-Nady, 2011; Sibomana et al., 2013; Zhou et al., 2019a). Changes in plant height, stem diameter, and internodes length were also reported for tomato plants under similar conditions (Sibomana et al., 2013; Al Hassan et al., 2015; Zhou et al., 2017). In conclusion, drought stress strongly impacts overall tomato plant growth and developmental traits, beginning with germination and impacting traits related to LA, root morphology, and biomass partitioning.

12.2.1.2 *Yield and fruit quality*

As a consequence of the impact on plant dry weight, drought affects crop productivity, reducing 50%–70% of crop yield worldwide (Siracusa et al., 2018). The flowering and fruit development stages are the most sensitive to water deficits in tomato (Chen et al., 2013; Khapte et al., 2019). A deficitary irrigation impacts flower number, which consequently results in fewer fruits produced and/or in fruits with decreased size, and ultimately to less marketable yield (Giardini et al., 1988). Further, drought during seed development decreases seed yield by shortening the final seed-filling periods, thus impacting yield (Pervez et al., 2009). Chen et al. (2013) showed that a reduction in irrigation supply (one-third or two-thirds of full irrigation) at the seedling stage has no negative significant influence on yield and quality of tomato, which suggests that a certain water saving can be achieved at the seedling stage. Similarly, Sivakumar and Srividhya (2016) observed that a moderate drought stress can be seen as a positive factor since it speeds up flower initiation. Neverthless, Hao et al. (2019) concluded that compared with the seedling and flowering stages, the application of mild drought stress while fruit is expanding was considered to be the optimal period to provide satisfactory fruit quality and yield while conserving water resources. On the opposite side, fruit ripening showed to be a highly sensitive stage and any heterogeneous distribution of irrigation, even if temporary, may induce physiological disorders such as fruit cracking (Srinivasa Rao et al., 2001) and blossom end rot (BER). The latter is a frequent physiological disorder caused by the limited capacity of tomato plants to internally translocate calcium toward fruits under drought conditions (Tabatabaie et al., 2004; Djangsou et al., 2019; Millones-Chanamé et al., 2019).

Fresh fruit weight is an important quality feature of tomato that is cultivar-dependent and also highly affected by drought (Wang et al., 2011). Madrid et al. (2009) showed that the requirements of tomato fruit for photosynthesis assimilates and water are drastically increased during the fruit maturation stage, which greatly increases the total water consumption. If the irrigation is reduced, the rate of water absorption will be lower than that of crop transpiration, which induces an internal water deficit affecting photosynthesis and resulting in reduced LA, cell size, and intercellular volume. Thus, fruit water accumulation, and consequently fruit weight, are decreased (Madrid et al., 2009). Sivakumar and Srividhya (2016) also observed that drought stress led to reduction of soluble protein content, sucrose phosphate synthase (involved in sucrose biosynthesis) activity, fruit volume, diameter, yield, and to an increased flower abscission percentage. Nevertheless, water management methods, such as irrigation deficit in which plants are irrigated with smaller amounts of water during the cultivation period, have been proposed as efficient tools to reduce stomatal conductance without signs of drought stress (Ćosić et al., 2015), while potentially improving fruit quality. The use of irrigation deficit was recently reviewed by Khapte et al. (2019) and it can apparently benefit tomato fruit quality by increasing the level of total soluble solids (sugars, amino acids, and organic acids), while avoiding the negative impact on yield. In the case of tomato, higher levels of lycopene, carotenoids, and total soluble solids content (TSS) are desired traits. A rise in TSS increases the value of the fresh fruits and improves their quality by affecting flavor, taste, and water content (Klunklin and Savage, 2017). Irrigation deficit also promotes the ripeness rate in tomato and increases fruit color, as demonstrated by Wang et al. (2011). The authors related this phenomenon to an increase in ethylene content, which in turn increases carotenoids concentration. Qi et al. (2003) demonstrated that water deficit during fruit maturation increases the activity of sucrose synthase and sucrose phosphate synthase, thus enlarging the gradient of sucrose concentration between leaves and fruits. Consequently, this potentiates the transport of more assimilates into the fruits, increasing the rate and amount of fructose and glucose transformation from sucrose, thus improving fruit TSS (Kan, 2008; Klunklin and Savage, 2017). Nangare et al. (2016) also observed improved fruit quality parameters like firmness, TSS, lycopene, ascorbic acid, acidity, and color with only a marginal decrease in fruit shape and size, under 60% ETc (evapotranspiration). Patanè et al. (2011) also reported an improvement in content of TSS, vitamin C, and acidity in tomato fruits under 80% of ETc without significant losses on marketable yield. Significant increases in TSS and lycopene were also observed in drought-stressed tomatoes (Sivakumar and Srividhya, 2016). Additionally, improvement in soluble sugars, vitamin C content, firmness (Hao et al., 2019), and lycopene content seem to be positively affected by deficit irrigation (Patanè et al., 2011; Nangare et al., 2016; Sivakumar and Srividhya, 2016; Takács et al., 2020). Takács et al. (2020) compared three different water supply levels (100%, 75%, and 50% of ETc) and concluded that, although 100% of ETc produced the best marketable yield, 75% ETc increased lycopene content, which could compensate for the lower marketable yield. However, results are sometimes inconsistent (Khapte et al., 2019), and as such this strategy must be further explored to better understand its potential interactions with other growing conditions.

Barbagallo et al. (2012) investigated the yield, chemical composition, and sensorial quality of processing tomato cultivar 'Brigade' grown in the field without irrigation after plant establishment in comparison with those fully irrigated (100% ETc), and concluded that although regular irrigation is important throughout the growing season to achieve satisfactory yields, under drought stress conditions there is an increase in antioxidant activity due to a rise in ascorbic acid and total phenolic contents and a decline in polyphenol oxidase (PPO) activity, thus improving the nutraceutical quality of the product. Djurović et al. (2016) also observed that 50% of ETc increased the organic acid and antioxidant content of tomatoes.

In both the fresh and processing tomato markets, the improvement of organoleptic and functional quality and the reduction of the impact of agriculture on environment are of major relevance. The abovementioned studies suggest that management of drought stress by applying water-saving irrigation strategies at the right moments might improve the quality and nutritional properties of tomato. Thus, an optimization of deficit-irrigation techniques, possibly combined with other sustainable agronomical practices (e.g., grafting, elicitors, biostimulants, and other growth-promoting substances) that have demonstrated an improvement of plant responses to drought stress conditions, is of utmost importance.

12.2.2 Physiological and biochemical mechanisms underlying drought stress and tolerance

12.2.2.1 Photosynthesis and stomatal behavior

Photosynthesis is a key metabolic pathway responsible for growth and development, thus playing a central role in plant performance (Lawlor and Tezara, 2009; Pinheiro and Chaves, 2011). The effects of drought stress on photosynthesis have already been extensively reviewed (Chaves et al., 2009; Pinheiro and Chaves, 2011; Flexas et al., 2012; Zargar et al., 2017). Decreased transpiration rate, CO_2 assimilation rate, stomatal conductance, and net photosynthesis were all reported under drought stress (Raja et al., 2020). The reduced photosynthetic rate seems to be mainly caused by stomatal closure (as a short-term response), by metabolic impairments of the photosynthetic machinery (as a long-term response), or both (Tezara et al., 1999). Recently, Liang et al. (2020) showed that the response of photosynthesis to drought stress transits from stomatal limitation to nonstomatal limitation, as the stress severity increases.

Adaptation of plants to long-term drought stress leads to adjustments in some leaf anatomical characteristics (Franks and Beerling, 2009; Galmés et al., 2013). For instance, stomatal size often decreases to allow a faster response to excessive water losses and to enhance WUE (Franks and Beerling, 2009). Results have demonstrated that the higher tolerance of 'Ramellet', a tomato landrace adapted to summer drought conditions in the Mediterranean region, is a consequence of a lower leaf mesophyll conductance to CO_2, and stomata conductance (Galmés et al., 2011). Galmés et al. (2013) have demonstrated that approximately 90% of the impairment in photosynthesis was caused by diffusion limitations to CO_2 transport from the atmosphere to the sites of carboxylation. By contrast, the biochemical limitations to CO_2 assimilation (RuBisCO activity) seem to be a minor component of photosynthesis limitation (Galmés et al., 2013; Parry et al., 2002; Flexas et al., 2006; Galmés et al., 2011). Nevertheless, accessions with higher concentration of RuBisCO-activated sites have higher rates of CO_2 assimilation (Galmés et al., 2011; Galmés et al., 2013). Drought-adapted tomato plants present optimal distribution, density, size, and area index of chloroplasts, which could influence leaf gas exchange and be partially responsible for the improved WUE (Galmés et al., 2011; Galmés et al., 2013).

Research has shown that there is considerable variation among tomato accessions regarding other key leaf structural and anatomical traits associated with mesophyll conductance (Galmés et al., 2013). Thick leaves with denser cuticles and cell walls also reduce leaf water loss and improve tolerance to drought stress (Galmés et al., 2011; Kamanga, 2020). A major role for trichomes in gas exchange and the determination of WUE was also established (Kamanga, 2020). The ratio of trichomes to stomata (T/S) seems to have a positive relationship with WUE, suggesting that this trait could be used to develop and screen drought-tolerant varieties (Galdon-Armero et al., 2018).

Drought stress also impairs photosynthesis by damaging the chloroplasts and by inactivating the electron transfer reaction, which can potentially reduce the amount of available oxygen, resulting in the production of reactive oxygen species (ROS) that can lead to serious photo-oxidative damages (Lawlor and Cornic, 2002). Chlorophyll fluorescence parameters (Fv/Fo and Fv/Fm), efficiency of excitation transfer (Fv′/Fm′), instantaneous quantum yield of photosystem II (ΦPSII), electron transport rate (ETR) of ΦPSII, and photochemical quenching coefficient (qP), as well as light harvesting efficiency, photochemical conversion, and photosynthetic electron transport (Pervez et al., 2009; Zhang et al., 2018) are all negatively affected by drought.

Oppositely, nonphotochemical quenching (NPQ) is increased in tomato exposed to drought (Yuan et al., 2016; Raja et al., 2020). Interestingly, Liang et al. (2020) found that the NPQ increase is transient suggesting that tomato leaves dissipate the excess of excitation energy by heat dissipation under mild drought stress. The authors suggested that as drought stress increases, heat dissipation capacity is reduced and PSII is damaged (Liang et al., 2020).

Besides the photosynthesis reduction, drought also leads to a deterioration of the thylakoid membranes, which causes substantial damage of photosynthetic pigments that control the absorption of energy (chlorophyll) and dissipate excess energy (carotenoids) (Baquedano and Castillo, 2006; Tamburino et al., 2017; Sivakumar et al., 2018; Raja et al., 2020). However, in the majority of cases, carotenoids are known to be less sensitive to drought stress than chlorophyll, which has been demonstrated for several crops (Salehi-Lisar et al., 2012).

12.2.2.2 Water relations and osmotic adjustment

Osmotic adjustment and cellular-compatible solute accumulation are widely recognized as key mechanisms to tolerate drought, as the increased solute concentration helps to maintain water potential gradients required to ensure continued uptake of water under these conditions (Chen and Jiang, 2010; Wani et al., 2013). In tomato, several studies have reported a significant increase in the accumulation of organic solutes such as glucose, fructose, sucrose, and proline, during drought stress (Hayat et al., 2008; Nahar et al., 2011; Tapia et al., 2016; Nahar and Ullah, 2017; Nahar and Ullah, 2018; Raja et al., 2020), which are cultivar-dependent (Tapia et al., 2016; Montesinos-Pereira et al., 2014; Moles et al., 2018). Although the increase in particular solutes is not directly related to tolerance/sensitivity to drought stress, some studies appear to demonstrate the relevance of this process (Patanè et al., 2016; Sánchez-Rodríguez et al., 2010a,b; Montesinos-Pereira et al., 2014). Polyamine synthesis and accumulation, for example, as a result of proline degradation, was higher in moderate drought-tolerant cherry tomato cultivar 'Zarina', than in 'Josefina', which is a more sensitive cultivar (Montesinos-Pereira et al., 2014).

Although previous results on solute accumulation point to interesting mechanisms that might contribute to a higher WUE, more studies are required to better clarify these processes.

12.2.2.3 Oxidative stress and scavenging system

Drought induces oxidative stress and increases the production of ROS, including antioxidant substrates and antioxidant enzymes (Sánchez-Rodríguez et al., 2010a,b, Zhou et al., 2019b), being the ability of plants to control oxidant levels highly correlated with stress tolerance (Laxa et al., 2019).

Importantly, ROS play a dual role in plant metabolism during abiotic stress. Under high concentrations they function as a toxic byproduct while under low concentrations they function as a signal transduction molecule triggering stress response. The major source of ROS in plants is the chloroplast, where stress-induced changes in redox status of electron transport lead to the formation of singlet oxygen ($^1O_2$), hydrogen peroxide (H_2O_2), superoxide anion (O_2^-), and hydroxyl radical (HO^-). They can directly act on membrane lipids, inactivate metabolic enzymes, and damage the nucleic acids, leading to cell death (Laxa et al., 2019; Mittler et al., 2011). The effects of drought stress on increased malondialdehyde (MDA, related to lipid peroxidation) content in tomato plants have been reported, indicating that the balance of the antioxidant enzymatic system is disrupted, resulting in ROS accumulation and increased lipid peroxidation, causing membrane damage (Zhou et al., 2019b; Liang et al., 2020; Raja et al., 2020). Yuan et al. (2016) has shown that this increase occurs at all developmental stages in tomato.

Among the ROS, H_2O_2 works as an oxidative burst signal due to its high permeability across the membranes and relatively long half-life (Mittler et al., 2011). In tomato, the activity of superoxide dismutase (SOD), an ubiquitous antioxidant enzyme that catalytically converts O_2^- to H_2O_2, gradually increased in 14 different cultivars exposed to drought stress (Aghaie et al., 2018; Hayat et al., 2008; Yuan et al., 2016; Zhou et al., 2019b, Raja et al., 2020; Liang et al., 2020). Besides SOD, the activity of peroxidase (POD) and catalase (CAT) also increases in tomato under prolonged drought. However, the activity of these enzymes sharply drops at the maturity stage (Yuan et al., 2016). As drought stress becomes more severe, the activities of SOD, POD, and CAT start decreasing (Liang et al., 2020).

The activities of glutathione reductase (GR) and ascorbate peroxidase (APX) increase in drought-stressed tomatoes. Oppositely, the activity of the dehydroascorbate reductase (DHAR), monodehydroascorbate reductase (MDHAR), PPO, and GPX (glutathione peroxidase) decline under drought stress (Raja et al., 2020; Sánchez-Rodríguez et al., 2011a). Thipyapong et al. (2004) proposed that a decrease in PPO lowers H_2O_2 concentration, which could improve tomato tolerance against drought. Sánchez-Rodríguez et al. (2011b) demonstrated that tomato cultivar 'Zarina' increased the synthesis of phenylpropanoids and flavonoids, suggesting that polyphenols play a significant physiological role in drought-stress tolerance in tomato.

Little is known about these responses in fruits subjected to drought. Murshed et al. (2013), while analyzing the fruits, found a relationship between drought and oxidative stress, with changes in ROS-scavenging enzymes (increase in SOD, CAT, and APX activity) and in oxidative stress biomarkers (increase in H_2O_2 and MDA). Regarding the enzymes from the ascorbate-glutathione cycle, the degree to which the AsA and DHA concentrations changed under drought stress was extremely variable among different drought stress treatments and different fruit development stages (Murshed et al., 2013).

In sum, drought stress leads to an increased production of ROS in tomato affecting proteins, lipids, and DNA (leaves and fruits), which ultimately compromises drought responses. An increased content on polyphenols in drought-tolerant cultivars appears to indicate that the antioxidant system can be a useful tool for breeding programs.

12.2.2.4 Nutrient uptake

Drought stress affects plant mineral nutrition, not only by decreasing nutrients uptake at the root level but also the mineral translocation to the shoots due to lower transpiration rate and membrane permeability (Rouphael et al., 2012).

Under drought conditions, N uptake diminishes (reviewed by Gonzalez-Dugo et al., 2010). Interestingly, tolerant cultivars seem to present increased amounts of N as compared to sensitive ones, which intensifies the synthesis of amino acids and proteins (Sánchez-Rodríguez et al., 2011a). (Sánchez-Rodríguez et al., 2010b) found lower N uptake and accumulation under water deficit in four sensitive cherry tomato cultivars. However, in the tolerant cultivar 'Zarina', an increase was found, suggesting a cultivar dependency for this trait. The authors suggested that these mechanisms in the cultivar 'Zarina' are essential for drought-stress tolerance, by acting on N metabolism as well as by helping to maintain or augment biomass. Additionally, Sánchez-Rodríguez et al. (2011a) reported that moderate water deficit also decreased the NO^- content in sensitive cultivars, as well as nitrate reductase (NR) activity. The decrease in tomato NR activity was previously reported by Hayat et al. (2008) and may be considered a biochemical adaptation to conserve energy by halting nitrate assimilation. Similar results were found for other macronutrients (P, K, Ca, Mg, S) and micronutrients (Fe, Cu, Mn, Zn, Cl, B), where lower concentrations were found in sensitive cultivars.

Oppositely, cultivar 'Zarina' showed increased concentrations and uptake of N, P, Mg, K, Fe, and Cl. Genotypic variability in nutrient uptake suggests that these different responses against drought might play an important role in drought-stress tolerance.

In short, although it is clear that drought stress hampers mineral uptake in tomato, more research should be done regarding the mechanisms involved, as the imbalances in nutrients ultimately have serious effects on various growth and developmental processes.

12.2.3 Molecular mechanisms underlying drought tolerance

Water deficit imposes selective pressure in the evolution of plant morphology and physiology. Breeding for drought-tolerant crops is a pressing issue due to the increasing frequency and duration of droughts caused by climate change. The activation of molecular pathways modulates the expression level of specific genes responsible for the accumulation of various osmolytes, improvement of the antioxidant system, reduction of transpiration, inhibition of shoot growth, and decrease of tillering, including hormonal regulation (Foolad, 2007; Joshi et al., 2016). Although ABA is reported to be abundant under drought conditions, inducing the expression of various stress-related genes, it has been shown that drought-inducible gene expression is also headed by ABA-independent regulatory systems (Joshi et al., 2016).

The genetic determinants of tomato response to drought are still poorly understood. Many quantitative trait loci (QTL) have been identified (Fooland et al., 2003; Gur and Zamir, 2004). However, the genomic regions covered by the QTL usually cover many genes, limiting the understanding of those involved in drought tolerance. Foolad et al. (2003) identified four drought-tolerant QTLs highly relevant for seed germination under drought stress, while Gur and Zamir (2004) reported three QTLs regulating increased yield under drought conditions. More recently, high-throughput analysis has identified thousands of genes whose expression is altered by drought stress (Candar-Cakir et al., 2016; Iovieno et al., 2016; Liu et al., 2017; Liu et al., 2018), but few studies have deeply analyzed the relevance of downregulated or upregulated genes on tomato drought-stress tolerance improvement (summarized in Table 12.1).

After a long and high-pressure selection process mostly based on yield traits, modern tomato cultivars have a very narrow genetic basis, and the vast majority are now drought-stress sensitive (Shirasawa et al., 2010). However, elite genes for stress tolerance appear to be well conserved in some wild tomato species, as *Solanum pennellii* and accessions from this species show high levels of drought tolerance (Bolger et al., 2014). A model of the different processes and genes involved in the *S. pennellii* drought tolerance response was proposed by Egea et al. (2018). Several genes linked to different physiological and metabolic processes have been identified, especially those involved in N

TABLE 12.1 Summary of genes in which overexpression has been associated with improved tolerance to drought stress in tomato.

Gene name	Encoded protein	Protein function on drought tolerance	Reference
AREB1	Abscisic acid (ABA)-insensitive-5-like protein	• *AREB1*-overexpressing transgenic tomato plants showed increased tolerance to drought stress as assessed by physiological responses such as higher relative water content and chlorophyll fluorescence and lower damage by lipoperoxidation. • Involved in ABA signals and in the upregulation of gene encoding oxidative stress-related proteins, lipid transfer proteins (LTPs), transcription regulators, and late embryogenesis abundant proteins.	(Orellana et al., 2010)
CML44	Hypothetical calcium-binding protein CML44	• Transgenic plants overexpressing *CML44* gene showed increased germination and better seedling growth, as well as higher antioxidant enzyme activity, gas exchange, and water-retention capacity and lower MDA accumulation and membrane damage. • Greater stress tolerance of transgenic plants was further reflected by the upregulation/downregulation of stress-related genes including SOD, GST, CAT, POD, LOX, PR, and ERD.	Munir et al. (2016)
ERF5	Ethylene-responsive transcription factor 5	• Increased tolerance to drought as well as the levels of relative water content.	Pan et al. (2012)

(*Continued*)

TABLE 12.1 (Continued)

HsfA1a	Heat stress transcription factor A-1a	• Silencing *HsfA1a* leads to hypersensitive response to drought stress with increased reduction of the stomatal aperture, increased level of ubiquitinated proteins, and compromised authophagosome formation. • Overexpression of *HsfA1a* increases plant drought tolerance, in a mechanism independent of the phytohormone ABA. • HsfA1a acts as a positive regulator of the expression of the autophagy-related genes *Atg10* and *Atg18f.*	Wang et al. (2015)
JUB1	Transcription factor JUNGBRUNNEN 1	• *SlJUB1* knockout plants showed increased ion leakage and H_2O_2 levels and decreased expression of various drought-responsive genes. • Overexpression of *SlJUB1* leads to activation of *DELLA* and of the stress-related genes *DREB2* and *DREB1*, responsible for decreasing ROS levels, increasing thereby drought tolerance.	Thirumalaikumar et al. (2018)
MAPK1-3	Mitogen-activated protein kinases	• Mapk1, Mapk2, and Mapk3 play a crucial role in enhancing tomato plants' drought tolerance by influencing stomatal activity and H_2O_2 production via the ABA-H_2O_2 pathway.	Li et al. (2013)
		• The *mapk3* knockout mutants exhibit severe wilting symptoms, higher H_2O_2 content, lower antioxidant enzyme activity, and severe	Wang et al. (2017)

(*Continued*)

TABLE 12.1 (Continued)

		membrane damages under drought stress. The *mapk3* knockout was shown to lead to upregulation or downregulation of drought stress-responsive genes including *SlLOX, SlGST,* and *SlDREB*, suggesting that *mapk3* is involved in drought response in tomato plants by protecting cell membranes from oxidative damage and modulating transcription of stress-related genes.	
		• Transgenic plants overexpressing *mapk3* showed an increased photosynthetic activity as well as content of chlorophyll, proline, and sugars under drought stress.	Muhammad et al. (2019)
MX1	MIXTA-like MYB transcription factor	• Substantially increases drought tolerance, by inducing trichome formation and cuticle deposition. • Improved fruit quality due to higher accumulation of carotenoids (including lycopene, β-carotene, phytoene, and lutein), and upregulation of *PSY, PDS, ZDS, PTOX,* and *LCY* that encode key carotenoid biosynthesis genes.	Ewas et al. (2016)
NPR1	Regulatory protein NPR1	• Increased tolerance to drought stress. • *SlNPR1* knockout plants showed severe oxidative damage and cell membrane damage and downregulation activity of antioxidant enzymes (APX, CAT, POD, and SOD). Downregulated expression of drought-related key	Li et al. (2019a)

(*Continued*)

TABLE 12.1 (Continued)

		genes, including *GST*, *DHN*, and *DREB*.	
PIP	Aquaporins (plasma membrane intrinsic proteins)	• Overexpression of *PIP2;1*, *PIP2;7*, and *PIP2;5* resulted in better water status and survival rates than wild-type plants.	Li et al. (2016)
PKE1		• Significantly improved the accumulation of chlorophyll and proline and limited the MDA content. • Upregulation of several ion transporter genes, abiotic-related transcriptional factors, and ROS-scavenging genes in *PKE1* overexpression lines.	Li et al. (2018); Li et al. (2019b)
SHN1	Ethylene-responsive transcription factor WIN1	• Overexpression of *SHN1* showed mild growth retardation phenotype with shiny and dark green leaves. SHN1 was shown to modulate wax accumulation, as demonstrated by higher cuticular wax deposition on leaf epidermal tissue and the induction of several wax-related synthesis genes.	Al-Abdallat, et al. (2014)
WRKY8	WRKY transcription factor 8	• Alleviated wilting or chlorosis phenotype, with higher levels of stress-induced osmotic substances like proline and higher transcript levels of the stress-responsive genes *AREB*, *DREB2A*, and *RD29*. • Smaller stomatal aperture, maintaining higher water content in leaves and reduced H_2O_2 and MDA, which was then followed by higher levels of	Gao et al., 2020

(Continued)

TABLE 12.1 (Continued)

		antioxidant enzyme activities under stress.	
ZIP1	Basic leucine zipper 1	• ZIP1 exerts an essential role in salt- and drought-stress tolerance by modulating an ABA-mediated pathway. • Silencing of *ZIP1* resulted in reduced expression of multiple ABA biosynthesis- and signal transduction-related genes in transgenic plants, as demonstrated by decreased ABA, chlorophyll contents and CAT activity, and increased MDA content.	Zhu et al. (2018)

assimilation, GOGAT/GS cycle, and GABA shunt. An upregulation of genes linked to jasmonic acid/ethylene (JA/ET) biosynthesis was also observed under drought stress in drought-tolerant species. Gong et al. (2010) studied the gene expression of two drought-tolerant introgression lines of *S. pennellii*, and one drought-sensitive cultivar (*S. lycopersicum* cultivar 'M82') showing that the drought-tolerant lines induced around 400 drought-responsive genes encoding for transcription factors and signaling proteins involved in the cell-wall structure, wax biosynthesis, and plant height. Key signaling molecular pathways regulating gluconeogenesis, purine and pyrimidine nucleotide biosynthesis, tryptophan degradation, starch degradation, methionine biosynthesis, and the removal of superoxide radicals were also specifically affected by drought stress, showing that tomato plants can adapt to drought conditions through decreasing energy dissipation, increasing ATP energy provision, and reducing oxidative damage (Gong et al., 2010). Also, genome-wide studies have identified genetic variation on fruit quality components upon drought stress (Albert et al., 2016). The authors proposed 41 putative candidate genes under three constitutive QTLs and 15 interactive or specific QTLs as being implicated on drought stress response on tomato. Regarding the interactive and specific QTLs, genes related to protein protection (chaperone and heat/cold shock proteins), water and solute transport (aquaporins and other transporters), sugar metabolism (sucrose phosphate synthase and invertases), and hormonal signaling (auxin, gibberellin, and ethylene) were identified (Albert et al., 2016). However, improved genetic tools for construction of knockout mutants on tomato are needed to validate the relevance of such genes on overall drought stress response. Therefore, further studies to stablish genetic determinants of drought stress tolerance in tomato are of utmost importance to globally understand the complex regulatory network(s) controlling this process.

12.3 Tomato plant responses to nitrogen stress

Tomato plants have high demands for fertilizers and are quite sensitive to N deficits (Scholberg et al., 2000; Du et al., 2018). Several reports have indicated that open-field grown tomato plants require high amounts of N, ranging from 100 to 574 kg/ha (with levels oftentimes higher than when grown in greenhouse conditions), which can be affected by different environmental and management factors, one of them being the N form supplied (Siddiqi et al., 2002; Du et al., 2018). Nitrate (NO_3^-) and ammonium (NH_4^+) are the most common forms of N supply in agriculture which have different uptake mechanisms and impacts on plant physiological and biochemical processes (Guo et al., 2007). It is commonly accepted that tomato plants have a better performance on NO_3^- or on a mix of NO_3^- and NH_4^+, than with NH_4^+ supply (Siddiqi et al., 2002).

The precise N fertilization that should be supplied in soil production systems is difficult to quantify particularly in open-field because part of these fertilizers can be lost through leaching (depending on soil type and rainfall) and denitrification. Consequently, it is difficult to adjust the exact N fertilizer supply to the plant demand, often making N availability vary from the N-deficient state to the N-excess state, depending on the agricultural practices (Jangam and Raghuram, 2015). While an excess of N can lead to damaging effects on the environment and human health, plants under N-deficit conditions can also have their growth, development, and yield compromised (Jangam and Raghuram, 2015; Okunlola et al., 2015). Therefore, new sustainable agricultural practices and strategies to improve NUE are needed (Anjana and Iqbal, 2007). Understanding the molecular, biochemical, and physiological changes occurring in tomato upon N deficiency will be a step forward for NUE improvement. In the following section, we analyze how N stress affects tomato plants, and we summarize the existing morphological and molecular traits conferring a higher NUE (Fig. 12.2).

12.3.1 Impact of nitrogen on plant growth, yield, and fruit quality

12.3.1.1 Plant growth

N encourages vegetative growth and it is a key element for plant growth, constituting the majority of plant dry matter, along with carbon (C) (Nawaz et al., 2012; Cohen et al., 2019). Several studies have demonstrated that while N availability favors tomato growth N deficiency strongly disturbs plant development (Scholberg et al., 2000; Tei et al., 2002; Nawaz et al., 2012; Luna et al., 2014; Okunlola et al., 2015). The effect of N deficiency on tomato plants is described by different mechanisms. First, N and carbohydrates are required for

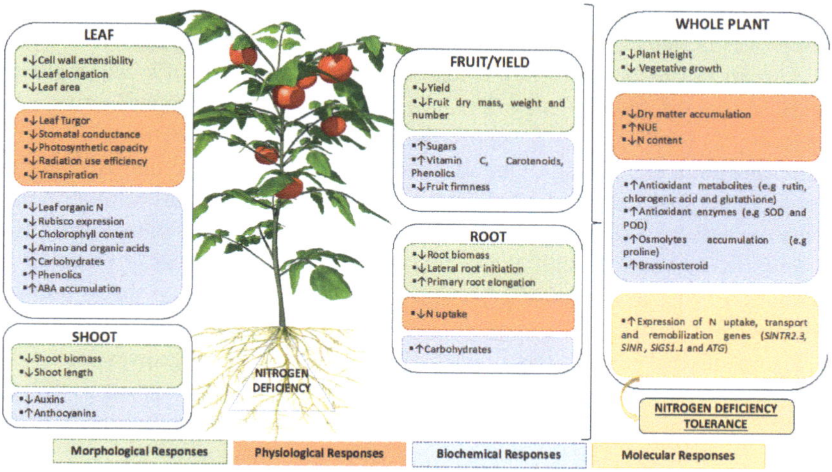

FIGURE 12.2 Schematic overview of the morphological, physiological, biochemical, and molecular processes occuring at specific tomato plant organs or at the whole-plant level in response to nitrogen deficiency.

protein synthesis; therefore the availability of N promotes plant growth by increasing protein synthesis. On the other hand, when N supply is insufficient, an excess of carbohydrate deposits in vegetative cells affects apical growth (Okunlola et al., 2015). Secondly, N availability influences auxin levels at the shoot apical meristem, affecting all aerial plant parts (Vernoux et al., 2010). As changes in root architecture help plants to cope with N deficiency by improving NUE, there is often a relocation of auxin from shoots to roots in plants under N stress for lateral root growth, decreasing the amount of auxin found on the apical meristem, and consequently affecting the shoot development (Rajagopal and Rao, 1974; Kiba et al., 2011). Chapin et al. (1988) reported that in tomato plants growing in hydroponic conditions, the shoot water content does not decline with N stress and authors have pointed out that reduced growth may involve a "root signal" that is able to induce the synthesis of ABA in leaves leading to reduced cell-wall extensibility. Finally, N supply influences photosynthetic activity (Guidi et al., 1997; Tei et al., 2002; De Groot et al., 2003). This effect is mainly a consequence of the high number of N-rich substances used during this process, which can lead to an even greater deposition of carbohydrates on vegetative cell walls (De Groot et al., 2003; Okunlola et al., 2015). An in-depth analysis on the effect of N stress on photosynthesis is given on Section 12.3.2.1 of this chapter.

The duration and intensity of the N stress influences the severity of growth impairment. For instance, tomato seedlings under N deficiency (one-eighth of optimum N) or N starvation conditions for 20, 30, and 40 days after sowing, showed that at 20 days only minor growth changes were detected among treatments. However, at 40 days control plants had almost four times the height of plants subjected to N starvation and more than double the height of plants subjected to N deficiency (Rajagopal and Rao, 1974).

The impact of N deficiency is also dependent on the plant's development stage. In tomato, N deficiency does not seem to impact the reproductive system as optimum N or 1/3 N supply did not directly influence truss development and number (Suniaga Quijada et al., 1992). Authors found that at the reproductive stage, under N deficiency the priority destination of assimilates is the reproductive organs, followed by roots, and finally the shoot vegetative system. In agreement with these findings, Guidi et al. (1997) reported a reduction of root dry mass when N deficiency was employed at the reproductive stage but not at the vegetative stage.

As mentioned above, the impact of N stress on plant growth is highly dependent on the form of N supply. Tomato plants are very prone to NH_4^+ toxicity when it is the sole source of N (Gerendás and Sattelmacher, 1990; Siddiqi et al., 2002). NH_4^+ supply can cause acidification of the root zone, constricting root growth, leading to cation imbalances (especially Ca, K, and Mg), uncoupling of chloroplastic electron transport from photophosphorylation, and reducing LA (Siddiqi et al., 2002; Guo et al., 2007; Hoopen et al., 2010). A 67% vegetative growth reduction was reported on tomato supplied exclusively with NH_4^+, when compared to a mixture of NO_3^- and NH_4^+ (Siddiqi et al., 2002). In fact, supplying NH_4^+ at levels above 0.10 of the total N restricts stem and leaf fresh and dry weight, as well as yield, being progressively more severe as the NH_4^+ supply fraction is increased (Akl et al., 2003).

NUE depends on two main steps, that is, how efficiently plants uptake N from the soil (uptake efficiency) and how plants can use the N more efficiently (utilization efficiency) (Garnett et al., 2009). Root morphology is highly affected by soil N availability, particularly with the NO_3^- form (Gruber et al. 2013; Kant et al., 2011). Generally, low N availability decreases tomato root growth and suppresses lateral root initiation (Guidi et al., 1997; Kant et al., 2011). Root architecture has been pointed out as a crucial breeding trait to improve NUE and to reduce N fertilizer losses (Garnett et al., 2009). Accordingly, some tolerant cultivars display more vigorous roots than sensitive ones, implying an association between root development and NUE (Liu et al., 2009; Fan et al., 2010). On tomato, two cultivars with contrasting NUE, 'Regina Ostuni' ('RO') (higher) and 'UC82' (lower), were identified and characterized (Abenavoli et al., 2016). Root length and fineness were significantly higher in 'RO' compared to 'UC82', regardless of NO_3^- supply dose. Thinner roots are related to higher root-soil surface, increasing uptake efficiency (Goss et al., 1993).

12.3.1.2 Yield and fruit quality

Crop yield and fruit quality are closely related to N availability. Several studies have reported that fruit yield is favored by a wide range of N fertilization rates, ranging from around 100 kg N/ha (Fátima et al., 2012) up to 600 kg N/ha

(Montagu and Goh, 1990). These differences are dependent on the developmental stage, cultivar, N source, and environmental conditions (Warner et al., 2004; Bénard et al., 2009; Elia and Conversa, 2012). Excessive N fertilization can also lead to a remobilization of N from reproductive organs to the leaves and favors vegetative growth to the detriment of fruit growth, resulting in lower yields (Scholberg et al., 2000; Hernández et al., 2020). Elia and Conversa (2012) showed that increasing N fertilization from 0 to 200 kg N/ha increased fruit yield by 28.1% whereas from 0 to 300 kg N/ha only increased 23.3%. Similarly, Du et al. (2017) concluded that 250 kg N/ha increased yields by 4.8% compared to a 350 kg N/ha N fertilization, reinforcing that excessive N fertilization can have detrimental effects on tomato yield. Recently, a study by Du et al. (2019) showed that aerated fertigation, an innovative method, promoted N uptake and improved yield of tomato plants grown in soil under protected cultivation even when N application was low.

It is important to mention that higher yield does not necessarily leads to a higher fruit quality, because it often leads to a decline of the organoleptic quality traits, as a result of a decrease in sugars or organic acids content (Bénard et al., 2009; Hernández et al., 2020). For instance, Wang et al. (2007) reported enhanced titratable acidity and soluble sugars in cherry tomato (*Lycopersicon esculentum* var. *carasiforme*) with higher N supplies, while maximum yields were found at lower N levels. Conversely, Bénard et al. (2009) concluded that increasing N from 4 to 12 mM produced higher yields but reduced soluble sugars.

The impact of N on the synthesis of bioactive compounds with antioxidant capacity (vitamin C, phenolics, and carotenoids) is an emerging area (Bénard et al., 2009; Hernández et al., 2020), as due to the high content of these antioxidants, tomato exhibits antimicrobial, antiinflammatory, antiallergic, and antimutagenic properties (Hernández et al., 2020; Tavallali et al., 2018). Interestingly, data have pointed that N availability directly affects vitamin C and phenolics compounds in fruits (Bénard et al., 2009; Hernández et al., 2020). However, the effect of N on carotenoids accumulation appears to be more ambiguous. While high N dosages appear to favor carotenoid accumulation in the fruit, nonsignificant changes of lycopene and β-carotene content were observed when N supply was reduced (Bénard et al., 2009; Poiroux-Gonord et al., 2010). The differences found in the literature could be largely due to different doses of mineral N applied and the phenological time at which the reduction was implemented (Hernández et al., 2020).

12.3.2 Physiological and biochemical mechanisms underlying N stress

12.3.2.1 *Photosynthesis*

N is involved in the synthesis of several components of the photosynthetic machinery, including pigments and RuBisCO enzyme, which alone encompasses about 50% of leaf N content and plays a major role in C assimilation

(Shangguan et al., 2000; Bassi et al., 2018; Cohen et al., 2019). Therefore, N deficiency is often associated with declines in RuBisCO concentration and leaf gas exchange, with a significant decrease on the leaf photosynthetic CO_2 assimilation capacity (Jin et al., 2015; Cohen et al., 2019). N deficiency also affects photosynthesis by triggering leaf senescence, a reduction in the new LA expansion, and a lack of sink to the photoassimilates produced in such processes, as lowering N supply decreases transport and/or utilization of assimilates (De Groot et al., 2003; Wei et al., 2016).

The impact of N stress on the photosynthetic activity of tomato has been extensively studied (Chapin et al., 1988; Cohen et al., 2019; De Groot et al., 2003; Halpern et al., 2018; Tei et al., 2002; Suniaga Quijada et al., 1992). Suniaga Quijada et al. (1992) noted that low N supply, during the early developmental stages of tomato, resulted in lower leaf elongation and lower leaf chlorophyll content, as well as abnormal starch accumulation in the chloroplasts. Further, Tei et al. (2002) observed that N availability is positively correlated with the growth rate of tomato plants, mainly due to an increase in the light interception as a consequence of a higher LA index. In addition, De Groot et al. (2003) showed that the photosynthetic capacity of tomato plants grown hydroponically following 2 days of N starvation was reduced due to a decrease in light absorption and assimilates use. Recently, some studies have been addressing the impacts of photosynthetic acclimation to elevated CO_2 (PAC) on N uptake and availability (Halpern et al., 2018; Cohen et al., 2019). It is well know that short-term high CO_2 enhances leaf photosynthetic rate. On the other hand, PAC arises under long-term exposure to elevated CO_2, where the positive effect on the photosynthetic activity becomes weaker as a consequence of (1) lower stomatal conductance and, therefore, reduced leaf gas exchange, and (2) downregulation of the concentration and expression of internal photosynthetic intermediates, particularly RuBisCO (Fig. 12.3). Moreover, long-term elevated CO_2 reduces NO_3 uptake due stomatal closure and reduced transpiration rate (Halpern et al., 2018). These authors also demonstrated that low NO_3^- aggravates photosynthetic performance regardless of the CO_2 concentration, but the negative effect was greater when it was combined with elevated CO_2. Furthermore, they demonstrated that higher NO_3^- fertilization can alleviate the acclimation effect of elevated CO_2 on the photosynthetic machinery of tomato plants, N status and sink to source imbalance. Finally, it has been shown that despite the negative effect that elevated CO_2 can have on photosynthetic parameters and on NO_3^- uptake, it can also generally improve the plant soil microbiome through a lower pH, resulting in higher N bacterial fixation, and hence, a greater N soil availability (Halpern et al., 2018; He et al., 2019; Wang et al., 2020) (Fig. 12.3).

12.3.2.2 Oxidative stress and scavenging system

An effect of N availability on oxidative stress response has been described for several plants (Tewari et al., 2007; Rubio-Wilhelmi et al., 2011; Hsieh et al., 2018). N deficiency diminishes the rate of the Calvin cycle and consequently

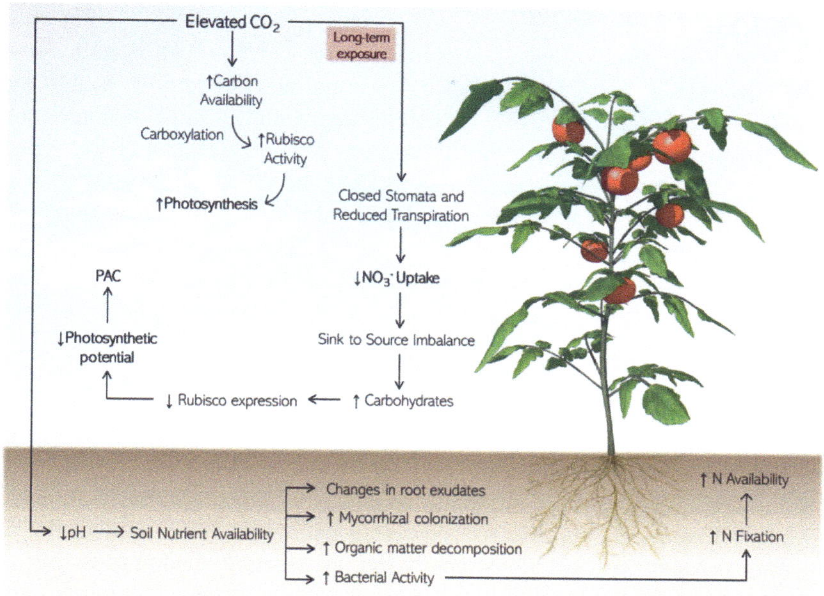

FIGURE 12.3 Interactive effects of elevated CO_2 on N uptake and availability, photosynthesis, photosynthetic acclimation (PAC) and soil microbiome.

prevents oxidation of NADPH and restoration of $NADP^+$ (Rubio-Wilhelmi et al., 2011). N deficiency promotes the generation of ROS, particularly of H_2O_2 and O_2^-, through excitation of the photosystem II that intensifies the Mehler reaction, consequently enhancing the activity of the enzymes POD and SOD (Koleška et al., 2017). In turn, ROS accumulation triggers leaf senescence and inhibition of the photosynthetic capacity by increasing lipid peroxidation, pigment loss, and protein degradation (Tewari et al., 2007; Rubio-Wilhelmi et al., 2011; Koleška et al., 2017).

Several studies in tomato have looked at the effect of N on the oxidative burst and on the scavenging system, particularly on phenolic compound content (Larbat et al., 2012, 2014; Koleška et al., 2017; Groher et al., 2018; Singh et al., 2019). Koleška et al. (2017) tested the effect of biostimulants on prevention of the oxidative damage produced by nitrogen-phosphate-potassium (NPK) deficiency and reported an increase of enzymatic antioxidant activity. Singh et al. (2019) demonstrated that high N supply could partly alleviate salt toxicity by the production of osmolytes, including proline.

N deficiency also triggers the accumulation of flavonoids via increases in phenylalanine ammonia lyase activity (Koleška et al., 2017). Several studies revealed that N deficiency or starvation induces an increase in rutin and chlorogenic acid (two antioxidant flavonoids and the main phenolic compounds found on tomato leaves) (Bénard et al., 2011; Larbat et al., 2012, 2014; Groher et al., 2018).

12.3.2.3 Nutrient uptake

N is a major macronutrient and its ionic form controls cation and anion uptake in order to maintain a charge balance and electroneutrality within the plant (Borgognone et al., 2013). In tomato, many studies have evaluated the effect of different N supplies on the uptake of other nutrients (Kirkby and Knight, 1977; Huett and Dettmann, 1988; Errebhi Wilcox, 1990; Smart and Bloom, 1993; Gunes et al., 1998). The understanding of these nutrient interactions is an important step towards achieving an efficient and sustainable strategy for integrated nutrient management (Aulakh and Malhi, 2005). The source of N is also a determining factor on the uptake of other nutrients. Under NO_3^- deficiency and/or higher NH_4^+ supplementation, cations like potassium (K), magnesium (Mg), and calcium (Ca) presented lower uptake levels and anions like phosphorous (P), sulfate (S), and most micronutrients had higher uptake rates, while with significantly higher NO_3^-/NH_4^+ ratio the inverse response was reported, e.g., (Borgognone et al., 2013; Smart and Bloom, 1993; Yoneyama et al., 2012). NH_4^+ can induce nutrient deficiency of cations like Ca, K, and Mg by competing with these cations for assimilation and transport sites and by acidifying the rooting medium (as NH_4^+ transport to the roots implies release of protons), altering the mineral nutrient bioavailability (Rayar and van Hai, 1977; Haynes and Goh, 1978; Gerendás et al., 1997; Hoopen et al., 2010). On the contrary, rhizospheric acidification can lead to increased uptake of some anions, while NO_3^-, associated with the proton consumption via $2H^+/NO_3^-$ symport, and consequently increase in the pH of the outer solution, can result in a decrease of anions uptake (Fageria, 2001; Borgognone et al., 2013). P uptake specifically is associated with a lower pH as it induces higher solubility of Ca-phosphates in the rooting medium and increases the phosphate ion that is preferentially absorbed by roots (Hoffmann et al., 1994) (Ashraf and Sultana, 2000).

12.3.3 Molecular mechanisms underlying nitrogen stress and tolerance

Improving NUE in crop plants is now, more than ever, an urgent environmental and agricultural goal that, despite extensive research over the past decade, remains one of the scientific "Grand Challenges" of the 21st century (Hsieh et al., 2018). To face this challenge and to be able to identify important sites of intervention for higher NUE, we need to have a broad understanding of the genetic machinery behind N uptake, transport, metabolism, and remobilization in crop plants (Jangam and Raghuram, 2015; Hsieh et al., 2018).

Depending on internal and external NO_3^- concentrations, plants have different uptake systems including two high-affinity transport systems (HATS), mainly activated under low NO_3^- availability (levels below 0.1 mM) and one low-affinity transport system (LATS) for cases of higher availability of NO_3^-

(Wang, 2001; Hildebrandt et al., 2002; Lea and Azevedo, 2006). Although the expression of both HATS is highly enhanced by NO_3^- treatment, cHATS is constitutively expressed even in the absence of NO_3^-, while iHATS is NO_3^- inducible (Hildebrandt et al., 2002). This root transport systems allows the plant to maximize NO_3^- uptake depending on soil N availability. Once inside the plant roots, NO_3^- may be transported to the shoot for assimilation (storage in the vacuole), released again into the rhizosphere, or reduced to nitrite or NH_4^+ via NR or nitrite reductases (NiR), respectively (Wang, 2001).

NH_4^+ is taken up by the plant directly from the rhizosphere by NH_4^+ transporters (AMT) and is further assimilated into amino acids via the glutamine synthetase/glutamate synthase (GS/GOGAT) cycle and/or the asparagine synthetase (AS) or the glutamate dehydrogenase (GDH) (Lam et al., 1994; Kant et al., 2011; Liu et al., 2016).

Several groups of genes related to N uptake, transport, and assimilation were identified including genes encoding NO_3^- transporters (*NRTs*), genes encoding *AMTs*, genes involved in the transport of NO_3^- to the vacuole (*CLC*), and genes related to the enzymes NR, NiR, of GS/GOGAT, GDH, and AS pathway for NH_4^+ assimilation (Lam et al., 1994; Wang, 2001; Liu et al., 2016). *NRTs* are further divided into two major groups, one mainly associated with LATS (*NRT1*) and the other associated with HATS (*NRT2*) (Wang, 2001).

12.3.3.1 N uptake and transport

Low N availability strongly induces the expression of genes associated with N uptake, particularly those associated with HATS such as the *NRT2* and the high-affinity *AMT* gene family (Peng et al., 2007; Hsieh et al., 2018). A study using an *Arabidopsis thaliana Atnrt2.1-nrt2.2* mutant evidenced that this NRT inhibits iHATS expression by up to 80% (Li et al., 2007). N starvation also induced an upregulation of genes belonging to the *NRT2* gene family in tomato (*LeNRT2-1* and *LeNRT2-2*) (Ono et al., 2000).

Tomato plants grown without N, with 200 μM NO_3^- or 100 μM NH_4^+ as their sole N source, showed that expression of *LeAMT1* and *LeNRT1-1* was independent of the N availability/source, while *LeNRT1-2* expression was only induced in the presence of NO_3^- (Lauter et al., 1996). This can be explained by the fact that (1) LATS can have two or even more differentially regulated transport systems in tomato, cLATS or NO_3^- iLATS (Huang, 1999); and that (2) *LeNRT1-1* can encode a dual-affinity transporter (Lauter et al., 1996; Ono et al., 2000). Indeed, it was already reported that if NRT1-1 is posttranscriptionally phosphorylated, it functions as a HATS, whereas if dephosphorylated it functions as a LATS (Masclaux-Daubresse et al., 2010). This last hypothesis was further explored in *A. thaliana*, but it was concluded that *NRT1-1* alone had a minor contribution to NO_3^- uptake at low external NO_3^- concentration. Curiously, the role of *AMT* genes appears to be more controversial. Von Wiren et al. (2000) evaluated the expression of three

AMT genes in hydroponically grown tomato (*LeAMT1-1*, *LeAMT1-2*, *LeAMT1-3*) and observed that *LeAMT1-1* expression was induced by N absence, whereas *LeAMT-2* expression was strongly induced by either NH_4^+ or NO_3^- sources suggesting that these genes encode an inducible and constitutive transport system, respectively. On the other hand, the expression of *LeAMT1-3* demonstrated little similarity with the other two genes and was identified only in leaves and when NO_3^- was present, having hypothetically a different mechanism of regulation in tomato plants.

Further studies in tomato plants aiming for functional characterization of LATS and HATS gene families are needed to fully understand the mechanism of regulation of these transporters in order to design new strategies to improve N uptake according to N availability.

12.3.3.2 N assimilation

N assimilation is initiated with the reduction of NO_3^- to NH_4^+, followed by NH_4^+ assimilation. As mentioned above, two enzymes, NR and NiR, have a crucial role in this process by reducing NO_3^- to nitrite and nitrite to NH_4^+, respectively (Masclaux-Daubresse et al., 2010). Studies concerning the effect of N deficiency on the genetic regulation of *NR* and *NiR* in tomato plants are scarce and suggest that the expression of NR can be affected by different temporal harvest time points and light conditions, or even by the plant organ analyzed, because they differ in the basal amount of NR enzyme (Galangau et al., 1988; Krapp et al., 2011). Accordingly, Galangau et al. (1988) stated no significant changes on the expression level of *NR* in tomato leaves after N starvation, in contrast to a decrease observed in the protein level of this transporter, suggesting that N starvation controls NR activity on tomato at least in part, through a posttranscriptional regulation.

The GS/GOGAT cycle is responsible for the assimilation of N into amino acids but it also plays a significant role in the reassimilation of NH_4^+ produced during N cycling (Liu et al., 2016). Besides the GS/GOGAT cycle, GDH is another alternative pathway responsible for the production of glutamate from NH_4^+. A study by Liu et al. (2016) allowed the identification of six GS (*SlGSI*, *SlGS2*, and *SlGS1.1–1.4*), two GOGAT (*SlNADH-GOGAT* and *SlFd-GOGAT*), and five GDH (*SlNAD-GDHA1-A3*, *SlNAD-GDHB1*, and *SlNADP-GDH*) genes in tomato plants. Importantly, *SlGS1.2*, *SlGS1.3*, *SlNADHGOGAT*, and *SlGDHB* had decreased transcriptional levels after N starvation, oppositely to *SlGS1.1* whose expression was increased by this condition. *SlGS1.1* may display an important function on N remobilization when plants go through a severe N starvation, playing an important role in plant survival to N stress (Liu et al., 2016). Interestingly, the expression of GDH was also induced by an excess of NH_4^+ (Vega-Mas et al., 2019).

Asparagine synthetase (AS) is also involved in NH_4^+ assimilation as it enables an ATP-dependent transfer of the amino group of glutamine to aspartate to generate glutamate and asparagine and it is considered an ideal amino acid for N transport and storage because of its high N:C ratio and its stability (Lam et al., 1994; Masclaux-Daubresse et al., 2010). A study by Wang (2001) demonstrated that the expression of an *AS* gene was induced by NO_3^- while no evidence on the effect of N starvation on *AS* gene regulation was collected so far.

12.3.3.3 N remobilization

Under severe N starvation, and depending on the developmental stage, plants may reuse stored N by degrading nitrogenous compounds and remobilizing the newly available N (Bassham, 2007; Lemaître et al., 2008). Therefore N remobilization is considered time- and environmental-dependent as it is more frequent during the vegetative and reproductive developmental stages and favored by limiting NO_3^- conditions (Lemaître et al., 2008; Masclaux-Daubresse et al., 2010). During leaf senescence, N remobilization is induced as several proteins are degraded, namely those involved in photosynthesis, providing great amounts of reutilizable N (Masclaux-Daubresse et al., 2010). GS1 isoforms are well-known markers of N remobilization and, particularly upon N deficiency, they have been described as senescence-related markers (Lemaître et al., 2008). In tomato, the expression of the *GS1* gene was upregulated by N starvation, suggesting that GS might have a role in the utilization of recycled N (Liu et al., 2016).

Autophagy is a highly regulated and conserved process that is crucial for N remobilization inside the cell. Under starvation, activation of autophagy induces the transport of cellular contents to the vacuole for subsequent digestion in order to recycle nutrients, thus optimizing plant growth, development, and stress responses (Bassham, 2007; Avin-Wittenberg et al., 2018; Wang et al., 2019). A downregulation of autophagy-related genes (*ATGs*) is associated with increased susceptibility of plants to N and C starvation and inhibits breakdown of mitochondria-resident proteins during senescence (Wang et al., 2019). The expression of *ATG* genes was reportedly increased upon N starvation in several plants (Zhou et al., 2015; Wang et al., 2016; Bedu et al., 2020). Brassinosteroid (BR) is a phytohormone that plays a critical role in autophagy (Wang et al., 2019). Data on tomato plants have revealed that in early stages of N starvation, the accumulation of *BZR1*, a BR-activated transcription factor, was enhanced to promote N recycling by inducing the expression of *ATGs*. Overall, these results appear to indicate that tomato plants are also dependent on ATGs for N remobilization upon N starvation in a process likely regulated by BR.

Despite the importance of understanding genetic regulation upon N deficiency in order to improve NUE, these mechanisms need further characterization in tomato plants especially regarding N remobilization. Given that tolerance to low N has become a highly relevant breeding goal, it is surprising that only few studies have compared the genetic variation in N-utilization traits among the accessions of tomato or compared measures of NUE for the same germplasm growing in different environments (Abenavoli et al., 2016). However, this slow breeding progress can be, at least in part, explained by the complexity of abiotic-stress tolerance associated with polygenes (Parajuli et al., 2018). Nevertheless, Abenavoli et al. (2016), reported that in the N stress tolerant 'RO' tomato genotype grown at low NO_3^- supply (0.1 mM NO_3^-), a higher shoot NO_3^- concentration and higher activity and gene expression of NR, as well as higher root *SlNTR2.3* expression, are likely associated with the increased NUE in this cultivar. Furthermore, the authors observed a high significant difference in the expression of *SlNTR2.3* between contrasting cultivars (at low NO_3^- supply) suggesting that this gene may function as a marker for the identification of improved NUE tomato genotypes, especially because when grown at high NO_3^- supply, the expression was similar for both genotypes. When compared with 'UC82' (a sensitive cultivar), 'RO' showed higher *SlNRT2.3* expression and shoot NO_3^- content at high NO_3^- supply but similar NR activity, suggesting an improved capacity to store NO_3^- on the leaves that may also be a potential phenotypic marker for improved NUE. Some of the NRT2 transporter activity is dependent on the high-affinity nitrate transporter-activating protein 2.1 (NAR2.1). The expression of the two-component high-affinity uptake system *SlNRT2.1/NAR2.1* was significantly higher in UC82 when compared with RO at low NO_3^- supply, but not at 5 mM NO_3^- supply. Further, while *SlNRT2.1* expression did not show any significant difference between genotypes, the *SlNAR2.1* expression was significantly higher in RO, suggesting that it might also have a role on NUE. The authors also concluded that their results may indicate that NUE components may play a greater role than N uptake efficiency in conferring higher NUE in tomato (Abenavoli et al., 2016). Recently, an interesting study by Domínguez-Figueroa et al. (2020) reported that the *A. thaliana* transcription factor CDF3 is involved in N responses and improves NUE in tomato, with *CDF3* expression being transiently induced after 1 day of N starvation. Further analyses of gain-of-function and loss-of-function lines allowed the authors to confirm the role of CDF3 as an important factor in N-regulated processes, specifically that CDF3 regulates (1) the expression of an important set of NO_3^- responsive genes, including *GS-1*, *GS-2*, *NR-1*, and NO_3^- transporters *NRT2.1*, *NRT2.4*, and *NRT2.5*; (2) the expression of carbon assimilation genes like *PK1* and *PEPC1*; (3) a rise in the content of major amino acids (glutamate, glutamine, proline, and asparagine) and total protein; and (4) an increase in enzyme activities of N metabolism (GS, NR, and GDH) in response to N availability (Domínguez-Figueroa et al., 2020). The authors also found that the overexpression of *CDF3* increased N accumulation, improved primary and lateral root elongation, enhanced the amounts of photoassimilates partitioned to the fruits (resulting from increased C and N metabolism), as well as the photosynthetic rate, biomass accumulation, and fruit yield. The recent results from Domínguez-Figueroa et al. (2020) suggest that CDF3 is an important regulatory factor for the NO_3^- response, and its potential role for improving NUE in tomato plants.

12.4 Conclusions and future prospects

Drought stress negatively affects tomato growth at several stages of development (from seed germination until fruit development) reducing fruit yield and quality. This is mostly due to a lower photosynthetic rate (leading to lower yield) and reduced transpiration rate (that affects Ca mobility to the fruits resulting in physiological disorders). N deficiency leads to plants with shorter vegetative growth periods and to reduced yields, mainly caused by lower photosynthetic capacity, carbohydrate accumulation, and hormonal readjustments. This literature review also presents and overview of the genes that are differently regulated to aid plants to adapt to both stresses, which might be of interest for future tomato drought and/or N-deficit-tolerant breeding programs. However, it is concluded that so far only few studies have been conducted to confirm the molecular regulatory networks and the key genes involved in the tolerance responses to water and N deficit, with the latter ones being almost uncharacterized. Still regarding N stress tolerance, promising new studies have pointed out that N remobilization might be crucial for increased tolerance to N stress on tomato but more studies are needed to confirm that hypothesis.

In the environment, plants are often subjected to several abiotic stresses, including water and N deficiencies, which are closely related. However, only few studies have addressed the complex interaction between them, and new insights are required in the future in order to improve WUE and NUE in tomato plants, especially in light of global climate change. Moreover, some studies here reviewed suggest that it is possible to manage drought stress by applying water-saving irrigation strategies in order to promote the quality and nutritional properties without compromising tomato yield, while also contributing to water conservation. Thus, continuing to explore and improve the use of deficit-irrigation techniques, especially when integrated with other novel sustainable agronomical tools (e.g. grafting, elicitors, biostimulants, and other growth-promoting substances) that have demonstrated to improve plant response to drought conditions, is of utmost importance.

Acknowledgments

The authors would like to thank Fundação para a Ciência e a Tecnologia (FCT) for JM and APGF the PhD scholarships (SFRH/BD/116147/2016 andUI/BD/151039/2021, respectively). This research was also supported by national funds through FCT (within the scope of UIDB/05748/2020, UIDP/05748/2020, UIDB/50016/2020) and through Norte2020 (NORTE-01-0145-FEDER-000041).

References

Abenavoli, M.R., Longo, C., Lupini, A., Miller, A.J., Araniti, F., Mercati, F., et al., 2016. Phenotyping two tomato genotypes with different nitrogen use efficiency. Plant Physiol. Biochem. 107, 21–32. Available from: https://doi.org/10.1016/j.plaphy.2016.04.021.

Aghaie, P., Tafreshi, S.A.H., Ebrahimi, M.A., Haerinasa, M., 2018. Tolerance evaluation and clustering of fourteen tomato cultivars grown under mild and severe drought conditions. Sci. Hortic. 232, 1–12. Available from: https://doi.org/10.1016/j.scienta.2017.12.041.

Akl, I.A., Savvas, D., Papadantonakis, N., Lydakis-Simantiris, N., Kefalas, P., 2003. Influence of ammonium to total nitrogen supply ratio on growth, yield and fruit quality of tomato grown in a closed hydroponic system. Eur. J. Hortic. Sci. 68 (5), 204–211.

Alaoui, S.M., Salghi, R., Abouatallah, A., Jaouhari, N., Hammouti, B., 2014. Impact of drip irrigation scheduling on vegetative parameters in tomato (*Lycopersicon esculentum*) under unheated greenhouse. Int. J. Eng. Res. Appl. 4 (1), 71–76.

Albert, E., Segura, V., Gricourt, J., Bonnefoi, J., Derivot, L., Causse, M., 2016. Association mapping reveals the genetic architecture of tomato response to water deficit: focus on major fruit quality traits. J. Exp. Bot. 67 (22), 6413–6430. Available from: https://doi.org/10.1093/jxb/erw411.

Al-Abdallat, A.M., Al-Debei, H.S., Ayad, J.Y., Hasan, S., 2014. Over-expression of SLSHN1 gene improves drought tolerance by increasing cuticular wax accumulation in tomato. Int. J. Mol. Sci. 15 (11), 19499–19515. Available from: https://doi.org/10.3390/ijms151119499.

Al Hassan, M., Fuertes, M.M., Sánchez, F.J.R., Vicente, O., Boscaiu, M., 2015. Effects of salt and water stress on plant growth and on accumulation of osmolytes and antioxidant compounds in cherry tomato. Notulae Botanicae Horti Agrobotanici 43 (1), 1–11. Available from: https://doi.org/10.15835/nbha4319793.

Anjana, S.U., Iqbal, M., 2007. Nitrate accumulation in plants, factors affecting the process, and human health implications. A review. Agronomy for Sustainable Development 27, 45–57. Available from: https://doi.org/10.1051/agro:2006021.

Ashraf, M., Sultana, R., 2000. Combination effect of NaCL salinity and nitrogen form on mineral composition of sunflower plants. Biol. Plant. 43 (4), 615–619. Available from: https://doi.org/10.1023/A:1002860202032.

Aulakh, M.S., Malhi, S.S., 2005. Interactions of nitrogen with other nutrients and water: effect on crop yield and quality, nutrient use efficiency, carbon sequestration, and environmental pollution. Adv. Agron. 86, 341–409. Available from: https://doi.org/10.1016/S0065-2113(05)86007-9.

Avin-Wittenberg, T., Baluška, F., Bozhkov, P.V., Elander, P.H., Fernie, A.R., Galili, et al., 2018. Corrigendum: autophagy-related approaches for improving nutrient use efficiency and crop yield protection. J. Exp. Bot. 69 (12), 3173. Available from: https://doi.org/10.1093/jxb/ery113.

Bassham, D.C., 2007. Plant autophagy-more than a starvation response. Curr. Opin. Plant Biol. 10, 587–593. Available from: https://doi.org/10.1016/j.pbi.2007.06.006.

Baquedano, F.J., Castillo, F.J., 2006. Comparative ecophysiological effects of drought on seedlings of the mediterranean water-saver *Pinus halepensis* and water-spenders *Quercus coccifera* and *Quercus ilex*. Trees 20 (6), 689–700. Available from: https://doi.org/10.1007/s00468-006-0084-0.

Barbagallo, R.N., Di Silvestro, I., Patanè, C., 2012. Yield, physicochemical traits, antioxidant pattern, polyphenol oxidase activity and total visual quality of field-grown processing tomato cv. Brigade as affected by water stress in Mediterranean climate. J. Sci. Food Agric. 93, 1449–1457. Available from: https://doi.org/10.1002/jsfa.5913.

Bassi, D., Menossi, M., Mattiello, L., 2018. Nitrogen supply influences photosynthesis establishment along the sugarcane leaf. Sci. Rep. 8, 2327. Available from: https://doi.org/10.1038/s41598-018-20653-1.

Bedu, M., Marmagne, A., Masclaux-Daubresse, C., Chardon, F., 2020. Transcriptional plasticity of autophagy-related genes correlates with the genetic response to nitrate starvation in *Arabidopsis thaliana*. Cells 9 (4), 1021. Available from: https://doi.org/10.3390/cells9041021.

Bénard, C., Gautier, H., Bourgaud, F., Grasselly, D., Navez, B., Caris-Veyrat, C., et al., 2009. Effects of low nitrogen supply on tomato (*Solanum lycopersicum*) fruit yield and quality with special emphasis on sugars, acids, ascorbate, carotenoids, and phenolic compounds. J. Agric. Food Chem. 57 (10), 4112–4123. Available from: https://doi.org/10.1021/jf8036374.

Bénard, C., Bourgaud, F., Gautier, H., 2011. Impact of temporary nitrogen deprivation on tomato leaf phenolics. Int. J. Mol. Sci. 12 (11), 7971–7981. Available from: https://doi.org/10.3390/ijms12117971.

Bhatt, R.M., Srinivasa Rao, N.K., 1987. Seed germination and seedling growth responses of tomato cultivars to imposed water stress. J. Hortic. Sci. 62 (2), 221–225. Available from: https://doi.org/10.1080/14620316.1987.11515772.

Bolger, A., Scossa, F., Bolger, M.E., Lanz, C., Maumus, F., Tohge, et al., 2014. The genome of the stress-tolerant wild tomato species *Solanum pennellii*. Nat. Genet. 46, 1034–1038. Available from: https://doi.org/10.1038/ng.3046.

Borgognone, D., Colla, G., Rouphael, Y., Cardarelli, M., Rea, E., Schwarz, D., 2013. Effect of nitrogen form and nutrient solution pH on growth and mineral composition of self-grafted and grafted tomatoes. Sci. Hortic. 149, 61–69. Available from: https://doi.org/10.1016/j.scienta.2012.02.012.

Candar-Cakir, B., Arican, E., Zhang, B., 2016. Small RNA and degradome deep sequencing reveals drought-and tissue-specific micrornas and their important roles in drought-sensitive and drought-tolerant tomato genotypes. Plant Biotechnol. J. 14 (8), 1727–1746. Available from: https://doi.org/10.1111/pbi.12533.

Chapin, F.S., Walter, C.H.S., Clarkson, D.T., 1988. Growth response of barley and tomato to nitrogen stress and its control by abscisic acid, water relations and photosynthesis. Planta 173, 352–366. Available from: https://doi.org/10.1007/BF00401022.

Chatterjee, A., Solankey, S., 2015. Functional physiology in drought tolerance of vegetable crops: an approach to mitigate climate change impact. In: Choudhary, M.L., Patel, V.B., Siddiqui, M.W., Mahdi, S.S. (Eds.), Climate Dynamics in Horticultural Science, Principles and Applications, vol. 1. Apple Academic Press, Oakville, ON, pp. 149–171. , CRC Press Taylor & Francis Group: Boca Raton, FL.

Chaves, M.M., Maroco, J.P., Pereira, J.S., 2003. Understanding plant responses to drought – from genes to the whole plant. Funct. Plant Biol. 30 (3), 239–264. Available from: https://doi.org/10.1071/FP02076.

Chaves, M.M., Flexas, J., Pinheiro, C., 2009. Photosynthesis under drought and salt stress: regulation mechanisms from whole plant to cell. Ann. Bot. 103 (4), 551–560. Available from: https://doi.org/10.1093/aob/mcn125.

Chen, H., Jiang, J.-G., 2010. Osmotic adjustment and plant adaptation to environmental changes related to drought and salinity. Environ. Rev. 18, 309–319. Available from: https://doi.org/10.1139/a10-014.

Chen, J., Kang, S., Du, T., Qiu, R., Guo, P., Chen, R., 2013. Quantitative response of greenhouse tomato yield and quality to water deficit at different growth stages. Agric. Water Manag. 129, 152–162. Available from: https://doi.org/10.1016/j.agwat.2013.07.011.

Cicco, A.D., 2019. The fruit and vegetable sector in the EU—a statistical overview Eurostat2019. Available online: https://ec.europa.eu/eurostat/statisticsexplained/index.php/The_fruit_and_vegetable_sector_in_the_EU_-_a_statistical_overview (accessed 12.09.2020).

Cohen, I., Halpern, M., Yermiyahu, U., Bar-Tal, A., Gendler, T., Rachmilevitch, S., 2019. CO_2 and nitrogen interaction alters root anatomy, morphology, nitrogen partitioning and photosynthetic acclimation of tomato plants. Planta 250 (5), 1423–1432. Available from: https://doi.org/10.1007/s00425-019-03232-0.

Ćosić, M., Djurović, N., Todorović, M., Maletić, R., Zečević, B., Stričević, R., 2015. Effect of irrigation regime and application of kaolin on yield, quality and water use efficiency of sweet pepper. Agric. Water Manag. 159, 139–147. Available from: https://doi.org/10.1016/j.agwat.2015.05.014.

De Groot, C.C., Boogaard, R.,V.D., Marcelis, L.F., Harbinson, J., Lambers, H., 2003. Contrasting effects of N and P deprivation on the regulation of photosynthesis in tomato plants in relation to feedback limitation. J. Exp. Bot. 54 (389), 1957–1967. Available from: https://doi.org/10.1093/jxb/erg193.

Deak, K.I., Malamy, J., 2005. Osmotic regulation of root system architecture. Plant J. 43 (1), 17–28. Available from: https://doi.org/10.1111/j.1365-313X.2005.02425.x.

Djangsou, H., Enrico, F., Domenico, R., Matteo, B., 2019. Blossom end-rot in tomato (*Solanum lycopersicum* L.): a multi-disciplinary overview of inducing factors and control strategies. Sci. Hortic. 249, 49–58. Available from: https://doi.org/10.1016/j.scienta.2019.01.042.

Djurović, N., Ćosíć, M., Stričević, R., Savić, S., Domazet, M., 2016. Effect of irrigation regime and application of kaolin on yield, quality and water use efficiency of tomato. Sci. Hortic. 201, 271–278. Available from: https://doi.org/10.1016/j.scienta.2016.02.017.

Domínguez-Figueroa, J., Carrillo, L., Renau-Morata, B., Yang, L., Molina, R.-V., Marino, D., et al., 2020. The *Arabidopsis* transcription factor CDF3 is involved in nitrogen responses and improves nitrogen use efficiency in tomato. Front. Plant Sci. 11, 601558. Available from: https://doi.org/10.3389/fpls.2020.601558.

Du, Y., Scheres, B., 2018. Lateral root formation and the multiple roles of auxin. J. Exp. Bot. 69 (2), 155–167. Available from: https://doi.org/10.1093/jxb/erx223.

Du, Y., Cao, H., Liu, S., Gu, X., Cao, Y., 2017. Response of yield, quality, water and nitrogen use efficiency of tomato to different levels of water and nitrogen under drip irrigation in Northwestern China. J. Integr. Agric. 16 (5), 1153–1161. Available from: https://doi.org/10.1016/s2095-3119(16)61371-0.

Du, Y.-D., Niu, W.-Q., Gu, X.-B., Zhang, Q., Cui, B.-J., 2018. Water-and nitrogen-saving potentials in tomato production: a meta-analysis. Agric. Water Manag. 210, 296–303. Available from: https://doi.org/10.1016/j.agwat.2018.08.035.

Du, Y.-D., Gu, X.-B., Wang, J.-W., Niu, W.-Q., 2019. Yield and gas exchange of greenhouse tomato at different nitrogen levels under aerated irrigation. Sci. Total Environ. 668, 1156–1164. Available from: https://doi.org/10.1016/j.scitotenv.2019.03.098.

Egea, I., Albaladejo, I., Meco, V., Morales, B., Sevilla, A., Bolarin, M.C., et al., 2018. The drought-tolerant *Solanum pennellii* regulates leaf water loss and induces genes involved in amino acid and ethylene/jasmonate metabolism under dehydration. Sci. Rep. 8, 2791. Available from: https://doi.org/10.1038/s41598-018-21187-2.

Elia, A., Conversa, G., 2012. Agronomic and physiological responses of a tomato crop to nitrogen input. Eur. J. Agron. 40, 64–74. Available from: https://doi.org/10.1016/j.eja.2012.02.001.

Errebhi, M., Wilcox, G.E., 1990. Tomato growth and nutrient uptake pattern as influenced by nitrogen form ratio. J. Plant Nutr. 13 (8), 1031–1043. Available from: https://doi.org/10.1080/01904169009364133.

Esan, V.I., Ayanbamiji, T.A., Adeyemo, J.O., Oluwafemi, S., 2018. Effect of drought on seed germination and early seedling of tomato genotypes using polyethylene glycol 6000. Int. J. Sci. 2, 36–43. Available from: https://doi.org/10.18483/ijSci.1533.

Ewas, M., Gao, Y., Wang, S., Liu, X., Zhang, H., Nishawy, E.M.E., Ali, F., Shahzad, R., Ziaf, K., Subthain, H., Martin, C., Luo, J., 2016. Manipulation of SLMXL for enhanced carotenoids accumulation and drought resistance in tomato. Sci. Bull. 61, 1413–1418. Available from: https://doi.org/10.1007/s11434-016-1108-9.

Fageria, V.D., 2001. Nutrient interactions in crop plants. J. Plant Nutr. 24 (8), 1269–1290. Available from: https://doi.org/10.1081/PLN-100106981.

Fageria, N.K., Baligar, V.C., 2005. Enhancing nitrogen use efficiency in crop plants. Adv. Agron. 88, 97–185. Available from: https://doi.org/10.1016/s0065-2113(05)88004-6.

Fan, J.-B., Zhang, Y.-L., Turner, D., Duan, Y.-H., Wang, D.-S., Shein, Q.-R., 2010. Root physiological and morphological characteristics of two rice cultivars with different nitrogen-use efficiency. Pedosphere 20 (4), 446–455. Available from: https://doi.org/10.1016/s1002-0160(10)60034-3.

FAO, 2015. Climate Change and Food Systems: Global Assessments and Implications for Food Security and Trade. Available online: Available from: http://www.fao.org/3/a-i4332e.pdf (accessed 20.09.2020.).

FAO, 2020. Statistical Database: Production Statistics. Available online:. Available from: http://www.fao.org/faostat/en/#data/QC/visualize (accessed 20.09.2020.).

Fátima, T., Teasdale, J.R., Bunce, J., Mattoo, A.K., 2012. Tomato response to legume cover crop and nitrogen: differing enhancement patterns of fruit yield, photosynthesis and gene expression. Funct. Plant Biol. 39 (3), 246. Available from: https://doi.org/10.1071/fp11240.

Flexas, J., Ribas-Carbo, M., Bota, J., Galmes, J., Henkle, M., Martinez-Canellas, S., et al., 2006. Decreased RuBisCo activity during water stress is not induced by decreased relative water content but related to conditions of low stomatal conductance and chloroplast CO_2 concentration. New Phytol. 172, 73–82. Available from: https://doi.org/10.1111/j.1469-8137.2006.01794.x.

Flexas, J., Gallé, A., Galmés, J., Ribas-Carbo, M., Medrano, H., 2012. The response of photosynthesis to soil water stress. In: Aroca, R. (Ed.), Plant Responses to Drought Stress. Springer, Berlin, Heidelberg, pp. 129–144. Available from: https://doi.org/10.1007/978-3-642-32653-0_5.

Foolad, M.R., 2007. Current status of breeding tomatoes for salt and drought tolerance. In: Jenkes, M.A., Hasegawa, P.M., Jain, S.M. (Eds.), Advances in Molecular Breeding Toward Drought and Salt Tolerant Crops. Springer, Dordrecht, pp. 669–700. Available from: https://doi.org/10.1007/978-1-4020-5578-2_27.

Foolad, M.R., Zhang, L.P., Subbiah, P., 2003. Genetics of drought tolerance during seed germination in tomato: inheritance and QTL mapping. Genome 46 (4), 536–545. Available from: https://doi.org/10.1139/g03-035.

Franco, J.A., Bañón, S., Vicente, M.J., Miralles, J., Martínez-Sánchez, J.J., 2011. Review article: root development in horticultural plants grown under abiotic stress conditions – a review. J. Hortic. Sci. Biotechnol. 86 (6), 543–556. Available from: https://doi.org/10.1080/14620316.2011.11512802.

Franks, P.J., Beerling, D.J., 2009. Maximum leaf conductance driven by CO_2 effects on stomatal size and density over geologic time. Proc. Natl. Acad. Sci. U S A 106, 10343–10347. Available from: https://doi.org/10.1073/pnas.0904209106.

Galangau, F., Daniel-Vedele, F., Moureaux, T., Dorbe, M.-F., Leydecker, M.-T., Caboche, M., 1988. Expression of leaf nitrate reductase genes from tomato and tobacco in relation to light-dark regimes and nitrate supply. Plant Physiol. 88 (2), 383–388. Available from: https://doi.org/10.1104/pp.88.2.383.

Galdon-Armero, J., Fullana-Pericas, M., Mulet, P.A., Conesa, M.A., Martin, C., Galmés, J., 2018. The ratio of trichomes to stomata is associated with water use efficiency in *Solanum lycopersicum* (tomato). Plant J. 96 (3), 607–619. Available from: https://doi.org/10.1111/tpj.14055.

Galmés, J., Conesa, M.À., Ochogavía, J.M., Perdomo, J.A., Francis, D.M., Ribas-Carbó, M., et al., 2011. Physiological and morphological adaptations in relation to water use efficiency in mediterranean accessions of *Solanum lycopersicum*. Plant Cell Environ. 34 (2), 245–260. Available from: https://doi.org/10.1111/j.1365-3040.2010.02239.x.

Galmés, J., Ochogavía, J.M., Gago, J., Roldán, E.J., Cifre, J., Conesa, M.À., 2013. Leaf responses to drought stress in Mediterranean accessions of *Solanum lycopersicum*: anatomical adaptations in relation to gas exchange parameters. Plant Cell Environ. 36 (5), 920–935. Available from: https://doi.org/10.1111/pce.12022.

Gao, Y.-F., Liu, J.-K., Yang, F.-M., Zhang, G.-Y., Wang, D., Zhang, L., et al., 2020. The WRKY transcription factor WRKY8 promotes resistance to pathogen infection and mediates drought and salt stress tolerance in *Solanum lycopersicum*. Physiol. Plant. 168 (1), 98–117. Available from: https://doi.org/10.1111/ppl.12978.

Garnett, T., Conn, V., Kaiser, B.N., 2009. Root based approaches to improving nitrogen use efficiency in plants. Plant Cell Environ. 32 (9), 1272–1283. Available from: https://doi.org/10.1111/j.1365-3040.2009.02011.x.

Gerendás, J., Sattelmacher, B., 1990. Influence of nitrogen form and concentration on growth and ionic balance of tomato (*Lycopersicon esculentum*) and potato (*Solanum tuberosum*). In: van Beusichem, M.L. (Ed.), Plant Nutrition—Physiology and Applications. Developments in Plant and Soil Sciences, vol. 41. Springer, Dordrecht, pp. 33–37. Available from: https://doi.org/10.1007/978-94-009-0585-6_6.

Gerendás, J., Zhu, Z., Bendixen, R., Ratcliffe, R.G., Sattelmacher, B., 1997. Physiological and biochemical processes related to ammonium toxicity in higher plants. Zeitschrift Für J. Planta Nutr. Soil Sci. 160 (2), 239–251. Available from: https://doi.org/10.1002/jpln.19971600218.

Giardini, L., Giovanardi, R., Borin, M., 1988. Water consumption and yield response of tomato in relation to water availability at different soil depth. Acta Hortic. 228, 119–126. Available from: https://doi.org/10.17660/ActaHortic.1988.228.12.

Gong, P., Zhang, J., Li, H., Yang, C., Zhang, C., Zhang, X., et al., 2010. Transcriptional profiles of drought-responsive genes in modulating transcription signal transduction, and biochemical pathways in tomato. J. Exp. Bot. 61 (13), 3563–3575. Available from: https://doi.org/10.1093/jxb/erq167.

Gonzalez-Dugo, V., Durand, J.-L., Gastal, F., 2010. Water deficit and nitrogen nutrition of crops. A review. Agron. Sustainable Develop. 30, 529–544. Available from: https://doi.org/10.1051/agro/2009059.

Goss, M.J., Miller, M.H., Bailey, L.D., Grant, C.A., 1993. Root growth and distribution in relation to nutrient availability and uptake. Eur. J. Agron. 2 (2), 57–67. Available from: https://doi.org/10.1016/s1161-0301(14)80135-4.

Groher, T., Schmittgen, S., Noga, G., Hunsche, M., 2018. Limitation of mineral supply as tool for the induction of secondary metabolites accumulation in tomato leaves. Plant Physiol. Biochem. 130, 105–111. Available from: https://doi.org/10.1016/j.plaphy.2018.06.033.

Gruber, B.D., Giehl, R.F.H., Friedel, S., von Wiren, N., 2013. Plasticity of the arabidopsis root system under nutrient deficiencies. Plant Physiol. 163 (1), 161–179. Available from: https://doi.org/10.1104/pp.113.218453.

Guidi, L., Lorefice, G., Pardossi, A., Malorgio, F., Tognoni, F., Soldatini, G.F., 1997. Growth and photosynthesis of *Lycopersicum esculentum* (L.) plants as affected by nitrogen deficiency. Biol. Plant. 40 (2), 235–244. Available from: https://doi.org/10.1023/a:1001068603778.

Gunes, A., Alpaslan, M., Inal, A., 1998. Critical nutrient concentrations and antagonistic and synergistic relationships among the nutrients of NFT-grown young tomato plants. J. Plant Nutr. 21 (10), 2035–2047. Available from: https://doi.org/10.1080/01904169809365542.

Guo, S., Zhou, Y., Shen, Q., Zhang, F., 2007. Effect of ammonium and nitrate nutrition on some physiological processes in higher plants – growth, photosynthesis, photorespiration, and water relations. Plant Biol. 9 (1), 21–29. Available from: https://doi.org/10.1055/s-2006-924541.

Gur, A., Zamir, D., 2004. Unused natural variation can lift yield barriers in plant breeding. PLoS Biol. 2 (10), e245. Available from: https://doi.org/10.1371/journal.pbio.0020245.

Halpern, M., Bar-Tal, A., Lugassi, N., Egbaria, A., Granot, D., Yermiyahu, U., 2018. The role of nitrogen in photosynthetic acclimation to elevated [CO_2] in tomatoes. Plant Soil 434, 397–411. Available from: https://doi.org/10.1007/s11104-018-3857-5.

Hao, S.X., Cao, H.X., Wang, H.B., Pan, X.Y., 2019. Effects of water stress at different growth stages on comprehensive fruit quality and yield in different bunches of tomatoes in greenhouses. Int. J. Agric. Biol. Eng. 12 (3), 67–76. Available from: https://doi.org/10.25165/j.ijabe.20191203.4468.

Hayat, S., Hasan, S.A., Fariduddin, Q., Ahmad, A., 2008. Growth of tomato (*Lycopersicon esculentum*) in response to salicylic acid under water stress. J. Plant Interact. 3 (4), 297–304. Available from: https://doi.org/10.1080/17429140802320797.

Haynes, R.J., Goh, K.M., 1978. Ammonium and nitrate nutrition of plants. Biol. Rev. 53, 465–510. Available from: https://doi.org/10.1111/j.1469-185X.1978.tb00862.x.

Hsieh, P.-H., Kan, C.-C., Wu, H.-Y., Yang, H.-C., Hsieh, M.-H., 2018. Early molecular events associated with nitrogen deficiency in rice seedling roots. Sci. Rep. 8, 12207. Available from: https://doi.org/10.1038/s41598-018-30632-1.

Hernandez-Espinoza, L.H., Barrios-Masias, F.H., 2020. Physiological and anatomical changes in tomato roots in response to low water stress. Sci. Hortic. 265, 109208. Available from: https://doi.org/10.1016/j.scienta.2020.109208.

Hernández, V., Hellín, P., Fenoll, J., Flores, P., 2020. Impact of nitrogen supply limitation on tomato fruit composition. Sci. Hortic. 264, 109173.

Heuvelink, E., Okello, R.C.O., Peet, M., Giovannoni, J.J., Dorais, M., 2020. Tomato. In: Wien, H. C., Stützel, H. (Eds.), The Physiology of Vegetable Crops. CABI, United Kingdom, p. 512.

He, W., Yoo, G., Moonis, M., Kim, Y., Chen, X., 2019. Impact assessment of high soil CO_2 on plant growth and soil environment: a greenhouse study. PeerJ, 7: e6311. Available from: https://doi.org/10.7717/peerj.6311.

Hildebrandt, U., Schmelzer, E., Bothe, H., 2002. Expression of nitrate transporter genes in tomato colonized by an arbuscular mycorrhizal fungus. Physiol. Plant. 115 (1), 125–136. Available from: https://doi.org/10.1034/j.1399-3054.2002.1150115.x.

Hoffmann, C., Ladewig, E., Claassen, N., Jungk, A., 1994. Phosphorus uptake of maize as affected by ammonium and nitrate nitrogen – measurements and model calculations. J. Plant Nutr. Soil Sci. 157 (3), 225–232. Available from: https://doi.org/10.1002/jpln.1994 1570310.

Hooker, T.S., Thorpe, T.A., 1997. Effects of water deficit stress on the developmental growth of excised tomato roots cultured in vitro. In Vitro Cell. Dev. Biol. Plant 33 (3), 245–251. Available from: https://doi.org/10.1007/s11627-997-0031-9.

Hoopen, F.T., Cuin, T.A., Pedas, P., Hegelund, J.N., Shabala, S., Schjoerring, J.K., et al., 2010. Competition between uptake of ammonium and potassium in barley and arabidopsis roots: molecular mechanisms and physiological consequences. J. Exp. Bot. 61 (9), 2303–2315. Available from: https://doi.org/10.1093/jxb/erq057.

Hsiao, T.C., Xu, L.K., 2000. Sensitivity of growth of roots versus leaves to water stress: biophysical analysis and relation to water transport. J. Exp. Bot. 51, 1595–1616. Available from: https://doi.org/10.1093/jexbot/51.350.1595.

Hsieh, P.-H., Kan, C.-C., Wu, H.-Y., Yang, H.-C., Hsieh, M.-H., 2018. Early molecular events associated with nitrogen deficiency in rice seedling roots. Sci. Rep. 8, 12207. Available from: https://doi.org/10.1038/s41598-018-30632-1.

Huang, N.-C., 1999. Cloning and functional characterization of an arabidopsis nitrate transporter gene that encodes a constitutive component of low-affinity uptake. Plant Cell Online 11 (8), 1381–1392. Available from: https://doi.org/10.1105/tpc.11.8.1381.

Huett, D.O., Dettmann, E.B., 1988. Effect of nitrogen on growth, fruit quality and nutrient uptake of tomatoes grown in sand culture. Aust. J. Exp. Agric. 28 (3), 391–399. Available from: https://doi.org/10.1071/EA9880391.

Iovieno, P., Punzo, P., Guida, G., Mistretta, C., Oosten, M.J.V., Nurcato, R., et al., 2016. Transcriptomic changes drive physiological responses to progressive drought stress and rehydration in tomato. Front. Plant Sci. 7, 371. Available from: https://doi.org/10.3389/fpls.2016.00371.

Jaleel, C.A., Manivannan, P., Wahid, A., Farooq, M., Somasundaram, R., Panneerselvam, R., 2009. Drought stress in plants: a review on morphological characteristics and pigments composition. Int. J. Agric. Biol. 11, 100–105.

Jangam, A.P., Raghuram, N., 2015. Nitrogen and stress. In: Pandey, G. (Ed.), Elucidation of Abiotic Stress Signaling in Plants. Springer, New York, NY, pp. 323–339. Available from: https://doi.org/10.1007/978-1-4939-2540-7_12.

Jin, X., Yang, G., Tan, C., Zhao, C., 2015. Effects of nitrogen stress on the photosynthetic CO_2 assimilation, chlorophyll fluorescence and sugar-nitrogen ratio in corn. Sci. Rep. 5, 9311. Available from: https://doi.org/10.1038/srep09311.

Joshi, R., Wani, S.H., Singh, B., Bohra, A., Dar, Z.A., Lone, A.A., et al., 2016. Transcription factors and plants response to drought stress: current understanding and future directions. Front. Plant Sci. 7, 1029. Available from: https://doi.org/10.3389/fpls.2016.01029.

Kamanga, R.M., 2020. Screening and differential physiological responses of tomato (*Solanum lycopersicum* L.) to drought stress. Plant Physiol. Rep. 25, 472–482. Available from: https://doi.org/10.1007/s40502-020-00532-6.

Kan, I., 2008. Yield quality and irrigation with saline water under environmental limitations: the case of processing tomatoes in California. Agric. Econ. 38, 57–66. Available from: https://doi.org/10.1111/j.1574-0862.2007.00281.x.

Kant, S., Bi, Y.-M., Rothstein, S.J., 2011. Understanding plant response to nitrogen limitation for the improvement of crop nitrogen use efficiency. J. Exp. Bot. 62 (4), 1499–1509. Available from: https://doi.org/10.1093/jxb/erq297.

Kapoor, D., Bhardwaj, S., Landi, M., Sharma, A., Ramakrishnan, M., Sharma, A., 2020. The impact of drought in plant metabolism: how to exploit tolerance mechanisms to increase crop production. Appl. Sci. 10, 5692. Available from: https://doi.org/10.3390/app10165692.

Khan, S.H., Khan, A., Litaf, U., Shah, A.S., Khan, M.A., Bilal, M., et al., 2015. Effect of drought stress on tomato cv. bombino. J. Food Process. Technol. 6, 465. Available from: https://doi.org/10.4172/2157-7110.1000465.

Khapte, P.S., Kumar, P., Burman, U., Kumar, P., 2019. Deficit irrigation in tomato: agronomical and physio-biochemical implications. Sci. Hortic. 248, 256–264. Available from: https://doi.org/10.1016/j.scienta.2019.01.006.

Kiba, T., Kudo, T., Kojima, M., Sakakibara, H., 2011. Hormonal control of nitrogen acquisition: roles of auxin, abscisic acid, and cytokinin. J. Exp. Bot. 62 (4), 1399–1409. Available from: https://doi.org/10.1093/jxb/erq410.

Kirkby, E.A., Knight, A.H., 1977. Influence of the level of nitrate nutrition on ion uptake and assimilation, organic acid accumulation, and cation-anion balance in whole tomato plants. Plant Physiol. 60 (3), 349–353. Available from: https://doi.org/10.1104/pp.60.3.349.

Klunklin, W., Savage, G., 2017. Effect on quality characteristics of tomatoes grown under well-watered and drought stress conditions. Foods 6 (8), 56. Available from: https://doi.org/10.3390/foods6080056.

Koevoets, I.T., Venema, J.H., Elzenga, J.T.M., Testerink, C., 2016. Roots withstanding their environment: exploiting root system architecture responses to abiotic stress to improve crop tolerance. Front. Plant Sci. 7, 1335. Available from: https://doi.org/10.3389/fpls.2016.01335.

Koleška, I., Hasanagić, D., Todorović, V., Murtić, S., Klokić, I., Parađiković, N., et al., 2017. Biostimulant prevents yield loss and reduces oxidative damage in tomato plants grown on reduced NPK nutrition. J. Plant Interact. 12 (1), 209–218. Available from: https://doi.org/10.1080/17429145.2017.1319503.

Krapp, A., Berthome, R., Orsel, M., Mercey-Boutet, S., Yu, A., Castaings, L., et al., 2011. Arabidopsis roots and shoots show distinct temporal adaptation patterns toward nitrogen starvation. Plant Physiol. 157 (3), 1255–1282. Available from: https://doi.org/10.1104/pp.111.179838.

Larbat, R., Le Bot, J., Bourgaud, F., Robin, C., Adamowicz, S., 2012. Organ-specific responses of tomato growth and phenolic metabolism to nitrate limitation. Plant Biol. 14 (5), 760–769. Available from: https://doi.org/10.1111/j.1438-8677.2012.00564.x.

Larbat, R., Paris, C., Le Bot, J., Adamowicz, S., 2014. Phenolic characterization and variability in leaves, stems and roots of micro-tom and patio tomatoes, in response to nitrogen limitation. Plant Sci. 224, 62–73. Available from: https://doi.org/10.1016/j.plantsci.2014.04.010.

Lam, H.M., Peng, S., Coruzzi, G.M., 1994. Metabolic regulation of the gene encoding glutamine-dependent asparagine synthetase in *Arabidopsis thaliana*. Plant Physiol. 106 (4), 1347–1357. Available from: https://doi.org/10.1104/pp.106.4.1347.

Lauter, F.R., Ninnemann, O., Bucher, M., Riesmeier, J.W., Frommer, W.B., 1996. Preferential expression of an ammonium transporter and of two putative nitrate transporters in root hairs of tomato. Proc. Natl. Acad. Sci. U S A 93 (15), 8139–8144. Available from: https://doi.org/10.1073/pnas.93.15.8139.

Laxa, M., Liebthal, M., Telman, W., Chibani, K., Dietz, K.-J., 2019. The role of the plant antioxidant system in drought tolerance. Antioxidants 8 (4), 94. Available from: https://doi.org/10.3390/antiox8040094.

Lawlor, D.W., Cornic, G., 2002. Photosynthetic carbon assimilation and associated metabolism in relation to water deficits in higher plants. Plant Cell Environ. 25, 275–294. Available from: https://doi.org/10.1046/j.0016-8025.2001.00814.x.

Lawlor, D.W., Tezara, W., 2009. Causes of decreased photosynthetic rate and metabolic capacity in water-deficient leaf cells: a critical evaluation of mechanisms and integration of processes. Ann. Bot. 103 (4), 561–579. Available from: https://doi.org/10.1093/aob/mcn244.

Lea, P.J., Azevedo, R.A., 2006. Nitrogen use efficiency. 1. Uptake of nitrogen from the soil. Ann. Appl. Biol. 149 (3), 243–247. Available from: https://doi.org/10.1111/j.1744-7348.2006.00101.x.

Lemaître, T., Gaufichon, L., Boutet-Mercey, S., Christ, A., Masclaux-Daubresse, C., 2008. Enzymatic and metabolic diagnostic of nitrogen deficiency in *Arabidopsis thaliana* Wassileskija accession. Plant Cell Physiol. 49 (7), 1056–1065. Available from: https://doi.org/10.1093/pcp/pcn081.

Li, W., Wang, Y., Okamoto, M., Crawford, N.M., Siddiqi, M.Y., Glass, A.D.M., 2007. Dissection of the *AtNRT2.1*:*AtNRT2.2* inducible high-affinity nitrate transporter gene cluster. Plant Physiol. 143 (1), 425–433. Available from: https://doi.org/10.1104/pp.106.091223.

Li, C., Yan, J.M., Li, Y.Z., Zhang, Z.C., Wang, Q.L., Liang, Y., 2013. Silencing the SpMPK1, SpMPK2, and SpMPK3 genes in tomato reduces abscisic acid-mediated drought tolerance. Int. J. Mol. Sci. 14 (11), 21983–21996. Available from: https://doi.org/10.3390/ijms141121983.

Li, R., Wang, J., Li, S., Zhang, L., Qi, C., Weeda, S., et al., 2016. Plasma membrane intrinsic proteins SlPIP2;1, SlPIP2;7 And SlPIP2;5 conferring enhanced drought stress tolerance in tomato. Sci. Rep. 6, 31814. Available from: https://doi.org/10.1038/srep31814.

Li, J., Wang, Y., Wei, J., Pan, Y., Su, C., Zhang, X., 2018. A tomato proline-, lysine-, and glutamic-rich type gene SpPKE1 positively regulates drought stress tolerance. Biochem. Biophys. Res. Commun. 499 (4), 777–782. Available from: https://doi.org/10.1016/j.bbrc.2018.03.222.

Li, R., Liu, C., Zhao, R., Wang, L., Chen, L., Yu, W., et al., 2019a. CRISPR/Cas9-Mediated SlNPR1 mutagenesis reduces tomato plant drought tolerance. BMC Plant Biol. 19, 38. Available from: https://doi.org/10.1186/s12870-018-1627-4.

Li, J., Chen, C., Wei, J., Pan, Y., Su, C., Zhang, X., 2019b. SpPKE1, a multiple stress-responsive gene confers salt tolerance in tomato and tobacco. Int. J. Mol. Sci. 20 (10), 2478. 10.3 https://doi.org/390/ijms20102478.

Liang, L., Ridoutt, B.G., Lal, R., Wang, D., Wu, W., Peng, P., et al., 2019. Nitrogen footprint and nitrogen use efficiency of greenhouse tomato production in North China. J. Cleaner Prod. 208, 285–296. Available from: https://doi.org/10.1016/j.jclepro.2018.10.149.

Liang, G., Liu, J., Zhang, J., Guo, J., 2020. Effects of drought stress on photosynthetic and physiological parameters of tomato. J. Am. Soc. Hortic. Sci. 145 (1), 12–17. Available from: https://doi.org/10.21273/JASHS04725-19.

Liu, H.S., Li, F.M., 2005. Root respiration, photosynthesis and grain yield of two spring wheat in response to soil drying. Plant Growth Regul. 46, 233–240. Available from: https://doi.org/10.1007/s10725-005-8806-7.

Liu, J., Chen, F., Olokhnuud, C., Glass, A.D.M., Tong, Y., Zhang, F., et al., 2009. Root size and nitrogen-uptake activity in two maize (*Zea mays*) inbred lines differing in nitrogen-use efficiency. J. Plant Nutr. Soil Sci. 172 (2), 230–236. Available from: https://doi.org/10.1002/jpln.200800028.

Liu, L., Wang, J., Han, Z., Sun, X., Li, H., Zhang, J., et al., 2016. Molecular analyses of tomato GS, GOGAT And GDH gene families and their response to abiotic stresses. Acta Physiol. Plant. 38, 229. Available from: https://doi.org/10.1007/s11738-016-2251-2.

Liu, M., Yu, H., Zhao, G., Huang, Q., Lu, Y., Ouyang, B., 2017. Profiling of drought-responsive microrna and mrna in tomato using high-throughput sequencing. BMC Genomics 18 (1), 481. Available from: https://doi.org/10.1186/s12864-017-3869-1.

Liu, M., Yu, H., Zhao, G., Huang, Q., Lu, Y., Ouyang, B., 2018. Identification of drought-responsive micrornas in tomato using high-throughput sequencing. Funct. Integr. Genomics 18, 67–78. Available from: https://doi.org/10.1007/s10142-017-0575-7.

Luna, A.M., Garcia, E.R., Servín, J.L., Herrera, A.L., Arellano, J.S., 2014. Evaluation of different concentrations of nitrogen for tomato seedling production (*Lycopersicon esculentum* Mill.). Univ. J. Agric. Res. 2 (8), 305–312. Available from: https://doi.org/10.13189/ujar.2014.020804.

Madrid, R., Barba, E.M., Sánchez, A., García, A.L., 2009. Effects of organic fertilis- ers and irrigation level on physical and chemical quality of industrial tomato fruit (cv. Nautilus). J. Sci. Food Agric. 89, 2608–2615. Available from: https://doi.org/10.1002/jsfa.3763.

Manoj, K., Uday, D., 2007. In vitro screening of tomato genotypes for drought resistance using polyethylene glycol. Afr. J. Biotechnol 5, 1488–1493.

Masclaux-Daubresse, C., Daniel-Vedele, F., Dechorgnat, J., Chardon, F., Gaufichon, L., Suzuki, A., 2010. Nitrogen uptake, assimilation and remobilization in plants: challenges for sustainable and productive agriculture. Ann. Bot. 105 (7), 1141–1157. Available from: https://doi.org/10.1093/aob/mcq028.

Millones-Chanamé, C.E., Souza de Oliveiras, A.M., Castro, E.M., Maluf, W.R., 2019. Inheritance of blossom end rot resistance induced by drought stress and of associated stomatal densities in tomatoes. Euphytica 215, 120. Available from: https://doi.org/10.1007/s10681-019-2444-z.

Mingo, D.M., Theobald, J.C., Bacon, M.A., Davies, W.J., Dodd, I.C., 2004. Biomass allocation in tomato (*Lycopersicon esculentum*) plants grown under partial rootzone drying: enhancement of root growth. Funct. Plant Biol. 31, 971–978. Available from: https://doi.org/10.1071/FP04020.

Mittler, R., Vanderauwera, S., Suzuki, N., Miller, G., Tognetti, V.B., Vandepoele, K., et al., 2011. ROS signaling: the new wave? Trends Plant Sci. 16 (6), 300–309. Available from: https://doi.org/10.1016/j.tplants.2011.03.007.

Moles, T.M., Mariotti, L., De Pedro, L.F., Guglielminetti, L., Picciarelli, P., Scartazza, A., 2018. Drought induced changes of leaf-to-root relationships in two tomato genotypes. Plant Physiol. Biochem. 128, 24–31. Available from: https://doi.org/10.1016/j.plaphy.2018.05.008.

Montagu, K.D., Goh, K.M., 1990. Effects of forms and rates of organic and inorganic nitrogen fertilisers on the yield and some quality indices of tomatoes (*Lycopersicon esculentum* Miller). New Zealand J. Crop Hortic. Sci. 18 (1), 31–37. Available from: https://doi.org/10.1080/01140671.1990.10428067.

Montesinos-Pereira, D., Barrameda-Medina, Y., Romero, L., Ruiz, J.M., Sánchez-Rodríguez, E., 2014. Genotype differences in the metabolism of proline and polyamines under moderate drought in tomato plants. Plant Biol. (Stuttg.) 16 (6), 1050–1057. Available from: https://doi.org/10.1111/plb.12178.

Muhammad, T., Zhang, J., Ma, Y., Li, Y., Zhang, F., Zhang, Y., et al., 2019. Overexpression of a mitogen-activated protein kinase SlMAPK3 positively regulates tomato tolerance to cadmium and drought stress. Molecules 24 (3), 556. Available from: https://doi.org/10.3390/molecules24030556.

Munir, S., Liu, H., Xing, Y., Hussain, S., Ouyang, B., Zhang, Y., et al., 2016. Overexpression of calmodulin-like (ShCML44) stress-responsive gene from *Solanum Habrochaites* enhances tolerance to multiple abiotic stresses. Sci. Rep. 6, 31772. Available from: https://doi.org/10.1038/srep31772.

Murshed, R., Lopez-Lauri, F., Sallanon, H., 2013. Effect of water stress on antioxidant systems and oxidative parameters in fruits of tomato (*Solanum Lycopersicon* L, cv. Micro-tom). Physiol. Mol. Biol. Plants 19 (3), 363–378. Available from: https://doi.org/10.1007/s12298-013-0173-7.

Nadeem, M., Li, J., Yahya, M., Sher, A., Ma, C., Wang, X., et al., 2019. Research progress and perspective on drought stress in legumes: a review. Int. J. Mol. Sci. 20 (10), 2541. Available from: https://doi.org/10.3390/ijms20102541.

Nahar, K., Gretzmache, R., 2011. Response of shoot and root development of seven tomato cultivars in hydrophonic system under water stress. Acad. J. Plant Sci. 4 (2), 57–63.

Nahar, K., Ullah, S.M., Islam, N., 2011. Osmotic adjustment and quality response of five tomato cultivars (*Lycopersicon esculentum* Mill) following water deficit stress under subtropical climate. Asian J. Plant Sci. 10 (2), 153–157. Available from: https://doi.org/10.3923/ajps.2011.153.157.

Nahar, K., Ullah, S.M., 2017. Fruit quality and osmotic adjustment of four tomato cultivars under drought stress. Asian J. Soil Sci. Plant Nutr. 2 (1), 1–8. Available from: https://doi.org/10.9734/AJSSPN/2017/36861.

Nahar, K., Ullah, S.M., 2018. Drought stress effects on plant water relations, growth, fruit quality and osmotic adjustment of tomato (*Solanum Lycopersicum*) under subtropical condition. Asian J Agric. Hortic. Res. 1 (2), 1–14. Available from: https://doi.org/10.9734/AJAHR/2018/39824.

Nangare, D.D., Singh, Y., Kumar, P.S., Minhas, P.S., 2016. Growth, fruit yield and quality of tomato (*Lycopersicon esculentum* Mill.) as affected by deficit irrigation regulated on phenological basis. Agric. Water Manag. 171, 73–79. Available from: https://doi.org/10.1016/j.agwat.2016.03.016.

Nawaz, H., Zubair, M., Derawadan, H., 2012. Interactive effects of nitrogen, phosphorus and zinc on growth and yield of tomato (*Solanum lycopersicum*). African J. Agric. Res. 7 (26), 3792–3799. Available from: https://doi.org/10.5897/ajar12.1027.

Nonami, H., 1998. Plant water relations and control of cell elongation at low water potentials. J. Plant Res. 111, 373–382. Available from: https://doi.org/10.1007/BF02507801.

Nuruddin, M.M., 2001. Effects of water stress on tomato at different growth stages, MSc thesis, McGill University, Macdonald Campus, Montreal, Canada.

Okunlola, G.O., Olatunji, O.A., Afolabi, A.M., Gbadegesin, K.K., 2015. Effects of nitrogen nutritional stress on the morphological and yield parameters of tomato (*Solanum lycopersicum* L.). Sci. Cold Arid Reg. 7 (2), 137–145.

Ono, F., Frommer, W.B., Wirén, N., 2000. Coordinated diurnal regulation of low- and high-affinity nitrate transporters in tomato. Plant Biol. 2 (1), 17–23. Available from: https://doi.org/10.1055/s-2000-297.

Orellana, S, Yañez, M., Espinoza, A., Verdugo, I, González, E., Ruiz-Lara, S., et al., 2010. The transcription factor SlAREB1 confers drought, salt stress tolerance and regulates biotic and abiotic stress-related genes in tomato. Plant Cell Environ 33 (12), 2191–2208. Available from: https://doi.org/10.1111/j.1365-3040.2010.02220.x.

Osakabe, Y., Osakabe, K., Shinozaki, K., Tran, L.-S.P., 2014. Response of plants to water stress. Front. Plant Sci. 5 (86), 1–8. Available from: https://doi.org/10.3389/fpls.2014.00086.

Pan, Y., Seymour, G.B., Lu, C., Hu, Z., Chen, X., Chen, G., 2012. An ethylene response factor (ERF5) promoting adaptation to drought and salt tolerance in tomato. Plant Cell Rep. 31, 349–360. Available from: https://doi.org/10.1007/s00299-011-1170-3.

Parajuli, S., Ojha, B.R., Ferrara, G.O., 2018. Quantification of secondary traits for drought and low nitrogen stress tolerance in inbreds and hybrids of maize (*Zea mays* L.). J. Plant Genet. Breed. 2 (1), 106.

Parry, M.A.J., Andralojc, P.J., Khan, S., Lea, P.J., Keys, A.J., 2002. Rubisco activity: effects of drought stress. Ann. Bot. 89 (7), 833–839. Available from: https://doi.org/10.1093/aob/mcf103.

Patanè, C., Tringali, S., Sortino, O., 2011. Effects of deficit irrigation on biomass, yield, water productivity and fruit quality of processing tomato under semi-arid Mediterranean climate conditions. Sci. Hortic. 129 (4), 590–596. Available from: https://doi.org/10.1016/j.scienta.2011.04.030.

Patanè, C., Scordia, D., Testa, G., Cosentino, S.L., 2016. Physiological screening for drought tolerance in mediterranean long-storage tomato. Plant Sci. 249, 25–34. Available from: https://doi.org/10.1016/j.plantsci.2016.05.006.

Peng, M., Bi, Y.-M., Zhu, T., Rothstein, S.J., 2007. Genome-wide analysis of *Arabidopsis* responsive transcriptome to nitrogen limitation and its regulation by the ubiquitin ligase gene NLA. Plant Mol. Biol. 65 (6), 775–797. Available from: https://doi.org/10.1007/s11103-007-9241-0.

Pervez, M.A., Ayub, C.M., Khan, H.A., Shahid, M.A., Ashraf, I., 2009. Effect of drought stress on growth, yield and seed quality of tomato (*Lycopersicon esculentum* L.). Pak. J. Agric. Sci. 46 (3), 174–178.

Pinheiro, C., Chaves, M.M., 2011. Photosynthesis and drought: can we make metabolic connections from available data? J. Exp. Bot. 62 (3), 869–882. Available from: https://doi.org/10.1093/jxb/erq340.

Poiroux-Gonord, F., Bidel, L.P.R., Fanciullino, A.-L., Gautier, H., Lauri-Lopez, F., Urban, L., 2010. Health benefits of vitamins and secondary metabolites of fruits and vegetables and prospects to increase their concentrations by agronomic approaches. J. Agric. Food Chem. 58 (23), 12065–12082. Available from: https://doi.org/10.1021/jf1037745.

Pokluda, R., Petříková, K., Abdelaziz, M.E., Jezdinský, A., 2010. Effect of water stress on selected physiological characteristics of tomatoes. Acta Univ. Agric. et Silvic. Mendelianae Brun. 58 (1), 131–138. Available from: https://doi.org/10.11118/actaun201058010131.

Qi, H.Y., Li, T.L., Zhang, J., Wang, L., Chen, Y.H., 2003. Effects on sucrose metabolism, dry matter distribution and fruit quality of tomato under water deficit. Agric. Sci. China 2, 1253–1258.

Raja, V., Qadir, S.U., Alyemeni, M.N., Ahmad, P., 2020. Impact of drought and heat stress individually and in combination on physio-biochemical parameters, antioxidant responses, and gene expression in *Solanum lycopersicum*. 3 Biotech 10 (5), 208. Available from: https://doi.org/10.1007/s13205-020-02206-4.

Rajagopal, V., Rao, I.M., 1974. Changes in the endogenous level of auxins and gibberellin-like substances in the shoot apices of nitrogen-deficient tomato plants (*Lycopersicon esculentum* Mill). Aust. J. Bot. 22 (3), 429–435. Available from: https://doi.org/10.1071/BT9740429.

Rayar, A.J., van Hai, T., 1977. Effect of ammonium on uptake of phosphorus, potassium, calcium and magnesium by intact soybean plants. Plant Soil 48, 81–87. Available from: https://doi.org/10.1007/BF00015159.

Rouphael, Y., Cardarelli, M., Schwarz, D., Franken, P., Colla, G., 2012. Effects of drought on nutrient uptake and assimilation in vegetable crops. In: Aroca, R. (Ed.), Plant Responses to Drought Stress. Springer. Available from: https://doi.org/10.1007/978-3-642-32653-0_7.

Rubio-Wilhelmi, M.M., Sanchez-Rodriguez, E., Rosales, M.A., Begoña, B., Rios, J.J., Romero, L., et al., 2011. Effect of cytokinins on oxidative stress in tobacco plants under nitrogen deficiency. Environ. Exp. Bot. 72 (2), 167–173. Available from: https://doi.org/10.1016/j.envexpbot.2011.03.005.

Salehi-Lisar, S.Y., Rahman, I.M., Hossain, M.M., Motafakkerazad, R. (2012). Water stress in plants: causes, effects and responses. In: IntechOpen. https://doi.org/10.5772/39363 Available from https://www.intechopen.com/books/water-stress/water-stress-in-plants-causes-effects-and-responses

Sánchez-Rodríguez, E., Rubio-Wilhelmi, M.M., Cervilla, L.M., Blasco, B., Rios, J.J., Rosales, M.A., et al., 2010a. Genotypic differences in some physiological parameters symptomatic for oxidative stress under moderate drought in tomato plants. Plant Sci. 178 (1), 30–40. Available from: https://doi.org/10.1016/j.plantsci.2009.10.001.

Sánchez-Rodríguez, E., Rubio-Wilhelmi, Md.M., Cervilla, L.M., Blasco, B., Rios, J.J., Leyva, R., et al., 2010b. Study of the ionome and uptake fluxes in cherry tomato plants under moderate water stress conditions. Plant Soil 335 (1-2), 339–347. Available from: https://doi.org/10.1007/s11104-010-0422-2.

Sánchez-Rodríguez, E., Rubio-Wilhelmi, Md.M., Ríos, J.J., Blasco, B., Rosales, M.A., Melgarejo, R., et al., 2011a. Ammonia production and assimilation: its importance as a tolerance mechanism during moderate water deficit in tomato plants. J. Plant Physiol. 168 (8), 816–823. Available from: https://doi.org/10.1016/j.jplph.2010.11.018.

Sánchez-Rodríguez, E., Moreno, D.A., Ferreres, F., Rubio-Wilhelmi, Md.M., Ruiz, J.M., 2011b. Differential responses of five cherry tomato varieties to water stress: changes on phenolic metabolites and related enzymes. Phytochemistry 72 (8), 723–729. Available from: https://doi.org/10.1016/j.phytochem.2011.02.011.

Scholberg, J., McNeal, B.L., Boote, K.J., Jones, J.W., Locascio, S.J., Olson, S.M., 2000. Nitrogen stress effects on growth and nitrogen accumulation by field-grown tomato. Agron. J. 92 (1), 159. Available from: https://doi.org/10.2134/agronj2000.921159x.

Selim, A.-F.H., El-Nady, M.F., 2011. Physio-anatomical responses of drought stressed tomato plants to magnetic field. Acta Astronaut. 69 (7-8), 387–396. Available from: https://doi.org/10.1016/j.actaastro.2011.05.025.

Senthilkumar, M., Sadashiva, A.T., Laxmanan, V., 2017. Impact of water stress on root architecture in tomato (*Solanum lycopersicum* Mill). Int. J. Curr. Microbiol. Appl. Sci. 6 (7), 2095–2103. Available from: https://doi.org/10.20546/ijcmas.2017.607.247.

Shangguan, Z.P., Shao, M.A., Dyckmans, J., 2000. Nitrogen nutrition and water stress effects on leaf photosynthetic gas exchange and water use efficiency in winter wheat. Environ. Exp. Bot. 44 (2), 141–149. Available from: https://doi.org/10.1016/S0098-8472(00)00064-2.

Shao, H.-B., Chu, L.-Y., Jaleel, C.A., Zhao, C.-X., 2008. Water-deficit stress-induced anatomical changes in higher plants. C. R. Biol. 331 (3), 215–225. Available from: https://doi.org/10.1016/j.crvi.2008.01.002.

Shirasawa, K., Isobe, S., Hirakawa, H., Asamizu, E., Fukuoka, H., Just, D., et al., 2010. SNP discovery and linkage map construction in cultivated tomato. DNA Res. 17 (6), 381–391. Available from: https://doi.org/10.1093/dnares/dsq024.

Sibomana, I.C., Aguyoh, J.N., Opiyo, A.M., 2013. Water stress affects growth and yield of container grown tomato (*Lycopersicon esculentum* Mill) plants. Global Journal of BioScience and Biotechnology 2 (4), 461–466.

Siddiqi, M.Y., Malhotra, B., Min, X., Glass, A.D.M., 2002. Effects of ammonium and inorganic carbon enrichment on growth and yield of a hydroponic tomato crop. J. Plant Nutr. Soil Sci. 165 (2), 191. https://doi.org/10.1002/1522-2624(200204)165:2 < 191::aid-jpln191 > 3.0.co;2-d.

Singh, M., Singh, V.P., Prasad, S.M., 2019. Nitrogen alleviates salinity toxicity in *Solanum lycopersicum* seedlings by regulating ROS homeostasis. Plant Physiol. Biochem. 141, 466–476. Available from: https://doi.org/10.1016/j.plaphy.2019.04.004.

Sinclair, T.R., Muchow, R.C., 2001. System analysis of plant traits to increase grain yield on limited water supplies. Agron. J. 93 (2), 263. Available from: https://doi.org/10.2134/agronj2001.932263x.

Siracusa, L., Patanè, C., Rizzo, V., Cosentino, S.L., Ruberto, G., 2018. Targeted secondary metabolic and physico-chemical traits analysis to assess genetic variability within a germplasm collection of "long storage" tomatoes. Food Chem. 244, 275–283. Available from: https://doi.org/10.1016/j.foodchem.2017.10.043.

Sivakumar, R., Srividhya, S., 2016. Impact of drought on flowering, yield and quality parameters in diverse genotypes of tomato (*Solanum lycopersicum* L.). Adv. Hortic. Sci. 30, 3–11. Available from: https://doi.org/10.13128/ahs-18696.

Sivakumar, R., Nandhitha, G.K., Nithila, S., 2018. Impact of drought on chlorophyll, soluble protein, abscisic acid, yield and quality characters of contrasting genotypes of tomato (*Solanum lycopersicum*). Curr. J. Appl. Sci. Technol. 21 (5), 1–10. Available from: https://doi.org/10.9734/BJAST/2017/34347.

Smart, D.R., Bloom, A.J., 1993. Relationships between the kinetics of NH_4^+ and NO_3^- absorption and growth in the cultivated tomato (*Lycopersicon esculentum* Mill, cv T-5). Plant Cell Environ. 16 (3), 259–267. Available from: https://doi.org/10.1111/j.1365-3040.1993.tb00868.x.

Spollen, W.G., Sharp, R.E., 1991. Spatial distribution of turgor and root growth at low water potentials. Plant Physiol. 96 (2), 438–443. Available from: https://doi.org/10.1104/pp.96.2.438.

Srinivasa Rao, N.K., Bhatt, R.M., Sadasivan, A.T., 2001. Tolerance to water stress in tomato cultivars. Photosynthetica 38 (3), 465–467. Available from: https://doi.org/10.1023/A:1010902427231.

Sun, Y., Wang, C., Chen, H.Y.H., Ruan, H., 2020. Response of plants to water stress: a meta-analysis. Front. Plant Sci. 11 (978), 1--18. Available from: https://doi.org/10.3389/fpls.2020.00978.

Suniaga Quijada, J., Bonafous, M., Dumas, Y., 1992. Growth and development of young tomato plants under nitrogen deficiency. In: International Symposium on Processing Tomatoes, Mendoza, Argentina, pp. 159–164.

Tabatabaie, S.J., Gregory, P.J., Hadley, P., 2004. Uneven distribution of nutrients in the root zone affects the incidence of blossom end rot and concentration of calcium and potassium in fruits of tomato. Plant Soil Dordrecht 258 (2), 169–178. Available from: https://doi.org/10.1023/B:PLSO.0000016548.84566.62.

Takács, S., Pék, Z., Csányi, D., Daood, H.G., Szuvandzsiev, P., Palotás, G., et al., 2020. Influence of water stress levels on the yield and lycopene content of tomato. Water 12 (8), 2165. Available from: https://doi.org/10.3390/w12082165.

Tamburino, R., Vitale, M., Ruggiero, A., Sassi, M., Sannino, L., Arena, S., et al., 2017. Chloroplast proteome response to drought stress and recovery in tomato (*Solanum lycopersicum* L.). BMC Plant Biol. 17 (40), 1–14. Available from: https://doi.org/10.1186/s12870-017-0971-0.

Tapia, G., Mendez, J., Inostroza, L., 2016. Different combinations of morpho-physiological traits are responsable for tolerance to drought in wild tomatoes *Solanum chilense* and *Solanum peruvianum*. Plant Biol. 18 (3), 406–416. Available from: https://doi.org/10.1111/plb.12409.

Tavallali, V., Esmaili, S., Karimi, S., 2018. Nitrogen and potassium requirements of tomato plants for the optimization of fruit quality and antioxidative capacity during storage. Journal of Food Measurement and Characterization 12, 755–762. Available from: https://doi.org/10.1007/s11694-017-9689-9.

Tei, F., Benincasa, P., Guiducci, M., 2002. Effect of N availability on growth, N uptake, light interception and photosynthetic activity in processing tomato. Acta Hortic. 571, 209–216. Available from: https://doi.org/10.17660/ActaHortic.2002.571.25.

Tewari, R.K., Kumar, P., Sharma, P.N., 2007. Oxidative stress and antioxidant responses in young leaves of mulberry plants grown under nitrogen, phosphorus or potassium deficiency. J. Integr. Plant Biol. 49 (3), 313–322. Available from: https://doi.org/10.1111/j.1744-7909.2007.00358.x.

Tezara, W., Mitchell, V.J., Driscoll, S.D., Lawlor, D.W., 1999. Water stress inhibits plant photosynthesis by decreasing coupling factor and ATP. Nature 401, 914–917. Available from: https://doi.org/10.1038/44842.

Tian, J., Guo, S.R., Sun, J., Wang, L.-P., Yang, Y., Li, B., 2011. Effects of exogenous spermidine on nitrogen metabolism of cucumber seedlings under high temperature stress. Chin. J. Ecol. 30 (10), 2197–2202. Available from: https://doi.org/10.21273/JASHS.141.4.381.

Thipyapong, P., Melkonian, J., Wolfe, D.W., Steffens, J.C., 2004. Suppression of polyphenol oxidases increases stress tolerance in tomato. Plant Sci. 167, 693–703. Available from: https://doi.org/10.1016/j.plantsci.2004.04.008.

Thirumalaikumar, V.P., Devkar, V., Mehterov, N., Ali, S., Ozgur, R., Turkan, I., et al., 2018. NAC transcription factor JUNGBRUNNEN1 enhances drought tolerance in tomato. Plant Biotechnol. J. 16 (2), 354–366. Available from: https://doi.org/10.1111/pbi.12776.

Vega-Mas, M.T., Rossi, K.J., Gupta, C., González-Murua, R.G., Ratcliffe, J.M., Estavillo, et al., 2019. Tomato roots exhibit in vivo glutamate dehydrogenase aminating capacity in response to excess ammonium supply. J. Plant Physiol. 239, 83–91. Available from: https://doi.org/10.1016/j.jplph.2019.03.009.

Vernoux, T., Besnard, F., Traas, J., 2010. Auxin at the shoot apical meristem. Cold Spring Harb. Perspect. Biol. 2 (4), 1487. Available from: https://doi.org/10.1101/cshperspect.a001487.

Von Wiren, N., Lauter, F.-R., Ninnemann, O., Gillissen, B., Walch-Liu, P., Engels, C., et al., 2000. Differential regulation of three functional ammonium transporter genes by nitrogen in root hairs and by light in leaves of tomato. Plant J. 21 (2), 167–175. Available from: https://doi.org/10.1046/j.1365-313x.2000.00665.x.

Wang, Y.-H., 2001. Nitrate-induced genes in tomato roots. array analysis reveals novel genes that may play a role in nitrogen nutrition. Plant Physiol. 127 (1), 345–359. Available from: https://doi.org/10.1104/pp.127.1.345.

Wang, Y.-T., Huang, S.-W., Liu, R.-L., Jin, J.-Y., 2007. Effects of nitrogen application on flavor compounds of cherry tomato fruits. J. Plant Nutr. Soil Sci. 170 (4), 461–468. Available from: https://doi.org/10.1002/jpln.200700011.

Wang, F., Kang, S., Du, T., Li, F., Qiu, R., 2011. Determination of comprehensive quality index for tomato and its response to different irrigation treatments. Agric. Water Manag. 98, 1228–1238. Available from: https://doi.org/10.1016/j.agwat.2011.03.004.

Wang, H., Fan, H., Yao, H., 2020. Effects of elevated CO_2 on tomato (Lycopersicon esculentum Mill.) growth and rhizosphere soil microbial community structure and functionality. Agronomy 10, 1752. Available from: https://doi.org/10.3390/agronomy10111752.

Wang, Y., Cai, S., Yin, L., Shi, K., Xia, X., Zhou, Y., et al., 2015. Tomato HsFA1a plays a critical role in plant drought tolerance by activating ATG genes and inducing Autophagy. Autophagy 11 (11), 2033–2047. Available from: https://doi.org/10.1080/15548627.2015.1098798.

Wang, P., Sun, X., Jia, X., Wang, N., Gong, X., Ma, F., 2016. Characterization of an autophagy-related gene MdATG8i from Apple. Front. Plant Sci. 7, 720. Available from: https://doi.org/10.3389/fpls.2016.00720.

Wang, L., Chen, L., Li, R., Zhao, R., Yang, M., Sheng, J., et al., 2017. Reduced drought tolerance by CRISPR/Cas9-mediated SlMAPK3 mutagenesis in tomato plants. J. Agric. Food Chem. 65 (39), 8674–8682. Available from: https://doi.org/10.1021/acs.jafc.7b02745.

Wang, Y., Cao, J.-J., Wang, K.-X., Xia, X.-J., Shi, K., Zhou, et al., 2019. BZR1 mediates brassinosteroid-induced autophagy and nitrogen starvation tolerance in tomato. Plant Physiol. 179 (2), 671–685. Available from: https://doi.org/10.1104/pp.18.01028.

Wani, S.H., Singh, N.B., Haribhushan, A., Mir, J.I., 2013. Compatible solute engineering in plants for abiotic stress tolerance – role of glycine betaine. Curr. Genomics 14 (3), 157–165. Available from: https://doi.org/10.2174/1389202911314030001.

Warner, J., Zhang, T.Q., Hao, X., 2004. Effects of nitrogen fertilization on fruit yield and quality of processing tomatoes. Canadian J. Plant Sci. 84 (3), 865–871. Available from: https://doi.org/10.4141/p03-099.

Wei, S., Wang, X., Shi, D., Li, Y., Zhang, J., Liu, P., et al., 2016. The mechanisms of low nitrogen induced weakened photosynthesis in summer maize (*Zea Mays* L.) under field conditions. Plant Physiol. Biochem. 105, 118–128. Available from: https://doi.org/10.1016/j.plaphy.2016.04.007.

Wu, Y., Cosgrove, D.J., 2000. Adaptation of roots to low water potentials by changes in cell wall extensibility and cell wall proteins. J. Exp. Bot. 51 (350), 1543–1553. Available from: https://doi.org/10.1093/jexbot/51.350.1543.

Yoneyama, K., Xie, X., Kim, H., Nomura, T., Sekimoto, H., et al., 2012. How do nitrogen and phosphorus deficiencies affect strigolactoneproduction and exudation? Planta 235, 1197–1207. Available from: https://doi.org/10.1007/s00425-011-1568-8.

Yuan, X.K., Yang, Z.Q., Li, Y.X., Liu, Q., Han, W., 2016. Effects of different levels of water stress on leaf photosynthetic characteristics and antioxidant enzyme activities of greenhouse tomato. Photosynthetica 54 (1), 28–39. Available from: https://doi.org/10.1007/s11099-015-0122-5.

Zargar, S.M., Gupta, N., Nazir, M., Mahajan, R., Malik, F.A., Sofi, N.R., et al., 2017. Impact of drought on photosynthesis: molecular perspective. Plant Gene 11, 154–159. Available from: https://doi.org/10.1016/j.plgene.2017.04.003.

Zhang, Y., Shi, Y., Gong, H.-J., Zhao, H.-l, Li, H.-l, Hu, Y.-h, et al., 2018. Beneficial effects of silicon on photosynthesis of tomato seedlings under water stress. J. Integr. Agric. 17 (10), 2151–2159. Available from: https://doi.org/10.1016/S2095-3119(18)62038-6.

Zhou, X., Zhao, P., Wang, W., Zou, J., Cheng, T., Peng, X., et al., 2015. A comprehensive, genome-wide analysis of autophagy-related genes identified in tobacco suggests a central role of autophagy in plant response to various environmental cues. DNA Res. 22 (4), 245–257. Available from: https://doi.org/10.1093/dnares/dsv012.

Zhou, R., Yu, X., Ottosen, C.-O., Rosenqvist, E., Zhao, L., Wang, Y., et al., 2017. Drought stress had a predominant effect over heat stress on three tomato cultivars subjected to combined stress. BMC Plant Biol. 17, 24. Available from: https://doi.org/10.1186/s12870-017-0974-x.

Zhou, R., Kong, L., Wu, Z., Rosenqvist, E., Wang, Y., Zhao, L., et al., 2019a. Physiological response of tomatoes at drought, heat and their combination followed by recovery. Physiol. Plant. 165, 144–154. Available from: https://doi.org/10.1111/ppl.12764.

Zhou, R., Kong, L., Yu, X., Ottosen, C.-O., Zhao, T., Jiang, F., et al., 2019b. Oxidative damage and antioxidant mechanism in tomatoes responding to drought and heat stress. Acta Physiol. Plant. 41, 20. Available from: https://doi.org/10.1007/s11738-019-2805-1.

Zhu, M., Meng, X., Cai, J., Li, G., Dong, T., Li, Z., 2018. Basic leucine zipper transcription factor SlbZIP1 mediates salt and drought stress tolerance in tomato. BMC Plant Biol. 18 (1), 83. Available from: https://doi.org/10.1186/s12870-018-1299-0.

Chapter 13

Arsenic stress and mineral nutrition in plants

Santanu Samanta and Aryadeep Roychoudhury
Department of Biotechnology, St. Xavier's College (Autonomous), Kolkata, India

13.1 Introduction

In the periodic table, arsenic (As) is probably the only element that has been the center of controversy for thousands of years. Arsenic is one of the most toxic element on earth, with several social impacts on public health from ancient times to modern times (Bhowmick et al., 2018). Arsenic and its compounds have been categorized as nonthreshold group I carcinogen according to the US Environmental Protection Agency (EPA) and the International Agency for Research on Cancer (IARC) (Niazi et al., 2018; Rosas-Castor et al., 2014). It is the 20th most common basic element in the earth's crust, a frequent contaminant of groundwater, and can be assimilated into plants destined for human consumption (Singh et al., 2015). Human exposure to As also comes from contaminated water.

Arsenic is considered nonessential for plants and has no known metabolic function (Zhao et al., 2010; Marin et al., 1993). Plant species vary in their tolerance to As, with toxicity threshold levels ranging from 5 to 100 mg As kg^{-1} dry weight for most plants (Meharg, 1994; Kabata-Pendias and Pendias, 1992). Arsenic does not break down during the natural metabolism of the biota. The oxidation states of As generally existing in soil, water, and environment are As^{-III}, As^{0}, As^{III}, and As^{V} (Panda et al., 2010). Both inorganic and organic species (e.g., monomethyl arsenic acid, dimethyl arsenic acid, arsenobetaine, and arsenosugars) of As are present, with the former being more toxic and mobile than organic As-species (Martinez et al., 2011). Arsenate (As^{V}) and arsenite (As^{III}) prevail in the natural environment; As^{III} is predominant in reducing environments such as submerged soils, whereas As^{V} is more abundant in oxidizing environments (Srivastava et al., 2016). These two forms are readily interconvertible depending upon environmental conditions (Mukhopadhyay et al., 2014). The abundance of organic species

Plant Nutrition and Food Security in the Era of Climate Change. DOI: https://doi.org/10.1016/B978-0-12-822916-3.00002-0

of As is relatively low in most soil and their phytotoxicity has not been closely examined to date.

Diverse reactions occurring in soil affect As speciation and regulate uptake of As through plant roots as well as its biogeochemical behavior in the soil-plant system. It has been generally considered that the inorganic forms of As are more lethal compared to their organic forms (Finnegan and Chen, 2012). Further, amongst the inorganic forms, As^{III} is more soluble and 60 times more toxic and mobile than As^{V}, since the former exerts detrimental effects on general protein metabolism by reacting with sulfhydryl (−SH) groups existing in cysteine residues, thereby affecting protein structures and/or catalytic functions and eventually cell death (Li et al., 2015; Pandey et al., 2017; Singh et al., 2017). The uptake of As through plant roots is dependent upon its concentration and more importantly on the As speciation in the soil and rhizosphere (Rafiq et al., 2017a; 2017b). The two dominant inorganic species of As, viz., arsenate and arsenite, enter into the plant systems through different transport systems. Due to the structural resemblance between the oxyanion structures of arsenate and phosphate, arsenate readily enters into cells nonspecifically via phosphate transporters as a charged anion. Arsenite is taken up by the plants through aquaporin channels because it is predominantly present as a neutral species in the environment (Samanta and Roychoudhury, 2021; Niazi et al., 2017, 2015; Kumar et al., 2015).

13.2 Effects of arsenic on plants

Arsenic availability in soil can affect normal functioning of plant metabolism, which can cause stunted growth and losses of crop productivity and yield. Although low doses of As ($4\ mg\ kg^{-1}$ soil) improved growth of *Helianthus annuus* seedlings (Imran et al., 2013), As hampers biochemical and metabolic functions at sufficiently higher concentrations, which can result in plant death. Upon exposure to As stress, the number of leaves, leaf area, and fresh and dry weight of plants were suppressed. After entry into the plants, As also induces several physiological alterations in plant tissues including changes in redox potential, resulting in excess formation of reactive oxygen species (ROS), which leads to cellular oxidative damages (Samanta et al., 2020a; Shahid et al., 2017; Abbas et al., 2018). These ROS are quite dangerous for plant metabolism, sometimes leading to membrane damages, nonspecific oxidation of proteins and membrane lipids, and DNA injury (Zhao et al., 2010; Sharma, 2012; Talukdar, 2013). At high concentration ($150\ \mu M$), As diminishes photosynthetic efficiency (Samanta et al., 2020a), perturbs hormonal (Samanta et al., 2021) and nutrient balance, and eventually suppresses growth. Arsenic toxicity in plants is governed by its speciation chemistry, its eventual bioaccumulation in particular tissues, and also its concentration in the medium. In a hydroponic study by Liu et al. (2008), it has been observed that As significantly influences potassium (K), calcium (Ca), magnesium (Mg), manganese (Mn), and zinc (Zn) concentration in *Triticum aestivum*. On the contrary, Farnese et al. (2014) showed no impact of As on the uptake of K, Ca, and Zn concentrations in *Pistia stratiotes*. Liu et al. (2008) have however demonstrated changes in the uptake of these nutrients in other plants during exposure to this metalloid.

13.3 Mineral nutrition in plants

Mineral elements are essential for normal plant growth and development and play a pivotal role in the accumulation of biomass. They can be categorized as (1) macronutrients or macroelements that are essential for plant nutrition and the composition ranges from 1000 to 15,000 mg kg^{-1} dry mass, for example, nitrogen (N), phosphorus (P), potassium (K), sulfur (S), calcium (Ca), and magnesium (Mg); and (2) micronutrient or microelement associated with various enzyme systems and believed to be intermediate in mobility. These are required in smaller amounts than macronutrients, ranging from 0.1 to 100 mg kg^{-1} dry mass, for example boron (B), manganese (Mn), iron (Fe), copper (Cu), zinc (Zn), and molybdenum (Mo). Plants generally maintain micronutrients in a stringent manner because concentration of most micronutrients is quite narrow in the plant body (Marschner, 1995). There are several parameters such as soil pH, form of nutrient, type of crops, extent of organic materials in soil, relative quantities of other nutrients as well as microbial activities that altogether regulate the optimal amount of each micronutrient and can affect the mineral species or forms and their efficiency of uptake, for example Fe is more soluble at pH 6.0 than pH 7.0, although Mo becomes more soluble toward alkalinization over the same pH range.

Plants maintain nutrient levels in the optimal range in a highly fascinating manner. When the nutrient level shifts toward the low or deficiency end, plants trigger the activity of selective transporters to enhance uptake and extrude protons or small molecules into the soil to increase the bioavailability of the respective nutrient (Marschner, 1983). Additionally, to enhance uptake of nutrients, root architecture is also modified, for example, by increasing cell division in root hair cells or lateral roots. When plants face nutrient demand greater than supply, the growth rate of plants is slowed down in order to survive the nutrient-limiting environment (Ramage and Williams, 2002). Plants respond by switching to alternative metabolic pathways that rely less on limiting nutrients, for example, by repressing the synthesis of nonessential metalloproteins in favor of those that are essential. Nutrient limitation can also lead to reallocation of nutrients, so that they can be used optimally to support the metabolism, growth, and defense needs of plants, whereas at the higher end or luxury end of the optimal range, excess nutrients are stored for future purpose. When a nutrient is present in very high levels, it can cause toxicity, which ultimately affects plant growth (Niedz and Evens, 2007).

Membrane-bound nutrient transporters also play a pivotal role in nutrient homeostasis. The nutrients are imported through the membrane-localized transporters and pass beyond the impermeable endodermis and/or exodermis (Park et al., 2008). To enter or exit the vacuole or other organelles for functional purposes, transporters are much needed. Translocation of nutrients through the xylem pool requires efflux of nutrients from symplast to the apoplast via transporters and subsequent influx through transporters back into the symplast of distal tissues. Any particular nutrient that is translocated nonselectively through the xylem pool tends to be accumulated within the highly transpiring tissues such as leaves (Clark and Duncan, 1991). Xylem and phloem are the important control points for nutrient translocation and homeostasis. Several reports have highlighted that leaf nodes in grasses act as hubs for nutrient redistribution between xylem and phloem (Yamaji and Ma, 2014).

13.4 Effect of arsenic on mineral nutrition in plants

During its route from soil to plants, As interacts with a number of elements and also influences nutrient uptake by competing directly with nutrients and altering metabolic processes. Arsenic has been found to alter macronutrient and micronutrient balances in rice grains (Hossain et al., 2008). In addition to this, Williams et al. (2009) reported that As toxicity limits microelements in rice grains, such as zinc (Zn), nickel (Ni), and selenium (Se). An imbalance in the composition of nutrient elements can induce As stress-induced injuries in rice (Shaibur et al., 2006, 2008a), barley (Shaibur et al., 2008b), sorghum (Shaibur et al., 2008a), and bush bean (Wallace et al., 1980). It has been reported in a hydroponic culture that the addition of As significantly influences the concentration of K, Ca, Mg, Mn, and Zn in winter wheat (*T. aestivum*) (Liu et al., 2008). Therefore a detailed knowledge about the consequences of As stress on the distribution, uptake, and translocation of mineral elements in different plant species is much required. The physiological roles of major macronutrients and micronutrients and how these elements are modulated during exposure to As in different plant species are discussed below.

13.4.1 Nitrogen

Nitrogen (N) is a primary component of proteins, amino acids, nucleic acids, purines, and pyrimidines as well as chlorophyll. Arsenic interrupts nutrient absorption, which indirectly affect plant growth. There are very few studies about the relationship between As and N in plants. Previously, Ghosh et al. (2013) reported that As might affect the key enzymes of nitrogen assimilation pathways, decreases nitrate (NO_3^-) uptake, and subsequently decline the concentration of N in leaves. Recently Li et al. (2019) have reported that

As exposure increased N content in the mature leaves but reduced its content in the expanded and emerging leaves of annual ryegrass. Although a reverse trend was observed in the case of perennial ryegrass, where treatment of As enhanced the N content slightly in emerging leaves, it did not show any effect in mature leaves of perennial ryegrass. This could be related to the different varieties of the same plants used for experiment, which also indicated different strategies adopted by different plants to counteract stress. Liu et al. (2008) found that As did not significantly affect the translocation coefficients of N, and the correlation between As and N was not significant, which suggested that As does not have significant effect on the absorption and metabolism of N in winter wheat. It has been observed by several researchers that arsenate was more toxic for N uptake (Garg and Kashyap, 2017; Wang et al., 2010).

13.4.2 Phosphorus

Phosphorus (P) is a component of high-energy substances such as adenosine triphosphate (ATP), adenosine diphosphate, and adenosine monophosphate, which are important components of nucleic acids and phospholipids. Arsenate is a phosphate analog and they share a transport channel in the root plasmalemma and are also controlled by the same gene. Because of the similar electron configuration and chemical properties of As and P, they compete with each other for soil sorption sites in soil (Rahman et al., 2008). Gusman et al. (2013) demonstrated that both As^{V} and As^{III} at low concentration (6.6 μmol L^{-1}) enhanced the P content in the roots and shoots of lettuce plant, whereas at higher concentrations (13.2, 26.4, and 52.8 μmol L^{-1}), a decreased trend of P accumulation was noted. Farnese et al. (2014) also showed that P uptake was higher up to 13 μM of As treatment, which was then decreased in *P. stratiotes*. The increase in P uptake up to a certain concentration of As was also reported in *Pteris vittata* (Singh and Ma, 2006). Some studies have reported low As concentration responsible for an induction of P uptake by plants, probably due to P deficiency, since arsenate replaces P in plants (Carbonell et al., 1998). It has been shown that the presence of P can strongly suppress the absorption of arsenate (Peryea, 1991; Hossain et al., 2009). Arsenate and P interaction depends on concentration as well as growth conditions. Reduction of P accumulation occurs at very high concentration of arsenate, due to As phytotoxicity within the plants or reduced uptake of P due to competition by As for the same transporter across the membrane (Gilberti et al., 2014). Similarly, Mohamed et al. (2016) demonstrated that high concentrations of arsenate (100, 200, and 400 μM) led to a deficiency of P concentration in shoots of *Vicia faba* plants. Several studies have postulated that the molar ratio of P:As influences plant growth (Gulz, 2002).

13.4.3 Potassium

Potassium (K) is essential for carbon and nitrogen metabolism as well as protein synthesis and enzyme activities. It regulates photosynthesis by influencing the opening and closing of stomata in plants. In addition, K maintains cell electronegativity and cell turgor pressure within the cells. Shaibur et al. (2015) found that concentration of K in the xylem sap composition was almost constant irrespective of treatment of As in growth medium. Arsenic uptake occurs through anion forms and it is widely known that K is a "cation for anion" (Clarkson and Hanson, 1980), and it is probably the factor governing stable concentration of K in the xylem sap composition. A contrasting result was reported by Carbonell and Mataix-Beneyto (1997), in which As declined the content and uptake of K through the roots, whereas the content increased in aboveground tissues in bean plants. Arsenic is known to inhibit key enzymatic activity by reacting with sulfhydryl (−SH) groups of protein (Karabal et al., 2003) and repressing root physiology (Orwick et al., 1976). Arsenic can block the K absorption site in roots (Shaibur et al., 2009). Li et al. (2019) showed that K content increased significantly in emerging leaves of two ryegrass species, perennial ryegrass (*Lolium perene*) and annual ryegrass (*Lolium multiflorum*) under As stress (Cakmak, 2005). Under stress situations, plants translocate more K to young portions of leaves to modulate stress-induced injury. Lombi et al. (2002) have found a positive correlation between As and K, while investigating As distribution in *P. vittata*. A significant reduction in K content in rice was also observed upon applying both inorganic and organic species of As (Marin, 1992). The decrease in K uptake could be the result of a competition with Na present in solution, since in all the above studies, As was applied as sodium salts.

13.4.4 Calcium

Calcium (Ca) is of fundamental importance for maintaining the integrity of plasma membranes and acts as a second messenger in metabolic regulation (Tu and Ma, 2005). It plays a pivotal role in cell elongation and division and transfer of carbohydrates in plants. Generally, higher plants contain 5−30 mg Ca kg^{-1} dry weight (Mengel and Kirkby, 2001) or 1−50 mg Ca kg^{-1} dry weight (Marschner, 1995). Arsenic toxicity reduced Ca content in shoots of *Hordeum vulgare* (Shaibur et al., 2009). The reduction of Ca translocation was possibly due to reduced transportation of water inside the plant tissues, which ultimately limits the rate of xylem flow under As stress (Stoeva et al., 2005). On the contrary, As treatment increased Ca content in mature and expanded leaves of two ryegrass species, although As did not impact Ca content in the emerging leaves of both ryegrass species (Li et al., 2019). Similar to this result, Mallick et al. (2011) and Li et al. (2006) also noticed increased Ca content in *Zea mays* and *P. vittata* respectively under As stress. This could be due to the long-distance translocation and distribution of Ca in the plants, which rely upon both transpiration rates and duration of transpiration. Ca^{2+} ion is absorbed only through the young root tips of plants (Clarkson and Sanderson, 1971) by the negative charge of the phosphate groups within the membrane lipids (Caldwell and Haug, 1982).

13.4.5 Sulfur

Sulfur (S) is an essential microelement in plants, which is required for the biosynthesis of amino acids such as cysteine and methionine (Capaldi et al., 2015). In addition, it also acts as a precursor for glutathione (GSH) and other thiol compounds in plants like nonprotein thiols (metabolite antioxidant), metallothioneins, and phytochelatins (Samanta et al., 2020b; Dixit et al., 2016), which help to detoxify As by sequestration of (As^{III}) with thiol-rich peptides and subsequently store them in the vacuole. By forming complexes, it also protects plants from oxidative damage and restricts As in roots and its translocation to the aboveground tissues. Campos et al. (2015) showed that As toxicity reduced translocation of S in non-*Pteris* species, *Pityrogramma calomelanos*, but increased nonprotein thiol content in pinnae, which suggested the remobilization of organic S. Similarly, Wei et al. (2010) also reported that As enhanced GSH biosynthesis in fronds of Chinese brake fern without modulating concentration of S. An increase in both S and S-containing compounds has been reported by several researchers with an increase in As in castor bean (Melo et al., 2009), *Chilopsis linearis* (Castillo-Michel et al., 2009), and coontail (*Ceratophyllum demersum*) (Mishra et al., 2008). Reed et al. (2015) found that uptake of S tended to spike at low concentrations of As (0.75 mg L^{-1}) and then decline with higher concentrations of As (2.25 and 3.75 mg L^{-1}) in a few selected ornamental plants like *Coreopsis lanceolata*, *Iris savannarum*, *Tagetes erecta*, *H. annuus*, and *Panicum virgatum*. The increased production of GSH and phytochelatins used in detoxification and sequestration of As would limit the concentration of S for general metabolic regulation. The initial increase in S concentration was probably due to subsequent reduction in tissue S when uptake could not supply enough S for both detoxification and metabolic function.

13.4.6 Magnesium

Magnesium (Mg) is the central atom of chlorophyll molecules in plants. One major role of Mg is to serve as cofactor of almost all enzymes that activate the phosphorylation process. Similar to Ca, Mg content was also decreased in plants under As stress. Shaibur et al. (2009) also reported that Mg concentration in plants was significantly decreased in As-treated plants as compared to control plants. Mg plays a crucial role in photosynthesis, but chlorosis induced by As severely impacts plant growth, chlorophyll content, and rate of photosynthesis (Jutamanee and Ngennoy, 2013; Stoeva et al., 2005). This decrease in Mg content occurred due to uncoupling activity of As in oxidative phosphorylation in the electron transport chain and decreased chlorophyll content in plants. Marschner (1995) stated that the toxic effects of As in plant roots is also a reason behind the reduced Mg uptake in stressed plants.

13.4.7 Manganese

Manganese (Mn) is very abundant in the earth's crust and is primarily taken up by plants as Mn^{2+}. The availability of Mn in soil is pH dependent and Mn^{2+} is more abundant in acidic soil. It acts as cofactor for several enzymes of carbohydrate metabolism, the Krebs cycle, as well as some other enzymes like glutamyl transferase, cysteine desulphydrase, and various Golgi-localized glycosyltransferases (Clarkson and Hanson, 1980). It plays a crucial role in photosystem II of photosynthesis. In plants, Mn participates in maintaining the structure of photosynthetic proteins and enzymes (Schmidt and Husted, 2019) and plays a role in ATP synthesis, in ribulose 1,5-bisphosphate carboxylase reactions, and the biosynthesis of proteins, fatty acids, and acyl lipids (Millaleo et al., 2010). In an earlier study by Päivöke and Simola (2001), Mn concentration was shown to be increased in both root and shoot of *Pisum sativum* under arsenate stress.

13.4.8 Boron

Boron (B) is the only nonmetal among micronutrients in plants. The best characterized physiological role played by B is to crosslink pectin polysaccharides in cell walls. The functions of B in plants are still unraveled. It is well established that many conditions that add B to fields can increase yields, so that B is now a common component of fertilizers. B uptake is mediated by the plants with the flow of water through the roots and is relatively immobile in plants (Mengel and Kirkby, 1987). It has been observed that the plants susceptible to As are compromised in B uptake and translocation, while plants not susceptible to As maintain B content similar to control plants (Reed et al., 2015).

13.4.9 Iron

Iron (Fe) plays an indispensable role in the biogeochemical cycle of As. On the particulate surfaces of soil, iron exists as Fe hydroxides; the root surfaces of wetland plants act as strong adsorbent for As. Under the reducing environment, reductive dissolution of Fe hydroxides releases the absorbed As which enhances As availability to plants. Fe is an important micronutrient, which has several functions in plants (Connorton et al., 2017). It acts as an important constituent of several antioxidative enzymes such as superoxide dismutase (SOD) and several peroxidases and catalase. Thus, an induction in Fe concentration is directly related to the enhancement of antioxidative enzymes in plants. Farnese et al. (2014) reported that a trend of increased Fe content in *P. stratiotes* roots was noted up to 13 μM of As, indicating activation of antioxidant machineries involved in elimination of excess ROS generated during exposure to this toxic metalloid. At concentration higher than 13 μM, Fe content was shown to be decreased because of diminished antioxidant capacity above this range. Membrane damage caused by this toxic metalloid severely affected the nutrient uptake systems. An induction in Fe content was also noted in both roots and shoots of castor bean plants grown in an As-enriched nutrient solution (Melo et al., 2009).

13.4.10 Copper

Copper (Cu) acts as a structural element in regulatory proteins and participates in photosynthetic electron transport, mitochondrial respiration, oxidative stress responses, cell wall metabolism, and hormone signaling (Gusman et al., 2013). Cu^{2+} ions also act as cofactors of several enzymes like Cu/Zn SOD, plastocyanin, cytochrome c oxidase, polyphenol oxidase, amino oxidase, and laccase (Yruela, 2005). Campos et al. (2015) reported that Cu content was reduced in pinna, stipes, and roots of *P. calomelanos* (a non-*Pteris* species) under 1 mM arsenate stress with respect to control plants. Similarly, Reed et al. (2015) showed that Cu content was gradually decreased when concentration of As increased in some ornamental plants, such as *C. lanceolata*, iris, marigold, sunflower, switchgrass, and *Tithonia rotundiflora*.

13.4.11 Zinc

Zinc (Zn) is the second most utilized d-group element in biological systems. It is essential for tryptophan biosynthesis, which acts as a precursor for the phytohormone, auxin. Upon exposure to As, reductions in plant dry mass also reduced Zn uptake. Additionally, As toxicity diminished the percentage of Zn translocation as well. Shaibur et al. (2009) demonstrated that Zn content decreased significantly in roots and shoots of As-treated barley seedlings. On the contrary, Zn concentration was shown to be increased in both roots and shoots of *Ricinus communis* during As treatment (Melo et al., 2009).

13.5 Conclusion and future perspectives

Arsenic toxicity disturbs the mineral-nutrient relations in plants by affecting nutrient availability, transport, and partitioning. Mineral nutrients can play a crucial role in determining plant resistance to As toxicity. The effect of As on nutrient uptake and allocation is differentially regulated and dependent upon the concentration of As, its speciation, and the sensitivity of plant species to As. Arsenic stress causes ion deficiency or imbalance due to competition of nutrients within the plants, which causes severe impact on plant growth and metabolism. Micronutrient deficiencies in plants due to As toxicity cause a human health concern because humans obtain micronutrients from dietary sources. Deficiencies cause a variety of health problems, including disease susceptibility, slow growth in children, and anemia in cases of iron deficiency. Understanding the changes in mineral profile in response to As stress is crucial for attenuating the adaptation strategy and availability in As-contaminated soil. Another approach is to breed plants that accumulate more nutrients even during As stress. The increasing supply of nutrients to the growth medium under As toxicity conditions can alleviate the adverse effects of stress-induced injuries on plant growth. More research should focus on alternative strategies of increasing plant resistance to As stress, including the exploration of the genetic potential in the form of conventional breeding and/or molecular technologies to introduce appropriate genes and regulatory signaling molecules.

Acknowledgments

Financial assistance from Science and Engineering Research Board, Government of India through the grant (EMR/2016/004799) and Department of Higher Education, Science and Technology and Biotechnology, Government of West Bengal, through the grant [264 (Sanc.)/ST/P/S&T/1G-80/2017] to Dr. Aryadeep Roychoudhury is gratefully acknowledged.

References

Abbas, G., Murtaza, B., Bibi, I., Shahid, M., Niazi, N.K., Khan, M.I., et al., 2018. Arsenic uptake, toxicity, detoxification, and speciation in plants: physiological, biochemical, and molecular aspects. Int. J. Environ. Res. Public Health 15, 59.

Bhowmick, S., Pramanik, S., Singh, P., Mondal, P., Chatterjee, D., Nriagu, J., 2018. Arsenic in groundwater of West Bengal, India: a review of human health risks and assessment of possible intervention options. Sci. Total Environ. 612, 148–169.

Cakmak, I., 2005. The role of potassium in alleviating detrimental effects of abiotic stresses in plants. J. Plant Nutr. 168, 521–530.

Caldwell, C.R., Haug, A., 1982. Divalent cation inhibition of barley root plasma membmne-bound Ca^{2+}-ATPase activity and its reversal by monovalent calcium. Physiol. Plant. 54, 112–118.

Campos, N.V., Arcanjo-Silva, S., Viana, I.B., Batista, B.L., Barbosa, F., Loureiro, M.E., et al., 2015. Arsenic-induced responses in *Pityrogramma calomelanos* (L.) link: arsenic speciation, mineral nutrition and antioxidant defenses. Plant Physiol. Biochem. 97, 28–35.

Capaldi, F.R., Gratão, P.L., Reis, A.R., Lima, L.W., Azevedo, R.A., 2015. Sulfur metabolism and stress defense responses in plants. Trop. Plant Biol 8, 60–73.

Carbonell, F., Mataix-Beneyto, J., 1997. Effect of sodium arsenite and sodium chloride on bean plant nutrition (macronutrients). J. Plant Nutr. 20, 1617–1633.

Carbonell, A.A., Aarabi, M.A., Delaune, R.D., Gambrell, R.P., Patrick, W.H., 1998. Arsenic in wetland vegetation: availability, phytotoxicity, uptake and effects on plant growth and nutrition. Sci. Total Environ. 217, 189–199.

Castillo-Michel, H.A., Zuverza-Mena, N., Parsons, J.G., Dokken, K.M., Duarte-Gardea, M., Peralta-Videa, J.R., et al., 2009. Accumulation, speciation, and coordination of arsenic in an inbred line and a wild type cultivar of the desert plant species *Chilopsis linearis* (desert willow). Phytochemistry 70, 540–545.

Clark, R.B., Duncan, R.R., 1991. Improvement of plant mineral nutrition through breeding. Field Crops Res 27, 219–240.

Clarkson, D.T., Hanson, J.B., 1980. The mineral nutrition of higher plants. Annu. Rev. Plant Physiol. 31, 239–298.

Clarkson, D.T., Sanderson, J., 1971. Inhibition of the uptake and long-distance transport of calcium by aluminium and other polyvalent cations. J. Exp. Bot. 22, 837–851.

Connorton, J.M., Balk, J., Rodríguez-Celma, J., 2017. Iron homeostasis in plants – a brief overview. Metallomics 9, 813–823.

Dixit, G., Singh, A.P., Kumar, A., Mishra, S., Dwivedi, S., Kumar, S., et al., 2016. Reduced arsenic accumulation in rice (*Oryza sativa* L.) shoot involves sulfur mediated improved thiol metabolism, antioxidant system and altered arsenic transporters. Plant Physiol. Biochem. 99, 86–96.

Farnese, F.S., Oliveira, J.A., Farnese, M.S., Gusman, G.S., Silveira, N.M., Siman, L.I., 2014. Uptake arsenic by plants: effects on mineral nutrition, growth and antioxidant capacity. IDESIA (Chile) 32, 99–106.

Finnegan, P.M., Chen, W., 2012. Arsenic toxicity: the effects on plant metabolism. Front. Physiol. 3, 182.

Garg, N., Kashyap, L., 2017. Silicon and *Rhizophagus irregularis*: potential candidates for ameliorating negative impacts of arsenate and arsenite stress on growth, nutrient acquisition and productivity in *Cajanus cajan* (L.) Millsp. genotypes. Environ. Sci. Pollut. Res. 24, 18520–18535.

Ghosh, S., Saha, J., Biswas, A.K., 2013. Interactive influence of arsenate and selenate on growth and nitrogen metabolism in wheat (*Triticum aestivum* L.) seedlings. Acta Physiol. Plant. 35, 1873–1885.

Gilberti, L., Menezes, A., Rodrigues, A.C., Fernandes, G.W., Berbara, R.L.L., Marota, H.B., 2014. Effects of arsenic on the growth, uptake and distribution of nutrients in the tropical species *Baccharis dracunculifolia* DC (Asteraceae). Euro. J. Toxicol. Sci. 2014, 1–18.

Gulz, P.A., 2002. Arsenic uptake of common crop plants from contaminated soil sand interaction with phosphate. Swiss Fed. Inst. Technol. Zurich 108p.

Gusman, G.S., Oliveira, J.A., Farnese, F.S., Cambraia, J., 2013. Mineral nutrition and enzymatic adaptation induced by arsenate and arsenite exposure in lettuce plants. Plant Physiol. Biochem. 71, 307–314.

Hossain, M., Islam, M.R., Jahiruddin, M., Abedin, A., Islam, S., Meharg, A.A., 2008. Effects of arsenic-contaminated irrigation water on growth, yield, and nutrient concentration in rice. Commun. Soil Sci. Plant Anal. 39, 302–313.

Hossain, M.B., Jahiruddin, M., Loeppert, R.H., Panaullah, G.M., Islam, M.R., Duxbury, J.M., 2009. The effects of iron plaque and phosphorus on yield and arsenic accumulation in rice. Plant Soil 317, 167–176.

Imran, M.A., Khan, R.M., Ali, Z., Mahmood, T., 2013. Toxicity of arsenic (As) on seed germination of sunflower (*Helianthus annuus* L.). Int. J. Phys. Sci. 8, 840–847.

Jutamanee, K., Ngennoy, S., 2013. Effect of magnesium and manganese sprays on SPAD readings and chlorophyll content of chlorotic leaves of jackfruit. Acta Hortic 984, 163–169.

Kabata-Pendias, A., Pendias, H., 1992. Trace Elements in Soils and Plants, second ed. CRC Press, Boca Raton, FL.

Karabal, E., Yucel, M., Öktem, H.V., 2003. Antioxidant responses of tolerant and sensitive barley cultivars to boron toxicity. Plant Sci 164, 925–933.

Kumar, S., Dubey, R.S., Tripathi, R.D., Chakrabarty, D., Trivedi, P.K., 2015. Omics and biotechnology of arsenic stress and detoxification in plants: current updates and prospective. Environ. Int. 74, 221–230.

Li, W., Chen, T., Huang, Z., Lei, M., Liao, X., 2006. Effect of arsenic on chloroplast ultrastructure and calcium distribution in arsenic hyperaccumulator *Pteris vittata* L. Chemosphere 62, 803–809.

Li, N., Wang, J., Song, W.-Y., 2015. Arsenic uptake and translocation in plants. Plant Cell Physiol 57, 4–13.

Li, J., Zhao, Q., Xue, B., Wu, H., Song, G., Zhang, X., 2019. Arsenic and nutrient absorption characteristics and antioxidant response in different leaves of two ryegrass (*Lolium perenne*) species under arsenic stress. PLoS One 14, e0225373.

Liu, Q., Hu, C., Tan, Q., Sun, X., Su, J., Liang, Y., 2008. Effects of As on As uptake, speciation, and nutrient uptake by winter wheat (*Triticum aestivum* L.) under hydroponic conditions. J. Environ. Sci 20, 326–331.

Lombi, E., Zhao, F.-J., Fuhrmann, M., Ma, L.Q., McGrath, S.P., 2002. Arsenic distribution and speciation in the fronds of the hyperaccumulator *Pteris vittata*. New Phytol 156, 195–203.

Mallick, S., Sinam, G., Sinha, S., 2011. Study on arsenate tolerant and sensitive cultivars of *Zea mays* L.: differential detoxification mechanism and effect on nutrients status. Ecotoxicol. Environ. Safety 74, 1316–1324.

Marin, A.R., 1992. Effect of Soil Redox Potential and pH on Nutrient Uptake by Rice with Special Reference to Arsenic Forms and Uptake (Ph.D. diss). Louisiana State University, Baton Rouge, LA, p. 108.

Marin, A.R., Masscheleyn, P.H., Patrick, W.H., 1993. Soil redox-pH stability of arsenic species and its influence on arsenic uptake by rice. Plant Soil 152, 245–253.

Marschner, H., 1983. General introduction to the mineral nutrition of plants. In: Läuchli, A., Bieleski, R.L. (Eds.), Inorganic Plant Nutrition. Springer-Verlag, Berlin, pp. 5–60. Encyclopaedia of Plant Physiology.

Marschner, H., 1995. Mineral Nutrition of Higher Plants, secon ed. Academic Press, Harcourt Brace and Company, London, San Diego, New York, Boston, Sydney, Tokyo, Toronto, pp. 1–889.

Martinez, V.D., Vucic, E.A., Becker-Santos, D.D., Gil, L., Lam, W.L., 2011. Arsenic exposure and the induction of human cancers. J. Toxicol 2011, 1–13.

Meharg, A.A., 1994. Integrated tolerance mechanisms: constitutive and adaptive plant responses to elevated metal concentrations in the environment. Plant Cell Environ. 17, 989–993.

Melo, E.E.C., Costa, E.T.S., Guilherme, L.R.G., Faquin, V., Nascimento, C.W.A., 2009. Accumulation of arsenic and nutrients by castor bean plants grown on an As-enriched nutrient solution. J. Hazard. Mater. 168, 479–483.

Mengel, K., Kirkby, E.A., 1987. Principles of Plant Nutrition. International Potash Institute, Worblaufen-Bern, pp. 559–572.

Mengel, K., Kirkby, E.A., 2001. Principles of Plant Nutrition, fifth ed. Kluwer Academic Publishers, Dordrecht.

Millaleo, R., Reyes-Diaz, M., Ivanov, A.G., Mora, M.L., Alberdi, M., 2010. Manganese as essential and toxic element for plants: transport, accumulation and resistance mechanisms. J. Soil Sci. Plant Nutr. 10, 470–481.

Mishra, S., Srivastava, S., Tripathi, R.D., Trivedi, P.K., 2008. Thiol metabolism and antioxidant systems complement each other during arsenate detoxification in *Ceratophyllum demersum* L. Aquat. Toxicol. 86, 205–215.

Mohamed, H.I., Latif, H.H., Hanafy, R.S., 2016. Influence of nitric oxide application on some biochemical aspects, endogenous hormones, minerals and phenolic compounds of *Vicia faba* plant grown under arsenic stress. Gesunde Pflanz 68, 99–107.

Mukhopadhyay, R., Bhattacharjee, H., Rosen, B.P., 2014. Aquaglyceroporins: generalized metalloid channels. Biochim. Biophys. Acta Gen. Subjects 1840, 1583–1591.

Niazi, N.K., Bibi, I., Fatimah, A., Shahid, M., Javed, M.T., Wang, H., et al., 2017. Phosphate-assisted phytoremediation of arsenic by *Brassica napus* and *Brassica juncea*: morphological and physiological response. Int. J. Phytoremed. 19, 670–678.

Niazi, N.K., Singh, B., Minasny, B., 2015. Mid-infrared spectroscopy and partial least-squares regression to estimate soil arsenic at a highly variable arsenic-contaminated site. Int. J. Environ. Sci. Technol. 12, 1965–1974.

Niazi, N.K., Bibi, I., Shahid, M., Ok, Y.S., Burton, E.D., Wang, H., et al., 2018. Arsenic removal by perilla leaf biochar in aqueous solutions and groundwater: an integrated spectroscopic and microscopic examination. Environ. Pollut. 232, 31–41.

Niedz, R.P., Evens, T.J., 2007. Regulating plant tissue growth by mineral nutrition. In Vitro Cell. Dev. Biol. 43, 370–381.

Orwick, P.L., Schreiber, M.M., Hodges, T.K., 1976. Absorption and efflux of chloro-s-triazines by *Setaria* roots. Weed Res 16, 139–144.

Päivöke, A.E.A., Simola, L.K., 2001. Arsenate toxicity to *Pisum sativum*: mineral nutrients, chlorophyll content, and phytase activity. Ecotoxicol. Environ. Saf. 49, 111–121.

Panda, S.K., Upadhyay, R.K., Nath, S., 2010. Arsenic stress in plants. J. Agron. Crop Sci 196, 161–174.

Pandey, C., Augustine, R., Panthri, M., Zia, I., Bisht, N.C., Gupta, M., 2017. Arsenic affects the production of glucosinolate, thiol and phytochemical compounds: a comparison of two *Brassica* cultivars. Plant Physiol. Biochem. 111, 144–154.

Park, J., Kim, Y.Y., Martinoia, E., Lee, Y., 2008. Long-distance transporters of inorganic nutrients in plants. J. Plant Biol. 51, 240–247.

Peryea, F.J., 1991. Phosphate-induced release of arsenic from soils contaminated with lead arsenate. Soil Sci. Soc. Am. J. 55, 1301–1306.

Rafiq, M., Shahid, M., Abbas, G., Shamshad, S., Khalid, S., Niazi, N.K., et al., 2017a. Comparative effect of calcium and EDTA on arsenic uptake and physiological attributes of *Pisum sativum*. Int. J. Phytoremed 19, 662–669.

Rafiq, M., Shahid, M., Shamshad, S., Khalid, S., Niazi, N.K., Abbas, G., et al., 2017b. A comparative study to evaluate efficiency of EDTA and calcium in alleviating arsenic toxicity to germinating and young *Vicia faba* L. seedlings. J. Soils Sedim. 18, 2271–2281.

Rahman, M.A., Hasegawa, H., Ueda, K., Maki, T., Rahman, M.M., 2008. Arsenic uptake by aquatic macrophyte *Spirodela polyrhiza* L.: interactions with phosphate and iron. J. Hazard. Mater. 160, 356–361.

Ramage, C.M., Williams, R.R., 2002. Mineral nutrition and plant morphogenesis. In Vitro Cell. Dev. Biol. 38, 116–124.

Reed, S.T., Ayala-Silva, T., Dunn, C.B., Gordon, G.G., 2015. Effects of arsenic on nutrient accumulation and distribution in selected ornamental plants. Agric. Sci. 6, 1513–1531.

Rosas-Castor, J., Guzmán-Mar, J., Hernández-Ramírez, A., Garza-González, M., Hinojosa-Reyes, L., 2014. Arsenic accumulation in maize crop (*Zea mays*): a review. Sci. Total Environ. 488, 176–187.

Samanta, S., Banerjee, A., Roychoudhury, A., 2020a. Melatonin application differentially modulates the enzymes associated with antioxidative machinery and ascorbate-glutathione cycle during arsenate exposure in indica rice varieties. Plant Biol. Available from: https://doi.org/10.1111/plb.13181.

Samanta, S., Banerjee, A., Roychoudhury, A., 2021. Exogenous melatonin regulates endogenous phytohormone homeostasis and thiol-mediated detoxification in two indica rice cultivars under arsenic stress. Plant Cell Rep. Available from: https://doi.org/10.1007/s00299-021-02711-7.

Samanta, S., Roychoudhury, A., 2021. Transporters involved in arsenic uptake, translocation, and efflux in plants. In: Roychoudhury, A., Tripathi, D.K., Deshmukh, R. (Eds.), Metal and nutrient transporters in abiotic stress. Academic Press, Cambridge, pp. 77–86.

Samanta, S., Singh, A., Roychoudhury, A., 2020b. Involvement of sulfur in the regulation of abiotic stress tolerance in plants. In: Roychoudhury, A., Tripathi, D.K. (Eds.), Protective Chemical Agents in the Amelioration of Plant Abiotic Stress: Biochemical and Molecular Perspectives. Wiley-Blackwell, pp. 437–466.

Schmidt, S.B., Husted, S., 2019. The biochemical properties of manganese in plants. Plants 8, 381.

Shahid, M., Dumat, C., Khalid, S., Schreck, E., Xiong, T., Niazi, N.K., 2017. Foliar heavy metal uptake, toxicity and detoxification in plants: a comparison of foliar and root metal uptake. J. Hazard. Mater. 325, 36–58.

Shaibur, M.R., Kitajima, N., Sugawara, R., Kondo, T., Imamul Huq, S.M., Kawai, S., 2006. Physiological and mineralogical properties of arsenic-induced chlorosis in rice seedlings grown hydroponically. Soil Sci. Plant Nutr. 52, 691–700.

Shaibur, M.R., Kitajima, N., Sugawara, R., Kondo, T., Imamul Huq, S.M., Kawai, S., 2008a. Physiological and mineralogical properties of arsenic-induced chlorosis in barley seedlings grown hydroponically. J. Plant Nutr. 31, 333–353.

Shaibur, M.R., Kitajima, N., Sugawara, R., Kondo, T., Alam, S., Imamul Huq, S.M., et al., 2008b. Critical toxicity level of arsenic and elemental composition of arsenic-induced chlorosis in hydroponic sorghum. Water Air Soil Pollut 191, 279–292.

Shaibur, M.R., Kitajima, N., Huq, S.M.I., Kawai, S., 2009. Arsenic-iron interaction: Effect of additional iron on arsenic-induced chlorosis in barley grown in water culture. Soil Sci. Plant Nutr 55, 739–746.

Shaibur, M.R., Sera, K., Kawai, S., 2015. Effect of arsenic on concentrations and translocations of mineral elements in the xylem of rice. J. Plant Nutr. 39, 365–376.

Sharma, I., 2012. Arsenic induced oxidative stress in plants. Biologia 67, 447–453.

Singh, N., Ma, L.Q., 2006. Arsenic speciation, and arsenic and phosphate distribution in arsenic hyperaccumulator *Pteris vittata* L. and nonhyperaccumulator *Pteris ensiformis* L. Environ. Pollut. 141, 238–246.

Singh, R., Singh, S., Parihar, P., Singh, V.P., Prasad, S.M., 2015. Arsenic contamination, consequences and remediation techniques: a review. Ecotoxicol. Environ. Saf. 112, 247–270.

Singh, A.P., Dixit, G., Kumar, A., Mishra, S., Kumar, N., Dixit, S., et al., 2017. Protective role for nitric oxide and salicylic acid for arsenite phytotoxicity in rice (*Oryza sativa* L.). Plant Physiol. Biochem. 115, 163–173.

Srivastava, S., Upadhyay, M.K., Tripathi, R.D., Dhankher, O.P., 2016. Arsenic transport, metabolism and toxicity in plants. Int. J. Plant Environ. 2, 17–28.

Stoeva, N., Berova, M., Zlatev, Z., 2005. Effect of arsenic on some physiological parameters in bean plants. Biol. Plant. 49, 293–296.

Talukdar, D., 2013. Arsenic-induced changes in growth and antioxidant metabolism of fenugreek. Russ. J. Plant Physiol. 60, 652–660.

Tu, C., Ma, L.Q., 2005. Effects of arsenic on concentration and distribution of nutrients in the fronds of the arsenic hyperaccumulator *Pteris vittata* L. Environ. Pollut. 135, 333–340.

Wallace, A., Mueller, R.T., Wood, R.A., 1980. Arsenic phytotoxicity and interactions in bush bean plants grown in solution culture. J. Plant Nutr. 2, 111–113.

Wang, H.B., He, H.B., Yang, G.D., Ye, C.Y., Niu, B.H., Lin, W.X., 2010. Effects of two species of inorganic arsenic on the nutrient physiology of rice seedlings. Acta Physiol. Plant. 32, 245–251.

Wei, S., Ma, L.Q., Saha, U., Mathews, S., Sundaram, S., Rathinasabapathi, B., et al., 2010. Sulfate and glutathione enhanced arsenic accumulation by arsenic hyperaccumulator *Pteris vittata* L. Environ. Pollut. 158, 1530–1535.

Williams, P.N., Islam, S., Islam, R., Jahiruddin, M., Adomako, E., Soliaman, A.R., et al., 2009. Arsenic limits trace mineral nutrition (selenium, zinc, and nickel) in Bangladesh rice grain. Environ. Sci. Technol. 43, 8430–8436.

Yamaji, N., Ma, J.F., 2014. The node, a hub for mineral nutrient distribution in graminaceous plants. Trends Plant Sci 19, 556–563.

Yruela, I., 2005. Copper in plants. Braz. J. Plant Physiol. 17, 145–156.

Zhao, F.-J., McGrath, S.P., Meharg, A.A., 2010. Arsenic as a food chain contaminant: mechanisms of plant uptake and metabolism and mitigation strategies. Annu. Rev. Plant Biol. 61, 535–559.

Further reading

Meharg, A.A., Naylor, J., Macnair, M.R., 1994. Phosphorus nutrition of arsenate-tolerant and nontolerant phenotypes of velvetgrass. J. Environ. Qual 23, 234–238.

Chapter 14

Recent advances in micronutrient foliar spray for enhancing crop productivity and managing abiotic stress tolerance

Ratnakumar Pasala[1], Ramesh Kulasekaran[1], Brij Bihari Pandey[1], C.H.L.N. Manikanta[1], K. Gopika[1], P.S. John Daniel[2], Sonia Elthury[1] and Praduman Yadav[1]

[1]*ICAR-Indian Institute of Oilseeds Research, Hyderabad, India,* [2]*International Crops Research Institute for the Semi-Arid Tropics, Patancheru, India*

14.1 Micronutrients

In the recent past, improper management of soil resources, intensive cultivation, limited crop rotation, and reductions of organic matter from soil have resulted in soil degradation. The soil illness causes a decline in yield and quality aspects of crops together with the appearance of nutrient deficiency in both soils and crops. The total available nutrient content in the soil is a function of pedogenic processes. Indian soils are moderately acceptable in terms of total soil micronutrient content. Despite this, the deficiency of nutrients has been largely observed in soils as well as plants (Shukla and Tiwari, 2016), and it will no longer be possible to achieve higher yields from crops even by surplus supplies of nutrients (Singh and Alloway, 2008). Besides, a reduction in water dynamics in the plant would affect the uptake of micronutrients from the soil, through lowering concentrations of zinc, iron, and manganese (Neenu and Ramesh, 2020).

Several factors including geochemical composition (total micronutrient contents of the soil parent material), type of soil (clay mineralogy, particle size distribution, soil horizon, soil age, soil formation processes), soil pH, redox potential, total soluble salts (EC), soil organic matter and calcium content, supply of trace elements (through atmospheric deposition, pesticides,

Plant Nutrition and Food Security in the Era of Climate Change. DOI: https://doi.org/10.1016/B978-0-12-822916-3.00008-1

manures, fertilizers), availability of micronutrients, and their interactions determine the micronutrient content in soil (Shukla et al., 2016). Other than soil application, the foliar spray method of nutrient application has been adapted to improve the performance of crops and reduce the consumption of chemical fertilizers and associated environmental risks. In other words, improper soil fertilization can cause leaching and could aggravate soil pollution. Furthermore, a lack of irrigation under dryland conditions may hinder nutrient uptake from the soil when crops experience droughts due to climate change; when crop productivity is impaired, irrigation water is a constraint for nutrient uptake from the soil. Therefore the micronutrient foliar spray is effective with minimum water (per hectare, 200 L required for each foliar spray) and cost-effective, to maximize on-farm crop yields under abiotic stress. For example, the irrigation required immediately after soil application of any nutrient is about 5−7 cm rainfall or irrigation water/hectare; comparatively, the amount and cost of irrigation water use for foliar application is only 200 L/hectare. The foliar sprays are efficient with the minimal quantity of nutrients received by the plant at a specific crop stage, with specific formulation and pH concentration including with targeted plant tissue (Srivastva et al., 2016; Ratnakumar et al., 2016a,b, 2017; Wackchaure et al., 2016a,b). The role of micronutrients that are being used for enhancing crop yields is documented in Table 14.1 and the combinations of micronutrients used are summarized in Table 14.2.

14.2 Micronutrients as foliar sprays

14.2.1 Boron

Soils serve as primary suppliers of trace nutrients, and among them boron (B) is known as one of the micronutrients for plants. Generally, plants must maintain a range of 0.3−1 ppm to perform normal functions. Any deviation either higher or lower from this concentration leads to impaired plant processes (Blevins and Lukaszewski, 1998). The boron is acquired in undissociated/uncharged boric acid form (H_3BO_3). Boron is required for optimum growth and development (Miwa et al., 2007) and plays several key functions in plants including biosynthesis of the cell wall and lignifications, tissue differentiation, vegetative growth, phenolic metabolism, membrane integrity, etc. Along with that, it is also considered as the key element for nitrogen fixation and nitrate assimilation (Reguera et al., 2010), oxidative damage (Kobayashi et al., 2004), and root growth (Martın-Rejano et al., 2011). Boron foliar application enhanced the growth of the vine, the number of fruits, and the size of cucurbitaceous crops. For the release of growth hormones and fruit enlargement besides boric acid (17%), borax (11%) was used. Earlier, in cucumber (*Cucumis sativus* L.) a significant response to three foliar sprays of 25 ppm of boric acid was observed (IIHR report, 2020). In sugarbeets, boron foliar sprays are used to cure young leaves' curl and stunted growth. In

TABLE 14.1 Micronutrient foliar sprays enhanced crop yields under normal and abiotic conditions.

Micronutrient	Crop	Dose	Role	References
Iron	Cotton	$FeSO_4$ (5 kg/ha) followed by boron (soil), $CuSO_4$ (soil), $ZnSO_4$ (foliar), and $CuSO_4$ (foliar)	Higher yield	Basavarajappa et al. (1997)
	Citrus	0.1% $FeSO_4$ in 0.05% citric acid solution	Controls iron chlorosis and increases fruit yield	El-Kassas (1984)
	Cupressus sempervirens	$FeSO_4$	Iron content is positively correlated with chlorophyll content	Farahat et al. (2007)
	Sugarcane			Rakkiyappan et al. (2002)
	Tree crops	2.27 kg of $FeSO_4$ in 378.5 L of water	Correction of chlorosis	Kansas State Universtiy, Kansas Forest Service Publication, (2012, MF-718)
Manganese	Barley, corn, oats, onion, peanut, safflower, soybean, and fruit trees	$MnSO_4$ (0.3–5.4 kg Mn ha^{-1}) in 31–230 L of water	Overcomes Mn deficiency in acidic pH soils	Nayyar et al. (1985)
	Soybean	$MnSO_4$ (32%) and chelated Mn (5%–12%)	Increases seed yield	Heckman (2010)
	Sweet orange		Increases leaf Mn content, fruit yield, and intensifies red color of skin and juice	Perveen and Rehman (2000)
	Beans	2 kg of $MnSO_4$ per hectare in 200 L of water	Increases grain yield	Bohner and Reid (2007)
Copper	Wheat	1 kg/ha of copper sulfate or copper oxychloride or copper chelates	Enhances maximum yield	Brennan (2005)

(*Continued*)

TABLE 14.1 (Continued)

Micronutrient	Crop	Dose	Role	References
Zinc	*Mentha piperita*	Zinc chloride (3 ppm)	Enhances the essential oil content by 28.2%	Akhtar et al. (2009)
	Blue sage (*Salvia farinacea* L.)	100 ppm of Zn	Enhances the length of peduncle and main inflorescence, number of inflorescence and florets, and fresh and dry weight of inflorescence per plant	Nahed and Balbaa (2007)
	Maize	300 ppm of Zn	Increases grain yield under optimum water availability	Bashir (2012)
		1.0 kg of Zn ha^{-1}	Increases the total N uptake and grain yield and also increases the length of cob and shape	Potarzycki and Grzebisz (2009)
	Cotton	0.5% $ZnSO_4$	Increases 100 seed weight, root length, shoot length, and vigor index	Rathinavel et al. (1999)
Boron	Brinjal	0.5% borax	Enhanced number of flowers per plant, number of productive flowers per plant, number of fruits per plant, individual fruit weight and yield	Kuruppaiah (2005)
	Chili	0.25% boron	Maximizes the fruit yield per plant and ha^{-1}	Dongre et al. (2000)
		0.1% H3BO3	Increases average fruit length and fruit diameter	
	Safflower	0.2% boron	Increases the 1000 kernel weight, seed and oil yield	Sangale et al. (1981)
	Rapeseed	0.4 kg/ha	Spraying at bud stage increases the seed yield and seed oil content	Bowszys (1996)

(*Continued*)

TABLE 14.1 (Continued)

Micronutrient	Crop	Dose	Role	References
Molybdenum	Grapevine	Two foliar sprays of Mo (118 g/ha)	Reduces the effects of fruit-set disorders, increases yield, and Mo concentration in petioles	Williams et al. (2004)
Chlorine	Wheat	NH_4Cl or $MgCl_2$	The plant grew well but showed some phytotoxicity and only little injury was caused from prevailing fungal diseases	Miller (1994)

TABLE 14.2 Combination of micronutrients on various crops.

Combination	Crop	Dose	Role	References
Mn and B	Sunflower	Along with recommended doses of N, P, K	Increases seed yield and 1000 seed weight	Mishra et al. (1985)
Fe/Mn	Soybean		Prevents reduction of yield and nutrient imbalance in calcareous soil	Moosavi and Ronaghi (2011)
Mn and Fe	Safflower		Increases the palmitic acid and oleic acid contents whereas percent of linoleic acid gets decreased; also compensates for the adverse effects of drought	Dehnavi (2008)
B and Cu	Cotton	Varying concentrations	Increases seed yield and number of bolls per plant but boll weight remains unaffected	Hosny et al. (1984)
Fe and Zn	Chamomile	Sprayed together	Improves flower and essential oil yields	Nasiri et al. (2010)
Zn, K, P	Cotton	Combined spray	Increases yield of lint and seed by 40%	Zakaria (2008)
Zn and B	Cotton	0.1% $ZnSO_4$ + 0.1% boron with 50 ppm GA3	Increase the number of squares, flowers, and matured bolls per plant and highest seed yield, shoot length, root length, and vigor index	Ullagaddi (2000)
B, Mn, Mo	Soybean	Combined spray	Increases the number of pods, seed per plant, 100 seed weight, and seed yield	Ziolek and Ziolek (1987)
Mo, urea, BA	Lettuce	Urea (1%) + Mo (1 mg/ dm^3) + BA (benzyladenine, 5 mg/ dm^3)	Decreases nitrates and increases soluble sugars	Wojciechowska and Kowalska (2011)

sunflower, boron foliar spray with 46 μm concentration enhanced capitulum size and weight, but excess boron caused leaf burn (Asad et al., 2003).

14.2.2 Copper

Copper (Cu) is a unique element due to its magnificent role in plants. The micronutrient Cu plays a significant role in plant growth and development, as these processes are highly dependent on the availability of copper. The element takes part in regulating several biochemical reactions in plants as a cofactor and also regulates the various physiological and metabolic functions such as respiration, photosynthesis, lignifications, phenol metabolism, protein synthesis, regulation of auxins, seed production, and chlorophyll formation (Rehm and Schmitt, 2002). In plants, Cu is present in oxidized (Cu^{+2}) and reduced form (Cu^{+1}) (Burkhead et al., 2009a,b) and involved in oxidation-reduction reactions. The element Cu serves as an electron carrier in reactions of chloroplasts and mitochondria in response to oxidative damage (Kramer and Clemens, 2006). Cu helps in cell wall metabolism, as the signal pathway of transcription, in oxidative phosphorylation, and in iron mobilization. In plants, Cu concentration of 2–50 ppm is present; 6 ppm is considered to be suitable for proper growth and development of shoots (Epstein, Bloom, 2005). Concentrations below 5 ppm can result in deficiency symptoms, and concentrations above 20 ppm lead to toxicity symptoms (Burkhead et al., 2009a,b). In dry beans, Cu fertilization influenced the growth, root growth, root fineness yield, and yield components and enhanced activity of COPT and other like COPT, ATP, EDTA, ABA, NADPH, ROS, SOD, CAT, POD, MDA, APX, DAS transporters, phloem transport, and copper remobilization during senescence. Cu foliar application less than 0.03% increased grain yield and applications higher than 0.03% induced leaf damage in wheat (Fouad et al., 2020).

14.2.3 Iron

Iron (Fe) is regarded as the fourth most plentiful and essential micronutrient found in the earth's crust after silicon, oxygen, and aluminum. Iron is a vital element for plants and required in trace amounts (Curie and Briat, 2003); however, it is less soluble and the availability of Fe to plants (inorganic form) is limited (Chatterjee et al., 2006). The essential role of Fe in the growth and optimal development of plants was first reported by Gris (1843). The chloroplasts, mitochondria, and vacuoles are the localized sites of Fe where it is found abundantly (Vigani et al., 2013) and functions as a cofactor for various metabolic reactions (Puig et al., 2005). These metabolic processes include reactions related to respiration, the tricarboxylic acid cycle, oxygen transport, synthesis of metabolic intermediates, DNA biosynthesis, chlorophyll synthesis, lipid metabolism, and gene regulation (Adamski et al.,

2012). The research in the past indicated that both excessive and limiting amounts of iron in plants can lead to impaired plant growth and processes (Wang et al., 2012). The deficiency of Fe is generally identified as impairment of several physiological and metabolic processes, which ultimately leads to severe damages to final yield (Donnini et al., 2010). The most visualized symptoms due to deficiency of Fe appear on young leaves as interveinal chlorosis between veins and chlorotic areas (Wang et al., 2012). Fe fertilizers, Fe-EDTA treatments of 1−1.5 g/L particularly with the foliar method of application caused increased growth characteristics such as height and chlorophyll of leaves and nutrient concentration of *Spathyphyllum* leaves (Mohamadipoor et al., 2013). Higher concentrations of iron also interfere with the uptake of nutrients *viz.*, potassium (K), phosphorus (P), calcium (Ca), and magnesium (Mg) (Zhang et al., 1999). Plants grown in calcareous soils generally experience Fe chlorosis; chlorosis-correcting sprays have been widely used in fruit crops (Àlvarez-Fernàndez et al., 2006).

14.2.4 Manganese

After iron, manganese (Mn) is considered as the most important element as it is involved in various metabolic processes of plants. Manganese is also known to play a crucial role in stress tolerance mechanisms either directly or indirectly by acting as a cofactor for several antioxidant enzymes (Burnell, 1988). Apart from this, it regulates the process of photolysis of water during the photosynthesis process. Furthermore, Mn is actively involved in the biosynthesis of lipids, fatty acids, protein, and ATP and also contributed to carboxylation reactions of the C3 cycle. The excess concentration of Mn in plant cells leads to several toxicity symptoms including a negative effect on the growth and yield of plants. The toxicity of Mn is also responsible for ROS that accelerates the damages to various plant levels (Millaleo et al., 2010a). On the other hand, soil with elevated pH (above 6.5) and high organic matter content is prone to Mn deficiency. The chloroplasts are highly susceptible to Mn deficiency (Mengel and Kirkby, 2001); therefore the primary and most observable symptoms appear on leaves as interveinal chlorosis. Apart from this, severe deficiency also caused brown necrotic spots, premature leaf fall, and delayed maturity. Two chemical forms such as inorganic ($MnSO_4$) and organic (Mn EDTA) were mostly used for the correction of Mn deficiency of fruit trees. Foliar application of manganese sulfate was more effective than manganese chelate in increasing leaf Mn concentrations and foliar sprays of 0.6% $MnSO_4$ and 0.3% $ZnSO_4$ together enhanced the yield and quality of pomegranate (Hasani et al., 2012).

14.2.5 Zinc

The essential role of zinc (Zn) as a significant micronutrient was elucidated in maize and barley. The optimal concentration of Zn required by plants is 15–20 mg/kg of dry weight (Marschner, 1995). Zinc plays several key roles in plants and is an important constituent of various cofactors for enzymes and proteins. Among the several functions, Zn played a crucial role in internodal elongation along with regulation of pollen growth and promotion of metabolism of auxins, proteins, and carbohydrates (Marschner, 1995). Leafy vegetable crops such as spinach and beet are known to accumulate a much higher concentration of Zn (Chaney, 1993). Zinc has intermediary mobility in plants; therefore the first symptoms appear on middle leaves as interveinal chlorosis that is concentrated between midrib and leaf margins, which later turns into pale green, yellow, or even white. The severe deficiency of Zn leads to gray-white discoloration, inward leaf curling, and even premature leaf fall and death. Moreover, the higher concentrations of Zn (above 0.2 mg/g of dry matter) caused leaf deterioration, impaired plant growth, and restricted development. Foliar application of Zn as Zn sulfate and Zn EDTA to wheat crop enhanced the grain concentration of Zn (Doolette et al., 2018). Foliar spray of Zn in combination with B and Fe enhanced productivity of guava (Bhoyar and Ramdevputra, 2016). Foliar-applied zinc chelate was 1.4–1.7 times more effective than zinc sulfate applied in wheat crop (Brennan, 1991).

14.2.6 Molybdenum

In a plant system, molybdenum (Mo) is required in trace amounts and is essential for the growth of plants. In plants, it participates in many oxo-reduction reactions. The element is mostly found as an important part of the organic protein complex, also known as the molybdenum cofactor (Moco). The complex binds to enzymes and catalyzes various kinds of metabolic reactions. The element involves various kinds of assimilation reactions where it acts as an enzyme cofactor. The enzymes are nitrate reductase, nitrogenase (present in nitrogen-fixing bacteria), xanthine dehydrogenase/oxidase (helps in the catabolism of purine and biosynthesis of ureide), aldehyde oxidase (active during ABA biosynthesis), and sulfite oxidase (sulfur metabolism) (Mendel and Robert, 2002). In soils, the uptake of Mo generally depends on the concentration of Fe oxides, soil pH (pH <5.5), and the presence of organic compounds. The Mo deficit in plants expresses as poor growth, leaves turning pale color, withered leaving midrib, and susceptibility to many diseases. Mo foliar application enhanced the yields in the plants belongs to the *Brassicaceae* family. Foliar Mo application of 75–100 g/hectare enhanced the nitrogen (N) acquisition by the plants and increased the yield and quality of vegetables like lettuce (Steiner et al., 2018).

Furthermore, the activity of several enzymes especially nitrate reductase caused restricted nitrogen assimilation and development of leaf symptoms in many horticulture, legume, and cereal crops (Chatterjee and Nautiyal, 2001).

14.2.7 Silica

Silica (Si) is an essential micronutrient and abundantly available in nature. Plants mostly absorb the Si in the form of silicates or orthosilicic acid. Si foliar sprays have been widely used for yield enhancement in crops like wheat (Ratnakumar et al., 2016a,b, 2017), soybean (Minhas et al., 2014), rice (Ma, 2007), and watermelon (Ratnakumar et al., 2015). Foliar application along with fertigation of Si encouraged significant fruit yields and crop recovery in a watermelon after a hailstorm (Ratnakumar et al., 2015; Bal et al., 2017); Si and B together in rice enhanced antioxidant enzyme activity (Farooq et al., 2015).

14.3 Micronutrient use efficiency

Use efficiency (UE) is the ability of plants to utilize and convert the foliar-sprayed micronutrient into yield. The micronutrient UE can be defined as the yield of biomass per unit input of micronutrient foliar spray. The effective use of micronutrients by plants is efficient even though they receive very minimal in the form of foliar spray. From the total amount of nutrient foliar spray, the plant may receive only 10%–20% after a portion of the formulation is lost into the soil or via evaporation. The formulation can be absorbed either by stomata or plasma-membrane intrinsic proteins or chelates. These nutrients may enhance the inbuilt resistance mechanism through redox-mediated mechanism to abiotic and biotic stress (Srivastva et al., 2016).

Zn UE can also be influenced by the size of the Zn source particle (Montalvo et al., 2016). In agriculture, water-soluble sources such as Zn sulfate ($ZnSO_4$, 20% Zn) act as the main source of Zn, which can be applied by broadcasting or in a furrow or as foliar spray close to grain filling. Significant variations in boron efficiency were reported in 24 wheat cultivars. Fang 60 cultivar showed an improved yield when grown under low boron, paralleled with higher partitioning of boron to ears nearly double that of other genotypes (Nachiangmai et al., 2004). During cold wet springs, manganese efficiency was associated with soil chemical and microbial processes. A higher percentage of plant molybdenum is allocated to upper leaves and spikes further enhanced higher phloem mobility in winter wheat (Yu et al., 2002).

14.4 Nano-form of micronutrients

Rectification of micronutrient deficiencies is routinely carried out through foliar application by the farming community, and hence the use of nano-forms of micronutrients as foliar spray would be advantageous in terms of UE as well as an environmental concern. Foliar applications of nano-micronutrients are reported to overcome these limitations. But the results of the use of nano-micronutrients are inconclusive. Plants fertilized with the recommended dose of Fe in the nano-form (54 mM) aided in the diminution of antioxidant enzyme activities more than a suboptimal dose of Fe in the macro form as salts (Elanchezhian et al., 2017). For example, foliar application of zinc nano-fertilizer (Mn, Fe, and Zn oxides) at 50 mg/L at 21 days after seeding and $10-20$ days after the first increased vegetative growth, improved fresh pod yield, pod physical quality, and nutritional value of snap bean cultivars in sandy soils (Marzouk et al., 2019). Nano-micronutrients foliar sprayed at 10% of the conventional fertilizer doses (i.e., Fe: 0.2%; Mn: 0.1%, Zn: 0.3%, B: 0.05%) along with soil application of macronutrients was inferior to conventional micronutrient sprays (Kailas et al., 2019).

14.5 Micronutrients for mitigation of abiotic stresses

14.5.1 Drought

The exact mechanism of how foliar sprays of micronutrients contribute to drought tolerance is yet to be defined; however, it was hypothesized that micronutrients trigger certain physiological and biochemical pathways in response to drought. As for Zn, it is evident that the application of Zn under water stress situations will increase wheat crop yield and quality. Yield-contributing traits such as height, spike length, and filling percentage were found to be influenced by treatment with zinc under drought conditions (Yavas and Unay, 2016). Zn also regulates the stomatal behavior and also ionic balance under stress thereby reducing the detrimental effects. Zn triggers the activity of NADPH oxidase, reduces the photo-oxidative damage, and through the activities of mentioned antioxidants, it reduces ROS activity (Hajiboland, 2012). It is known that drought limits the availability of micronutrients; the application of Mn changes the nutritional status of the plant and contributes to defense against drought-induced ROS. Mn application under water stress also improves photosynthesis, stomatal conductance, leaf area, pollen fertility, and water UE of the crop (Karim et al., 2006). Fe under drought stress may increase the assimilate production ultimately increasing the yield. Zn and Fe together improve the leaf characteristics. Iron also increases the oil content of crops like sunflower and sesame (unpublished data).

14.5.2 Heat stress

Temperature stress (high or low) inhibits the translocation of water and nutrients by damaging the cell and induces many morphological and physiological disorders in plants. Micronutrients play an important role in the proper growth and development of plants under heat stress (Khalil et al., 2018). The role of foliar application of micronutrients under high temperature is crop-specific. Se in cucumber enhances the heat resistance and helps the plants improve their growth and yield by maintaining the stomatal conductance and transpiration rate (Balal et al., 2016). In wheat, exogenous application of Si increases the activity of antioxidants, chlorophyll contents, and osmotic adjustment capacity but decreases H_2O_2 contents, MDA, and electrolyte leakage (Ratnakumar et al., 2016a,b). Si in rice prevents the weakening of function and structure of cell membranes (Nelson, 1944). Exogenous application of Ca in wheat reduces oxidative stress and increases the activity of SOD, CAT, POD, amino acid contents, and photosynthetic rate (Goswami et al., 2015). Mn application in pepper increases photosynthesis and N metabolism to reduce the adverse effects of heat stress on the plant. It also reduces premature leaf drop and brown necrotic spots. The involvement of Mn activates many enzymes in decarboxylation and redox reaction in plants grown under heat stress (Aktas et al., 2005). Application of sulfur in cluster bean showed increased growth and plant biomass (Mobin et al., 2017). Zn application in wheat enhances the defense system against heat stress by maintaining the turgidity of the cell (Graham, 2004). Mg in maize and wheat increases the biomass, growth, and heat tolerance at a higher concentration of Mg (540 μM); whereas in beans, maize, pepper, and mulberry to avoid oxidative damage, the concentration and activities of oxidative molecules and oxidative enzymes were increased (Mengutay et al., 2013).

14.5.3 Salinity stress

During salt stress, it is evident that plants uptake higher quantities of salts which brings the osmotic potential down, and to cope with the condition they are known to produce different compatible solutes. Micronutrients are known to trigger the production of these osmoprotectants, such as Fe, Zn, and Mn, to escalate the production of proline (Hussain et al., 2018). Zn under lower concentrations acts synergistically with NaCl increasing biomass as well as yield. Rose plants treated with a foliar spray of boron 40–60 μM showed maximum resistance to exposed levels of salt stress, that is 1.8 and 2.5 dSm (Habiballah et al., 2020). Soil application of boron at moderate amounts conferred tolerance toward salt stress. Fe as mentioned previously activates compatible solutes and biosynthesis of proteins needed for salt tolerance; it also triggers the activity of RuBisCO by three times (Yasmeen et al., 2016). A salt-tolerant genotype of *Setariaitalica* showed increased levels of Cu-Zn SOD when exposed to salt stress (Sreenivasulu et al., 2000). Application of Mb to wheatgrass increases the activity of enzymes aldehyde oxidase, nitrate oxidase, and xanthine dehydrogenase that confers tolerance to salinity stress (Babenko et al., 2015).

14.5.4 Cold stress

Micronutrients play an important role in uplifting the harmful effects of cold stress by eliminating ROS via activation of defense compounds such as SOD and others. A study on an external supply of Mn and Zn in combination with seaweed on cold-stressed maize plants showed beneficial effects by reducing leaf necrosis, reducing shoot growth, and increasing root growth two-fold, which helps in nutrient acquisition. Also, this external supply of Mn and Zn resulted in increased levels of Cu-Zn SOD (Bradáčová et al., 2016). N fertilization in ammoniacal form amplified the Zn and Mn nutritional status increasing the shoot level ABA content by 33% that moderately detoxifies the ROS generated (Moradtalab et al., 2020).

14.5.5 Light stress

Photo-oxidative damage, which is the light-dependent generation of ROS in chloroplasts, is higher when the absorption of light energy exceeds the capacity of photosynthetic electron transport rate (Foyer and Noctor, 2005). High light intensity combined with various stress conditions will aggravate the photo-oxidative damage and ultimately affects the yield. The mineral nutritional status of the plants will influence the photosynthetic electron transport rate and CO_2 fixation in many ways (Mengel and Kirkby, 2001). Therefore, the deficiencies of mineral nutrients will enhance the sensitivity of plants to photo-oxidative damage. Micronutrient foliar sprays for mitigation of high light stress need to be tested to understand their mechanism.

Apart from regulating and changing the functions of a plant, micronutrients under optimum concentrations are known to influence the plant's inbuilt tolerance to abiotic stress situations by regulating cellular redox-homeostasis mechanisms (Srivastava et al., 2016). On-farm crops experience severe stress situations that affect nutrient uptake and availability. As a mitigation strategy, foliar applications of different bioregulators, biomolecules, and micronutrients have shown enhanced crop yields under abiotic stresses (Srivastava et al., 2016; Ratnakumar et al., 2016a,b; Wackchaure et al., 2016a,b). Either by oxidative—reduction reactions or through forming ligands with the organic compounds and also forming metalloenzymes that act as a catalyst, micronutrients help plants in conferring tolerance to stress.

Under abiotic stress situations, plants are vulnerable to utilizing the nutrients, which may lead to the formation of ROS that destroy membrane lipids, proteins, and DNA. Micronutrients under these circumstances act as protectors by activating antioxidative enzymes needed for the scavenging of these ROS molecules. Micronutrients are not only meant for growth and development in plants, some of them such as Fe, Mn, Zn, and Cu function as cofactors in activating several antioxidant enzymes. The activity of antioxidant enzymes varies with the availability of these micronutrient foliar applications in plant systems, which ultimately determines the plant sensitivity toward the environmental stress incurred (Jyoti Ranjan, 2019).

Zn as an activator of nearly 200 enzymes by serving as a cofactor is also important in activating antioxidant enzymes in defense against ROS. Zn along with Cu activates an enzyme Cu-Zn SOD (Sajedi et al., 2011). Apart from activating the enzyme needed for ROS scavenging, Mn was found to activate nearly 35 enzymes, either by serving as a cofactor or as an activator. Mn contributes to abiotic stress tolerance by an important enzyme, Mn-SOD, which quenches ROS produced at mitochondria (Khan et al., 2018). In vivo interchange of Cu, that is Cu + 2 and Cu + 1, results in the formation of OH* radicals which demands the existence of Cu chaperones in plant systems. Cu activates nearly 30 enzymes needed for various developmental and growth functions (Harrison et al., 2000). Similar to Cu, Fe is a redox-active element and interchange between Fe + 2 and Fe + 3 governs various metabolic activities of the plant. Iron in its free state has the potential to trigger ROS and results in oxidative damage to plants, so it always exists in chelated forms and gets transported in the plant system on an as-needed basis (Khan et al., 2018). Iron is a principal constituent of important antioxidant enzymes, that is ferro-superoxide dismutase, catalase, ascorbate peroxidase, and guaiacol peroxidase (Becana et al., 1998). Five of fifty enzymes activated by Mo are present in the plant system governing various metabolic functions. Experimental results of Wu et al. (2014) suggest that the activities of SOD, POD, and APX increase in stressed wheat plants upon external supply of Mo.

14.6 Cell homeostasis and crosstalk between micronutrients

Micronutrients are considered essential for plants and play vital roles in a plant's life cycle, yet there are many impediments to proper homeostasis. For the growth and development of plants, micronutrient uptake and homeostasis are critical. Cellular homeostasis can be maintained by micronutrient foliar sprays by enhancing the activity of scavenging enzymes such as antioxidative enzymes. The role of each micronutrient under each abiotic stress condition has been documented in the previous sections of this chapter. Crosstalk between the micronutrients to maintain the balance in cell homeostasis has been reported properly. Fig. 14.1 represents the schematic diagram of crosstalk among Fe, Zn, and P to maintain the redox states in the electron transport chain of the mitochondrion and the chloroplast (Fig. 14.1). P, Zn, and Fe fertilizers are widely used for agricultural production. Foliar

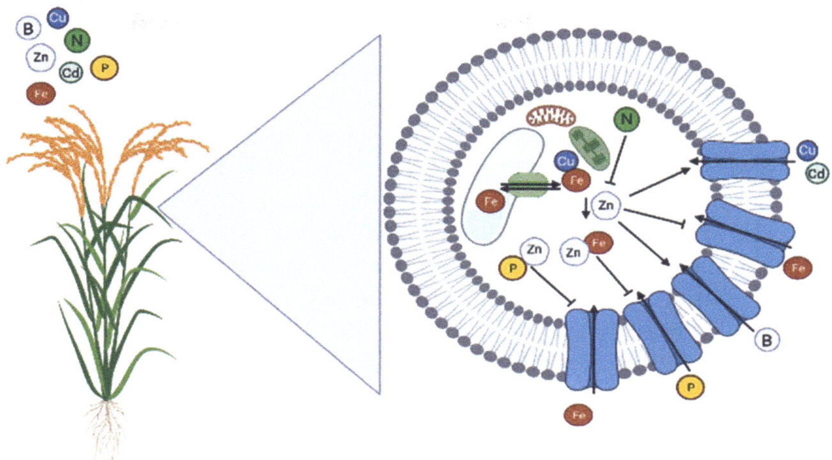

FIGURE 14.1 Schematic representation of micronutrient foliar application to enhance cellular homeostasis by crosstalk among Fe, Zn, and P.

applications of micronutrients are more suitable than soil applications, due to their rapid overcoming of deficiencies, ease of use, reductions in toxicity caused by accumulation, and prevention of element stabilization in the soil.

14.7 Enhanced crop yield and productivity

Micronutrient applications in soil, foliar sprays, and seed priming are known to enhance the yield of agricultural, horticultural, and floricultural crops. Sunflower (*Helianthus annuus* L. cv. Shumos) seed yield increased in response to boron with 200–250 mg/L application (Al-Amery et al., 2011). In most of the studies involving sunflower 0.20% borax at ray floret initiation stage enhanced yield in association with $ZnSO_4$ spray (Patil et al., 2006). A concentration of 0.5% each of Zn and Fe at 30 and 65 DAS of safflower enhanced the number of the capitula, seed weight per head, 1000 seed weight, and seed yield (Ravi et al., 2008), and this could probably be through synergetic interaction between iron and zinc over single applications of micronutrients. Boric acid at 0.2% at 60 DAS in Niger was found to enhance seed as well as oil yield (Lalitha et al., 2008). Other than oilseeds, the micronutrient foliar application in cereals, pulses, and horticulture crops was encouraging and well documented in Table 14.1.

14.8 Way forward

Micronutrient foliar sprays are essential, effective, eco-friendly, and advanced tools of low-input sustainable agriculture. The nano-form of micronutrients is available in the market at cheaper prices with a combination of other micronutrients to enhance the production and productivity of agricultural and horticultural crops. The role of these micronutrients in maintaining cell homeostasis and crosstalk between micronutrients has been documented in this chapter including the mitigation strategies of each micronutrient under abiotic stress and their combinations. The dose, timing, and number of foliar sprays and stage of the crop need to be standardized before the recommendation of any micronutrient. The number of foliar sprays should not be excessive and create other micronutrient imbalances in the crop and soil microbiomes. The combination of micronutrients should lead to a synergetic effect rather antagonistic effect. The foliar sprays of micronutrients are cost effective when they are applied to on-farm crops with appropriate dose and frequency. Micronutrient foliar spray is environmentally protective and essential to enhance plant fitness and productivity.

References

Adamski, J.M., Danieloski, R., Deuner, S., Braga, E.J., de Castro, L.A., Peters, J.A., 2012. Responses to excess iron in sweet potato: impacts on growth, enzyme activities, mineral concentrations and anatomy. Acta Physiol. Plant. 34 (5), 1827–1836.

Akhtar, N., Abdul, M.S.M., Akhter, H., Katrun, N.M., 2009. Effect of planting time and micronutrient as zinc chloride on the growth, yield and oil content of *Mentha piperita*. Bangladesh J. Sci. Ind. Res. 44 (1), 125–130.

Aktas, H., Karni, L., Chang, D.C., Turhan, E., Bar-Tal, A., Aloni, B., 2005. The suppression of salinity associated oxygen radicals production, in pepper (*Capsicum annuum*) fruit, by manganese, zinc and calcium in relation to its sensitivity to blossom-end rot. Physiol. Plant. 123, 67–74.

Al-Amery, M.M., Hamza, J.H., Fuller, M.P., 2011. Effect of boron foliar application on reproductive growth of sunflower (*Helianthus annuus* L.). Int. J. Agron. 230712. Article ID.

Àlvarez-Fernàndez, A., Abadía, J., Abadía, A., 2006. Iron deficiency, fruit yield and fruit quality. In: Barton, L.L., Abadia, J. (Eds.), Iron Nutrition in Plants and Rhizospheric Microorganisms. Dordrecht, Springer. Available from: https://doi.org/10.1007/1-4020-4743-6_4.

Asad, A., Blamey, F.P.C., Edwards, D.G., 2003. Effects of boron foliar applications on vegetative and reproductive growth of sunflower. Ann. Bot. 92 (4), 565–570.

Babenko, O.N., Brychkova, G., Sagi, M., Alikulov, Z.A., 2015. Molybdenum application enhances adaptation of crested wheatgrass to salinity stress. Acta Physiol. Plant. 37, 1.

Bal, S.K., Minhas, P.S., Singh, Y., Kumar, M., Patel, D.P., Rane, J., et al., 2017. Coping with hailstorm in vulnerable deccan plateau region of India: technological interventions for crop recovery. Curr. Sci. 2021–2027.

Balal, R.M., Shahid, M.A., Javaid, M.M., 2016. The role of selenium in amelioration of heat-induced oxidative damage in cucumber under high temperature stress. Acta Physiol. Plant 38, 158. Available from: https://doi.org/10.1007/s11738-016-2174-y.

Basavarajappa, R., Koraddi, V.R., Kamath, K.S., Doddamani, M.B., 1997. Response cotton cv. abadhita (*Gossypium hirsutum*) to soil and foliar application of micro-nutrients under rainfed condition. Karnataka J. Agric. Sci. 10 (2), 287–291.

Bashir, F., Muhammad, M., Sarwar, N., Ali, H.M., Muhammad, K., Shehzad, A., 2012. Effect of foliar application of zinc on yield and radiation use efficiency (RUE) of maize (*Zea mays* L.) under reduced irrigation conditions. Asian J. Pharm. Biol. Res. 2 (1), 33–39.

Becana, M., Moran, J.F., Iturbe-Ormaetxe, I., 1998. Iron-dependent oxygen free radical generation in plants subjected to environmental stress: toxicity and antioxidant protection. Plant Soil. 201 (1), 137–147.

Bhoyar, M.G., Ramdevputra, M.V., 2016. Effect of foliar spray of zinc, iron and boron on the growth, yield and sensory characters of guava (*Psidium guajava L.*) Cv. Sardar L-49. J. Appl. Nat. Sci. 8 (2), 701–704.

Blevins, D.G., Lukaszewski, K.M., 1998. Boron in plant structure and function. Annu. Rev. Plant. Biol. 49 (1), 481–500.

Bohner, H. and Reid, K., 2007. Manganese deficiency, crop pests, Ontario, <http://www.omafra.gov.on.ca/english/crops/field/news/croppest/2007/12cpo07a2.html>

Bowszys, T., 1996. Response of winter rape to foliar application of boron fertilizer. Zesyty Problemowe Portepow Nauk. Rolniczych 434 (1), 71–76.

Bradáčová, K., Weber, N.F., Morad-Talab, N., Asim, M., Imran, M., Weinmann, M., et al., 2016. Micronutrients (Zn/Mn), seaweed extracts, and plant growth-promoting bacteria as cold-stress protectants in maize. Chem. Biol. Technol. Agric. 3 (1), 19.

Brennan, R., 2005. Overcoming copper deficiency in wheat with foliar sprays, Department of Agriculture and Food, Govt. of Australia, <http://www.agric.wa.gov.au/PC_92056.html>.

Brennan, R.F., 1991. Effectiveness of zinc sulfate and zinc chelate as foliar sprays in alleviating zinc deficiency of wheat grown on zinc-deficient soils in Western Australia. Aust. J. Exp. Agric. 31 (6), 831–834.

Burkhead, J.L., Gogolin-Reynolds, K.A., Abdel-Ghany, S.E., Cohu, C.M., Pilon, M., 2009a. Copper homeostasis. New Phytol. 182 (4), 799–816.

Burkhead, J.L., Gogolin-Reynolds, K.A., Abdel-Ghany, S.E., Cohu, C.M., 2009b. Pilon M. Copperhomeostasis. New Phytol. 182 (4), 799–816.

Burnell, J.N., 1988. The biochemistry of manganese in plants. Manganese in Soils and Plants. Springer Netherlands, pp. 125–137.

Chaney, R.L., 1993. Zinc phytotoxicity. In: Robson, A.D. (Ed.), Zinc in Soil and Plants. Kluwer Academic Publishers, Dordrecht, pp. 135–150.

Chatterjee, C., Nautiyal, N., 2001. Molybdenum stress affects viability and vigour of wheat seeds. J. Plant. Nutr. 24, 1377–1386.

Chatterjee, S., Chattopadhyay, B., Mukhopadhyay, S.K., 2006. Trace metal distribution in tissues of Cichlids (Oreochromis niloticus and O. mossambicus) collected from waste water fed fish ponds in East Calcutta Wetlands, a Ramsar site. Acta Ichthyologica et. Piscat. 36 (2), 119–125.

Curie, C., Briat, J.F., 2003. Iron transport and signalling in plants. Annu. Rev. Plant. Biol. 54, 183–206.

Dehnavi, M.M., Mohammad, S.A., Sanavy, M., 2008. Effects of withholding irrigation and foliar application of zinc and manganese on fatty acid composition and seed oil content in winter safflower. In: 7th International safflower conference, Wagga Wagga Australia.

Dongre, S.M., Mahorkar, V.K., Joshi, P.S., Deo, D.D., 2000. Effect of micronutrients spray on yield and quality of chilli (*Capsicum annum* L.) cv. Jayanti. Agric. Sci. Dig. 20, 106–107.

Donnini, S., Prinsi, B., Negri, A.S., Vigani, G., Espen, L., Zocchi, G., 2010. Proteomic characterization of iron deficiency responses in *Cucumis sativus* L. roots. BMC Plant. Biol. 10 (1), 268.

Doolette, C.L., Read, T.L., Li, C., Scheckel, K.G., Donner, E., Kopittke, P.M., Schjoerring, J.K., Lombi, E., 2018. Foliar application of zinc sulphate and zinc EDTA to wheat leaves: differences in mobility, distribution, and speciation. J. Exp. Bot. 69 (18), 4469–4481.

Elanchezhian, R., Dameshwar, Kumar, Kulasekaran, Ramesh, et al., 2017. Morpho-physiological and biochemical response of maize (*Zea mays* L.) plants fertilized with nano-iron (Fe_3O_4) micronutrient. J. Plant Nutr. 40 (14), 1969–1977.

El-Kassas, S.E., 1984. Effect of iron nutrition on the growth, yield, fruit quality, and leaf composition of ceded balady lime trees grown on sandy calcareous soils. J. Plant Nutr. 7, 301–311.

Epstein, E., Bloom, A.J., 2005. second ed. Mineral Nutrition of Plants: Principles and Perspectives, 17–39. Sinauer Associates, Sunderland, MA.

Farahat, M.M., Ibrahim, S.S., Lobna, T., Fatma, E.M.El-Quesni, 2007. Respons of Vegetative Growth and Some Chemical Constituents of *Cupressus Sempervirens* L. to Foliar Application of Ascorbic Acid and Zinc at Nubaria to Respons of vegetative growth and some chemical constituents of *Cupressus Sempervirens* L. to foliar application of ascorbic acid and zinc at Nubaria. World J. Agric. Sci. 3 (4), 496–502. In this issue.

Farooq, M.A., Saqib, Z.A., Akhtar, J., Bakhat, H.F., Ratnakumar, P., Dietz, K., 2015. Protective role of silicon (Si) against combined stress of salinity and boron (B) toxicity by improving antioxidant enzymes activity in rice. Silicon 11, 2193–2197. Available from: https://doi.org/10.1007/s12633-015-9346-z.

Fouad, A., Saad, D., Kacem, M., Abdelwahed, M., Khalid, D., Abderrahim, R., et al., 2020. Efficacy of copper foliar spray in preventing copper deficiency of rainfed wheat (*Triticum aestivum* L.) grown in a calcareous soil. J. Plant. Nutr. 43 (11), 1617–1626.

Foyer, C.H., Noctor, G., 2005. Oxidant and antioxidant signalling in plants; a re-evaluation of the concept of oxidative stress in physiological concept. Plant Cell Environ. 28, 1056–1071.

Goswami, S., Kumar, R.R., Sharma, S.K., Kala, Y.K., Singh, K., Gupta, R., et al., 2015. Calcium triggers protein kinases-induced signal transduction for augmenting the thermotolerance of developing wheat (Triticum aestivum) grain under the heat stress. J. Plant Biochem. Biotechnol. 24 (4), 441–452.

Graham, A.W., 2004. Effects of zinc nutrition and high temperature on the growth, yield and grain quality of wheat (*Triticum aestivum* L.). School of Agriculture, Food and Wine, The University of, Adelaide, p. 297–p.

Gris, E., 1843. Mémoir relatifá l'action des composés soluble ferrugineux sur la végétation. Compt Rend. Acad. Sci. (Paris.) 17, 679.

Habiballah, Y., Neda, D., Rasoul, R., Alireza, B., 2020. The alleviation of salinity-induced stress by using boron in soilless grown rose. J. Plant. Nutr. 43 (4), 526–537.

Hajiboland, R., 2012. Effect of micronutrient deficiencies on plants stress responses. Abiotic Stress Responses in Plants. Springer, New York, NY, pp. 283–329.

Harrison, M.D., Jones, C.E., Solioz, M., Dameron, C.T., 2000. Intracellular copper routing: the role of copper chaperones. Trends Biochem. Sci. 25 (1), 29–32.

Hasani, M., Zamani, Z., Savaghebi, G., Fatahi, R., 2012. Effects of zinc and manganese as foliar spray on pomegranate yield, fruit quality and leaf minerals. J. Soil. Sci. Plant. Nutr. 12 (3), 471–480.

Heckman, J. (2010). Effective Treatment of Mn deficiency in soybean: Comparing Fertilizer Sources, Plant & pest advisory, A Rutgers Cooperative Extension Publication, 16(7), 3.

Hosny, A.A., Kadry, W., Mohamad, H.M.H., 1984. Effect of foliar application of boron and copper on growth, yield and yield components of Giza 75 cotton variety. An. Agric. Sc. 21, 25–35.

Hasanuzzaman M., Fujita M., Oku H., et al., 2018. Role of micronutrients in salt stress tolerance to plants. In: Plant Nutrients and Abiotic Stress Tolerance. pp. 363–376.

IIHR. 2020. Foliar Application of Boron Increased the Productivity of Cucurbitaceous Crops at Farmer's Field. <https://icar.org.in/node/297>.

Iron Chlorosis in Trees, 2012. Kansas State. Universtiy, Kansas Forest Service Publication, p. MF-718. Available from: http://www.hfrr.ksu.edu/doc1649.ashx.

Jyoti Ranjan, R., 2019. Micronutrients induced alteration of antioxidant enzymes in plants. Int. J. Nutr. Sci. 4 (2), 1034.

Kailas, Veeresh, H., Narayana Rao, K., Balanagoudar, S.R., Sharanagouda, H., 2019. Effect of conventional and nano micronutrient fertilizers on yield and economics of pigeonpea (*Cajanus cajan* (L.) Mill sp.). Int. J. Curr. Microbiol. App Sci. 8 (9), 185–193.

Karim, M.R., Islam, M.R., Jahiruddin, M., Islam, M.R., 2006. Effects of Manganese, sulphur, zinc, boron and molybdenum on the yield and nutrient uptake of BRRI Dhan 30. J. Ban. Soc. Agric. Sci. Technol. 3, 141–144.

Khalil, U., Ali, S., Rizwan, M., et al., 2018. Role of mineral nutrients in plant growth under extreme temperatures. Plant Nutrients and Abiotic Stress Tolerance. Springer Nature Singapore., pp. 499–524.

Khan, M., Ahmad, R., Khan, M.D., Rizwan, M., Ali, S., Khan, M.J., et al., 2018. Trace elements in abiotic stress tolerance. In: Plant Nutrients and Abiotic Stress Tolerance, Springer, Singapore, pp. 137–151.

Kobayashi, M., Mutoh, T., Matoh, T., 2004. Boron nutrition of cultured tobacco BY-2 cells. IV. Genes induced under low boron supply. J. Exp. Bot. 55, 1441–1443.

Kramer, U., Clemens, S., 2006. Functions and homeostasis of zinc, copper, and nickel in plants. In: Tamás, M., Martinoia, E. (Eds.), Molecular Biology of Metal Homeostasis and Detoxification from Microbes to Man. Springer, Berlin, pp. 214–272.

Kuruppaiah, P., 2005. Foliar application of micro-nutrients on growth, flowering and yield characters of brinjal cv. Annamalai. Plant. Arch. 5 (2), 605–608.

Lalitha, B.H., Yeledhalli, N.A., Ravi, M.V., 2008. Effect of foliar application of potassium sulphate and boric acid on the yield, nutrient uptake and quality of niger (*Guizotiaa byssinica* L.). Karnataka J. Agric. Sci. 21, 282–283.

Ma, J.F., Yamaji, N., Tamai, K., Mitani, N., 2007. Genotypic difference in silicon uptake and expression of silicon transporter genes in rice. Plant. Physiol. 145 (3), 919–924.

Marschner H. Mineral Nutrition of Higher Plants, second ed. Elsevier, Academic Press, 1995;. 889.

Martın-Rejano, E.M., Camacho-Cristobal, J.J., Herrera-Rodrıguez, M.B., Rexach, J., Navarro-Gochicoa, M.T., González-Fontes, A., 2011. Auxin and ethylene are involved in the responses of root system architecture to low boron supply in *Arabidopsis* seedlings. Physiol. Plant. 142, 170–180.

Marzouk, N.M., Abd-Alrahman, H.A., EL-Tanahy, A.M.M., Mahmoud, S.H., 2019. Impact of foliar spraying of nano micronutrient fertilizers on the growth, yield, physical quality, and nutritional value of two snap bean cultivars in sandy soils. Bull. Natl. Res. Cent. 43 (1), 1–9.

Mendel, R., Robert, H., 2002. Molybdo-enzymes and molybdenum cofactor in plants. J. Exp. Bot 53, 1689–1698.

Mengel, K., Kirkby, E.A., 2001. Principles of Plant Nutrition, fifth ed. Kluwer Academic Publishers, Dordrecht, p. 848.

Mengutay, M., Ceylan, Y., Kutman, U.B., Cakmak, I., 2013. Adequate magnesium nutrition mitigates adverse effects of heat stress on maize and wheat. Plant. Soil. 368, 57–72.

Millaleo, R., Reyes-Dıaz, M., Ivanov, A.G., Mora, M.L., Alberdi, M., 2010a. Manganese as essential and toxic element for plants: transport, accumulation and resistance mechanisms. J. Soil. Sci. Plant. Nutr. 10 (4), 470–481.

Miller, T.D., 1994. The effect of chloride fertilizers on yield and disease progress in Texas wheat, Intensive Wheat Management Conference Proceedings.

Minhas P.S., Rane J., Ratnakumar P., Kumar P.S., Singh Y., Singh N.P., 2014. Annual Report of NIASM 2013-14. pp. 80.

Mishra, N.M., Ramkrishna Reddy, M.G., Subramanyam, M.V.K., 1985. Effect of micro-nutrients on sunflower. Andhra Agric. J. 32, 149–159.

Miwa, K., Takano, J., Omori, H., Seki, M., Shinozaki, K., Fujiwara, T., 2007. Plants tolerant of high boron levels. Science 318 (5855), 1417.

Mobin, M., Khan, M.N., Abbas, Z.K., Ansari, H.R., Al-Mutairi, K.A., 2017. Significance of sulfur in heat stressed cluster bean (*Cymopsiste tragonoloba* L. Taub) genotypes: responses of growth, sugar and antioxidative metabolism. Arch. Agron. Soil. Sci. 63, 288–295.

Mohamadipoor, R., Sedaghathoor, S., Mahboub-Khomami, A., 2013. Effect of application of iron fertilizers in two methods 'foliar and soil application' on growth characteristics of Spathyphyllum illusion. Eur. J. Exp. Biol. 3 (1), 232–240.

Montalvo, D., Degryse, F., Da Silva, R.C., Baird, R., McLaughlin, M.J., 2016. Agronomic effectiveness of zinc sources as micronutrient fertilizer. Adv. Agron 139, 215–267.

Moosavi, A.A., Ronaghi, A., 2011. Influence of foliar and soil applications of iron and manganese on soybean dry matter yield and iron-manganese relationship in a Calcareous soil. Asian J. Crop. Sc. 5 (12), 1550–1556.

Moradtalab, N., Ahmed, A., Geistlinger, J., Walker, F., Höglinger, B., Ludewig, U., et al., 2020. Synergisms of microbial consortia, N forms, and micronutrients alleviate oxidative damage and stimulate hormonal cold stress adaptations in maize. Front. Plant. Sci. 11, 396.

Nachiangmai, D., Dell, B., Bell, R., Huang, L., Rerkasem, B., 2004. Enhanced boron transport into the ear of wheat as a mechanism for boron efficiency. Plant. soil. 264 (1), 141–147.

Nahed, G.A.E., Balbaa, L.K., 2007. Influence of tyrosine and zinc on growth, flowering and chemical constituents of *Salvia farinacea* plants. J. Appl. Sci. Res. 3 (11), 1479–1489.

Nasiri, Y., Zehtab-Salmasi, S., Nasrullahzadeh, S., Najafi, N., Ghassemi-Golezani, K., 2010. Effects of foliar application of micro-nutrients (Fe and Zn) on flower yield and essential oil of chamomile (*Matricaria chamomilla* L.). J. Med. Plants Res. 4 (17), 1733–1737.

Nayyar, V.K., Sadanas, U.S., Takkar, P.N., 1985. Methods and rates of application of Mn and its critical levels for wheat following rice on coarse textured soils. Fert. Res. 8, 173–178.

Neenu, S., Ramesh, K., 2020. Weather – micronutrient interactions in soil and plants – a critical review. Chem. Sci. Rev. Lett. 9 (33), 205–219.

Nelson, N., 1944. A photometric adaptation of the Somogyi method for the determination of glucose. J. Biol. Chem. 153, 375–380.

Patil, I., Ramesh, H., Patil, M., 2006. Effect of boron and zinc on yield parameters of sunflower. Karnataka. J. Agric. Sci. 16, 243–250.

Perveen, S., Rehman, H., 2000. Effect of foliar application of zinc, manganese and boron in combination with urea on the yield of sweet orange. Pak. J. Agric. Res. 16 (2), 135–141.

Potarzycki, J., Grzebisz, W., 2009. Effect of zinc foliar application on grain yield of maize and its yielding components. Plant. Soil. Env. 55 (12), 519–527.

Puig, S., Askeland, E., Thiele, D.J., 2005. Coordinated remodeling of cellular metabolism during iron deficiency through targeted mRNA degradation. Cell 120 (1), 99–110.

Rakkiyappan, P., Thangavelu, S., Radhamani, R., 2002. Effect of ferrous sulphate on sugarcane varieties grown in iron deficient soil. Sugar Tech 4 (1-2), 33–37. Available from: https://doi.org/10.1007/BF02956877. In this issue.

Rathinavel, K., Dharmalingam, C., Paneersewam, S., 1999. Effect of micronutrient on the productivity and quality of cotton seed Cv. TCB 209 (*Gossypium barbadense* L.). Madras Agric. J. 86, 313–316.

Ratnakumar, P., Khan, M.I.R., Minhas, P.S., Sultana, R., Per, T.S., Deokate, P.P., et al., 2016a. Can plant bioregulators minimize the crop productivity losses caused by drought, heat stress and salinity? An integrated review. J. Appl. Bot. Food Qual. 89, 13–125.

Ratnakumar, P., Deokate, P.P., Rane, J., Jain, N., Kumar, V., Berhe, D.V., et al., 2016b. Effect of stabilized silicic acid foliar sprays on wheat under water stress conditions. J. Funct. Environ. Bot. 6, 17–19.

Ratnakumar, P., Minhas, P.S., Wackchaure, G.C., Chaudary, R.L., Deokate, P.P., 2015. Yield and water production functions of wheat (*Triticum aestivum L.*) cultivars and response to exogenous application of thiourea and ortho-silicic acid. J. Environ. Agric. Sci. 2 (6), 1228–1250.

Ratnakumar, P., Minhas, P.S., Wakchaure, G.C., 2017. Plant bioregulators : A mitigation strategies for resilient agriculture. *In* Minhas et al 2017. Abiotic Stress Management for Resilience Agriculture. Springer Nature Publishers, India, pp. 221–235.

Ravi, S., Channal, H.T., Hebsur, N.S., Dharmatti, P.R., 2008. Effect of sulpher, zinc, iron nutrition on growth, yield and nutrient uptake of safflower (*Carthanus tinctorious* L.). Karnataka J. Agric. Sci. 21, 382–385.

Reguera, M., Wimmer, M., Bustos, P., Goldbach, H.E., Bolaños, L., Bonilla, I., 2010. Ligands of boron in Pisum sativum nodules are involved in regulation of oxygen concentration and rhizobial infection. Plant Cell Environ. 33 (6), 1039–1048.

Rehm G., Schmitt M., 2002. Copper for crop production. <http://www.extension.umn.edu>

Sajedi, N.A., Ardakani, M.R., Madani, H., Naderi, A., Miransari, M., 2011. The effects of selenium and other micronutrients on the antioxidant activities and yield of corn (*Zea mays* L.) under drought stress. Physiol. Mol. Biol. Plants 17, 215–222.

Sangale, P.B., Patil, G.D., Daftardar, S.Y., 1981. Effect of foliar application of zinc, iron and boron on yield of safflower. J. Maharashtra Agric. Univ. 6, 65–66.

Sawan, Z.M., Mahmoud, M.H., El-Guibali, A.H., 2008. Influence of potassium fertilization and foliar application of zinc and phosphorus on growth, yield components, yield and fiber properties of Egyptian cotton (*Gossypium barbadense* L.). J. Plant. Ecol. 1 (4), 259–270.

Shukla, A.K., Tiwari, P.K., 2016. Micro and secondary nutrients and pollutant elements research in India. Coordinator's Report. AICRP on Micro- and Secondary Nutrients and Pollutant Elements in Soils and Plants. ICAR-IISS, Bhopal.

Shukla, A.K., Behera, S.K., Lenka, N.K., Tiwari, P.K., Prakash, C., Malik, R.S., et al., 2016. Spatial variability of soil micronutrients in the intensively cultivated trans-gangetic plains of India. Soil Tillage Res. 163, 282–289.

Singh, M.V., 2008. Micronutrients deficiency in Indian soils and crops. In: Alloway, B.J. (Ed.), Micronutrient Deficiencies in Global Crop Production. Springer, Dordrecht, pp. 93–125.

Sreenivasulu, N., Grimm, B., Wobus, U., Weschke, W., 2000. Differential response of antioxidant compounds to salinity stress in salt-tolerant and salt-sensitive seedlings of foxtail millet (*Setaria italica*). Physiol. Plant. 109, 435–442.

Srivastva, A.K., Ratnakumar, P., Minhas, P.S., Suprasanna, P., 2016. Plant bioregulators for sustainable agriculture: integrating redox signaling as a possible unifying mechanism. Adv. Agron. 137, 237–278. Available from: https://doi.org/10.1016/bs.agron.2015.12.002.

Steiner, F., Zoz, T., Zuffo, A.M., Machado, P.P., Zoz, J., 2018. Foliar application of molybdenum enhanced quality and yield of crispleaf lettuce (Lactuca sativa L., cv. Grand Rapids). Acta Agronómica 67. Available from: https://doi.org/10.15446/acag.v67n1.59272.

Ullagaddi, M.S., 2000. Effect of Dates of Sowing and Chemicals on Seed Yield and Quality of Male Parent SB (YF)-425 Cotton (*Gossypium barbadonse* L.). Thesis, University of Agricultural Sciences, Dharwad.

Vigani, G., Zocchi, G., Bashir, K., Philippar, K., Briat, J.F., 2013. Cellular iron homeostasis and metabolism in plant. Front. plant. Sci. 4 (490), 1–3.

Wackchaure, G.C., Minhas, P.S., Ratnakumar, P., Chaudary, R.L., 2016a. Effect of plant bioregulators on growth, yield and water production functions of sorghum [*Sorghum bicolor* (L.) Moench]. Agric. Water Manage. 177, 138–145.

Wackchaure, G.C., Minhas, P.S., Ratnakumar, P., Chaudary, R.L., 2016b. Optimizing supplemental irrigation for wheat (*Triticum aestivum* L.) and the impact of plant bio-regulators in a semi-arid region of deccan plateau in India. Agric. Water Manage. 172, 9–17.

Wang, Y.Y., Hsu, P.K., Tsay, Y.F., 2012. Uptake, allocation and signaling of nitrate. Trends Plant. Sci. 17 (8), 458–467.

Williams, C.M.J., Maier, N.A., Bartlett, L., 2004. Effect of molybdenum foliar sprays on yield, berry size, seed formation, and petiolar nutrient composition of "Merlot" grapevines. J. Plant Nutr. 27, 1891–1916.

Wojciechowska, R., Kowalska, I., 2011. The effect of foliar application of urea, Mo and BA on nitrate metabolism in lettuce leaves in the spring and summer, autumn seasons. Folia Hortic. 23 (2), 119–123.

Wu, S., Hu, C., Tan, Q., Nie, Z., Sun, X., 2014. Effects of molybdenum on water utilization, antioxidativedefense system and osmotic-adjustment ability in winter wheat (*Triticum aestivum*) under drought stress. Plant. Physiol. Biochem. 83, 365–374.

Yasmeen, F., Raja, N.I., Razzaq, A., Komatsu, S., 2016. Gel-free/label-free proteomic analysis of wheat shoot in stress tolerant varieties under iron nanoparticles exposure. Biochim. Biophys. Acta 1864, 1586–1598.

Yavas, I., Unay, A., 2016. Effects of zinc and salicylic acid on wheat under drought stress. J. Anim. Plant. Sci. 26 (4), 1012–1101.

Yu, M., Hu, C.X., Wang, Y.H., 2002. Molybdenum efficiency in winter wheat cultivars as related to molybdenum uptake and distribution. Plant. Soil. 245 (2), 287–293.

Zhang, X., Zhang, F., Mao, D., 1999. Effect of iron plaque outside roots on nutrient uptake by rice (*Oryza sativa* L.): phosphorus uptake. Plant Soil. 209, 187–192.

Ziolek, E., Ziolek, W., 1987. The effect of trace element application on yield and quality of soybean seeds. Agraria 26, 195–207.

Chapter 15

Biotechnological tools for manipulating nutrient homeostasis in plants

Jorge González-Villagra[1,2], Marjorie Reyes-Díaz[3,4], Zed Rengel[5] and Claudio Inostroza-Blancheteau[1,2]

[1]*Department of Agricultural Sciences and Aquaculture, Faculty of Natural Resources, Universidad Católica de Temuco, Temuco, Chile,* [2]*Nucleus of Research in Food Production, Faculty of Natural Resources, Universidad Católica de Temuco, Temuco, Chile,* [3]*Department of Chemical Sciences and Natural Resources, Faculty of Engineering and Sciences, Universidad de La Frontera, Temuco, Chile,* [4]*Center of Plant, Soil Interaction and Natural Resources Biotechnology, Scientific and Technological Bioresource Nucleus (BIOREN-UFRO), Universidad de La Frontera, Temuco, Chile,* [5]*UWA School of Agriculture and Environment, University of Western Australia, Perth, WA, Australia*

15.1 Introduction

The growth of the human population has forced development of nutritious and high-yielding crops in an attempt to ensure food production for the future (Ali et al., 2018; Wang et al., 2020). Food production and high-yielding crops have also significantly increased the use of fertilizers (López-Arredondo et al., 2013; Nguyen and Kant, 2018). Currently, global fertilizer consumption is estimated at about 201 million tons, including 119 million tons of nitrogen (N) fertilizers, 45 million of phosphorus (P) fertilizers, and 37 million of potassium (K) fertilizers (Food and Agriculture Organization of The United Nations, 2020). However, only 30%–50% of the applied fertilizers are actually absorbed by crops, and the remainder is lost to the environment, causing not only severe damage such as soil acidification, air pollution, and aquatic eutrophication, but also increasing the costs of agricultural production (Li et al., 2017; Ali et al., 2018).

Environmental factors such as soil, water, and temperature influence plant nutrient uptake, transport, and assimilation (Fischer et al., 2019). It has

Plant Nutrition and Food Security in the Era of Climate Change. DOI: https://doi.org/10.1016/B978-0-12-822916-3.00009-3

been widely reported that drought stress, which is the most severe manifestation of climate change around the world, decreases the quality and quantity of macronutrients and micronutrients in plants, reducing plant nutrition and crop yields (Marschner and Rengel, 2012; Savage et al., 2016; Sevanto et al., 2018; Soares et al., 2019). Therefore, low nutrient uptake, transport, and assimilation in plants under climate change are huge problems jeopardizing food and nutrition security because plants are the major part of the human diet and the main source of nutrients (Thompson et al., 2010; Cai et al., 2014; Fischer et al., 2019). Currently, about 40% of the earth's surface is (at least periodically) subjected to drought stress, and it is predicted that this proportion will increase up to 50% by 2025 (Topcu and Kirda, 2013). Therefore, it is important to develop strategies to improve nutrient use efficiency in crop plants in the context of climate change.

Numerous studies have been conducted to improve nutrient use efficiency by manipulating genes directly involved in nutrient uptake, transport, and assimilation (e.g., López-Arredondo et al., 2013; Ning et al., 2015; Gojon, 2017; Wang et al., 2020). Biotechnological tools have been used to modify regulatory pathways in nutrient uptake, overexpression of nutrient transporters, and introgression of novel genes to improve nutrient uptake (Feng et al., 2017; Gao et al., 2019; Wang et al., 2020). In this chapter, we have reviewed the main advances in the use of biotechnological tools to manipulate nutrient uptake, transport, and assimilation in plants in the context of climate change.

15.2 Biotechnological tools for macronutrient manipulation

15.2.1 Nitrogen

Nitrogen (N) is an essential nutrient for many physiological and biochemical processes in plants, playing a structural role in proteins and nucleic acids, seed dormancy, flowering time, leaf expansion, root development, and control of multiple N-responsive genes (Bouguyon et al., 2012). The N metabolism regulation in plants is highly complex and influenced by physiological and metabolic processes such as amino acid synthesis, carbon metabolism, and NO_3^- concentration in plants (López-Arredondo et al., 2013). Nitrogen is taken up in inorganic forms as nitrate (NO_3^-) and ammonium (NH_4^+) from soil, with ammonium-based fertilizers being a major N source (Krouk et al., 2010; Marschner and Rengel, 2012). Plants have root-specific transporters to take up N from soil (Masclaux-Daubresse et al., 2010). The NITRATE TRANSPORTER1 (NRT1)/PEPTIDE TRANSPORTER (PRT) family (NPF) includes the low-affinity transporters (LATs) that operate at relatively high nitrate supply, whereas the NITRATE TRANSPORTER2 (NRT2) family comprises the high-affinity transporters under low N levels in soil (Williams and Miller, 2001). In addition, the CHLORIDE CHANNEL family and SLOWLY ACTIVATING ANION CHANNEL (SLAC) are also involved in nitrate uptake, whereas NH_4^+ is taken up by AMMONIUM/METHYLAMMONIUM TRANSPORTERS (AMTs) (Fan et al., 2016; Feng et al., 2011; Jacquot et al., 2017; Li et al., 2012; Marschner and Rengel, 2012; Yan et al., 2011). The N absorbed by AMTs is assimilated into amino acids via the glutamine synthetase/glutamine-2-oxoglutarate aminotransferase cycle and translocated to developing organs (Li et al., 2017).

For the last 2 decades, biotechnological tools for improving N use efficiency have focused mainly on the N transport from soil to plants (Gojon et al., 2017). For example, the overexpression of *OsNRT1.1A* enhanced grain yield and N utilization in *Oryza sativa*, showing constitutive expression in the root epidermis and root hairs (Lin et al., 2000; Wang et al., 2018a). Likewise, He et al. (2020) reported that the VvNPF6.5, a member of nitrate transporter 1/peptide transporter family (NRT1/PRT/NPF), regulates nitrate uptake in *Vitis vinifera* plants. Under N limitation, the NRT2 transporter has an important role in NO_3^- uptake from soil (Wang et al., 2012; O'Brien et al., 2016). Overexpression of *OsNRT2.1*, the high-affinity nitrate transporter, increased nitrate uptake in *O. sativa* plants (Luo et al., 2018). Similarly, NRT2.1 has an important role in adaptation to low nitrate levels in *Manihot esculenta* (Zou et al., 2019). In *Arabidopsis*, many NRT2 transporters, such as NRT2.1, NRT2.2, NRT2.4, and NRT2.5, require the nitrate transport accessory protein 2.1 (NAR2.1) as a partner protein, which targets NRT protein to the plasma membrane, increasing nitrate uptake (Feng et al., 2011; Wang et al., 2018a). Thus, Chen et al. (2016) showed that transgenic plants overexpressing *OsNAR2.1* gene enhanced *OsNRT2.1* expression, exhibiting higher biomass, yield (38%), and nitrogen use efficiency (about 130%) compared with wild-type plants. Interestingly, the NRT2.1 transporter may also improve water uptake by regulating plasma membrane aquaporins (plasma membrane intrinsic protein: PIP) in *Arabidopsis* plants subjected to drought stress (Li et al., 2016). Therefore, NRT2.1 transporter could be an important biotechnological tool to enhance drought resistance in crop plants.

Using knockout mutant plants and gene overexpression analysis, Lou et al. (2020) reported that the osmotic stress/ABA-activated protein kinase 2, a member of sucrose nonfermenting 1-related kinases 2, promotes NO_3^- influx rate and nitrate concentration in *O. sativa* under drought stress, indicating a connection with nitrate transporters.

Transcription factors have also been used as a potential target to improve N use efficiency. Ge et al. (2020) showed that the transcription factor NIN-like protein ZmNIN-LIKE PROTEIN 5 (*ZmNPL5*) had a central role in N assimilation in *Zea mays* plants. Defective *ZmNPL5* plants accumulated less N in the roots; in contrast, overexpression of *ZmNLP5* resulted in higher N accumulation compared to wild-type plants. Recently, Wang et al. (2020) showed that the MYB (myeloblastosis) transcription factor *OsMYB305* participates in NO_3^- uptake and assimilation, upregulating the expression of *OsNRT2.1*, *OSNRT2.2* and *OsNAR2.1* in *O. sativa* roots. In addition,

OsMYB305 overexpression resulted in increased tiller number, shoot dry weight, and total N concentration in *O. sativa*. Likewise, Yu et al. (2016) have shown that the overexpression of *NIN-LIKE PROTEIN 7* transcription factor increased nitrogen uptake, total nitrogen content, photosynthesis rate, carbon assimilation, and plant growth in *Arabidopsis thaliana* and *Nicotiana tabacum* plants. Therefore, these transcription factors could be key candidates for improving N use efficiency in the future.

The AMT transporter mediates ammonium uptake and is expressed mainly in root hairs (Lauter et al., 1996; Ludewig et al., 2007; Nath and Tuteja, 2016). AMT transporters can be subdivided into two subfamilies, AMT1 and AMT2 (Loque and von Wiren, 2004). Bu et al. (2013) reported that *Puccinellia tenuflora* plants overexpressing the ammonium transporter (*PutAMT1;1*) gene exhibited greater root length compared to wild-type under low NH_4^+ supply. The *AMT1;1* is mainly expressed in the root tip and in the epidermal cells and is upregulated under low-N supply. Interestingly, the overexpression of *PutAMT1;1* increased *P. tenuflora* tolerance to drought stress (Bu et al., 2019). The *AMT* genes were upregulated in several other species such as *Z. mays*, *O. sativa*, and *Solanum lycopersicum* under low-NH_4^+ supply (Sonoda et al., 2003; Gu et al., 2013). In summary, N uptake could be improved by regulating NO_3^- and NH_4^+ transporters such as NRT and AMT in crop species.

15.2.2 Phosphorus

Phosphorus is involved in several physiological and metabolic processes in plants, and is a key element in DNA, RNA, and plasma membrane structures (Marschner, 1995). Total P content is between 200 and 3000 mg P kg^{-1} in soil; however, it is poorly available to plants due to chemical fixation and slow diffusion properties, being taken up as inorganic monohydrogen- and dihydrogen-phosphate (Pi) that represents less than 1% of the total P content in soil (Marschner, 1995; Raghothama, 1999; Richardson et al., 2009). Thus, plants have developed complex mechanisms to improve P availability and thus Pi uptake (Ai et al., 2009; Gu et al., 2016).

The Pi transporters and Pi starvation genes are key biotechnological tools to improve Pi uptake, its assimilation, and Pi use efficiency in plants (Wang et al., 2018b; Levi et al., 2019). Plants have high- and low-affinity Pi transporters located in the plasma membrane of root cells (Remy et al., 2012). Until now, the PHOSPHATE TRANSPORTER 1 (PHT1) family has been the most studied; it includes five subfamilies characterized in *Arabidopsis* and *O. sativa* (Nussaume et al., 2011; Wang et al., 2017).

Five *PHT1* genes were functionally characterized in *Arabidopsis* (Remy et al., 2012). The PHT1 was upregulated by P starvation (Secco et al., 2013; Ayadi et al., 2015). The Pi taken up is loaded into the xylem by *PHOSPHATE1* (*PHO1*) transporter that is expressed in the root stele cells, distributing Pi within the plant (Hamburger et al., 2002; Wege et al., 2016). Zhang et al. (2016) showed that the expression of *PtPHT1.2* and *PtPHO9* genes was significantly increased under drought conditions (e.g., 90-fold for *PtPHT1.2* gene compared to wild-type). The authors reported that the expression of *PtPHT1* and *PtPHO* was upregulated under drought stress and low supply of phosphate and increased only slightly when no phosphate was supplied.

The *PHT2* is expressed in green tissues in *Arabidopsis* and via AtPHT2-GFP fusion protein in the chloroplast and is involved in Pi/H^+ symport (Versaw and Harrison, 2002; Zhao et al., 2003). Interestingly, mutant *AtPHT2* reduced Pi transport into chloroplast (about 20 times compared to wild-type) (Versaw and Harrison, 2002). In *O. sativa*, *PHT2* is expressed in leaves and induced by phosphate deficiency (Shi et al., 2013).

The PHT3 has been reported as a Pi transporter in mitochondria, playing an important role in ATP production in plants (Zhu et al., 2012). Zhang et al. (2016) studied *PHT* expression in different tissues in *Populus simonii*. They reported six homologous *PsPHT3* genes, including *PsPHT3.1a*, *PsPHT3.1b*, *PsPHT3.2a*, *PsPHT3.2b*, *PsPHT3.3a*, and *PsPHT3.3b*. The same authors reported that *PsPHT3.1b* and *PsPHT3.3b* were highly expressed in *P. simonii* roots under Pi deficiency.

The PHT4 has six members in *Arabidopsis* located in different plant tissues and organelles, with PHT4;1 and PHT4;4 found in leaf chloroplasts, PHT4;2 in root plastids, PHT4;3 and PHT4;5 in phloem plastids, and PHT4;6 in the Golgi (Hassler et al., 2012). According to Karlsson et al. (2015), among the PHT4 members, the PHT4;1 is key in providing phosphate for ATP synthesis and plant growth in *Arabidopsis*. They showed that *PHT4;1* mutant plant had decreased stromal phosphate, limiting ATP synthesis and impairing CO_2 assimilation in plants, which in turn reduced plant growth.

The PHT5 transporters are also known as vacuolar phosphate transporters involved in sequestration of Pi into vacuoles (Remy et al., 2012). According to Wang et al. (2015), three (SPX)-Major Facility Superfamily proteins (OsSPX-MFS1, OsSPX-MFS2, and OsSPX-MFS3) belonging to the PHT5 family have been identified as Pi transporters in the tonoplast of *O. sativa*. Thus, PHT5 transporters help plants maintain Pi homeostasis inside the cells.

In summary, characterizing Pi transporters is an important tool to improve Pi uptake by plants and could be used as a strategy to cope with limited phosphate availability in soil, and to decrease phosphorus fertilization.

15.2.3 Potassium

Potassium (K^+) is an important macronutrient involved in stomatal regulation and as cofactor for enzymes, regulating photosynthesis, intracellular pH, and cation-anion balance (Maathuis, 2009). The two types of potassium uptake are the low-affinity transport system (LATS) and high-affinity transport system (HATS) (Epstein et al., 1963; Maathuis and Sanders, 1994, 1996). The LATS mediates K uptake in the millimolar range, and HATS in the micromolar range (Glass, 1976; Maathuis and Sanders, 1997; Chérel et al., 2014).

The K transport can be mediated by K^+ channels and K^+ transporters (Li et al., 2018). The K^+ transporters are classified into four families: K^+ transporter/high-affinity K^+ transporter/K uptake permease (KT/HAK/KUP), high-affinity K^+/Na^+ transporter (HKT), K^+ exchange antiporters (KEA), and cation/H^+ exchangers (CHX) (Aranda-Sicilia et al., 2016; Wang and Wu, 2017). Channel proteins involved in K^+ uptake belong to three families, including Shaker, Tandem-Pore K^+ (TPK), and K^+ inward rectifier (KIR) (Lebaudy et al., 2007; Li et al., 2018). According to Véry et al. (2014), the Shaker channel and the KT/HAK/KUP transporters play a crucial role in K^+ uptake. In *Arabidopsis*, 13 members of the KT/HAK/KUP family have been reported (Gierth and Mäser, 2007; Véry et al., 2014).

The KT/HAK/KUP transporters in *Hordeum vulgare*, *Arabidopsis*, *O. sativa*, and *Z. mays* are localized in the plasma membrane, chloroplasts, and other endomembranes, reflecting diverse roles in K^+ homeostasis (Zhang et al., 2012; Han et al., 2016). The HAK5 from *Arabidopsis*, HAK1 from *H. vulgare*, and HAK1 from *O. sativa* have been reported as high-affinity K^+ transporters in roots and are positively induced under low K^+ concentration in soil (Santa-María et al., 1997; Qi et al., 2008; Chen et al., 2015).

The guard cell outward rectifying K (GORK) and stelar K1 outward rectifier (SKOR), the Shaker channels, are involved in K^+ transport in roots from peripheral root cells to the root stele tissues and translocated to shoot via xylem (Gaymard et al., 1998; Ivashikina et al., 2001). In addition, AKT2 was involved in the K^+ loading into phloem tissues in *Arabidopsis* (Lacombe et al., 2000).

It has been reported that K^+ transporters have an important role in drought-stress tolerance, especially *OsHAK1* transporter (Osakabe et al., 2013; Han et al., 2016; Chen et al., 2017). The *OsHAK1* mediates K^+ acquisition and its transport in *O. sativa* under different K^+ supplies (Chen et al., 2015). Recently, Chen et al. (2017) showed *OsHAK1* mutants had low drought-stress tolerance, whereas *OsHAK1* overexpression in *O. sativa* plants was associated with better drought-stress tolerance and 35% greater grain yield under drought than wild-type. Therefore, K^+ transporters could be used in molecular breeding for improving K^+ uptake and its assimilation in crop plant species.

15.2.4 Calcium

Calcium (Ca) is an essential macronutrient required for cell-wall structure and as a secondary messenger in signal transduction pathways (Dodd et al., 2010; Hedrich, 2012; Demidchik and Shabala, 2018). Calcium deficiency symptoms are usually observed in the young tissues due to lack of Ca remobilization from old tissues (Thor et al., 2019). In soil, Ca is in the form of a divalent cation (Ca^{2+}) that is taken up by Ca^{2+}-permeable ion channels located in the plasma membrane of root cells (González-Fontes et al., 2017; Tan and Luan, 2017). White (2001) suggested that Ca^{2+} is transported apoplastically from the root epidermis through the cortex to reach the Casparian strip of the endodermis. In the Casparian strip, Ca^{2+} moves into cytosol of the endodermal cells via channel proteins and is then exported into the stele apoplast and loaded into the xylem via Ca^{2+}-ATPases or Ca^{2+}/H^+ antiporters (White, 2001; Demidchik et al., 2002). However, the identity of Ca^{2+}-permeable channels mediating Ca^{2+} uptake into endodermal cells in roots is currently unknown (Thor, 2019).

Earlier studies classified Ca^{2+} channels into three groups according to their voltage dependence: depolarization-activated Ca^{2+}-permeable channels (DACCs), hyperpolarization-activated Ca^{2+}-permeable channels (HACCs), and voltage-independent Ca^{2+}-permeable channels (VICCs) (Miedema et al., 2001). The DACCs are involved in rapid Ca^{2+} uptake in response to external stimuli, and HACCs are responsible for nutritional Ca^{2+} uptake in root hairs (Thion et al., 1998; Miedema et al., 2001; Miedema et al., 2008). On the other hand, the VICCs play a critical role in maintaining the basal Ca^{2+} activity in the cytosol (White and Davenport, 2002). Advanced studies based on tissue expression, electrophysical characteristics, and plasma membrane localization have proposed annexins to contribute to the HACC currents, whereas cyclic nucleotide-gated (CNGV) proteins and glutamate receptor homologs (GLRs) could account for VIC-induced currents (Demidchik et al., 2002; White et al., 2002; Demidchik et al., 2018). Thus, plants have highly redundant systems to take up Ca^{2+} from soil.

At the cellular level, genetic analyses have shown that Ca^{2+}-ATPases and tonoplast Ca^{2+}/H^+ exchangers (CAX) contribute to Ca^{2+} sequestration into vacuoles, maintaining Ca^{2+} activity in the cytosol of leaf cells (Conn et al., 2011). Graham et al. (2014) identified four potential Ca^{2+} transporters (CAX1, CAX3, ACA8, and Eca4) involved in Ca^{2+} sequestration into vacuoles in *Arabidopsis* leaves. Biotechnological tools could be an important factor in increasing our understanding of the genes and mechanisms involved in Ca^{2+} uptake and its transport in plants to improve nutritional value of crops in the future.

15.2.5 Magnesium

Magnesium (Mg^{2+}) plays a crucial role in physiological and biochemical processes such as chlorophyll synthesis, as enzyme cofactor, and membrane stability (Ceylan et al., 2016; Marschner and Rengel, 2012). Magnesium deficiency reduces photosynthesis, carbohydrate synthesis, and plant growth (Yang et al., 2012; Farhat et al., 2016). Therefore, Mg^{2+} is an important macronutrient in plants. However, Mg^{2+} uptake is frequently depressed by other cations such as H^+, Mn, and Al^{3+} in soil, triggering Mg^{2+} deficiency in plants (Farhat et al., 2016; Marschner and

Rengel, 2012). Mg^{2+} uptake is performed by the functionally characterized Mg^{2+} transporter (*MGT*) genes in plants (Liu et al., 2019). Initially, the *MGT* gene family was named AtMRS2 considering its structural similarity to bacterial CorA and yeast protein MRS (Bui et al., 1999; Lunin et al., 2006). The *MGT* gene family has 10 members in *Arabidopsis* expressed in a broad range of plant tissues, and the proteins are localized in the plasma membrane, mitochondria, and tonoplast (Schock et al., 2000; Gebert et al., 2009). The *AtMGT1* is expressed in the plasma membrane of the root hairs, vascular tissues, and old leaves (Li et al., 2001; Gebert et al., 2009). Therefore, *AtMGT1* could be a key transporter in Mg^{2+} uptake and translocation from roots to shoot.

It has been reported that *AtMGT6* is a highly expressed gene in the plasma membrane of roots during Mg^{2+} deficiency in different species such as *Arabidopsis, Solanum tuberosum, S. lycopersicum, Brachypodium distachyon, Fragaria vesca,* and *Brassica juncea* (Mao et al., 2014; Regon et al., 2019; Koch et al., 2020), suggesting a key role for *AtMGT6* under Mg^{2+} deficiency.

Recently, Liu et al. (2019) identified eight *MGT* genes in *Poncirus trifoliata* through genome-wide analysis. They reported that *PtrMGT* genes were classified into five subfamilies, mainly expressed in vegetative plant tissues. Interestingly, the authors also showed that *PtrMGT5* overexpression enhanced Mg^{2+} deficiency tolerance in *Arabidopsis* transgenic plants, improving fresh weight and Mg^{2+} content in plant tissues (Liu et al., 2019). According to Li et al. (2008) MGT5 is a mitochondria Mg transporter, regulating pollen development.

Other *MGT* genes have been suggested to play an important role in Mg^{2+} homeostasis, such as MGT2 and MGT3 that are the tonoplast-located Mg^{2+} transporters (Conn et al., 2011; Gebert et al., 2009). Drummond et al. (2006) showed that the *MGT10* was localized to the envelope membrane of chloroplasts, suggesting a role in Mg^{2+} translocation in this organelle. Sun et al. (2017) reported that *MGT10 Arabidopsis* mutant plants had dramatically altered thylakoid structure, decreasing photosynthesis and plant growth, suggesting that *MGT10* is essential for normal chloroplast development and photosynthesis. In summary, Mg transporters (MGT) could be an interesting biotechnological tool for improving Mg^{2+} uptake and translocation, plant nutrition, photosynthesis, and plant growth (Table 15.1).

15.3 Biotechnological tools for micronutrient manipulation

15.3.1 Boron

Climatic change advances rapidly to reduce plant growth and productivity, directly affecting distribution, quality, and nutrient supply (Soares et al., 2019). These changes in climatic variables, such as low and high temperature, extreme rainfall, drought, high humidity, and high irradiance, can affect the availability and uptake of nutrients from the soil solution (Brouder and

TABLE 15.1 Overexpression of biotechnological tools for manipulating nutrient homeostasis in plants.

Species	Biotechnological tools	Effects	References
Arabidposis thaliana Oryza sativa Manihot esculenta	Overexpressing the high-affinity nitrate transporter *NRT2.1*	Improves water and nitrogen uptake by regulating plasma membrane aquaporins (plasma membrane intrinsic protein: PIP) under drought stress	Li et al. (2016), Luo et al. (2018), Zou et al. (2019)
S. tuberosum	Overexpressing the nitrate transporter *NPF1.11*	Improves plant height, leaf chlorophyll, and tuber protein content	Klaassen et al (2020)
O. sativa	Overexpressing the nitrate transporter *NRT2.3b*	Improves nitrogen and phosphorus uptake, root length, and grain yield	Feng et al. (2017)
O. sativa	Overexpressing the nitrate transporter *NRT2.1*	Enhances nitrate uptake and root length	Naz et al. (2019)
A. thaliana and *N. tabacum*	Overexpressing the transcription factor *NLP7*	Improves nitrogen uptake, photosynthesis rate, and plant growth	Yu et al. (2016)
A. thaliana	Overexpressing the *PHT1* and *PHOSPHATE 1* (*PHO*) families	Increases phosphorus uptake under drought-stress conditions	Zhang et al. (2016)
O. sativa	Overexpressing the high-affinity K^+ (*HAK1*) transporter	Mediates drought-stress tolerance and K^+ acquisition and its transport	Chen et al. (2015), Chen et al. (2017)
O. sativa	Overexpressing the natural resistance-associated macrophage protein (*NRAMP6*)	Increases Mn uptake and tolerance to abiotic stress	Peris-Peris et al. (2017)

Volenec, 2008). Boron (B) can be leached easily in the soil, reducing its amount in the soil solution. In contrast, another factor affecting B nutrition of plants is drought (Bulut, 2019). For example, boron uptake by roots is reduced in arid and semiarid conditions because B is transported mostly by mass water flow in soil (Hajiboland and Farhanghi, 2010; Bulut, 2019).

Hence, transpiration has great importance in the translocation of B from soil to roots and then to leaves (Miwa and Fujiwara, 2010; Bulut, 2019). Boron is taken up by plants from the soil as a charge-neutral molecule boric acid (H_3BO_3) (a major chemical form of B in soil solution) and as borate ($H_3BO^-_4$) (Marschner, 1995; Yoshinari and Takano, 2017).

Boron is necessary for normal growth and development of plants, implicated in several physiological functions, such as cell wall formation, membrane integrity, cell divisions, phenol, and protein metabolism, as well as nucleic acid metabolism (Cakmak and Römheld, 1997; Yang et al., 2013; Julkowska, 2018). The structural role in the cell wall is the primary function of B in plants (Brown et al., 2002). In fact, more than 90% of B in the cell wall is found as pectin, forming borate ester cross-linkages with rhamnogalacturonan II (RG-II), which is a highly conserved structural component in a wide range of plant species (Kobayashi et al., 1996; O'Neill et al., 2004).

Boron deficiency generally causes defects in growth and development of plants, being one of the most widespread micronutrient deficiencies in agriculture (Loomis and Durst, 1992). For example, plants exposed to low B concentrations in soil solution may show a wide range of symptoms such as decreased root elongation, leaf expansion, and loss of fertility, depending on the genotype, plant species, and phenological stages (Gupta, 1979; Dell and Huang, 1997; Goldbach et al., 2007).

Boron deficiency reduces not only yield but also crop quality (Shorrocks, 1997; Brown et al., 2002). In trees, it is reported that B deficiency depresses growth of specific parts such as root tips, buds, flowers, and young leaves and deformation of the organs such as roots, shoots, leaves, and fruits. At present, there is no evidence that B plays a direct role in photosynthesis (Dell and Huang, 1997; Wang et al., 2015). However, B deficiency limits root growth, resulting in weak vascular tissue, which may restrain water uptake and transport within the plant and thus alter leaf function, reducing the stomata number and making abnormal shapes (Wimmer and Eichert, 2013). Boron deficiency in trees reduces the chlorophyll content, photosynthesis, and stomatal conductance, as well as the activities of photosynthetic enzymes and catalase, but enhances the production of reactive oxygen species (ROS) and intercellular CO_2 concentration, resulting in a decrease in photosynthesis (Han et al., 2008; Wojcik et al., 2008; Tewari et al., 2010; Yoshinari and Takano, 2017). Moreover, the soluble sugar accumulation in B-deficient leaves of trees may also produce feedback inhibition of photosynthesis (Han et al., 2008; Ruuhola et al., 2011; Hajiboland et al., 2013; Lu et al., 2014).

Boron deficiency also increases lignin and phenol content in woody plant species (Wang et al., 2015), negatively influencing quality and quantity of wood, fruits, and yield. Boron requirements are greater in the reproductive than vegetative structures because boron is involved in flowering, pollen tube growth, and fruit ripening (Herrera-Rodríguez et al., 2010).

In the early stage of *V. vinifera* cultured under low-B conditions, diffuse yellowing of the young leaves, brownish areas of apical tendrils, and cupping of the third and fourth leaves from the shoot tips were observed (Christensen et al., 2006). Another report in *V. vinifera* showed that flower and fruit cluster necrosis, and small fruits appear due to B deficiency (Christensen et al., 2006).

Reduced length of the xylem vessel in both leaf and fruit vascular bundles as well as the reduced diameter of the xylem vessels in the leaf vascular bundle were found in *Citrus grandis* under B deficiency (Liu et al., 2013). In agreement with this report, an increase in vascular cross-sectional areas in *Picea abis* and *Citrus sinensis* in leaf veins were observed under B deficiency (Sutinen et al., 2006, 2007; Yang et al., 2013).

Five B transporters (*AtBOR6*, *AtBOR7*, *OsBOR4*, *VvBOR1*, and *AtNIP6;1*) may be associated with B retranslocation due to their expression in nodal regions of shoots or flowers (Tanaka et al., 2008, 2013; Fujiwara et al., 2010; Pérez-Castro et al., 2012). However, only the *AtNIP6;1* and *OsBOR4* genes have been functionally characterized (Tanaka et al., 2008, 2013). The *AtNIP6;1* gene is highly expressed in nodal regions of shoots, particularly the phloem, and NIP6;1 mutant showed a reduced expansion and low B concentration in young rosette leaves under B-limited conditions (Tanaka et al., 2008). *OsBOR4* is expressed in pollen, and *BOR4* mutants showed a reduced pollen tube elongation as a consequence of low B supply (Tanaka et al., 2013). These findings suggest that *AtNIP6;1* and *OsBOR4* have important roles in B distribution into growing leaves or flowers (Tanaka et al., 2008, 2013). In addition, *AtNIP6;1* gene may be responsible for xylem-to-phloem transfer of B due to its predominant expression in the phloem of the nodes (Wang et al., 2015).

15.3.2 Manganese

The changes in climatic variables (IPCC, 2014) are contributing to increased manganese availability, placing the productivity of important ecosystems at risk (Fernando and Lynch, 2015). In many regions around the world, climate change has affected rainfall distribution, provoking longer dry periods during the main growing season of crops, becoming an important yield-limiting factor (Raza et al., 2019).

Manganese (Mn) is an essential micronutrient for plant growth but is phytotoxic under certain edaphic and climatic conditions (Millaleo et al., 2010; Fernando and Lynch, 2015; Schmidt and Husted, 2019). Manganese in soil is found in three oxidation states: Mn(I), Mn(III), and Mn(IV); nevertheless only the divalent form (Mn^{2+}) is available for plant uptake, whereas Mn(III) is unstable and Mn(IV) is highly insoluble as oxides that precipitate in soil solution (Millaleo et al., 2010; Schmidt and Husted, 2019).

In plants, Mn is part of the structure of photosynthetic proteins and some metalloenzymes. Its deficiency provokes negative effects on the water-splitting part of photosystem II (PSII) that provides the electrons in photosynthesis

(Buchanan et al., 2000). In addition, an excess of Mn also can produce damage in the proteins of the photosynthetic apparatus, disrupting the photosynthetic electron flow in chloroplasts (Fernando and Lynch, 2015). Also, excessive Mn can alter several metabolic processes, such as the activity of enzymes; uptake and translocation of other nutrients such as calcium (Ca), magnesium (Mg), iron (Fe), and phosphorus (P); and provoking oxidative damage (Ducic and Polle, 2005; Lei et al., 2007). Besides, Mn plays an important role in diverse processes related to ATP synthesis, purine, and urea catabolism, phospholipid biosynthesis, and DNA repair (Millaleo et al., 2010; Alejandro et al., 2020).

Mn uptake from soil and translocation to shoot is regulated by a range of proteins and genes. In *Arabidopsis*, Mn^{2+} uptake is mediated by *AtNRAMP1* protein, which is mainly localized in the plasma membrane of the epidermis and cortex cell in roots (Cailliatte et al., 2010; Castaings et al., 2016). Under Mn deficiency, the expression of *AtNRAMP1* gene is moderately upregulated, and the *NRAMP1* mutant accumulated less Mn in shoots, suggesting a function of this protein in high-affinity Mn^{2+} uptake (Cailliatte et al., 2010). In *O. sativa*, the first identified transporter related to Mn uptake and translocation was *OsNRAMP5* localized in the plasma membrane of the distal side of the exodermis and endodermis and was involved in uptake of Mn^{2+} from the soil solution (Sasaki et al., 2012). *OsMTP9* is another similar Mn transporter identified in *O. sativa* associated with Mn uptake and translocation (Ueno et al., 2015). Both transporter genes (*OsNRAMP5* and *OsMTP9*) are involved in Mn transport from soil to the xylem, corroborated by using knockout mutants of these genes (Sasaki et al., 2012; Ueno et al., 2015; Chang et al., 2020).

More recently, the gene *HvIRT1* that contributed to Mn uptake and translocation in *H. vulgare* plants has been characterized (Long et al., 2018). Other genes such as *OsNRAMP6* are localized in the plasma membrane and expressed only in shoots of *O. sativa* regulating the Mn and Fe transport and could be involved in Mn uptake and biotic stress resistance (Peris-Peris et al., 2017).

In a tropical legume *Stylosanthes hamata* tolerant to acid soil conditions with high Mn concentrations, Delhaize et al. (2003) identified tonoplast Mn transporter *ShMTP8*. This protein is involved in Mn detoxification by sequestering into the vacuole. Studies of overexpression of *ShMTP8* gene in *Arabidopsis* conferred Mn tolerance (Delhaize et al., 2003). In *Arabidopsis*, *AtMTP8* has been characterized as an Mn and Fe transporter localized in the tonoplast (Eroglu et al., 2016, 2017; Chu et al., 2017).

Plant lines overexpressing *OsNRT2.1* under alternating wet and dry (AWD) and waterlogged (WL) conditions showed an increase of Mn content in *OsNRT2.1* transgenic seeds, which could be explained by the increased expression of *OsNRAMP* family genes (Luo et al., 2018). Proteomic analysis of roots revealed that *Stylosanthes guianensis* malate dehydrogenase1 (*SgMDH1*) protein is accumulating in response to Mn excess, with the overexpression of *SgMDH1* conferring Mn tolerance in *Arabidopsis* based on increased malate exudation and decreased plant Mn concentrations, suggesting that this organic acid could ameliorate Mn toxicity in plants (Chen et al., 2015). All biotechnological tools represent a key strategy for maintaining yield and improved food quality in a sustainable agricultural system under new climatic scenarios.

15.3.3 Zinc

Zinc is a crucial micronutrient involved in a number of cellular functions in plants, with its deficiency decreasing plant growth, productivity, and quality (Kabir et al., 2014). Zinc enters plants through root absorption of divalent Zn^{2+} from the soil solution; the uptake depends on ion concentration at the root surface, plant demand, and root absorption capacity, and is driven by transpiration (Sadeghzadeh, 2013). Even though Zn deficiency significantly reduces the yield, the main constraint is the reduction in Zn concentration in grain crops that is associated with deficiency in human nutrition (Krithika and Balachandar, 2016). The deficiency of this micronutrient in different agroclimatic regions worldwide causes significant damage in many crop species (Broadley et al., 2007). Zinc deficiency is accentuated particularly in arid and semiarid climatic conditions, where alkaline soils dominate (Cakmak, 2000).

At physiological levels Zn is required for a wide range of vital processes, including CO_2 fixation in the Calvin cycle (Hajiboland and Amirazad, 2010), nitrogen metabolism, and the maintenance of vacuolar homeostasis (Lan et al., 2013; Siemianowski et al., 2013). It can also reduce the activity of superoxide dismutase, whilst the peroxidase activity is increased (Pandey et al., 2002; Kosesakal and Unal, 2009). On the other hand, high Zn concentration can produce negative effects in plants, such as inhibition of root growth, changes in root and leaf morphologies, difficulties in reproductive growth, disintegration of organelles, and chromatin condensation (Rout and Das, 2003).

Zn^{2+} is taken up by roots via Zn and Fe transporters, such as ZIP family protein by ZRT, IRT-like Protein, IRT1 (iron-regulated transporter). In *O. sativa* several ZIPs have been described as responsible for Zn uptake from soil, translocation from roots to shoot, as well as the storage in grains (Krithika and Balachandar, 2016). Nonetheless, some *ZIP* genes, such as *OsZIP1*, *OsZIP4*, and *OsZIP5* are regulated not only by Zn but also by Fe, Cu, and Mn (Bughio et al., 2002; Ishimaru et al., 2005; Lee et al., 2010).

Some transcription factors such as basic region leucine-zipper transcription factor (*bZIP19*) are regulated by Zn depletion in soil solution (Cabot et al., 2019). In plants, *bZIPs* are involved in the response to biotic and abiotic stresses. In fact, bZIP transcription factors are key to plant immunity, as well as in the hormonal signaling networks (Noman et al., 2017, 2019). Recently, it has been reported that the overexpression of a member of the natural resistance-associated macrophage protein (NRAMP) family *AhNRAMP1* gene of *Arachis hypogaea* contributes to Mn and Zn transport and may be a candidate for improving plant biofortification (Wang et al., 2019).

15.4 Future directions

Biotechnological tools are an important strategy that allows the acceleration of some biological processes, which in recent years have allowed us to identify and characterize at molecular and functional levels a number of genes involved in crucial physiological and biochemical processes in plants. Thorough understanding of the processes of absorption, translocation, remobilization, and storage of nutrients, especially within the context of climate change, underpins the efforts to identify and develop plant genotypes resistant to various types of biotic and abiotic stresses that have been exacerbated by climate change. At the nutritional level, using these molecular tools, such as overexpression of specific genes or master genes as well as the use of the editing gene with new tools (e.g., CRISPR-Cas9), will enable us to obtain plants with efficiency in nutrient use, allowing sustainable optimization of agricultural production and avoiding negative environmental effects.

References

Aranda-Sicilia, M.N., Aboukila, A., Armbruster, U., Cagnac, O., Schumann, T., Kunz, H.H., et al., 2016. Envelope K^+/H^+ antiporters AtKEA1 and AtKEA2 function in plastid development. Plant Physiol. 172, 441–449.

Ai, P., Sun, S., Zhao, J., Fan, X., Xin, W., Guo, Q., et al., 2009. Two rice phosphate transporters, *OsPht1;2* and *OsPht1;6*, have different functions and kinetic properties in uptake and translocation. Plant J. 57, 798–809.

Alejandro, S., Höller, S., Meier, B., Peiter, E., 2020. Manganese in plants: from acquisition to subcellular allocation. Front. Plant Sci. 11, 300.

Ali, J., Jewel, Z.A., Mahender, A., Anandan, A., Hernandez, J., Li, Z., 2018. Molecular genetics and breeding for nutrient use efficiency in rice. Int. J. Mol. Sci. 19 (6), 1762.

Ayadi, A., David, P., Arrighi, J.F., Chiarenza, S., Thibaud, M.C., Nussaume, L., et al., 2015. Reducing the genetic redundancy of *Arabidopsis* PHOSPHATE TRANSPORTER1 transporters to study phosphate uptake and signaling. Plant Physiol. 167, 1511–1526.

Bouguyon, E., Gojon, A., Nacry, P., 2012. Nitrate sensing and signaling in plants. Semin. Cell Dev. Biol. 23, 648–654.

Broadley, M.R., White, P.J., Hammond, J.P., Zelko, I., Lux, A., 2007. Zinc in plants. New Phytol. 173 (4), 677–702.

Brouder, S.M., Volenec, J.J., 2008. Impact of climate change on crop nutrient and water use efficiencies. Physiol. Plant 133, 705–724.

Brown, P.H., Bellaloui, N., Wimmer, M.A., Bassil, E.S., Ruiz, J., Hu, H., et al., 2002. Boron in plant biology. Plant Biol. 4, 205–223.

Bu, Y., Sun, B., Zhou, A., Zhang, X., Lee, I., Liu, S., 2013. Identification and characterization of a PutATM1;1 gene from *Puccinellia tenuiflora*. PLoS One 8 (12), e83111.

Bu, Y., Takano, T., Liu, S., 2019. The role of ammonium transporter (AMT) against salt stress in plants. Plant Signal. Behav. 14, 8.

Buchanan, B., Grusen, W., Jones, R., 2000. Biochemistry and Molecular Biology of Plants, second ed. American Society of Plant Physiologists, Rockville, Maryland, USA, pp. 1367.

Bughio, N., Yamaguchi, H., Nishizawa, N.K., Nakanishi, H., Mori, S., 2002. Cloning an iron-regulated metal transporter from rice. J. Exp. Bot. 53, 1677–1682.

Bui, D.M., Gregan, J., Jarosch, E., Ragnini, A., Schweyen, R.J., 1999. The bacterial magnesium transporter CorA can functionally substitute for its putative homologue Mrs2p in the yeast inner mitochondrial membrane. J. Biol. Chem. 274, 20438–20443.

Bulut, H., 2019. The effect of climatic factors on boron nutrition of plants. In: 9th International Symposium on Atmospheric Sciences ATMOS 2019, Istambul, Turkey.

Cabot, C., Martos, S., Llugany, M., Gallego, B., Tolrà, R., Poschenrieder, C., 2019. A role for zinc in plant defense against pathogens and herbivores. Front. Plant Sci. 10, 1171.

Cai, W., Borlace, S., Lengaigne, M., Rensch, P., Collins, M., Vecchi, G., et al., 2014. Increasing frequency of extreme El Niño events due to greenhouse warming. Nature Clim. Change 4, 111–116.

Cailliatte, R., Schikora, A., Briat, J.-F., Mari, S., Curie, C., 2010. High-affinity manganese uptake by the metal transporter *NRAMP1* is essential for Arabidopsis growth in low manganese conditions. Plant Cell 22, 904–917.

Cakmak, I., 2000. Role of zinc in protecting plant cells from reactive oxygen species. New Phytol. 146, 185–205.

Cakmak, I., Römheld, V., 1997. Boron deficiency induced impairments of cellular functions in plants. Plant Soil. 193, 71–83.

Castaings, L., Caquot, A., Loubet, S., Curie, C., 2016. The high-affinity metal transporters NRAMP1 and IRT1 team up to take up iron under sufficient metal provision. Sci. Rep. 6, 37222.

Ceylan, Y., Kutman, U.B., Mengutay, M., Cakmak, I., 2016. Magnesium applications to growth medium and foliage affect the starch distribution, increase the grain size and improve the seed germination in wheat. Plant Soil. 406, 145–156.

Chang, J., Huang, S., Yamaji, N., Zhang, W., Ma, J.F., Zhao, F., 2020. OsNRAMP1 contributes to cadmium and manganese uptake in rice. Plant Cell Environ. 43, 2476–2491.

Chen, G., Hu, Q., Luo, L.E., Yang, T., Zhang, S., Hu, Y., et al., 2015. Rice potassium transporter OsHAK1 is essential for maintaining potassium-mediated growth and functions in salt tolerance over low and high potassium concentration ranges. Plant Cell Environ. 38, 2747–2765.

Chen, G., Liu, C., Gao, Z., Zhang, Y., Jiang, H., Zhu, L., et al., 2017. OsHAK1, a high-affinity potassium transporter, positively regulates responses to drought stress in rice. Front. Plant Sci. 8, 1885.

Chen, J., Zhang, Y., Tan, Y., Zhang, M., Zhu, L., Xu, G., et al., 2016. Agronomic nitrogen-use efficiency of rice can be increased by driving OsNRT2.1 expression with the OsNAR2.1 promoter. Plant Biotechnol. J. 14, 1705–1715.

Chérel, I., Lefoulon, C., Boeglin, M., Sentenac, H., 2014. Molecular mechanisms involved in plant adaptation to low K^+ availability. J. Exp. Bot. 65, 833–848.

Christensen, L., Beede, R., Peacock, W., 2006. Fall foliar sprays prevent boron-deficiency symptoms in grapes. Calif. Agric. 60, 100–103.

Conn, S.J., Gilliham, M., Athman, A., Schreiber, A.W., Baumann, U., Moller, I., et al., 2011. Cell-specific vacuolar calcium storage mediated by CAX1 regulates apoplastic calcium concentration, gas exchange, and plant productivity in *Arabidopsis*. Plant Cell 23 (1), 240–257.

Conn, S.J., Conn, V., Tyerman, S.D., Kaiser, B.N., Leigh, R.A., Gilliham, M., 2011. Magnesium transporters, MGT2/MRS2-1 and MGT3/MRS2-5, are important for magnesium partitioning within *Arabidopsis thaliana* mesophyll vacuoles. New Phytol. 190, 583–594.

Delhaize, E., Kataoka, T., Hebb, D.M., White, R.G., Ryan, P.R., 2003. Genes encoding proteins of the cation diffusion facilitator family that confer manganese tolerance. Plant Cell 15, 1131–1142.

Dell, B., Huang, L., 1997. Physiological responses of plants to boron. Plant Soil. 193, 103–120.

Demidchik, V., Bowen, H.C., Maathuis, F.J.M., Shabala, S.N., Tester, M.A., White, P.J., et al., 2002. *Arabidopsis thaliana* root non-selective cation channels mediate calcium uptake and are involved in growth. Plant J. 32, 799–808.

Demidchik, V., Shabala, S., 2018. Mechanisms of cytosolic calcium elevation in plants: the role of ion channels, calcium extrusion systems and NADPH oxidase-mediated "OS-Ca^{2+} Hub". Funct. Plant Biol. 45, 9–27.

Demidchik, V., Shabala, S., Isayenkov, S., Cuin, T., Pottosin, I., 2018. Calcium transport across plant membranes: mechanisms and functions. New Phytol. 220 (1), 49–69.

Dodd, A.N., Kudla, J., Sanders, D., 2010. The language of calcium signaling. Annu. Rev. Plant Biol. 61, 593–620.

Drummond, R.S.M., Tutone, A., Li, Y.C., Gardner, R.C., 2006. A putative magnesium transporter AtMRS2-11 is localized to the plant chloroplast envelope membrane system. Plant Sci. 170, 78–89.

Ducic, T., Polle, A., 2005. Transport and detoxification of manganese and copper in plants. Braz. J. Plant Physiol. 17, 103–112.

Epstein, E., Rains, D.W., Elzam, O.E., 1963. Resolution of dual mechanisms of potassium absorption by barley roots. Proc. Natl. Acad. Sci. USA 49, 684–692.

Eroglu, S., Giehl, R.F.H., Meier, B., Takahashi, M., Terada, Y., Ignatyev, K., et al., 2017. Metal tolerance protein 8 mediates manganese homeostasis and iron reallocation during seed development and germination. Plant Physiol. 174, 1633–1647.

Eroglu, S., Meier, B., von Wirén, N., Peiter, E., 2016. The vacuolar manganese transporter MTP8 determines tolerance to iron deficiency-induced chlorosis in Arabidopsis. Plant Physiol. 170, 1030–1045.

Fan, X.R., Tang, Z., Tan, Y.W., Zhang, Y., Luo, B.B., Yang, M., et al., 2016. Overexpression of a pH-sensitive nitrate transporter in rice increases crop yields. Proc. Natl. Acad. Sci. USA 113, 7118–7123.

Feng, H., Li, B., Zhi, Y., Chen, J., Li, R., Xia, X., et al., 2017. Overexpression of the nitrate transporter, OsNRT2.3b, improves rice phosphorus uptake and translocation. Plant Cell Rep. 36, 1287–1296.

Feng, H.M., Yan, M., Fan, X.R., Li, B.Z., Shen, Q.R., Miller, A.J., et al., 2011. Spatial expression and regulation of rice high-affinity nitrate transporters by nitrogen and carbon status. J. Exp. Bot. 62, 2319–2332.

Food and Agriculture Organization of the United Nations (2020) World Fertilizer Trends and Outlook to 2020. <http://www.fao.org/3/a-i6895e.pdf> (accessed 04.06.20.).

Farhat, N., Elkhouni, A., Zorrig, W., Smaoui, A., Abdelly, C., Rabhi, M., 2016. Effects of magnesium deficiency on photosynthesis and carbohydrate partitioning. Acta Physiol. Plant 38, 145.

Fernando, D.R., Lynch, J.P., 2015. Manganese phytotoxicity: new light on an old problem. Ann. Bot. 116, 313–319.

Fischer, S., Hilger, T., Piepho, H.-P., Jordan, I., Cadish, G., 2019. Do we need more drought for better nutrition? The effect of precipitation on nutrient concentration in East African food crops. Sci. Total Environ. 658, 405–415.

Fujiwara, T., Tanaka, M., Miwa, K., 2010. Optimisation of nutrient transport processes by plants-boron transport as an example. In: R. J. Gilkes, R.J., Prakongkep, N. (Eds.),

Proceedings of the 19th World Congress of Soil Science: Soil Solutions for a Changing World, IUSS, Brisbane, QLD, pp. 26–29.

Gao, Z., Wang, Y., Chen, G., Zhang, A., Yang, S., Shang, L., et al., 2019. The indica nitrate reductase gene OsNR2 allele enhances rice yield potential and nitrogen use efficiency. Nat. Commun. 10, 5207.

Gaymard, F., Pilot, G., Lacombe, B., Bouchez, D., Bruneau, D., Boucherez, J., et al., 1998. Identification and disruption of a plant shaker-like outward channel involved in K^+ release into the xylem sap. Cell 94, 647–655.

Ge, M., Wang, Y., Liu, Y., Jiang, L., He, B., Ning, L., et al., 2020. The NIN-like protein 5 (ZmNLP5) transcription factor is involved in modulating the nitrogen response in maize. Plant J. 102 (2), 353–368.

Gebert, M., Meschenmoser, K., Svidova, S., Weghuber, J., Schweyen, R., Eifler, K., et al., 2009. A root-expressed magnesium transporter of the MRS2/MGT gene family in *Arabidopsis thaliana* allows for growth in low-Mg^{2+} environments. Plant Cell 21, 4018–4030.

Gierth, M., Mäser, P., 2007. Potassium transporters in plants-involvement in K^+ acquisition, redistribution and homeostasis. FEBS Lett. 581, 2348–2356.

Glass, A.D.M., 1976. Regulation of potassium absorption in barley roots: an allosteric model. Plant Physiol. 58, 33–37.

Gojon, A., 2017. Nitrogen nutrition in plants: rapid progress and new challenges. J. Exp. Bot. 68, 2457–2462.

Goldbach, H., Huang, L., Wimmer, M., 2007. Boron functions in plants and animals: recent advances in boron research and open questions. In: Xu, F., Goldbach, H.E., Brown, P.H., Bell, R.W., Fujiwara, T., Hunt, C.D., et al., (eds) Advances in plant and animal boron nutrition. Springer, Dordrecht, pp. 3–25.

González-Fontes, A., Navarro-Gochicoa, C., Ceacero, C., et al., 2017. Understanding calcium transport and signaling, and its use efficiency in vascular plants. Plant Macronutrient Use Efficiency. pp. 165–180.

Graham, N.S., Hammond, J.P., Lysenko, A., Mayes, S., O'Lochlainn, S., Blasco, B., et al., 2014. Genetical and comparative genomics of *Brassica* under altered Ca supply identifies *Arabidopsis* Ca-transporter orthologs. Plant Cell 26, 2818–2830.

Gu, M., Chen, A., Sun, S., Xu, G., 2016. Complex regulation of plant phosphate transporters and the gap between molecular mechanisms and practical application: what is missing? Mol. Plant 9, 396–416.

Gu, R., Duan, F., An, X., Zhang, F., von Wiren, N., Yuan, L., 2013. Characterization of AMT-mediated high-affinity ammonium uptake in roots of maize (*Zea mays* L.). Plant, Cell Physiol. 54, 1515–1524.

Gupta, U.C., 1979. Boron nutrition of crops. Adv. Agron. 31, 273–307.

Hajiboland, R., Amirazad, F., 2010. Drought tolerance in Zn-deficient red cabbage (*Brassica oleracea* L. var. *capitata* f. *rubra*) plants. Hort. Sci. (Prague) 37, 88–98.

Hajiboland, R., Bahrami-Rad, S., Bastani, S., Tolrà, R., Poschenrieder, C., 2013. Boron re-translocation in tea (*Camellia sinensis* (L.) O. Kuntze) plants. Acta Physiol. Plant 35, 2373–2381.

Hajiboland, R., Farhanghi, F., 2010. Remobilization of boron, photosynthesis, phenolic metabolism and anti-oxidant defense capacity in boron-deficient turnip (*Brassica rapa* L.) plants. Soil Sci. Plant Nutr. 56, 427–437.

Hamburger, D., Rezzonico, E., MacDonald-Comber Petetot, J., Somerville, C., Poirier, Y., 2002. Identification and characterization of the *Arabidopsis* PHO1 gene involved in phosphate loading to the xylem. Plant Cell 14, 889–902.

Han, S., Chen, L.S., Jiang, H.X., Smith, B.R., Yang, L.T., Xie, C.Y., 2008. Boron deficiency decreases growth and photosynthesis, and increases starch and hexoses in leaves of citrus seedlings. J. Plant Physiol. 165, 1331–1341.

Han, M., Wu, W., Wu, W.H., Wang, Y., 2016. Potassium transporter KUP7 is involved in K + acquisition and translocation in *Arabidopsis* root under K^+-limited conditions. Mol. Plant 9, 437–446.

Hassler, S., Lemke, L., Jung, B., Möhlmann, T., Krüger, F., Schumacher, K., et al., 2012. Lack of the Golgi phosphate transporter PHT4;6 causes strong developmental defects, constitutively activated disease resistance mechanisms and altered intracellular phosphate compartmentation in *Arabidopsis*. Plant J. 72, 732–744.

He, Y., Xi, X., Zha, Q., Lu, Y., Jiang, A., 2020. Ectopic expression of a grape nitrate transporter VvNPF6.5 improves nitrate content and nitrogen use efficiency in Arabidopsis. BMC Plant Biol. 20 (1), 549.

Hedrich, R., 2012. Ion channels in plants. Physiol. Rev. 92, 1777–1811.

Herrera-Rodríguez, M.B., González-Fontes, A., Rexach, J., Camacho-Cristóbal, J.J., Maldonado, J.M., Navarro-Gochicoa, M.T., 2010. Role of boron in vascular plants and response mechanisms to boron stresses. Plant Stress 4 (Special Issue 2), 115–122.

Ishimaru, Y., Suzuki, M., Kobayashi, T., Takahashi, M., Nakanishi, H., Mori, S., et al., 2005. OsZIP4, a novel zinc-regulated zinc transporter in rice. J. Exp. Bot. 56, 3207–3214.

IPCC, 2014. Climate change 2014: synthesis report. In: Core Writing Team, Pachauri, R.K., Meyer, L.A. (Eds.), Proceedings of the Contribution of Working Groups I, II and III to the Fifth Assessment Report of the Intergovernmental Panel on Climate Change, IPCC, Geneva.

Ivashikina, N., Becker, D., Ache, P., Meyerhoff, O., Felle, H.H., Hedrich, R., 2001. K + channel profile and electrical properties of *Arabidopsis* root hairs. FEBS Lett. 508, 463–469.

Jacquot, A., Li, Z., Gojon, A., Schulze, W., Lejay, L., 2017. Post-translational regulation of nitrogen transporters in plants and microorganisms. J. Exp. Bot. 68 (10), 2567–2580.

Julkowska, M., 2018. Adjusting boron transport by two-step tuning of levels of the efflux transporter BOR1. Plant Physiol. 177 (2), 439–440.

Kabir, A.H., Swaraz, A.M., Stangoulis, J., 2014. Zinc-deficiency resistance and biofortification in plants. J. Plant Nutr. Soil Sci. 177, 311–319.

Karlsson, P., Herdean, A., Adolfsson, L., Beebo, A., Nziengui, H., Irigoyen, S., et al., 2015. The *Arabidopsis* thylakoid transporter PHT4;1 influences phosphate availability for ATP synthesis and plant growth. Plant J. 84 (1), 99–110.

Klaassen, M., Dees, D., Garrido, R., Báez, J.A., Schrijen, M., Mendoza, P., et al., 2020. Overexpression of a putative nitrate transporter (StNPF1.11) increases plant height, leaf chlorophyll content and tuber protein content of young potato plants. Funct. Plant Biol. 47 (5), 464–472.

Kobayashi, M., Matoh, T., Azuma, J., 1996. Two chains of rhamnogalacturonan II are cross-linked by borate-diol ester bonds in higher plant cell walls. Plant Physiol. 110, 1017–1020.

Koch, M., Winkelmann, M., Hasler, M., Pawelzik, E., Naumann, M., 2020.) Root growth in light of changing magnesium distribution and transport between source and sink tissues in potato (*Solanum tuberosum* L.). Sci. Rep. 10, 8796.

Kosesakal, T., Unal, M., 2009. Role of zinc deficiency in photosynthetic pigments and peroxidase activity of tomato seedlings. IUFS J. Biol. 68 (2), 113–120.

Krithika, S., Balachandar, D., 2016. Expression of zinc transporter genes in rice as influenced by zinc-solubilizing enterobacter cloacae strain ZSB14. Front. Plant Sci. 7, 446.

Krouk, G., Crawford, N.M., Coruzzi, G.M., Tsay, Y.F., 2010. Nitrate signaling: adaptation to fluctuating environments. Curr. Opin. Plant Biol. 13 (3), 265–272.

Lacombe, B., Pilot, G., Michard, E., Gaymard, F., Sentenac, H., Thibaud, J.B., 2000. A Shakerlike $K.^+$ channel with weak rectification is expressed in both source and sink phloem tissues of *Arabidopsis*. Plant Cell 12, 837–851.

Lan, H.X., Wang, Z.F., Wang, Q.H., Wang, M.M., Bao, Y.M., Huang, J., et al., 2013. Characterization of a vacuolar zinc transporter OZT1 in rice (*Oryza sativa* L.). Mol. Biol. Rep. 40, 1201–1210.

Lauter, F.R., Ninnemann, O., Bucher, M., Riesmeier, J., Frommer, W.B., 1996. Preferential expression of an ammonium transporter and two putative nitrate transporters in root hairs of tomato. Proc. Natl. Acad. Sci. USA 93, 8139–8144.

Lebaudy, A., Véry, A.A., Sentenac, H., 2007. K^+ channel activity in plants: genes, regulations and functions. FEBS Lett. 581, 2357–2366.

Lee, S., Jeong, H.J., Kim, S.A., Lee, J., Guerinot, M.L., An, G., 2010. OsZIP5 is a plasma membrane zinc transporter in rice. Plant Mol. Biol. 73, 507–517.

Lei, Y., Korpelainen, H., Li, C., 2007. Physiological and biochemical responses to high Mn concentrations in two contrasting *Populus cathayana* populations. Chemosphere 68, 686–694.

Levi, M., Gratton, E., Forster, I., Hernando, N., Wagner, C., Biber, J., et al., 2019. Mechanisms phosphate transport. Nat. Rev. Nephrol. 15, 482–500.

Li, L., Tutone, A.F., Drummond, R.S., Gardner, R.C., Luan, S., 2001. A novel family of magnesium transport genes in *Arabidopsis*. Plant Cell 13, 2761–2775.

Li, G., Tillard, P., Gojon, A., Maurel, C., 2016. Dual regulation of root hydraulic conductivity and plasma membrane aquaporins by plant nitrate accumulation and high-affinity nitrate transporter NRT2.1. Plant Cell Physiol. 57, 733–742.

Li, H., Hu, B., Chu, C.C., 2017. Nitrogen use efficiency in crops: lessons from *Arabidopsis* and rice. J. Exp. Bot. 68, 2477–2488.

Li, S.M., Li, B.Z., ShiWM, 2012. Expression patterns of nine ammonium transporters in rice in response to N status. Pedosphere 22, 860–869.

Li, W., Xu, G., Alli, A., Yu, L., 2018. Plant HAK/KUP/KT K^+ transporters: function and regulation. Semin. Cell Dev. Biol. 74, 133–141.

Li, L.G., Sokolov, L.N., Yang, Y.H., Li, D.P., Ting, J., Pandy, G.K., et al., 2008. A mitochondrial magnesium transporter functions in *Arabidopsis* pollen development. Mol. Plant 1, 675–685.

Lin, C.M., Koh, S., Stacey, G., Yu, S.M., Lin, T.Y., Tsay, Y.F., 2000. Cloning and functional characterization of a constitutively expressed nitrate transporter gene, OsNRT1, from rice. Plant Physiol. 122, 379–388.

Liu, X., Guo, L.X., Luo, L.J., Liu, Y.Z., Peng, S.A., 2019. Identification of the magnesium transport (MGT) family in *Poncirus trifoliata* and functional characterization of PtrMGT5 in magnesium deficiency stress. Plant Mol. Biol. 101, 551–560.

Liu, Y.Z., Li, S., Yang, C.Q., Peng, S.A., 2013. Effects of boron-deficiency on anatomical structures in the leaf main vein and fruit mesocarp of pummelo [*Citrus grandis* (L.) Osbeck]. Korean J. Hortic. Sci. 88, 693–700.

Long, L., Persson, D.P., Duan, F., Jørgensen, K., Yuan, L., Schjoerring, J.K., et al., 2018. The iron-regulated transporter 1 plays an essential role in uptake, translocation and grain-loading of manganese, but not iron, in barley. New Phytol. 217, 1640–1653.

Loomis, W., Durst, R., 1992. Chemistry and biology of boron. BioFactors 3, 229–239.

López-Arredondo, D.L., Leyva-González, M.A., Alatorre-Cobos, F., Herrera-Estrella, L., 2013. Biotechnology of nutrient uptake and assimilation in plants. Int. J. Dev. Biol. 57, 595–610.

Loque, D., von Wiren, N., 2004. Regulatory levels for the transport ammonium in plant roots. J. Exp. Bot. 55, 1293–1305.

Lou, D., Chen, Z., Yu, D., Yang, X., 2020. SAPK2 contributes to rice yield by modulating nitrogen metabolic processes under reproductive stage drought stress. Rice 13, 35.

Lu, Y.B., Yang, L.T., Li, Y., Xu, J., Liao, T.T., Chen, Y.B., et al., 2014. Effects of boron deficiency on major metabolites, key enzymes and gas exchange in leaves and roots of *Citrus sinensis* seedlings. Tree Physiol. 34, 608–618.

Ludewig, U., Neuhduser, B., Dynowski, M., 2007. Molecular mechanisms of ammonium transport and accumulation in plants. FEBS Lett. 581, 2301–2308.

Lunin, V.V., Dobrovetsky, E., Khutoreskaya, G., Zhang, R., Joachimiak, A., Doyle, D.A., et al., 2006. Crystal structure of the CorA Mg^{2+} transporter. Nature 440, 833.

Luo, B., Chen, J., Zhu, L., Liu, S., Li, B., Lu, H., et al., 2018. Overexpression of a high-affinity nitrate transporter OsNRT2.1 increases yield and manganese accumulation in rice under alternating wet and dry condition. Front. Plant Sci. 9, 1192.

Marschner, H., 1995. Mineral nutrition of higher plants. J. Ecol. 76 (4), 681–861.

Marschner, P., Rengel, Z., 2012. Nutrient availability in soils. In: Marschner, P. (Ed.), Marschner's Mineral Nutrition of Higher Plants, third ed. Academic Press, pp. 315–330.

Masclaux-Daubresse, C., Daniel-Vedele, F., Dechorgnat, J., Chardon, F., Gaufichon, L., Suzuki, A., 2010. Nitrogen uptake, assimilation and remobilization in plants: challenges for sustainable and productive agriculture. Ann. Bot. 105, 1141–1157.

Maathuis, F.J.M., 2009. Physiological functions of mineral macronutrients. Curr. Opin. Plant Biol. 12, 250–258.

Maathuis, F.J.M., Sanders, D., 1997. Regulation of K^+ absorption in plant root cells by external K + : interplay of different plasma membrane K^+ transporters. J. Exp. Bot. 48, 451–458.

Maathuis, F.J.M., Sanders, D., 1994. Mechanism of high-affinity potassium uptake in roots of *Arabidopsis thaliana*. Proc. Natl. Acad. Sci. USA 91, 9272–9276.

Maathuis, F.J.M., Sanders, D., 1996. Mechanisms of potassium absorption by higher plant roots. Physiol. Plant 96, 158–168.

Mao, D., Chen, J., Tian, L., Liu, Z., Yang, L., Tang, R., et al., 2014. *Arabidopsis* transporter MGT6 mediates magnesium uptake and is required for growth under magnesium limitation. Plant Cell 26, 2234–2248.

Miedema, H., Bothwell, J.H.F., Brownlee, C., Davies, J.M., 2001. Calcium uptake by plant cells—channels and pumps acting in concert. Trends Plant Sci. 6, 514–519.

Miedema, H., Demidchik, V., Véry, A.A., Bothwell, J.H.F., Brownlee, C., Davies, J.M., 2008. Two voltage-dependent calcium channels co-exist in the apical plasma membrane of *Arabidopsis thaliana* root hairs. New Phytol. 179, 378–385.

Millaleo, R., Reyes-Diaz, M., Ivanov, A.G., Mora, M.L., Alberdi, M., 2010. Manganese as essential and toxic element for plants: transport, accumulation and resistance mechanisms. J. Soil Sci. Plant Nutr. 10, 476–494.

Miwa, K., Fujiwara, T., 2010. Boron transport in plants: co-ordinated regulation of transporters. Ann. Bot. 105, 1103–1108.

Nath, M., Tuteja, N., 2016. NPKS uptake, sensing, and signaling and miRNAs in plant nutrient stress. Protoplasma 253, 767–786.

Naz, M., Luo, B., Guo, X., Chen, J., Fan, X., 2019. Overexpression of the nitrate transporter OsNRT2.1 enhances nitrate dependent root elongation. Genes 10, 290.

Nguyen, G., Kant, S., 2018. Improving nitrogen use efficiency in plants: effective phenotyping in conjunction with agronomic and genetic approaches. Funct. Plant Biol. 45 (6), 606–619.

Ning, L., Sun, P., Wang, Q., Ma, D., Hu, Z., Zhang, D., et al., 2015. Genetic architecture of biofortification traits in soybean (*Glycine max* L. Merr.) revealed through association analysis and linkage mapping. Euphytica 204, 353–369.

Noman, A., Aqeel, M., Khalid, N., Islam, W., Sanaullah, T., Anwar, M., et al., 2019. Zinc finger protein transcription factors: integrated line of action for plant antimicrobial activity. Microb. Pathog. 132, 141–149.

Noman, A., Liu, Z., Aqeel, M., Zainab, M., Khan, M.I.K., Hussain, A., 2017. Basic leucine zipper domain transcription factors: the vanguards in plant immunity. Bioch. Let. 39, 1779–1791.

Nussaume, L., Kanno, S., Javot, H., Marin, E., Pochon, N., Ayadi, A., et al., 2011. Phosphate import in plants: focus on the PHT1 transporters. Front. Plant Sci. 2, 83.

O'Brien, J.A., Vega, A., Bouguyon, E., Krouk, G., Gojon, A., Coruzzi, G., et al., 2016. Nitrate transport, sensing, and responses in plants. Mol. Plant 9, 837–856.

O'Neill, M.A., Ishii, T., Albersheim, P., Darvill, A.G., 2004. Rhamnogalacturonan II: structure and function of a borate cross-linked cell wall pectic polysaccharide. Annu. Rev. Plant Biol. 55, 109–139.

Osakabe, Y., Arinaga, N., Umezawa, T., Katsura, S., Nagamachi, K., Tanaka, H., et al., 2013. Osmotic stress responses and plant growth controlled by potassium transporters in *Arabidopsis*. Plant Cell 25, 609–624.

Pandey, N., Pathac, G.C., Singh, A.K., Sharma, C.P., 2002. Enzymic changes in response to zinc nutrition. J. Plant Physiol. 159, 1151–1153.

Pérez-Castro, R., Kasai, K., Gainza-Cortés, F., Ruiz-Lara, S., Casaretto, J.A., Peña-Cortés, H., et al., 2012. *VvBOR1*, the grapevine ortholog of *AtBOR1*, encodes an efflux boron transporter that is differentially expressed throughout reproductive development of *Vitis vinifera* L. Plant Cell Physiol. 53, 485–494.

Peris-Peris, C., Serra-Cardona, A., Sánchez-Sanuy, F., Campo, S., Ariño, J., San Segundo, B., 2017. Two NRAMP6 isoforms function as iron and manganese transporters and contribute to disease resistance in rice. Mol. Plant Microbe Interact. 30, 385–398.

Qi, Z., Hampton, C.R., Shin, R., Barkla, B.J., White, P.J., Schachtman, D.P., 2008. The high affinity K^+ transporter AtHAK5 plays a physiological role in planta at very low K^+ concentrations and provides a calcium uptake pathway in *Arabidopsis*. J. Exp. Bot. 59, 595–607.

Raghothama, K.G., 1999. Phosphate acquisition. Annu. Rev. Plant Physiol. Plant Mol. Biol. 50, 665–693.

Raza, A., Razzaq, A., Mehmood, S.S., Zou, X., Zhang, X., Lv, Y., et al., 2019. Impact of climate change on crops adaptation and strategies to tackle its outcome: a review. Plants 8 (2), 34.

Regon, P., Chowra, U., Prakash, J., Awasthi, P., Borgohain, P., Panda, S., 2019. Genome-wide analysis of magnesium transporter genes in *Solanum lycopersicum*. Comput. Biol. Chem. 80, 498–511.

Remy, E., Cabrito, T.R., Batista, R.A., Teixeira, M.C., Sá-Correia, I., Duque, P., 2012. The Pht1;9 and Pht1;8 transporters mediate inorganic phosphate acquisition by the *Arabidopsis thaliana* root during phosphorus starvation. New Phytol. 195, 356–371.

Richardson, A.E., Barea, J.M., McNeill, A.M., Prigent-Combaret, C., 2009. Acquisition of phosphorus and nitrogen in the rhizosphere and plant growth promotion by microorganisms. Plant Soil. 321, 305–339.

Rout, G.R., Das, P., 2003. Effect of metal toxicity on plant growth and metabolism. I. Zinc Agron. 23, 51–64.

Ruuhola, T., Keinänen, M., Keski-Saari, S., Lehto, T., 2011. Boron nutrition affects the carbon metabolism of silver birch seedlings. Tree Physiol. 31, 1251–1261.

Sadeghzadeh, B., 2013. A review of zinc nutrition and plant breeding. J. Soil Sci. Plant Nutr. 13 (4), 905–927.

Santa-María, G.E., Rubio, F., Dubcovsky, J., RodrÃguez-Navarro, A., 1997. The HAK1 gene of barley is a member of a large gene family and encodes a high-affinity potassium transporter. Plant Cell 9, 2281–2289.

Sasaki, A., Yamaji, N., Yokosho, K., Ma, J.F., 2012. Nramp5 is a major transporter responsible for manganese and cadmium uptake in rice. Plant Cell 24, 2155–2167.
Savage, J.A., Clearwater, M.J., Haines, D.F., Klein, T., Mencuccini, M., Sevanto, S., et al., 2016. Allocation, stress tolerance and carbon transport in plants: how does phloem physiology affect plant ecology? Plant Cell Environ. 39, 709–725.
Secco, D., Jabnoune, M., Walker, H., Shou, H., Wu, P., Poirier, Y., et al., 2013. Spatiotemporal transcript profiling of rice roots and shoots in response to phosphate starvation and recovery. Plant Cell 25, 4285–4304.
Sevanto, S., 2018. Drought impacts on phloem transport. Curr. Opin. Plant Biol. 43, 76–81.
Shi, S., Wang, D., Yan, Y., Zhang, F., Wang, H., Gu, M., et al., 2013. Function of phosphate transporter OsPHT2; 1 in improving phosphate utilization in rice. Chin. J. Rice Sci. 27, 457–465.
Schmidt, S.B., Husted, S., 2019. The biochemical properties of manganese in plants. Plants 8, 381.
Schock, I., Gregan, J., Steinhauser, S., Schweyen, R., Brennicke, A., Knoop, V., 2000. A member of a novel *Arabidopsis thaliana* gene family of candidate Mg2 + ion transporters complements a yeast mitochondrial group II intron-splicing mutant. Plant J. 24, 489–501.
Shorrocks, V.M., 1997. The occurrence and correction of boron deficiency. Plant Soil. 193, 121–148.
Siemianowski, O., Barabasz, A., Weremczuk, A., Ruszczyñska, A., Bulska, E., Williams, L.E., et al., 2013. Development of Zn-related necrosis in tobacco is enhanced by expressing *AtHMA4* and depends on the apoplastic Zn levels. Plant Cell Environ. 36, 1093–1104.
Soares, J.C., Santos, C.S., Carvalho, S.P., Pintado, M., Vasconcelos, M., 2019. Preserving the nutritional quality of crop plants under a changing climate: importance and strategies. Plant Soil. 443, 1–26.
Sonoda, Y., Ikeda, A., Yamay, T., Yamaguchi, J., 2003. Feedback regulation of the ammonium transporter gene family AMT1 by glutamine in rice. Plant Cell Physiol. 44, 1396–1402.
Sun, Y., Yang, R.A., Li, L.G., Huang, J.R., 2017. The magnesium transporter MGT10 is essential for chloroplast development and photosynthesis in *Arabidopsis thaliana*. Mol. Plant 10, 1584–1587.
Sutinen, S., Aphalo, P.J., Lehto, T., 2007. Does timing of boron application affect needle and bud structure in Scots pine and Norway spruce seedlings? Trees 21, 661–670.
Sutinen, S., Vuorinen, M., Rikala, R., 2006. Developmental disorders in buds and needles of mature Norway spruce, *Picea abies* (L.) Karst., in relation to needle boron concentrations. Trees 20, 559–570.
Tan, R.J., Luan, S., 2017. Regulation of calcium and magnesium homeostasis in plants: from transporters to signaling network. Curr. Opin. Plant Biol. 39, 97–105.
Tanaka, N., Uraguchi, S., Saito, A., Kajikawa, M., Kasai, K., Sato, Y., et al., 2013. Roles of pollen-specific boron efflux transporter, OsBOR4, in the rice fertilization process. Plant Cell Physiol 54, 2011–2019.
Tanaka, M., Wallace, I.S., Takano, J., Roberts, D.M., Fujiwara, T., 2008. NIP6;1 is a boric acid channel for preferential transport of boron to growing shoot tissues in *Arabidopsis*. Plant Cell 20, 2860–2875.
Tewari, R.K., Kumar, P., Sharma, P.N., 2010. Morphology and oxidative physiology of boron-deficient mulberry plants. Tree Physiol. 30, 68–77.
Thion, L., Mazars, C., Nacry, P., Bouchez, D., Moreau, M., Ranjeva, R., et al., 1998. Plasma membrane depolarization-activated calcium channels, stimulated by microtubule-depolymerizing drugs in wild-type *Arabidopsis thaliana* protoplasts, display constitutively

large activities and a longer half-life in ton 2 mutant cells affected in the organization of cortical microtubules. Plant J. 13, 603–610.

Thompson, H.E., Berrang-Ford, L., Ford, J.D., 2010. Climate change and food security in Sub-Saharan Africa: a systematic literature review. Sustainability 2, 2719–2733.

Thor, K., 2019. Calcium-nutrient and messenger. Front. Plant Sci. 10, 440.

Topcu, S., Kirda, C. 2013. In irrigation: environmental effects. In: Reference Module in Systems and Environmental Sciences. https://doi:10.1016/B978-0-12-409548-9.05194-0.

Ueno, D., Sasaki, A., Yamaji, N., Miyaji, T., Fujii, Y., Takemoto, Y., et al., 2015. A polarly localized transporter for efficient manganese uptake in rice. Nat. Plants 1, 15170.

Versaw, W.K., Harrison, M.J., 2002. A chloroplast phosphate transporter, PHT2;1, influences allocation of phosphate within the plant and phosphate-starvation responses. Plant Cell 14, 1751–1766.

Véry, A.A., Nieves-Cordones, M., Daly, M., Khan, I., Fizames, C., Sentenac, H., 2014. Molecular biology of K^+ transport across the plant cell membrane: what do we learn from comparison between plant species? J. Plant Physiol. 171, 748–769.

Wang, Y.Y., Hsu, P.K., Tsay, Y.F., 2012. Uptake, allocation and signaling of nitrate. Trends Plant Sci. 17, 458–467.

Wang, C., Yue, W., Ying, Y., Wang, S., Secco, D., Liu, Y., et al., 2015. OsSPX-MFS3, a vacuolar phosphate efflux transporter, is involved in maintaining Pi homeostasis in rice. Plant Physiol. 169, 2822–2831.

Wang, W., Hu, B., Yuan, D., Liu, Y., Che, R., Hu, Y., et al., 2018a. Expression of the nitrate transporter gene OsNRT1.1A/OsNPF6.3 confers high yield and early maturation in rice. Plant Cell 30 (3), 638–651.

Wang, F., Deng, M., Xu, J., Zhu, X., Mao, C., 2018b. Molecular mechanism of phosphate transport and signaling in higher plants. Semin. Cell Dev. Biol. 74, 114–122.

Wang, D., Xu, T., Yin, Z., Wu, W., Geng, H., Li, L., et al., 2020. Overexpression of *OsMYB305* in rice enhances the nitrogen uptake under low-nitrogen condition. Front. Plant Sci. 11, 369.

Wang, N., Qiu, W., Dai, J., Guo, X., Lu, Q., Wang, T., et al., 2019. AhNRAMP1 enhances manganese and zinc uptake in plants. Front. Plant Sci. 10, 415.

Wang, Y., Wu, W.H., 2017. Regulation of potassium transport and signaling in plants. Curr. Opin. Plant Biol. 39, 123–128.

Wege, S., Khan, G., Jung, J., Vogiatzaki, E., Pradervand, S., Aller, I., et al., 2016. The EXS domain of PHO1 participates in the response of shoots to phosphate deficiency via a root-to-shoot signal. Plant Physiol. 170 (1), 385–400.

White, P.J., 2001. The pathways of calcium movement to the xylem. J. Exp. Bot. 52, 891–899.

White, P.J., Davenport, R.J., 2002. The voltage-independent cation channel in the plasma membrane of wheat roots is permeable to divalent cations and may be involved in cytosolic Ca^{2+} homeostasis. Plant Physiol. 130, 1386–1395.

White, P.J., Bowen, H.C., Demidchik, V., Nichols, C., Davies, J.M., 2002. Genes for calcium-permeable channels in the plasma membrane of plant root cells. Biochim. Biophys. Acta Biomembr. 1564, 299–309.

Williams, L., Miller, A., 2001. Transporters responsible for the uptake and partitioning of nitrogenous solutes. Annu. Rev. Plant Physiol. Plant Mol. Biol. 52, 659–688.

Wimmer, M.A., Eichert, T., 2013. Review: mechanisms for boron deficiency-mediated changes in plant water relations. Plant Sci. 203, 25–32.

Wojcik, P., Wojcik, M., Klamkowski, K., 2008. Response of apple trees to boron fertilization under conditions of low soil boron availability. Sci. Hortic. 116, 58–64.

Yan, M., Fan, X.R., Feng, H.M., Miller, A.J., Shen, Q.R., Xu, G.H., 2011. Rice OsNAR2.1 interacts with *OsNRT2.1*, *OSNRT2.2* and OsNRT2.3a nitrate transporters to provide uptake over high and low concentration ranges. Plant Cell Environ. 34, 1360–1372.

Yang, C.Q., Liu, Y.Z., An, J.C., Li, S., Jin, L.F., Zhou, G.F., et al., 2013. Digital gene expression analysis of corky split vein caused by boron deficiency in "Newhall" navel orange (*Citrus sinensis* Osbeck) for selecting differentially expressed genes related to vascular hypertrophy. PLoS ONE 8e65737.

Yang, G.H., Yang, L.T., Jiang, H.X., Li, Y., Wang, P., Chen, L.S., 2012. Physiological impacts of magnesium-deficiency in citrus seedlings: photosynthesis, antioxidant system and carbohydrates. Trees 26, 1237–1250.

Yoshinari, A., Takano, J., 2017. Insights into the mechanisms underlying boron homeostasis in plants. Front. Plant Sci. 8, 1951.

Yu, L.H., Wu, J., Tang, H., Yuan, Y., Wang, S.M., Wang, Y.P., et al., 2016. Overexpression of *Arabidopsis* NLP7 improves plant growth under both nitrogen-limiting and -sufficient conditions by enhancing nitrogen and carbon assimilation. Sci. Rep. 6, 27795.

Zou, L., Qi, D., Sun, J., Zheng, X., Peng, M., 2019. Expression of the cassava nitrate transporter NRT2.1 enables *Arabidopsis* low nitrate tolerance. J. Genet. 98, 74.

Zhang, Z., Zhang, J., Chen, Y., Li, R., Wang, H., Wei, J., 2012. Genome-wide analysis and identification of HAK potassium transporter gene family in maize (*Zea mays* L.). Mol. Biol. Rep. 39, 8465–8473.

Zhang, C., Meng, S., Li, M., Zhao, Z., 2016. Genomic identification and expression analysis of the phosphate transporter gene family in poplar. Front. Plant Sci. 7, 1398.

Zhao, L., Versaw, W., Liu, J., Harrison, M., 2003. A phosphate transporter from *Medicago truncatula* is expressed in the photosynthetic tissues of the plant and located in the chloroplast envelope. New Phytol. 157, 291–302.

Zhu, W., Miao, Q., Sun, D., Guodong, Y., Wu, C., Huang, J., et al., 2012. The mitochondrial phosphate transporters modulate plant responses to salt stress via affecting ATP and gibberellin metabolism in *Arabidopsis thaliana*. PLoS One 7, e43530.

Chapter 16

Crop biofortification and food security

Erum Shoeb and Kathleen Hefferon
Department of Cell & Systems Biology, University of Toronto, Toronto, ON, Canada

16.1 Introduction

Crops generated through genetic modification, often referred to as genetically modified organisms (GMOs), have been available commercially for more than 20 years, yet they remain as controversial as ever. According to the International Service for the Acquisition of Agri-biotech Applications, the number of smallholder farmers in developing countries that grow genetically modified (GM) crops exceeds the number in industrialized countries. While first-generation GM crops were developed based on traits that increased yields and saved costs by providing resistance traits to pathogens, second-generation GM crops have targeted abiotic stresses caused by climate change such as heat, drought, and flooding, as well as traits to improve the nutritional outcomes of populaces who are food insecure. These include biofortified, nutritionally enhanced food crops or crops to reduce food waste. This chapter describes new breeding technologies associated with agricultural biotechnology, including GM and genome-edited crops, and how they can be used to address food insecurity. The chapter describes the development of crops biofortified in iron, zinc, and vitamin A, then ends with a discussion of the challenges that remain regarding consumer acceptance of these technologies, and the future of biofortified crops for addressing global health issues such as food security and malnutrition.

16.2 New plant-breeding technologies

Many of the varieties of crops that are available today have had their genomes altered by a technology that existed long before the advent of recombinant DNA. Mutagenesis breeding, for example, originates from mutation research that originated in the 1930s, and includes the introduction of random mutations to plant cuttings using chemical or irradiation mutagenesis.

Plant Nutrition and Food Security in the Era of Climate Change. DOI: https://doi.org/10.1016/B978-0-12-822916-3.00018-4

Explants propagated from these mutation events can then be developed into new crop varieties. More than 3000 varieties of crops, including the popular ruby red grapefruit, have been generated through mutagenesis breeding according to the Mutant Variety Database (https://mvd.iaea.org).

Genetic engineering of crops for improved nutritional and agronomic traits is necessary to combat food insecurity and climate change. This can include creating transgenic plants through the introduction of a novel trait into a crop by either adding new genes or making small changes to preexisting genes that are already part of the crop genome. Novel forms of genetic material are incorporated into the plant genome via agrobacterium-mediated transformation or particle bombardment (gene gun). Cisgenic crops, on the other hand, express genes from closely related plants rather than other species, thus allowing farmers to regain useful traits that have been lost over years of crop breeding. RNAi, or RNA interference technology, represents another form of genetic engineering in which the antisense RNA version of a specific plant or pathogen gene is expressed to block the gene targeted via gene silencing.

By far the most recent and interesting technology used in agricultural biotechnology is genome editing. Genome editing does not require the introduction of new gene sequences; rather, it may direct only one or a few nucleotide changes within a plant genome. Genome-editing technologies such as CRISPR-Cas9 make use of various defense strategies derived by bacteria to target specific sequences of DNA and cleave them at these sites with Cas nucleases. The technology is then able to take advantage of DNA repair mechanisms found already in the cells of all organisms. By repairing the sites of cleavage, specialized changes that will be carried through the genome of the "edited" organism to subsequent generations can be produced. Genome editing is an extremely promising new tool for plant breeding, as many complex traits such as crop yield, photosynthesis, and drought resistance involve multiple genes, and thus would be difficult to achieve through conventional breeding alone.

16.3 Agricultural biotechnology to address food insecurity and poverty

GM-biofortified crops can help meet nutrition targets of the sustainable development goals, to end hunger, improve nutrition, and improve livelihoods by increasing crop yields, thereby ending poverty. One avenue to improve yield has been to make use of miRNAs, transcription factors and proteins involved in signaling cascades (Garcia, 2008). Another way to improve plant yield is to increase levels of plant proteins responsible for stimulating cell proliferation or expansion (Ciura and Kruk, 2018). Manipulating the levels of phytohormones, especially cytokinins, auxins, and gibberellins, can also improve plant growth (Mezzetti et al., 2004). Other approaches including the regulation of plant senescence can increase yield (Zalabák et al., 2013). Overexpression of phosphatases has brought about faster growth in plants and increased seed yield (Sun et al., 2012). Similar results were found for *Camelina sativa* plants, which overexpressed a phosphatase from *Arabidopsis* (Zhang et al., 2012; Chhikara et al., 2018). The authors were able to increase oil yield up to 13% higher seed oil content and increase seed mass up to 52%, in comparison with wild-type plants. Improvements in yield have also been explored using CRISPR-Cas9 technologies (Ricroch and Henard-Damave, 2016). Li et al. (2017) edited rice (Zhonghua 11) genes responsible for grain number, panicle architecture, grain size, and plant architecture.

Besides hunger, approximately 2 billion people suffer from one or more micronutrient deficiencies, also known as hidden hunger (Hay et al., 2017). Transdisciplinary strategies are needed to provide micronutrients, such as vitamin A or iron supplementation (Wesseler et al., 2017). Biofortification of crops is an alternative solution to alleviate the burden of malnutrition (Zhao and Shewry, 2011; Masset et al., 2012; Ruel and Alderman, 2013). Progress in conventional biofortification has been made over the past century, with many micronutrient-enriched crops now available in more than 50 countries (HarvestPlus, 2015). The evidence is accumulating that indicates such crops are a cost-effective strategy (Meenakshi et al., 2010) to improve micronutrient status among target populations (Low et al., 2007; Haas et al., 2013).

Other GM foods with improved nutritional traits are starting to become commercialized; for example low-acrylamide GM potatoes (Rommens, 2008) and GM oilseed crops. Consumers are willing to pay on average 24% more for GM-biofortified food than conventional food (De Steur, 2017). Consumers can be significantly influenced by misinformation as well (Wesseler and Zilberman, 2014). This form of negative publicity for GM crops threatens the lives of malnourished populations in developing regions.

Besides nutritional supplements, micronutrient deficiencies in crops can be addressed by using fertilizers and foliar sprays, or via the biofortification of plant varieties using modern breeding techniques (Ricroch and Henard-Damave, 2016). Staple crops such as rice, banana, cassava, and maize are low in micronutrients and sufficient quantities of each must also be bioavailable in order to contribute to a healthy diet. The production of genetically modified, biofortified food crops is potentially advantageous for populations who find it difficult to change their dietary habits, who do not have access to fruits and vegetables, or who lack access to nutritional programs.

16.4 Crops biofortified with iron and zinc

Phytate is found to vary greatly among plant products, with grains (such as rice) and legumes (such as beans) having particularly high levels. Plant-produced phytic acids can bind to zinc and form insoluble complexes that result in reduced zinc absorption in the human digestive tract. The greater the amount of phytate present in a particular food crop, the less zinc accumulation takes place in the digestive tract. Besides poor soils, plant-based diets focusing on a single staple grain such as rice, with high phytate-to-zinc molar ratios, are a major factor contributing to zinc deficiency worldwide.

Reducing inhibitors of zinc absorption such as plant phytates is one way to address worldwide zinc deficiency (Rooney et al., 2020). Reducing the level of phytates using RNAi technology, for example, would increase the bioavailability of zinc and other trace metal elements, thus increasing the nutritional benefits of many crop species (Luan et al., 2020). RNA silencing to lower phytate levels has led to an increase in the iron and zinc concentration in milled rice seeds (Pottinger et al., 2020, Yang et al., 2020).

An important goal of micronutrient biofortification is to enhance the amount of bioavailable zinc found in the edible seed of cereals (Hefferon, 2019). Because rice (*Oryza sativa*) is an important grain to many of the world's rural poor, substantial efforts have been made to create transgenic rice with improved zinc accumulation in the endosperm. Plant metal transporter proteins can use multiple metal substrates, including iron, zinc, and even cadmium as substrates for uptake from soil into the roots. Mutations to create loss of function of these transporter proteins result in loss of uptake of all three of these metals into plant cells (Zang et al., 2017).

Masuda et al. (2012) increased the accumulation of the iron storage protein ferritin as well as enhanced iron translocation by overexpressing the iron(II)-nicotianamine transporter OsYSL2 in rice endosperm. While yield remained similar to conventional rice, the transgenic lines produced higher levels of iron (6-fold in the greenhouse and 4.4-fold in the paddy) and zinc (1.6-fold). The authors concluded that the introduction of multiple genes involved in iron and zinc homeostasis is more effective for iron biofortification than the introduction of a single gene. The same research group increased iron and zinc accumulation even more by enhancing the uptake and transport of iron using the ferric iron chelator, mugineic acid. In this case, the authors generated transgenic plants expressing the soybean ferritin gene (SoyferH2) driven by two endosperm-specific promoters, along with the barley nicotianamine synthase gene (HvNAS1), two nicotianamine aminotransferase genes (HvNAAT-A and -B), and a mugineic acid synthase gene (IDS3) to increase mugineic acid production in rice plants. Transgenic lines grown under iron-sufficient conditions increased iron accumulation by 4-fold as much as lines that were grown in both commercially supplied soil (iron-sufficient conditions) and calcareous soil (iron-deficient conditions). Transgenic plants that expressed both ferritin and mugineic acid biosynthetic genes showed signs of iron-deficiency tolerance in calcareous soil. By increasing iron content in cereal crops such as rice, an increase in zinc accumulation can also be achieved.

Aung et al. (2013) generated a transgenic line of rice that overexpressed the nicotianamine synthase gene HvNAS1 to enhance iron transport, the Fe (II)-nicotianamine transporter gene OsYSL2 to transport iron to the endosperm, and the Fe storage protein gene SoyferH2 to increase iron accumulation in the endosperm. Transgenic rice plants accumulated over 3.4-fold higher Fe concentrations and 1.3-fold higher zinc concentrations compared to conventional, nontransformed rice.

A transgenic high-yielding indica rice cultivar that expressed the ferritin gene from soybeans exhibited over a 2.6-fold higher level of ferritin (Paul et al., 2014). The plants, when milled, provided a 2.54-fold and 1.54-fold increase in iron and zinc content, respectively. Similarly, Tan et al. (2015) used the iron transporter gene MxIRT1 from apple trees to generate transgenic rice. Rice plants expressing MxIRT1 exhibited a 3-fold increase in iron and zinc accumulation, yet a decrease in cadmium concentration. Trijatmiko et al. (2016) expressed rice nicotianamine synthase (OsNAS2) and soybean ferritin (SferH-1) genes in plants that exhibited an endosperm Fe and Zn enrichment. Banakar et al. (2017) developed transgenic rice plants that expressed nicotianamine and 2′-deoxymugenic acid to accumulate iron and zinc in rice endosperm. Transgenic rice plants were capable of accumulating up to 4-fold more iron (Fe) and 2-fold more zinc (Zn) in rice endosperm, as well as lower levels of cadmium compared to wild-type plants, thus increasing the nutritional quality of rice plants for human consumption.

16.5 Golden rice and golden bananas

The annual loss of life for children due to the lack of vitamin A is estimated to be around 100,000 worldwide. Major causes of damage are due to deficiency of vitamin A in underdeveloped fetal organs such as the lungs, and a weakened immune system that affects children and underdeveloped fetuses. Vitamin A deficiency not only causes death but also blindness in young children. Dryness of mucus membranes is also reported, along with effects on the cornea, leading to blindness. As many as 250,000–500,000 children lose vision each year due to vitamin A deficiency (Cetin et al., 2019; Stevens et al., 2015). The best sources of vitamin A include high-priced foodstuffs, which are often out of reach for poor people. Since β-carotene efficiently transforms to vitamin A in the presence of starches and reduced fiber, staple crops such as rice and bananas were selected by scientists to produce genetically modified β-carotene rich varieties.

A well-known example is golden rice (GR). The latest version of GR produces enough beta-carotene provitamin A (PVA) to provide vitamin A to the consumer in developing countries where rice is being used as staple food (Moghissi et al., 2016). Beta-carotene in genetically modified GR is found to be an effective source of vitamin A in humans (Tang et al., 2009). When administered to children of 6–8 years of age, 50 g (dry weight) of GR was found to be equally effective as β-carotene in providing approximately 60% of the recommended dose of vitamin A (Tang et al., 2012).

The story of GR begins in 1990, when Ingo Potrykus started to focus on genetic manipulation of rice to improve quality of food and to fight malnutrition (Potrykus, 2001, 2012). Before introducing GR for public use, regulatory checkpoints were met through backcross breeding and field tests of the transfer of trait into popular rice varieties (Mallikarjuna Swamy and Hefferon, 2019). Ye et al. (2000) developed GR-1 with phytoene synthase (Psy) gene of daffodil in transgenic rice expressed under the control of rice glutelin promoter along with the bacterial phytoene desaturase (CrtI) gene expressed under the constitutive promoter CaMV 35S. One of the selected endosperm line showed carotenoid dry weight of 1.6 μg/g. Paine et al. (2005) later reported the production of GR-2, which was developed with a maize Psy gene instead of daffodil, exhibited a carotenoid dry weight of up to 37 μg/g (Paul et al., 2017; Tang et al., 2009). GR is very well supported by the scientific and agriculture research community, who considered it a brilliant example of using plant genetic engineering for the benefit of humankind, but it is equally and very strongly rejected by the challengers of GMOs, calling it a fake attempt to open the developing countries to GMO technology and GMO food.

Bananas are one of the most important fruit crops and are widely grown in the world as a dietary component taken both in the raw and cooked form. It is the staple food in a number of African countries, including Uganda, with consumption levels of 0.5−1.0 kg/person per day and serving as the primary starch source of the people living in these countries (Komarek, 2010; Smale and Tushemereirwe, 2007). Even for the rest of the world, bananas are a perfect food for young children because of their pleasant taste, low cost, and convenience in availability. Naturally, bananas are not rich in PVA and iron, thus making banana-based diets inadequate in these micronutrients. PVA carotenoid-rich banana cultivars may offer a possible solution in developing countries to combat vitamin A deficiency and stunting in children. Acceptable carotenoid-rich banana cultivars have been identified and the "Banana21" project was started in 2005 to address the issues of malnutrition through production of genetically modified nutrient-rich banana, with increased levels of PVA carotenoids (Paul et al., 2018). "Cavendish" bananas were produced in Australia with significantly elevated levels of β-carotene banana cultivars with edible yellow- and orange-fleshed fruit. There was a positive significant correlation between carotenoid content and deeper yellow/orange coloration indicators (Englberger et al., 2006). As the maize phytoene synthase 1 (ZmPsy1) gene was used to develop GR-2, banana-derived phytoene synthase 2a (MtPsy2a) gene expressed high levels of PVA (Paul et al., 2017). Another banana cultivar, "Asupina," with high-carotenoid-accumulating capability was also introduced. The carotenoid biosynthesis in banana appears to be a complex process involving a network of signals and the regulatory mechanisms of the two cultivars seems to differ significantly. Both the cultivars are acceptable for taste and other attributes, but the higher expression of carotenoid cleavage dioxygenase 4 in Cavendish indicating enzymatic degradation in the cultivar and the conversion of amyloplasts to chromoplasts is observed in Asupina only. Chromoplast development causes reduction in dry fruit content and firmness suggesting greater potential of the cultivar to accumulate high levels of carotenoids (Buah et al., 2016).

16.6 Biofortified maize and cassava

Different maize varieties can produce β-carotene at levels that are highly variable. β-carotene content can be improved via conventional breeding or using biotechnology. Transgenic maize biofortified with β-carotene improved health in clinical trials held in North America and Africa. Cassava is a staple of 250 million sub-Saharan Africans, but is high in cyanide and low in protein, iron, and β-carotene. However, the fact that biofortified maize is orange in color may be a challenge for public perception. For example, orange maize is traditionally associated with animal feed in Africa, while white maize is considered to be used for human consumption. Transgenic cassava expressing high levels of β-carotene has been demonstrated to successfully increase vitamin A levels in feeding studies, indicating that nutritionally enhanced cassava can be generated.

16.7 Nutritionally enhanced tomatoes

Tomatoes are attractive as a crop as they can be grown worldwide and are also easy to metabolically engineer. For example, two transcription factors isolated from snapdragon have been shown to increase anthocyanin, found also in blueberries and cranberries, in tomato plants. Anthocyanin is an antioxidant with clearly demonstrable health benefits, and tomato fruits metabolically engineered to produce high levels of anthocyanins were able to extend the life spans of cancer-susceptible mice by up to 30% (Li et al., 2014; Kiferle et al., 2015; Su et al., 2016). Similarly, transgenic tomatoes that express the gene-encoding grape resveratrol (stilbene synthase) accumulated high concentrations and transgenic "designer oilseed" plants such as false flax (*C. sativa*) have been metabolically engineered to synthesize omega-3 fatty acids and are used to feed farmed fishes (Ruiz-Lopez et al., 2014; Betancor et al., 2015). Two other examples of traits that have been added to designer oil crops include higher levels of acetyl glyceride and enriched glucoraphanin content, the latter example having preventative effect against cancer (Augustine and Bisht, 2015).

16.8 Future of biofortified crops and importance to global food security

Concerns regarding GM crops concentrate on potential harmful effects to human health and/or the environment, as well as some ethical issues regarding how the terms *natural* or *unnatural* are defined. In addition to this, there are concerns regarding patent rights and control by large multinational corporations. These problems facing the use of plant biotechnology in the food space continue to impact policy, regulation, and international trade.

The problems become accentuated for countries with underdeveloped policies, research programs, and regulatory infrastructures, such as sub-Saharan Africa (Paarlberg, 2008). The fact that many of these countries have poorly developed research programs has greatly reduced their opportunity for GMO development. Many GM products, including GM banana and GM maize (under R&D programs and confined field trials) must be sent to industrialized countries (e.g., United States, Australia) for laboratory analysis where there is an ongoing collaboration (Adenle et al., 2017; Wesseler et al., 2017). Consequently, many African countries are unable to regulate and develop GM crops.

The United States follows a "substantial equivalence" policy toward regulating innovations in crop biotechnology, whereas the EU follows the "precautionary principle," initially designed by the United Nations of the Convention on Biological Diversity under the Cartenega Protocol on Biosafety. These differences in regulatory framework concerning product versus process have led to serious obstacles in terms of food trade, particularly for those in developing countries. NGOs collectively spent over $10 billion USD in 2015 to prevent the spread of use of GM crops around the globe (Smyth, 2017; Wesseler and Zilberman, 2014). This can be compared to $8.5 billion spent by multinational companies on crop biotechnology (Adenle et al., 2017; Wesseler et al., 2017). The result of this incentive has been bleak for postcolonial Africa and Southeast Asia, where regulation of crops such as BT cotton, GM eggplant, and GR has been prohibitively restrictive.

In the case of GR, for example, release has been hindered in countries such as Bangladesh and the Philippines, with field trials destroyed by anti-GMO activists on more than one occasion in both countries (Potrykus, 2010). GR could make a world of difference by contributing to the reduction of malnutrition in the developing world. After 20 years, GR has finally been approved in several countries (Wesseler and Zilberman, 2017). Over the past 10 years, 1.4 billion life years have been lost in India due to vitamin A deficiency. More recently, a genome-edited version of GR has also been developed. Whether anti-GM activists will block its release as well remains to be seen (Dong et al., 2020) (Figs. 16.1 and 16.2).

Today, the number of chronically malnourished people in the world exceeds 1 billion, and inequity between wealthy and poor countries continues

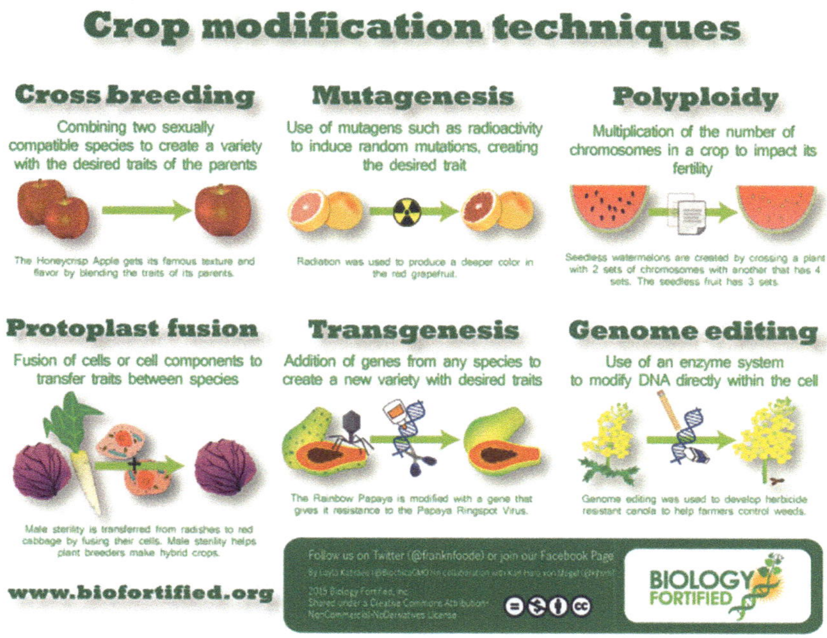

FIGURE 16.1 Crop-breeding strategies. *Reproduced with permission from Biofortified.com.*

FIGURE 16.2 Golden rice and golden bananas. *Reproduced with permission from goldenrice.org, GatesNotes.com, respectively.*

to grow. More recently, the advent of climate change has necessitated a revision of these goals to incorporate overriding sustainability objectives that can be addressed alongside poverty-reduction objectives. Rising incomes, along with predicted population growth, will increase food demand particularly with respect to meat consumption, and in turn are expected to exert more pressure on ecosystems. New breeding technologies, such as the generation of biofortified crops, could offer significant and viable solutions to these challenges.

References

Adenle, A.A., Morris, E.J., Murphy, D.J., 2017. Genetically Modified Organisms in Developing Countries: Risk Analysis and Governance. Cambridge University Press.

Augustine, R., Bisht, N.C., 2015. Biofortification of oilseed *Brassica juncea* with the anti-cancer compound glucoraphanin by suppressing *GSL-ALK* gene family. Sci Rep 5, 18005. Available from: https://doi.org/10.1038/srep18005.

Aung, M.S., Masuda, H., Kobayashi, T., Nakanishi, H., Yamakawa, T., Nishizawa, N.K., 2013. Iron biofortification of Myanmar rice. Front. Plant Sci. 4, 158. Available from: https://doi.org/10.3389/fpls.2013.00158.

Banakar, R., Alvarez Fernandez, A., Díaz-Benito, P., Abadia, J., Capell, T., Christou, P., 2017. Phytosiderophores determine thresholds for iron and zinc accumulation in biofortified rice endosperm while inhibiting the accumulation of cadmium. J. Exp. Bot. 68, 4983–4995. Available from: https://doi.org/10.1093/jxb/erx304.

Betancor, M.B., Sprague, M., Sayanova, O., Usher, S., Campbell, P.J., Napier, J.A., et al., 2015. Evaluation of a high-EPA oil from transgenic *Camelina sativa* in feeds for Atlantic salmon (*Salmo salar* L.): effects on tissue fatty acid composition, histology and gene expression. Aquaculture 444, 1–12.

Buah, S., Mlalazi, B., Khanna, H., Dale, J.L., Mortimer, C.L., 2016. The quest for golden bananas: investigating carotenoid regulation in a Fe'i group musa cultivar. J. Agric. Food Chem. (16), 3176–3185. Available from: https://doi.org/10.1021/acs.jafc.5b05740.

Cetin, I., Bühling, K., Demir, C., Kortam, A., Prescott, S.L., Yamashiro, Y., et al., 2019. Impact of micronutrient status during pregnancy on early nutrition programming. Ann. Nutr. Metab. 74 (4), 269–278. Available from: https://doi.org/10.1159/000499698.

Chhikara, S., Abdullah, H.M., Akbari, P., Schnell, D., Dhankher, O.P., 2018. Engineering *Camelina sativa* (L.) Crantz for enhanced oil and seed yields by combining diacylglycerol acyltransferase1 and glycerol-3-phosphate dehydrogenase expression. Plant Biotechnol. J. 16 (5), 1034–1045. Available from: https://doi.org/10.1111/pbi.12847. Epub 2017 Nov 19.

Ciura, J., Kruk, 2018. Phytohormones as targets for improving plant productivity and stress tolerance. J. Plant Physiol. 229, 32–40.

De Steur, H., Wesana, J., et al., 2017. Methods matter: a meta-regression on the determinants of willingness-to-pay studies on biofortified foods. Ann. N Y Acad. Sci. 1390 (1), 34–46.

Dong, O.X., Yu, S., Jain, R., Zhang, N., Duong, P.Q., Butler, C., et al., 2020. Marker-free carotenoid-enriched rice generated through targeted gene insertion using CRISPR-Cas9. Nat. Commun. 11 (1). Available from: https://doi.org/10.1038/s41467-020-14981-y.

Englberger, L., Wills, R.B., Blades, B., Dufficy, L., Daniells, J.W., Coyne, T., 2006. Carotenoid content and flesh color of selected banana cultivars growing in Australia. Food Nutr. Bull. 27 (4), 281–291. Available from: https://doi.org/10.1177/156482650602700401.

Garcia, D., 2008. A miRacle in plant development: role of microRNAs in cell differentiation and patterning. Semin. Cell Dev. Biol. 19 (6), 586–595. Available from: https://doi.org/10.1016/j.semcdb.2008.07.013. Epub 2008 Jul 30. Review.

Haas, J.D., Finkelstein, J.L., et al., 2013. Iron biofortified pearl millet improves iron status in Indian school children: results of a feeding trial. FASEB J. 27 (1 Suppl.), 355.2-355.2.

HarvestPlus, 2015. Nutritious Staple Food Crops: Who is Growing What? available from <bit.ly/CropMap2015>.

Hay, S.I., Abajobir, A.A., et al., 2017. Global, regional, and national disability-adjusted life-years (DALYs) for 333 diseases and injuries and healthy life expectancy (HALE) for 195 countries and territories, 1990–2016: a systematic analysis for the global burden of disease study 2016. Lancet 390 (10100), 1260–1344.

Hefferon, K., 2019. Biotechnological approaches for generating zinc-enriched crops to combat malnutrition. Nutrients. 11 (2), 253.

Kiferle, C., Fantini, E., Bassolino, L., Povero, G., Spelt, C., Buti, S., et al., 2015. Tomato R2R3-MYB proteins SlANT1 and SlAN2: same protein activity, different roles. PLoS One 10 (8), e0136365.

Komarek, A., 2010. The determinants of banana market commercialisation in Western Uganda. Afr. J. Agric. Res. 5, 775–784. Available from: https://doi.org/10.5897/AJAR09.022.

Li, H., Deng, Z., Liu, R., Loewen, S., Tsao, R., 2014. Bioaccessibility, in vitro antioxidant activities and in vivo anti-inflammatory activities of a purple tomato (*Solanum lycopersicum* L.). Food Chem 159, 353–360.

Li, H., Hu, B., Chu, C., 2017. Nitrogen use efficiency in crops: lessons from *Arabidopsis* and rice. J. Exp. Bot. 68, 2477–2488.

Low, J., Arimond, M., et al., 2007. A food-based approach introducing orange-fleshed sweet potatoes increased vitamin a intake and serum retinol concentrations in young children in rural mozambique. J. Nutr. 137 (5), 1320–1327.

Luan, H., Liao, W., Song, Y., Niu, H., Hu, T., Zhi, H., 2020. Transgenic plant generated by RNAi-mediated knocking down of soybean *Vma12* and *soybean mosaic virus* resistance evaluation. AMB Express 10 (1), 62. Available from: https://doi.org/10.1186/s13568-020-00997-6.

Mallikarjuna Swamy, B.P., Hefferon, K., 2019. Biotechnological approaches for generating zinc-enriched crops to combat malnutrition. Nutrients. 11 (2), 253. Available from: https://doi.org/10.3390/nu11020253. PMID: 30678136; PMCID: PMC6413068.

Masset, E., Haddad, L., et al., 2012. Effectiveness of agricultural interventions that aim to improve nutritional status of children: systematic review. BMJ 344.

Masuda, H., Ishimaru, Y., Aung, M.S., Kobayashi, T., Kakei, Y., Takahashi, M., et al., 2012. Iron biofortification in rice by the introduction of multiple genes involved in iron nutrition. Sci. Rep. 2, 543. Available from: https://doi.org/10.1038/srep00543.

Meenakshi, J.V., Johnson, N., et al., 2010. How cost-effective is biofortification in combating micronutrient malnutrition? An ex-ante assessment. World Dev. 38 (1), 64–75.

Mezzetti, B., Landi, L., Pandolfini, T., Spena, A., 2004. The *defH9-iaaM* auxin-synthesizing gene increases plant fecundity and fruit production in strawberry and raspberry. BMC Biotechnol 4 (4).

Moghissi, A.A., Pei, S., Liu, Y., 2016. Golden rice: scientific, regulatory and public information processes of a genetically modified organism. Crit. Rev. Biotechnol 36 (3), 535–541. Available from: https://doi.org/10.3109/07388551.2014.993586.

Paarlberg, R., 2008. Africa's organic farms. The New York Times. Feb. 2008. <http://www.nytimes.com/2008/02/29/opinion/29iht-edpaarlberg.1.10576543.html>.

Paine, J.A., Shipton, C.A., Chaggar, S., Howells, R.M., Kennedy, M.J., Vernon, G., et al., 2005. Improving the nutritional value of golden rice through increased pro-vitamin A content. Nat. Biotechnol. 23 (4), 482–487. Available from: https://doi.org/10.1038/nbt1082.

Paul, S., Ali, N., Datta, S.K., Datta, K., 2014. Development of an iron-enriched high-yieldings indica rice cultivar by introgression of a high-iron trait from transgenic iron-biofortifiedrice. Plant Foods Hum. Nutr. 203–208. Available from: https://doi.org/10.1007/s11130-014-0431-z.

Paul, J.Y., Khanna, H., Kleidon, J., Hoang, P., Geijskes, J., Daniells, J., et al., 2017. Golden bananas in the field: elevated fruit pro-vitamin A from the expression of a single banana transgene. Plant Biotechnol. J. 15 (4), 520–532. Available from: https://doi.org/10.1111/pbi.12650.

Paul, J.Y., Harding, R., Tushemereirwe, W., Dale, J., 2018. Banana21: from gene discovery to deregulated golden bananas. Front. Plant Sci. 9, 558. Available from: https://doi.org/10.3389/fpls.2018.00558.

Potrykus, I., 2001. Golden rice and beyond. Plant Physiol. 125 (3), 1157. Available from: https://doi.org/10.1104/pp.125.3.1157.

Potrykus, I., 2010. Lessons from the "Humanitarian Golden Rice" project: regulation prevents development of public good genetically engineered crop products. New Biotechnol. 27 (5), 466–472.

Potrykus, I., 2012. "Golden Rice", a GMO-product for public good, and the consequences of GE-regulation. J. Plant Biochem. Biotechnol. 21 (1), 68–75. Available from: https://doi.org/10.1007/s13562-012-0130-5.

Pottinger, S.E., Bak, A., Margets, A., Helm, M., Tang, L., Casteel, C., et al., 2020. Optimizing the PBS1 decoy system to confer resistance to potyvirus infection in arabidopsis and soybean. Mol. Plant Microbe Interact . Available from: https://doi.org/10.1094/MPMI-07-19-0190-R.

Ricroch, A.E., Henard-Damave, M.C., 2016. Next biotech plants: new traits, crops, developers and technologies for addressing global challenges. Crit. Rev. Biotechnol. 36 (4), 675–690.

Rommens, C.M., Yan, H., et al., 2008. Low-acrylamide french fries and potato chips. Plant Biotechnol. J. 6 (8), 843–853.

Rooney, W.M., Grinter, R.W., Correia, A., Parkhill, J., Walker, D.C., Milner, J.J., 2020. Engineering bacteriocin-mediated resistance against the plant pathogen *Pseudomonas syringae*. Plant Biotechnol. J. 18 (5), 1296–1306. Available from: https://doi.org/10.1111/pbi.13294. Epub 2019 Dec 3.

Ruel, M.T., Alderman, H., 2013. Nutrition-sensitive interventions and programmes: how can they help to accelerate progress in improving maternal and child nutrition? Lancet. 382 (9891), 536–551.

Ruiz-Lopez, N.1, Haslam, R.P., Napier, J.A., Sayanova, O., 2014. Successful high-level accumulation of fish oil omega-3 long-chain polyunsaturated fatty acids in a transgenic oilseed crop. Plant J 77 (2), 198–208. Available from: https://doi.org/10.1111/tpj.12378. Epub 2013 Dec 6.

Smale, M., Tushemereirwe, W., 2007. An Economic Assessment of Banana Genetic Improvement and Innovation in the Lake Victoria Region of Uganda and Tanzania. Research Report. International Food Policy Research Institute (IFPRI).

Smyth, S.J., 2017. Genetically modified crops, regulatory delays, and international trade. Food Energy Secur. 6 (2), 78–86.

Stevens, G.A., Bennett, J.E., Hennocq, Q., Lu, Y., De-Regil, L.M., Rogers, L., et al., 2015. Trends and mortality effects of vitamin A deficiency in children in 138 low-income and middle-income countries between 1991 and 2013: a pooled analysis of population-based surveys. Lancet. Glob. Health 3 (9), e528–e536. Available from: https://doi.org/10.1016/s2214-109x(15)00039-x.

Sun, F.P.K., Suen, Y., Zhang, C., Liang, C., Carrie, J., Whelan, et al., 2012. A dual-targeted purple acid phosphatase in *Arabidopsis thaliana* moderates carbon metabolism and its over expression leads to faster plant growth and higher seed yield. New Phytol. 194, 206–219.

Su, X., Xu, J., Rhodes, D., Shen, Y., Song, W., Katz, B., et al., 2016. Identification and quantification of anthocyanins in transgenic purple tomato. Food Chem. 202, 184–188.

Tan, S., Han, R., Li, P., Yang, G., Li, S., Zhang, P., et al., 2015. Over-expression of the MxIRT1 gene increases iron and zinc content in rice seeds. Transgenic Res. 24, 109–122. Available from: https://doi.org/10.1007/s11248-014-9822-z.

Tang, G., Qin, J., Dolnikowski, G.G., Russell, R.M., Grusak, M.A., 2009. Golden rice is an effective source of vitamin A. Am. J. Clin. Nutr. 89 (6), 1776–1783. Available from: https://doi.org/10.3945/ajcn.2008.27119.

Tang, G., Hu, Y., Yin, S.A., Wang, Y., Dallal, G.E., Grusak, M.A., et al., 2012. β-Carotene in Golden Rice is as good as β-carotene in oil at providing vitamin A to children. Am. J. Clin. Nutr. 96 (3), 658–664. Available from: https://doi.org/10.3945/ajcn.111.030775.

Trijatmiko, K.R., Dueñas, C., Tsakirpaloglou, N., Torrizo, L., Arines, F.M., Adeva, C., et al., 2016. Biofortified indica rice attains iron and zinc nutrition dietary targets in the field. Sci. Rep 6, 19792. Available from: https://doi.org/10.1038/srep19792.

Wesseler, J., Smart, R.D., et al., 2017. Foregone benefits of important food crop improvements in Sub-Saharan Africa. PLoS One 12 (7), e0181353. Available from: https://doi.org/10.1371/journal.pone.0181353.

Wesseler, J., Zilberman, D., 2014. The economic power of the golden rice opposition. Environ. Dev. Econ. 19 (06), 724–742.

Wesseler, J., Zilberman, D., 2017. Golden Rice: no progress to be seen. Do we still need it? Environ. Dev. Econ. 22 (2), 107–109.

Wesseler, J., et al., 2017. Foregone benefits of important food crop improvements in Sub-Saharan Africa. PLoS One 12 (7), e0181353. Available from: https://doi.org/10.1371/journal.pone.0181353.

Yang, X., Lu, Y., Wang, F., Chen, Y., Tian, Y., Jiang, L., et al., 2020. Involvement of the chloroplast gene ferredoxin 1 in multiple responses of *Nicotiana benthamiana* to potato virus X infection. J. Exp. Bot. 71 (6), 2142–2156. Available from: https://doi.org/10.1093/jxb/erz565.

Ye, X., Al-Babili, S., Klöti, A., Zhang, J., Lucca, P., Beyer, P., et al., 2000. Engineering the provitamin A (beta-carotene) biosynthetic pathway into (carotenoid-free) rice endosperm. Science 287 (5451), 303–305. Available from: https://doi.org/10.1126/science.287.5451.303.

Zalabák, D., Pospíšilová, H., Šmehilová, M., Mrízová, K., Frébort, I., Galuszka, P., 2013. Genetic engineering of cytokinin metabolism: prospective way to improve agricultural traits of crop plants. Biotechnol. Adv. 31, 97–117.

Zang, X., Geng, X., Wang, F., Liu, Z., Zhang, L., Zhao, Y., et al., 2017. Overexpression of wheat ferritin gene TaFER-5B enhances tolerance to heat stress and other abiotic stresses associated with the ROS scavenging. BMC Plant Biol 17, 14. Available from: https://doi.org/10.1186/s12870-016-0958-2.

Zhao, F.-J., Shewry, P.R., 2011. Recent developments in modifying crops and agronomic practice to improve human health. Food Policy 36, S94–S101.

Zhang, Y., Yu, L., Yung, K.F., Leung, D.Y., Sun, F., Lim, B.L., 2012. Overexpression of AtPAP2 in *Camelina sativa* leads to faster plant growth and higher seed yield. Biotechnol. Biofuels 5, 19.

Further reading

Barabaschi, D., Tondelli, A., Desiderio, F., Volante, A., Vaccino, P., Valè, G., Cattivelli, L., 2016. Next generation breeding. Plant Sci. 242, 3–13.

James, C., 2017. Global Status of Commercialized Biotech/GM Crops: 2016. ISAAA, New York, ISAAA briefs: No. 49-199.

Mao, Y., Botella, J.R., Zhu, J.K., 2016. Heritability of targeted gene modifications induced by plant-optimized CRISPR systems. Cell Mol Life Sci.

Marundan Jr., S., Samia, M., Ordonio, R.L., Rebong, D.B., Miranda, R., Alibuyog, A., et al., 2021. Development and characterization of GR2E golden rice introgression lines. Sci Rep 11 (1), 2496. Available from: https://doi.org/10.1038/s41598-021-82001-0.

Rani, R., Yadav, P., Barbadikar, K.M., Baliyan, N., Malhotra, E.V., Singh, B.K., et al., 2016. CRISPR/Cas9: a promising way to exploit genetic variation in plants. Biotechnol Lett 38 (12), 1991–2006. Available from: https://doi.org/10.1007/s10529-016-2195-z. Epub 2016 Aug 29.

Chapter 17

Biotechnological approaches for generating iron-rich crops

Samriti Mankotia, Jagannath Swain and Santosh B. Satbhai
Department of Biological Sciences, Indian Institute of Science Education and Research (IISER) Mohali, SAS Nagar, India

17.1 Introduction

Due to anticipated growth, the world population is estimated to exceed 9.7 billion by 2050. One of the huge challenges is to ensure adequate food and nutritional security to this population especially under increasingly challenging environmental conditions. Adequate food containing essential micronutrients is required for humans to maintain healthy life. Iron is an essential micronutrient for almost every living organism including plants and humans. Iron is required for the biosynthesis of oxygen transport proteins, hemoglobin and myoglobin, and for the formation of heme enzymes and other iron-containing enzymes involved in electron transfer and oxidation reductions (McDowell, 2003). Iron deficiency leads to anemia, lower immunity, impaired cognitive development, and reduced work productivity resulting from tiredness (Murata et al., 2010). Almost 50% of anemia cases are assumed to be because of iron deficiency, and it has been found that anemia in pregnant women increases chances of maternal and perinatal mortality. Iron-deficiency anemia is estimated to cause 90,000 deaths and around 800 million children and women are estimated to be affected by anemia in the year 2011 (Murata et al., 2010). According to the World Health Organization, iron deficiency is the only micronutrient deficiency that is prevalent in both developing and developed countries (Murata et al., 2010).

Iron deficiency results from low iron intake from diets based on staple crops such as white rice, corn meal, wheat flour, potatoes, or cassava, which are rich in starch and phytic acid (an iron absorption inhibitor) but nutrient poor (Hurrell, 2002). The daily iron requirements for people of different age groups are shown in Table 17.1. Iron biofortification of staple crops is considered a powerful tool to alleviate iron deficiency. The iron content of

Plant Nutrition and Food Security in the Era of Climate Change. DOI: https://doi.org/10.1016/B978-0-12-822916-3.00011-1

TABLE 17.1 Average daily iron requirement for people of different age groups (Trumbo et al., 2001).

Group	Daily iron requirement (average) (mg/day)
Children (4–8 years of age)	10
Adult men (31–50 years of age)	8
Adult women (19–50 years of age)	18

TABLE 17.2 The iron content of various crops and the estimated target level to be achieved through biofortification.

Crop	Iron content (μg/g)	Recommended iron levels (μg/g)
Wheat	30	52
Rice	2	13
Beans	50	94
Pearl millet	47	77
Maize	30	52
Sweet potato	6	28
Cassava	4	15

Source: Modified from Bouis, H.E., Hotz, C., McClafferty, B., Meenakshi, J.V., Pfeiffer, W.H., 2011. Biofortification: a new tool to reduce micronutrient malnutrition. Food Nutr. Bull. 32 (Suppl. 1), S31–S40. https://doi.org/10.1177/15648265110321S105.

various crops and the estimated target level of iron content of biofortified crops is presented in Table 17.2 (Bouis et al., 2011).

Crop plants mainly absorb iron from the rhizosphere through roots. It is absorbed either in the form of Fe(II) or bound to an organic acid, which is then transported to shoots and leaves through xylem. Though iron is abundant in the earth's crust, it is not bioavailable for absorption by plants. Iron availability for absorption by plants depends on the pH of the soil (Morrissey and Guerinot, 2009). At basic and neutral pH, it tends to form insoluble crystals of ferric oxide under aerobic conditions, which cannot be absorbed by plants (Kim and Guerinot, 2007). To overcome this problem, plants have adopted two strategies for the uptake of iron.

The chapter describes the molecular mechanisms behind iron-uptake strategy, transport, and storage into plants and use of this knowledge for biotechnological applications to produce iron-biofortified staple crops. This chapter also describes new possible genetic modifications that can improve the iron content of plants and make them resilient to iron-deficient soils.

17.2 Biofortification methods

Conventional breeding approaches have been used for centuries to produce crops with desirable traits. For conventional breeding, natural genetic variation for the character or trait of interest should be present in the germplasm. In many cases, because of absence of natural variation, conventional breeding cannot be used for increasing the nutrient content of crops (Amiri et al., 2015). In these cases, genetic engineering has been successfully used for biofortification of food crops.

The iron content in plants can be increased by enhancing processes of iron uptake, transport, and storage in the edible parts of plants (Carvalho and Vasconcelos, 2013). These processes can be enhanced by identifying the candidate genes that are involved in regulating them. In addition to this, transcription factors involved in controlling the expression of a number of genes involved in iron signaling can be genetically modified for improving the iron content of plants. Genome-wide association studies (GWAS) and quantitative trait loci (QTL) mapping have identified many loci associated with high iron. For example, it has been found using GWAS that natural allelic variants of Ferric reductase oxidase 2 (FRO2) (ferric chelate reductase), which is involved in iron uptake, differ in their transcript levels as well as in their activity (Satbhai et al., 2017). This information can be used for improvement of iron content in crop species grown in low-iron soils or in calcareous soils where iron availability is limited. GWAS of a multiparent advanced generation intercross plus population in rice has identified loci that are associated with iron content of grains. The genes identified by this study include nicotianamine (NA) synthase OsNAS3 and vacuolar iron transport OsVIT1, which are involved in iron homeostasis (Descalsota et al., 2018). A meta-QTL analysis from seven studies in Andean and Middle American intragene and intergene pool populations has identified two QTL that are specifically associated with high iron and eight QTL that are associated with both iron and zinc (Izquierdo et al., 2018). Both the QTL and GWAS approaches depend on natural diversity, which may take a longer time for significantly increasing the micronutrient content. Transgenic approaches on the other hand represent a faster way for improving micronutrient content in the edible parts of a plant. These approaches have targeted genes involved in iron uptake, transport, and storage for improving the iron content of crops.

17.3 Transgenic approaches for improving iron content of plants

17.3.1 Increasing iron content by modifying iron-uptake strategy

For iron uptake, plants utilize two different pathways that are activated under iron deficiency: reduction-based strategy and chelation-based strategy (Thomine and Lanquar, 2011). Dicotyledons and monocotyledons plants, except graminaceous plants, use the reduction-based strategy (strategy I), which consists of three main steps: (1) Protons are released into the rhizosphere by the proton pump (H^+-ATPase), which lowers the soil pH making Fe^{3+} soluble, (2) The solubilized Fe^{3+} is then reduced to Fe^{2+} by the activity of root membrane FRO2, and (3) The Fe^{2+} is then transported into the root epidermal cell by the Iron-regulated transporter 1 (IRT1) (Thomine and Lanquar, 2011) (Fig. 17.1).

In addition, iron availability is also enhanced by the secondary metabolites with iron-mobilizing properties, which are secreted by pleiotropic drug resistance9 into the rhizosphere making Fe^{3+} more available for reduction and uptake by plants (Tsai, et al., 2017) (Fig. 17.1). Graminaceous plants use the chelation-based strategy (strategy II), which consists of three main steps: (1) Mugineic acid (MA) family phytosiderophores (PS) are synthesized and secreted into the rhizosphere, (2) PS binds Fe^3 + forming Fe(III)-PS complexes, and (3) Fe(III)-PS complexes are then transported into the root by the Yellow Stripe 1 transporter (Thomine and Lanquar, 2011; Schaaf et al., 2004) (Fig. 17.2).

Takahashi et al. (2001) showed that introduction of barley genomic DNA fragment consisting of two nicotianamine aminotransferase (NAAT) genes resulted in the production of transgenic rice with enhanced tolerance to low

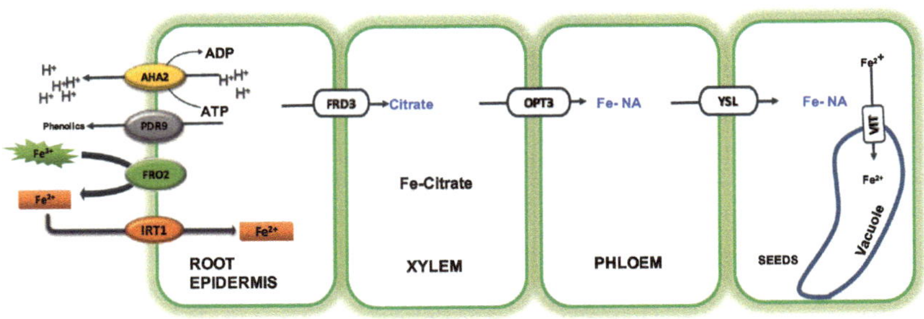

FIGURE 17.1 Process of iron uptake, transport, and storage in dicotyledon plants. Iron is absorbed in the root epidermal cells from the soil in the form of Fe2 + with the help of AHA2, FRO2, and IRT1. From the roots iron is transported to the shoots through xylem in the form of Fe-citrate. The Fe chelation to citrate is facilitated by FRD3. Iron is then transported to the phloem from xylem in the form of Fe (II)-NA complexes with the help of OPT3 transporters. In the developing seeds, iron is transported from the phloem with the help of YSL transporters and finally iron is transported in vacuoles by VIT1.

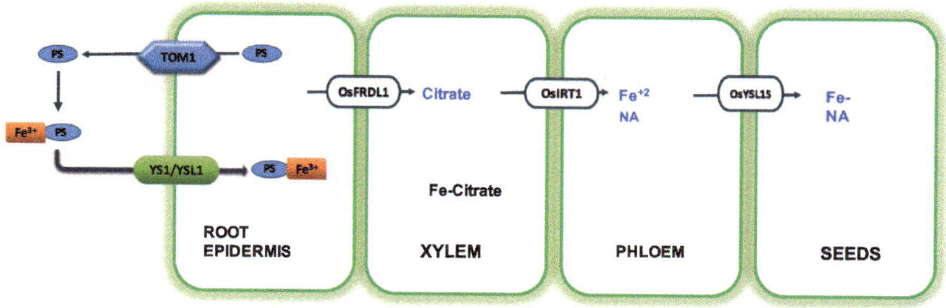

FIGURE 17.2 Process of iron uptake, transport, and storage in graminaceous plants. In strategy II plants such as rice, the PS are initially released into the rhizosphere by the TOM1 (transporter of MAs) and chelate Fe^{3+}. The PS-Fe(III) complex is taken inside the root epidermis by the YS/YSL family. From the root epidermis iron is transported to the xylem where it is bound by citrate which is transported by OsFRDL1 in rice. From the xylem it is then transported to the phloem by the OsIRT1 where it forms a complex with NA. From the phloem, the Fe-NA complex is finally transported to the seeds by the OsYSL15.

iron levels in alkaline soil (Takahashi et al., 2001). Expression of NAAT results in the improvement of iron uptake in the plants by stimulating the production of PS (Takahashi et al., 2001). In another study (Lee and An, 2009) overexpressed IRT1 resulted in increased iron content in rice leaves by 1.7-fold and only 1.1-fold in grains, suggesting that iron accumulated in the vegetative tissues in the absence of an extra sink in the seeds (Lee and An, 2009). Boonyaves et al. (2017) overexpressed IRT1 with *Phaseolus vulgaris* Ferritin 1 (PvFER1) in the endosperm, which increased the iron concentration up to 4-fold in polished rice (Boonyaves et al., 2017). The overexpression of *Arabidopsis* IRT1 together with Ferritin 1 (FER1) in cassava, resulted in increase in iron content by 5.5-fold in tubers (Narayanan et al., 2015). It has been found that when nicotianamine synthase genes (NAS) including OsNAS1, OsNAS2, or OsNAS3 were constitutively overexpressed in three rice populations, respectively, the iron content increased in the grains of all three transgenic populations (Johnson et al., 2011).

17.3.2 Facilitating iron transport and distribution

The proper iron distribution and transport within each plant organ is important for healthy growth and development of the plant. Ferric Reductase Domain (FRD), Nicotianamine Synthases (NAS), and Yellow Stripe (YSL) genes play an important role in iron transport and distribution within the plant (Ishimaru et al., 2010) (Figs. 17.1 and 17.2). Iron is transported through xylem from roots to shoots in a chelated form, mainly citrate and malate that is facilitated by FRD3 in *Arabidopsis* and OsFRDL1 in rice (Durrett et al., 2007; Yokosho et al., 2009). Iron is transported from xylem to phloem within developing seeds and young leaf tissues in the form of Fe (II)-NA complexes by Fe transporters such as YSLs and Oligopeptide Transporter (OPT), which are located in the xylem parenchyma (Klatte et al., 2009; Koike et al., 2004; Curie et al., 2009) (Figs. 17.1 and 17.2).

The overexpression of NAS genes that are involved in the synthesis of NA, increased iron levels up to 2-fold in wheat and rice grains (Lee et al., 2009). In case of grasses, the NAAT and 2-deoxymugineic acid (DMA) synthase converts NA to DMA, which is the precursor for all MA PS, secreted by roots into the rhizosphere facilitating uptake of Fe^{3+} (Bashir et al., 2006). The overexpression of NAS1 together with NAAT in rice increased DMA content up to 29-fold and iron content up to 4-fold in polished rice (Banakar et al., 2017). Ishimaru et al. (2010) showed that the expression of the OsYLS2 gene using sucrose transporter 1 (SUT1) promoter results in 4.4-fold enhancement in Fe content in rice grains (Ishimaru et al., 2010). Similarly, the overexpression of native ferritin (MusaFer1) in banana cultivar *Rasthali* leads to increase in iron content and makes it more tolerant to oxidative stress (Yadav et al., 2017). It was also reported that when soybean ferritin cDNA was expressed in embryonic cells of banana cv. *Rasthali* (AAB) the iron content increases by 6.32-fold (Kumar et al., 2011).

17.3.3 Facilitating iron storage

Ferritins are the major iron storage proteins in plants. These proteins play an important role in preventing oxidative stress and maintaining Fe homeostasis (Briat et al., 2010). Iron is also stored in plants in vacuoles. Iron concentration increased up to 3-fold in unpolished and polished grains when soybean Ferritin (expressing H1 subunit) (FERH1) gene was expressed in rice, while in maize it was not as effective; however, in the case of wheat, iron concentration increased up to 1.5-fold when FER1 was expressed from common bean (*Phaseolus vulgaris*) or from wheat itself (Borg et al., 2012; Neal et al., 2013; Singh et al., 2017a). Connorton et al. (2017) reported that the iron concentration in rice increases more than 2-fold when vacuolar iron transporter TaVIT2 was overexpressed under the endosperm specific promoter GLU-1D-1, which was used for ferritin overexpression (Connorton et al., 2017). The Fe levels increased up to 6-fold when NAS and ferritin genes were concomitantly expressed in rice endosperm (Wirth et al., 2009). When *Arabidopsis* VIT1 was overexpressed in cassava tubers, iron levels increased up to 3–4-fold (Narayanan et al., 2015). The expression of *Chlamydomonas reinhardtii* iron assimilatory 1 (FEA1) gene in cassava lines results in increased iron levels in the roots (Ihemere et al., 2012). The transcription factors that are involved in regulating the expression of a number of genes involved in iron homeostasis have been targeted successfully for increasing iron content of staple crops. For example, when the transcription factor OsIRO2, which is involved in activation of a number of genes for iron uptake, is overexpressed the iron concentration increases up to 3-fold in brown rice (Ogo et al., 2011).

17.4 Role of microbes in the biofortification of iron

Several plant growth-promoting microorganisms play an important role in the fortification of macronutrients and micronutrients in crops by increasing the availability of nutrients in soil through various mechanisms like production of PS, secretion of organic acids, and phenolics (Khan et al., 2019; Singh et al., 2018). Therefore microorganisms can be utilized for increasing the micronutrient content of food crops, and there are some successful results in wheat but this must be tested in different farming conditions and other crops. Singh et al. reported that inoculation of wheat grains with *Arthrobacter sulfonivorans* (DS-68) and *Enterococcus hirae* (DS-163) leads to increase in Fe content in wheat grains (Singh et al., 2017c; 2018). The Fe content in wheat grains also increases significantly on inoculation of *Providencia* sp. PW_5 (Rana, 2012). Similarly, in chickpea grains, Fe content was increased after inoculation of *Pseudomonas* (Khalid et al., 2015).

17.5 Conclusions and future perspectives

Approximately 2 billion of the world's population is suffering from iron deficiency. Biofortification is a method of increasing the micronutrient content in the edible parts of staple crops to alleviate micronutrient deficiency. This chapter has described biotechnological tools that have been used for producing iron-biofortified crops. The mechanism of iron uptake, transport, and storage has been described. Most of the genes involved in these processes have been discovered and the iron content in various crops has been successfully increased by modifying the expression of these genes alone or simultaneously. Most of the research has focused on the overexpression of genes involved in iron uptake, transport, or storage in plants. Another method for improving iron content would be to target genes that are master regulators of iron-deficiency responses controlling the expression of a number of genes. IMA/FEPs genes that are positive regulators of transcriptional response to iron deficiency can be targeted for producing iron-biofortified crops (Grillet et al., 2018; Hirayama et al., 2018). Some of the genes involved in iron homeostasis are still being discovered and this information can be used for biotechnological applications to produce iron-rich crops. CRISPR/Cas9 can also be used as a method of gene editing for producing iron-biofortified crops. This method is more efficient and faster compared to conventional breeding methods, and it can be used to replace existing alleles with alleles associated with high iron content at specific sites in the genome. Iron nanoparticles can also be used for improving the iron content of crops. It has been found that priming of wheat seeds with iron nanoparticles (conc. 25

TABLE 17.3 List of iron-biofortified staple crops.

Process enhanced	Promoter	Gene	Crop	Increase in Fe content	References
Uptake and transport	OsACT1	OsYSL15	Rice/*Japonica* cv. Dongjin	1.3	Lee et al. (2009)
	2 × 35S	OsNAS1	Rice/*Japonica* cv. Nipponbare	2.2-fold	Johnson et al. (2011)
	2 × 35S	OsNAS2	Rice/*Japonica* cv. Nipponbare	4.2-fold	Johnson et al. (2011)
	2 × 35S	OsNAS3	Rice/*Japonica* cv. Nipponbare	2.2-fold	Johnson et al. (2011)
	ZmUBQ	OsNAS2	Wheat/Bobwhite	Up to 2.1-fold	Singh et al. (2017b)
	OsSUT1	OsYSL2	Rice/*Japonica* cv. Tsukinohikari	4.4-fold	Ishimaru et al. (2010)
	HvIDS3	HvIDS3	Rice/*Japonica* cv. Tsukinohikari	1.4-fold	Masuda et al. (2008)
	StPATATIN	CrFEA1	Cassava/TMS60444	>3-fold	Ihemere et al. (2012)
Iron Storage	OsGluB-1	GmFERRITIN	Rice/*Japonica* cv. Kitaake	3-fold	Goto et al. (1999)
	OsGluB-1	GmFERRITIN	Rice/*Indica* cv. IR68144	3.7-fold	Vasconcelos et al. (2003)

	OsGluA2	OsFERRITIN 2	Rice/*Indica* cv. Pusa-Sugandhi II	2.09-fold	Paul et al. (2012)
	GLUB4/GLUB1	GmFERRITIN H1	Rice/*Indica* cv. IR64	3.4-fold	Oliva et al. (2014)
	TaGLU-1D-1	TaFERRITIN 1-A	Wheat/Bobwhite	50%–85%	Borg et al. (2012)
	Zm27gZEIN	GmFERRITIN	Maize/Hill (A188 × B73)	0.2-fold	Kanobe et al. (2013)
	CaMV 35S	GmFERRITIN	Maize/Hill (A188 × B73) and Jubilee	20%–70%	Drakakaki et al. (2005)
	CaMV 35S	GmFERRITIN	Lettuce (cv. Green Leaf)	1.7-fold	Goto et al. (2000)
Iron deficiency signaling pathway	CaMV 35S	OsIRO2	Rice/*Japonica* cv. Tsukinohikari	>2-fold	Ogo et al. (2011)
		AtFRO2	Soybean	10%	Vasconcelos et al. (2014)
Storage and transport		OsVIT1 or OsVIT2 functional disruption (mutants)	Rice/*Japonica* cv. Zhonghua11 and cv. Dongjin genetic background, respectively	Upto 1.5-fold	Zhang et al. (2012)
		OsVIT2 T-DNA insertion line	Rice	>1.5-fold	Bashir et al. (2013)
	StPATATIN	AtVIT1	Cassava/TME 204	3–4-fold	Narayanan et al. (2015)

(*Continued*)

TABLE 17.3 (Continued)

Process enhanced	Promoter	Gene	Crop	Increase in Fe content	References
Both transport and storage	OsGLUB and 35S	PvFERRITIN, AfPHYTASE and AtNAS1,	Rice/*Japonica* cv. Taipei 309	Up to 6.3-fold	Wirth et al. (2009)
	OsGLUB1, OsACT1 and OsSUT1, OsGlb1	GmFERRITIN H2 × 2, HvNAS1 and OsYSL2 × 2	Rice/*Japonica* cv. Tsukinohikari	Up to 6-fold	Masuda et al. (2012)
	OsGlb1, OsACT1 and OsSUT1	GmFERRITIN H2 × 2, HvNAS1 and OsYSL2 × 2	Rice/*Japonica* cv. Paw San Yin	Up to 3.4-fold	Aung et al. (2013)
	MsENOD12B, OsGLUA2 and CaMV35S	AtIRT1, PvFERRITIN AtNAS1	Rice/*Japonica* cv. Taipei 309	Up to 4.24-fold in polished seeds	Boonyaves et al. (2016)
	GluA2 and 35S	GmFERRITIN and OsNAS2	Rice/IR64	Up to 6-fold in polished seeds	Trijatmiko et al. (2016)

Abbreviations for plant species: At, *Arabidopsis thaliana*; Gm, *Glycine max* (soybean); Hv, *Hordeum vulgare* (barley); Ms, *Medicago sativa*; Os, *Oryza sativa* (rice); St, *Solanum tuberosum* (potato); Ta, *T. aestivum* (wheat).

ppm) leads to increase in grain iron content up to 45.7% (Sundaria et al., 2019). This research will become more valuable in coming years because of climate change and changes in micronutrient levels in the most important staple crops (Table 17.3).

Acknowledgments

The research in SBS's lab is supported by the grants from Department of Biotechnology (Ramalingaswami Fellowship) and SERB (ECR/2018/001580) DST, Govt. of India. SBS acknowledges intramural funding support from IISER Mohali. SM acknowledges IISER Mohali and JS acknowledges CSIR, Govt. of India for their Ph.D. fellowships.

References

Amiri, R., Bahraminejad, S., Sasani, S., Jalali-Honarmand, S., Fakhri, R., 2015. Bread wheat genetic variation for grain's protein, iron and zinc concentrations as uptake by their genetic ability. Eur. J. Agron 67, 20–26. Available from: https://doi.org/10.1016/j.eja.2015.03.004.

Aung, M.S., Masuda, H., Kobayashi, T., Nakanishi, H., Yamakawa, T., Nishizawa, N.K., 2013. Iron biofortification of Myanmar rice. Front. Plant Sci. 4, 158. Available from: https://doi.org/10.3389/fpls.2013.00158.

Banakar, R., Alvarez Fernandez, A., Díaz-Benito, P., Abadia, J., Capell, T., Christou, P., 2017. Phytosiderophores determine thresholds for iron and zinc accumulation in biofortified rice endosperm while inhibiting the accumulation of cadmium. J. Exp. Bot. 68 (17), 4983–4995. Available from: https://doi.org/10.1093/jxb/erx304.

Bashir, K., et al., 2006. Cloning and characterization of deoxymugineic acid synthase genes from graminaceous plants. J. Biol. Chem. 281 (43), 32395–32402. Available from: https://doi.org/10.1074/jbc.M604133200.

Bashir, K., Takahashi, R., Akhtar, S., Ishimaru, Y., Nakanishi, H., Nishizawa, N.K., 2013. The knockdown of OsVIT2 and MIT affects iron localization in rice seed. Rice (N Y) 6 (1), 31. Available from: https://doi.org/10.1186/1939-8433-6-31.

Boonyaves, K., Gruissem, W., Bhullar, N.K., 2016. NOD promoter-controlled AtIRT1 expression functions synergistically with NAS and FERRITIN genes to increase iron in rice grains. Plant Mol. Biol. 90 (3), 207–215. Available from: https://doi.org/10.1007/s11103-015-0404-0.

Boonyaves, K., Wu, T.Y., Gruissem, W., Bhullar, N.K., 2017. Enhanced grain iron levels in iron-regulated metal transporter, nicotianamine synthase, and ferritin gene cassette. Front. Plant Sci. 8, 130. Available from: https://doi.org/10.3389/fpls.2017.00130.

Borg, S., et al., 2012. Wheat ferritins: improving the iron content of the wheat grain. J. Cereal Sci 56 (2), 204–213. Available from: https://doi.org/10.1016/j.jcs.2012.03.005.

Bouis, H.E., Hotz, C., McClafferty, B., Meenakshi, J.V., Pfeiffer, W.H., 2011. Biofortification: a new tool to reduce micronutrient malnutrition. Food Nutr. Bull. 32 (Suppl. 1), S31–S40. Available from: https://doi.org/10.1177/15648265110321S105.

Briat, J.F., et al., 2010. New insights into ferritin synthesis and function highlight a link between iron homeostasis and oxidative stress in plants. Ann. Bot. 105 (5), 811–822. Available from: https://doi.org/10.1093/aob/mcp128.

Carvalho, S.M.P., Vasconcelos, M.W., 2013. Producing more with less: Strategies and novel technologies for plant-based food biofortification. Food Res. Int. 54 (1), 961–971. Available from: https://doi.org/10.1016/j.foodres.2012.12.021.

Connorton, J.M., Jones, E.R., Rodríguez-Ramiro, I., Fairweather-Tait, S., Uauy, C., Balk, J., 2017. Wheat vacuolar iron transporter TaVIT2 transports Fe and Mn and is effective for biofortification. Plant Physiol 174 (4), 2434–2444. Available from: https://doi.org/10.1104/pp.17.00672.

Curie, C., et al., 2009. Metal movement within the plant: contribution of nicotianamine and yellow stripe 1-like transporters. Ann. Bot. 103 (1), 1–11. Available from: https://doi.org/10.1093/aob/mcn207.

Descalsota, G.I.L., et al., 2018. Genome-wide association mapping in a rice magic plus population detects qtls and genes useful for biofortification. Front. Plant Sci 9, 1347. Available from: https://doi.org/10.3389/fpls.2018.01347.

Drakakaki, G., et al., 2005. Endosperm-specific co-expression of recombinant soybean ferritin and *Aspergillus* phytase in maize results in significant increases in the levels of bioavailable iron. Plant Mol. Biol. 59 (6), 869–880. Available from: https://doi.org/10.1007/s11103-005-1537-3.

Durrett, T.P., Gassmann, W., Rogers, E.E., 2007. The FRD3-mediated efflux of citrate into the root vasculature is necessary for efficient iron translocation. Plant Physiol 144 (1), 197–205. Available from: https://doi.org/10.1104/pp.107.097162.

Goto, F., Yoshihara, T., Shigemoto, N., Toki, S., Takaiwa, F., 1999. Iron fortification of rice seed by the soybean ferritin gene. Nat. Biotechnol. 17 (3), 282–286. Available from: https://doi.org/10.1038/7029.

Goto, F., Yoshihara, T., Saiki, H., 2000. Iron accumulation and enhanced growth in transgenic lettuce plants expressing the iron- binding protein ferritin. Theor. Appl. Genet. 100 (5), 658–664. Available from: https://doi.org/10.1007/s001220051336.

Grillet, L., Lan, P., Li, W., Mokkapati, G., Schmidt, W., 2018. IRON MAN is a ubiquitous family of peptides that control iron transport in plants. Nat. Plants 4 (11), 953–963. Available from: https://doi.org/10.1038/s41477-018-0266-y.

Hirayama, T., Lei, G.J., Yamaji, N., Nakagawa, N., Ma, J.F., 2018. The putative peptide gene FEP1 regulates iron deficiency response in *Arabidopsis*. Plant Cell Physiol 59 (9), 1739–1752. Available from: https://doi.org/10.1093/pcp/pcy145.

Hurrell, R.F., 2002. Fortification: overcoming technical and practical barriers. J. Nutr. 132 (Suppl. 4). Available from: https://doi.org/10.1093/jn/132.4.806s.

Ihemere, U.E., Narayanan, N.N., Sayre, R.T., 2012. Iron biofortification and homeostasis in transgenic cassava roots expressing the algal iron assimilatory gene, FEA1. Front. Plant Sci. 3, 171. Available from: https://doi.org/10.3389/fpls.2012.00171.

Ishimaru, Y., et al., 2010. Rice metal-nicotianamine transporter, OsYSL2, is required for the long-distance transport of iron and manganese. Plant J 62 (3), 379–390. Available from: https://doi.org/10.1111/j.1365-313X.2010.04158.x.

Izquierdo, P., Astudillo, C., Blair, M.W., Iqbal, A.M., Raatz, B., Cichy, K.A., 2018. Meta-QTL analysis of seed iron and zinc concentration and content in common bean (*Phaseolus vulgaris* L.). Theor. Appl. Genet. 131 (8), 1645–1658. Available from: https://doi.org/10.1007/s00122-018-3104-8.

Johnson, A.A.T., et al., 2011. Constitutive overexpression of the OsNAS gene family reveals single-gene strategies for effective iron- and zinc-biofortification of rice endosperm. PLoS One 6 (9), e24476. Available from: https://doi.org/10.1371/journal.pone.0024476.

Kanobe, M.N., Rodermel, S.R., Bailey, T., Paul Scott, M., 2013. Changes in endogenous gene transcript and protein levels in maize plants expressing the soybean ferritin transgene. Front. Plant Sci. 4, 196. Available from: https://doi.org/10.3389/fpls.2013.00196.

Khalid, S., Asghar, H.N., Akhtar, M.J., Aslam, A., Zahir, Z.A., 2015. Biofortification of iron in chickpea by plant growth promoting rhizobacteria. Pakistan J. Bot. 47 (3), 1191–1194.

Khan, A., Singh, J., Upadhayay, V.K., Singh, A.V., Shah, S., 2019. Microbial biofortification: A green technology through plant growth promoting microorganisms. Sustainable Green Technologies for Environmental Management.

Kim, S.A., Guerinot, M.L., 2007. Mining iron: iron uptake and transport in plants. FEBS Lett. 581 (12), 2273–2280. Available from: https://doi.org/10.1016/j.febslet.2007.04.043.

Klatte, M., Schuler, M., Wirtz, M., Fink-Straube, C., Hell, R., Bauer, P., 2009. The analysis of *Arabidopsis* nicotianamine synthase mutants reveals functions for nicotianamine in seed iron loading and iron deficiency responses. Plant Physiol 150 (1), 257–271. Available from: https://doi.org/10.1104/pp.109.136374.

Koike, S., et al., 2004. OsYSL2 is a rice metal-nicotianamine transporter that is regulated by iron and expressed in the phloem. Plant J. 39 (3), 415–424. Available from: https://doi.org/10.1111/j.1365-313X.2004.02146.x.

Kumar, G.B.S., Srinivas, L., Ganapathi, T.R., 2011. Iron fortification of banana by the expression of soybean ferritin. Biol. Trace Elem. Res. 142 (2), 232–241. Available from: https://doi.org/10.1007/s12011-010-8754-6.

Lee, S., An, G., 2009. Over-expression of OsIRT1 leads to increased iron and zinc accumulations in rice. Plant, Cell Environ 32 (4), 408–416. Available from: https://doi.org/10.1111/j.1365-3040.2009.01935.x.

Lee, S., et al., 2009. Iron fortification of rice seeds through activation of the nicotianamine synthase gene. Proc. Natl. Acad. Sci. U S A. 106 (51), 22014–22019. Available from: https://doi.org/10.1073/pnas.0910950106.

Masuda, H., et al., 2008. Increase in iron and zinc concentrations in rice grains via the introduction of barley genes involved in phytosiderophore synthesis. Rice 1 (1), 100–108. Available from: https://doi.org/10.1007/s12284-008-9007-6.

Masuda, H., et al., 2012. Iron biofortification in rice by the introduction of multiple genes involved in iron nutrition. Sci. Rep. 2, 543. Available from: https://doi.org/10.1038/srep00543.

McDowell, L.R., 2003. Minerals in Animal and Human Nutrition, second ed.

Morrissey, J., Guerinot, M.L., 2009. Iron uptake and transport in plants: the good, the bad, and the ionome. Chem. Rev. 109 (10), 4553–4567. Available from: https://doi.org/10.1021/cr900112r.

Murata, Y., Yamamoto, K., Yamaguchi, Y., Morishita, H., 2010. The expression method of the spacecraft operations procedure. SpaceOps 2010 Conference. pp. 1–48. Available from: http://doi.org/10.2514/6.2010-2232.

Narayanan, N., et al., 2015. Overexpression of *Arabidopsis* VIT1 increases accumulation of iron in cassava roots and stems. Plant Sci. 240, 170–181. Available from: https://doi.org/10.1016/j.plantsci.2015.09.007.

Neal, A.L., et al., 2013. Iron and zinc complexation in wild-type and ferritin-expressing wheat grain: implications for mineral transport into developing grain. J. Biol. Inorg. Chem. 18 (5), 557–570. Available from: https://doi.org/10.1007/s00775-013-1000-x.

Ogo, Y., Itai, R.N., Kobayashi, T., Aung, M.S., Nakanishi, H., Nishizawa, N.K., 2011. OsIRO2 is responsible for iron utilization in rice and improves growth and yield in calcareous soil. Plant Mol. Biol. 75 (6), 593–605. Available from: https://doi.org/10.1007/s11103-011-9752-6.

Oliva, N., et al., 2014. Large-scale production and evaluation of marker-free *indica rice* IR64 expressing phytoferritin genes. Mol. Breed. 33 (1), 23–37. Available from: https://doi.org/10.1007/s11032-013-9931-z.

Paul, S., Ali, N., Gayen, D., Datta, S.K., Datta, K., 2012. Molecular breeding of Osfer 2 gene to increase iron nutrition in rice grain. GM Crops Food 3 (4), 310–316. Available from: https://doi.org/10.4161/gmcr.22104.

Rana, A., Joshi, M., Prasanna, R., Shivay, Y.S., Nain, L., 2012. Biofortification of wheat through inoculation of plant growth promoting rhizobacteria and cyanobacteria. Eur. J. Soil Biol. 50, 118–126. Available from: https://doi.org/10.1016/j.ejsobi.2012.01.005.

Satbhai, S.B., Setzer, C., Freynschlag, F., Slovak, R., Kerdaffrec, E., Busch, W., 2017. Natural allelic variation of *FRO2* modulates *Arabidopsis* root growth under iron deficiency. Nat. Commun. 8. Available from: https://doi.org/10.1038/ncomms15603. Article Number: 15603.

Schaaf, G., Ludewig, U., Erenoglu, B.E., Mori, S., Kitahara, T., Von Wirén, N., 2004. ZmYS1 functions as a proton-coupled symporter for phytosiderophore- and nicotianamine-chelated metals. J. Biol. Chem. 279 (10), 9091–9096. Available from: https://doi.org/10.1074/jbc.M311799200.

Singh, S.P., Gruissem, W., Bhullar, N.K., 2017a. Single genetic locus improvement of iron, zinc and β-carotene content in rice grains. Sci. Rep. 7 (1). Available from: https://doi.org/10.1038/s41598-017-07198-5. Article Number: 6883.

Singh, S.P., Keller, B., Gruissem, W., Bhullar, N.K., 2017b. Rice *Nicotianamine synthase 2* expression improves dietary iron and zinc levels in wheat. Theor. Appl. Genet. 130 (2), 283–292. Available from: https://doi.org/10.1007/s00122-016-2808-x.

Singh, D., Rajawat, M.V.S., Kaushik, R., Prasanna, R., Saxena, A.K., 2017c. Beneficial role of endophytes in biofortification of Zn in wheat genotypes varying in nutrient use efficiency grown in soils sufficient and deficient in Zn. Plant Soil 416, 107–116. Available from: https://doi.org/10.1007/s11104-017-3189-x.

Singh, D., et al., 2018. Prospecting endophytes from different Fe or Zn accumulating wheat genotypes for their influence as inoculants on plant growth, yield, and micronutrient content. Ann. Microbiol. 68, 815–833. Available from: https://doi.org/10.1007/s13213-018-1388-1.

Sundaria, N., Singh, M., Upreti, P., Chauhan, R.P., Jaiswal, J.P., Kumar, A., 2019. Seed priming with iron oxide nanoparticles triggers iron acquisition and biofortification in wheat (*Triticum aestivum* L.) grains. J. Plant Growth Regul 38 (1), 122–131. Available from: https://doi.org/10.1007/s00344-018-9818-7.

Takahashi, M., Nakanishi, H., Kawasaki, S., Nishizawa, N.K., Mori, S., 2001. Enhanced tolerance of rice to low iron availability in alkaline soils using barley nicotianamine aminotransferase genes. Nat. Biotechnol. 19 (5), 466–469. Available from: https://doi.org/10.1038/88143.

Thomine, S., Lanquar, V., 2011. Iron transport and signaling in plants. In: Geisler, M., Venema, K. (Eds.), Transporters and Pumps in Plant Signaling. pp. 99–131.

Trijatmiko, K.R., et al., 2016. Biofortified indica rice attains iron and zinc nutrition dietary targets in the field. Sci. Rep. 6, 19792. Available from: https://doi.org/10.1038/srep19792.

Trumbo, P., Yates, A.A., Schlicker, S., Poos, M., 2001. Dietary reference intakes: vitamin A, vitamin K, arsenic, boron, chromium, copper, iodine, iron, manganese, molybdenum, nickel, silicon, vanadium, and zinc. J. Am. Diet Assoc. 101 (3), 294–301. Available from: https://doi.org/10.1016/S0002-8223(01)00078-5.

Tsai, H.H., Schmidt, W., 2017. Mobilization of iron by plant-borne coumarins. Trends Plant Sci. 22 (6), 538–548. Available from: https://doi.org/10.1016/j.tplants.2017.03.008.

Vasconcelos, M., et al., 2003. Enhanced iron and zinc accumulation in transgenic rice with the ferritin gene. Plant Sci. 164 (3), 371–378. Available from: https://doi.org/10.1016/S0168-9452(02)00421-1.

Vasconcelos, M.W., Clemente, T.E., Grusak, M.A., 2014. Evaluation of constitutive iron reductase (*AtFRO2*) expression on mineral accumulation and distribution in soybean (*Glycine max.* L). Front. Plant Sci. 5, 112. Available from: https://doi.org/10.3389/fpls.2014.00112.

Wirth, J., et al., 2009. Rice endosperm iron biofortification by targeted and synergistic action of nicotianamine synthase and ferritin. Plant Biotechnol. J. 7 (7), 631–644. Available from: https://doi.org/10.1111/j.1467-7652.2009.00430.x.

Yadav, K., Patel, P., Srivastava, A.K., Ganapathi, T.R., 2017. Overexpression of native ferritin gene MusaFer1 enhances iron content and oxidative stress tolerance in transgenic banana plants. PLoS One 12 (11), e0188933. Available from: https://doi.org/10.1371/journal.pone.0188933.

Yokosho, K., Yamaji, N., Ueno, D., Mitani, N., Jian, F.M., 2009. OsFRDL1 is a citrate transporter required for efficient translocation of iron in rice. Plant Physiol 149 (1), 297–305. Available from: https://doi.org/10.1104/pp.108.128132.

Zhang, Y., Xu, Y.H., Yi, H.Y., Gong, J.M., 2012. Vacuolar membrane transporters OsVIT1 and OsVIT2 modulate iron translocation between flag leaves and seeds in rice. Plant J 72 (3), 400–410. Available from: https://doi.org/10.1111/j.1365-313X.2012.05088.x.

Chapter 18

Use of nanomaterials in plant nutrition

Adalberto Benavides-Mendoza
Horticulture Department, Autonomus Agricultural University Antonio Narro, Saltillo, Mexico

18.1 Introduction

Soils provide a wide range of environmental services, from the more tangible such as the provision of food and water to other less tangible like the regulation of atmosphere composition. With the accelerated human population growth and changing lifestyles, soils and their environmental services are being exposed to great pressure. Approximately 98.8% (an average of 2849 kcal) of the daily calories consumed by humans come from soil and only 1.2% (an average of 35 kcal) from aquatic sources. In order to reach food security for the global population, food production must increase in 2050 by 70% from the level of 2005. In the case of continuing to apply the same production techniques such as tillage, fertilization, irrigation, pest and pathogen control, as well as their environmental outcomes such as overfertilization, overtillage, pesticide leaching, and soil decarbonization, this increasing pressure will undoubtedly put the sustainability of the soils used for agricultural, livestock, and forestry production into question (Kopittke et al., 2019).

Among the production techniques mentioned, fertilization occupies a central place. On the one hand, without the use of fertilizers, it is currently impossible to produce the food, fiber, timber, flowers, medicinal plants, and biofuels necessary to meet human needs. On the other hand, excessive or poorly planned use leads to soil degradation through decarbonization and loss of organic matter, salinization, acidification, and modifications in soil microbiomes (FAO and ITPS, 2015; Kopittke et al., 2019). These degradation processes can be accelerated by environmental factors, such as increases in temperature and changes in precipitation patterns, which are already occurring as a result of climate change at this time. In 2015 it was estimated that 33% of soils were subject to one or more types of degradation (Fig. 18.1) (FAO and ITPS, 2015). Some problems associated with soil degradation are the slow recovery of soils (in times measured in several human

Plant Nutrition and Food Security in the Era of Climate Change. DOI: https://doi.org/10.1016/B978-0-12-822916-3.00016-0

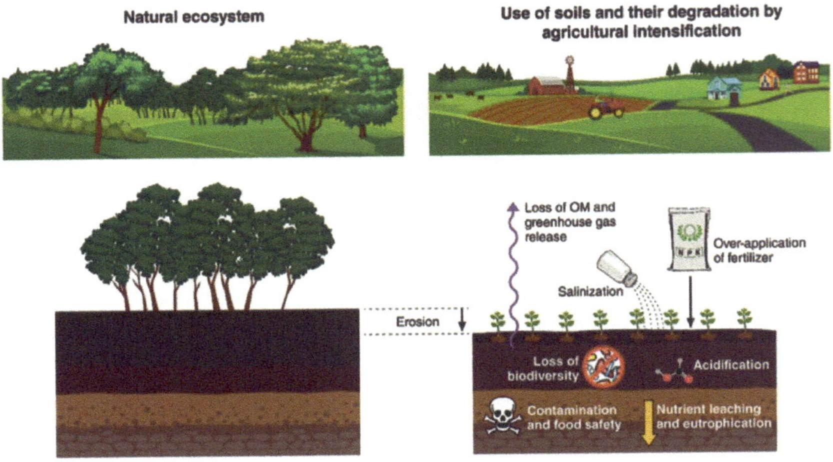

FIGURE 18.1 Different factors that trigger soil degradation under intensive agriculture. *From Kopittke, P.M., Menzies, N.W., Wang, P., McKenna, B.A., Lombi, E., 2019. Soil and the intensification of agriculture for global food security. Environ. Int. 132, 105078. https://doi.org/10.1016/j.envint.2019.105078.*

generations) and the concomitant decrease in food-production capacity, which may lead to regional food security crises in the near future. Until now there has been no agreement as to the adequate strategies to resolve the challenges on recovery or stabilization of soil degradation (Fraser et al., 2016).

In the scope of fertilizers, great efforts are made to improve its use, form, and time of application, such as the generation and dissemination of The International Code of Conduct for the Sustainable Use and Management of Fertilizers (FAO and ITPS, 2019). A central message of this document is the need to consider fertilization practices in a systemic context, taking into account all the components of agroecosystems, as well as their connections with biosphere processes.

Likewise, from a scientific, technical, and innovation point of view, much work is carried out to substantially increase the extrinsic and intrinsic efficiency of fertilizers, which, in the case of N, the most widely used fertilizer, takes values as low as 47% (Kopittke et al., 2019). For P, it has been estimated that only 15% of the total P destined for food production is eventually present in human foods and the rest is lost to the ecosystems (Suh and Yee, 2011). In the case of Zn, bioavailability can vary from 61.4% to 43.3% in flooded-dried and unflooded conditions, respectively (Mandal et al., 2000); and the crop's use efficiency for the micronutrients Fe, Zn, Cu, and B is estimated at <5% (Monreal et al., 2016). This low efficiency is the result of (1) factors extrinsic to the fertilizer formulation, such as inadequate soil management and irrigation, use of less than optimal fertilization planning and application times, and inadequate fertilizer application techniques; (2)

intrinsic restrictions of the edaphic system and of the capacity of the radical systems of plants to access fertilizers; and (3) inherent restrictions of the formulation or of the additives with which the formulation is accompanied, such as very slow or very fast solubilization, not coordinated with plant growth, that causes low yield and losses by leaching, volatilization, and precipitation-fixation by soil colloids (Ojeda-Barrios et al., 2020).

The mentioned efficiency of 47% of nitrogen fertilizers implies that more than half of the applied nitrogen fertilizer is not used by crops, causing economic losses, potential productivity, soil degradation, and contamination of water and atmosphere. Another way of looking at it is that, if it had highly efficient techniques, plant genotypes, and materials, it would be possible to produce the same amount of food by applying a much smaller amount of the fertilizers currently used, significantly mitigating the environmental impact of the food-production activity and fiber (Cui et al., 2014; FAO and ITPS, 2019).

Projections of climate models and economic models that determine the environmental impact indicate that the resilience of agriculture, and the resulting food security, depends on the efficiency in the use of natural resources and inputs necessary for food production: soil, water, energy, and fertilizers, among others. In this sense, among the essential improvements in the efficiency of agricultural activity is the increase in the efficiency of fertilizers and the use of nutrients by plants (Celikkol Erbas and Guven Solakoglu, 2017).

Among the innovations proposed years ago to improve the intrinsic efficiency of fertilizers is the use of nanofertilizers and/or the use of nanomaterials (NMs) that function as fertilizer additives and that increase the effectiveness of conventional fertilizers (Chhipa, 2017; Dimkpa and Bindraban, 2018; Monreal et al., 2016; Ojeda-Barrios et al., 2020).

This chapter presents updated information about the application of different NMs to increase efficiency in plant nutrition techniques, either directly with the use of nanofertilizers, or indirectly with the employment of NMs that increase the bioavailability and stability of conventional fertilizers.

18.2 Fertilizers and their efficiency

According to their composition, form, and time of application, fertilizers supply plants with 14 of the 17 elements recognized as essential for plants: N, P, K, S, Mg, Ca, Fe, Zn, Mn, Cu, B, Cl, Ni, and Mo. The remaining three elements, H, O, and C are contributed by the water and CO_2 used in the photosynthetic process.

The definition of *fertilizer* is "a chemical or natural substance or material that is used to provide nutrients to plants, usually via application to the soil, but also to foliage or through water in rice systems, fertigation or hydroponics or aquaculture operations" (FAO and ITPS, 2019). In the above definition, the term "nutrients" refers to any of the 14 elements listed previously.

When comparing with nanofertilizers, the conventional fertilizers are termed *bulk fertilizers*, for containing the essential elements that they provide in a nonnanometric form, or bulk form.

Ideally, as part of a comprehensive vision, fertilizer sources and formulations should depend on the type of soil, quality and application system of irrigation water, characteristics of the species or plant variety, and the stage of development of the plant (FAO and ITPS, 2019). In this ideal case, using a very efficient fertilizer (without loss of nutrients by volatilization, leaching, or chemical fixation) the volume of nutrients applied to a crop should equal the crop extractions by subtracting the contributions of the irrigation water. Extractions are estimated from the total dry biomass harvested, which is lost by the agroecosystem when the harvest is obtained, assuming that the total biomass that is not part of the yield is returned to the soil. Under this scheme, using highly efficient fertilizers, the applied amounts should be much less than what is normally used. For example, 155 tons of tomato fruit represents an extraction of 2.3 kg of Fe in the fruits and 1.7 kg of Fe in stems and leaves. However, the reality is that agricultural producers apply much greater amounts of fertilizers since a substantial part of it is lost due to volatilization, leaching, and fixing processes. These same processes are what originates the soil, water, and air pollution linked to fertilizers.

Different systems have been proposed to make the use of fertilizers more effective and sustainable, such as the 4R System, based on the selection of the right source of nutrients, at the right rate, at the right time, and in the right place. The procedure takes into account a series of performance indicators (e.g., yield, quality, yield stability, biodiversity, and soil erosion) together with the local characteristics of the soil, water, and climate to integrate a 4R process of soil and plant nutrition management (Mikkelsen, 2011). The preceding process requires solid technical and diverse-incentive support and the participation of different stakeholders—growers, extensionists, fertilizer manufacturers, governments, consumers of agricultural products—during the process of using the fertilizers (FAO and ITPS, 2019). The objective of a system for the sustainable use of fertilizers is to balance the production benefits of fertilizers versus the environmental and economic risks (Mikkelsen, 2011).

However, the application of a system for the sustainable use of fertilizers is not the typical situation. In many cases, the use of fertilizers, unfortunately, is accompanied by an environmental cost in the form of soil decarbonization, modification of edaphic microbiomes, groundwater pollution, and emissions to the atmosphere. The aim of the fertilizers' stakeholders is then to expand the available alternatives to mitigate the fertilizers' ecological damage, and one of them is the use of fertilizers that show greater efficiency in terms of their contribution of nutrients to plants, presenting lower rates of loss due to fixation, leaching, or volatilization. Having more efficient fertilizers is part of the 4R process.

The available fertilizers have different characteristics, cost, and accessibility in the market. Granulated fertilizers are the most economical and available in the market. Granulated fertilizers can be easily applied to the soil and provide the primary elements N-P-K and secondary Ca-Mg-S. They are followed by high-solubility fertilizers, more costly, aimed at a more specialized market, and relatively less available to all farmers. The use of high-solubility fertilizers depends on pressurized irrigation systems or foliar spraying and provides the primary and secondary elements, and the microelements Cl, Fe, Zn, Mn, B, Cu, Mn, and Ni. In the end, the most expensive fertilizers are those with more controlled nutrient release: organic fertilizers, chelated fertilizers, slow-release fertilizers (SRFs), and controlled-release fertilizers (CRFs) (Fig. 18.2). Compared with the other fertilizer categories, the use of these highly specialized fertilizers is minimal (Morgan et al., 2009).

For fertilizer use, the cost, ease of access for acquisition, application mode, and efficiency should be considered. The efficiency of the fertilizer is the quotient between the amount of plant absorption and the amount applied to the soil. Different physical, chemical, and biological features of the soil such as composition of the soil parent material, porosity or compaction, organic matter and inorganic soil colloids, cation exchange capacity, interactions between different nutrient ions, pH, ORP (Oxidation Reduction Potential), the edaphic microbiome, extremes of temperature, and humidity, make the fertilizer efficiency very low (Baligar et al., 2001).

The diverse soil factors mentioned induce leaching or volatilization of nutrients contained in the fertilizer or transform it into nonbioavailable chemical forms by speciation to nonabsorbable forms, precipitation, complexation, or formation of chelated forms. The leaching and volatilization, especially for N and P, represent a significant source of water and atmospheric pollution (Foley et al., 2011).

Granulated fertilizers are the less efficient category, followed by those with high solubility. Considering the low efficiency of nutrient supply of the granular fertilizers, it is necessary to apply great quantities and spread over diverse plant development times to make up for moderate losses due to fixation, leaching, and volatilization. Beyond the economic costs due to the larger use of machinery and labor, the surplus of fertilizers and excessive tillage produce soil degradation and atmospheric pollution, mainly due to the oxidation of organic matter that produces huge quantities of CO_2 (Baligar et al., 2001; FAO and ITPS, 2015).

High-solubility fertilizers are more efficient than granulated because of the application of technical high-pressure irrigation systems or the use of mechanized foliar spray for foliar fertilizers, growth regulators, and pesticides (Baligar et al., 2001). Pressurized irrigation techniques increase the water and nutrient use efficiency through the continual application of small amounts of water and fertilizers according to the crop needs, thus diminishing the exposure of fertilizers to the previously mentioned edaphic factors

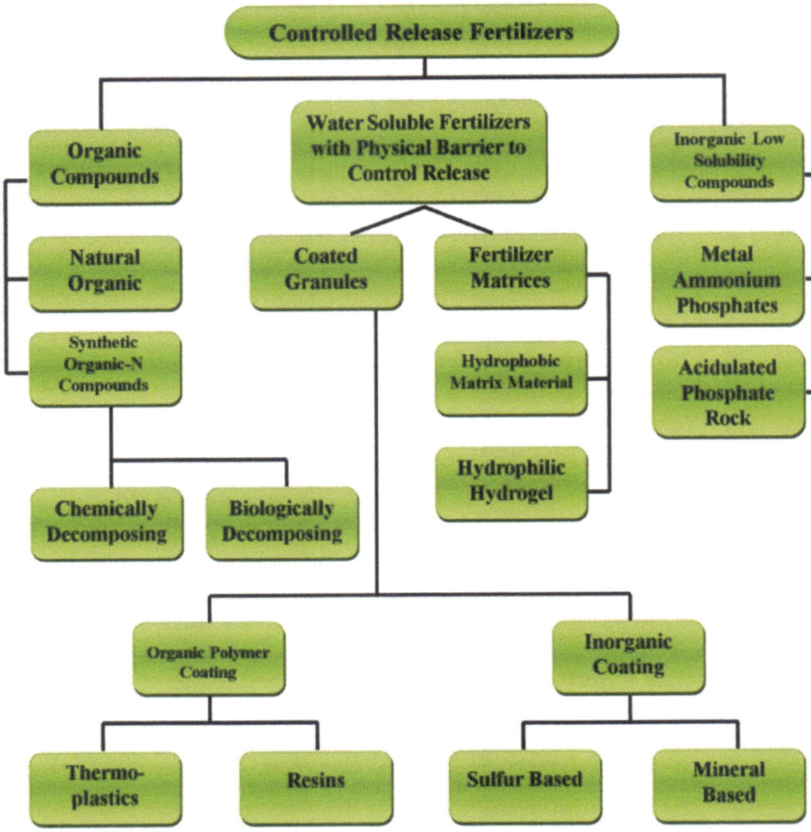

FIGURE 18.2 Scheme illustrating the classification of different types of fertilizers with controlled release of nutrients. This category includes slow-release fertilizers (SRFs) and controlled-release fertilizers (CRFs). *From Azeem, B., KuShaari, K., Man, Z.B., Basit, A., Thanh, T.H., 2014. Review on materials & methods to produce controlled release coated urea fertilizer. J. Control. Release 181, 11–21. https://doi.org/10.1016/j.jconrel.2014.02.020.*

that make the nutrients nonbioavailable. On the other hand, as water is used efficiently, and nutrients can be applied at the right time and in the right amount, manual labor is reduced. The fertilizer is distributed uniformly, and there is less danger of environmental pollution. The result is higher productivity and yield and better crop quality (Pinto et al., 2002).

There are some inconveniencies with the use of pressurized irrigation. These include the high initial cost of irrigation systems, the nonuniform application of water and fertilizers when the system is not properly designed, and the probability of precipitation due to incompatibility of the fertilizer preparations, producing blockage of the emitters and substantial economic losses to cultivators (Silva et al., 1996).

The fertilizers that presently have the higher efficiency are chelated fertilizers, SRFs, CRFs, and organic fertilizers. In the chelated fertilizers, some nutrients form a complex (called a chelate) with the ligands of a synthetic or natural organic molecule. The chelate is chemically stable versus the losses associated with the diverse edaphic conditions and is absorbed by the roots or the leaf or fruit surfaces of plants. The main disadvantage of chelates is the high cost and availability only for some elements, mostly microelements. The SRFs and organic fertilizers provide the nutrient elements in a formula that is not immediately soluble but liberated through the microbial breakdown or chemical hydrolysis. The CRFs include soluble fertilizers embedded in low-solubility materials or coated with low-solubility materials such as sulfur, some kind of minerals (e.g., phosphate rock), or polymers and hydrogels (Fig. 18.3) (Souri, 2016). The term CRF is usually applied to fertilizers in which the features governing the rate, pattern, and time of release are well known and manageable during CRF fabrication. On the other hand, the SRFs are characterized by the liberation of the nutrients at a slower rate than in a conventional fertilizer, but

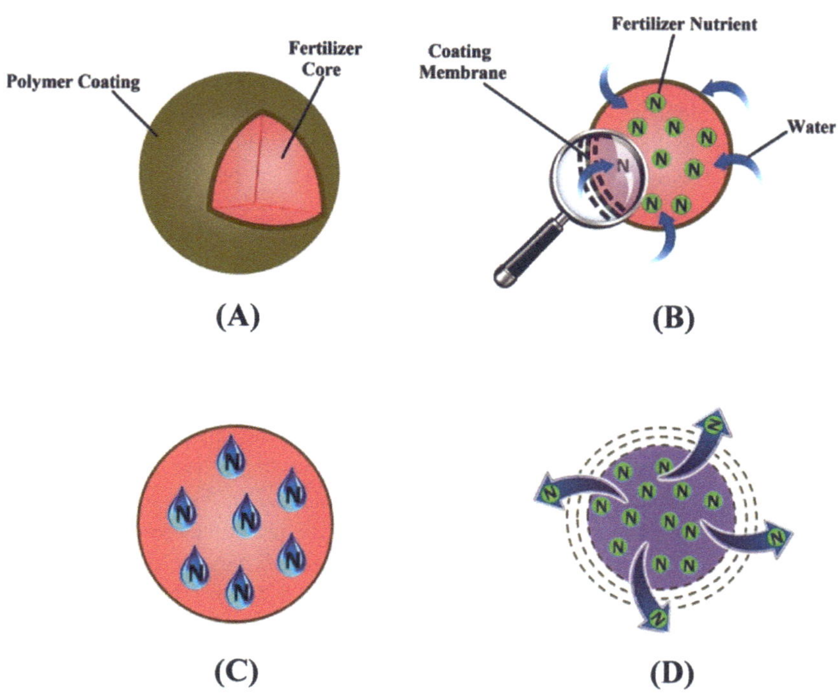

FIGURE 18.3 The diffusion process in the controlled release of nutrients. (A) Fertilizer material covered with a polymer layer, (B) water penetrates into the layer and dissolves the fertilizer material, (C) the dissolved fertilizer raises the osmotic pressure, (D) the controlled release of plant nutrient through swelled cover layer. *From Azeem, B., KuShaari, K., Man, Z.B., Basit, A., Thanh, T.H., 2014. Review on materials & methods to produce controlled release coated urea fertilizer. J. Control. Release 181, 11–21. https://doi.org/10.1016/j.jconrel.2014.02.020.*

the rate, pattern, and time of release are not well controlled. The release in SRFs can be dependent on handling and storage conditions or by soil conditions and edaphic biological activity (Sempeho et al., 2014).

Both CRF and SRF deliver necessary elements for plants at a low rate, which increases the chance for plants to absorb the dissolved nutrients, reducing soil fixation, leaching, and volatilization. The main weaknesses of SRFs are the cost and the dependence on temperature and microbial activity for the release of fertilizer elements (Morgan et al., 2009). This temperature dependence can reduce its applicability in regions with low temperatures; the microbial activity-dependence would cause a lower efficiency in soils with a small amount of organic matter (Souri, 2016).

18.3 Nanofertilizers and fertilizer nanoadditives

Following the previous definition of fertilizer, a *nanofertilizer* is defined as any material with a minimum dimension of between 1 and 100 nm, and that contains and directly contributes at least one of the 14 elements listed in the previous section (Ojeda-Barrios et al., 2020). The above definition includes the nanochelates. The direct contribution occurs if at least one essential element is contained in the nanofertilizer so that the reaction with the soil components will make it accessible in ionic form (e.g., Fe^{2+}), chelates (in the nanometric regime), or as a nanoparticle (e.g., Fe or Fe_2O_3) that can enter the root and other plant tissues.

A concept associated with that of fertilizer is that of *fertilizer additives*, defined as "substances added to or modifications of fertilizers, or products added to the soil, designed to increase fertilizer use efficiency" (FAO and ITPS, 2019). That is, fertilizer additives are substances that can function as fertilizer or soil conditioners or plant biostimulants, including humic substances, zeolites, and beneficial elements such as silicon, selenium, sodium, and iodine, that do not directly provide essential nutrients for plants, but that through indirect processes make nutrients more bioavailable in soil or the plant root and its microbiome more efficient to their absorption (Hajiboland et al., 2020). As a consequence of the above, an increase in the efficiency occurs in the use of nutrients.

Analogously to nanofertilizers, *fertilizer nanoadditives* are defined as any material with at least a dimension between 1 and 100 nm that does not contribute directly to essential plant nutrients but increases the efficiency of fertilizers or nanofertilizers. An example is the indirect inputs of essential nutrients that occurs in the presence of NMs formulated with nonessential elements, such as TiO_2, silica, and carbon NMs (Andelkovic et al., 2018; Boroumand et al., 2020; Ditta and Arshad, 2016; Raliya et al., 2018).

The indirect impact of fertilizer nanoadditives (Fig. 18.4) occurs by means of two processes: (1) one or more components of the nanoadditive cause a physicochemical reaction in the solid matrix of soil or in the soil pore water,

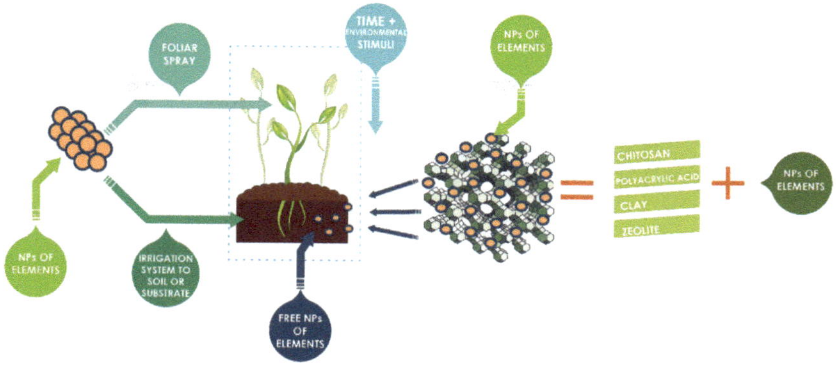

FIGURE 18.4 Nanofertilizers can provide the essential elements for plants directly (left side). Additives or nanoadditives (right side) (e.g., chitosan or nanochitosan, silica or nanosilica) do not directly provide essential elements, but rather absorb or adsorb bulk fertilizers or nanofertilizers, allowing their better absorption by plants and mitigating losses by soil fixation, leaching, and volatilization. An additional advantage of additives or nanoadditives is that they biostimulate plants and increase their ability to absorb nutrients from the soil. *From Morales-Díaz, A.B., Ortega-Ortíz, H., Juárez-Maldonado, A., Cadenas-Pliego, G., González-Morales, S., Benavides-Mendoza, A., 2017. Application of nanoelements in plant nutrition and its impact in ecosystems. Adv. Nat. Sci Nanosci. Nanotechnol. 8, 013001. https://doi.org/10.1088/2043-6254/8/1/013001.*

causing a higher bioavailability of an essential element and (2) one or more components of the nanoadditive change the metabolic capabilities of the plants and their microbiomes, resulting in supplementary capacity to take essential nutrients from the soil (Morales-Díaz et al., 2017; Raliya et al., 2018).

Nanofertilizers constitute a relatively new category of high-efficiency fertilizers used to provide essential elements for crop plants. As a consequence of their higher efficiency, lesser amounts of fertilizers formulated with NMs are expected to be applied to meet the needs of the plants compared to bulk fertilizers, and the anticipated ecological impact is less. On the other hand, the direct use of nanofertilizers or combinations of nanofertilizers and additives or nanoadditives is expected to decrease the impact of climatic anomalies, especially temperature, and factors such as water deficit and salinity (Ojeda-Barrios et al., 2020; Sharma et al., 2020). Similarly, the biostimulant effect of additives, such as humic substances, chitosan, silicon, or their nanometric equivalents, is a way to increase the resilience of crops against climatic anomalies (Abdel-Aziz et al., 2016; Al Murad et al., 2020; Jindo et al., 2020; Juárez-Maldonado et al., 2019).

Mineral nutrients in the form of nanofertilizers can be provided in three ways that correspond to the classification shown by Ojeda-Barrios et al. (2020) (Table 18.1).

1. The first category of nanofertilizers includes the bulk ionic form of elements adsorbed or absorbed in a nanocomplex such as carbon nanotubes,

TABLE 18.1 Classification of nanofertilizers, some examples, and comparative advantages compared to conventional fertilizers.

Categories of nanofertilizers and nanoadditives	Elements involved and references	Comparative advantages reported
(1) Nanostructured materials incorporating the bulk ionic form of essential elements	N. The zeolite clinoptilolite was used as carrier for NH_4 (Millán et al., 2008)	Greater yield and efficiency in the use of N in *Lolium multiflorum*
	N. NPs of hydroxyapatite modified with urea and encapsulated in wood (Kottegoda et al., 2011)	The stability for N-release of the fertilizer was twice (60 days) that of the conventional fertilizer (30 days)
	N. Nanocarbon material incorporated in a SRF and applied to rice (Wu, 2013)	More yield and N use efficiency and less N runoff loss when the carbon nanomaterial was added to the SRF
	N. Urea–hydroxyapatite nanohybrid (Raguraj et al., 2020)	In *Camellia sinensis* the nanohybrid showed positive effects on yield (more pronounced during unfavorable climatic conditions). Fewer applications and 50% of N resulted in the same yield obtained applying more N as conventional fertilizer
	N-P-K + nanochitosan (Abdel-Aziz et al., 2016)	Fertilizer with more efficiency, increase in yield
	N-P-K (500, 60, and 400 mg/L, respectively) + nanochitosan (0.2%) and N-P-K + carbon nanotubes (20 μg/L) (Abdel-Aziz et al., 2019)	Foliar application with both NMs improved yield, seed quality, plant growth, and reduced the days to harvest
	P-nanohydroxyapatite and nanozeolite (Mikhak et al., 2017)	Decrease in the P fixation in soil
	P. Graphene oxide-Fe(III) composite containing	Compared with commercial phosphorus fertilizer

(*Continued*)

TABLE 18.1 (Continued)

Categories of nanofertilizers and nanoadditives	Elements involved and references	Comparative advantages reported
	PO_4^{3-} (Andelkovic et al., 2018)	Monoammonium phosphate (MAP), the nanocomposite loaded with phosphate reduced the possibility for leaching or runoff of soluble P
	K. The zeolite clinoptilolite with K 160 meq/g as a SRF (Hershey et al., 1980)	50 g clinoptilolite in 1.5 L of substrate produced 3-month chrysanthemum production equivalent to that obtained with daily irrigation using a nutritive solution with 234 mg/L K
	P and K release from the zeolite phillipsite (Notario del Pino et al., 1995)	The zeolite showed a more stable, controlled, and over a longer time (more than 70 days) of P and K, compared to KH_2PO_4 (50 days)
	Zn encapsulated in a Mn nano hollow core shells of polymers (Yuvaraj and Subramanian, 2015)	70% more bioavailability of Zn^{2+} in a percolation reactor; more Zn uptake in rice plants
	Zn incorporated in nanozeolite (clinoptilolite) (Yuvaraj and Subramanian, 2018)	544% more bioavailability of Zn^{2+} in a percolation reactor
	Zn encapsulated in nanochitosan (Choudhary et al., 2019)	The Zn-chitosan NPs improved the growth, yield, and stress tolerance of maize plants
	Zn. Zinc-loaded graphene oxide in barley and *Medicago truncatula* (Watts-Williams et al., 2020)	Increased growth and yield in plants treated with graphene oxide + Zn compared to bulk Zn ($ZnSO_4$)

(Continued)

TABLE 18.1 (Continued)

Categories of nanofertilizers and nanoadditives	Elements involved and references	Comparative advantages reported
(2) Nanostructured essential elements incorporated in the carrier material. The carrier may or may not be a nanomaterial	P-nano-hydroxyapatite (Liu and Lal, 2015)	Higher efficiency of P fertilization of the nano-hydroxyapatite versus Ca $(H_2PO_4)_2$
	Cu NPs enveloped in chitosan (Juarez-Maldonado et al., 2016)	More yield and better fruit quality
	Cu NPs enveloped in chitosan-PVA (Polyvinyl alcohol) hydrogels (Hernández et al., 2017)	More yield and better fruit quality
	Cu NPs encapsulated in chitosan-PVA hydrogels (González Gómez et al., 2017)	In grafted watermelon, the Cu NPs with chitosan-PVA improved the stem length and root length
(3) NPs or nanochelates of essential elements for direct application in soil, substrate, leaves, fruits, or seeds	N. NH_4NO_3 NPs in pomegranate (Davarpanah et al., 2017)	The N NPs applied by foliar spraying (1.8 kg/ha) yielded more fruit with better quality compared to bulk urea (16.3 kg/ha)
	N. Nanochelated N by foliar application (1.02–1.36 gN/L) in olive (Rohi Vishekaii et al., 2019)	The nanochelated N raised the fruit set and the oil percentage. The bulk urea presented the highest yield and quality of fruit and the highest oil yield
	N, P, K nanofertilizers in liquid formulations were applied to soil and compared with equivalent quantities of bulk N, P, K fertilizers in potato (Abd El-Azeim et al., 2020)	Compared with control, the plants with nanofertilizers showed higher tuber yield, starch rates, N-P-K nutrient use efficiency, and harvest index
	Mg and Fe NPs applied by leaf aspersion (0.5 g/L and 0.25–0.5 g/L, respectively) in *Vigna unguiculata* (Delfani et al., 2014)	Mg NPs significantly increased yield. Fe (in both nano and bulk forms) significantly improved the impact of Mg NPs

(*Continued*)

TABLE 18.1 (Continued)

Categories of nanofertilizers and nanoadditives	Elements involved and references	Comparative advantages reported
	Ca. Nano-Ca (1.5%–2.5%) applied in preharvest by spraying the apple trees (Ranjbar et al., 2018)	Compared with $CaCl_2$, the nano-Ca was more effective at increasing the quality and postharvest life of the fruits
	Ca-Mg-Fe-Si nanosize calcite (0.5 g/L) (Sabir et al., 2014)	Higher berry yield and quality in grapevines
	Fe. Superparamagnetic iron oxide NPs applied in the nutritive solution (30–60 mg/L) of soybean (Ghafariyan et al., 2013)	Fe NPs increased the chlorophyll level in plants, compared to Fe EDTA fertilizer
	Fe. Comparison of Fe_2O_3 NPs (5–40 μmol) and Fe-EDTA in *Cataranthus roseus* (Askary et al., 2017)	Fe_2O_3 NPs increased plant growth and quality in comparison with Fe-EDTA
	Fe. Iron nanochelates applied to the soil (8 kg/ha) and by foliar spraying (2.5 g/L) (Fakharzadeh et al., 2020)	In rice the nanochelated Fe yielded 27% more, with 13% more protein
	Zn. Nanozinc oxide NPs to coating seeds (25–50 mg Zn/g seed) of maize, soybean, pigeon pea, and *Abelmoschus esculentus* (Adhikari et al., 2016)	The requirement of Zn of the plants can be applied entirely with the seed coat without exerting any osmotic potential
	Zn (60–120 mg/L) and B (3.25–6.5 mg/L) NPs in pomegranate (Davarpanah et al., 2016)	Foliar application of nano-Zn and nano-B improved yield and quality contrasted to the conventional fertilizer
	Zn. Comparison of ZnO NPs (40–160 mg/kg soil) and bulk $ZnCl_2$ applied to soybean cultivated in soil (Yusefi-Tanha et al., 2020)	In comparison to $ZnCl_2$ the yield increased with the use of ZnO NPs

(*Continued*)

TABLE 18.1 (Continued)

Categories of nanofertilizers and nanoadditives	Elements involved and references	Comparative advantages reported
	Mn. Comparison of Mn NPs and $MnSO_4$ fertilizers (0.05–0.5 mg/L) in *Vigna radiata* (Pradhan et al., 2013)	A higher speed of energy transfer in photosynthetic systems with the use of Mn NPs
	Cu. CuO NPs (200–400 mg/kg soil) used as fertilizer for lettuce plants (Wang et al., 2019)	Compared with microparticulate CuO, the nanoparticulate CuO increased biomass and more Cu accumulation without causing adverse effects
	Cu. Application on CuO NPs, bulk CuO, and $CuSO_4$ (75–600 mg Cu/kg soil) in *Allium fistulosum* (Wang et al., 2020)	Better absorption of Cu, and increased bulb Ca and Mg, in plants treated with CuO NPs. Leaf allicin was higher with the application of CuO NPs and $CuSO_4$
	Mo NPs (a solution with not less than 8 mg/L of Mo NPs was applied at 1 μL/g seed) mixed with *Bradyrhizobium japonicum* for seed inoculation (Taran et al., 2014)	The application of bacteria and Mo NPs increased the number and mass of the nodules in chickpea plants. Mo favored the abundance of nitrifying microorganisms. Plants had higher catalase activity

graphene, nanohydroxyapatite, nanozeolites, nanoclays, or nanochitosan that regulates the release and availability of nutrients.

2. The second nanofertilizer category uses the essential elements in nano form but incorporated in a carrier substance, which in turn may or may not be a nanomaterial.
3. The third nanofertilizer category uses the essential element as nanoparticles (NPs) in suspension or nanochelates. The NPs or nanochelates can be applied directly to the soil or substrate using drench or the irrigation system; a second alternative is sprinkling the NP suspension or nanochelates to the leaves, fruits, or seeds.

An example of the first category is nanochitosan, which forms complexes between the ions of metals such as Mg, Ca, Cu, and Fe, among others, and their functional groups amino and hydroxyl. Nanochitosan is a potent biostimulant and a potential alternative for the development of agricultural chelates (Abdel-Aziz et al., 2016; Khati et al., 2017). Other examples of the first category are the carbon nanotubes loaded with N, P, and K (Hasaneen et al., 2017), biochar NPs composites (Joseph et al., 2013), and the urea-hydroxyapatite nanofertilizer with a high N content and a slow-release rate (Kottegoda et al., 2017).

Some examples for the second category are the NPs of hydroxyapatite (source of P) stabilized in carboxymethyl cellulose (Liu and Lal, 2015), and the NPs of Cu and Fe incorporated by absorption or adsorption in materials such as clays, zeolites, chitosan, or polymeric hydrogels (Golbashy et al., 2017; González Gómez et al., 2017).

Examples of the third category are the NPs of Fe, Cu, Zn, and their oxides (Juarez-Maldonado et al., 2016; Juárez-Maldonado et al., 2018; López-Vargas et al., 2018; Morales-Díaz et al., 2017).

Table 18.1 shows the three categories explained above of the different combinations of bulk fertilizers, nanofertilizers, fertilizer additives, and fertilizer nanoadditives that together make up various alternatives of nanofertilizers. Table 18.1 includes examples of each of the three categories for different essential elements and some examples of the differential impact of nanofertilizers versus conventional fertilizers.

A wealth of literature is available on the subject of NMs and their impact on plants; however many cases do not necessarily refer to the use of NMs as fertilizer. In many cases, NMs were applied to plants in levels above or below the nutrient recommendations. For example, for a nutrient solution for the production of guava seedlings, it is suggested 1.8 mg/L of Fe and 0.11 mg/L of Zn (Monteiro Corrêa et al., 2012), while in many studies it is common to find concentrations of up to 60 mg/L of NPs of Fe or Zn (Juárez-Maldonado et al., 2018). On the other hand, in some studies amounts as small as 0.03 mg/L substrate can induce biostimulation in plants (Juarez-Maldonado et al., 2016). Additionally, many studies have been carried out under laboratory conditions or on a small or microcosm scale. However, in cultivation fields, on a larger scale (>1 ha), many climatic, edaphic, and biological factors occur that modify the response of plants to NMs, which makes many results described in the literature challenging to extrapolate to situations encountered in the field (Morales-Díaz et al., 2017; Reddy et al., 2016). A substantial part of the literature, especially in the case of metals, focuses on toxicity or biostimulation, rather than on fertilizer usefulness.

Different studies Table 18.1 show specific properties that provide advantages to nanofertilizers over conventional fertilizers. These properties raise the efficiency of the nanofertilizers, promote yield and quality, and decrease the quantity of the element applied as nanofertilizer, as well as the number of applications. Some examples of the higher fertilization efficiency of the nanofertilizers will be mentioned next.

18.3.1 Minor soil fixation, precipitation, or speciation in nonbioavailable forms

Fertilizers with N, P, S, Fe, Mn, Cu, Zn, and Mo are susceptible to fixation in the soil. The process of complexation of nutrients with inorganic and organic soil components results in nonbioavailable forms for plant roots (Pilbeam, 2015).

In comparison with conventional fertilizers, nanofertilizers have a lower rate of nutrient release and higher stability against aggregation or adsorption; on the other hand, nanofertilizers do not change the chemical speciation of their components due to soil redox potential, pH-dependent phenomena, and presence of ligands (Raliya et al., 2018).

Examples of the above are the use of nano rock phosphate that increases the stability and bioavailability of P for plants, since it inhibits the fixation of P by Ca, Fe, or silicic acid; the use of modified zeolites, with micro or nanopores that develop a high internal surface, to regulate the ion exchange of N, P, K, and other elements with the soil solution and plant roots (Singh et al., 2017).

18.3.2 Less volatilization and leaching

Volatilization and leaching are processes that lead to the loss of nutrients from soil horizons. Volatilization is critical for nitrogen fertilizers, intensifies by increasing the amount of N applied, and does not diminish with split applications. N volatilization reduces with the use of urease or nitrification inhibitors, deep incorporation of fertilizers, and use of coated fertilizers for controlled release (Pan et al., 2016). The leaching happens in N and K fertilizers, but it occurs similarly in fertilizers containing other elements when applied in light textured soils (Lal, 2015). Leaching is mitigated by the presence in the soil matrix of inorganic and organic ion exchange complexes, by the chemical speciation phenomena that change the solubility of the elements, and as result of the recycling of nutrients between soil and plants. The leaching of plant nutrients can result from the excess application of water and fertilizers in agricultural systems; however, it is also a phenomenon associated with soils under degradation (Lal, 2015).

In order to reduce volatilization and leaching losses, nanofertilizers can furnish some advantages, such as decreased solubility, enabling the slow release or controlled release of the nutrients, and an enhanced capability to stabilize by aggregation or by adsorption in edaphic components (Raliya et al., 2018).

To diminish volatilization and leaching of nutrients, it has been recommended to decrease their release rate by combining them with nanofertilizer additives or fertilizer additives such as hydrogels, films, or other biopolymers such as C nanotubes (Chen et al., 2015; Liu et al., 2009), chitosan (Abdel-Aziz et al., 2016; Hasaneen et al., 2014; Kashyap et al., 2015), montmorillonite-urea clay nanocomposites (Golbashy et al., 2017) or other types of ceramic materials (Choy et al., 2007), NPs of SiO_2, and natural or synthetic zeolites (Ditta and Arshad, 2016; Lehman and Larsen, 2014; Eroglu et al., 2017; Nakhli et al., 2017). Zeolites positively regulate the bioavailability of nutrients, decrease leaching, and are inexpensive and innocuous materials (Eroglu et al., 2017).

Another alternative to reduce the leaching of essential nutrients such as Ca, Fe, Zn, Cu, and Mn is to apply them directly to the substrate or soil as NPs. Due to their large surface/volume ratio, NPs have high surface energy. The natural tendency of this high-energy unstable system is to decrease the surface energy of the NPs in the nutritive solution, soil, or soil solution. The above produces the aggregation of the NPs in the nutritive solution or soil solution, or their adsorption with the solid components of the substrate or soil (Ojeda-Barrios et al., 2020).

Aggregation in the substrate or soil stabilizes the NPs and reduces leaching or mass flow that results in the pollution of soils and water or other ecosystem components. However, it also reduces the absorption by the roots and can even provoke unfavorable effects by excess deposition of aggregates in the root surfaces (Morales-Díaz et al., 2017).

To reduce the aggregation process, the engineered NPs are functionalized before being utilized. The functionalization diminishes the surface free energy of the NPs and, in consequence, the natural tendency to form aggregates. However, when exposed to natural media or biological fluids, the surface of the NPs is covered with a new corona, or layer of organic molecules that substantially modify their stability and bioavailability. Corona formation occurs in all NPs. Its composition and properties are specific for each type of NP and the environment in which they are found. Likewise, the corona modifies the impact that NPs have on organisms, creating a certain level of uncertainty about the environmental impact resulting from the use of nanofertilizers (Bakshi, 2020; Morales-Díaz et al., 2017).

On the other hand, adsorption on substrate components or soil colloids stabilizes NPs, diminishes leaching, and keeps them available for plants. The equilibrium between adsorption/desorption is dynamic and rests positively on the concentration of salts in the nutritive or soil solution and negatively on the amount of dissolved organic matter (Bae et al., 2013). In sandy soils, the adsorption of NPs represents an advantage (Rui et al., 2016) with an enhanced availability of Fe using Fe_2O_3 NPs, in comparison with Fe-EDTA chelate. In loam or clayey soils, larger adsorption of the NPs on the superficial components or colloids of substrate or soil would be expected, as the salt concentration increases; the opposite result is observed by the presence of humic substances (Bae et al., 2013) and possibly exudates from the roots.

18.3.3 Greater productivity and yield

Nanofertilizers can provide nutrients in two forms: by releasing ions after the reaction of the materials (nanofertilizers + additives) with the substrate or soil components, or by direct absorption of the NPs or nanochelates in plant roots (Morales-Díaz et al., 2017). The ions of essential elements liberated from the NMs or NMs + additives in the nutritive solution or soil solution are absorbed by the typical mechanisms that involve cell membrane integral proteins that act as channels and transporters. On the other hand, the NPs applied to the substrate or soil (e.g., Fe or Cu NPs or carbon NPs loaded with nutrients) can be absorbed by the root cells and transported by the apoplastic or symplastic path, directly through preexisting pores in the cell walls or by the creation of new pores resulting from the surface interactions between the cells and the NPs (Juárez-Maldonado et al., 2019).

It is also possible to add nanofertilizers by leaf spray. Depending on the higher or slighter size of the nanofertilizers and additives used, the absorption can occur through stomata or the cuticle (Juárez-Maldonado et al., 2019). After the nutrients from the nanofertilizers enter the plant through the stomata or the epidermal surfaces, they are transported to the different leaf tissues (Eichert et al., 2008).

For both substrate, soil, and foliar use, nanofertilizers exhibit a faster dissolution, absorption, and assimilation by the plants compared to conventional fertilizers. The above advantages have been reported for the elements N, P, K, Ca, Mg, Fe, Mn, Zn, Cu, and Mo (Ditta and Arshad, 2016), including the nanochelates (Al-juthery et al., 2019), and maybe explicate why the application of nanofertilizers generally shows better results in crop growth and yield when compared to conventional fertilizers (Abdel-Aziz et al., 2016; Kabir et al., 2018; Liu and Lal, 2015).

18.4 Pros and cons of nanofertilizers and nanomaterials

The use of NMs in plant nutrition, although it has been extensively studied on an experimental or pilot scale, still requires a lot of basic and applied research work to be delivered to the crop fields at a full scale. It is necessary to define specific critical issues, such as the safe and profitable application on a commercial scale, the safety of foods produced under these fertilization systems, and the long-term effects of the potential trophic and between-ecosystems transfer of NMs used as plant nutrients (Kah, 2015; Morales-Díaz et al., 2017).

In all the conventional fertilizers, the essential elements are released as ionic species whose fate in ecosystems is relatively well known; therefore, although there is concern about their environmental impact on soil, water, and the atmosphere, there is no concern about the health of the human consumer. For nanofertilizers, on the other hand, the subject still has knowledge gaps and conflicting results about the short- and long-term effects on ecosystems and food innocuity (Chun et al., 2020; Kah et al., 2018; Morales-Díaz et al., 2017; Reddy Pullagurala et al., 2019). For example, it is only partially known the complete life cycle of NMs in the ecosystems. Equally only partly known is whether the nanofertilizers are completely transformed to ionic forms in the plant organs, or if a certain quantity of the NMs can be conserved and potentially be transferred to the consumers (Chun et al., 2020; Morales-Díaz et al., 2017).

NMs are found naturally in ecosystems. They result from natural processes such as the weathering of materials by climatic action; from chemical reactions that occur in soils, water, and the atmosphere; and from processes such as volcanism and biological activity. However, human activity has caused NMs to be discharged into ecosystems in unprecedented amounts, which are likely to rise as time passes and the number and class of applications of NMs increases (Kranjc and Drobne, 2019).

The responses of plants and their microbiome, as well as those of other organisms, are variable depending on the species, the NMs involved and their characteristics, the environmental modifications of the corona of the NMs, as well as the environmental conditions where the interaction between organism-nanomaterial occurs (Bandeppa et al., 2019; Reddy Pullagurala et al., 2019). The complexity of the multitude of interactions in ecosystems is one of the factors that make it difficult to define legal regulations regarding the use, disposition, and destination in ecosystems of the different products that contain NMs (Lai et al., 2018). As a result of the above points, there is great controversy over the long-term safety of the use of NMs. Unfortunately, not enough information is available, in the relevant time and space scales, to reach a well-supported conclusion (Kranjc and Drobne, 2019; Morales-Díaz et al., 2017; Reddy Pullagurala et al., 2019).

In the case of nanofertilizers, the same dilemma about available information arises. There are not sources with sufficient information to predict with certainty the results of the use of nanofertilizers in terms of ecological safety and food innocuity (Reddy Pullagurala et al., 2019). In the study by Kranjc and Drobne (2019), which included an exhaustive search for information in publications from 2017 to 2018, it was concluded that the exposure of the plants to the NMs of C, Ag, Ce, and Ti does not seem to represent a risk for crops because these materials are present in low concentration in the soil and water; that the easily dissolving NMs (e.g., Ag, Cu, and Zn) pose a risk to plants from ions released from the NM, rather than from the NM itself; but that there is no information about the responses of the plants in the long term, including against exposure to low-dose NMs.

However, certain aspects of the use of nanofertilizers should not be equated with the unintentional release events of NMs, as in the case of environmental pollution. In the ideal situation, the higher efficiency of nanofertilizers should mean a decrease in the volume of materials applied to the soil as fertilizers, with less environmental impact resulting from a lower rate of soil degradation, less volatilization into the atmosphere, and less water pollution by leaching and runoff.

A type of regulated scheme for nanofertilizer usage should consider the available information on the effect of different NMs used as fertilizer materials. The amounts and types of nanofertilizer to be used would be defined, taking into account local information on the composition of soil and water and the crop's nutritional needs. Under that scheme, it could be expected that a transition from the use of fertilizers to nanofertilizers would indeed result in an environmental benefit. On the other hand, even thinking about the maximum amounts of nanofertilizers that have been reported in the literature, they would not be expected to constitute a short-term danger for crops (Kranjc and Drobne, 2019; Ojeda-Barrios et al., 2020).

A zinc nanofertilizer in the form of ZnO NPs can be used as an example of the above. Under certain environmental conditions, ZnO NPs concentrations >100 mg/kg soil can negatively impact the edaphic microbiome. In most cases, ZnO NPs in concentrations >500 mg/kg soil are expected to be detrimental to crop growth. However, levels <50 mg/kg soil and <20 mg/L by foliar spray are generally beneficial (Reddy Pullagurala et al., 2018). Tentatively, the limits for the use of ZnO as a nanofertilizer could be set at 50 mg/kg soil unless there was an extremely deficient level of Zn in the soil.

If we consider the application of 50 mg/kg of ZnO NPs in a loam soil (in a profile 0–0.2 m, with a bulk density of 1.2 g/mL), this is equivalent to about 120 kg/ha of ZnO NPs, that is, about 96 kg Zn/ha. The above is a very high amount of Zn if we take into account that an adequate corrective application of Zn for crops in calcareous soil is 100 kg/ha $ZnSO_4{\bullet}7H_2O$, which represents 23 kg/ha of Zn (Yilmaz et al., 1997). That means that in the case of the ZnO NPs, an application of 29 kg/ha (12 mg/kg) of ZnO NPs to the soil could be enough to equal the 23 kg of bulk Zn mentioned. This amount could even be decreased, taking into account the higher efficiency of the ZnO NPs-based nanofertilizer compared to $ZnSO_4$.

A factor associated with the use of nanofertilizers that can indirectly modify food innocuity is the ability of NMs to induce higher tolerance in plants against biotic and abiotic stresses. One of the consequences of increased use of nanofertilizers would be less use of pesticides, which would mean potentially innocuous foods. This potential greater innocuity depends on certainty about the nonoccurrence of the trophic transfer.

A first mechanism that explains the ability of nanofertilizers to induce stress tolerance is that presenting higher stability and bioavailability of nutrients, improves plant nutrition, allowing biofortified crops with greater vigor and therefore greater ability to tolerate environmental stress, in addition to presenting better nutritional quality, especially in the case of microelements such as Fe, Zn, and Cu. The nutritional attributes and the stress tolerance of plants are correlated characteristics: the adequate equilibrium of mineral nutrients, osmolytes, antioxidants, and metabolites related with stress signaling and the cellular responses results on the one hand in plants with more capability to adapt to the changing environment and, on the other hand, in foods that provide more significant benefits for the health of the consumer (Cabrera-De la Fuente et al., 2018).

The second mechanism is probably associated with changes in metabolism and adjustments in the profiles of microorganism species that shape the different microbiomes of plants. The use of NMs can cause changes in the growth rate of certain microorganisms, affecting more certain groups compared to others, resulting in changes in the relative abundance of different classes of microorganisms. This change in the composition of the microbial community, due to the complexity of the biotic interactions in the soil and the resulting difficulty in its control and prediction, is considered an adverse effect. Furthermore, these changes in the edaphic microbial community can spread to other groups of organisms such as protozoa and soil mesofauna, with unpredictable long-term effects (González-Morales et al., 2020). It should be noted that changes in microbial communities, which modify the global metabolism of the soil, would not necessarily be reflected in the short term in the growth or yield of crops. Soil degradation processes are not immediately noticeable, but in the medium or long term, they modify the soil health and its ability to provide environmental services, including crop production (FAO and ITPS, 2015). The same can apply to the microbiome of humans and other animals: not observing dysfunctions in the short term does not automatically imply their absence in the long term (Morales-Díaz et al., 2017), and can be illustrated with certain food additives (Cao et al., 2020). Until now, no information indicates how nanofertilizers should be used to mitigate the impact on beneficial soil microorganisms (Reddy Pullagurala et al., 2019). For this reason, it is recommended to take care to avoid the inhibitory effect of the NMs on the soil's beneficial microorganisms (Reddy Pullagurala et al., 2018).

Finally, the third mechanism by which nanofertilizers are thought to induce stress tolerance in plants is associated with the biostimulant capacity of these materials (Juárez-Maldonado et al., 2019). The biostimulation occurs due to the interactions of NMs with cell components: walls, membranes, proteins, membranes, and metabolites, or by the release of metal ions able to induce oxidative stress, and by the existence in the NM of unpaired electrons that provoke redox signals (Hernández et al., 2017; Juárez-Maldonado et al., 2018; López-Vargas et al., 2018). When the concentration of the NMs is below some threshold, the level of oxidative stress does not exceed the toxicity limit, and stimulates in the plant cells the induction of defense genes and metabolites, defense proteins, and more activity and concentration of enzymatic and nonenzymatic antioxidants (Van Aken, 2015).

It can be seen that the balance between the favorable and unfavorable aspects of the use of nanofertilizers seems still uncertain. As with other technologies, such as the use of fertilizers, pesticides, and other techniques for agriculture, the final balance appears to depend on the generation of more knowledge, and the transformation of these into new techniques aimed at increasing productivity with the least possible environmental impact. The set of new techniques must be well supported by regulations at the global and local levels. An example of this kind of regulation is The International Code of Conduct for the Sustainable Use and Management of Fertilizers (FAO and ITPS, 2019). The indications in this document are transferable to situations where nanofertilizers are used for crop production.

18.5 Conclusions

The use of nanofertilizers promises functional, environmental, and maybe economic advantages compared to conventional fertilizers. The benefits have been seen for practically all the essential elements for plants.

The use of NMs, both in the form of nanofertilizers to provide essential nutrients directly and in the form of nanoadditives to increase the efficiency of nutrient use, has been proven in different plant species. Studies indicate that such materials are safe, at least in the short term, for crops when care is taken in handling the concentrations and doses to be used.

The toxicology of NMs has been well studied and constitutes a broad and relatively well-established area of knowledge. On the other hand, less information is available about the ecological life cycle of nanofertilizers. Little information is available to predict the impacts of nanofertilizers on ecosystems and food innocuity.

Some research needs that are mentioned by the authors of the papers consulted are:

1. Determine the long-term impact of repeated applications of nanofertilizers on the soil health, soil microbiome, and different crop species in different edaphoclimatic regions.
2. Carry out more comparisons in different edaphoclimatic conditions of the effectiveness of the application of the elements contained in nanofertilizers versus those of conventional fertilizers.
3. More epidemiological studies on the transfer of NMs to the food chain and food consumed by humans are needed.

References

Abd El-Azeim, M.M., Sherif, M.A., Hussien, M.S., Tantawy, I.A.A., Bashandy, S.O., 2020. Impacts of nano- and non-nanofertilizers on potato quality and productivity. Acta Ecol. Sin. 40 (5), 388–397. Available from: https://doi.org/10.1016/j.chnaes.2019.12.007.

Abdel-Aziz, H.M.M., Hasaneen, M.N.A., Omer, A.M., 2016. Nano chitosan-NPK fertilizer enhances the growth and productivity of wheat plants grown in sandy soil. Spanish J. Agric. Res. 14, 0902. Available from: https://doi.org/10.5424/sjar/2016141-8205.

Abdel-Aziz, H.M.M., Hasaneen, M.N.A., Omer, A.M., 2019. Impact of engineered nanomaterials either alone or loaded with NPK on growth and productivity of French bean plants: seed priming vs foliar application. S. Afr. J. Bot. 125, 102–108. Available from: https://doi.org/10.1016/j.sajb.2019.07.005.

Adhikari, T., Kundu, S., Rao, A.S., 2016. Zinc delivery to plants through seed coating with nano-zinc oxide particles. J. Plant Nutr. 39, 136–146. Available from: https://doi.org/10.1080/01904167.2015.1087562.

Al Murad, M., Khan, A.L., Muneer, S., 2020. Silicon in horticultural crops: cross-talk, signaling, and tolerance mechanism under salinity stress. Plants 9, 460. Available from: https://doi.org/10.3390/plants9040460.

Al-juthery, H.W.A., Al-taee, R.A.H.G., Al-Obaidi, Z.H.H., Ali, E.A.H.M., NAl-Shami, Q.M., 2019. Influence of foliar application of some nano-fertilizers in growth and yield of potato under drip irrigation. J. Phys. Conf. Ser. 1294, 092024. Available from: https://doi.org/10.1088/1742-6596/1294/9/092024.

Andelkovic, I.B., Kabiri, S., Tavakkoli, E., Kirby, J.K., McLaughlin, M.J., Losic, D., 2018. Graphene oxide-Fe(III) composite containing phosphate – a novel slow release fertilizer for improved agriculture management. J. Clean. Prod. 185, 97–104. Available from: https://doi.org/10.1016/j.jclepro.2018.03.050.

Askary, M., Amirjani, M.R., Saberi, T., 2017. Comparison of the effects of nano-iron fertilizer with iron-chelate on growth parameters and some biochemical properties of *Catharanthus roseus*. J. Plant Nutr. 40, 974–982. Available from: https://doi.org/10.1080/01904167.2016.1262399.

Azeem, B., KuShaari, K., Man, Z.B., Basit, A., Thanh, T.H., 2014. Review on materials & methods to produce controlled release coated urea fertilizer. J. Control. Release 181, 11–21. Available from: https://doi.org/10.1016/j.jconrel.2014.02.020.

Bae, S., Hwang, Y.S., Lee, Y.-J., Lee, S.-K., 2013. Effects of water chemistry on aggregation and soil adsorption of silver nanoparticles. Environ. Health Toxicol. 191, 28. Available from: https://doi.org/10.5620/eht.2013.28.e2013006.

Bakshi, M.S., 2020. Impact of nanomaterials on ecosystems: mechanistic aspects in vivo. Environ. Res. 182, 109099. Available from: https://doi.org/10.1016/j.envres.2019.109099.

Baligar, V.C., Fageria, N.K., He, Z.L., 2001. Nutrient use efficiency in plants. Commun. Soil Sci. Plant Anal. 32, 921–950. Available from: https://doi.org/10.1081/CSS-100104098.

Bandeppa, G.S., Gobinath, R., Latha, P.C., Manasa, V., Chavan, S., 2019. Soil ecological pros and cons of nanomaterials: impact on microorganisms and soil health. In: Panpatte, D.G., Jhala, Y.K. (Eds.), Nanotechnology for Agriculture: Advances for Sustainable Agriculture. Springer, Singapore, pp. 145–159. Available from: https://doi.org/10.1007/978-981-32-9370-0_10.

Boroumand, N., Behbahani, M., Dini, G., 2020. Combined effects of phosphate solubilizing bacteria and nanosilica on the growth of land cress plant. J. Soil Sci. Plant Nutr. 20, 232–243. Available from: https://doi.org/10.1007/s42729-019-00126-8.

Cabrera-De la Fuente, M., González-Morales, S., Juárez-Maldonado, A., Leija-Martínez, P., Benavides-Mendoza, A., 2018. Plant nutrition and agronomic management to obtain crops with better nutritional and nutraceutical quality. In: Holban, A.M., Grumezescu, A.M. (Eds.), Therapeutic Foods, Handbook of Food Bioengineering. Academic Press, pp. 99–140. Available from: https://doi.org/10.1016/B978-0-12-811517-6.00004-0.

Cao, X., Han, Y., Gu, M., Du, H., Song, M., Zhu, X., et al., 2020. Foodborne titanium dioxide nanoparticles induce stronger adverse effects in obese mice than non-obese mice: gut microbiota dysbiosis, colonic inflammation, and proteome alterations. Small 16, 2001858. Available from: https://doi.org/10.1002/smll.202001858.

Celikkol Erbas, B., Guven Solakoglu, E., 2017. In the presence of climate change, the use of fertilizers and the effect of income on agricultural emissions. Sustainability 9, 1989. Available from: https://doi.org/10.3390/su9111989.

Chen, G., Qiu, J., Liu, Yan, Jiang, R., Cai, S., et al., 2015. Carbon nanotubes act as contaminant carriers and translocate within plants. Sci. Rep. 5, 1–9. Available from: https://doi.org/10.1038/srep15682.

Chhipa, H., 2017. Nanofertilizers and nanopesticides for agriculture. Environ. Chem. Lett. 15, 15–22. Available from: https://doi.org/10.1007/s10311-016-0600-4.

Choudhary, R.C., Kumaraswamy, R.V., Kumari, S., Sharma, S.S., Pal, A., Raliya, R., et al., 2019. Zinc encapsulated chitosan nanoparticle to promote maize crop yield. Int. J. Biol. Macromol. 127, 126–135. Available from: https://doi.org/10.1016/j.ijbiomac.2018.12.274.

Choy, J.-H., Choi, S.-J., Oh, J.-M., Park, T., 2007. Clay minerals and layered double hydroxides for novel biological applications. Appl. Clay Sci. Clays Health 36, 122–132. Available from: https://doi.org/10.1016/j.clay.2006.07.007.

Chun, S., Muthu, M., Gopal, J., 2020. Nanotoxic impacts on staple food crops: there's plenty of room for the unpredictables. Crit. Rev. Food Sci. Nutr. 60 (22), 1–12. Available from: https://doi.org/10.1080/10408398.2019.1707158.

Cui, Z., Wang, G., Yue, S., Wu, L., Zhang, W., Zhang, F., et al., 2014. Closing the N-use efficiency gap to achieve food and environmental security. Environ. Sci. Technol. 48, 5780–5787. Available from: https://doi.org/10.1021/es5007127.

Davarpanah, S., Tehranifar, A., Davarynejad, G., Abadía, J., Khorasani, R., 2016. Effects of foliar applications of zinc and boron nano-fertilizers on pomegranate (*Punica granatum* cv. Ardestani) fruit yield and quality. Sci. Hortic. 210, 57–64. Available from: https://doi.org/10.1016/j.scienta.2016.07.003.

Davarpanah, S., Tehranifar, A., Davarynejad, G., Aran, M., Abadía, J., Khorassani, R., 2017. Effects of foliar nano-nitrogen and urea fertilizers on the physical and chemical properties of pomegranate (*Punica granatum* cv. Ardestani) fruits. HortScience 52, 288–294. Available from: https://doi.org/10.21273/HORTSCI11248-16.

Delfani, M., Firouzabadi, M.B., Farrokhi, N., Makarian, H., 2014. Some physiological responses of black-eyed pea to iron and magnesium nanofertilizers. Commun. Soil Sci. Plant Anal. 45, 530–540. Available from: https://doi.org/10.1080/00103624.2013.863911.

Dimkpa, C.O., Bindraban, P.S., 2018. Nanofertilizers: new products for the industry? J. Agric. Food Chem. 66, 6462–6473. Available from: https://doi.org/10.1021/acs.jafc.7b02150.

Ditta, A., Arshad, M., 2016. Applications and perspectives of using nanomaterials for sustainable plant nutrition. Nanotechnol. Rev. 5, 209–229. Available from: https://doi.org/10.1515/ntrev-2015-0060.

Eichert, T., Kurtz, A., Steiner, U., Goldbach, H.E., 2008. Size exclusion limits and lateral heterogeneity of the stomatal foliar uptake pathway for aqueous solutes and water-suspended nanoparticles. Physiol. Plant 134, 151–160. Available from: https://doi.org/10.1111/j.1399-3054.2008.01135.x.

Eroglu, N., Emekci, M., Athanassiou, C.G., 2017. Applications of natural zeolites on agriculture and food production. J. Sci. Food Agric. 97, 3487–3499. Available from: https://doi.org/10.1002/jsfa.8312.

Fakharzadeh, S., Hafizi, M., Baghaei, M.A., Etesami, M., Khayamzadeh, M., Kalanaky, S., et al., 2020. Using nanochelating technology for biofortification and yield increase in rice. Sci. Rep. 10, 4351. Available from: https://doi.org/10.1038/s41598-020-60189-x.

FAO and ITPS, 2015. Status of the World's Soil Resources (SWSR). Food & Soils, Rome, Italy.

FAO and ITPS, 2019. The International Code of Conduct for the Sustainable Use and Management of Fertilizers. FAO. Available from: https://doi.org/10.4060/CA5253EN.

Foley, J.A., Ramankutty, N., Brauman, K.A., Cassidy, E.S., Gerber, J.S., Johnston, M., et al., 2011. Solutions for a cultivated planet. Nature 478, 337–342. Available from: https://doi.org/10.1038/nature10452.

Fraser, E., Legwegoh, A., Kc, K., CoDyre, M., Dias, G., Hazen, S., et al., 2016. Biotechnology or organic? Extensive or intensive? Global or local? A critical review of potential pathways to resolve the global food crisis. Trends Food Sci. Technol. 48, 78–87. Available from: https://doi.org/10.1016/j.tifs.2015.11.006.

Ghafariyan, M.H., Malakouti, M.J., Dadpour, M.R., Stroeve, P., Mahmoudi, M., 2013. Effects of magnetite nanoparticles on soybean chlorophyll. Environ. Sci. Technol. 47, 10645–10652. Available from: https://doi.org/10.1021/es402249b.

Golbashy, M., Sabahi, H., Allahdadi, I., Nazokdast, H., Hosseini, M., 2017. Synthesis of highly intercalated urea-clay nanocomposite via domestic montmorillonite as eco-friendly slow-release fertilizer. Arch. Agron. Soil Sci. 63, 84–95. Available from: https://doi.org/10.1080/03650340.2016.1177175.

González Gómez, H., Ramírez Godina, F., Ortega Ortiz, H., Benavides Mendoza, A., Robledo Torres, V., Cabrera De la Fuente, M., 2017. Use of chitosan-PVA hydrogels with copper nanoparticles to improve the growth of grafted watermelon. Molecules 22, 1031. Available from: https://doi.org/10.3390/molecules22071031.

González-Morales, S., Parera, C.A., Juárez-Maldonado, A., la Fuente, M.C.D., Benavides-Mendoza, A., 2020. Chapter 13 – The ecology of nanomaterials in agroecosystems. In: Husen, A., Jawaid, M. (Eds.), Nanomaterials for Agriculture and Forestry Applications, Micro and Nano Technologies. Elsevier, pp. 313–355. Available from: https://doi.org/10.1016/B978-0-12-817852-2.00013-5.

Hajiboland, R., Sadeghzadeh, N., Bosnic, D., Bosnic, P., Tolrà, R., Poschenrieder, C., et al., 2020. Selenium activates components of iron acquisition machinery in oilseed rape roots. Plant Soil 452 (1-2), 569–586. Available from: https://doi.org/10.1007/s11104-020-04599-w.

Hasaneen, M.N.A., Abdel-Aziz, H.M.M., El-Bialy, D.M.A., Omer, A.M., 2014. Preparation of chitosan nanoparticles for loading with NPK fertilizer. Afr. J. Biotechnol. 13 (31), 3158–3164. Available from: https://doi.org/10.5897/AJB2014.13699.

Hasaneen, M.N.A.-G., Abdel-Aziz, H.M.M., Omer, A.M., 2017. Characterization of carbon nanotubes loaded with nitrogen, phosphorus and potassium fertilizers. Am. J. Nano Res. Appl. 5 (2), 12–18. Available from: https://doi.org/10.11648/j.nano.20170502.11.

Hernández, H.H., Benavides-Mendoza, A., Ortega-Ortiz, H., Hernández-Fuentes, A.D., Juárez-Maldonado, A., 2017. Cu nanoparticles in chitosan-PVA hydrogels as promoters of growth, productivity and fruit quality in tomato. Emir. J. Food Agric. 29 (8), 573–580. Available from: https://doi.org/10.9755/ejfa.2016-08-1127.

Hershey, D.R., Paul, J.L., Carlson, R.M., 1980. Evaluation of potassium-enriched clinoptilolite as a potassium source for potting media. HortScience 15, 87–89.

Jindo, K., Olivares, F.L., Malcher, D.J., da, P., Sánchez-Monedero, M.A., Kempenaar, C., et al., 2020. From lab to field: role of humic substances under open-field and greenhouse conditions as biostimulant and biocontrol agent. Front. Plant Sci. 11, 426. Available from: https://doi.org/10.3389/fpls.2020.00426.

Joseph, S., Graber, E.R., Chia, C., Munroe, P., Donne, S., Thomas, T., et al., 2013. Shifting paradigms: development of high-efficiency biochar fertilizers based on nano-structures and soluble components. Carbon Manage. 4, 323–343. Available from: https://doi.org/10.4155/cmt.13.23.

Juárez-Maldonado, A., González-Morales, S., Cabrera-De la Fuente, M., Medrano-Macías, J., Benavides-Mendoza, A., 2018. Nanometals as promoters of nutraceutical quality in crop plants. In: Grumezescu, A.M., Holban, A.M. (Eds.), Impact of Nanoscience in the Food Industry, Handbook of Food Bioengineering. Academic Press, pp. 277–310. Available from: https://doi.org/10.1016/B978-0-12-811441-4.00010-8.

Juarez-Maldonado, A., Ortega-Ortíz, H., Pérez-Labrada, F., Cadenas-Pliego, G., Benavides-Mendoza, A., 2016. Cu nanoparticles absorbed on chitosan hydrogels positively alter morphological, production, and quality characteristics of tomato. J. Appl. Bot. Food Qual. 89. Available from: https://doi.org/10.5073/JABFQ.2016.089.023.

Juárez-Maldonado, A., Ortega-Ortíz, H., Morales-Díaz, A.B., González-Morales, S., Morelos-Moreno, Á., Cabrera-De la Fuente, M., et al., 2019. Nanoparticles and nanomaterials as plant biostimulants. Int. J. Mol. Sci. 20, 162. Available from: https://doi.org/10.3390/ijms20010162.

Kabir, E., Kumar, V., Kim, K.-H., Yip, A.C.K., Sohn, J.R., 2018. Environmental impacts of nanomaterials. J. Environ. Manage. 225, 261–271. Available from: https://doi.org/10.1016/j.jenvman.2018.07.087.

Kah, M., 2015. Nanopesticides and nanofertilizers: emerging contaminants or opportunities for risk mitigation? Front. Chem. 3, 64. Available from: https://doi.org/10.3389/fchem.2015.00064.

Kah, M., Kookana, R.S., Gogos, A., Bucheli, T.D., 2018. A critical evaluation of nanopesticides and nanofertilizers against their conventional analogues. Nat. Nanotechnol. 13, 677–684. Available from: https://doi.org/10.1038/s41565-018-0131-1.

Kashyap, P.L., Xiang, X., Heiden, P., 2015. Chitosan nanoparticle based delivery systems for sustainable agriculture. Int. J. Biol. Macromol. 77, 36–51. Available from: https://doi.org/10.1016/j.ijbiomac.2015.02.039.

Khati, P., Chaudhary, P., Gangola, S., Bhatt, P., Sharma, A., 2017. Nanochitosan supports growth of *Zea mays* and also maintains soil health following growth. 3 Biotech 7, 81. Available from: https://doi.org/10.1007/s13205-017-0668-y.

Kopittke, P.M., Menzies, N.W., Wang, P., McKenna, B.A., Lombi, E., 2019. Soil and the intensification of agriculture for global food security. Environ. Int. 132, 105078. Available from: https://doi.org/10.1016/j.envint.2019.105078.

Kottegoda, N., Munaweera, I., Madusanka, N., Karunaratne, V., 2011. A green slow-release fertilizer composition based on urea-modified hydroxyapatite nanoparticles encapsulated wood. Curr. Sci. 101, 73–83.

Kottegoda, N., Sandaruwan, C., Priyadarshana, G., Siriwardhana, A., Rathnayake, U.A., Berugoda Arachchige, D.M., et al., 2017. Urea-hydroxyapatite nanohybrids for slow release of nitrogen. ACS Nano 11, 1214–1221. Available from: https://doi.org/10.1021/acsnano.6b07781.

Kranjc, E., Drobne, D., 2019. Nanomaterials in plants: a review of hazard and applications in the agri-food sector. Nanomaterials (Basel) 9, 1094. Available from: https://doi.org/10.3390/nano9081094.

Lai, R.W.S., Yeung, K.W.Y., Yung, M.M.N., Djurišić, A.B., Giesy, J.P., Leung, K.M.Y., 2018. Regulation of engineered nanomaterials: current challenges, insights and future directions. Environ. Sci. Pollut. Res. 25, 3060–3077. Available from: https://doi.org/10.1007/s11356-017-9489-0.

Lal, R., 2015. Restoring soil quality to mitigate soil degradation. Sustainability 7, 5875–5895. Available from: https://doi.org/10.3390/su7055875.

Lehman, S.E., Larsen, S.C., 2014. Zeolite and mesoporous silica nanomaterials: greener syntheses, environmental applications and biological toxicity. Environ. Sci. Nano 1, 200–213. Available from: https://doi.org/10.1039/C4EN00031E.

Liu, Q., Chen, B., Wang, Q., Shi, X., Xiao, Z., Lin, J., et al., 2009. Carbon nanotubes as molecular transporters for walled plant cells. Nano Lett 9, 1007–1010. Available from: https://doi.org/10.1021/nl803083u.

Liu, R., Lal, R., 2015. Synthetic apatite nanoparticles as a phosphorus fertilizer for soybean (*Glycine max*). Sci. Rep. 4, 5686. Available from: https://doi.org/10.1038/srep05686.

López-Vargas, E.R., Ortega-Ortíz, H., Cadenas-Pliego, G., De Alba Romenus, K., Cabrera de la Fuente, M., Benavides-Mendoza, A., et al., 2018. Foliar application of copper nanoparticles increases the fruit quality and the content of bioactive compounds in tomatoes. Appl. Sci. 8 (7), 1020. Available from: https://doi.org/10.3390/app8071020.

Mandal, B., Hazra, G.C., Mandal, L.N., 2000. Soil management influences on zinc desorption for rice and maize nutrition. Soil Sci. Soc. Am. J. 64, 1699–1705. Available from: https://doi.org/10.2136/sssaj2000.6451699x.

Mikhak, A., Sohrabi, A., Kassaee, M.Z., Feizian, M., 2017. Synthetic nanozeolite/nanohydroxyapatite as a phosphorus fertilizer for German chamomile (*Matricariachamomilla* L.). Ind. Crops Prod. 95, 444–452. Available from: https://doi.org/10.1016/j.indcrop.2016.10.054.

Mikkelsen, R.L., 2011. The "4R" nutrient stewardship framework for horticulture. HortTechnology 21, 658–662. Available from: https://doi.org/10.21273/HORTTECH.21.6.658.

Millán, G., Agosto, F., Vázquez, M., 2008. Use of clinoptilolite as a carrier for nitrogen fertilizers in soils of the Pampean regions of Argentina. Int. J. Agric. Nat. Resour. 35, 293–302.

Monreal, C.M., DeRosa, M., Mallubhotla, S.C., Bindraban, P.S., Dimkpa, C., 2016. Nanotechnologies for increasing the crop use efficiency of fertilizer-micronutrients. Biol. Fertil. Soils 52, 423–437. Available from: https://doi.org/10.1007/s00374-015-1073-5.

Monteiro Corrêa, R., Carmo Pinto, S.I., do, Soares, É., Mendes de Carvalho, V.A., 2012. Hydroponic production of fruit tree seedlings in Brazil. In: Asao, T. (Ed.), Hydroponics – A Standard Methodology for Plant Biological Researches. InTech, Rijeka, Croatia, pp. 225–244.

Morales-Díaz, A.B., Ortega-Ortíz, H., Juárez-Maldonado, A., Cadenas-Pliego, G., González-Morales, S., Benavides-Mendoza, A., 2017. Application of nanoelements in plant nutrition and its impact in ecosystems. Adv. Nat. Sci Nanosci. Nanotechnol. 8, 013001. Available from: https://doi.org/10.1088/2043-6254/8/1/013001.

Morgan, K.T., Cushman, K.E., Sato, S., 2009. Release mechanisms for slow- and controlled-release fertilizers and strategies for their use in vegetable production. HortTechnology 19, 10–12. Available from: https://doi.org/10.21273/HORTSCI.19.1.10.

Nakhli, S.A.A., Delkash, M., Bakhshayesh, B.E., Kazemian, H., 2017. Application of zeolites for sustainable agriculture: a review on water and nutrient retention. Water Air Soil Pollut. 228, 464. Available from: https://doi.org/10.1007/s11270-017-3649-1.

Notario del Pino, J.S., Arteaga Padrón, I.J., González Martin, M.M., Garcia Hernández, J.E., 1995. Phosphorus and potassium release from phillipsite-based slow-release fertilizers. J. Control. Release 34, 25–29. Available from: https://doi.org/10.1016/0168-3659(94)00116-C.

Ojeda-Barrios, D.L., Morales, I., Juárez-Maldonado, A., Sandoval-Rangel, A., Fuentes-Lara, L. O., Benavides-Mendoza, A., 2020. Importance of nanofertilizers in fruit nutrition. In: Srivastava, A.K., Hu, C. (Eds.), Fruit Crops. Elsevier, pp. 497–508. Available from: https://doi.org/10.1016/B978-0-12-818732-6.00035-6.

Pan, B., Lam, S.K., Mosier, A., Luo, Y., Chen, D., 2016. Ammonia volatilization from synthetic fertilizers and its mitigation strategies: a global synthesis. Agric. Ecosyst. Environ. 232, 283–289. Available from: https://doi.org/10.1016/j.agee.2016.08.019.

Pilbeam, D.J., 2015. Breeding crops for improved mineral nutrition under climate change conditions. J. Exp. Bot. 66, 3511–3521. Available from: https://doi.org/10.1093/jxb/eru539.

Pinto, J.M., Silva, D.J., Borges, A.L., Coelho, E.F., Feitosa, F.J., 2002. Fertirrigação. In: Genu, P.J.C., Pinto, A.C. (Eds.), A Cultura Da Mangueira. Embrapa Informação Tecnológica, Brasilia, pp. 225–241.

Pradhan, S., Patra, P., Das, S., Chandra, S., Mitra, S., Dey, K.K., et al., 2013. Photochemical modulation of biosafe manganese nanoparticles on vigna radiata: a detailed molecular, biochemical, and biophysical study. Environ. Sci. Technol. 47, 13122–13131. Available from: https://doi.org/10.1021/es402659t.

Raguraj, S., Wijayathunga, W.M.S., Gunaratne, G.P., Amali, R.K.A., Priyadarshana, G., Sandaruwan, C., et al., 2020. Urea–hydroxyapatite nanohybrid as an efficient nutrient source in *Camellia sinensis* (L.) Kuntze (tea). J. Plant Nutr. 43 (15), 2383–2394. Available from: https://doi.org/10.1080/01904167.2020.1771576.

Raliya, R., Saharan, V., Dimkpa, C., Biswas, P., 2018. Nanofertilizer for precision and sustainable agriculture: current state and future perspectives. J. Agric. Food Chem. 66, 6487–6503. Available from: https://doi.org/10.1021/acs.jafc.7b02178.

Ranjbar, S., Rahemi, M., Ramezanian, A., 2018. Comparison of nano-calcium and calcium chloride spray on postharvest quality and cell wall enzymes activity in apple cv. Red Delicious. Sci. Hortic. 240, 57–64. Available from: https://doi.org/10.1016/j.scienta.2018.05.035.

Reddy, P.V.L., Hernandez-Viezcas, J.A., Peralta-Videa, J.R., Gardea-Torresdey, J.L., 2016. Lessons learned: are engineered nanomaterials toxic to terrestrial plants? Sci. Total Environ. 568, 470–479. Available from: https://doi.org/10.1016/j.scitotenv.2016.06.042.

Reddy Pullagurala, V.L., Adisa, I.O., Rawat, S., Kim, B., Barrios, A.C., Medina-Velo, I.A., et al., 2018. Finding the conditions for the beneficial use of ZnO nanoparticles towards plants-a review. Environ. Pollut. 241, 1175–1181. Available from: https://doi.org/10.1016/j.envpol.2018.06.036.

Reddy Pullagurala, V.L., Adisa, I.O., Rawat, S., White, J.C., Zuverza-Mena, N., Hernandez-Viezcas, J.A., et al., 2019. Chapter 4 – Fate of engineered nanomaterials in agroenvironments and impacts on agroecosystems. In: Marmiroli, N., White, J.C., Song, J. (Eds.), Exposure to Engineered Nanomaterials in the Environment, Micro and Nano Technologies. Elsevier, pp. 105–142. Available from: https://doi.org/10.1016/B978-0-12-814835-8.00004-2.

Rohi Vishekaii, Z., Soleimani, A., Ghasemnezhad, M., Hasani, A., 2019. The feasibility for replacement of urea with nitrogen nano-chelated fertilizer in olive (*Olea europaea* L.) orchards. Iran. J. Plant Physiol. 10, 3047–3058. Available from: https://doi.org/10.22034/ijpp.2019.670790.

Rui, M., Ma, C., Hao, Y., Guo, J., Rui, Y., Tang, X., et al., 2016. Iron oxide nanoparticles as a potential iron fertilizer for peanut (*Arachis hypogaea*). Front. Plant Sci. 7, 815. Available from: https://doi.org/10.3389/fpls.2016.00815.

Sabir, A., Yazar, K., Sabir, F., Kara, Z., Yazici, M.A., Goksu, N., 2014. Vine growth, yield, berry quality attributes and leaf nutrient content of grapevines as influenced by seaweed extract (*Ascophyllum nodosum*) and nanosize fertilizer pulverizations. Sci. Hortic. 175, 1–8. Available from: https://doi.org/10.1016/j.scienta.2014.05.021.

Sempeho, S.I., Kim, H.T., Mubofu, E., Hilonga, A., 2014. Meticulous overview on the controlled release fertilizers. Adv. Chem. 2014, . Available from: https://doi.org/10.1155/2014/363071Article ID 363071.

Sharma, G., Kumar, A., Devi, K.A., Prajapati, D., Bhagat, D., Pal, A., et al., 2020. Chitosan nanofertilizer to foster source activity in maize. Int. J. Biol. Macromol. 145, 226–234. Available from: https://doi.org/10.1016/j.ijbiomac.2019.12.155.

Silva, E.M., da, Pinto, A.C., de, Q., Azevedo, J.A., de, 1996. Manejo da Irrigação e Fertirrigação na Cultura da Mangueira. EMBRAPA, Brasilia.

Singh, M.D., Chirag, G., Prakash, P.O., Mohan, M.H., Prakasha, G., Vishwajith, 2017. Nano fertilizers is a new way to increase nutrients use efficiency in crop production. Int. J. Agric. Sci. 9, 3831–3833.

Souri, M.K., 2016. Aminochelate fertilizers: the new approach to the old problem; a review. Open Agric. 1, 118–123. Available from: https://doi.org/10.1515/opag-2016-0016.

Suh, S., Yee, S., 2011. Phosphorus use-efficiency of agriculture and food system in the US. Chemosphere 84, 806–813. Available from: https://doi.org/10.1016/j.chemosphere.2011.01.051.

Taran, N.Y., Gonchar, O.M., Lopatko, K.G., Batsmanova, L.M., Patyka, M.V., Volkogon, M.V., 2014. The effect of colloidal solution of molybdenum nanoparticles on the microbial composition in rhizosphere of *Cicer arietinum* L. Nanoscale Res. Lett. 9, 289. Available from: https://doi.org/10.1186/1556-276X-9-289.

Van Aken, B., 2015. Gene expression changes in plants and microorganisms exposed to nanomaterials. Curr. Opin. Biotechnol. 33, 206–219. Available from: https://doi.org/10.1016/j.copbio.2015.03.005.

Wang, Y., Lin, Y., Xu, Y., Yin, Y., Guo, H., Du, W., 2019. Divergence in response of lettuce (*var. ramosa Hort.*) to copper oxide nanoparticles/microparticles as potential agricultural fertilizer. Environ. Pollut. Bioavail. 31, 80–84. Available from: https://doi.org/10.1080/26395940.2019.1578187.

Wang, Y., Deng, C., Cota-Ruiz, K., Peralta-Videa, J.R., Sun, Y., Rawat, S., et al., 2020. Improvement of nutrient elements and allicin content in green onion (*Allium fistulosum*) plants exposed to CuO nanoparticles. Sci. Total Environ. 725, 138387. Available from: https://doi.org/10.1016/j.scitotenv.2020.138387.

Watts-Williams, S.J., Nguyen, T.D., Kabiri, S., Losic, D., McLaughlin, M.J., 2020. Potential of zinc-loaded graphene oxide and arbuscular mycorrhizal fungi to improve the growth and zinc nutrition of Hordeum vulgare and Medicago truncatula. Appl. Soil Ecol. 150, 103464. Available from: https://doi.org/10.1016/j.apsoil.2019.103464.

Wu, M.-Y., 2013. Effects of incorporation of nano-carbon into slow-released fertilizer on rice yield and nitrogen loss in surface water of paddy soil. In: Proceedings of the 2013 Third International Conference on Intelligent System Design and Engineering Applications, pp. 676–681. https://doi.org/10.1109/ISDEA.2012.161.

Yilmaz, A., Ekiz, H., Torun, B., Gultekin, I., Karanlik, S., Bagci, S.A., et al., 1997. Effect of different zinc application methods on grain yield and zinc concentration in wheat cultivars grown on zinc-deficient calcareous soils. J. Plant Nutr. 20, 461–471. Available from: https://doi.org/10.1080/01904169709365267.

Yusefi-Tanha, E., Fallah, S., Rostamnejadi, A., Pokhrel, L.R., 2020. Zinc oxide nanoparticles (ZnONPs) as a novel nanofertilizer: influence on seed yield and antioxidant defense system in soil grown soybean (*Glycine* max cv. Kowsar). Sci. Total Environ. 738, 140240. Available from: https://doi.org/10.1016/j.scitotenv.2020.140240.

Yuvaraj, M., Subramanian, K.S., 2015. Controlled-release fertilizer of zinc encapsulated by a manganese hollow core shell. Soil Sci. Plant Nutr. 61, 319–326. Available from: https://doi.org/10.1080/00380768.2014.979327.

Yuvaraj, M., Subramanian, K.S., 2018. Development of slow release Zn fertilizer using nano-zeolite as carrier. J. Plant Nutr. 41, 311–320. Available from: https://doi.org/10.1080/01904167.2017.1381729.

Chapter 19

Plant beneficial microbes in mitigating the nutrient cycling for sustainable agriculture and food security

Richa Salwan[1] and Vivek Sharma[2]
[1]*College of Horticulture and Forestry, Dr YS Parmar University of Horticulture and Forestry, Neri, Hamirpur, India,* [2]*University Centre for Research and Development, Chandigarh University, Gharuan, Mohali, India*

19.1 Introduction

The rising population, urbanization, and available incomes are causing a serious threat and raising concerns for sustainable agriculture and food security (El Mujtar et al., 2019). Agriculture is largely dependent on the nutrient availability in soil for efficient production and healthy food crops. Use of inadequate management practices and climatic factors are some of the major concerns leading to the availability and stability of food supply (Gourdji et al., 2015). To meet current demand, farmers are using excessive amounts of external inputs and heavy machinery for more intensified agriculture. The overall production has increased, but serious impacts such as reduction in available natural resources as well as environmental and other food safety issues have also been raised (Zhang et al., 2015). Moreover, soil erosion, decreased organic matter content, nutrient availability, salinization, acidification, and loss of biodiversity have negatively affected the agricultural productivity and nutritional quality (Kihara et al., 2017; Vanlauwe et al., 2015). High doses of agrochemicals have destroyed the native beneficial microbes and affected nutrient cycling, crop yield, and ecosystem sustenance (Naik et al., 2019; Calderon et al., 2017). However, in the last decade, efforts have been made to restore the soil health by adopting natural agricultural practices and ensuring healthy and sustainable crop production (Chappell, 2018; Smith and Haddad, 2015).

Plant Nutrition and Food Security in the Era of Climate Change. DOI: https://doi.org/10.1016/B978-0-12-822916-3.00010-X

Plant growth-promoting microbes are of paramount importance in regulating the flux of processes like degradation of organic matter, integrated nutrient management, and nutrient availability for promoting sustainability in agriculture (Lalitha, 2017). These microorganisms can improve agronomic efficiency by decreasing input cost through reduced usage of chemical fertilizers (De Souza et al., 2015). Microbes including members of fungi, bacteria, actinomycetes, and yeasts have been explored for promoting plant growth through potassium and phosphate solubilization, siderophore production, nitrogen fixation, and exopolysaccharide secretion (Naik et al., 2019). For example, *Azospirillum* spp., *Pseudomonas* spp., and *Rhizobium* spp. are involved in N_2 fixation, and phosphate solubilization ability (De Souza et al., 2015; Lin et al., 2015). Therefore considering these points scientific studies based on biological practices involving the use of beneficial microbes should be undertaken for enhanced production (Vassilev et al., 2015). Bioformulations containing such effective microbes could induce photosynthesis; provide nutrients like P, K, and N, indole-3-acetic acid (IAA)-like growth hormones, and others for improved plant growth; as well as various secondary metabolites to combat the effect of soilborne plant pathogens. In this context, "-omics" technologies including metabolomics, proteomics, and transcriptomics are being used to understand the complex interactions underlying plants and beneficial microbes including plant growth-promoting bacteria (PGPB) and mycorrhiza residing in soil. In this chapter, we have described the role of plant beneficial microorganisms in nutrient recycling for sustainable agriculture and food security.

19.2 Plant beneficial microorganisms

Soil is the promising nutrient center and bioresource reservoir for the production of agricultural crops. Plants live in association with diverse rhizosphere communities, which support ecological processes. Among these diverse communities, microbes are promising candidates in influencing plant as well as soil health and act as ecological barriers to overcome abiotic stress. The interactive behavior of soil microorganisms and plants also plays an important role in regulating plant growth, quality of produce, and safety of food. Association of microbes can be epiphytic, endophytic, and rhizospheric, depending upon the plant species.

Application of plant beneficial microbes is a promising tool to overcome abiotic and biotic stresses and to obtain high crop yields. Such agricultural practices can also help in strengthening ecosystem services like restoration/remediation of degraded soils, biological control of soilborne pests and diseases, and mitigation and adaptation for climatic factors (Kaul et al., 2018; El Mujtar et al., 2019). For the sustainable growth of plants, beneficial microbes have gained considerable importance because of their positive effects on physiological processes of plants (Sharma et al., 2020a). These beneficial microbes promote nutrient acquisition, siderophore production, and nitrogen fixation in the rhizosphere, and modulate plant hormones such as auxin, cytokinin, gibberellin, ethylene, and abscisic acid (ABA), which ultimately improve the crop production (Fig. 19.1). Various bacteria including *Azotobacter*, *Pseudomonas*, and *Rhizobium*, and fungi including mycorrhiza have been reported as plant beneficial microorganisms in absorbing nutrients under stressed conditions (Table 19.1).

In the rhizosphere, microbes live in close association with plant roots by establishing symbiotic interactions, and negative interactions occur between parasitic plants and pathogens. These interactions are determinative for plant health and microbes can be referred as commensal, mutualistic, and pathogenic (Yu et al., 2019). Mutualism occurs when microbes provide nutritional benefits to the plants and in turn, get carbon from the plants (Basu et al., 2018). For example, mutualistic interactions take place between N-fixing rhizobia and legume plants, and arbuscular mycorrhiza fungi to provide N and P to the plants under nutrient-deficit environments. Mycorrhiza including arbuscular mycorrhizal (AM) and vesicular mycorrhizal (VM) fungi are found associated with plants. AM forms extensive hyphal networks near the plant roots to absorb water and minerals, and in turn, provides minerals N, P, and K, and water and protects the plants from stress (Clark and Zeto, 2000; Bernardo et al., 2019). Hence plants living in association with AM could be

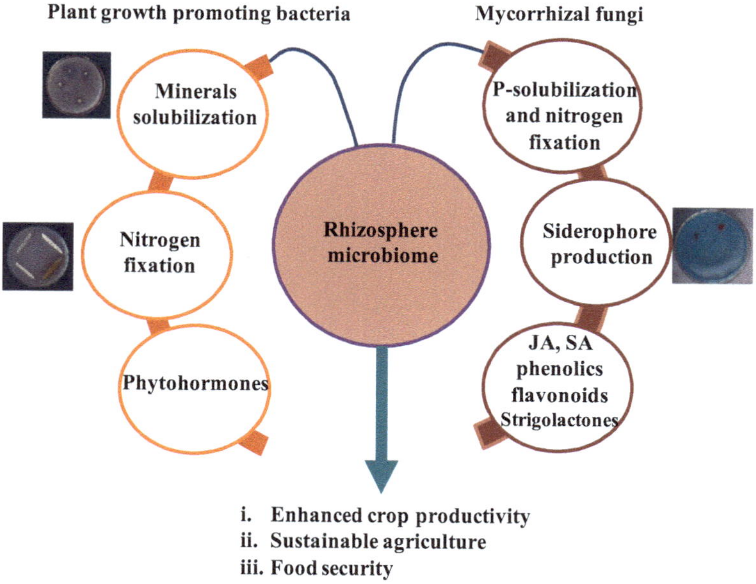

FIGURE 19.1 Beneficial attributes of plant growth-promoting bacteria and mycorrhiza in soil for promoting crop production and protection, with application in sustainable agriculture and food security.

TABLE 19.1 Plant beneficial microbes with respective beneficial attributes and ability of biotic and abiotic stress tolerance.

Bacteria	Beneficial attributes	Host	Stress tolerance	References
Achromobacter xylosoxidans,	Cytokinin	*Helianthus annuus*	Drought	Forchetti et al. (2010)
Acinetobacter sp.	IAA, phosphate solubilization, siderophores			Rokhbakhsh-Zamin et al. (2011)
Agrobacterium radiobacter	Agrocin 84	Stone fruit, rose		Bending (2017)
Azospirillum lipoferum	Gibberellin	*Triticum aestivum*	Drought	Creus et al. (2004)
Bacillus amyloliquefaciens	ABA	*Oryza sativa*	Salt	Shahzad et al. (2017)
Bacillus aryabhattai	IAA, GA, ABA	*Glycine max*	Heat	Park et al. (2017)
Bacillus licheniformis,	ABA	*Vitis vinifera*	Water	Salomon et al. (2014)
Bacillus pumilus	Cytokinin	*H. annuus*	Drought	Forchetti et al. (2010)
Bacillus sp	IAA	*Sulla carnosa*	Salt	Hidri et al. (2016)
Bacillus sp. SBA12	P solubilizer	—	Biotic and abiotic	Salwan and Sharma (2020)

Bacillus subtilis	Picrocrocin, crocetin, and safranal compounds	*Crocus sativus*	Biotic elicitor	Sharaf-Eldin et al. (2008)
Burkholderia cepacia	P solubilizer IAA, GA			Martínez-Gallegos et al. (2018)
Claroideoglomus etunicatum	—	*Aeluropus littoralis*	Salt	Hajiboland et al. (2015)
Curtobacterium flaccumfaciens	IAA	Barley	Salt	Cardinale et al. (2015)
Funneliformis mosseae	—	*Poncirus trifoliata*	Drought	Zhang et al. (2019)
Glomus fasciculatum	Phosphatases	*Zea mays*	Abiotic	Usharani et al. (2014)
Glomus mosseae	—	*T. aestivum*	Drought	Rani (2016)
Hartmannibacter diazotrophicus E19	ACC deaminase	*Hordeum vulgare*	Salinity	Suarez et al. (2015)
Pseudomonas fluorescens	ABA	*V. vinifera* L.,	Water	Salomon et al. (2014)
P. fluorescens	Phenazine-1-carboxylate, 2,4-diacetylphloroglucinol HCN, 2,4-diacetylphloroglucinol, pyoluteorin, pyrrolnitrin, pyoluteorin	Wheat, *Beta vulgaris*, *Nicotiana tabacum*, *Gossypium* spp	Biocontrol	Bending (2017)
Paenibacillus polymyxa E681	Antibiotics/jasmonic acid	*Arabidopsis* and other host	Biotic/ environmental stress	Park et al. (2013)
Piriformospora indica	Acid phosphatases biofertilizer, bioregulator, and bioprotector	Wide-host range	Biotic and abiotic	

(*Continued*)

TABLE 19.1 (Continued)

Bacteria	Beneficial attributes	Host	Stress tolerance	References
Pseudomonas putida UW4	ACC deaminase 19			
Pseudomonas sp.	Phytohormone	*S.* carnosa	Salt	Hidri et al. (2016)
Pseudomonas sp. A3R3 P	IAA, siderophores, heavy metal mobilization		Heavy metal phytoremediation	Ma et al. (2011)
Rhizophagus irregularis	—	Maize	Improve soil productivity	Moreira et al. (2016)
Serratia marcescens	P solubilization	Maize	Salt	Lavania and Nautiyal (2013)
Sphingomonas paucimobilis	P solubilizer IAA	—	—	Martínez-Gallegos et al. (2018)
Streptomyces sp. EN28	Induce defense pathway	*Arabidopsis*	Biotic	Conn et al. (2008)
Streptomyces sp. AC30	P solubilizer	—	Abiotic and biotic	Salwan et al. (2020)
Streptomyces sp. AC40	P solubilizer	—	Abiotic and biotic	Salwan et al. (2020)

used as an important approach for sustainable food production. VM also produces filamentous hyphae and helps the plant in providing more minerals to exchange carbon as an energy source (Peterson et al., 2004).

19.3 Microorganisms in nutrient recycling

Microbes have wide interactions with the ecosystem and are involved in the recycling of carbon, enhancing water-holding capacity of soil, and maintaining carbon–nitrogen ratio. Soil structure, fertility, nutrient availability, and turnover processes are also affected by environmental factors and microbes residing in soil (Dijkstra and Cheng, 2008; Phillips et al., 2006). To assimilate nutrients from soil, it is important to understand the response of soil microbes with regards to environmental factors and agricultural practices.

Microorganisms play important roles in agriculture as they help in nutrient exchange and reduce usage of chemical fertilizers. Progressive farmers supply heavy doses of N, P, and K for enhanced crop yield and to maintain their economy. But the majority of these fertilizers remain unused and run off in water streams, ponds, and rivers. These unused agricultural effluents cause harm to biodiversity, creating environmental as well as health hazards. The introduction of these unutilized nutrients again into the environment disturbs the natural biogeochemical cycles of nutrients and causes eutrophication. Therefore advanced technologies such as increased nutrient use efficiency are required for intensive agriculture, to manage food security, and to overcome environmental impact. It is also necessary to make use of underutilized and nonrenewable resources to manage nutrients with the help of biological sources/indigenous microbes. It is well known that plants are unable to use dinitrogen and organic phosphorus but these are important for enhanced crop production and yield. The microbes thriving in soil have the ability to fix dinitrogen and solubilize complex mineral sources and perform symbiotically with the plants. In order to absorb these minerals, microbes assimilate nitrogen into nitrate (NO_3^-) and ammonium (NH_4^+) by the process of nitrogen fixation and make them available to the plants (Xu et al., 2012). Accordingly, microbes with promising N_2-fixing ability help the plants in revitalizing growth, disease resistance, and maintaining the amount of nitrogen in the soil (Damam et al., 2016).

19.4 Nitrogen fixation

Soil microbes play an important role in C and N recycling by involving N_2 fixation through breakdown of organic matter and immobilization. The flux of nitrogen involves deposition, fixation, adsorption, leaching, volatilization, and immobilization of nitrogenous compounds. Nitrogen is found in organic components like amino and nucleic acids, enzymes, chlorophyll, and proteins. Various microbes such as mycorrhizae produce extracellular enzymes to degrade nitrogenous compounds and thereby affect N turnover in soils (Jones et al., 2005). Microbes like bacteria and fungi can fix atmospheric dinitrogen while other organisms can absorb nitrogen only in the form of ammonium and nitrate ions. The fixation of nitrogen by microbes involves processes: ammonification, nitrification, denitrification, and anaerobic ammonium oxidation, arranged in an order. The organisms are known as ammonifiers, nitrifiers, denitrifiers, and nitrogen fixers for carrying out ammonification, nitrification, denitrification, and nitrogen fixation, respectively. Nitrogen fixation and denitrification are simultaneously carried out by different groups of microorganisms (Kuypers et al., 2018; Stein et al., 2016). There are some microorganisms which are considered as nitrite oxidizers and degrade formate, hydrogen, and sulfide. Recently, reactions involving oxidation of hydroxylamine to nitric oxide and formation of dinitrogen gas and hydrazine have also been reported. Besides this, microorganisms including ammonia-oxidizing archaea, denitrifying eukaryotic foraminifera, and symbiotic heterotrophic nitrogen-fixing cyanobacteria have been identified for carrying oxidation of nitrite and ammonia (Daims et al., 2015; van Kessel et al., 2015; Thompson et al., 2012; Griffin et al., 2007; Risgaard-Petersen et al., 2006; Konneke et al., 2005). Eukaryotes have not been reported for nitrogen fixation but they establish symbiotic relationships with nitrogen-fixing microorganisms. For example, *Candidatus atelocyanobacterium* thalassa, a cyanobacterium, lives in symbiotic association with algae *Braarudosphaera bigelowii* and species of *Rhizobium* living in association with root nodules of legume crops (Burris et al., 1993; Thompson et al., 2012; Martinez-Perez et al., 2016).

Nitrogen-fixing microorganisms contain a variety of enzymes such as ammonia monooxygenase, nitrogenases, hydroxylamine oxidoreductase, hydroxylamine oxidase, nitrite oxidoreductase, nitrate reductase, and nitrite reductase to fix biological nitrogen. The transformation of nitrogen involves fixing of dinitrogen to ammonia, which further assimilates into organic nitrogen as biomass. The organic nitrogen is degraded into ammonia by ammonification, and further oxidation of ammonia leads to nitrate through a process known as nitrification. Finally, nitrate is again converted into dinitrogen by involving denitrification and annamox reaction (Fig. 19.2).

Aerobic bacteria and archaea can oxidize ammonia to hydroxylamine in the presence of ammonia monooxygenase, and, further, ammonia-oxidizing bacteria can also oxidize ammonia to nitrite (Arp et al., 2003). The oxidation of ammonia involves two groups of microorganisms. One group involves ammonia-limited environments that completely oxidize ammonia to nitrate *Candidatus Nitrospira inopinata* and the other group involves oxygen-limited conditions and partially oxidizes ammonia to nitrite (Kuypers et al., 2018). The reaction involving complete oxidation of ammonia to nitrate is called the comammox reaction while partial oxidation of ammonia to nitrite is called the anammox reaction. Methanotrophs such as *Candidatus*

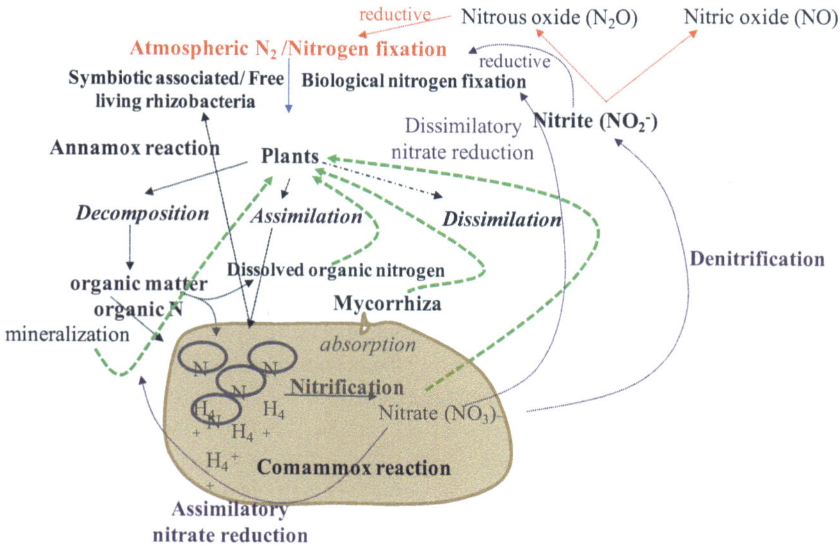

FIGURE 19.2 Role of microorganisms in the dynamics of nitrogen. These organisms can fix atmospheric dinitrogen in the form of ammonium and nitrate ions. The microbes-mediated nitrogen fixation involves ammonification, nitrification, denitrification, and anaerobic ammonium oxidation.

Methylomirabilis oxyfera can also oxidize ammonia to hydroxylamine by involving methane monooxygenase (Stein et al., 2011). Ammonia oxidizers oxidize hydroxylamine to produce nitric oxide in the presence of hydroxylamine oxidoreductase and further oxidize to nitrite in the presence of nitrite reductase (Kuypers et al., 2018). *Nitrospira* spp. which is an aerobic ammonia oxidizer contains both ammonia oxidoreductase and hydroxylamine oxidoreductase (Daims et al., 2015; van Kessel et al., 2015). Nitric oxide serves as a signaling molecule and an intermediate of the anammox, denitrification, and nitrification processes. Nitric oxide is converted into nitrous oxide in presence of nitric oxide reductases, which are found in *Paracoccus denitrificans* and *Pseudomonas stutzeri*. In *Pseudomonas chlororaphis*, nitrous oxide can be the end product of denitrification (Zumft et al., 1997). Similarly, bacteria including *Nitrosospira* sp. and *Nitrososphaera viennensis* degrade urea in the presence of urease, *Nitrososphaera gargensis* degrades cyanate in the presence of cyanase and produces ammonia (Tourna et al., 2011; Palatinszky et al., 2015).

After nitrite formation from ammonia, nitrite oxidoreductase present in aerobic nitrite-oxidizing bacteria converts nitrite to nitrate and is related to energy conservation. Anaerobes including *Thiocapsa* sp. KS1 and *Rhodopseudomonas* sp. LQ17 oxidize nitrite anaerobically by linking carbon fixation to phototrophy. Nitrite oxidoreductase is also related to nitrate reductases in bacteria and archaea. Nitrate reduction is a major source of nitrite for nitrogen-cycling processes like nitrite oxidation and anammox (Tsementzi et al., 2016; Bristow et al., 2017). Nitrate is a major nitrogen source for eukaryotes, bacteria, and archaea that contain assimilatory nitrate reductases together with assimilatory nitrite reductases produce ammonia, which is incorporated into biomass (Moreno-Vivian et al., 1999). Nitrite reduction to ammonium is used for both dissimilatory and assimilatory purposes.

Known as dissimilatory nitrate reduction, the reduction of nitrate to nitrite is catalyzed by nitrate reductase, which may be present in cytoplasm or periplasm. In cytoplasm, nitrate reductase reduces nitrate contributions to energy conservation by releasing protons, whereas periplasmic nitrate reductase reduces nitrate to nitrite and does not contribute to the proton motive force (Moreno-Vivian et al., 1999; Kuypers et al., 2018). Bacteria such as *Paracoccus denitrificans* contain both cytoplasmic and periplasmic nitrate reductase. Dissimilatory nitrate reduction to nitrite is not merely the first step in denitrification. For example, sulfur-oxidizing *Beggiatoa* sp. reduce nitrate to nitrite to ammonium while others can convert nitrate to nitrite only (Preisler et al., 2007). This process is also linked with the degradation of organic matter (Zumft et al., 1997), methane (Ettwig et al., 2016), sulfur compounds (Cardoso et al., 2006), hydrogen, and iron (Weber et al., 2006). Dissimilatory nitrite reduction also leads to the formation of ammonium as reported in case of *Pyrolobus fumarii*, archaea, diatoms, and fungi (Kuypers et al., 2018). Further, degradation of two molecules of nitric oxide leads to the formation of a dinitrogen and an oxygen molecule (Ettwig et al., 2010). Microbial nitrous oxide reduction to nitrogen gas is catalyzed by nitrous oxide reductase (Zumft et al., 2006).

Mycorrhiza is also involved in transferring N to the roots of plants. Hyphae absorb NH_4^+ and NO_3^- which are transported by antiport and symport mechanism, respectively. Sometimes, the availability of carbon becomes deficient for mycorrhiza under increased NH_4^+ and limited NO_3^- concentration. Therefore mycorrhiza makes use of amino acids such as arginine for transfer of nitrogen (Basu et al., 2018). Besides this, other mechanisms include storage of inorganic forms of nitrogen, absorption of N by the urea cycle and asparagine synthase, transfer of stored nitrogen to mycelium, and transfer of NH_4^+ to the plants (Basu et al., 2018). The nitrogen uptake also involves multiple transporters and also helps in recycling organic biomass, which affects the soil microbial community.

19.5 Mineral solubilization by beneficial microbes

Plant beneficial microbes use various minerals as a source of energy and nutrition. Mineral solubilization is accomplished by involving transfer of electrons over the surface of minerals to degrade them. Bacteria have several cytochromes on the surface of their outer membranes to facilitate electron transfer while degrading mineral oxides (Uroz et al., 2009). Such microbes act on complex sources of phosphorus, potassium, nitrogen, iron, and others to facilitate nutrient availability to plants. Various bacteria including *Arthrobacter* sp. (Zarjani et al., 2013), *Burkholderia*, (Uroz et al., 2009), *Paenibacillus glucanolyticus* (Sangeeth et al., 2012), *Paenibacillus mucilaginosus* (Liu et al., 2012), *Penicillium frequentans*, *Pseudomonas*, and *Sphingomonas* can solubilize both K and P. By utilizing promising metal solubilizers, the agricultural productivity that has been degraded can be again restored (Basak and Biswas, 2009). These microbes can be utilized for sustainable agriculture to enhance crop productivity as well as food security.

Various urease-producing bacteria (Helmi et al., 2016), nitrate-reducing bacteria (Hu et al., 2019), sulfate-reducing bacteria (Braissant et al., 2007), Fe-reducing bacteria (Liu et al., 2017), Fe-oxidizing bacteria (Emerson et al., 2010), and phosphate-solubilizing bacteria (Salwan et al., 2020) that play important roles in biogeochemical cycling of N, S, Fe, and P have been reported for inducing mineralization (Han et al., 2020). Mycorrhiza also accomplishes nutrient transfer by forming arbuscules and in turn, plants release strigolactones to establish symbiotic association by involving novel transporters or receptor kinases (Basu et al., 2018).

19.6 Phosphorus solubilization

Phosphorus is a vital element in various metabolic processes like biosynthesis of macromolecules, energy transfer processes, photosynthesis, respiration, and signal transduction. The major portion of P is biologically unavailable as it is available in immobilized, insoluble, and precipitated forms and hence is difficult to be absorbed by the plants. The supply of phosphorus for plant growth at critical stages is essential for optimal production (Lukowiak et al., 2016). Therefore involvement of phosphorus-solubilizing bacteria (PSB) is beneficial for plants as these produce extracellular phosphatases and organic acids to solubilize phosphorus. These PSB also direct various soil processes by facilitating decomposition, nutrient mobilization and mineralization, and release of nutrients in the P cycle. Phosphate-solubilizing bacteria *Acinetobacter*, *Bacillus*, *Burkholderia*, *Pseudomonas*, *Rhizobium*, *Serratia*, *Sphingomonas*, and *Streptomyces* have been explored for improved plant growth and production (Table 19.1). A nematofungus *Arthrobotrys oligospora* also has the ability to solubilize rock phosphate (Cordero et al., 2012).

In order to withstand P limitation, plants develop strategies such as increased root hairs, extensive root branching, and production of phosphatase and organic acids. Plants also develop mutual association with AM surrounding plant roots as an absorption strategy to enhance Pi and nutrient acquisition. This association is an evolutionary successful strategy as $>80\%$ plants live symbiotically with AM. Such plants can use two different pathways

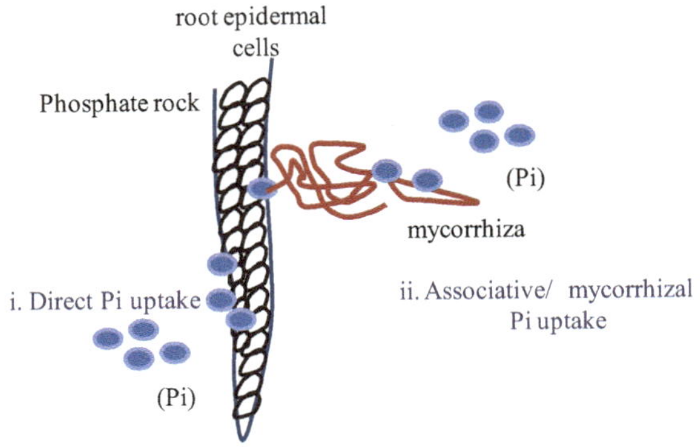

FIGURE 19.3 Mechanism of phosphate transport: (1) Direct pathway involves direct absorption of Pi from the rhizosphere through plant Pi transporters in root hairs and (2) Second pathway involves associative mycorrhizal Pi uptake and in return gets carbohydrates from the host plant. Here, Pi is first transferred into fungal Pi transporters located away from the root surface and then subsequently translocated to arbuscules in root cortical cells. *Adapted from Smith, S.E., Jakobsen, I., Grønlund, M., Smith, F.A., 2011. Roles of arbuscular mycorrhizas in plant phosphorus nutrition: interactions between pathways of phosphorus uptake in arbuscular mycorrhizal roots have important implications for understanding and manipulating plant phosphorus acquisition. Plant Physiol. 156, 1050–1057.*

including P uptake by root epidermal cells and root hairs, and another through the AM fungal pathway (Fig. 19.3). P uptake by symbiotic interaction involves the uptake and translocation of P from the soil to the root and subsequent transfer to the plants. Both plants and fungi uptake negatively charged Pi ions, which causes thermodynamic issues as the cell membrane carries negative electric potential and higher concentration inside the cell than the soil. Therefore Pi transport across the plasma membrane requires metabolic energy and involves high-affinity Pi transporters and P-type H^+-ATPase. Mycorrhiza-specific plant phosphate transporters are required for P transfer as well as for maintaining symbiosis. Plants living in association with mycorrhiza are able to survive in low Pi concentration, maintaining nutrient homeostasis by enhancing altered nutrient use efficiency (Ning et al., 2001).

19.7 Potassium solubilization

Potassium (K) is one of the major macronutrients and plays an important role in plant growth promotion. Potassium deficiency causes chlorosis, leaf falling, slow growth rate, poor root development, reduced production of seeds, and reduced yield in plants. Therefore it is necessary to apply an alternate potassium source such as bioformulations containing potassium solubilizers for improved plant growth and sustainability in agricultural crops (Prasad et al., 2019). Potassium mainly exists in three forms including soil minerals, nonexchangeable, and available form. About 90%–98% of the total K exists as insoluble rock and silicate minerals, micas, or feldspars in the rooting zone, which is relatively unavailable for plant uptake (Scheffer, 2002). The nonexchangeable form of K constitutes 10% of total K and exists as reserve to manage loss from the soil. Only 1%–2% of total K is available and found either in the solution or as part of the exchangeable cation on clay mineral. Further constraints such as imbalanced fertilizers, intensive cropping, soil erosion, leaching, and introduction of hybrids and other high-yielding crop varieties also cause difficulties to the plants. Different species of bacteria *Bacillus subtilis*, *Bacillus mucilaginosus*, *Bacillus edaphicus*, *Bacillus circulans*, *Burkholderia* spp, *Paenibacillus* spp., and *Pseudomonas* spp. are able to solubilize potassium by producing enzymes and organic acids (Saha et al., 2016; Liu et al., 2012; Hafeez and Hassan, 2011; Basak and Biswas, 2009). Other microbes including fungi, arbuscular mycorrhiza, and yeast also solubilize complex K sources including illite, micas, and orthoclase into soluble forms with the production of organic acid (Zeng et al., 2012). Members of fungi including *Aspergillus terreus*, *Aspergillus niger*, *Glomas mosseae*, *Glomas intraradices*, and *Penicillium* sp. have been reported promising in solubilizing complex sources of K (Rajawat et al., 2016; Meena et al., 2015; Lian et al., 2002).

These microbes adopt different approaches including direct and indirect mechanism, polysaccharide secretion, and biofilm formation on mineral surfaces to solubilize complex forms of K into simple ions. A direct method of solubilization is accomplished by the production of organic acids like oxalic, tartaric and citric acids, acidolysis, and enhanced solubility of minerals in the rhizosphere and chemical weathering based on carbonic acid (Mendes et al., 2013; Gerke, 1992; Park et al., 2009; Gadd, 2007). An indirect method of K solubilization includes chelation of the cations, followed by exchange and solubilization on mineral surfaces, formation of metal-ligand complexes, and release of plant hormones (Uroz et al., 2009; Sattar et al., 2019).

In addition to direct and indirect methods, beneficial microbes also adopt K solubilization by releasing exopolysaccharides (EPS), which helps in adhering of microbes over the surface of minerals to enhance the production of organic acids (Liu et al., 2012). These EPS are biodegradable, high-molecular-weight polymers made up of monosaccharides, and play an important role in aggregating soil particles, maintaining water potential, ensuring strict contact with bacteria and roots of plant, and protecting against phytopathogens (Pawar et al., 2016). EPS also adsorb organic acids, maintain the equilibrium between soil and minerals, and enhance dissolution and release of K^+ (Lian et al., 2002). Bacteria including *Bacillus*, *Clostridium*, and *Thiobacillus* secrete capsules made of polysaccharides for degrading feldspar and illite to release K^+ (Sheng and He, 2006).

Another important approach to solubilize K is the formation of biofilms. Bacteria produce biofilms to adhere to the surfaces of minerals and release various metabolites and organic acids, which lower the pH and help in solubilization of complex minerals and facilitate uptake by plants. The biofilms also help to protect, to adapt, and to survive in extremities of the environment by extracting nutrients through the release of extracellular polymers, polysaccharides exudates, and enzymes and mobilization and weathering of complex minerals.

19.8 Siderophore production

Iron plays an important role as cofactor in various metabolic processes and is considered as a vital element for living organisms. It is abundant on earth and forms insoluble complexes in soils of neutral to basic pH. Both microbes and plants need an appropriate amount of iron for their survival but soluble forms of iron are limited. In this regard, microorganisms use different approaches to overcome the availability of low iron in the environment. Among these approaches, microbes produce iron-chelating agents, popularly known as siderophores, to scavenge Fe, forming complexes with metals including divalents, trivalents, and actinides and making them easily available to microbes (Braud et al., 2009). Siderophores also bind tightly to available Fe^{3+} in the rhizosphere and prevent phytopathogens by creating iron-limiting conditions (Shen et al., 2013). Sometimes these microbes also acquire iron from heterologous siderophores produced by other inhabiting microorganisms in soil. Three classes of siderophores have been characterized: catecholates, hydroxamates, and carboxylate (Salwan and Sharma, 2019; Sharma et al., 2020b). Hydroxamate siderophores form Fe(III)-EDTA complex (10_{30}), whereas catecholates and carboxylates form Fe(III)-EDDHA (10_{40}) (Robert and Chenu, 1992). Approximately 270 siderphores have been structurally characterized among all known siderophores (Boukhalfa et al., 2003). Bacterial siderophores including enterobactin, rhizobactin, and ferrioxamine B and pyoverdine possess a mixture of functional groups (Ahmed and Holmström, 2014; Sharma et al., 2020b). *Pseudomonas* is well known for siderophores and can produce >50 pyoverdine siderophores (Boukhalfa and Crumbliss, 2002). The structure of fluorescent pyoverdine is variable and also used to determine relatedness among species (Chincholkar et al., 2005). Studies revealed affiliation of fluorescent and nonfluorescent species into different taxa and well-defined species based on the type of siderophore (Meyer et al., 2002). Further, siderotyping of fluorescent *Pseudomonas* strains revealed prevalence of pyoverdines that can be employed as a potential strategy for diversity and taxonomic studies of microbes (Meyer, 2010). Siderophores have been reported as chemotaxonomic markers depending upon ornibactins in the identification of *Burkholderia* spp. and mycobactin in *Mycobacterium* spp. (Bultreys et al., 2006; Mokracka et al., 2004). Siderophores produced in species of actinobacteria including *Arthrobacter* and *Streptomyces* revealed higher dissolution of Fe compared to synthetic siderophores (desferrioxamine B) (Kalinowski et al., 2000). The fate of iron is variable among gram-positive and gram-negative bacteria. In gram-negative bacteria, outer membrane receptors sense the binding of Fe(III)−siderophore complexes to the receptor at the cell surface and then moved to the periplasmic space by involving outer membrane receptor proteins, periplasmic binding proteins, and inner membrane transport proteins. The complex reaches the cytoplasm by passing through the cytoplasmic membrane with the help of an ATP-binding cassette (ABC) transport. At the end, the complex releases Fe(II) by reducing Fe(III). But in gram-positive bacteria, the outer membrane and its receptors are absent. Therefore Fe(III)−siderophore complexes bind to the membrane proteins and are transported to the cytoplasm by involving ABC transporters similar to gram-negative bacteria (Fukushima et al., 2013; Braun and Hantke, 2011).

Fungi are generally known to produce hydroxamate-type siderophores like ferrichrome and these are subdivided further based on the side chain (Saha et al., 2016; Glick, 2015; Winkelmann, 2007; Renshaw et al., 2002). *Suillus granulatus* produce hydroxamate siderophores to solubilize goethite formed of iron(III) oxide-hydroxide (Ahmed and Holmström, 2014). Further, siderphores of the ferrichrome family have also been reported to alter the structure of biotite by dissolution (Sokolova et al., 2010). The plants living in association with mycorrhizal fungi have the capacity to take up more Fe and hence play important roles in nutrition (Van Schöll et al., 2008). Fungi use shuttle system, taxicab, hydrolytic and reductive mechanisms to facilitate the flux of iron. The shuttle system involves transport of ferrichrome in fungi like *Ustilago maydis*. The Fe(III)−siderophore complex first moves across the cell membrane, followed by release of Fe(III) and reduction and recycling of free siderophores. The taxicab mechanism involves transport of Fe (III) across the cell membrane in *Rhodotorula* species (Ahmed and Holmström, 2014). Further, the hydrolytic mechanism involves first release of Fe(III) from Fe(III)−siderophore complex and reduction of Fe(III) to Fe (II) in the cell. The reductive mechanism involves reduction of Fe(III) to Fe (II) at the cell membrane and reduced Fe is taken up by the cell without the involvement of the cell membrane. The hydrolytic and reductive mechanisms have been reported in *Mycelia sterilia* and *Ustilago sphaerogena*, respectively (Ahmed and Holmström, 2014). Plants also contain siderophores of hydroxy- and amino-substituted iminocarboxylic acids including mugineic acid and avenic acid (Olanrewaju et al., 2017; Glick, 2015). Some phytosiderophores have the ability to dissolute ferrihydrate and goethite.

19.9 Zinc solubilization

Zinc is one of the vital micronutrients required for promoting growth and reproduction in plants. The deficiency of Zn leads to nutritional disorders causing reduced synthesis of auxins, chlorophyll, cytochromes, and nucleotides, which makes plants susceptible to high temperature (Cakmak, 2008). Fertilizers including organic and inorganic composition are being employed in agriculture to overcome the deficiency of Zn in plants (Cakmak, 2008). Biofertilizers including Zn-solubilizing microbes also serve as attractive agents in providing Zn to the plants. These microbes solubilize complexes ZnO and $ZnCO_3$ and other sources of Zn in soil and enhance uptake, which leads to grain fortification. Various studies have reported species of *Acinetobacter*, *Bacillus*, *Gluconacetobacter*, and *Pseudomonas* with potential for solubilizing Zn in aqueous conditions (Saravanan et al., 2004; Fasim et al., 2002; Whiting et al., 2001; Di Simine et al., 1998). Bacteria *Thiobacillus thioxidans*, *Thiobacillus ferroxidans*, and facultative thermophilic iron oxidizers also have potential to solubilize Zn from sulfide ore sphalerite. Besides Zn solubilizers, Zn mobilizers have also been reported for enhancing uptake and thus increasing yield in agricultural crops (Hafeez et al., 2013). Zn mobilizers also follow P solubilizers and Fe mobilizers by producing chelating agents and organic acids to enhance soil nutritional value, crop production, and yield.

19.10 Phytohormone production

Plant hormones are the substances or factors found in cells, tissues, intracellular spaces, and even vascular bundles. These hormones are capable of influencing physiological, biochemical, and molecular responses to enhance tolerance against abiotic and biotic stress (Vessey, 2003). These are capable of promoting growth by influencing cell division, cell expansion, shoot and root elongation, development of lateral roots, photosynthesis, and stress tolerance by improving nitrogen use efficiency. Based on their biochemical action, these plant hormones are divided into auxins for cell elongation, ABA for seed dormancy, cytokinins for cell division, ethylene for fruit ripening, gibberellins for shoot growth, and jasmonic acids for pathogen resistance (Rademacher, 2015). It is reported that exogenous augmentation of phytohormone is also suitable for improving plant growth and development under abiotic stress. Accordingly, the beneficial microbes enhance plant growth and stress tolerance by either producing or regulating phytohormones including cytokinins, ethylene, gibberellins, and IAA (Sharma and Salwan, 2018). The AM fungi also produce compounds similar to plant hormones and are involved in plant growth.

Among these phytohormones, auxins such as IAA, indole-3-butyric acid, and indole-3-propionic acid are capable of regulating growth and development, improving plant breeding, and increasing grain production (Zhao, 2010). Likewise, the production of IAA, an active auxin, regulates cell division, differentiation, and elongation by enhancing osmotic pressure, cell permeability, reduced wall pressure, and enhanced synthesis of cell walls. IAA activates the radicular system to elongate roots and develop lateral roots for enhanced uptake. Auxins are also capable of promoting leaf development and played important roles in gravitropism and phototropism. IAA is also produced by various species of genus *Arthrobacter Enterobacter, Frankia, Kitasatospora, Nocardia, Pseudomonas, Rhizobium, Stenotrophomonas*, and *Streptomyces* (Kumar et al., 2019; Piccoli et al., 2011; Khan and Doty, 2009; Lawongsa et al., 2008; Péret et al., 2008; Shrivastava et al., 2008; da Silva Sousa et al., 2008; Yanni et al., 2001; Manulis et al., 1994).

Besides auxins, gibberellins are also produced in the nodules of plants and modified by various microorganisms (Numan et al., 2018). Gibberellins are capable of inducing elongation in shoots, germination of seeds, as well as maturation of flowers and fruits. Gibberellins (GAs) also has the ability to break seed dormancy and induce germination as well as hyperelongation phenotype in stems. This helps the plants in adapting to climatic conditions and protects the plants from high temperatures and early sprouting. Gibberellic acid, belonging to class E of phytohormones, positively influences plant growth and regulates off-seasonal processes through reserve mobilization, leaf and stem elongation, and promotion of flowering. Various gibberellins GA1, GA19, GA20, and GA44 produced by bacterial strains have shown enhanced growth among plants (Dobert et al., 1992). Different strains of *Rhizobium* are known to produce auxin and gibberellin (GA_7). In similar cases, production of gibberellins from *Azospirillum* sp. and *Bacillus* sp. has been related to enhanced uptake of N in wheat as well as maize roots (Fulchieri et al., 1993; Numan et al., 2018). Another phytohormone such as cytokinin is involved in differentiation of roots and shoots and helps the plant to maintain totipotency of cells. Plant beneficial microbes also produce cytokinins for promoting plant growth. Another hormone, ABA helps the plants in combatting environmental stress such as drought and has the ability to reverse the effect of cytokinins and gibberellins by blocking seed germination and enhancing formation of storage proteins. Similarly in other studies, bacteria *Acetobacter* sp., *Achromobacter xylosoxidans, Arthrobacter koreensis, Azospirillum Halomonas desiderata, Bacillus megaterium, Bacillus cereus, B. subtilis, Bacillus licheniformis, Escherichia coli, Proteus mirabilis, Pseudomonas fluorescens* G20-18 have been reported for ABA, cytokinins, gibberellins, IAA, and jasmonic acid (Großkinsky et al., 2016; Salomon et al., 2014; Piccoli et al., 2011; Naz et al., 2009; Forchetti et al., 2007; Karadeniz et al., 2006; Bottini et al., 2004; García de Salamone and Ocampo, 2002; Gutiérrez-Mañero et al., 2001).

Mycorrhizal association with plants causes immense changes in the signaling mechanism. The AM fungi also produce compounds like strigolactones, flavonoids such as formononetin, salicylic acid, jasmonic acid, and phenolic compounds *p*-coumaric acid, *p*-hydroxybenzoic acid, or quercetin help in establishing symbiotic association by fungal colonization and hyphal branching through increased sporulation (Basu et al., 2018).

19.11 Application of plant beneficial microorganisms in agriculture and food security

Agriculture is the biggest economic and financial source in developing countries (Gindling and Newhouse, 2012). Therefore an approach toward sustainable agriculture for improved crop production and protection is necessary. On the other hand, due to rising population, the demands for food production are increasing and estimated to be doubled by 2050 (Prieto et al., 2017). So it is predicted that major portions of food production must be accomplished by utilizing organic nutrients like crop residues and green manures, composted biomass, humic and fulvic acid-containing preparations, vermicasts, and seaweed emulsions and extracts (Cavagnaro, 2015). Modern agriculture relies upon biologically regulated nutrient supply systems to drive nutrients by transforming organic nutrients available in soil. Plant beneficial microbes are preferred as biofertilizers for maintaining sustainable growth as well as protection of plants in an ecofriendly manner. These microbes living in association with the plant roots use direct and indirect processes to manage nutrients and maintain soil fertility (Lalitha, 2017; Philippot et al., 2013). Due to the high diversity of plant beneficial microbes and their practical application in agricultural systems, research on such microbes, particularly mycorrhizal fungi and PGPRs, has been accumulating progressively. These microbes are preferred not only for crop production but also to maintain the quality and yield of the products. Mycorrhiza also promotes belowground functions by extending hyphae and offering nutritional as well as nonnutritional benefits (Moreno-Espandola et al., 2007). The extraradical hyphal networks formed by mycorrhiza act as a channel for exchanging nutrients between the environment and roots and taking carbon from the plants. Mycorrhiza is involved in various ecological processes like carbon sequestration and reducing emissions of greenhouse gases. Mycorrhizal association can act as an integral link in global carbon cycling as it alters the carbon metabolism of the plant by supplying carbon to the roots. Mycorrhiza also produces various exudates including amino acids, organic acids, sugars, polysaccharides, and glomalin, which are assimilated by the soil microbial biomass, thereby improving soil structure and biological heterogeneity by creating hotspots of microbial activity (Rillig, 2004; Zhu et al., 2003). Mycorrhiza also acts as a promising fertilizer as it maintains nutrient use efficiency by decreasing the leaching process and enhancing plant nutrient uptake, maintaining soil structure and fertility, and encouraging microbe-mediated immobilization of biomass (Cavagnaro et al., 2015). Further, mycorrhiza is also responsible for stabilizing ecological efficiency and in achieving nutrition and food security as the mycelium extends deep in the soil, enhancing mineralization and immobilization of nutrients and thus maintaining plant growth even in stressed conditions. Hence as an applied fertilizer, mycorrhiza can be an effective and sustainable part of efficient agricultural practices to harness the nutrient supply and reduce production costs.

The association of AM also promotes biological diversity by establishing further interactions with plant growth-promoting rhizobacteria (PGPR) (Rooney et al., 2009). Additionally, bacterial species including *P. fluorescens* SBW25 and *Paenibacillus brailensis* PB177 use exudates of AM as substrates and some microbes attach themselves to dead hyphae for nutrition (Toljander et al., 2006). Some bacteria have been known as mycorrhiza helper bacteria, which can enhance colonization of mycorrhiza and suppression of pathogens (Bending et al., 2006). Similarly, PGPB have also gained importance for sustainable agriculture as they are involved in promoting plant growth by establishing various mechanisms. These microbes also secrete lytic enzymes, growth factors such as vitamins, and antimicrobial substances in order to enhance nutrient use efficiency and stress tolerance for improved growth of plants and yield (Liu et al., 2016). These beneficial microbes also release reactive oxygen species, which activates the secretion of hormones, ABA, ethylene, jasmonic acid, and salicylic acid and elicits various pathways to combat stress (Egamberdieva et al., 2017). Various plant beneficial microbes including *Arthrobacter*, *Azotobacter*, *Azospirillum*, *Bacillus*, *Microbacterium*, *Paenibacillus*, *Pseudomonas*, and *Streptomyces* have been reported for enhancing plant growth, soil fertility, and nutrient uptake, as well as resistance toward stress factors (Numan et al., 2018; Paul and Lade, 2014; Vurukonda et al., 2016). Some studies have reported a combination of rhizobia and *Pseudomonas* spp. for improving negative effects of salt stress in *Galega officinalis* (Salwan et al., 2019). Therefore PGPRs play an important role in agriculture as they are not only gaining attention for enhanced plant growth, soil improvement, and stress but also preferred to manage soilborne plant diseases (Li et al., 2020; Lally et al., 2017; Brooke et al., 2017; Turan et al., 2011).

19.12 Conclusion and future prospects

Sustainable agriculture is important for improved health and economic benefits as well as environmental protection and to meet food requirements for the growing population. The present agriculture system is facing problems because of insufficient production, reduced economics, and inappropriate use of agrochemicals. Therefore the plant beneficial microbes are of considerable importance to improve crop production and protection as well as ecofriendly natures. Farmers should be trained for using biological agents and biofertilizers to manage nutrients for improved plant growth. As mentioned, a variety of microbes are known for their biological properties but commercialization of biofertilizers is in the infancy stage and needs further studies. Research is needed to explore management of nutrient supplements, viability, colonization, interaction mechanisms, and soil fertility by the biological fertilizers in the field for future sustainable agricultural production. Mycorrhiza is among the most important keystones in maintaining crop productivity and crop security to belowground functioning. However, lack of information regarding the underlying functional relationships between plants, mycorrhizas, pests, and microorganisms represents a major challenge in the attempt to achieve sustainability. It is suggested that management of these beneficial organisms is important to exploit or recycle soil inorganic nutrients to overcome the depletion of finite nutrient resources. Integration of mycorrhizal systems with other carbon sequestration management practices could also be an important future practice. Modern technologies including genomics, proteomics, and metabolomics can help in finding underexplored genes, proteins, and enzymes as well as metabolites that can also be employed as potential candidates to manage mineral availability in plants.

Acknowledgments

The authors are thankful to SEED Division, DST for providing financial assistance under the projects SP/YO/125/2017 and SEED-TIASN-023–2018. RS is also thankful to College of Horticulture and Forestry for providing infrastructural facilities to complete this work.

References

Ahmed, E., Holmström, S.J.M., 2014. Siderophores in environmental research: roles and applications. Microb. Biotechnol. 7, 196–208. Available from: https://doi.org/10.1111/1751-7915.12117.

Arp, D.J., Stein, L.Y., 2003. Metabolism of inorganic N compounds by ammonia-oxidizing bacteria. Crit. Rev. Biochem. Mol. Biol. 38, 471–495.

Basak, B.B., Biswas, B.D.R., 2009. Influence of potassium solubilizing microorganism (*Bacillus mucilaginosus*) and waste mica on potassium uptake dynamics by sudan grass (*Sorghum vulgare* Pers.) grown under two Alfisols. Plant Soil 317, 235–255.

Basu, S., Rabara, R.C., Negi, S., 2018. AMF: the future prospect for sustainable agriculture. Physiol. Mol. Plant Pathol. 102, 36–45. Available from: https://doi.org/10.1016/j.pmpp.2017.11.007.

Bending, G.D., 2017. The rhizopshere and its microorganisms, second ed.Encyclopedia of Applied Plant Sciences, Vol. 3. pp. 347–351. Available from: https://doi.org/10.1016/B978-0-12-394807-6.00165-9.

Bending, G.D., et al., 2006. Significance of microbial interactions in the mycorrhizosphere. Adv. Appl. Microbiol. 60, 97–132.

Bernardo, L., Carletti, P., Badeck, F.W., Rizza, F., Morcia, C., Ghizzoni, R., et al., 2019. Metabolomic responses triggered by arbuscular mycorrhiza enhance tolerance to water stress in wheat cultivars. Plant Physiol. Biochem. 137, 203–212. Available from: https://doi.org/10.1016/j.plaphy.2019.02.007.

Bottini, R., Cassán, F., Piccoli, P., 2004. Gibberellin production by bacteria and its involvement in plant growth promotion and yield increase. Appl. Microbiol. Biotechnol. 65, 497–503. Available from: https://doi.org/10.1007/s00253-004-1696-1.

Boukhalfa, H., Crumbliss, A.L., 2002. Chemical aspects of siderophores mediated iron transport. Biometals 15, 325–339.

Boukhalfa, H., Lack, J., Reilly, S.D., Hersman, L., Neu, M.P., 2003. Siderophore production and facilitated uptake of iron and plutonium in *P. putida*. AIP Conf. Proc. 673, 343–344.

Braissant, O., Decho, A.W., Dupraz, C., Glunk, C., Przekop, K.M., Visscher, P.T., 2007. Exopolymeric substances of sulfate-reducing bacteria: interactions with calcium at alkaline pH and implication for formation of carbonate minerals. Geobiology 5, 401–411. Available from: https://doi.org/10.1111/j.1472-4669.2007.00117.x.

Braud, A., Hoegy, F., Jezequel, K., Lebeau, T., Schalk, I.J., 2009. New insights into the metal specificity of the *Pseudomonas aeruginosa* pyoverdine–iron uptake pathway. Environ. Microbiol. 11, 1079–1091.

Braun, V., Hantke, K., 2011. Recent insights into iron import by bacteria. Curr. Opin. Chem. Biol. 15, 328–334.

Bristow, L.A., et al., 2017. N_2 production rates limited by nitrite availability in the Bay of Bengal oxygen minimum zone. Nat. Geosci. 10, 24–29.

Brooke, J.S., Di Bonaventura, G., Berg, G., Martinez, J.L., 2017. Editorial: a multi- disciplinary look at *Stenotrophomonas maltophilia*: an emerging multi-drug-resistant global opportunistic pathogen. Front. Microbiol. 8, 1511. Available from: https://doi.org/10.3389/fmicb.2017.01511.

Bultreys, A., Gheysen, I., de Hoffmann, E., 2006. Yersiniabactin production by *Pseudomonas syringae* and *Escherichia* coli, and description of a second yersiniabactin locus evolutionary group. Appl. Environ. Microbiol. 72, 3814–3825.

Burris, R.H., Roberts, G., 1993. Biological nitrogen fixation. Annu. Rev. Nutr. 13, 317–335.

Cakmak, I., 2008. Enrichment of cereal grains with zinc: agronomic or genetic biofortification? Plant Soil 302, 1–17.

Calderon, K., Spor, A., Breuil, M.C., Bru, D., Bizouard, F., Violle, C., et al., 2017. Effectiveness of ecological rescue for altered soil microbial communities and functions. ISME J. 11, 272–283. Available from: https://doi.org/10.1038/ismej.2016.86.

Cardinale, M., Ratering, S., Suarez, C., Zapata Montoya, A.M., Geissler-Plaum, R., Schnell, S., 2015. Paradox of plant growth promotion potential of rhizobacteria and their actual promotion effect on growth of barley (*Hordeum vulgare* L.) under salt stress. Microbiol. Res. 181, 22–32.

Cardoso, R.B., et al., 2006. Sulfide oxidation under chemolithoautotrophic denitrifying conditions. Biotechnol. Bioeng. 95, 1148–1157.

Cavagnaro, T.R., 2015. Biologically regulated nutrient supply systems. Compost and arbuscular mycorrhizas – a review. Adv. Agron. 129, 293–321. Available from: https://doi.org/10.1016/bs.agron.2014.09.005.

Chappell, M., 2018. Beginning to End Hunger Food and the Environment in Horizonte, Brazil, and Beyond. University of California Press, Berkeley, CA.

Chincholkar, S.B., Rane, M.R., Chaudhari, B.L., 2005. Siderophores: their biotechnological applications. In: Varma, A., Podila, G.K. (Eds.), Biotechnological Application of Microbes. IK International, New Delhi, pp. 177–198.

Clark, R.B., Zeto, S.K., 2000. Mineral acquisition by arbuscular mycorrhizal plants. J. Plant Nutr. 23, 867–902.

Conn, V.M., Walker, A.R., Franco, C.M., 2008. Endophytic actinobacteria induce defense pathways in *Arabidopsis thaliana*. Mol. Plant Microbe Interact. 21, 208–218.

Cordero, O.X., Ventouras, L.A., DeLong, E.F., Polz, M.F., 2012. Public good dynamics drive evolution of iron acquisition strategies in natural bacterioplankton populations. Proc. Natl. Acad. Sci. USA 109, 20059–20064. Available from: https://doi.org/10.1073/pnas.1213344109.

Creus, C.M., Sueldo, R.J., Barassi, C.A., 2004. Water relations and yield in *Azospirillum* inoculated wheat exposed to drought in the field. Can. J. Bot. 82, 273–281.

Daims, H., et al., 2015. Complete nitrification by *Nitrospira* bacteria. Nature 528, 504–509.

Damam, M., Kaloori, K., Gaddam, B., Kausar, R., 2016. Plant growth promoting substances (phytohormones) produced by rhizobacterial strains isolated from the rhizosphere of medicinal plants. Int. J. Pharm. Sci. Rev. 37, 130–136.

da Silva Sousa, C., Cristina Fermino Soares, A., da Silva Garrido, M., 2008. Characterization of *Streptomycetes* with potential to promote plant growth and biocontrol. Sci. Agric. 65, 50–55.

De Souza, R., Ambrosini, A., Passaglia, L.M.P., 2015. Plant growth-promoting bacteria as inoculants in agricultural soils. Genet. Mol. Biol. 38, 401–419. Available from: https://doi.org/10.1590/S1415-475738420150053.

Di Simine, C.D., Sayer, J.A., Gadd, G.M., 1998. Solubilization of zinc phosphate by a strain of *Pseudomonas fluorescens* isolated from a forest soil. Biol. Fertil. Soils 28, 87–94.

Dijkstra, F., Cheng, W., 2008. Increased soil moisture content increases plant N uptake and the abundance of 15N in plant biomass. Plant Soil 302, 263–271.

Dobert, R.C., Rood, S.B., Blevins, D.G., 1992. Gibberellins and the legume-*Rhizobium* symbiosis I. Endogenous gibberellins of Lima Bean (*Phaseolus lunatus* L.) stems and nodules. Plant Physiol. 98, 221–224.

Egamberdieva, D., Wirth, S.J., Alqarawi, A.A., Abd Allah, E.F., Hashem, A., 2017. Phytohormones and beneficial microbes: essential components for plants to balance stress and fitness. Front. Microbiol. 8, 2104. Available from: https://doi.org/10.3389/fmicb.2017.02104.

El Mujtar, V., Muñoz, N., Prack Mc Cormick, B., Pulleman, M., Tittonell, P., 2019. Role and management of soil biodiversity for food security and nutrition; where do we stand? Glob. Food Sec. 132–144. Available from: https://doi.org/10.1016/j.gfs.2019.01.007.

Emerson, D., Fleming, E.J., McBeth, J.M., 2010. Iron-oxidizing bacteria: an environmental and genomic perspective. Annu. Rev. Microbiol. 64, 561–583. Available from: https://doi.org/10.1146/annurev.micro.112408.134208.

Ettwig, K.F., et al., 2010. Nitrite-driven anaerobic methane oxidation by oxygenic bacteria. Nature 464, 543–548.

Ettwig, K.F., et al., 2016. Archaea catalyze iron-dependent anaerobic oxidation of methane. Proc. Natl. Acad. Sci. USA 113, 12792–12796.

Fasim, F., Ahmed, N., Parsons, R., Gadd, G.M., 2002. Solubilization of zinc salts by a bacterium isolated from the air environment of a tannery. FEMS Microbiol. Lett. 213, 1–6.

Forchetti, G., Masciarelli, O., Alemano, S., Alvarez, D., Abdala, G., 2007. Endophytic bacteria in sunflower (*Helianthus annuus* L.) isolation, characterization, and production of jasmonates and abscisic acid in culture medium. Appl. Microbiol. Biotechnol. 76, 1145–1152. Available from: https://doi.org/10.1007/s00253-007-1077-7.

Forchetti, G., Masciarelli, O., Izaguirre, M.J., Alemano, S., Alvarez, D., Abdala, G., 2010. Endophytic bacteria improve seedling growth of sunflower under water stress, produce salicylic acid, and inhibit growth of pathogenic fungi. Curr. Microbiol. 61, 485–493.

Fukushima, T., Allred, B.E., Sia, A.K., Nichiporuk, R., Andersen, U.N., Raymond, K.N., 2013. Gram-positive siderophore-shuttle with iron-exchange from Fe-siderophore to apo-siderophore by *Bacillus cereus* YxeB. Proc. Natl. Acad. Sci. USA 110, 13821–13826.

Fulchieri, M., Lucangeli, C., Bottini, R., 1993. Inoculation with *Azospirillum lipoferum* affects growth and gibberellin status of corn seedling roots. Plant. Cell. Physiol. 34, 1305–1309.

Gadd, G.M., 2007. Geomycology: biogeochemical transformations of rocks, minerals, metals and radionuclides by fungi, bioweathering and bioremediation. Mycol. Res. 11, 3–49.

García-Garrido, J.M., Ocampo, J.A., 2002. Regulation of the plant defence response in arbuscular mycorrhizal symbiosis. J. Exp. Bot. 53, 1377–1386.

Gerke, J., 1992. Phosphate, aluminium and iron in the soil solution of three different soils in relation to varying concentrations of citric acid. J. Plant Nutr. Soil Sci. 155 (4), 339–343. Available from: https://doi.org/10.1002/jpln.19921550417.

Gindling, T.H., Newhouse, D., 2012. Policy Resarch Working Paper Series 6201Self-Employment in the Developing World. The World Bank. Available from: https://doi.org/10.1596/1813-9450-6201.

Glick B.R. (ed)., 2015. Resource acquisition. Beneficial Plant-Bacterial Interactions. Springer, New York, pp. 29–63.

Gourdji, S., Läderach, P., Valle, A.M., Martinez, C.Z., Lobell, D.B., 2015. Historical climate trends, deforestation, and maize and bean yields in Nicaragua. Agric. For. Meteorol. 200, 270–281. Available from: https://doi.org/10.1016/j.agrformet.2014.10.002.

Griffin, B.M., Schott, J., Schink, B., 2007. Nitrite, an electron donor for anoxygenic photosynthesis. Science 316, 1870.

Großkinsky, D.K., Tafner, R., Moreno, M.V., Stenglein, S.A., García de Salamone, I.E., Nelson, L.M., et al., 2016. Cytokinin production by *Pseudomonas fluorescens* G20-18 determines biocontrol activity against *Pseudomonas syringae* in *Arabidopsis*. Sci. Rep. 6, 23310. Available from: https://doi.org/10.1038/srep23310.

Gutiérrez-Mañero, F.J., Ramos, B., Probanza, A., Mehouachi, J., Talon, M., 2001. The plant growth promoting rhizobacteria *Bacillus pumilus* and *Bacillus licheniformis* produce high amounts of physiologically active gibberellins. Physiol. Plant 111, 206–211. Available from: https://doi.org/10.1034/j.1399-3054.2001.1110211.x.

Hafeez, F.Y., Hassan, M.N., 2011. Utilization of plant associated rhizobacteria to improve plant nutrition and protection: a cost effective and ecofriendly strategy. WIT Trans. Ecol. Environ. 167, 231–238.

Hafeez, F.Y., Abaid-Ullah, M., Hassan, M.N., 2013. Plant growth- promoting rhizobacteria as zinc mobilizers: a promising approach for cereals biofortification. In: Maheshwari, D.K., Saraf, M., Aeron, A. (Eds.), Bacteria in Agrobiology: Crop Productivity. Springer, Berlin, pp. 217–235.

Hajiboland, R., Dashtebani, F., Aliasgharzad, N., 2015. Physiological responses of halophytic C_4 grass *Aeluropus littoralis* to salinity and arbuscular mycorrhizal fungi colonization. Photosynthetica 53, 572–584. Available from: https://doi.org/10.1007/s11099-015-0131-4.

Han, L., Li, J., Xue, Q., Chen, Z., Zhou, Y., Poon, C.S., 2020. Bacterial-induced mineralization (BIM) for soil solidification and heavy metal stabilization: a critical review. Sci. Total Environ. 746, 140967. Available from: https://doi.org/10.1016/j.scitotenv.2020.140967.

Helmi, F.M., Elmitwalli, H.R., Elnagdy, S.M., El-Hagrassy, A.F., 2016. Calcium carbonate precipitation induced by ureolytic bacteria *Bacillus licheniformis*. Ecol. Eng. 90, 367–371. Available from: https://doi.org/10.1016/j.ecoleng.2016.01.044.

Hidri, R., Barea, J.M., Mahmoud, O.M., Abdelly, C., Azcón, R., 2016. Impact of microbial inoculation on biomass accumulation by *Sulla carnosa* provenances, and in regulating nutrition, physiological and antioxidant activities of this species under non-saline and saline conditions. J. Plant Physiol. 201, 28–41. Available from: https://doi.org/10.1016/j.jplph.2016.06.013.

Hu, M., Chen, P.C., Sun, W.M., Li, F.B., Cui, J.H., 2019. Physiological and genomic characterization of a nitrate-reducing Fe(II)-oxidizing bacterium isolated from paddy soil. Geomicrobiol. J. 36, 433–442. Available from: https://doi.org/10.1080/01490451.2019.1566416.

Jones, D.L., Healey, J.R., Willett, V.B., Farrar, J.F., Hodge, A., 2005. Dissolved organic nitrogen uptake by plants—an important N uptake pathway. Soil Biol. Biochem. 37, 413–423.

Kalinowski, B.E., Liermann, L.J., Brantley, S.L., Barnes, A.S., Pantano, C.G., 2000. X-ray photoelectron evidence for bacteria-enhanced dissolution of hornblende. Geochim. Cosmochim. Acta 64, 1331–1343.

Karadeniz, A., Topcuoğlu, S., Inan, S., 2006. Auxin, gibberellin, cytokinin and abscisic acid production in some bacteria. World J. Microbiol. Biotechnol. 22, 1061–1064. Available from: https://doi.org/10.1007/s11274-005-4561-1.

Kaul, S., Gupta, S., Sharma, T., Dhar, M.K., 2018. Unfolding the role of rhizomicrobiome toward sustainable agriculture. In: Giri, B., Prasad, R., Varma, A. (Eds.), Root Biology Soil Biology, vol. 52.

Khan, Z., Doty, S.L., 2009. Characterization of bacterial endophytes of sweet potato plants. Plant Soil 322, 197–207.

Kihara, J., Sileshi, G.W., Nziguheba, G., Kinyua, M., Zingore, S., Sommer, R., 2017. Application of secondary nutrients and micronutrients increases crop yields in sub-Saharan Africa. Agron. Sustain. Dev. 37, 25. Available from: https://doi.org/10.1007/s13593-017-0431-0.

Konneke, M., Bernhard, A.E., de la Torre, J.R., Walker, C.B., Waterbury, J.B., Stahl, D.A., 2005. Isolation of autotrophic ammonia-oxidizing marine archaeon. Nature 437, 543–546.

Kumar, A., Patel, J.S., Meena, V.S., Srivastava, R., 2019. Recent advances of PGPR based approaches for stress tolerance in plants for sustainable agriculture. Biocatal. Agric. Biotechnol. 20, 101271. Available from: https://doi.org/10.1016/j.bcab.2019.101271.

Kuypers, M.M.M., Marchant, H.K., Kartal, B., 2018. The microbial nitrogen-cycling network. Nat. Rev. Microbiol. 16, 263–276. Available from: https://doi.org/10.1038/nrmicro.2018.9.

Lalitha, S., 2017. Plant growth–promoting microbes: a boon for sustainable agriculture. In: Dhanarajan, A. (Ed.), Sustainable Agriculture towards Food Security. Springer Singapore, Singapore, pp. 125–158. Available from: https://doi.org/10.1007/978-981-10-6647-4_8.

Lally, R.D., Galbally, P., Moreira, A.S., Spink, J., Ryan, D., Germaine, K.J., et al., 2017. Application of endophytic *Pseudomonas fluorescens* and a bacterial consortium to *Brassica napus* can increase plant height and biomass under greenhouse and field conditions. Front. Plant Sci. 8, 2193.

Lavania, M., Nautiyal, C., 2013. Solubilization of tricalcium phosphate by temperature and salt tolerant *Serratia marcescens* NBRI1213 isolated from alkaline soils. Afr. J. Microbiol. Res. 7, 4403–4413.

Lawongsa, P., Boonkerd, N., Wongkaew, S., O'Gara, F., Teaumroong, N., 2008. Molecular and phenotypic characterization of potential plant growth-promoting *Pseudomonas* from rice and maize rhizospheres. World J. Microbiol. Biotechnol. 24, 1877–1884.

Li, H., Qiu, Y., Yao, T., Ma, Y., Zhang, H., Yang, X., 2020. Effects of PGPR microbial inoculants on the growth and soil properties of *Avena sativa*, *Medicago sativa*, and *Cucumis sativus* seedlings. Soil Tillage Res. 199, 104577. Available from: https://doi.org/10.1016/j.still.2020.104577.

Lian, B., Fu, P., Mo, D., Liu, C., 2002. A comprehensive review of the mechanism of potassium releasing by silicate bacteria. Acta Mineral. Sin. 22, 179–183.

Lin, S.Y., Hameed, A., Liu, Y.C., Hsu, Y.H., Lai, W.A., Shen, F.T., et al., 2015. *Azospirillum soli* sp. nov., a nitrogen-fixing species isolated from agricultural soil. Int. J. Syst. Evol. Microbiol. 65, 4601–4607. Available from: https://doi.org/10.1099/ijsem.0.000618.

Liu, D., Lian, B., Dong, H., 2012. Isolation of *Paenibacillus* sp. and assessment of its po- tential for enhancing mineral weathering. Geomicrobiol. J. 29, 413–421.

Liu, K., Garrett, C., Fadamiro, H., Kloepper, J.W., 2016. Induction of systemic resistance in Chinese cabbage against black rot by plant growth-promoting rhizobacteria. Biol. Control. 99, 8–13.

Liu, X.Y., Zhang, M.J., Li, Y.B., Wang, Z.N., Wen, J.K., 2017. *In situ* bioremediation of tailings by sulfate reducing bacteria and iron reducing bacteria: lab- and field-scale remediation of sulfidic mine tailings. Solid State Phenom. 262, 651–655. Available from: https://doi.org/10.4028/www.scientific.net/SSP.262.651.

Lukowiak, R., Grzebisz, W., Sassenrath, G.F., 2016. New insights into phosphorus man- agement in agriculture-a crop rotation approach. Sci. Total Environ. 542, 1062–1107.

Ma, Y., Rajkumar, M., Luo, Y., Freitas, H., 2011. Inoculation of endophytic bacteria on host and non-host plants-effects on plant growth and Ni uptake. J. Hazard. Mater. 195, 230237.

Manulis, S., Shafrir, H., Epstein, E., Lichter, A., Barash, I., 1994. Biosynthesis of indole-3-acetic acid via the indole-3-acetamide pathway in Streptomyces spp. Microbiology 140 (Pt 5), 1045–1050.

Martínez-Gallegos, V., Bautista-Cruz, A., Martínez-Martínez, L., Soledad Sanchez-Medina, P., 2018. First report of phosphate-solubilizing bacteria associated with *Agave angustifolia*. Int. J. Agric. Biol. 20, 1298–1302.

Martinez-Perez, C., et al., 2016. The small unicellular diazotrophic symbiont, UCYN-A, is a key player in the marine nitrogen cycle. Nat. Microbiol. 1, 16163.

Meena, R.S., Meena, V.S., Meena, S.K., Verma, J.P., 2015. The needs of healthy soils for a healthy world. J. Cleaner Prod. 102, 560–561.

Mendes, R., Garbeva, P., Raaijmakers, J.M., 2013. The rhizosphere microbiome: Significance of plant beneficial, plant pathogenic, and human pathogenic microorganisms. FEMS Microbiol. Rev. 37, 634–663. Available from: https://doi.org/10.1111/1574-6976.12028.

Meyer, J.M., 2010. Pyoverdine siderophores as taxonomic and phylogenic markers. In: Ramos, J.L., Filloux, A. (Eds.), Pseudomonas, vol. 6, Molecular Microbiology, Infection and Biodiversity. Springer, London, pp. 201–233.

Meyer, J.M., Geoffroy, V.A., Baida, N., Gardan, L., Izard, D., Lemanceau, P., et al., 2002. Siderophore typing, a powerful tool for the identification of fluorescent and nonfluorescent *Pseudomonas*. Appl Environ Microbiol 68, 2745–2753.

Mokracka, J., Koczura, R., Kaznowski, A., 2004. Yersiniabactin and other siderophores produced by clinical isolates of *Enterobacter* spp. and *Citrobacter* spp. FEMS Immunol. Med. Microbiol. 40, 51–55.

Moreira, H., Pereira, S.I., Marques, A.P., Rangel, A.O., Castro, P.M., 2016. Mine land valorization through energy maize production enhanced by the application of plant growth-promoting rhizobacteria and arbuscular mycorrhizal fungi. Environ. Sci. Pollut. Res. 23, 6940–6950.

Moreno-Espandola, I.P., et al., 2007. Role of root-hairs and hyphae in adhesion of sand particles. Soil Biol Biochem. 39, 2520–2526.

Moreno-Vivian, C., Cabello, P., Martinez-Luque, M., Blasco, R., Castillo, F., 1999. Prokaryotic nitrate reduction: molecular properties and functional distinction among bacterial nitrate reductases. J. Bacteriol. 181, 6573–6584.

Naik, K., Mishra, S., Srichandan, H., Singh, P.K., Sarangi, P.K., 2019. Plant growth promoting microbes: potential link to sustainable agriculture and environment. Biocatal. Agric. Biotechnol. 21, 101326. Available from: https://doi.org/10.1016/j.bcab.2019.101326.

Naz, I., Bano, A., Ul-Hassan, T., 2009. Isolation of phytohormones producing plant growth promoting rhizobacteria from weeds growing in Khewra salt range, Pakistan and their implication in providing salt tolerance to *Glycine max* L. Afr. J. Biotechnol. 8, 5762–5766. Available from: https://doi.org/10.5897/AJB09.1176.

Ning, J., Cumming, J.R., 2001. Arbuscular mycorrhizal fungi alter phosphorus relations of broomsedge (*Andropogon virginicus* L.) plants. J. Exp. Bot. 52, 1883–1891.

Numan, M., Bashir, S., Khan, Y., Mumtaz, R., Shinwari, Z.K., Khan, A.L., et al., 2018. Plant growth promoting bacteria as an alternative strategy for salt tolerance in plants: a review. Microbiol. Res. 209, 21–32. Available from: https://doi.org/10.1016/j.micres.2018.02.003.

Olanrewaju, O.S., Glick, B.R., Babalola, O.O., 2017. Mechanisms of action of plant growth promoting bacteria. World J. Microbiol. Biotechnol. 33, 197. Available from: https://doi.org/10.1007/s11274-017-2364-9.

Palatinszky, M., et al., 2015. Cyanate as an energy source for nitrifiers. Nature 524, 105–108.

Park, H.B., Lee, B., Kloepper, J.W., Ryu, C.M., 2013. One shot-two pathogens blocked: exposure of Arabidopsis to hexadecane, a long chain volatile organic compound, confers induced resistance against both *Pectobacterium carotovorum* and *Pseudomonas syringae*. Plant Signal. Behav. 8, e24619.

Park, J., Sanford, R.A., Bethke, C.M., 2009. Microbial activity and chemical weathering in the Middendorf aquifer, South Carolina. Chem. Geol. 258, 232–241.

Park, Y.G., Mun, B.G., Kang, S.M., Hussain, A., Shahzad, R., Seo, C.W., et al., 2017. *Bacillus aryabhattai* SRB02 tolerates oxidative and nitrosative stress and promotes the growth of soybean by modulating the production of phytohormones. PLoS One 12, e0173203.

Paul, D., Lade, H., 2014. Plant-growth-promoting rhizobacteria to improve crop growth in saline soils: a review. Agron. Sustain. Dev. 34, 737–752.

Pawar, S.T., Bhosale, A.A., Gawade, T.B., Nale, T.R., 2016. Isolation, screening and optimization of exo-polysaccharide producing bacterium from saline soil. J. Microbiol. Biotechnol. Res. 3, 24–31.

Péret, B., Svistoonoff, S., Lahouze, B., Auguy, F., Santi, C., Doumas, P., et al., 2008. A role for auxin during actinorhizal symbioses formation. Plant Signal. Behav. 3, 34–35.

Peterson, R.L., Massicotte, H.B., Melville, L.H., 2004. Mycorrhizas: Anatomy and Cell Biology. NRC Research Press, Ottawa.

Philippot, L., Raaijmakers, J.M., Lemanceau, P., Van Der Putten, W.H., 2013. Going back to the roots: The microbial ecology of the rhizosphere. Nat. Rev. Microbiol. 11, 789–799. Available from: https://doi.org/10.1038/nrmicro3109.

Phillips, D.A., Fox, T.C., Six, J., 2006. Root exudation (net efflux of amino acids) may increase rhizodeposition under elevated CO_2. Glob. Change Biol. 12, 561–567.

Piccoli, P., Travaglia, C., Cohen, A., Sosa, L., Cornejo, P., Masuelli, R., et al., 2011. An endophytic bacterium isolated from roots of the halophyte *Prosopisstrom bulifera* produces ABA, IAA, gibberellins A1 and A3 and jasmonic acid in the chemically defined culture medium. Plant Growth Regul. 64, 207–210.

Prasad, M., Srinivasan, R., Chaudhary, M., Choudhary, M., Jat, L.K., 2019. Plant growth promoting rhizobacteria (PGPR) for sustainable agriculture: perspectives and challenges. PGPR Amelioration in Sustainable Agriculture. Woodhead Publishing (Elsevier), pp. 129–157. Available from: https://doi.org/10.1016/b978-0-12-815879-1.00007-0.

Preisler, A., et al., 2007. Biological and chemical sulfide oxidation in a *Beggiatoa* inhabited marine sediment. ISME J 1, 341–353.

Prieto, K.R., Echaide-Aquino, F., Huerta-Robles, A., Valerio, H.P., Macedo-Raygoza, G., Prado, F.M., et al., 2017. Endophytic bacteria and rare earth elements; promising candidates for nutrient use efficiency in plants. In: Plant macronutrient use efficiency, Molecular and Genomic Perspectives in Crop Plants, pp. 285–306.

Rademacher, W., 2015. Plant growth regulators: backgrounds and uses in plant production. J. Plant Growth Regul. 34, 845–872. Available from: https://doi.org/10.1007/s00344-015-9541-6.

Rajawat, M.V.S., Singh, S., Tyagi, S.P., Saxena, A.K., 2016. A modified plate assay for rapid screening of potassium-solubilizing bacteria. Pedosphere 26, 768–773.

Rani B., 2016. Effect of arbuscular mycorrhiza fungi on biochemical parameters in wheat *Triticum aestivum* L. under drought conditions. Doctoral dissertation, CCSHAU, Hisar.

Renshaw, J.C., Robson, G.D., Trinci, A.P.J., Wiebe, M.G., Livens, F.R., Collison, D., et al., 2002. Fungal siderophores: structures, functions and applications. Mycol. Res. 106, 1123–1142.

Rillig, M.C., 2004. Arbuscular mycorrhizae, glomalin and soil aggregation. Can. J. Soil Sci. 84, 355–363.

Risgaard-Petersen, N., et al., 2006. Evidence for complete denitrification in a benthic foraminifer. Nature 443, 93–96.

Robert, M., Chenu, C., 1992. Interactions between soil minerals and microorganisms. In: Stotzky, G., Bollag, J.-M. (Eds.), Soil Biochemistry, vol. 7. Marcel Dekker, New York, pp. 307–404.

Rokhbakhsh-Zamin, F., Sachdev, D., Kazemi-Pour, N., Engineer, A., Pardesi, K.R., Zinjarde, S., et al., 2011. Characterization of plant-growth-promoting traits of *Acinetobacter* species isolated from rhizosphere of *Pennisetum glaucum*. J. Microbiol. Biotechnol. 21, 556–566.

Rooney, D.C., Killham, K., Bending, G.D., Baggs, E., Weih, M., Hodge, A., 2009. Mycorrhizas and biomass crops: opportunities for future sustainable development. Trends Plant Sci. 14, 542–549. Available from: https://doi.org/10.1016/j.tplants.2009.08.004.

Saha, M., Maurya, B.M., Bahadur, I., Kumar, A., Meena, V.S., 2016. Can potassium- solubilising bacteria mitigate the potassium problems in India? In: Meena, V.S., Maurya, B.R., Prakash Verma, J., Meena, R.S. (Eds.), Potassium Solubilizing Microorganisms for Sustainable Agriculture, first ed. Springer, pp. 127–147. , Chapter 9.

Salomon, M.V., Bottini, R., de Souza Filho, G.A., Cohen, A.C., Moreno, D., Gil, M., et al., 2014. Bacteria isolated from roots and rhizosphere of *Vitis vinifera* retard water losses, induce abscisic acid accumulation and synthesis of defense-related terpenes in in vitro cultured grapevine. Physiol. Plant. 51, 359–374. Available from: https://doi.org/10.1111/ppl.12117.

Salwan, R., Sharma, A., Sharma, V., 2019. Microbes mediated plant stress tolerance in saline agricultural ecosystem. Plant Soil 442, 1–22. Available from: https://doi.org/10.1007/s11104-019-04202-x.

Salwan, R., Sharma, V., Sharma, A., Singh, A., 2020. Molecular imprints of plant beneficial *Streptomyces* sp. AC30 and AC40 reveal differential capabilities and strategies to counter environmental stresses. Microbiol. Res 235, 126449.

Sangeeth, K.P., Bhai, R.S., Srinivasan, V., 2012. *Paenibacillus glucanolyticus*, a promising potassium solubilizing bacterium isolated from black pepper (*Piper nigrum* L.) rhizosphere. J. Spices Aromat. Crops 21, 118–124.

Saravanan, V.S., Subramoniam, S.R., Raj, S.A., 2004. Assessing in vitro solubilization potential of different zinc solubilizing bacterial (ZSB) isolates. Braz. J. Microbiol. 35, 121–125.

Sattar, A., Naveed, M., Ali, M., Zahir, Z.A., Nadeem, S.M., Yaseen, M., et al., 2019. Perspectives of potassium solubilizing microbes in sustainable food production system: a review. Appl. Soil Ecol. 133, 146–159. Available from: https://doi.org/10.1016/j.apsoil.2018.09.012.

Scheffer, F., 2002. Soil Science, fifteenth ed. Spektrum Verlag, Heidelberg, pp. 123–125.

Shahzad, R., et al., 2017. Inoculation of abscisic acid-producing endophytic bacteria enhances salinity stress tolerance in *Oryza sativa*. Environ. Exp. Bot. 136, 68–77.

Sharaf-Eldin, M., Elkholy, S., Fernandez, J.A., et al., 2008. *Bacillus subtilis* FZB24 affects flower quantity and quality of Saffron (*Crocus sativus*). Planta. Med. 74, 1316–1320.

Sharma, V., Salwan, R., 2018. Biocontrol potential and applications of actinobacteria. Elsevier Book Volume on Actinobacteria: Diversity and Biotechnological Applications 93–108. Available from: https://doi.org/10.1016/B978-0-444-63994-3.00006-0.

Sharma, V., Sharma, A., Salwan, R., 2020a. Overview and challenges in the implementation of plant beneficial microbes. In: Vivek, Sharma, Richa, Salwan, Laith, Ani Al (Eds.), Molecular Aspects of Plant Beneficial Microbes in Agriculture, vol. 2. Elsevier. Available from: https://doi.org/10.1016/b978-0-12-818469-1.00001-8.

Sharma, V., Sharma, A., Malannavar, A.B., Salwan, R., 2020b. Molecular aspects of biocontrol species of *Streptomyces* in agricultural crops. In: Vivek, Sharma, Richa, Salwan, Laith, Ani Al (Eds.), Molecular Aspects of Plant Beneficial Microbes in Agriculture, Elsevier, pp. 89–109. Available from: https://doi.org/10.1016/b978-0-12-818469-1.00008-0.

Shen, J., Li, C., Mi, G., Li, L., Yuan, L., Jiang, R., et al., 2013. Maximizing root/rhizosphere efficiency to improve crop productivity and nutrient use efficiency on intensive agriculture of China. J. Exp. Bot. 64, 1181–1192.

Sheng, X.F., He, L.Y., 2006. Solubilization of potassium-bearing minerals by a wild-type strain of *Bacillus edaphicus* and its mutants and increased potassium uptake by wheat. Can. J. Microbiol. 52, 66–72.

Shrivastava, S., D'souza, S., Desai, P., 2008. Production of indole-3-acetic acid by immobilized actinomycete (*Kitasatospora* sp.) for soil applications. Curr. Sci. 94, 1595–1604.

Smith, L., Haddad, L., 2015. Reducing child undernutrition: past drivers and priorities for the post-MDG era. World Dev. 68, 180–204.

Smith, S.E., Jakobsen, I., Grønlund, M., Smith, F.A., 2011. Roles of arbuscular mycorrhizas in plant phosphorus nutrition: interactions between pathways of phosphorus uptake in arbuscular mycorrhizal roots have important implications for understanding and manipulating plant phosphorus acquisition. Plant Physiol. 156, 1050–1057.

Sokolova, T.A., Tolpeshta, I.I., Topunova, I.V., 2010. Biotite weathering in podzolic soil under conditions of a model field experiment. Eurasian Soil Sci. 43, 1150–1158.

Stein, L.Y., Klotz, M.G., 2011. Nitrifying and denitrifying pathways of methanotrophic bacteria. Biochem. Soc. Trans. 39, 1826–1831.

Stein, L.Y., Klotz, M.G., 2016. The nitrogen cycle. Curr. Biol. 26, R94–R98.

Suarez, C., Cardinale, M., Ratering, S., Steffens, D., Jung, S., Montoya, A.M.Z., et al., 2015. Plant growth-promoting effects of Hartmannibacter diazotrophicus on summer barley (*Hordeum vulgare* L.) under salt stress. Appl. Soil Ecol. 95, 23–30. Available from: https://doi.org/10.1016/j.apsoil.2015.04.017.

Thompson, A.W., et al., 2012. Unicellular cyanobacterium symbiotic with a single-celled eukaryotic alga. Science 337, 1546–1550.

Toljander, J.F., et al., 2006. Attachment of different soil bacteria to arbuscular mycorrhizal fungal extraradical hyphae is determined by hyphal vitality and fungal species. FEMS Microbiol. Lett. 254, 34–40.

Tourna, M., et al., 2011. Nitrososphaera viennensis, an ammonia oxidizing archaeon from soil. Proc. Natl. Acad. Sci. USA 108, 8420–8425.

Tsementzi, D., et al., 2016. SAR11 bacteria linked to ocean anoxia and nitrogen loss. Nature 536, 179–183.

Turan, M., Gulluce, M., Karadayi, M., Sahin, F., 2011. Role of soil enzymes produced by PGPR strains in wheat growth and nutrient uptake parameters in the filed conditions. Curr. Opin. Biotech. 22, 133–140.

Uroz, S., Calvaruso, C., Turpault, M.P., Frey-Klett, P., 2009. Mineral weathering by bacteria: ecology, actors and mechanisms. Trends Microbiol. 17, 378–387.

Usharani, G., Sujitha, D., Sivasakthi, S., Saranraj, P., 2014. Effect of arbuscular mycorrhizal (AM) fungi (*Glomus fasciculatum* L.) for the improvement of growth and yield of maize (*Zea mays* L.). Cent. Eur. J. Exp. Biol. 3, 19–25.

van Kessel, M.A., et al., 2015. Complete nitrification by a single microorganism. Nature 528, 555–559.

Van Schöll, L., Kuyper, T.W., Smits, M.M., Landeweert, R., Hoffland, E., van Breemen, N., 2008. Rock-eating mycorrhizas: their role in plant nutrition and biogeo- chemical cycles. Plant Soil 303, 35–47.

Vanlauwe, B., Descheemaeker, K., Giller, K.E., Huising, J., Merckx, R., Nziguheba, G., et al., 2015. Integrated soil fertility management in sub-Saharan Africa: unravelling local adaptation. Soil 1, 491–508. Available from: https://doi.org/10.5194/soil-1-491-2015.

Vassilev, N., Vassileva, M., Lopez, A., Martos, V., Reyes, A., Maksimovic, I., et al., 2015. Unexploited potential of some biotechnological techniques for biofertiliser production and formulation. Appl Microbiol. Biotechnol. 99, 4983–4996. Available from: https://doi.org/10.1007/s00253-015-6656-4.

Vessey, J.K., 2003. Plant growth promoting rhizobacteria as biofertilizer. Plant Soil 255, 571–586.

Vurukonda, S.S.K., Vardharajula, S., Shrivastava, A.S., 2016. Enhancement of drought stress tolerance in crops by plant growth promoting rhizobacteria. Microbiol. Res. 184, 13–24.

Weber, K.A., Achenbach, L.A., Coates, J.D., 2006. Microorganisms pumping iron: anaerobic microbial iron oxidation and reduction. Nat. Rev. Microbiol. 4, 752–764.

Whiting, S.N., De Souza, M.P., Terry, N., 2001. Rhizosphere bacteria mobilize Zn for hyperaccumulation by *Thlaspi caerulescens*. Environ. Sci. Technol. 35, 3144–3150.

Winkelmann, G., 2007. Ecology of siderophores with special reference to the fungi. Biometals 20, 379–392.

Xu, G., Fan, X., Miller, A.J., 2012. Plant nitrogen assimilation and use efficiency. Ann. Rev. Plant Biol. 63, 153–182.

Yanni, Y.G., Rizk, R.Y., El-Fattah, F.K.A., Squartini, A., Corich, V., Giacomini, A., et al., 2001. The beneficial plant growth-promoting association of *Rhizobium leguminosarum bv. trifolii* with rice roots. Funct. Plant Biol. 28, 845–870.

Yu, K., Pieterse, C.M.J., Bakker, P.A.H.M., Berendsen, R.L., 2019. Beneficial microbes going underground of root immunity. Plant Cell Environ. 42, 2860–2870. Available from: https://doi.org/10.1111/pce.13632.

Zarjani, K.J., Aliasgharzad, N., Oustan, S., Emadi, M., Ahmadi, A., 2013. Isolation and characterization of potassium solubilizing bacteria in some Iranian soils. Arch. Agron. Sol. Sci. 59, 1713–1723.

Zeng, X., Liu, X., Tang, J., Hu, S., Jiang, P., Li, W., et al., 2012. Characterization and potassium-solubilizing ability of *Bacillus circulans* Z1-3. Adv. Sci. Lett. 10, 173–176.

Zhang, X., Zhong, T., Liu, L., Ouyang, X., 2015. Impact of soil heavy metal pollution on food safety in China. PLoS One 10, 1–14. Available from: https://doi.org/10.1371/journal.pone.0135182.

Zhang, F., Wang, P., Zou, Y.N., Wu, Q.S., Kuča, K., 2019. Effects of mycorrhizal fungi on root-hair growth and hormone levels of taproot and lateral roots in trifoliate orange under drought stress. Arch. Agron. Soil Sci. 65 (9), 1316–1330. Available from: https://doi.org/10.1080/03650340.2018.1563780.

Zhao, L., 2010. The tale of our other genome. Nature 465, 879–880.

Zhu, Y.G., Miller, R.M., 2003. Carbon cycling by arbuscular mycorrhizal fungi in soil-plant systems. Trends Plant Sci. 8, 407–409.

Zumft, W.G., Kroneck, P.M., 2006. Respiratory transformation of nitrous oxide (N_2O) to dinitrogen by bacteria and archaea. Adv. Microb. Physiol. 52, 107–227.

Zumft, W.G., 1997. Cell biology and molecular basis of denitrification. Microbiol. Mol. Biol. Rev. 61, 533–616.

Further reading

Barbagelata, P.A., Mallarino, A.P., 2013. Field correlation of potassium soil test methods based on dried and field-moist soil samples for corn and soybean. Soil Sci. Soc. Am. J. 77, 318–327.

Barroso, C.V., Pereira, G.T., Nahas, E., 2006. Solubilization of $CaHPO_4$ and $AlPO_4$ by *Aspergillus niger* in culture media with different carbon and nitrogen sources. Braz. J. Microbiol. 37, 434–438.

Daims, H., Lücker, S., Wagner, M.A., 2016. New perspective on microbes formerly known as nitrite-oxidizing bacteria. Trends Microbiol. 24, 699–712.

Garcia-Seco, D., Zhang, Y., Gutierrez-Mañero, F.J., Martin, C., Ramos-Solano, B., 2015. Application of *Pseudomonas fluorescens* to blackberry under field conditions improves fruit quality by modifying flavonoid metabolism. PLoS One 10, e0142639. Available from: https://doi.org/10.1371/journal.pone.0142639.

George, P., Gupta, A., Gopal, M., Thomas, L., Thomas, G.V., 2013. Multifarious beneficial traits and plant growth promoting potential of *Serratia marcescens* KiSII and *Enterobacter* sp. RNF267 isolated from the rhizosphere of coconut palms (Cocosnucifera L.). World J. Microbiol. Biotechnol. 29, 109–117. Available from: https://doi.org/10.1007/s11274-012-1163-6.

Kumar, A., Maurya, B.R., Raghuwanshi, R., Meena, V.S., Islam, M.T., 2017. Co-inoculation with *Enterobacter* and rhizobacteria on yield and nutrient uptake by wheat (*Triticum aestivum* L.) in the alluvial soil under Indo-gangetic plain of India. J. Plant Growth Regul. 1–10.

Mal, B., Mahapatra, P., Mohanty, S., 2015. Effect of diazotrophs and chemical fertilizers on production and economics of okra (*Abelmoschus esculentus*, L.) cultivars. Am. J. Plant Sci. 1, 168–174. Available from: https://doi.org/10.4236/ajps.2014.51022.

Mariotte, P., Mehrabi, Z., Bezemer, T.M., De Deyn, G.B., Kulmatiski, A., et al., 2018. Plant-soil feedback: bridging natural and agricultural sciences. Trends Ecol. Evol. 33 (2), 129–142. Available from: https://doi.org/10.1016/j.tree.2017.11.005.

Oteino, N., Lally, R.D., Kiwanuka, S., Lloyd, A., Ryan, D., Germaine, K.J., et al., 2015. Plant growth promotion induced by phosphate solubilizing endophytic *Pseudomonas* isolates. Front. Microbiol. 6, 745.

Placea, F., Barrettb, C.B., Freemanc, H.A., Ramischd, J.J., Vanlauwed, B., 2003. Prospects for integrated soil fertility management using organic and inorganic inputs: evidence from smallholder African agricultural systems. Food Policy 28, 365–378.

Tenenboim, H., Brotman, Y., 2016. Omic relief for the biotically stressed: metabolomics of plant biotic interactions. Trends Plant Sci. 21, 781–791.

Chapter 20

Nutritional imbalance in plants under rising atmospheric CO_2

Pravesh Kundu[1,2], Komal Goel[1,2] and Gaurav Zinta[1,2]
[1]*CSIR-Institute of Himalayan Bioresource Technology (IHBT), Palampur, India,* [2]*Academy of Scientific and Innovative Research (AcSIR), CSIR-HRDC Campus, Ghaziabad, India*

20.1 Introduction

The concentration of atmospheric carbon dioxide (CO_2) is increasing steadily and predicted to double by 2100. The main reason for the elevation of CO_2 is industrialization and burning of fossil fuels. CO_2 is a substrate for photosynthesis, and rising atomspheric CO_2 levels alter plant growth and productivity. This growth-promoting effect of high CO_2 is called CO_2 fertilization, and particularly observed in C3 as compared to C4 plant species (Long et al., 2006). Ribulose-1,5-bisphosphate carboxylase/oxygenase (RuBisCO) catalyzes CO_2 fixation via photosynthetic reactions. High CO_2 favors C3 over C4 because C4 photosynthesis is nearly staurated under current atmospheric CO_2 (Drake et al., 1997; Makino and Mae, 1999). For example, rice being a C3 crop shows photosynthetic enhancement under high CO_2 that results in increase in leaf area and C-related resources (carbohydrates) (Long et al., 2005; Lin et al., 2015), while nitrogen (N) resources decline due to the dilution effect of carbohydrates, which eventually causes reduction in protein concentration (Jena et al., 2018; Loladze, 2014). Thus there is an emerging concern about the effects of high CO_2 on agriculture productivity and nutritional quality (Solomon et al., 2007). As elevated CO_2 increases crop yield and decreases nutritional quality, thus this qualitative change affects those populations that rely only on a single food crop. For example, many countries rely only on rice as a cereal, which would be affected negatively by elevated CO_2 at the nutrients level. This decline in the nutritients profile is expected to create nutritional deficiency in the human population, which eventually would affect gross domestic productivity (GDP) (Zhu et al., 2018).

The world population will be 2.3 billion by the end of 2050, and to feed the growing population the demand for food supply will increase (Alexandratos

Plant Nutrition and Food Security in the Era of Climate Change. DOI: https://doi.org/10.1016/B978-0-12-822916-3.00006-8

and Bruinsma, 2012). Elevated CO_2 is expected to increase agricultural productivity by increasing photosynthesis especially in C3 crops (Kimball, 2016; Makino and Mae, 1999). Studies show that elevated CO_2 increases photosynthetic rates initially, however, long-term exposure of elevated CO_2 leads to photosynthetic acclimation in C3 plants (Long et al., 2004). This could be linked with the nitrogen limitation (Seneweera et al., 2002), changes in the source-sink relations (Ruiz-Vera et al., 2017), and suppression of photosynthetic genes (Carbonero et al., 2017). Photosynthetic acclimation occurs mainly in nitrogen-deficient plants, as compared to well-fertilized ones (Seneweera et al., 2002; Vicente et al., 2015a). As vegetables are essential for human nutrition, and their yield also increases under elevated CO_2 (Pérez-López et al., 2015). The yield of cucumber increased by 73% under elevated CO_2 (Dong et al., 2017). However, species-specific and tissue-specific responses have also been reported. For example, leaves of lettuce and celery decrease their nitrate contents, whereas in lettuce stems it increases (Jin et al., 2009). In this chapter we discuss the effects of elevated CO_2 on C3 and C4 plants, and describe it on the level of growth, yield, nutrition, stoichiometry and impact on human health.

20.2 Effect of elevated CO_2 on C3 and C4 plants

C3 plants contribute to 90% of carbohydrate production of all the plants on earth (Still et al., 2003). The two main functional groups of C3 plants are grasses and legumes. Doubling of CO_2 increases sugar level from 19%–46% in C3 plants due to photosynthetic enhancement (Leakey et al., 2009). However, C4 plants are unable to the take the benefit of elevated CO_2 because of Kranz anatomy, which leads to the saturation of photosynthesis under current atmospheric CO_2 (Loladze, 2014). The rising atmospheric CO_2 also increase the content of nonstructural carbohydrates in C3 plants and decrease the levels of protein and nitrogen. The ionome decreases with the elevation of CO_2 in C3 plants, while no effect is observed in C4 plants (Loladze, 2014). In C3 grasses and legumes, concentration of zinc and iron decreases which increases the risk of disease related to nutritional deficiency. Of the global population, 60%–70% depend on C3 and legume plants for their need for food. For example, we can see wheat plant in elevated CO_2 condition, iron levels decrease 5.1% [95% confidence interval (CI) 26.5%–23.7%] and zinc decreases 9.5% (95% [CI] 212.7%–25.9%). The concentration of protein also gets lowered in rice by 7.8% (95% CI 28.9%–26.8%) and in wheat by 6.3% (95% CI 27.5%–25.2%), but there is no effect on the protein of the C4 plants (Myers et al., 2014). Legume plants do not have the same effect of elevated CO_2 as nonleguminous. Under elevated CO_2 conditions, the seed mass of legumes increases due to an increase of seed nitrogen, which is a positive effect of elevated CO_2. For plant growth, nitrogen and phosphorous should be present in adequate amounts. When nitrogen is present in the full amount, crops give a good biomass production. Enhanced growth of plants demands more nitrogen due to an acute limitation of nitrogen taking place. Due to deficiency of nitrogen in C3 plants, the process of photosynthesis gets affected because the activity of RuBisCO and other photosynthetic enzymes decreases and the number of proteins also decreases (Stitt and Krapp, 1999). One improvement in C3 crops is the increase in water conduction and use efficiency in the leaves (Ainsworth and Long, 2005). C4 crops also have a water-saving effect in leaves but there is no effect on production, for example, in sorghum (Conley et al., 2001), but Ainsworth and Long (2005) did not recognize the water-conduction effect in C4 plants. The growth of C3 plants increases; similarly the C4 crop sorghum also shows the same result in the early developmental stage (Conley et al., 2001) when at its maturity C4 photosynthesis is also expressed at a high level (Cousins et al., 2003). Elevated CO_2 with well-watered conditions elevates the canopy of C3 plants as well as their early maturation stage, but on the C4 plant there is no effect observed. Improved water and nutrient relations indirectly stimulate the growth of C3 and C4 plants (Wand et al., 1999; Kimball et al., 2002; Leakey et al., 2009). In the next 50 years under higher CO_2 conditions, the protein content of C3 crops like wheat and rice would decrease up to 15% while legumes and C4 crops may experience a lower effect on them (Myers et al., 2014). Leguminous crops benefit from the symbiosis effect; they have arbuscular fungi for nitrogen assimilation and phosphorous absorption (Smith et al., 2011). Plants with the association of arbuscular mycorrhizal (AM) reduce low phosphorus as compared to nonmycorrhizal; for example, the seedling of locust (*Robinia pseudoacacia* L.) loses 50% phosphorus while only 22% decline in the mycorrhizal seedling (Olesniewicz and Thomas, 1999). Nitrogen absorption is also done by AM in legumes in nitrate and ammonia form (Ames et al., 1983; Johansen et al., 1993). In the future, it must be in the form of ammonia because recent studies show that assimilation of nitrogen in the form of nitrate is inhibited by elevated CO_2 (Bloom et al., 2010). Similarly, a positive effect of mycorrhiza was also studied in tomatoes (*Solanum Lycopersicum*), where absorption of nitrogen increases under elevated CO_2 as compared to ambient (Cavagnaro et al., 2007). Chen et al. (2007) observed that under elevated CO_2 absorption of nitrogen by AM is species-specific, for example, absorption of nitrogen increases in *Plantago lanceolata* L, while in *Festuca arundinacea* Schreb it decreases. Other effects were also observed in C3 plants like the decline in iron, zinc, and other minerals, which disturbs the nutritional quality of crops like spinach (Jain et al., 2007). By changing the nutrient concentration, elevated CO_2 disturbs the ionome concentration of C3 plants. All minerals get diluted with the elevation of carbohydrates except carbon (Loladze, 2002). The researcher has to pay attention to the elevated concentration of CO_2, reasons for it, and its effect. Besides reducing the level of CO_2 emission, resilient and nutrition-rich crops have to be promoted.

20.3 Effect of elevated CO_2 on photosynthesis

Day by day CO_2 is getting elevated in the environment due to industrialization. It has increased from 208 to 408 mL/moll through March 2018 and continues to increase. Fossil fuel combustion and deforestation are the main reasons for the elevation of CO_2. We can see the direct effect of elevated CO_2 on the photosynthesis of plants. Photosynthesis gets promoted and thus yields and productivity also improve (Long et al., 2004). As photosynthesis increases in plants under elevated CO_2 nitrogen deficiency also occurs.

Simultaneously, deficiency of nitrogen causes a sugar-moderated suppression of crops' sink metabolism affecting enzymes, that is the tendency to produce carbon storage new sinks (Stitt and Krapp, 1999). Due to the increase of soluble sugars, plants get tolerant to environmental stress (Drake et al., 2011; Huang and Xu, 2015). In all the vegetable concentrations of sugar increase in the edible part by the elevation of CO_2 like glucose by 13.2%, fructose by 14.2%, sucrose by 3.7% (at $P = .07$) (Dong et al., 2018). Among all the parts, the leaf has a higher concentration of sugar because of the higher level of photosynthesis. For example, the sugar level of Chinese cabbage leaves can increase 38%–188% and 16%–53% for oily sowthistle leaves (Jin et al., 2009). Photosynthesis increases the level of soluble sugar, resulting in antioxidants beginning to accumulate in the plants because sugars are the precursor of the antioxidants (Wang et al., 2003; Jaafar et al., 2012; Becker and Kläring, 2016). For example, the lettuce leaf phenol synthesis starts with the accumulation of sugar when the control CO_2 level changes from 200 to 1000 mmol/mol (Becker and Kläring, 2016). Photosynthesis initially increases with the elevation of CO_2 but later on it gets acclimatized (with long-term downregulation of photosynthesis), which directly affects the quality of vegetables (Long et al., 2004; Kirschbaum, 2011). In rain-fed crops, elevated CO_2 plays a positive role. It increases photosynthesis in water-stress conditions with the closure of stomata, which reduces evapotranspiration and increases the water productivity of the plants. We can see more effect on photosynthesis in C3 plants because there is no saturation of carbon fixation; hence they are more productive in the elevated CO_2 condition (Drake et al., 1997). Many primary and secondary responses are directly affected by the elevated CO_2 because CO_2 is the primary substrate for photosynthesis and it also influences many other processes like growth, nutrient composition, and dry matter allocation (Stitt and Krapp, 1999; Schortemeyer et al., 1999; Uprety et al., 2003; Pal et al., 2005). With the elevation of CO_2 magnesium content also decreases in the leafy tissue of plants, which is essential for the formation of chlorophyll by which the concentration of chlorophyll starts to decline, so it indirectly affects the rate of photosynthesis (McGrath and Lobell, 2013b).

20.4 Effect of elevated CO_2 on yield and growth

As the level of CO_2 increases in the atmosphere, there is a positive effect on the yield and growth of the plants. There is a 46% increase in the production of sugar and other carbohydrates (Loladze, 2014). There are two types of carbohydrates in the plants: total structural carbohydrates (TSC; e.g., cellulose and fiber) that are not digestible for humans and total nonstructural carbohydrates (TNC; e.g., starch, glucose, fructose) that are easily digestible to the human gut. Production of TNC is supported by the elevated CO_2. TNC increases 10%–45% with elevated CO_2 (Stiling and Cornelissen, 2007; Robinson et al., 2012). CO_2 is the main source of photosynthesis, providing positive effects on yield, growth, and biomass due to the fertilization effect. For example, soybean (*Glycine max*), a Kharif crop, was grown under elevated CO_2 and temperatures to understand the impact of climate change on it. Soybean reached maximum height after 63 days of sowing in all climate conditions ($P < .05$) like high temperature, high CO_2, or both. The chamber effect is more on the crop because there is more growth in ambient OTC than in an open field. And there is a positive growth in the yield or soybean pods, which increased 36%–37% in number when compared with elevated and ambient conditions. The seed index changes from 11.28 to 14.20 g under elevated CO_2 which is measured through the weight of 100 seeds. An overall increase in the biomass of the crop is 47% under elevated CO_2 conditions. This all occurs because of the increase in the net rate of photosynthesis due to higher carboxylation (Lenka et al., 2017). As carboxylation increases, respiration is decreased and it improves the water and nutrient relationship. When enough nutrients are available, it promotes the growth of C3 and C4 plants (Wand et al., 1999; Kimball et al., 2002; Leakey et al., 2009). In tomato, elevated CO_2 causes a mismatch effect on the color and maturity of the fruit. The concentration of soluble sugars increases at the early fruiting stages of tomato while color pigments do not synthesize at the same speed and quantity. The collection of tomato fruits is done based on the maturity of their color, resulting in harvesting being delayed in elevated CO_2 (Islam et al., 1996; Lin et al., 2014). Since plant growth and yield are increased by the elevated CO_2 it possibly contributes to feeding the increasing population (Drake et al., 1997; Kimball et al., 2002; Ainsworth and Long, 2005). The effect of elevated CO_2 on plant growth can be measured by the interactivity of CO_2 and genotype. Dry matter of the wheat also increases in the condition of elevated CO_2 (Pal et al., 2005; Aranjuelo et al., 2013; Benlloch-Gonzalez et al., 2014). Durum wheat in conditions of low levels of nitrogen shows collection of the dry matter under elevated CO_2 condition (Vicente et al., 2015b). Overall elevated CO_2 benefits the crop by increasing biomass production, growth, and water production and decreases the evapotranspiration of the plants. It also increases the root:shoot ratio of the plants, which improves the underground biomass. In potato, the amount of sugar increases as well as growth but upon frying, it turns brown and produces acrylamide in higher amounts (Donnelly et al., 2001; Kumari and Agrawal, 2014). Elevated CO_2 also decreases citrate concentration, which gives a good taste to the potato but generates the risk of discoloration. One of the results of elevated CO_2 is the increase in the atmospheric temperature. There can be a 20%–40% decline in the yield of staple crops like rice and maize in higher CO_2 and temperature conditions. Because high-temperature conditions are risky for these cereal crops, the food supply for all the population will be at risk. It is necessary to understand the effect of elevated CO_2 on agriculture production (Gebregergis, 2016).

20.5 Effect of elevated CO_2 on leaf area

Plants grown under elevated CO_2 often show accelerated growth as observed in the form of leaf area increase that further enhance photosynthesis (Conroy, 1992; Centritto et al., 1999; Drake et al., 1997). For example, spinach and fenugreek grown under elevated CO_2 show a maximum increase in the leaf area at 80 days of high CO_2 exposure. Also, there is a significant increase in the leaf dry mass of spinach and fenugreek by 16.4% and 37%, respectively (Jain et al., 2007).

20.6 Effect of elevated CO_2 on proteins

Proteins are one of the micronutrients of plants. With the elevation of CO_2 quantity of proteins gets decreased in the plants. For example, the proteins of rice plants decrease when the rice plants grow in elevated CO_2 conditions (Zhu et al., 2018). Additionally, temperature also plays a role in the reduction of proteins in the rice. It is predicted that in the next 50 years protein content of rice and wheat will decrease up to 15% (Myers et al., 2014). Growth, tissue repair, and turnover get limited due to protein deficiency (Gropper et al., 2008). Generally, protein deficiency happens when micronutrients and energy get reduced (Millward and Jackson, 2004). The use of meta-regression on pea protein found that the content of protein decreases 4.1% (1.6%–6.7%) when the level of CO_2 increases 150 ppm in the atmosphere, but overall there is no remarkable change in the protein content on meta-analysis (Medek et al., 2017). Due to inadequate statistical power, meta-regression is not notable for other crops. In maize, there is not much difference in protein in both ambient and elevated CO_2. Proteins' availability to the organisms also decreases when elevated CO_2 increases the level of secondary metabolites and toxins (Cavagnaro et al., 2011). Stoichiometry of the necessary food also gets disturbed with the disturbance in the ratio of proteins and carbohydrates. Proteins of pollen grain also get affected by the elevation of CO_2, according to data collected from the Natural History Archives of the Smithsonian Institute. In the museum, there is a floral sample of *S. canadensis* plant that had been collected from different range biota of the United States and Canada since 1842–1998 and to the ancient data collected some samples have been added from Maryland (2008, 2012, 2014) and Texas (2012, 2014) in situ. The decline in protein concentration in pollen grain directly affects the bee diet because there is a one-third decrease in the proteins of pollen and there is no other source for excessive protein to recompense (Ziska et al., 2016). Vegetable protein concentration also decreases (9.5%) by the elevation of CO_2. Protein concentration especially decreases in fruit vegetable (10.5%), stem vegetable (12.6%), and root vegetable (20.5%) while leafy vegetables are unaffected because of the presence of high amounts of nitrate, possibly by elevated CO_2 promoting nitrogen uptake in the leaves (Stitt and Krapp, 1999). For example, lettuce's (Batavia Rubia Munguia) inner side leaves (without AM fungi) have more nitrogen concentration in elevated CO_2 as compared to outer leaves (Baslam et al., 2012). Also, vegetables are less affected by elevated CO_2 on the nitrogen assimilation and production of nitrogenous compounds than other crops (Taub and Wang, 2008; Loladze, 2014); mainly grain crops have low nitrogen in their cultivated soil. In wheat and *Arabidopsis*, nitrogen assimilation is limited on the elevation of CO_2 because it impedes nitrogen uptake (Bloom et al., 2010), which has a direct impact on the protein concentration of these crops. Plants grown under elevated CO_2 have low protein in the leaves as measured per mass (Cotrufo et al., 1998). Complete nitrogen gets translocated to the grain in the period of senescence (Martre et al., 2003), so grain protein also has the effect of low leaf protein. The reason for the low protein concentration is low nitrogen, which is reduced due to the stimulation of growth of plants in elevated CO_2 conditions (Coleman et al., 1993). It is predicted that by 2050 protein deficiency risk will reach its peak in countries such as India, Tajikistan, Bangladesh, Iran, and Afghanistan.

20.7 Effect of elevated CO_2 on stoichiometry

Stoichiometry is the comparative ratio of nutrients or chemical elements (Sterner and Elser, 2002). Plants' stoichiometry becomes disturbed due to elevation of CO_2 because elevated CO_2 disturbs the nutrient content of plants. The concentration of total nonstructural carbohydrates (TNC) increases with the elevated CO_2 while the level of protein decreases, which disturbs the stoichiometry of the TNC:protein ratio (Loladze, 2014). In plant tissue, only a few grams of TNC were boosted by elevated CO_2, and this concentration of carbohydrates dilutes supplementary nutrients of the plant by $\sim 1\% - 7.4\%$ (Poorter et al., 1997; Keutgen and Chen, 2001; Katny et al., 2005; Erbs et al., 2010; Azam et al., 2013). In the elevated CO_2 conditions demanded of the photosynthetic apparatus for leaves, nitrogen reduces while the amount of carbohydrates increases (Stitt and Krapp, 1999), which permits the plant to devote defense resources to recovery (i.e., roots) (Lambers, 1993; Coley et al., 2002). With the change in nitrogen content, concentration and place of phosphorus also change as this stoichiometry of P:C and P:C:N gets disturbed, which directly affects nutritional quality and yield of the crops (Loladze, 2002). Disturbed stoichiometry of plants also affects nutritional quality and defense systems (Lambers, 1993; Coley et al., 2002; Bidart-Bouzat and Imeh-Nathaniel, 2008). A quarter-century ago, the first proof was found for the low content of minerals in plants in elevated CO_2 conditions (Porter and Grodzinski, 1984; Peet et al., 1986; O'Neill et al., 1987). Physiological effects such as increased photosynthesis and reduced transpiration are responsible for the reduction of minerals of the plants (Loladze, 2002; McGrath and Lobell, 2013a). Nitrogen has twice the effect of elevated CO_2 as compared with minerals. Stoichiometry of the N:S and N:P reduces approximately 7% when elevated CO_2 creates a gap difference on nitrogen (15%), sulfur (9%), and phosphorus (9%). As nitrogen and minerals reduce, 6% of carbon increases with the elevation of CO_2. The resulting stoichiometric ratio of C:P increases by (16%), C:S (16%), and C:N (25%) (Poorter et al., 1997; Stiling and Cornelissen, 2007; Robinson et al., 2012). Imbalanced stoichiometry directly affects human nutrition. Populations with dependence only on a single type of food are highly affected. For example, people dependent only on rice get a low nutrition effect on health and the area suffers from low GDP (Loladze, 2014).

20.8 Effect of elevated CO_2 on photorespiration

Photosynthesis is stimulated with a higher concentration of CO_2 while photorespiration is suppressed. Photorespiration is a process of O_2 fixation by the RuBisCO enzyme. It decreases the assimilation of CO_2 if the level of O_2 rises. Photorespiration is a waste processor and a side mechanism of the RuBisCO enzyme (Kebeish et al., 2007; Maurino and Peterhansel, 2010). It decreases in plants when the level of CO_2 rises in the environment. At the present CO_2 level, photorespiratory mutant does not grow well (Timm and Bauwe, 2013). This suggests that carbon metabolism has a very important role in this process and growing evidence is that stress tolerance is provided by photorespiration in plants (Voss et al., 2013). Reduced photorespiration decreases the production of oxygen radicals by this antioxidant metabolism (Pérez-López et al., 2015). Photorespiration is necessary for nitrogen assimilation. Due to the reduction of photorespiration, the amount of nitrogen also decreases in plants because the level of photorespiration-derived reductants (NADPH) decreases. These reductants are useful for the conversion of nitrogen to nitrate, nitrite, and then ammonia for the formation of amino acids (Wujeska-Klause et al., 2019). The chloroplast vesiculation (CV) gene plays an important role in the regulation of photorespiration. In elevated CO_2 conditions, silencing of the CV gene increases photorespiration. As CO_2 is elevated, the interaction of OsCV protein with OsPEX11-1 protein occurs in the peroxisome. This interaction of proteins is maintained throughout trafficking to vacuoles where degradation of proteins takes place. Resulting OsCV protein-mediated peroxisomal turnover occurs, which regulates the uptake of nitrogen and photorespiration in rice plants in the condition of elevated CO_2 (Umnajkitikorn et al., 2020).

20.9 Effect of elevated CO_2 on nitrogen

Nitrogen is an essential component of plants because it is necessary for protein synthesis. Elevated CO_2 harms the nitrogen concentration of the plants. From all parts of the plants 10%–18% nitrogen reduces with the elevation of CO_2 (Cotrufo et al., 1998; Jablonski et al., 2002; Taub and Wang, 2008). In elevated CO_2 conditions, there are two main reasons for low nitrogen concentration: (1) reduction in RuBisCO (protein)—it has a large pool of nitrogen and is found in huge quantities on earth (Drake et al., 1997), and (2) inhibition of nitrogen uptake (Bloom et al., 2010). Due to the reduction in nitrogen concentration, protein also decreases in the edible part of the crops (Taub and Wang, 2008). An experiment was done on spinach and fenugreek in the elevated CO_2 conditions to see the effect on nitrogen concentration. After 40 days nitrogen concentration increases but after 80 days it decreases (Jain et al., 2007); this indicates that at the initial stage nitrogen uptake is increased in these leafy vegetables and the same results are also shown by barley seedlings (Robredo et al., 2011) but nitrogen uptake is limited at a later stage due to acclimatization of photosynthesis (Taub and Wang, 2008). Reduction in nitrogen also has been noticed in the wheat and berseem plant under elevated CO_2 (Pal et al., 2003) and also in soybean (Reeves et al., 1994). Reduction in nitrogen and increase in carbohydrates changes the C:N ratio in both leafy vegetables. The C:N ratio is highest after 60 days exposure because after 60 days nitrogen concentration is lowest (Gifford et al., 2000; Pal et al., 2004). Many plants, crop or noncrop, show a 4.9%–19% decline in nitrogen concentration under elevated CO_2 conditions (Ainsworth et al., 2004; McGrath and Lobell, 2013a; Bloom et al., 2014; Vicente et al., 2016; Dier et al., 2018). Different species have different effects on nitrogen like in wheat and *Arabidopsis* nitrogen uptake decreases (Bloom et al., 2010), in cucumber it also decreases (Tang et al., 2018), but in tomato it increases (Wei et al., 2018). In soybean, concentration of carbohydrates increases with the increase of CO_2 while nitrogen assimilation decreases (Bloom et al., 2014). For nitrogen reduction, some additional reasons are (1) dilution effect (elevation of biomass) (Reich et al., 2006), (2) low conduction of nutrients from the soil due to the reduction in transpiration (McGrath and Lobell, 2013a), (3) reduction in the proteins like RuBisCO (Long et al., 2004), (4) prevention in nitrate uptake (Bloom et al., 2012, 2014; Myers et al., 2014), and (5) reduction in nitrogen uptake with the change in rhizosphere environment (Cure et al., 1988). Due to the reduction in nitrogen concentration, the C:N ratio of plants disturbed by the cycling of C and N in the soil also gets disturbed (Li et al., 2017). Nitrogen deficiency created under elevated CO_2 (Cotrufo et al., 1998) can be balanced by mycorrhizal fungi, which increases the assimilation of nitrogen in plants (Tanaka and Yano, 2005).

20.10 Effect of elevated CO_2 on human nutrition

Elevated CO_2 is the main reason for the nutritional imbalance in plants as we studied. The nutritional imbalance in plants affects human health. Human health is directly (cognitive development, weak immunity, and metabolism) and indirectly (diabetes type 2 and obesity) affected by nutritionless food. The effect of elevated CO_2 on 130 different varieties of plants was observed by Loladze and he noticed a reduction in the level of all essential 25 minerals such as zinc, iron, and iodine at an average 8% (Loladze, 2002) based on which carbohydrates to minerals ratio increased. In effect, this diet became sufficient in calories but low in minerals, causing malnutrition. In Japan, an experiment was done on rice to know the elevated CO_2 effect on the nutritional quality of rice, and the result shows no decrease in minerals but a decreased level of nitrogen (Lieffering et al., 2004). In contrast, Seneweera and Conroy (1997) found a decline in the amount of Zn and Fe which badly affects poor countries. Lieffering et al. (2004) disagreed with Seneweera and Conroy (1997) and said this decrease of zinc and iron may be the result of the "pot effect." A minerals-poor diet causes stunted growth in childhood, weakens the immune system, and increases the chances of death in mothers and children. Due to the increase in photosynthesis under elevated CO_2, levels of starch increase in plants. When people eat starch-rich food, obesity increases; then people eat more and more to compensate for the lower mineral diet (Loladze, 2002). Due to the decline of minerals in main grain crops like wheat and rice, "hidden hunger" is generated in people who are eating in adequate quantities but consuming food that is not nutritional. A deficiency of iodine in the diet is one of the main reasons for brain damage and lower IQ in children. Iodized salt is beneficial to overcome iodine deficiency, but billions of people still do not have access to it (Welch and Graham, 1999; Welch, 2002). The risk of zinc deficiency is present in every third person globally, which causes stunted growth, weakened immune systems, and increases the death rate of children (Brown et al., 2009). In the world, approximately 2 billion people are affected by iron deficiency, which leads to anemia by which the rate of maternal mortality increases (Organization, 2002). Millions of people have Ca, Mg, and Se deficiency (White and Broadley, 2009). The "hunger-obesity paradox" is also the result of a mineral-deficit diet that causes obesity (Scheier, 2005). As an example, poor and homeless people in the United States are at great risk because they are dependent on cheap, calorie-dense food but that food does not contain nutrients (Koh et al., 2012). In the major 20 health risk problems of the world, obesity and undernutrition are ranked by WHO (Organization, 2002; Hill et al., 2003; Stein and Goldman, 2009), and every third person is affected by obesity (Wiggins and Keats, 2014). To compensate for the deficiency of minerals people eat 14%–16% more food from nutrition-deficient plants that are grown in elevated CO_2 (Stiling and Cornelissen, 2007). For a person who is calorie-deficient, taking 5% extra food is beneficial, but for a person who is not calorie-deficient but is deficient for minerals 5% extra food could be detrimental. The green revolution is an unintentional reason for mineral deficiency in crops that increases the yield of the crops but reduces their mineral content (Davis et al., 2004). A huge risk is that CO_2 is increasing with time so the upcoming generation will be solely dependent on nutrition-deficient food (Myers et al., 2014). Elevated CO_2 not only affects the nutritional concentration of crops, but it also increases a nutrition inhibitor phytate, which inhibits the zinc absorption in the gut of humans, sometimes leading to zinc deficiency. Wheat is the only exceptional crop in which the level of phytate decreases with the increase of CO_2 (Myers et al., 2014). We can overcome this problem with the selection of cultivar, by the use of minerals-rich fertilizers, genetic modification, postharvest biofortification, and by providing education to people regarding the effects of elevated CO_2 on the nutritional quality of crops. Physiologists and nutritionists also can monitor the effect of elevated CO_2 on crops and humans.

20.11 Effect of elevated CO_2 on secondary metabolites

Secondary metabolites are organic molecules that are not essential for plant growth and development, but play a crucial role in defense = against stresses such as herbivory (Agrawal and Fishbein, 2006). We will discuss the defensive function of secondary metabolites against herbivores concerning partitioning of assets like nitrogen and carbon because secondary metabolites contain carbon (CCSMs) and nitrogen (Rhoades, 1979). Consumption of the plant tissue is reduced by the phenolics and tannins, which are CCSMs (Agrawal and Fishbein, 2006). When in the elevated CO_2, the TNC pool is increased and levels of carbon-based secondary metabolites (tannin, phenolics, and terpenes) also increase (Bryant et al., 1983). For example, in the leaves of herbs and wooded plants, the level of carbon-containing secondary metabolites is high when the level of TNC is high and the level of nitrogen is low under elevated CO_2 conditions (Reichardt et al., 1991; Herms and Mattson, 1992). By the increase of atmospheric CO_2 and altered nitrogen cycle, the portioning of carbon in the secondary metabolites gets altered (Lambers, 1993). An increased amount of secondary metabolites also increases the protection of plants against herbivores and pathogens (Bernays and Chapman, 1987; Nicholson and Hammerschmidt, 1992), breakdown of litter decreases (Horner et al., 1988; Kuiters, 1990), and competitive interaction of plants also changes (Inderjit and Dakshini, 1994). With the increases of CO_2, the level of terpenes does not increase much as compared with phenolics because of different portioning of carbon in different secondary metabolites (Fajer et al., 1992). This might be due to genetic restrictions on the synthesis of terpenes through the mevalonic acid pathway (Lawler et al., 1996). More consumption of nitrogen takes place in C3 plants under elevated CO_2 (Drake et al., 1997; Ainsworth and Long, 2005) by which nitrogen incorporation starts in secondary metabolites instead of primary (Gleadow et al., 2009). Low amounts of nitrogen play a crucial role in the synthesis of secondary metabolites because in the low nitrogen condition, high amounts of phenylalanine amino acids participate in the production of phenolic compounds via the shikimic acid pathway, not in protein synthesis (Lambers, 1993; Lincoln et al., 1993). Compared with mature trees, juvenile or small plants typically have a high amount of water and nutrients so normally they are edible (Herms and Mattson, 1992; Bryant and Julkunen-Tiitto, 1995) but juvenile plants produce more quantities of secondary metabolites so they are more protected against herbivory than mature plants (Bryant and Julkunen-Tiitto, 1995). For example, in the young leaves of many species of oak, the total amount of phenolics was higher while in mature leaves, condensed tannin increased (Dawra et al., 1988). So mainly changes in the secondary metabolites are studied in the leaves of the plants (Bryant et al., 1983; Coley, 1983; Herms and Mattson, 1992), but in loblolly pine the amount of phenolics and tannins increased 35% in the belowground parts of the plant as compared with fascicular needles (Gebauer et al., 1997). Loblolly pine followed the carbon-nutrient balance (CNB) hypothesis for the production of secondary metabolites, not carbon and nitrogen as in other plants, so the CNB hypothesis was proven. After all, there is a powerful connection between the amount of C-based secondary metabolites and nonstructural carbohydrates because in the needles and lateral roots the amount of total phenolics also increases. Under elevated CO_2, the increment of secondary metabolites is independent of developmental stages but dependent on carbohydrate concentration, while production of tannin shows part-specific dependency on the CNB hypothesis. Without depending on developmental stages, the collection of tannin in different parts of loblolly pine, such as root, stem, and needles is dependent on low nitrogen concentration but not on CO_2 elevation (Reichardt et al., 1991).

Under elevated CO_2 α-pinene monoterpene increases (Heyworth et al., 1998) but other monoterpenes do not show the same response. Many hypotheses like CNB (Bryant et al., 1983) and nutrient limitation (Coley et al., 1985) are developed to understand the evolutionary model of the dispersal of secondary metabolites. All these hypotheses forecast the impact of elevated CO_2 on secondary metabolites of the plants (Peñuelas and Estiarte, 1998). According to the CNB hypothesis, the amount of secondary metabolites that are carbon-based increases when photosynthesis is favored by elevated CO_2 and low nutrients, while showing no effect on growth (Bryant et al., 1983). The overall result of increased secondary metabolites in the plant is that the defense system of plants becomes strong, but due to a decrease in protein

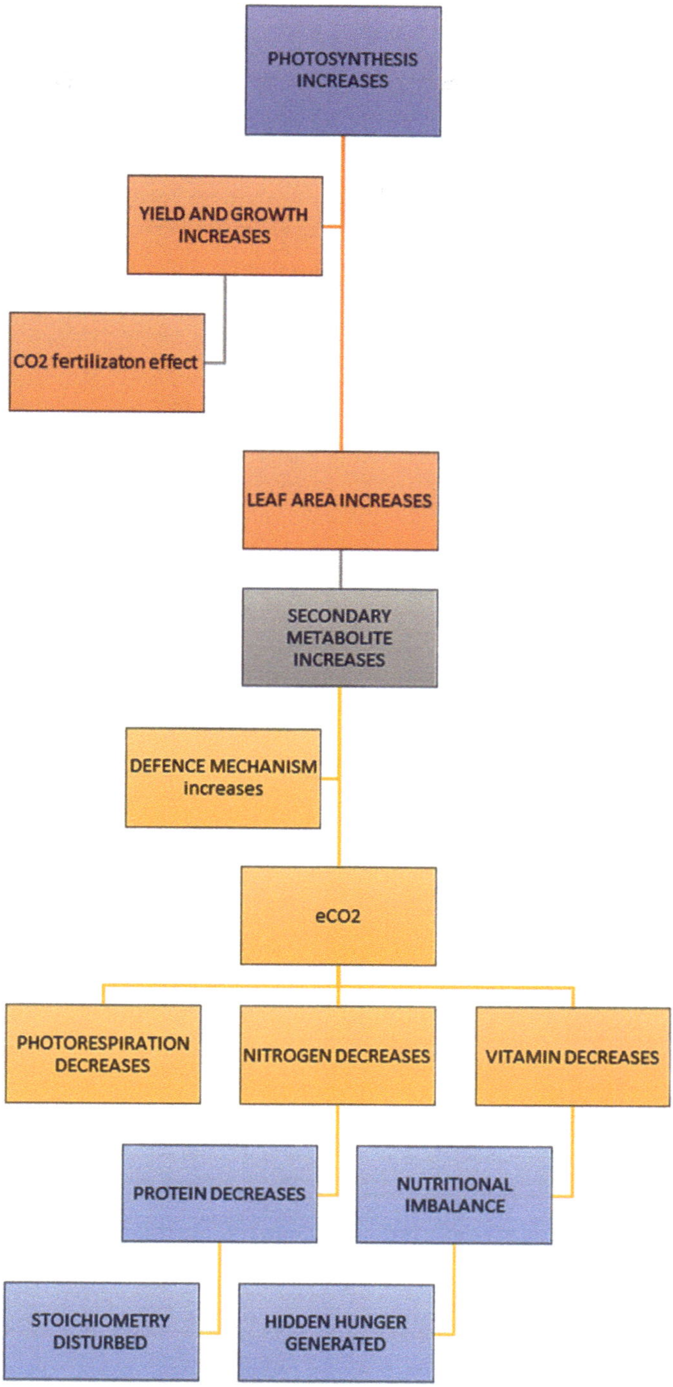

FIGURE 20.1 The effect of elevated CO_2 on the nutritional quality of crops under a changing climate.

TABLE 20.1 Effect of elevated CO_2 on different parameters of C3 and C4 crops and vegetables.

Parameters affected by the elevation of CO_2	C3 crops	Effect	Reference	C4 crops	Effect	References	Vegetables	Effect	Reference
Photosynthesis	Rice	Increases	Jena et al. (2018)	Sorghum	Saturated	Loladze (2014)	Chinese cabbage	Increases	Dong et al. (2018)
Yield and growth	Soybean	Increases	Lenka et al. (2017)	Sorghum	No effect	Conley et al. (2001)	Cucumber	Increases	Dong et al. (2017)
Leaf area	Fenugreek	Increases	Jain et al. (2007)	Maize	No effect	Silva et al. (2020)	Spinach	Increases	Jain et al. (2007)
Proteins	Rice	Decreases	Zhu et al. (2018)	Maize	No effect	Myers et al. (2014)	Radish	Decreases	Stitt and Krapp (1999)
Stoichiometry	Barley	Changes	Pérez-López et al. (2014)	Maize	No effect	Loladze (2014)	Tomato	Changes	Khan et al. (2013)
Photorespiration	*Arabidopsis*	Decreases	Kebeish et al. (2007)	–	–	–	Lettuce	Decreases	Pérez-López et al. (2018)

Nitrogen	Barley	Decreases	Robredo et al. (2011)	Maize	Decreases	Stitt and Krapp (1999)	Cucumber	Decreases	Tang et al. (2018)
Secondary metabolites	Oak trees	Increases	Dawra et al. (1988)	Maize	Increases	Xu et al. (2019)	Lettuce	Increases	Pérez-López et al. (2018)
Vitamins	Rice	Vitamin E increases; vitamin B complex decreases	Zhu et al. (2018)	–	–	–	Carrot	Vitamin C decreases	Azam et al. (2013)

concentration nutritional quality decreases which becomes a severe problem for herbivores and humans in the future (Gleadow and Woodrow, 2000).

20.12 Effect of elevated CO_2 on vitamins

Vitamins are essential for normal growth and development of human body. The effect of elevated CO_2 has been observed on the content of vitamins (vitamin B and E) in various crops. Elevated CO_2 reduced the contents of vitamin B complex, while vitamin E (α-tocopherol) increased in rice grains. The percent reduction observed in vitamin B complexes was B1 (thiamine) (17.1%), B2 (riboflavin) (16.6%), B5 (pantothenic acid) (12.7%), and B9 (folate) (30.3%), whereas vitamin E increased (Zhu et al., 2018). This is consistent with the carbon-nutrient balance (CNB) hypothesis, that suggests increase in the level of carbon-based vitamin (vitamin E) and decrease in nitrogen-based vitamins (vitamin B group) (Hamilton et al., 2001) (Fig., 20.1 Table 20.1).

20.13 Conclusion

This chapter discusses the effects of elevated CO_2 on nutritional quality of crops. The yield of crops is reported to increase with elevation of CO_2 but nutrients like minerals, proteins, and vitamins would decrease. Such decline in the nutrient profile of crops will affect human health and cause malnutrition. Thus nutrient-deficient crops may increase the prevelance of obesity, hidden hunger, and hypertension in the human population. Therefore, there is a need to enhance the nutritional quality of crops under elevated CO_2, for that several approaches can be opted which include: (1) cultivar selection that show less decline in nutrients under high CO_2; (2) modulation of environmental factors like light intensity and quality to increase nutrient uptake and counteract the CO_2 dilution effect; (3) early harvest of vegetables before maturation; and (4) optimization of fertilization procedures to increase the availability of N, Fe or Zn in the soil.

References

Agrawal, A.A., Fishbein, M., 2006. Plant defense syndromes. Ecology 87, S132–S149.

Ainsworth, E.A., Long, S.P., 2005. What have we learned from 15 years of free-air CO_2 enrichment (FACE)? A meta-analytic review of the responses of photosynthesis, canopy properties and plant production to rising CO_2. New Phytol. 165, 351–372.

Ainsworth, E.A., Rogers, A., Nelson, R., Long, S.P., 2004. Testing the "source–sink" hypothesis of down-regulation of photosynthesis in elevated (CO_2) in the field with single gene substitutions in Glycine max. Agric. For. Meteorol. 122, 85–94.

Alexandratos N., Bruinsma J., 2012. World agriculture towards 2030/2050: the 2012 revision.

Ames, R., Reid, C., Porter, L., Cambardella, C., 1983. Hyphal uptake and transport of nitrogen from two 15N-labelled sources by Glomus mosseae, a vesicular-arbuscular mycorrhizal fungus. New Phytol. 95, 381–396.

Aranjuelo, I., Sanz-Sáez, Á., Jauregui, I., et al., 2013. Harvest index, a parameter conditioning responsiveness of wheat plants to elevated CO_2. J. Exp. Bot. 64, 1879–1892.

Azam, A., Khan, I., Mahmood, A., Hameed, A., 2013. Yield, chemical composition and nutritional quality responses of carrot, radish and turnip to elevated atmospheric carbon dioxide. J. Sci. Food Agric. 93, 3237–3244.

Baslam, M., Garmendia, I., Goicoechea, N., 2012. Elevated CO_2 may impair the beneficial effect of arbuscular mycorrhizal fungi on the mineral and phytochemical quality of lettuce. Ann. Appl. Biol. 161, 180–191.

Becker, C., Kläring, H.-P., 2016. CO_2 enrichment can produce high red leaf lettuce yield while increasing most flavonoid glycoside and some caffeic acid derivative concentrations. Food Chem. 199, 736–745.

Benlloch-Gonzalez, M., Bochicchio, R., Berger, J., Bramley, H., Palta, J.A., 2014. High temperature reduces the positive effect of elevated CO_2 on wheat root system growth. Field Crops Res. 165, 71–79.

Bernays, E., Chapman, R., 1987. The evolution of deterrent responses in plant-feeding insects. In: Chapman, R.F., Bernays, E.A., Stoffolano, J.G. (Eds.), Perspectives in Chemoreception and Behavior. Proceedings in Life Sciences. Springer, New York, NY. Available from: https://doi.org/10.1007/978-1-4612-4644-2_10.

Bidart-Bouzat, M.G., Imeh-Nathaniel, A., 2008. Global change effects on plant chemical defenses against insect herbivores. J. Integr. Plant Biol. 50, 1339–1354.

Bloom, A.J., Burger, M., Asensio, J.S.R., Cousins, A.B., 2010. Carbon dioxide enrichment inhibits nitrate assimilation in wheat and *Arabidopsis*. Science 328, 899–903.

Bloom, A.J., Asensio, J.S.R., Randall, L., Rachmilevitch, S., Cousins, A.B., Carlisle, E.A., 2012. CO_2 enrichment inhibits shoot nitrate assimilation in C3 but not C4 plants and slows growth under nitrate in C3 plants. Ecology 93, 355–367.

Bloom, A.J., Burger, M., Kimball, B.A., Pinter Jr, P.J., 2014. Nitrate assimilation is inhibited by elevated CO_2 in field-grown wheat. Nat. Clim. Change 4, 477–480.

Brown, K.H., Peerson, J.M., Baker, S.K., Hess, S.Y., 2009. Preventive zinc supplementation among infants, preschoolers, and older prepubertal children. Food Nutr. Bull. 30, S12–S40.

Bryant, J.P., Julkunen-Tiitto, R., 1995. Ontogenic development of chemical defense by seedling resin birch: energy cost of defense production. J. Chem. Ecol. 21, 883–896.

Bryant, J.P., Chapin Iii, F.S., Klein, D.R., 1983. Carbon/nutrient balance of boreal plants in relation to vertebrate herbivory. Oikos 40, 357–368.

Carbonero, P., Iglesias-Fernández, R., Vicente-Carbajosa, J., 2017. The AFL subfamily of B3 transcription factors: evolution and function in angiosperm seeds. J. Exp. Bot. 68, 871–880.

Cavagnaro, T.R., Sokolow, S.K., Jackson, L.E., 2007. Mycorrhizal effects on growth and nutrition of tomato under elevated atmospheric carbon dioxide. Funct. Plant Biol. 34, 730–736.

Cavagnaro, T., Gleadow, R., Miller, R., 2011. Plant nutrient acquisition and utilisation in a high carbon dioxide world. Funct. Plant Biol. 38, 87–96.

Centritto, M., Magnani, F., Lee, H.S., Jarvis, P.G., 1999. Interactive effects of elevated [CO_2] and drought on cherry (*Prunus avium*) seedlings II. Photosynthetic capacity and water relations. New Phytol. 141, 141–153.

Chen, X., Tu, C., Burton, M.G., Watson, D.M., Burkey, K.O., Hu, S., 2007. Plant nitrogen acquisition and interactions under elevated carbon dioxide: impact of endophytes and mycorrhizae. Glob. Change Biol. 13, 1238–1249.

Coley, P.D., 1983. Herbivory and defensive characteristics of tree species in a lowland tropical forest. Ecol. Monogr. 53, 209–234.

Coley, P.D., Bryant, J.P., Chapin, F.S., 1985. Resource availability and plant antiherbivore defense. Science 230, 895–899.

Coleman, J., Mcconnaughay, K., Bazzaz, F., 1993. Elevated CO_2 and plant nitrogen-use: is reduced tissue nitrogen concentration size-dependent? Oecologia 93, 195–200.

Coley, P., Massa, M., Lovelock, C., Winter, K., 2002. Effects of elevated CO_2 on foliar chemistry of saplings of nine species of tropical tree. Oecologia 133, 62–69.

Conley, M.M., Kimball, B., Brooks, T., et al., 2001. CO_2 enrichment increases water-use efficiency in sorghum. New Phytol. 151, 407–412.

Conroy, J.P., 1992. Influence of elevated atmospheric CO_2 concentrations on plant nutrition. Aust. J. Bot. 40, 445–456.

Cotrufo, M.F., Ineson, P., Scott, A., 1998. Elevated CO_2 reduces the nitrogen concentration of plant tissues. Glob. Change Biol. 4, 43–54.

Cousins, A., Adam, N., Wall, G., et al., 2003. Development of C4 photosynthesis in sorghum leaves grown under free-air CO_2 enrichment (FACE). J. Exp. Bot. 54, 1969–1975.

Cure, J.D., Israel, D.W., Rufty Jr, T.W., 1988. Nitrogen stress effects on growth and seed yield of nonnodulated soybean exposed to elevated carbon dioxide. Crop Sci. 28, 671–677.

Davis, D.R., Epp, M.D., Riordan, H.D., 2004. Changes in USDA food composition data for 43 garden crops, 1950 to 1999. J. Am. Coll. Nutr. 23, 669–682.

Dawra, R.K., Makkar, H.P., Singh, B., 1988. Protein-binding capacity of microquantities of tannins. Anal. Biochem. 170, 50–53.

Dier, M., Meinen, R., Erbs, M., et al., 2018. Effects of free air carbon dioxide enrichment (FACE) on nitrogen assimilation and growth of winter wheat under nitrate and ammonium fertilization. Glob. Change Biol. 24, e40–e54.

Dong, J., Li, X., Chu, W., Duan, Z., 2017. High nitrate supply promotes nitrate assimilation and alleviates photosynthetic acclimation of cucumber plants under elevated CO_2. Sci. Hortic. 218, 275–283.

Dong, J., Gruda, N., Lam, S.K., Li, X., Duan, Z., 2018. Effects of elevated CO_2 on nutritional quality of vegetables: a review. Front. Plant Sci. 9, 924.

Donnelly, A., Craigon, J., Black, C.R., Colls, J.J., Landon, G., 2001. Elevated CO_2 increases biomass and tuber yield in potato even at high ozone concentrations. New Phytol. 149, 265–274.

Drake, B.G., Gonzàlez-Meler, M.A., Long, S.P., 1997. More efficient plants: a consequence of rising atmospheric CO_2? Ann. Rev. Plant Biol. 48, 609–639.

Drake, J.E., Gallet-Budynek, A., Hofmockel, K.S., et al., 2011. Increases in the flux of carbon belowground stimulate nitrogen uptake and sustain the long-term enhancement of forest productivity under elevated CO_2. Ecol. Lett. 14, 349–357.

Erbs, M., Manderscheid, R., Jansen, G., Seddig, S., Pacholski, A., Weigel, H.-J., 2010. Effects of free-air CO_2 enrichment and nitrogen supply on grain quality parameters and elemental composition of wheat and barley grown in a crop rotation. Agric. Ecosyst. Environ. 136, 59–68.

Fajer, E., Bowers, M., Bazzaz, F., 1992. The effect of nutrients and enriched CO_2 environments on production of carbon-based allelochemicals in Plantago: a test of the carbon/nutrient balance hypothesis. Am. Nat. 140, 707–723.

Gebauer, R.L., Strain, B.R., Reynolds, J.F., 1997. The effect of elevated CO_2 and N availability on tissue concentrations and whole plant pools of carbon-based secondary compounds in loblolly pine (Pinus taeda). Oecologia 113, 29–36.

Gebregergis, Z., 2016. Effect of elevated carbon dioxide, CO_2 and temperature on crop production: a review. J. Nat. Sci. Res. 6, 42–46.
Gifford, R.M., Barrett, D.J., Lutze, J.L., 2000. The effects of elevated [CO_2] on the C: N and C: P mass ratios of plant tissues. Plant Soil 224, 1–14.
Gleadow, R.M., Woodrow, I.E., 2000. Polymorphism in cyanogenic glycoside content and cyanogenic β-glucosidase activity in natural populations of Eucalyptus cladocalyx. Funct. Plant Biol. 27, 693–699.
Gleadow, R.M., Evans, J.R., Mccaffery, S., Cavagnaro, T.R., 2009. Growth and nutritive value of cassava (Manihot esculenta Cranz.) are reduced when grown in elevated CO_2. Plant Biol. 11, 76–82.
Gropper, S., Smith, J., Groff, J., 2008. Nutrition and Metabolism. Wadsworth, Belmont, CA.
Hamilton, J., Zangerl, A., Delucia, E., Berenbaum, M., 2001. The carbon–nutrient balance hypothesis: its rise and fall. Ecol. Lett. 4, 86–95.
Herms, D.A., Mattson, W.J., 1992. The dilemma of plants: to grow or defend. Quart. Rev. Biol. 67, 283–335.
Heyworth, C., Iason, G., Temperton, V., Jarvis, P., Duncan, A., 1998. The effect of elevated CO_2 concentration and nutrient supply on carbon-based plant secondary metabolites in *Pinus sylvestris* L. Oecologia 115, 344–350.
Hill, J.O., Wyatt, H.R., Reed, G.W., Peters, J.C., 2003. Obesity and the environment: where do we go from here? Science 299, 853–855.
Horner, J.D., Gosz, J.R., Cates, R.G., 1988. The role of carbon-based plant secondary metabolites in decomposition in terrestrial ecosystems. Am. Nat. 132, 869–883.
Huang, B., Xu, Y., 2015. Cellular and molecular mechanisms for elevated CO_2–regulation of plant growth and stress adaptation. Crop Sci. 55, 1405–1424.
Inderjit, D.K., Dakshini, K., 1994. Algal allelopathy. Bot. Rev. 60, 182–196.
Islam, M.S., Matsui, T., Yoshida, Y., 1996. Effect of carbon dioxide enrichment on physico-chemical and enzymatic changes in tomato fruits at various stages of maturity. Sci. Hortic. 65, 137–149.
Jaafar, H.Z., Ibrahim, M.H., Karimi, E., 2012. Phenolics and flavonoids compounds, phenylanine ammonia lyase and antioxidant activity responses to elevated CO_2 in *Labisia pumila* (Myrisinaceae). Molecules 17, 6331–6347.
Jablonski, L.M., Wang, X., Curtis, P.S., 2002. Plant reproduction under elevated CO_2 conditions: a meta-analysis of reports on 79 crop and wild species. New Phytol. 156, 9–26.
Jain, V., Pal, M., Raj, A., Khetarpal, S., 2007. Photosynthesis and nutrient composition of spinach and fenugreek grown under elevated carbon dioxide concentration. Biol. Plant. 51, 559.
Jena, U.R., Swain, D.K., Hazra, K.K., Maiti, M.K., 2018. Effect of elevated [CO_2] on yield, intra-plant nutrient dynamics, and grain quality of rice cultivars in eastern India. J. Sci. Food Agric. 98, 5841–5852.
Jin, C., Du, S., Wang, Y., Condon, J., Lin, X., Zhang, Y., 2009. Carbon dioxide enrichment by composting in greenhouses and its effect on vegetable production. J. Plant Nutr. Soil Sci. 172, 418–424.
Johansen, A., Jakobsen, I., Jensen, E., 1993. External hyphae of vesicular–arbuscular mycorrhizal fungi associated with Trifolium subterraneum L. 3. Hyphal transport of 32P and 15N. New Phytol. 124, 61–68.
Katny, M.A.C., Hoffmann-Thoma, G., Schrier, A.A., Fangmeier, A., Jäger, H.-J., Van Bel, A.J., 2005. Increase of photosynthesis and starch in potato under elevated CO_2 is dependent on leaf age. J. Plant Physiol. 162, 429–438.
Kebeish, R., Niessen, M., Thiruveedhi, K., et al., 2007. Chloroplastic photorespiratory bypass increases photosynthesis and biomass production in *Arabidopsis thaliana*. Nat. Biotechnol. 25, 593–599.

Khan, I., Azam, A., Mahmood, A., 2013. The impact of enhanced atmospheric carbon dioxide on yield, proximate composition, elemental concentration, fatty acid and vitamin C contents of tomato (*Lycopersicon esculentum*). Environ. Monit. Assess. 185, 205–214.

Keutgen, N., Chen, K., 2001. Responses of citrus leaf photosynthesis, chlorophyll fluorescence, macronutrient and carbohydrate contents to elevated CO_2. J. Plant Physiol. 158, 1307–1316.

Kimball, B.A., 2016. Crop responses to elevated CO_2 and interactions with H_2O, N, and temperature. Curr. Opin. Plant Biol. 31, 36–43.

Kimball, B., Kobayashi, K., Bindi, M., 2002. Responses of agricultural crops to free-air CO2 enrichment. Adv. Agron. 77, 293–368.

Kirschbaum, M.U., 2011. Does enhanced photosynthesis enhance growth? Lessons learned from CO_2 enrichment studies. Plant Physiol. 155, 117–124.

Koh, K.A., Hoy, J.S., O'connell, J.J., Montgomery, P., 2012. The hunger–obesity paradox: obesity in the homeless. J. Urban Health 89, 952–964.

Kuiters, A., 1990. Role of phenolic substances from decomposing forest litter in plant-soil interactions. Acta Bot. Neerl. 39, 329–348.

Kumari, S., Agrawal, M., 2014. Growth, yield and quality attributes of a tropical potato variety (*Solanum tuberosum* L. cv Kufri chandramukhi) under ambient and elevated carbon dioxide and ozone and their interactions. Ecotoxicol. Environ. Saf. 101, 146–156.

Lambers, H., 1993. Rising CO_2, secondary plant metabolism, plant-herbivore interactions and litter decomposition. CO_2 and Biosphere. Springer, pp. 263–271.

Lawler, I., Foley, W., Woodrow, I., Cork, S., 1996. The effects of elevated CO_2 atmospheres on the nutritional quality of Eucalyptus foliage and its interaction with soil nutrient and light availability. Oecologia 109, 59–68.

Leakey, A.D., Ainsworth, E.A., Bernacchi, C.J., Rogers, A., Long, S.P., Ort, D.R., 2009. Elevated CO_2 effects on plant carbon, nitrogen, and water relations: six important lessons from FACE. J. Exp. Bot. 60, 2859–2876.

Lenka, N.K., Lenka, S., Thakur, J., et al., 2017. Interactive effect of elevated carbon dioxide and elevated temperature on growth and yield of soybean. Curr. Sci. 113, 2305.

Li, Y., Yu, Z., Liu, X., et al., 2017. Elevated CO_2 increases nitrogen fixation at the reproductive phase contributing to various yield responses of soybean cultivars. Front. Plant Sci. 8, 1546.

Lieffering, M., Kim, H.-Y., Kobayashi, K., Okada, M., 2004. The impact of elevated CO_2 on the elemental concentrations of field-grown rice grains. Field Crops Res. 88, 279–286.

Lin, T., Zhu, G., Zhang, J., et al., 2014. Genomic analyses provide insights into the history of tomato breeding. Nat. Genet. 46, 1220–1226.

Lin, Y.-S., Medlyn, B.E., Duursma, R.A., et al., 2015. Optimal stomatal behaviour around the world. Nat. Clim. Change 5, 459–464.

Lincoln, D.E., Fajer, E.D., Johnson, R.H., 1993. Plant-insect herbivore interactions in elevated CO_2 environments. Trends Ecol. Evol. 8, 64–68.

Loladze, I., 2002. Rising atmospheric CO_2 and human nutrition: toward globally imbalanced plant stoichiometry? Trends Ecol. Evol. 17, 457–461.

Loladze, I., 2014. Hidden shift of the ionome of plants exposed to elevated CO_2 depletes minerals at the base of human nutrition. eLife 3, e02245.

Long, S.P., Ainsworth, E.A., Rogers, A., Ort, D.R., 2004. Rising atmospheric carbon dioxide: plants FACE the future. Annu. Rev. Plant Biol. 55, 591–628.

Long, S.P., Ainsworth, E.A., Leakey, A.D., Morgan, P.B., 2005. Global food insecurity. Treatment of major food crops with elevated carbon dioxide or ozone under large-scale fully open-air conditions suggests recent models may have overestimated future yields. Philos. Trans. R. Soc. B Biol. Sci. 360, 2011–2020.

Long, S.P., Ainsworth, E.A., Leakey, A.D., Nösberger, J., Ort, D.R., 2006. Food for thought: lower-than-expected crop yield stimulation with rising CO_2 concentrations. Science 312, 1918–1921.

Makino, A., Mae, T., 1999. Photosynthesis and plant growth at elevated levels of CO_2. Plant Cell Physiol. 40, 999–1006.

Martre, P., Porter, J.R., Jamieson, P.D., Triboï, E., 2003. Modeling grain nitrogen accumulation and protein composition to understand the sink/source regulations of nitrogen remobilization for wheat. Plant Physiol. 133, 1959–1967.

Maurino, V.G., Peterhansel, C., 2010. Photorespiration: current status and approaches for metabolic engineering. Curr. Opin. Plant Biol. 13, 248–255.

McGrath, J.M., Lobell, D.B., 2013a. Reduction of transpiration and altered nutrient allocation contribute to nutrient decline of crops grown in elevated CO_2 concentrations. Plant Cell Environ. 36, 697–705.

McGrath, J.M., Lobell, D.B., 2013b. Regional disparities in the CO_2 fertilization effect and implications for crop yields. Environ. Res. Lett. 8, 014054.

Medek, D.E., Schwartz, J., Myers, S.S., 2017. Estimated effects of future atmospheric CO_2 concentrations on protein intake and the risk of protein deficiency by country and region. Environ. Health Perspect. 125, 087002.

Millward, D.J., Jackson, A.A., 2004. Protein/energy ratios of current diets in developed and developing countries compared with a safe protein/energy ratio: implications for recommended protein and amino acid intakes. Public Health Nutr. 7, 387–405.

Myers, S.S., Zanobetti, A., Kloog, I., et al., 2014. Increasing CO_2 threatens human nutrition. Nature 510, 139–142.

Nicholson, R.L., Hammerschmidt, R., 1992. Phenolic compounds and their role in disease resistance. Ann. Rev. Phytopathol. 30, 369–389.

O'neill, E., Luxmoore, R., Norby, R., 1987. Elevated atmospheric CO_2 effects on seedling growth, nutrient uptake, and rhizosphere bacterial populations of *Liriodendron tulipifera* L. Plant Soil 104, 3–11.

Olesniewicz, K.S., Thomas, R.B., 1999. Effects of mycorrhizal colonization on biomass production and nitrogen fixation of black locust (*Robinia pseudoacacia*) seedlings grown under elevated atmospheric carbon dioxide. New Phytol. 142, 133–140.

Pal, M., Rao, L., Srivastava, A., Jain, V., Sengupta, U., 2003. Impact of CO_2 enrichment and variable nitrogen supplies on composition and partitioning of essential nutrients of wheat. Biol. Plant. 47, 227.

Pal, M., Karthikeyapandian, V., Jain, V., Srivastava, A., Raj, A., Sengupta, U., 2004. Biomass production and nutritional levels of berseem (*Trifolium alexandrium*) grown under elevated CO_2. Agric. Ecosyst. Environ. 101, 31–38.

Pal, D., Sahoo, M., Mishra, A.K., 2005. Analgesic and anticonvulsant effects of saponin isolated from the stems of Opuntia vulgaris Mill in mice. Eur. Bull. Drug Res. 13, 91–97.

Peet, M.M., Huber, S.C., Patterson, D.T., 1986. Acclimation to high CO_2 in monoecious cucumbers: II. Carbon exchange rates, enzyme activities, and starch and nutrient concentrations. Plant Physiol. 80, 63–67.

Peñuelas, J., Estiarte, M., 1998. Can elevated CO_2 affect secondary metabolism and ecosystem function? Trends Ecol. Evol. 13, 20–24.

Pérez-López, U., Miranda-Apodaca, J., Mena-Petite, A., Muñoz-Rueda, A., 2014. Responses of nutrient dynamics in barley seedlings to the interaction of salinity and carbon dioxide enrichment. Environ. Exp. Bot. 99, 86–99.

Pérez-López, U., Miranda-Apodaca, J., Lacuesta, M., Mena-Petite, A., Muñoz-Rueda, A., 2015. Growth and nutritional quality improvement in two differently pigmented lettuce cultivars grown under elevated CO_2 and/or salinity. Sci. Hortic. 195, 56–66.

Pérez-López, U., Sgherri, C., Miranda-Apodaca, J., et al., 2018. Concentration of phenolic compounds is increased in lettuce grown under high light intensity and elevated CO_2. Plant Physiol. Biochem. 123, 233–241.

Poorter, H., Van Berkel, Y., Baxter, R., et al., 1997. The effect of elevated CO_2 on the chemical composition and construction costs of leaves of 27 C3 species. Plant Cell Environ. 20, 472–482.

Porter, M.A., Grodzinski, B., 1984. Acclimation to high CO_2 in bean: carbonic anhydrase and ribulose bisphosphate carboxylase. Plant Physiol. 74, 413–416.

Reeves, D., Rogers, H., Prior, S., Wood, C., Runion, G., 1994. Elevated atmospheric carbon dioxide effects on sorghum and soybean nutrient status. J. Plant Nutr. 17, 1939–1954.

Reich, P.B., Hungate, B.A., Luo, Y., 2006. Carbon-nitrogen interactions in terrestrial ecosystems in response to rising atmospheric carbon dioxide. Annu. Rev. Ecol. Evol. Syst. 37, 611–636.

Reichardt, P., Chapin, F., Bryant, J., Mattes, B., Clausen, T., 1991. Carbon/nutrient balance as a predictor of plant defense in Alaskan balsam poplar: potential importance of metabolite turnover. Oecologia 88, 401–406.

Rhoades, D.F., 1979. Evolution of plant chemical defense against herbivores. In: Rosenthal G.A., Janzen D.H. (Eds.). Herbivores: Their Interaction with Secondary Plant Metabolites. Academic Press, New York, 3–54.

Robinson, E.A., Ryan, G.D., Newman, J.A., 2012. A meta-analytical review of the effects of elevated CO_2 on plant–arthropod interactions highlights the importance of interacting environmental and biological variables. New Phytol. 194, 321–336.

Robredo, A., Pérez-López, U., Miranda-Apodaca, J., Lacuesta, M., Mena-Petite, A., Muñoz-Rueda, A., 2011. Elevated CO_2 reduces the drought effect on nitrogen metabolism in barley plants during drought and subsequent recovery. Environ. Exp. Bot. 71, 399–408.

Ruiz-Vera, U.M., De Souza, A.P., Long, S.P., Ort, D.R., 2017. The role of sink strength and nitrogen availability in the down-regulation of photosynthetic capacity in field-grown *Nicotiana tabacum* L. at elevated CO_2 concentration. Front. Plant Sci. 8, 998.

Scheier, L.M., 2005. What is the hunger-obesity paradox? J. Acad. Nutr. Diet. 105, 883–885.

Schortemeyer, M., Atkin, O.K., Mcfarlane, N., Evans, J.R., 1999. The impact of elevated atmospheric CO_2 and nitrate supply on growth, biomass allocation, nitrogen partitioning and N2 fixation of Acacia melanoxylon. Funct. Plant Biol. 26, 737–747.

Seneweera, S.P., Conroy, J.P., 1997. Growth, grain yield and quality of rice (*Oryza sativa* L.) in response to elevated CO_2 and phosphorus nutrition. Soil Sci. Plant Nutr. 43, 1131–1136.

Seneweera, S.P., Ghannoum, O., Conroy, J.P., et al., 2002. Changes in source–sink relations during development influence photosynthetic acclimation of rice to free air CO_2 enrichment (FACE). Funct. Plant Biol. 29, 947–955.

Silva, R.G.D., Alves, R.D.C., Zingaretti, S.M., 2020. Increased [CO2] causes changes in physiological and genetic responses in C4 crops: a brief review. Plants 9, 1567.

Smith, S.E., Jakobsen, I., Grønlund, M., Smith, F.A., 2011. Roles of arbuscular mycorrhizas in plant phosphorus nutrition: interactions between pathways of phosphorus uptake in arbuscular mycorrhizal roots have important implications for understanding and manipulating plant phosphorus acquisition. Plant Physiol. 156, 1050–1057.

Solomon, S., Manning, M., Marquis, M., Qin, D., 2007. Climate Change 2007-The Physical Science Basis: Working Group I Contribution to the Fourth Assessment Report of the IPCC. Cambridge University Press.

Stein, P.D., Goldman, J., 2009. Obesity and thromboembolic disease. Clin. Chest Med. 30, 489–493.

Sterner, R.W., Elser, J.J., 2002. Ecological Stoichiometry: The Biology of Elements From Molecules to the Biosphere. Princeton University Press.

Stiling, P., Cornelissen, T., 2007. How does elevated carbon dioxide (CO_2) affect plant–herbivore interactions? A field experiment and meta-analysis of CO_2-mediated changes on plant chemistry and herbivore performance. Glob. Change Biol. 13, 1823–1842.

Still, C.J., Berry, J.A., Collatz, G.J., Defries, R.S., 2003. Global distribution of C3 and C4 vegetation: carbon cycle implications. Global. Biogeochem. Cycles 17, 6-1–6-14.

Stitt, M., Krapp, A., 1999. The interaction between elevated carbon dioxide and nitrogen nutrition: the physiological and molecular background. Plant Cell Environ. 22, 583–621.

Tanaka, Y., Yano, K., 2005. Nitrogen delivery to maize via mycorrhizal hyphae depends on the form of N supplied. Plant Cell Environ. 28, 1247–1254.

Tang, W., Arellano, A.F., Digangi, J.P., et al., 2018. Evaluating high-resolution forecasts of atmospheric CO and CO_2 from a global prediction system during KORUS-AQ field campaign. Atmos. Chem. Phys. 18, 11007–11030.

Taub, D.R., Wang, X., 2008. Why are nitrogen concentrations in plant tissues lower under elevated CO_2? A critical examination of the hypotheses. J. Integr. Plant Biol. 50, 1365–1374.

Timm, S., Bauwe, H., 2013. The variety of photorespiratory phenotypes–employing the current status for future research directions on photorespiration. Plant Biol. 15, 737–747.

Umnajkitikorn, K., Sade, N., Rubio Wilhelmi, M.D.M., Gilbert, M.E., Blumwald, E., 2020. Silencing of OsCV (chloroplast vesiculation) maintained photorespiration and N assimilation in rice plants grown under elevated CO_2. Plant Cell Environ. 43, 920–933.

Uprety, D., Dwivedi, N., Jain, V., et al., 2003. Responses of rice cultivars to the elevated CO_2. Biol. Plant. 46, 35–39.

Vicente, R., Pérez, P., Martínez-Carrasco, R., et al., 2016. Metabolic and transcriptional analysis of durum wheat responses to elevated CO_2 at low and high nitrate supply. Plant Cell Physiol. 57, 2133–2146.

Vicente, R., Pérez, P., Martínez-Carrasco, R., Gutiérrez, E., Morcuende, R., 2015a. Nitrate supply and plant development influence nitrogen uptake and allocation under elevated CO_2 in durum wheat grown hydroponically. Acta Physiol. Plant. 37, 1–13.

Vicente, R., Pérez, P., Martínez-Carrasco, R., Usadel, B., Kostadinova, S., Morcuende, R., 2015b. Quantitative RT–PCR platform to measure transcript levels of C and N metabolism-related genes in durum wheat: transcript profiles in elevated [CO_2] and high temperature at different levels of N supply. Plant Cell Physiol. 56, 1556–1573.

Voss, I., Sunil, B., Scheibe, R., Raghavendra, A., 2013. Emerging concept for the role of photorespiration as an important part of abiotic stress response. Plant Biol. 15, 713–722.

Wand, S.J., Midgley, G.F., Jones, M.H., Curtis, P.S., 1999. Responses of wild C4 and C3 grass (*Poaceae*) species to elevated atmospheric CO_2 concentration: a meta-analytic test of current theories and perceptions. Global Change Biol. 5, 723–741.

Wang, S.Y., Bunce, J.A., Maas, J., 2003. Elevated carbon dioxide increases contents of antioxidant compounds in field-grown strawberries. J. Agric. Food Chem. 51, 4315–4320.

Wei, Z., Du, T., Li, X., Fang, L., Liu, F., 2018. Interactive effects of elevated CO_2 and N fertilization on yield and quality of tomato grown under reduced irrigation regimes. Front. Plant Sci. 9, 328.

Welch, R.M., 2002. The impact of mineral nutrients in food crops on global human health. Plant Soil 247, 83–90.

Welch, R.M., Graham, R.D., 1999. A new paradigm for world agriculture: meeting human needs: productive, sustainable, nutritious. Field Crops Res. 60, 1–10.

White, P.J., Broadley, M.R., 2009. Biofortification of crops with seven mineral elements often lacking in human diets–iron, zinc, copper, calcium, magnesium, selenium and iodine. New Phytol. 182, 49–84.

Wiggins S., Keats S., 2014. Future diets: under-and over-nutrition in developing countries. https://www.commonwealthhealth.org.

World Health Organization, 2002. The World Health Report 2002: Reducing Risks, Promoting Healthy Life. World Health Organization.

Wujeska-Klause, A., Crous, K.Y., Ghannoum, O., Ellsworth, D.S., 2019. Lower photorespiration in elevated CO_2 reduces leaf N concentrations in mature Eucalyptus trees in the field. Glob Chang Biol 25, 1282–1295.

Xu, H., Xie, H., Wu, S., Wang, Z., He, K., 2019. Effects of elevated CO_2 and increased N fertilization on plant secondary metabolites and chewing insect fitness. Front. Plant Sci. 10, 739.

Zhu, C., Kobayashi, K., Loladze, I., et al., 2018. Carbon dioxide (CO_2) levels this century will alter the protein, micronutrients, and vitamin content of rice grains with potential health consequences for the poorest rice-dependent countries. Sci. Adv. 4, eaaq1012.

Ziska, L.H., Pettis, J.S., Edwards, J., et al., 2016. Rising atmospheric CO_2 is reducing the protein concentration of a floral pollen source essential for North American bees. Proc. R. Soc. B Biol. Sci. 283, 20160414.

Index

Note: Page numbers followed by "*f*" and "*t*" refer to figures and tables, respectively.

B

O

P

MIX
Papier | Fördert gute Waldnutzung
FSC® C083411

Kolibri 360 GmbH
Ferdinand-Jühlke-Straße 7
99095 Erfurt, Deutschland
produktsicherheit@kolibri360.de